Lecture Notes in Computer Science 5446

Commenced Publication in 1973
Founding and Former Series Editors:
Gerhard Goos, Juris Hartmanis, and Jan van Leeuwen

Qing Li   Ling Feng   Jian Pei
Sean X. Wang   Xiaofang Zhou
Qiao-Ming Zhu (Eds.)

Advances in Data and Web Management

Joint International Conferences, APWeb/WAIM 2009
Suzhou, China, April 2-4, 2009
Proceedings

 Springer

Volume Editors

Qing Li
City University of Hong Kong, Department of Computer Science
Kowloon, Hong Kong, China
E-mail: itqli@cityu.edu.hk

Ling Feng
Tsinghua University, Department of Computer Science & Technology
Beijing 100084, China
E-mail: fengling@tsinghua.edu.cn

Jian Pei
Simon Fraser University, School of Computing Science
Burnaby, BC V5A 1S6, Canada
E-mail: jpei@cs.sfu.ca

Sean X. Wang
University of Vermont, Department of Computer Science
Burlington, VT 05405, USA
E-mail: sean.wang@uvm.edu

Xiaofang Zhou
The University of Queensland, School of ITEE
Brisbane, QLD 4072, Australia
E-mail: zxf@itee.uq.edu.au

Qiao-Ming Zhu
Soochow University, School of Computer Science & Technology
Suzhou, Jiangsu 215006, China
E-mail: qmzhu@suda.edu.cn

Library of Congress Control Number: 2009922514
CR Subject Classification (1998): H.2-5, C.2, J.1, K.4
LNCS Sublibrary: SL 3 – Information Systems and Application,
incl. Internet/Web and HCI

ISSN   0302-9743
ISBN-10   3-642-00671-X Springer Berlin Heidelberg New York
ISBN-13   978-3-642-00671-5 Springer Berlin Heidelberg New York

springer.com

© Springer-Verlag Berlin Heidelberg 2009
Printed in Germany

Typesetting: Camera-ready by author, data conversion by Scientific Publishing Services, Chennai, India
Printed on acid-free paper   SPIN: 12632534   06/3180   5 4 3 2 1 0

Preface

APWeb and WAIM are two leading international conferences on the research, development, and applications of Web technologies, database systems, information management and software engineering, with a focus on the Asia-Pacific region. The previous APWeb conferences were held in Beijing (1998), Hong Kong (1999), Xi'an (2000), Changsha (2001), Xi'an (2003), Hangzhou (2004), Shanghai (2005), Harbin (2006), Huangshan (2007), and Shenyang (2008); and the previous WAIM conferences were held in Shanghai (2000), Xi'an (2001), Beijing (2002), Chengdu (2003), Dalian (2004), Hangzhou (2005), Hong Kong (2006), Huangshan (2007), and Zhangjiajie (2008). For the second time, APWeb and WAIM were combined to foster closer collaboration and research idea sharing, and were held immediately after IEEE ICDE 2009 in Shanghai.

This high-quality program would not have been possible without the authors who chose APWeb+WAIM as a venue for their publications. Out of 189 submitted papers from 21 countries and regions, including Australia, Austria, Bengal, Brazil, Canada, France, Greece, Hong Kong, India, Iran, Japan, Korea, Macau, Mainland China, Myanmar, The Netherlands, Norway, Taiwan, Thailand, UK, and USA, we selected 42 full papers and 26 short papers for publication. The acceptance rate for regular full papers is 22%. The contributed papers addressed a wide scope of issues in the fields of Web-age information management and advanced applications, including Web data mining, knowledge discovery from streaming data, query processing, multidimensional data analysis, data management support to advanced applications, etc.

A conference like this can only succeed as a team effort. We want to thank the Program Committee members and the reviewers for their invaluable efforts. Special thanks to the local Organizing Committee headed by Jiwen Yang and Liusheng Huang. Many thanks also go to the Workshop Co-chairs (Lei Chen and Chengfei Liu), Tutorial Co-chairs (Liu Wenyin and Vivek Gopalkrishnan), Publicity Co-chairs (Lei Yu and Baihua Zheng), and Finance Co-chairs (Howard Leung and Genrong Wang). Last but not least, we are grateful for the hard work and effort of our webmaster (Zhu Fei, Tony), and to the generous sponsors who enabled the smooth running of the conference.

We hope that you enjoy the proceedings of APWeb+WAIM 2009.

April 2009

Qing Li
Ling Feng
Jian Pei
Sean X. Wang
Xiaofang Zhou
Qiaoming Zhu

Organization

APWeb+WAIM 2009 was organized by Soochow University, China.

Organizing Committee

Conference Co-chairs

Sean X. Wang	University of Vermont, USA
Xiaofang Zhou	University of Queensland, Australia
Qiaoming Zhu	Soochow University, China

Program Committee Co-chairs

Qing Li	City University of Hong Kong, China
Ling Feng	Tsinghua University, China
Jian Pei	Simon Fraser University, Canada

Local Organization Co-chairs

Jiwen Yang	Soochow University, China
Liusheng Huang	University of Science and Technology, China

Workshop Co-chairs

Lei Chen	Hong Kong University of Science and Technology, China
Chengfei Liu	Swinburne University of Technology, Australia

Tutorial/Panel Co-chairs

Wenyin Liu	City University of Hong Kong, China
Vivek Gopalkrishnan	Nanyang Technological University, Singapore

Industrial Chair

Hui Xue	Soochow University, China

Publicity Co-chairs

Lei Yu	State University of New York at Binghamton, USA
Baihua Zheng	Singapore Management University, Singapore

Finance Co-chairs

Howard Leung   City University of Hong Kong, China
Genrong Wang   Soochow University, China

CCF DB Society Liaison

Xiaofeng Meng   Renmin University of China, China

WISE Society Liaison

Yanchun Zhang   Victoria University, Australia

Webmaster

Fei Zhu   Soochow University, China

Program Committee

Aijun An   York University, Canada
James Bailey   University of Melbourne, Australia
Ladjel Bellatreche   ENSMA - Poitiers University, France
Rafae Bhatti   IBM Almaden Research Center, USA
Sourav Bhowmick   Nanyang Technological University, Singapore
Haiyun Bian   Wright State University, USA
Klemens Boehm   University of Karlsruhe, Germany
Athman Bouguettaya   Virginia Polytechnic Institute and State
                              University, USA
Stephane Bressan   National University of Singapore, Singapore
Ji-Won Byun   ORACLE, USA
Jinli Cao   La Trobe University, Australia
Badrish Chandramouli   Duke University, USA
Akmal Chaudhri   IBM DeveloperWorks, USA
Enhong Chen   University of Science and Technology of China,
                              China
Jian Chen   South China University of Technology, China
Lei Chen   Hong Kong University of Science and
                              Technology, China
Qiming Chen   Hewlett-Packard Laboratories, USA
Yi Chen   Arizona State University, USA
Hong Cheng   University of Illinois at Urbana-Champaign,
                              USA
Reynold Cheng   Hong Kong Polytechnic University, China
David Cheung   The University of Hong Kong, China
Dickson Chiu   Dickson Computer Systems, Hong Kong, China
Byron Choi   Hong Kong Baptist University, China
Gao Cong   Microsoft Research Asia, China

Bin Cui	Peking University, China
Alfredo Cuzzocrea	University of Calabria, Italy
Guozhu Dong	Wright State University, USA
Wenliang Du	Syracuse University, USA
Xiaoyong Du	Renmin University of China, China
Jianhua Feng	Tsinghua University, China
Yaokai Feng	Kyushu University, Japan
Eduardo Fernandez	Florida Atlantic University, USA
Ada Fu	The Chinese University of Hong Kong, China
Benjamin Fung	Concordia University, Canada
Gabriel Fung	The University of Queensland, Australia
Byron Gao	University of Wisconsin,USA
Hong Gao	Harbin Institute of Technology, China
Jun Gao	Peking University, China
Yunjun Gao	Singapore Management University, Singapore
Zhiguo Gong	University of Macau, China
Anandha Gopalan	Imperial College, UK
Vivekanand Gopalkrishnan	Nanyang Technological University, Singapore
Guido Governatori	The University of Queensland, Australia
Madhusudhan Govindaraju	State University of New York at Binghamton, USA
Stephane Grumbach	INRIA, France
Ning Gu	Fudan University, China
Giovanna Guerrini	Università di Genova, Italy
Jingfeng Guo	Yanshan University, China
Mohand-Said Hacid	Université Claude Bernard Lyon 1, France
Weihong Han	National Laboratory for Parallel and Distributed Processing, China
Michael Houle	National Institute of Informatics, Japan
Ming Hua	Simon Fraser University, Canada
Xiangji Huang	York University, Canada
Yan Huang	University of North Texas, USA
Ela Hunt	University of Strathclyde, UK
Renato Iannella	National ICT Australia (NICTA), Australia
Yoshiharu Ishikawa	Nagoya University, Japan
Yan Jia	National University of Defence Technology, China
Daxin Jiang	Microsoft Research Asia, China
Ruoming Jin	Kent State University, USA
Panagiotis Kalnis	National University of Singapore, Singapore
Murat Kantarcioglu	UT Dallas, USA
Wee Keong	Nanyang Technological University, Singapore
Markus Kirchberg	Institute for Infocomm Research, A*STAR, Singapore
Hiroyuki Kitagawa	University of Tsukuba, Japan

Flip Korn	AT&T Labs Research, USA
Manolis Koubarakis	National and Kapodistrian University of Athens, Greece
Anne LAURENT	University of Montpellier 2, France
Chiang Lee	National Cheng-Kung University
Dik Lee	Hong Kong University of Science and Technology, China
Wang-Chien Lee	Pennsylvania State University,USA
Yoon-Joon Lee	Korea Advanced Institute of Science and Technology, Korea
Carson Leung	University of Manitoba, Canada
Chen Li	University of California, Irvine, USA
Jinyan Li	Nanyang Technological University, Singapore
Jiuyong Li	University of South Australia, Australia
Quanzhong Li	IBM, USA
Tao Li	Florida International University, USA
Xue Li	The University of Queensland, Australia
Zhanhuai Li	Northeastern Polytechnical University, China
Wenxin Liang	Japan Science and Technology Agency (JST) and CSIC of Tokyo Institute of Technology, Japan
Ee Peng Lim	Nanyang Technological University, Singapore
Chengfei Liu	Swinburne University of Technology, Australia
Hongyan Liu	Tsinghua University, China
Mengchi Liu	Carleton University, Canada
Qing Liu	CSIRO, Australia
Tieyan Liu	Microsoft Research Asia, China
Wei Liu	University of Western Australia, Australia
Weiyi Liu	Yunnan University, China
Jiaheng Lu	University of California, Irvine, USA
Liping Ma	Ballarat University, Australia
Xiaosong Ma	NC State University, USA
Lorenzo Martino	Purdue University, USA
Weiyi Meng	Binghamton University, USA
Xiaofeng Meng	Renmin University of China, China
Miyuki Nakano	University of Tokyo, Japan
Wilfred Ng	Hong Kong University of Science and Technology, China
Anne Ngu	Texas State University, USA
Junfeng Pan	Google, USA
Chaoyi Pang	CSIRO, Australia
Zhiyong Peng	Wuhan University, China
Evaggelia Pitoura	University of Ioannina, Greece
Marc Plantevit	University of Montpellier 2, France
Tieyun Qian	Wuhan University, China
Weining Qian	East China Normal University, China

Wenyu Qu	Tokyo University, Japan
Vijay Raghavan	University of Louisiana at Lafayette, USA
Cartic Ramakrishnan	Wright State University, USA
Keun Ryu	Chungbuk National University, Korea
Monica Scannapieco	ISTAT, Italy
Markus Schneider	University of Florida, USA
Derong Shen	Northeastern University, China
Dou Shen	Microsoft, USA
Heng Tao Shen	University of Queensland, Australia
Jialie Shen	Singapore Management University, Singapore
Timothy Shih	Tamkang University, Taiwan
Youngin Shin	Microsoft, USA
Adam Silberstein	Yahoo, USA
Peter Stanchev	Kettering University, USA
Xiaoping Sun	Chinese Academy of Science, China
Changjie Tang	Sichuan University, China
Jie Tang	Tsinghua University, China
Nan Tang	CWI, The Netherlands
Zhaohui Tang	Microsoft, USA
David Taniar	Monash University, Australia
Yufei Tao	Chinese University of Hong Kong, China
Maguelonne Teisseire	University of Montpellier 2, France
Yunhai Tong	Peking University, China
Yicheng Tu	University of South Florida, USA
Daling Wang	Northeastern University, China
Guoren Wang	Northeastern University, China
Haixun Wang	IBM T. J. Watson Research Center, USA
Hua Wang	University of Southern Queensland, Australia
Jianyong Wang	Tsinghua University, China
Junhu Wang	Griffith University, Australia
Min Wang	IBM T. J. Watson Research Center, USA
Shan Wang	Renmin University of China, China
Tengjiao Wang	Peking University, China
Wei Wang	Fudan University, China
Wei Wang	University of New South Wales, Australia
X. Sean Wang	University of Vermont, USA
Jirong Wen	Microsoft Research Asia, China
Raymond Wong	University of New South Wales, Australia
Raymond Chi-Wing Wong	HKUST, China
Fei Wu	Zhejiang University, China
Weili Wu	The University of Texas at Dallas, USA
Wensheng Wu	IBM, USA
Xintao Wu	University of North Carolina at Charlotte, USA
Yuqing Wu	Indiana University, USA
Zonghuan Wu	University of Louisiana at Lafayette, USA

Jitian Xiao	Edith Cowan University, Australia
Xiaokui Xiao	Chinese University of Hong Kong, China
Hui Xiong	Rutgers University, USA
Wei Xiong	National University of Defense Technology, China
Guandong Xu	Victoria University, Australia
Jianliang Xu	Hong Kong Baptist University, Hong Kong, China
Jun Yan	University of Wollongong, Australia
Xifeng Yan	IBM, China
Chunsheng Yang	NRC, Canada
Jian Yang	Macquarie University, Australia
Shuqiang Yang	National Laboratory for Parallel and Distributed, China
Xiaochun Yang	Northeastern University, China
Jieping Ye	Arizona State University, USA
Laurent Yeh	Prism, France
Ke Yi	Hong Kong University of Technology, China
Jian Yin	Sun Yat-Sen University, China
Cui Yu	Monmouth University, USA
Ge Yu	Northeastern University, China
Jeffrey Yu	Chinese University of Hong Kong, China
Lei Yu	State University of New York at Binghamton, USA
Philip Yu	University of Illinois at Chicago, USA
Xiaohui Yu	York University, Canada
Lihua Yue	Science and Technology University of China, China
Chengqi Zhang	University of Technology, Sydney, Australia
Donghui Zhang	Northeastern University, USA
Qing Zhang	CSIRO, Australia
Rui Zhang	University of Melbourne, Australia
Shichao Zhang	University of Technology, Sydney, Australia
Xiuzhen Zhang	RMIT University, Australia
Yanchun Zhang	Victoria University, Australia
Ying Zhang	University of New South Wales, Australia
Hongkun Zhao	Bloomberg, USA
Yanchang Zhao	University of Technology, Sydney, Australia
Sheng Zhong	State University of New York at Buffalo, USA
Aoying Zhou	East China Normal University, China
Lizhu Zhou	Tsinghua University, China
Shuigeng Zhou	Fudan University, China
Xuan Zhou	CSIRO ICT Centre, Australia
Yongluan Zhou	EPFL, Sweden
Qiang Zhu	University of Michigan, USA
Xingquan Zhu	Florida Atlantic University, USA

External Reviewers

Watve Alok
Mafruzzaman Ashrafi
Markus Bestehorn
Thorben Burghardt
Mustafa Canim
Cai Chen
Rui Chen
Shu Chen
Xiaoxi Du
Frank Eichinger
Shi Feng
Li zhen Fu
Qingsong Guo
Yi Guo
Christopher Healey
Raymond Heatherly
Ali Inan
Fangjiao Jiang
Fangjiao Jiang
Hideyuki Kawashima
Dashiell Kolbe
Lily Li
Huajing Li

Guoliang Li
Yukun Li
Zhisheng Li
Bing-rong Lin
Xingjie Liu
Xumin Liu
Yang Liu
Yintian Liu
Chao Luo
Zaki Malik
Massimo Mecella
Chibuike Muoh
Chibuike Muoh
Khanh Nguyen
Robert Nix
Bahadorreza Ofoghi
Yoann Pitarch
Yoann Pitarch
Gang Qian
Chedi Raissi
Mathieu Roche
Paola Salle
Guido Sautter

Heiko Schepperle
Mingsheng Shang
Damon
    Sotoudeh-Hosseini
Xiaoxun Sun
MingJian Tang
Fadi Tashtoush
Fadi Tashtoush
Yuan Tian
Hiroyuki Toda
Puwei Wang
Wei Wang
Yousuke Watanabe
Yousuke Watanabe
Qi Yu
Ting Yu
Kun Yue
Chengyang Zhang
Xiaohui Zhao
Da Zhou
Jing Zhou

Table of Contents

Topic-Based Techniques

Web Data Processing

Multidimensional Data Analysis

Stream Data Processing

Data Mining and Its Applications

Data Management Support to Advanced Applications

Miscellaneous

Short Papers

Distributed XML Processing

M. Tamer Özsu

University of Waterloo
Cheriton School of Computer Science
tozsu@cs.uwaterloo.ca

Abstract. XML is commonly used to store data and to exchange it between a variety of systems. While centralized querying of XML data is increasingly well understood, the same is not true in a scenario where the data is spread across multiple nodes in a distributed system. Since the size of XML data collections are increasing along with the heavy workloads that need to be evaluated on top of these collections, scaling a centralized solution is becoming increasingly difficult. A common method for addressing this issue is to distribute the data and parallelize query execution. This is well understood in relational databases, but the issues are more complicated in the case of XML data due to the complexity of the data representation and the flexibility of the schema definition. In this talk, I will introduce our new project to systematically study distributed XML processing issues. The talk will focus on data fragmentation and localization issues.

This is joint work with Patrick Kling.

Q. Li et al. (Eds.): APWeb/WAIM 2009, LNCS 5446, p. 1, 2009.

Towards Multi-modal Extraction and Summarization of Conversations

Raymond Ng

University of British Columbia, Canada
rng@cs.ubc.ca
www.cs.ubc.ca/~rng

Abstract. For many business intelligence applications, decision making depends critically on the information contained in all forms of "informal" text documents, such as emails, meeting summaries, attachments and web documents. For example, in a meeting, the topic of developing a new product was first raised. In subsequent follow-up emails, additional comments and discussions were added, which included links to web documents describing similar products in the market and user reviews on those products. A concise summary of this "conversation" is obviously valuable. However, existing technologies are inadequate in at least two fundamental ways. First, extracting "conversations" embedded in multi-genre documents is very challenging. Second, applying existing multi-document summarization techniques, where were designed mainly for formal documents, have proved to be highly ineffective when applied to informal documents like emails. In this talk, we first review some of the earlier works done on extracting email conversations. We also give an overview of email summarization and meeting summarization methods. We then present several open problems that need to be solved for multi-modal extraction and summarization of conversations to become a reality. Last but not least, we specifically focus on extraction and summarization of sentiments in emails and blogs.

Q. Li et al. (Eds.): APWeb/WAIM 2009, LNCS 5446, p. 2, 2009.
© Springer-Verlag Berlin Heidelberg 2009

Simple but Effective Porn Query Recognition by k-NN with Semantic Similarity Measure

Shunkai Fu[1], Michel C. Desmarais[2], Bingfeng Pi[1], Ying Zhou[1],
Weilei Wang[1], Gang Zou[1], Song Han[1], and Xunrong Rao[1]

[1] Roboo Inc.
[2] Ecole Polytechnique de Montreal

Abstract. Access to sexual information has to be given some restricts on commercial search engine. Compared with filtering porn contents directly, we prefer to recognize porn queries and recommend appropriate ones considering several potential advantages. However, how to recognize them in an automatic way is not a trivial job due that its short length, in most scenarios, doesn't allow enough information for machine to make correct decision. In this paper, a simple but effective solution is proposed to recognize porn queries as exist in very large query log. Instead of checking purely if there are sensitive words contained in the queries, which may work for some cases but has obvious limitations, we go a little further by collecting and studying the semantic content of queries. Our experiments with real data demonstrate that small cost in training a k-Nearest Neighbor classifier (k-NN) will bring us quite impressive classification performance, especially the recall.

Keywords: Porn-query recognition, semantic similarity, k-Nearest Neighbor, mobile search engine.

1 Introduction

Sexuality is part of human everyday life behaviors and human information behaviors. Human Internet sexuality is a growing area of research in the social science. In cyber space, the seeking of sexual information is known as an important human element of everyday life information seeking, as shown by the comprehensive study about the nature and characteristics of human information searching for sexual information [1,2,3]. Being a leading mobile search service provider in China, we agree with this point with high confidence based on our years of observation and experience.

However, this doesn't mean that sexual information should be accessible by all without any restrict. At least, we should screen the non-adult users, leaving them a cleaner cyber environment. To do so, there are two possible approaches:

1. Control the information source, i.e. sites found on Internet. It is a thorough solution, and it is believed the responsibility of some official regulation agencies;
2. Filter out the non-appropriate retrieved results given a query. This is what we can do as a search service provider, and we believe indeed something should be done as a responsible company.

Q. Li et al. (Eds.): APWeb/WAIM 2009, LNCS 5446, pp. 3–14, 2009.

Even with the second solution, there are two options at the search engine side:

1. Given a query, do the seeking as usual, but filter out any porn results contained in the retrieved set;
2. Recognize a submitted query as porn, rejecting the search request but recommending some non-porn choices.

Both solutions mentioned above may have to work with users' ID certification/recognition if better effect is expected; otherwise, we may filter out any sexual information, even for adult users. For mobile search in China, this is possible since most phone numbers are registered with formal certification, but this is not what we are going to discuss here.

On our search engine, we take the second option, i.e. suggest some non-porn choices if a query is recognized as porn type, based on the following consideration:

- Firstly, recognizing a porn query, if it is possible, will be more economic solution because no computing resource is spent to retrieve the related results, especially when the target repository is large;
- Secondly, based on the current relatedness measure model as applied in search engines today, if we choose to filter out sexual items, very likely most or even all results would be removed. Then, what remained in the seeking result may be empty or some pretty marginal results that users are not interested at all, which means poor usage experience though effort is spent to do the screening;
- Thirdly, blocking out porn results is based on the fact that we can classify each result correctly. This is possible for text page (or document) seeking since there are mature methods, e.g. SVM and Naïve Bayes, available for document classification [4], where enough terms are required for training a reliable model. However, till today, there is no known published effective approach for the recognition of porn image and video, but the seeking of these two types of resource contributes a large portion of visiting flow as observed from our query log. Even there are some limited labels affiliated with each image or video resource, which is the most common method to make them searchable before content-based seeking [5,6] becomes mature, these information can be subjective and prone to errors (as annotated in manual or automatic manner [7]). Besides, even all these labels are appropriate, they are still not enough to build a reliable classification model;
- Finally, recognizing the query as porn and recommending explicitly some non-porn one(s) is useful as well as meaningful. (a) It allows to response users with this detection and to explain why their search is prohibited. This information will prevent the possible fruitless repeating trials if users don't know what happened behind. (b) This applies to not only document seeking, but image, video, etc, because it is query-, instead of content-dependent. (c) Recommending a different one, no matter chosen automatically by algorithms or manually by editors, relieves the disappointed feeling of users if the recommendation is interesting or attractive as well, which is quite important for mobile search where any input on mobile equipment consumes much more time than on PC[8]. (d) Directing a minor searcher to some pre-defined healthy or positive content may be a helpful action for our community.

Even though we know the potential advantage of recognizing a porn query, how to classify a query into porn or non-porn category is not a trivial task. The study of [8] indicates that the average query (in English) on Google mobile search engine is 15 letters long, about two words, and this is in accordance with our own study [13]. Given such short text, it is difficult to judge if a query is porn or not. This problem becomes more severe on Chinese search engine as we are engaging with. For example, given the same word, "同志" (comrade), it carries very different meaning and different target when appear in different context: "解放军同志" refers to the Chinese army, but "同志图片", based on our operation experience, is about gay images in most cases. Therefore, we can't simply determine if a query is porn or not via simply checking if it contains someone specific word. A more challenging problem is that even though a query is all composed of non-porn or non-sensitive words, it still will result with lots of porn content, e.g. "狼盟" (wolf alliance) and"武滕兰"(one Japanese pornographic star). Manual checking of such queries is possible, but the cost will be terribly expensive considering the huge number of queries being issued every day.

The primary contribution of this paper is four-folder:

1. We point out the advantages of recognizing porn queries and recommending other appropriate ones on mobile search engine;
2. We propose to measure how related of two queries based on their semantic background information, and it is shown a reliable relatedness measure even when two queries share no terms in common but, in fact, are closely correlated;
3. Given the similarity measure proposed, k-Nearest Neighbor (k-NN) is used to vote if a query is porn or not;
4. Empirical studies with real web data and query log demonstrates the effectiveness and efficiency of this solution;
5. Some traditionally non-porn queries are categorized as porn by this method, which reflects the exact status of current usage and the promising aspect of this method.

In section 2, semantic similarity measure for short queries is introduced. Then, in section 3, how to classify a query by k-NN based on proposed similarity measure is explained. Empirical studies on the effectiveness and efficiency are covered in section 4. We end with brief conclusion and potential future work in section 5.

2   Similarity Measure with Semantic Knowledge Space

2.1   Motivation and Overall Introduction

How to measure the similarity between queries in a robust way is the basis of our solution. Given two porn queries, our intuition tells us that they should be measured as related even though no single term is shared between them, e.g. "同志图片" (gay image) and "色情小说" (porn novel). However, the traditional similarity measures, such as the vector support model (VSM) with cosine coefficient as the basic distance measure [10], fail to achieve this goal, yielding a similarity of 0. Even in cases where two queries may share terms, e.g. "同志图片" (gay image) and "解放军同志" (comrade army), they, in fact, refers to completely different meaning although their cosine coefficient is 0.5 due to the shared lexical term "同志". So, whether sharing term(s) or not is not a trustable indication of relatedness between queries.

To address this problem, we need a more reasonable as well as reliable measure about the closeness among queries, referring more comprehensive information about the queried targets, instead of simple term-wise happenings. Fortunately, the web provides a potential rich source of data that may be utilized to address this problem, and modern search engine helps us to leverage the large volume of pages on the web to determine the greater context for a query in an efficient and effective manner.

Our solution to this problem is quite simple, and it is based on the intuition that queries which are similar in concept are expected to have similar context. For each query, we depend on the search engine to seek for a group of related and ranked documents, from which a feature vector is abstracted. Such feature vector contains those words that tend to co-occur with the original query, so it can be viewed as the corresponding context of the query, providing us more semantic background information. Then, the comparison of queries becomes the comparison of their corresponding feature, or context, vectors. Actually, we found that there was similar application of this method to determine the similarity between more objects on the Web [9].

2.2 Formalizing the Approach

Let q represent a query, and we get its feature vector, denoted as $FV(q)$, with the following sequential steps:

1. Issue the query q to a search engine;

2. Let $D(q)$ be the set of retrieved documents. In practice, we only keep the top K documents, assuming that they contain enough information, so $D(q) = \{d_1, d_2, ..., d_K\}$;

3. Compute the TF-IDF term vector tv_i for each document $d_i \in D(q)$, with each element as

$$tv_i(j) = tf_{i,j} \times \log(\frac{N}{df_j})$$

where $tf_{i,j}$ is the frequency of the jth term in d_i, N is the total number of documents available behind the search engine, and df_j is the total number of documents that contain the jth term. This weighting scheme is chosen considering its wide application in information retrieval (IR) community and has been proved work well in many applications, including ours here;

4. Sum up tv_i, $i = 1..K$, to get a single vector. Here, for the same term, its IDF, i.e. $\log(\frac{N}{df})$, is known as identical in different tv_i, so their corresponding weights are addable;

5. Normalize and rank the features of the vector with respect to their weight (TF-IDF), and truncate the vector to its M highest weighted terms. What left is the target $FV(q)$;

After we get the feature vector for each query of interest, the measure of semantic similarity between two queries is defined as:

$$Sim(q_i, q_j) = FV(q_i) \bullet FV(q_j)$$

Our tests indicate that this is truly a reliable semantic similarity measure. For example, "儒豹" (Roboo Inc., the company providing mobile search service in China) and "手机搜索" (mobile search) are determined closely related, but they are known as obviously not related if look at the terms composed only. Reversely, "解放军同志" and "同志图片"are judged as non-related since they indeed refer to completely different concepts, though they share one same term, "同志".

3　Porn Query Classification by k-NN with Semantic Similarity Measure

3.1　Why k-NN?

Being a classical method, the k-Nearest Neighbor (k-NN) algorithm classifies objects based on closest training examples in the feature space. It is kind of instance-based learning where the function is only approximated locally and all computation is deferred until classification. [11,12].

Like any other classification approach, the application of k-NN is composed of two phases: training and classification. The training phase consists only of storing the feature vectors and class labels of the training samples. In the actual classification phase, the new sample (whose class is not known) is represented as a vector in the feature space. Distances from the new vector to all stored vectors are computed and k closest samples are selected. The most common class among the k-selected neighbors is assigned to this new instance, simulating the democratic voting.

Given the binary classification problem, porn or non-porn query, we choose k-NN, instead of other classifiers, based on the following points:

1. It is simple, with respect to both training and classification, which ensures a quick response, no matter offline or online application;
2. The proposed semantic distance measure completely complies the computation of k-NN; of course, more distance measures are supported by k-NN;
3. Few training samples are enough to get a satisfactory classification model based on our empirical studies. This reduces the cost to do time-consuming manual labeling job as required by the training phase of all classification models;
4. It is easy to update the trained model by just re-loading their feature vectors with up-to-date content as retrieved by search engine. This will ensure a stable classification performance although the Internet is known with constant changes, and the possible cost is affordable considering the small amount of training samples;

5.   More classes can be added and become effective right now, and this just requires adding their corresponding feature vectors.

We believe the potential merits of k-NN are not limited to these, and more discussion is left to interesting readers.

In the following two sub-sections, both training and classification phases will be covered with more details.

3.2 Training Phase

Figure 1 demonstrates the training stage of this proposed approach in sequence:

1.   Same number of porn and non-porn queries is selected manually from the query log respectively, and each is attached with the corresponding label;
2.   Issue a search for each query selected in the previous step, assuming a Web repository is ready;
3.   Get the top K documents from the retrieved results, ignoring the remaining ones;
4.   Do the feature vector extraction as explained in Section 2.2, and store the results in a format of <Query, porn/non-porn, feature vector>, which will be referred in the classification phase.

Queries with highest frequencies, both porn and non-porn, are selected with priority considering that they reflect the most typical query interest of users. How many documents to be analyzed and how many features (terms) to be kept in the feature vector are arbitrary. In our system, the top $K = 100$ results are referred, and the top 60% terms extracted and ranked in term of their weight remain in the final feature vector.

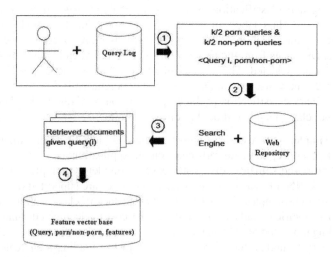

Fig. 1. Overview of the training phase: how the feature vectors are abstracted regarding to individual porn query

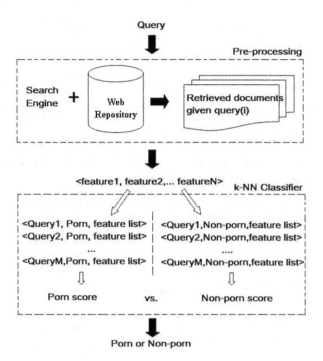

Fig. 2. The overall procedure of how to classify a query into porn/non-porn category

3.3 Classification Phase

There are a number of ways to classify the new vector to a particular class, and one of the most used techniques is to predict the new vector to the most common class amongst the k nearest neighbors. A major drawback to using this technique to classify a new vector to a class is that the more frequent examples tend to dominate the prediction of the new vector, as they tend to come up in the k nearest neighbors when the neighbors are computed due to their large number. One of the ways to overcome this problem is to take into account the distance of each k nearest neighbors with the new vector of the query being studied, and predict its class based on these summed relative distances.

How a query is classified into porn or non-porn category by k-NN with semantic measure is shown in Figure 2:

1. Given a query q, we depend on the search engine to retrieve a group of related documents, and the top K ones are considered as the semantic context;
2. A feature vector corresponding to this query is extracted from the K documents as shown in Section 2.2 and in the training phase;
3. The Euclidean distance between this new feature vector and all the stored feature vectors, as selected and stored in the training phase, is calculated independently;
4. Based on the known labels of training samples, the distance between this new vector and the porn/non-porn queries is summarized respectively:

$$PornScore = \sum_{q_i \in PornQueries} Sim(q, q_i) \text{, and}$$

$$NonPornScore = \sum_{q_i \in NonPornQueries} -Sim(q, q_i)$$

They are denoted as *PornScore* and *NonPornScore* for easy reference, and *NonPornScore* is negative in value;

5. The sum of *PornScore* and *NonPornScore* is the final score of a given query; if it is higher than a pre-defined threshold value, θ, it is known as porn query; otherwise, it is classified as non-porn.

So, we determine if a query is porn or not based on its relative distance with respect to the porn and non-porn feature sub-space (Figure 3).

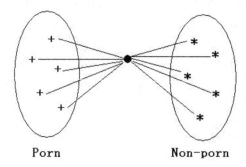

Porn   Non−porn

Fig. 3. Determine the class of a query based on its distance with respect to porn and non-porn feature sub-space

4 Empirical Study

4.1 Design of Experiments

The primary goal of our empirical study is on the effectiveness of the proposed solution. To do so, we crawled and indexed 19 millions of pages, using it as the Web repository.

To train the model, 200 queries were selected manually from the query log, including 100 most frequently appearing porn queries and 100 most frequent non-porn queries. Their corresponding feature vectors were abstracted and stored in advance, and the average length of the feature vector is 1,619 terms.

Then, a number of queries are selected randomly from the query log of page, image and video channels respectively, and they play as the test cases in this study.

4.2 Data Analysis

We select testing queries from three different repositories (channel), i.e. page, image and video, and the same trained k-NN model using data from page search log is used.

The corresponding summary information of selected testing queries is presented in Table 1, including the total amount, average length of query (in Chinese character), number of porn query, number of non-porn query and number of pseudo non-porn query. Here the so-called pseudo non-porn queries refer to those queries which are known as non-porn traditionally, but they are actually widely used in porn context currently, e.g. "美丽护士 (pretty nurse)".

Five persons are employed to label each query independently, and their opinion is summarized at the end. **The primary rule for such labeling is whether most results retrieved by this query were porn or not for a minor users**, and the decision is restricted to the mobile cyber environment of China (i.e. WAP sites only), which may be somewhat different from that on traditional (wired) Internet.

Table 1. Summary of queries for testing (Note: the length refers to the number of Chinese character)

Channel	Total Query	Average Length	Porn query	Non-porn Query	Pseudo-non-porn Query
Page	1989	4.23	1101	838	50
Image	1411	3.12	449	904	58
Video	870	3.27	407	407	56

In Table 1, it is noted that, (1) the average length of query on mobile search is quite short; and (2) high ratio of queries aims at porn content. These two facts exactly match our discussion in Section 1, reflecting both the challenge and significance of this project.

The threshold value of score, θ, to classify if a query is porn or not is set as 1.0, 3.0 and 2.0 for page, image and video channel respectively. Actually, this setting is arbitrary, and different value may result in different statistics. Our choices are based on a series of tests. The overall classification precision is shown with Table 2. To avoid any bias, we filter out the pseudo non-porn queries here to make the study easier.

$$\text{Overall Precision} = \frac{\text{\# of correctly classified queries}}{\text{total \# of queries being test}}$$

Table 2. Classification performance in terms of different search channel

Channel	Overall Precision
Page ($\theta = 1.0$)	94.87%
Image ($\theta = 3.0$)	93.64%
Video ($\theta = 2.0$)	92.75%

For each channel, we also collect the statistics of precision and recall for both porn and non-porn queries, as shown in Table 3. The definition of precision and recall are given below with porn query as example, but it applies to the calculation of non-porn query as well:

$$Precision = \frac{\#(true\ porn\ queries) \cap (queries\ classified\ as\ porn)}{total\ \#\ of\ queries\ classified\ as\ porn}$$

$$Recall = \frac{\#(of\ true\ porn\ queries) \cap (queries\ classified\ as\ porn)}{total\ \#\ of\ true\ porn\ queries}$$

Table 3. Precision and recall statistics about both porn and non-porn queries' classification performance on the channel of page, image and video

Channel	Query	Precision	Recall
Page	Porn	92.51%	98.73%
($\theta = 1.0$)	Non-porn	98.04%	89.50%
Image	Porn	88.87%	92.43%
($\theta = 3.0$)	Non-porn	96.16%	94.25%
Video	Porn	90.47%	95.58%
($\theta = 2.0$)	Non-porn	95.31%	89.93%

```
A new query comes;
If (it is classified before)
    Just retrieve its class label;
Else
    Do search and extract its feature vector;
    Calculate its distance relative to both 100 porn queries and 100 non-porn quries;
    If(the summary of distance to porn queries > that of non-porn queries)
        It is a porn queries;
    else
        It is a non-porn queries;
    Store it into recognized query repository
```

Fig. 4. Overall description of online classification

4.3 Discussion

From Table 2, and 3, the following conclusions can be made:

1. The overall classification accuracy is impressive for all three channels, which indicates that this semantic-based approach is not only trustable but truly content independent (same high performance for queries of image and video, as that of page search);
2. The k-NN classification model works pretty well with the semantic distance measure as the basis;
3. The recall performance about porn query recognition for all three channels is obviously better than the corresponding precision. It means that (1) most porn queries will be correctly classified, and (2) the primary error made is to category a non-porn query as porn, which, based on our experience, reflects the current characteristic of WAP sites, very young for this industry and preferring to use sensitive contents to attract users.

5 Conclusion

In this paper, a novel porn query recognition approach is proposed so that some necessary restrict can be employed on the seeking of sexual content via search engine. We discuss its potential merits, (1) reliable similarity measure based on semantic context; (2) content independent so that workable to non-text content, like image and video; (3) effective in recognizing a porn query, even it has no porn term at all apparently; (4) in-time reflection of the current initiative of issuing some query which, traditionally, is not porn, but widely to refer some porn object now. How it works is discussed with enough detail, and experiments with real Web and query data indicate that it is an effective, even though surprisingly simple, method.

Although there is no separate discussion about its online application, we still include a possible solution here:

If millions of queries are classified and stored in advance offline, the class of a submitted query normally can be found directly from the repository. When the hit is missing, the online waiting for the extraction of feature vector won't be long. Considering the result can be stored for later reference, without having to re-calculate each time, this procedure won't bring annoying effect to the users.

Several future work stay in our project list, including:

- If the selection of training samples would influence the performance;
- If the number of training samples has an influence on the performance;
- Is there any automatic way to choose the appropriate threshold value, θ, for each channel;
- Is there any other voting mechanism?

References

1. Spink, A., Ozmutlu, H.C., Lorence, D.P.: Web searching for sexual information: an exploratory study. Journal of Information Processing and Management 40, 113–123 (2004)
2. Cooper, A.: Sexuality and Internet: surfing into the new millennium. Journal of CyberPsychology and Behavior 1(2), 181–187 (1998)
3. Cooper, A. (ed.): Sex and the Internet: A Guidebook for Clinicians. Brunner-Routledge, New York (2002)
4. Yang, Y., Zhang, J., Kisiel, B.: A scalability analysis of classifiers in text categorization. In: Proceedings of 26th ACM SIGIR, pp. 96–103 (2003)
5. Lew, M., Sebe, N., Djeraba, C., Jain, R.: Content-based multimedia information retrieval: state of the art and challenges. ACM Transactions on Multimedia Computing, Communications and Applications, 1–19 (2006)
6. Datta, R., Joshi, D., Li, J., Wang, J.Z.: Image retrieval, ideas, influences and trends of the new age. ACM Computing Surveys 40(2) (2008)
7. Cramer, J.K., Hersh, W.: Medical image retrieval and automatic annotation: OHSU at ImageCLEF 2007. In: Working Notes for the Cross Language Evaluation Forum (CLEF) Workshop (2007)
8. Kamvar, M., Baluja, S.: Query suggestions for mobile search: understanding usage patterns. In: Proceedings of the SIGCHI conference on Human Factors in computing systems (CHI) (2008)

9. Sahami, M.: Mining the Web to determine similarity between words, objects, and communities. In: Proceedings of AAAI FLAIRS (2006)
10. Salton, G., McGill, M.J.: Introduction to Modern Information Retrieval. McGraw-Hill Book Company, New York (1983)
11. Shakhnarovich, G., Darrell, T., Indyk, P. (eds.): Nearest-Neighbor Methods in Learning and Vision: Theory and Practice. MIT Press, Cambridge (2006)
12. http://en.wikipedia.org/wiki/K-nearestneighbor_al-gorithm
13. Fu, S., Pi, B., Han, S., Zou, G., Guo, J., Wang, W.: User-centered solution to detect near-duplicate pages on mobile search engine. In: Proceedings of 31th ACM SIGIR Workshop on Mobile IR, Singapore (2008)

Selective-NRA Algorithms for Top-k Queries

Jing Yuan, Guang-Zhong Sun*, Ye Tian, Guoliang Chen, and Zhi Liu

MOE-MS Key Laboratory of Multimedia Computing and Communication,
Department of Computer Science and Technology,
University of Science and Technology of China, Hefei, 230027, P.R. China
yuanjing@mail.ustc.edu.cn, {gzsun,yetian,glchen}@ustc.edu.cn,
liuzhi10@mail.ustc.edu.cn

Abstract. Efficient processing of top-k queries has become a classical research area recently since it has lots of application fields. Fagin et al. proposed the "middleware cost" for a top-k query algorithm. In some databases there is no way to perform a random access, Fagin et al. proposed NRA (No Random Access) algorithm for this case. In this paper, we provided some key observations of NRA. Based on them, we proposed a new algorithm called Selective-NRA (SNRA) which is designed to minimize the useless access of a top-k query. However, we proved the SNRA is not instance optimal in Fagin's notion and we also proposed an instance optimal algorithm Hybrid-SNRA based on algorithm SNRA. We conducted extensive experiments on both synthetic and real-world data. The experiments showed SNRA (Hybrid-SNRA) has less access cost than NRA. For some instances, SNRA performed 50% fewer accesses than NRA .

1 Introduction

Assume there are a huge amount of objects and every object has m attributes, for each attribute the object has a score, these scores can be aggregated to a total score by an aggregate function, and we want to know which k objects have the largest total scores. This scenario is generalized as "top-k queries".

Top-k query has become a classical research area since it has many applications such as information retrieval[5],[7], multimedia databases[2],[9], data mining[6]. In top-k queries, we can do *random access* and *sorted access* to get an object's score. Under a "random access" we can get some object's score of a given attribute at one step; a "sorted access" means we proceed through an attribute list sequentially from the top of a list, i.e. if some object is the lth largest object in the ith list, we should do l sorted access to the ith list to obtain the object's score. As stated in [8], if there are s sorted accesses and r random accesses, the total access cost will be $sC_S + rC_R$. (C_S is the cost of a single sorted access and C_R is the cost of a single random access.) In some cases, random access is forbidden or restricted.[1] For example, a typical web search engine has no way to return a score of a document of our choice

---

* Corresponding author.

Q. Li et al. (Eds.): APWeb/WAIM 2009, LNCS 5446, pp. 15–26, 2009.
© Springer-Verlag Berlin Heidelberg 2009

under a query. On the assumption that random access is not supported by the database, in [1] the authors proposed the "No Random Access Algorithm" (NRA)[1].

In this paper we first demonstrate some observations of NRA algorithm and propose a "Selective-NRA" (SNRA) algorithm, which performs significantly better than NRA algorithm in terms of access cost and bookkeeping; secondly, we turn SNRA into an instance optimal algorithm which we call Hybrid-SNRA; thirdly, we carry out extensive experiments to compare SNRA (Hybrid-SNRA) algorithms with NRA both on synthetic and real-world data sets. The results also demonstrate that our algorithms have lower access cost than NRA algorithm.

The rest of this paper is organized as follows. In section 2, we define the problem and review related work. In section 3, we introduce SNRA algorithm and Hybrid-SNRA algorithm. In section 4, we show the experiment results. Finally in section 5, we conclude the paper.

2   Problem Definition and Related Work

Our model can be described as follows: assume there are m lists and n objects, the aggregation function is t which has m variables. For a given object R and for list i, R has a score x_i and $0 \leq x_i \leq 1$. R has a total score of $t(x_1, x_2, \ldots, x_m)$. We shall denote the lists as $L_1, L_2 \ldots, L_m$ which are sorted lists and can not be random accessed. We refer to L_i as list i. Each entry of L_i is (R, x_i) where x_i is the ith field of R. We assume there is an exact value for each object, so the length of each list is n. Since we consider only sorted access, the access cost of an algorithm will be $C_R \cdot r$ if r sorted accesses are performed.

There have been several algorithms which satisfy the assumption of no random access. The famous two algorithms are Güntzer et al.'s "Stream-Combine Algorithm"[2] and Fagin et al.'s "No Random Access Algorithm"[1]. As mentioned in [3], *Stream-Combine* algorithm considers only upper bounds on overall grades of objects, and cannot say that an object is in the top-k unless that object has been seen in every sorted list. In this sense, NRA is better than *Stream-Combine*. In [1], the authors proved that algorithm NRA is instance optimal with optimality ratio m, and no deterministic algorithm has a lower optimality ratio. (The definition of *instance optimal* and *optimality ratio* appeared in [1].) In [10], the authors studied NRA algorithm and proposed an algorithm which they call "LARA". In terms of run time cost, LARA is significantly faster than NRA algorithm. However, in terms of access cost, the advantage of LARA is sometimes marginally. Theobald et al. presented several probabilistic algorithms [4] which are also variants of the NRA algorithm .

The basic idea of NRA is to evaluate an object's exact value using upper bounds (*best value* $B^{(d)}(R)$) and lower bounds (*worst value* $W^{(d)}(R)$). The detail of algorithm NRA is given as follows. If not specified elsewhere, we will use the same notations (such as $W^d(R), T_k^{(d)}, M_k^{(d)}$) as NRA algorithm used.

Algorithm NRA (by Fagin et al.)[1]

- Do sorted access in parallel to $L_1, L_2, \ldots, L_m$. At each depth d (when d objects have been accessed under sorted access in each list) do:
  - Maintain the bottom values $(\underline{x}_1^{(d)}, \underline{x}_2^{(d)}, \ldots, \underline{x}_m^{(d)})$ encountered in each list.
  - For every object R with discovered fields $S = S^{(d)}(R) = \{i_1, i_2, \ldots, i_l\} \subseteq \{1, \ldots, m\}$ with values $x_{i1}, x_{i2}, \ldots, x_{il}$, compute $W^{(d)}(R) = W_S(R) = t(w_1, w_2, \ldots, w_m)$ where $w_i = \begin{cases} x_i & \text{if } i \in S \\ 0 & \text{else} \end{cases}$ and $B^{(d)}(R) = B_S(R) = t(b_1, b_2, \ldots, b_m)$ where $b_i = \begin{cases} x_i & \text{if } i \in S \\ \underline{x}_i^{(d)} & \text{else} \end{cases}$.
  - Let $T_k^{(d)}$, the current top k list, contain the k objects with the largest $W^{(d)}$ values seen so far (and their grades); if two objects have the same $W^{(d)}$ value, then ties are broken using the $B^{(d)}$ values, such that the object with the highest $B^{(d)}$ value wins (and arbitrarily among objects that tie for the highest $B^{(d)}$ value). Let $M_k^{(d)}$ be the kth largest $W^{(d)}$ value in $T_k^{(d)}$.
- Call an object R *viable* if $B^{(d)}(R) > M_k^{(d)}$. Halt when (a) at least k distinct objects have been seen (so that in particular $T_k^{(d)}$ contains k objects) and (b) there are no viable objects left outside $T_k^{(d)}$, that is, when $B^{(d)}(R) \leq M_k^{(d)}$ for all $R \notin T_k^{(d)}$. Return the objects in $T_k^{(d)}$.

3 Selective-NRA Algorithms

In this section we will propose our Selective-NRA algorithms. In the rest of this section we first give an equivalent form of algorithm NRA's stopping rule, and introduce some lemmas and observations which motivate us to propose algorithm SNRA; secondly we will show our algorithm SNRA and prove its correctness; finally we will propose an instance optimal algorithm based on algorithm SNRA.

3.1 Observations of NRA Algorithm

Definition 1. *Call an object R "best competitor" if R has the largest "best value" ($B^d(R)$) among all "viable"[1] objects which are not in the current top-k. If there is more than one best competitor, choose one which we have seen earliest (sorted accessed earliest) as the best competitor.*

The stopping rule(b) of NRA is :"there are no viable objects left outside $T_k^{(d)}$, that is, when $B^{(d)}(R) \leq M_k^{(d)}$ for all $R \notin T_k^{(d)}$" which is equivalent to

$$\text{The best competitor's best value} \leq M_k^{(d)}.$$

We now give two lemmas which motivate us to propose algorithm SNRA. In this paper, we suppose $m \geq 2$.

---

[1] The *viable* object and *best value* are well defined at the previous section.

Lemma 1. *If at depth d object R is the best competitor and algorithm NRA does not halt at depth d, then R has at least one missing (undiscovered) field.*

Proof. We assume that at depth d we have known all fields of R. Then $W^{(d)}(R) = B^{(d)}(R)$. Since R has the largest *best value*, the *best value* of all *viable* objects except objects in $T_k^{(d)}$ is not more than $B^{(d)}(R)$. Since algorithm NRA does not halt at depth d, there is some object $R' \in T_k^{(d)}$ such that $B^{(d)}(R) > W^{(d)}(R')$, then it follows that, R should be in $T_k^{(d)}$ because $B^{(d)}(R) = W^{(d)}(R) > W^{(d)}(R')$. It leads to a confliction. So the conclusion follows, as desired. □

Lemma 2. *If at depth d object R is the best competitor and algorithm NRA does not halt at depth d, $B^{(d)}(R)$ will decrease at depth $d + 1$ **only if** we sorted access R's missing field.*

Proof. If we sorted access ith list which is not R's missing field, R's *best value* will not decrease because we compute R's *best value* by substituting R's real value for the discovered field and *bottom value* for the missing field. So R's *best value* will decrease only if we sorted access R's missing field. (R's *best value* maybe won't decrease when the values at the next level are the same with this level. It is a necessary but not sufficient condition.) □

Another Observation: We note that the *best competitor* changes with the algorithm running. However, by doing experiments, we found that the last *best competitor* (which means the last one before algorithm NRA terminated) would occupy the position of *best competitor* for a long time. We speculate the results are most likely similar for common data sets. Table 1 gives an experiment result for a synthetic uniform data set. We set $n=100000$, $m=2,4,\ldots,12$, $k = 20$. The aggregation function is summation.

Suppose that the last *best competitor* has only one missing field when algorithm NRA is running. In this case, we still have to sorted access the *best competitor*'s all other known fields. This will incur lots of useless sorted accesses if the last *best competitor* continue for a long time before the top-k objects are obtained. These observations motivate us to propose Selective-NRA algorithm.

3.2   Selective-NRA Algorithm (SNRA)

We will show our algorithm in pseudo-code form.

As stated in section3.1, algorithm NRA performs some sorted accesses which will definitely not reduce the *best competitor*'s best value. Our approach can avoid these sorted accesses. To prove the correctness of SNRA algorithm, we

Table 1. Accessed depth for last *best competitor* vs. total accessed depth

m	2	4	6	8	10	12
depth of the last best competitor	315	11997	31008	41087	37126	34614
total depth of NRA	1987	23211	54029	80440	91492	97532

Algorithm SNRA-Initialize

1: bottom[j]:=1.0, $j = 1, 2, \ldots, m$
2: topk:=[dummy$_1$,dummy$_2$,...,dummy$_k$] dummy$_i$'s best:=worst:=0, missingfield:=$\emptyset$
3: **for** each R_i, $i = 1, 2, \ldots, n$ **do**
4:     R_i.best:=m
5:     R_i.worst:=0
6:     R_i.missingfield:=[1,2,...,m]
7: **end for**
8: candidates:=$\emptyset$
9: bestcompetitor:=dummy with missingfield=[1,2,...,m]
10: best:=m
11: min:=dummy$_1$ with mink:=0

Algorithm Selective-NRA

Call Algorithm SNRA-Initialize
while (mink < best) $\vee$ (less than k objects have been accessed) **do**
    for each $j \in$ bestcompetitor.missingfield **do**
        sorted access L_j obtain (p, x_j)
        if (p has not been seen before) **then**
            candidates:=candidates$\cup$\{p\}
        end if
        bottom[j]:=x_j //is used for updating an object's *best value*
        p.missingfield:=p.missingfield-\{j\}
        update p.worst
        mink:=min\{d.worst| d$\in$ topk\}// our "mink" is "$M_k^{(d)}$" in NRA
        min:=argmin$_d$\{d.worst | d$\in$ topk\}
        if (p.worst > mink) $\wedge$ (p$\in$ candidates) **then**
            candidates:=candidates-\{p\}
            candidates:=candidates$\cup$ \{min\}
            topk:=topk-\{min\}
            topk:=topk$\cup$ \{p\}
        end if
    end for
    for each R$\in$ (candidates $\cup$ topk) **do**
        update R.best
        if (R.best $\leq$ mink) **then**
            candidates:=candidates-\{R\}
        end if
    end for
    bestcompetitor:=argmax$_R$\{R.best | R$\in$ candidates\}
    best:=max$_R$\{R.best | R$\in$ candidates\}
end while

---

need to give a lemma first. This lemma indicates that SNRA algorithm will not lead to an "infinite loop".

Lemma 3. *Assume algorithm SNRA has sorted accessed d_j depth to L_j ($j = 1, 2, \ldots, m$), and the stopping rule has not been satisfied, algorithm SNRA will proceed to access the next object at least in one list.*

Proof. Let d be an array of $[d_1, d_2, \ldots, d_m]$, in the rest of the paper when we say **depth d**, it means depth d_j in L_j. We only need to prove that at depth d the *best competitor* has at least one missing field. Thus the remaining part of this lemma's proof is the same as Lemma 1's. □

Theorem 1. *If the aggregation function is monotone, then algorithm SNRA correctly finds the top k objects.*

Proof. According to Lemma 3, if algorithm SNRA doesn't halt, it will proceed to access the next level at least in one list until the stopping rule is satisfied. Assume that algorithm SNRA halts after d_j sorted access to L_j ($j = 1, 2, \ldots, m$) and the objects output by algorithm SNRA are $R_1, R_2, \ldots, R_k$. Let R be an object not among $R_1, R_2, \ldots, R_k$. We must show that $t(R) \le t(R_i)$ for each $i = 1, 2, \ldots, k$.

Let $d = [d_1, d_2, \ldots, d_m]$. Since algorithm SNRA halts at depth d, the *best competitor*'s best value is less than $M_k^{(d)}$, then, $B^{(d)}(R) \le B^{(d)}(best\ competitor) \le M_k^{(d)}$ and $t(R) \le B^{(d)}(R)$. Also for each of the k objects R_i we have $M_k^{(d)} \le W^{(d)}(R_i) \le t(R_i)$. Combining the inequalities we have shown, we have

$$t(R) \le B^{(d)}(R) \le M_k^{(d)} \le W^{(d)}(R_i) \le t(R_i)$$

for each i, as desired. □

Not only accesses fewer objects, our algorithm requires less bookkeeping than algorithm NRA. At step 2 of algorithm NRA, it will update all seen objects's worst values and best values. Our algorithm just updates an object's worst value when we sorted access it. This is reasonable because the other objects's worst values will not change if they are not accessed.

Now we give an example to show how our algorithm SNRA works.

Example 1. Assume m=3, n=5, $k = 1$, the aggregation function is summation, and the lists shown in Tab. 2 can only be sorted accessed. Table 3-6 show how our algorithm SNRA performs sorted accesses at each step on this database. At step(a) SNRA sorted accesses all the 3 lists. After that, object R_1 becomes the top 1 object and R_2 becomes the *best competitor*. Then at step(b) we only sorted access L_1 and L_3 since L_2 is R_2's missing field. Now R_1 is the *best competitor* and R_2 is the top 1 whose worst value is 1.8. At step(c) L_2 is sorted accessed. R_1 becomes the top 1 again and R_2 becomes the *best competitor*. At this time, R_2's missing field is L_3 so at step(d) we sorted access L_3. Now the *best competitor* is R_4 whose best value is not more than top 1's worst value. The algorithm terminates. Table 7 shows how the top 1 object and the *best competitor* update at each step.

On this database, algorithm NRA sorted accesses to depth 3 of each list and performs 9 sorted accesses in total while our SNRA's sorted access cost is 7.

Table 2. Sorted lists

$(R_1,0.9)$	$(R_2,0.9)$	$(R_1,0.6)$
$(R_2,0.9)$	$(R_1,0.8)$	$(R_4,0.6)$
$(R_3,0.6)$	$(R_3,0.7)$	$(R_3,0.4)$
$(R_4,0.5)$	$(R_4,0.6)$	$(R_2,0.3)$
$(R_5,0.2)$	$(R_5,0.2)$	$(R_5,0.3)$

Table 3. SNRA(a)

$(R_1,0.9)$	$(R_2,0.9)$	$(R_1,0.6)$
$(R_2,0.9)$	$(R_1,0.8)$	$(R_4,0.6)$
$(R_3,0.6)$	$(R_3,0.7)$	$(R_3,0.4)$
$(R_4,0.5)$	$(R_4,0.6)$	$(R_2,0.3)$
$(R_5,0.2)$	$(R_5,0.2)$	$(R_5,0.3)$

Table 4. SNRA(b)

$(\mathbf{R_1,0.9})$	$(\mathbf{R_2,0.9})$	$(\mathbf{R_1,0.6})$
$(\mathbf{R_2,0.9})$	$(R_1,0.8)$	$(\mathbf{R_4,0.6})$
$(R_3,0.6)$	$(R_3,0.7)$	$(R_3,0.4)$
$(R_4,0.5)$	$(R_4,0.6)$	$(R_2,0.3)$
$(R_5,0.2)$	$(R_5,0.2)$	$(R_5,0.3)$

Table 5. SNRA(c)

$(\mathbf{R_1,0.9})$	$(\mathbf{R_2,0.9})$	$(\mathbf{R_1,0.6})$
$(\mathbf{R_2,0.9})$	$(\mathbf{R_1,0.8})$	$(\mathbf{R_4,0.6})$
$(R_3,0.6)$	$(R_3,0.7)$	$(R_3,0.4)$
$(R_4,0.5)$	$(R_4,0.6)$	$(R_2,0.3)$
$(R_5,0.2)$	$(R_5,0.2)$	$(R_5,0.3)$

Table 6. SNRA(d)

$(\mathbf{R_1,0.9})$	$(\mathbf{R_2,0.9})$	$(\mathbf{R_1,0.6})$
$(\mathbf{R_2,0.9})$	$(\mathbf{R_1,0.8})$	$(\mathbf{R_4,0.6})$
$(R_3,0.6)$	$(R_3,0.7)$	$(\mathbf{R_3,0.4})$
$(R_4,0.5)$	$(R_4,0.6)$	$(R_2,0.3)$
$(R_5,0.2)$	$(R_5,0.2)$	$(R_5,0.3)$

Table 7. Each step of SNRA

top 1		best competitor	
object	worst	object	best
R_1	1.5	R_2	2.4
R_2	1.8	R_1	2.4
R_1	2.3	R_2	2.4
R_1	2.3	R_4	2.3

3.3   Turning SNRA into an Instance Optimal Algorithm

Fagin et al. defined "instance optimality" for a top-k query algorithm and proved instance optimality of algorithm NRA. Unfortunately our algorithm SNRA is not "instance optimal" in his notion. In this section we first show an example that demonstrates SNRA is not instance optimal and then we will modify algorithm SNRA and turn it to be an instance optimal algorithm.

Example 2. Assume $m=2$, $k=1$, the aggregation function is summation.

Table 8 shows a database over which algorithm NRA performs only 6 sorted accesses to depth 3, outputs R_1 as the top 1 object while algorithm SNRA sorted accesses the whole $L1$ list and totally does $n + 3$ sorted accesses. (After depth $[1, 1]$, R_2 is the *best competitor*, so we sorted access to depth $[2, 1]$, then R_2 is still the *best competitor*, then sorted access to depth $[3, 1]$, at this depth, R_3 is the *best competitor*, then sorted access to depth $[3, 2]$, R_2 becomes the *best competitor* since R_3's exact value is less than R_2's worst value, then R_2 will be the *best competitor* until the end of L_1 at depth $[n - 1, 2]$. After depth $[n, 2]$, R_2 becomes the top 1 and R_1 becomes the *best competitor*. Then sorted access L_2, at depth $[n, 3]$ R_1 becomes the top 1 and the algorithm terminates.) Since n could be arbitrary large, algorithm SNRA is not an instance optimal algorithm.   □

The reason SNRA is not instance optimal is that SNRA "selects" some of the lists instead of all the lists to sorted access. By doing this selection, SNRA may

Table 8. An example shows SNRA is not instance optimal

$(R_1,1.0)$	$(R_2,0.9)$
$(R_3,0.5)$	$(R_3,0.39)$
$(R_4,0.45)$	$(R_1,0.38)$
$(R_5,0.32)$	$(R_5,0.32)$
...	...
$(R_i,0.32\text{-}\frac{0.2}{n-i+1})$	$(R_i,0.32\text{-}\frac{0.2}{n-i+1})$
...	...
$(R_n,0.12)$	$(R_n,0.12)$
$(R_2,0.11)$	$(R_4,0.11)$

do fewer sorted access than NRA algorithm in most databases but may "miss" some important information in some particular databases like our example 2. Nevertheless, we can force SNRA to be an instance optimal algorithm by a little modifying of algorithm SNRA. We call the modified algorithm "Hybrid-SNRA", and we now show it as follows.

Algorithm Hybrid-SNRA

1: **Call** Algorithm Initialize
2: step:=0;
3: **while** (mink < best) ∨ (less than k objects have been accessed) **do**
4:     step++;
5:     **if** (step mod p =0) **then**
6:         field=[1,2,...,m]-{the fields which have been accessed to the bottom}
7:     **else**
8:         field=bestcompetitor.missingfield
9:     **end if**
10:    **for** each $j \in$ field **do**
11:        ...//the same as SNRA from line 4 to line 18
12:    **end for**
13:    ...//the same as SNRA from line 20 to line 27
14: **end while**

We note the sorted access cost of Hybrid-SNRA is at most p times as algorithm NRA where p is an constant. So algorithm Hybrid-SNRA is instance optimal. Since the optimal ratio of algorithm NRA is m, the optimal ratio of algorithm Hybrid-SNRA is pm under some natural assumption [1]. (In fact, if $p=1$, Hybrid-SNRA is equivalent to NRA and if $p=$INFINITY, Hybrid-SNRA becomes SNRA.)

4   Experiment Results

Our algorithms were implemented in C++. We performed our experiments on an AMD 1.9 GHz PC with 2GB of memory. In our experiments we used summation

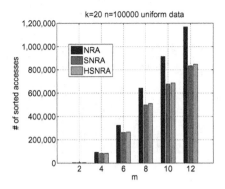
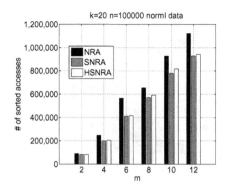

Fig. 1. Access cost over uniform database   **Fig. 2.** Access cost over normal database

as the aggregation function which was a most common one. For Hybrid-SNRA we set $p=11$ as default. We used both synthetic and real-world data to evaluate SNRA, Hybrid-SNRA and NRA algorithms.

4.1 Evaluation for Synthetic Data

We conducted experiments on three synthetic data sets with different distributions. They are uniform distributed, normal (Gaussian) distributed and exponential distributed. We set $n = 100000$, $m = 2, 4, \ldots, 12$, and $k = 20$.

Figure 1-3 show that on the uniform, normal and exponential distributed databases our algorithm SNRA as well as Hybrid-SNRA performs fewer sorted accesses than algorithm NRA, and algorithm SNRA performs the fewest sorted accesses among these three algorithms. When m is small, the difference is not so significant, but as m becomes larger, SNRA outperforms NRA more and more significantly in terms of sorted access cost.

4.2 Evaluation for Real-World Data

In addition of synthetic data, we carried out experiments on three different real-world data sets which were all downloaded from UCI KDD Archive[2].

- The first real-world data (CE) is IPUMS Census Database. This data set contains unweighted PUMS census data from the Los Angeles and Long Beach areas for the year 1990. It contains 88443 objects and we extracted 4 to 10 attributes from this data set. We normalized this data set with formula: $\frac{x_i - Min}{Max - Min}$ if an object's value is x_i.
- The next real-world data is KDD cup 1998 data set (cup98). It contains 95412 different objects. This is the data set used for The Second International Knowledge Discovery and Data Mining Tools Competition, which was held in conjunction with KDD-98 The Fourth International Conference on Knowledge Discovery and Data Mining. We extracted 10 attributes to perform our experiment. We normalized this data set with the same formula as CE data.

---

[2] http://kdd.ics.uci.edu

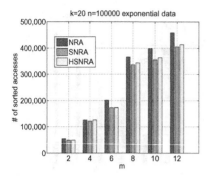
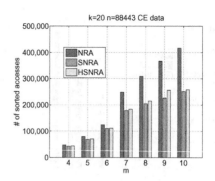

Fig. 3. Access cost over exponential database

Fig. 4. Access cost over IPUMS Census database

- The last data set "El Nino" data set contains oceanographic and surface meteorological readings taken from a series of buoys positioned throughout the equatorial Pacific. The data is expected to aid in the understanding and prediction of El Nino/Southern Oscillation (ENSO) cycles. We removed those objects which had some missing fields from the original data set. We normalized the data set with the same formula as CE data. The remaining data contains 93935 objects. We chose 7 attributes to test our algorithms.

Figure 4 shows the experiment result of SNRA, Hybrid-SNRA and NRA over IPUMS Census data set. We tested when $m = 4, 5, \ldots, 10$ algorithm SNRA, Hybrid-SNRA and NRA would do how many sorted accesses to the lists. The result demonstrates that our algorithms are more better in terms of access cost even though the degree of advantage is relevant to the specific database. In addition, this figure also shows that algorithm Hybrid-SNRA does a little more sorted accesses than algorithm SNRA, under our default set ($p=11$).

Figure 5 shows the experiment result of SNRA, Hybrid-SNRA and NRA over Cup98 data set. In this figure, it is also significant that our SNRA does fewer sorted accesses. Furthermore, as m goes larger, the advantage of SNRA is gradually obvious.

Figure 6 shows the experiment result of SNRA, Hybrid-SNRA and NRA over El Nino data set. On this data set our SNRA and Hybrid-SNRA also perform very well. Our algorithm saves 50% of NRA's sorted accesses, because of the "selecting" strategy.

4.3 Summarize

Our experiment illustrates that algorithm SNRA does fewer sorted accesses than algorithm NRA, both on synthesized data and real-world data. Algorithm Hybrid-SNRA does a little more sorted accesses than algorithm SNRA. As m becomes larger, the decrease of sorted accesses becomes more significant. (When k changes, we obtain similar results which are omitted due to space limitations.)

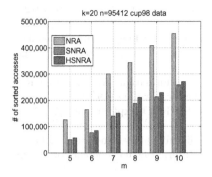

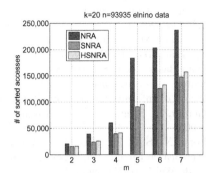

Fig. 5. Access cost over KDD Cup98 database

Fig. 6. Access cost over El Nino database

The reason why our algorithm does fewer sorted accesses is that it "selects" some lists and does useful sorted accesses instead of sorted accessing all the lists.

5   Conclusion and Future Work

In this paper, we analyzed algorithm NRA and gave some observations. We proposed a new algorithm which we called Selective-NRA and we turned it into an instance optimal algorithm Hybrid-SNRA. Extensive experiment results both on synthetic and real-world data show that our algorithms SNRA and Hybrid-SNRA perform significantly better than NRA in terms of sorted access cost.

Another interesting result according to our experiments is that algorithm NRA and algorithm Hybrid-SNRA are instance optimal but they perform fewer sorted accesses than a non-"instance optimal" algorithm on common data sets. This is an issue for our further research. In the future, we will also consider the run time cost of SNRA compared with algorithm NRA. We will design some techniques to lower down the run time cost of SNRA.

Acknowledgements. This work is supported by the National Science Foundation of China under the grant No. 60533020 and No. 60873210. This work is also supported by the Science Research Fund of MOE-Microsoft Key Laboratory of Multimedia Computing and Communication (Grant No. 06120801).

References

1. Fagin, R., Lotem, A., Naor, M.: Optimal Aggregation Algorithms for Middleware. In: Proceedings of the 20th ACM SIGACT-SIGMOD-SIGART Symposium on Principles of Database Systems, pp. 102–113 (2001)
2. Güntzer, U., Balke, W.T., Kie, W.: Towards Efficient Multi-Feature Queries in Heterogeneous Environments. In: Proceedings of the IEEE International Conference on Information Technology: Coding and Computing, pp. 622–628 (2001)

3. Fagin, R.: Combining Fuzzy Information: an Overview. SIGMOD Record 31(2), 109–118 (2002)
4. Theobald, M., Keikum, G., Schenkel, R.: Top-k Query Evaluation with Probabilistic Guarantees. In: Proceedings of the 30th International Conference on Very Large Data Bases, pp. 648–659 (2004)
5. Salton, G.: Automatic Text Processing: The Transformation, Analysis, and Retrieval of Information by Computer. Addison-Wesley, Reading (1989)
6. Getoor, L., Diehl, C.P.: Link Mining: a Survey. SIGKDD Explorations 7(2), 3–12 (2005)
7. Long, X., Suel, T.: Three-Level Caching for Efficient Query Processing in Large Web Search Engines. In: Proceedings of the 14th International Conference on World Wide Web, pp. 257–266 (2005)
8. Fagin, R.: Combining Fuzzy Information from Multiple Systems. J. Comput. Syst. Sci. 58(1), 83–99 (1999)
9. Nepal, S., Ramakrishna, M.V.: Query Processing Issues in Image (Multimedia) Databases. In: Proceedings of the 15th International Conference on Data Engineering, pp. 22–29 (1999)
10. Mamoulis, N., Yiu, M.H., Cheng, K.H., Cheung, D.W.: Efficient Top-k Aggregation of Ranked Inputs. ACM Transactions on Database Systems (TODS) 32(3), 19 (2007)

Continuous K-Nearest Neighbor Query over Moving Objects in Road Networks

Yuan-Ko Huang[1], Zhi-Wei Chen[1], and Chiang Lee[2]

Department of Computer Science and Information Engineering, National
Cheng-Kung University, Tainan, Taiwan, R.O.C.
[1]{hyk,howard1118}@dblab.csie.ncku.edu.tw,
[2]leec@mail.ncku.edu.tw

Abstract. Continuous K-Nearest Neighbor ($CKNN$) query is an important type of spatio-temporal queries. A $CKNN$ query is to find among all moving objects the K-nearest neighbors ($KNNs$) of a moving query object at each timestamp. In this paper, we focus on processing such a $CKNN$ query in road networks, where the criterion for determining the $KNNs$ is the shortest network distance between objects. We first highlight the limitations of the existing approaches, and then propose a cost-effective algorithm, namely the *Continuous KNN algorithm*, to overcome these limitations. Comprehensive experiments are conducted to demonstrate the efficiency of the proposed approach.

1 Introduction

With the advent of mobile and ubiquitous computing, processing continuous spatio-temporal queries over moving objects has become a necessity for many applications, such as traffic control systems, geographical information systems, and location-aware advertisement. One type of spatio-temporal queries is the *continuous K-nearest neighbor query* (or $CKNN$ query for short) that *finds among all moving objects the K-nearest neighbors ($KNNs$) of a moving query object at each timestamp.* An example of a $CKNN$ query is "finding two nearest cabs of a pedestrian based on his/her current moving speed and direction."

Early methods [1,2,3] proposed to efficiently process such a $CKNN$ query focus exclusively on Euclidean spaces, which are inapplicable to road networks. Recently, several studies [4,5,6,7] have investigated how to search the $KNNs$ in road networks. These techniques are designed to deal with $CKNN$ queries over static objects. Mouratidis et al. [8] first address the issue of continuous monitoring $KNNs$ on moving objects. They choose to re-evaluate snapshot KNN query at those time instants at which updates occur. However, due to the nature of discrete location updates, the $KNNs$ within two consecutive updates are unknown.

In this paper, our efforts on processing such a $CKNN$ query are devoted to overcoming the limitations of the previous works mentioned above. That is, we investigate the $CKNN$ problem under the following three conditions: (1) All objects (including the query object) move continuously in a road network. (2)

Q. Li et al. (Eds.): APWeb/WAIM 2009, LNCS 5446, pp. 27–38, 2009.

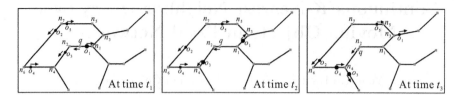

Fig. 1. Example of a $CKNN$ query

The distance between two objects is defined as the distance along the shortest path between them in the network. (3) The KNNs set of the query object at each timestamp should be completely determined. Let us consider an example in Figure 1, where a set of objects o_1 to o_5 and a query object q move in a road network which is represented as a graph consisting of nodes and edges. In this example, each object moves steadily towards the direction indicated by the corresponding arrow. Assume that the moving query object q wants to know within a certain period of time the nearest neighbor on its route (i.e., a C1NN query is issued). The query result will consist of tuples $< [t_1, t_2], \{o_1\} >$, $<[t_2, t_3], \{o_3\} >$, ..., where each tuple $< [t_a, t_b], \{o_i\} >$ represents that object o_i is the 1NN of q within the time interval $[t_a, t_b]$.

To efficiently process $CKNN$ query over moving objects in road networks, we first address the problem of how to significantly reduce the overhead of representing the network distance between moving objects at each timestamp. Although the network distance can be computed based on the Dijkstra's algorithm [9] or the A* algorithm [10] that have been shown to be simple and efficient for computing the network distance between stationary objects, recomputing the network distance whenever objects change locations would incur extremely high computational cost which makes the idea infeasible, especially for mobile environments in which objects move continuously. In order to greatly reduce the recomputation cost, we design a technique using the information about the relative speed of moving objects and the shortest path between network nodes. By exploiting this technique, the network distance between two objects at each timestamp is simply represented as a linear function of time, and thus can be easily computed. Another major problem we need to tackle is to avoid repetitively processing snapshot KNN queries at each timestamp. Let us consider again the example in Figure 1, where object o_3 replaces o_1 to be the 1NN at t_2 and is replaced by o_4 at t_3. We term these time points at which the KNNs set changes from one to another (e.g., t_2 and t_3) the KNNs-changing time points. An important characteristic of $CKNN$ query is that the KNNs in between two consecutive KNNs-changing time points remain the same. Based on this characteristic, the problem of performing repetitive queries can be greatly reduced to finding the KNNs-changing time points and their corresponding KNNs. In this paper, we design a $CKNN$ method combined with the proposed distance model to determine the KNNs-changing time points with corresponding KNNs.

The major contributions of this paper are summarized as follows.

- Our work remedies the major drawbacks of the past related work and provides a more practical and efficient solution for the CKNN problem.
- A technique is designed for representing the network distance between each moving object and the query object at each timestamp.
- We propose a main-memory algorithm, namely the *Continuous KNN (CKNN) algorithm*, to efficiently find the KNNs set of the moving query object at each timestamp.
- A comprehensive set of experiments is conducted. The performance results manifest the efficiency and the usefulness of the proposed CKNN approach.

The remainder of the paper is organized as follows. We review the related work for snapshot and continuous KNN queries in road networks in Section 2. Section 3 describes the data structures used for representing road networks. In Section 4, we present how to efficiently calculate the network distance as a preliminary to the proposed CKNN algorithm. The CKNN algorithm is presented in Section 5. Section 6 shows extensive experiments on the performance of our method. In Section 7, we conclude the paper with directions on future work.

2   Related Work

Processing KNN queries in road networks is an emerging research topic in recent years. In this section, we first overview the existing methods for processing snapshot KNN queries, and then discuss the related work that investigates how to answer continuous KNN queries.

2.1   Methods for Snapshot KNN Queries

Numerous algorithms have been proposed for efficient processing of snapshot KNN queries over static objects in road networks. Jensen *et al.* [5] propose a data model and a definition of abstract functionality required for KNN queries in spatial network databases. To compute network distances, they adopt the Dijkstra's algorithm for online evaluation of the shortest path. Papadias *et al.* [7] describe a framework that integrates network and Euclidean information, and answers snapshot KNN queries. They propose *incremental Euclidean restriction* (IER) to first process a query in the Euclidean space so as to obtain the results as the candidates, and then compute network distances of these candidates for the actual results. In addition, they present an alternative approach *incremental network expansion* (INE), which is based on generating a search region that expands from the query point. Kolahdouzan *et al.* [11] propose a solution-based approach for KNN queries in spatial networks. Their approach, called VN^3, precalculates the network Voronoi polygons (NVPs) and some network distances, and then processes the KNN queries based on the properties of the Network Voronoi diagrams. In summary, the above methods proposed for processing snapshot KNN queries in road networks are applicable to static objects. Once objects move continuously, the KNN result would change over time. As a result, the above methods must be repeatedly evaluated and thus their performance significantly degrades because of the high re-evaluation cost.

2.2   Methods for Continuous KNN Queries

Recently, efficient processing of CKNN queries in road networks has also been studied. Kolahdouzan *et al.* [6] propose a solution, called *upper bound algorithm* (UBA), that only performs snapshot KNN queries at the locations where they are required, and hence provides better performance by reducing the numbe of KNN evaluations. Cho *et al.* [4] develop a *unique continuous search algorithm* (UNICONS) to improve the search performance for CKNN queries. UNICONS first divides the path of the query object into subpaths by the intersections, and then the snapshot KNN queries are performed at two endpoints of each subpath. Finally, the KNNs for each subpath can be found from the union of the KNN sets at its two endpoints and the objects along it. The UBA and UNICONS methods are adequately designed to deal with CKNN queries over static objects (that is, only the query object moves continuously). When objects' locations change over time instead of being fixed, the performance of these techniques would be significantly degraded. Mouratidis *et al.* [8] first address the issue of continuous monitoring KNNs on moving objects. They propose an incremental monitoring algorithm (IMA) to re-evaluate query at those time instants at which updates occur. At each evaluation time, the query may get benefit from the previous result of the last evaluation so that the overhead incurred by processing repetitive queries can be reduced. However, due to the nature of discrete location updates, the KNNs of the query object within two consecutive updates are unknown. As such, IMA would return incorrect results as long as there exist some time points within two consecutive updates at which the KNNs set changes. In this paper, our efforts on processing CKNN query are devoted to overcoming the limitations of the above approaches. We develop an efficient method to determine the KNNs at each timestamp under the situation that all objects (including the query object) move continuously in a road network.

3   Data Structures

In our system, each object moves with fixed speed in a road network, which is represented as an undirected weighted graph consisting of a set of nodes and edges. Information about edges and nodes of the network and the moving objects is stored in three tables, respectively. The first is the *edge table* T_{edge} storing for each edge e_i: (1) its start node n_s and end node n_e (where $n_s < n_e$), (2) its length *len* (i.e., the distance between n_s and n_e), (3) the maximum speed limit s_{max}, and (4) a set S_{obj} of moving objects currently on e_i. The second one is the *node table* T_{node}, that maintains for each node n_i the set S_{adj} of edges connecting n_i. The last one is the *object table* T_{obj} maintaining the information of each moving object. T_{obj} stores for each object o_i: (1) the edge e_j containing it, (2) the update time t_u, (3) the distance *dist* between o_i and the start node n_s of edge e_j

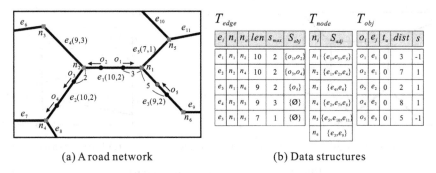

Fig. 2. Representation of road network

(i.e., $e_j.n_s$) at t_u, and (4) its moving speed s. The three tables are updated only when objects reach the network nodes.

To exemplify the information stored in the three tables, we give a road network in Figure 2(a), which consists of nodes n_1 to n_6 and edges e_1 to e_{11} (where an edge label is followed by its len and s_{max} enclosed in parentheses). In this figure, nodes at the boundary of the data space and len and s_{max} of most edges are omitted for clearness. Each of the objects o_1 to o_5 moves with speed 1 m/sec in the road network, and their moving directions are indicated by the corresponding arrows. The detailed information of the tables T_{edge}, T_{node}, and T_{obj} is shown in Figure 2(b). A positive speed value of an object in T_{obj} (e.g., object o_2) indicates that this object moves from the start node to the end node, and a negative value (e.g., object o_1) indicates that it moves in opposite direction.

4 Network Distance Calculation

In this section, we design a technique to calculate the network distance between objects at each timestamp. Let the distance along the shortest path between two stationary nodes n_i and n_j on the road network be $|SP_{n_i,n_j}|$. We choose to precompute $|SP_{n_i,n_j}|$ between any pairs of network nodes and then use them to speed up the calculation of the network distance.

Given a query object q and a moving object o, the calculation of the network distance between q and o at any time t involves two possible cases according to whether they are moving on the same edge or not. In the first case that q and o lie on the same edge e, the network distance between q and o at time t, denoted as $ND_{q,o}(t)$, is determined as follows.

$$ND_{q,o}(t)=|(q.dist+q.s\times(t-q.t_u))-(o.dist+o.s\times(t-o.t_u))|, \qquad (1)$$

where $q.s$ and $q.t_u$ are the moving speed and the update time of q respectively, and $q.dist$ is the distance between q and the start node of edge e at time $q.t_u$ (referring to Section 3). Similarly, $o.s$, $o.t_u$, and $o.dist$ are the information about object o. The obtained $ND_{q,o}(t)$ is valid during the time period in which q and

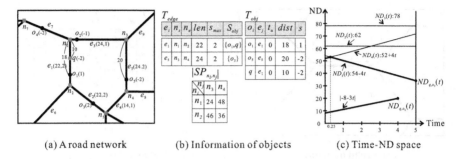

(a) A road network   (b) Information of objects   (c) Time-ND space

Fig. 3. Time-network-distance model

o move on the edge e simultaneously. That is, it needs to be recomputed only when q or o reaches the terminal node of edge e. For instance, the objects q and o_1 lie on the same edge e_1 in Figure 3(a). The corresponding information (i.e., $dist$, s, and t_u) of objects q and o_1 is described in Figure 3(b). By substituting the corresponding information into the above equation, we obtain $ND_{q,o_1}(t) = |-8-3t|$, which can be represented in the *time-network-distance* (*Time-ND*) *space* as shown in Figure 3(c). Note that object o_1 leaves from edge e_1 at time 4 (i.e., $ND_{q,o_1}(t)$ is changed after time 4).

The second case is that objects q and o move on the different edges e_i and e_j, respectively. In this case, the network distance $ND_{q,o}(t)$ between q and o at time t would be equal to the sum of the network distance from q to a node n_i, the network distance from o to a node n_j, and the distance along the shortest path between the two nodes n_i and n_j (i.e., $|SP_{n_i,n_j}|$), where node n_i (or n_j) is either a start node n_s or an end node n_e of edge e_i (or e_j). By using the four distances $|SP_{n_s,n_s}|$, $|SP_{n_s,n_e}|$, $|SP_{n_e,n_s}|$, and $|SP_{n_e,n_e}|$, the network distance $ND_{q,o}(t)$ can be represented as $\min\left(ND_1(t), ND_2(t), ND_3(t), ND_4(t)\right)$, where

$$ND_1(t)=q.dist+q.s\times(t-q.t_u)+|SP_{n_s,n_s}|+o.dist+o.s\times(t-o.t_u), \tag{2}$$

$$ND_2(t)=q.dist+q.s\times(t-q.t_u)+|SP_{n_s,n_e}|+e_j.len-(o.dist+o.s\times(t-o.t_u)),$$

$$ND_3(t)=e_i.len-(q.dist+q.s\times(t-q.t_u))+|SP_{n_e,n_s}|+o.dist+o.s\times(t-o.t_u),$$

$$ND_4(t)=e_i.len-(q.dist+q.s\times(t-q.t_u))+|SP_{n_e,n_e}|+e_j.len-(o.dist+o.s\times(t-o.t_u)).$$

Similar to the first case, $ND_{q,o}(t)$ remains valid until q or o reaches a network node. Consider again the road network in Figure 3(a), where objects q and o_5 are moving on the different edges e_1 and e_5. As nodes n_1 and n_3 are the start nodes of edges e_1 and e_5, respectively, $ND_1(t)$ is obtained by substituting the corresponding information (shown in Figure 3(b)) into Equation (4) and represented as $54-4t$. Similarly, $ND_2(t)$, $ND_3(t)$, and $ND_4(t)$ can also be obtained as 62, 78, and $52+4t$, respectively. As we can see from Figure 3(c), in the Time-ND space $ND_4(t)$ has the minimum value within the time interval $[0, 0.25]$, and then is replaced by $ND_1(t)$ within $[0.25, 5]$. Hence, we obtain

$$ND_{q,o}(t)=\begin{cases} 52+4t \ \forall t \in [0, 0.25], \\ 54-4t \ \forall t \in [0.25, 5], \end{cases} \tag{3}$$

where time 0.25 can be determined by solving the equation $ND_4(t) = ND_1(t)$ and time 5 is the time point at which the query object q reaches node n_1.

5   CKNN Algorithm

To achieve the goal of continuously monitoring the KNN result, the CKNN algorithm needs to divide the time interval into disjoint subintervals, and these subintervals are considered sequentially in finding the KNNs of the query object. Recall that the network distances $ND_{q,o}(t)$ between all objects and the query object are required to be updated once the query object reaches an intersection (i.e., a node) of the road network. Thus, we choose to divide the time interval into subintervals by the time points at which the query object reaches a network node. That is, each subinterval refers to the time period during which the query object moves on the same edge. As a result, each object's $ND_{q,o}(t)$ remains valid within a subinterval until it reaches a network node.

For each subinterval $[t_i, t_j]$, the procedure of determining the KNNs consists of two phases. One is the *filtering phase*, which is used to efficiently prune the non-qualifying objects. This phase is able to guarantee that a pruned object is impossible to be the KNN within the subinterval $[t_i, t_j]$ regardless of whether it reaches the network node. Another is the *refinement phase*, which examines whether the candidates are the KNNs or not. In the following, we present the two phases in details.

5.1   Filtering Phase

The main idea of the filtering phase is to find a pruning network distance, denoted as $ND_{pruning}$, to ensure that if an object o whose network distance at time t_i is greater than $ND_{pruning}$, then o is impossible to be the KNN of the query object q within the subinterval $[t_i, t_j]$. Let objects $NN_1, NN_2, ..., NN_k$ be the KNNs of q at time t_i (note that the KNNs can be retrieved using the previous approaches such as [11,7]). Then, the pruning network distance $ND_{pruning}$ can be represented as $ND_{q,KNN}([t_i, t_j]) + d_{add}$, where $ND_{q,KNN}([t_i, t_j])$ refers to the maximum of the network distances between these KNNs (i.e., NN_1 to NN_k) and q within $[t_i, t_j]$, and d_{add} is an additional distance (which will be explained later).

The network distance $ND_{q,KNN}([t_i, t_j])$ is used to guarantee that at least K objects whose network distances to q are less than or equal to $ND_{q,KNN}([t_i, t_j])$ can be found. As such, those objects whose $ND_{q,o}(t) > ND_{q,KNN}([t_i, t_j])$ at any time $t \in [t_i, t_j]$ can be safely pruned. The following three distances are used to determine $ND_{q,KNN}([t_i, t_j])$: (1) the network distance $ND_{q,NN_\alpha}(t_i)$ from the α-th NN to q at time t_i, (2) the moving distance $d_{NN_\alpha}([t_i, t_j])$ of the α-th NN within $[t_i, t_j]$, and (3) the moving distance $d_q([t_i, t_j])$ of q within $[t_i, t_j]$. With these distances, $ND_{q,KNN}([t_i, t_j])$ can be computed as follows.

$$ND_{q,KNN}([t_i, t_j]) = \max\{ND_{q,NN_\alpha}(t_i) + d_{NN_\alpha}([t_i, t_j]) + d_q([t_i, t_j]) \mid 1 \le \alpha \le k\}. \qquad (4)$$

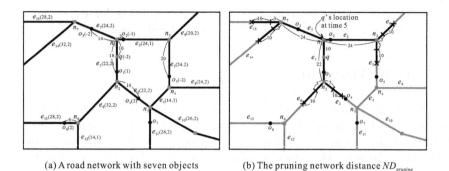

(a) A road network with seven objects   (b) The pruning network distance $ND_{pruning}$

Fig. 4. Example of the filtering phase

Consider Figure 4(a), where eight objects (including the query object q) are moving in a road network. In this figure, each object (edge) label is followed by its corresponding speed (length and maximum speed limit) enclosed in parentheses. Assume that two NNs (i.e., $K = 2$) are to be found within the subinterval $[0, 5]$ as the query object q reaches the node n_1 at time 5. Initially, objects o_1 and o_2 are the 2NNs, and their network distances $ND_{q,NN_1}(0)$ and $ND_{q,NN_2}(0)$ are 8 and 12, respectively. As object o_1 moves with speed 1 m/sec, the moving distance $d_{NN_1}([0,5])$ is equal to 5. Similarly, the moving distance of object o_2 can be computed as $d_{NN_2}([0,5]) = 5$. For the query object q, it reaches the node n_1 at time 5 so that its moving distance $d_q([0,5])$ is equal to 10. Hence, $ND_{q,KNN}([0,5])$ is represented as $\max(8 + 5 + 10, 12 + 5 + 10) = 27$. As we can see from Figure 4(b), the network distances between the 2NNs (i.e., o_1 and o_2) and q within $[0, 5]$ must be less than or equal to 27. That is, there are at least two objects in the region R that starts at q's location at time 5 (i.e., at node n_1) and ends at the vertical marks with distance 27. Clearly, an object which is outside the region R at any time $t \in [0, 5]$ can never be the 2NN.

The additional distance d_{add} is used to facilitate the determination of whether an object is outside the region R within the entire subinterval $[t_i, t_j]$ (i.e., whether this object can be pruned or not). Let the set of the boundaries of the region R be S_{bound} (e.g., the short vertical bars on the roads in Figure 4(b)). If an object outside the region R at time t_i moves with the maximum speed limit of the edge containing it within $[t_i, t_j]$ without being able to reach any of the boundaries in S_{bound}, then it is impossible to be the query result. Motivated by this, the distance d_{add} can be set to $\max\{d_b([t_i, t_j]) | b \in S_{bound}\}$, where $d_b([t_i, t_j])$ refers to the maximum moving distance of an object located on the boundary b within $[t_i, t_j]$ (i.e., this object moves with the maximum speed limit of each edge containing it). An object o whose network distance to each boundary of the region R at time t_i is greater than d_{add} can be safely pruned because o must be outside the region R within $[t_i, t_j]$.

Continuing the example in Figure 4(b), the distance d_{add} is computed as $2 \times 5 = 10$. With the distances $ND_{q,KNN}([0,5])$ and d_{add}, the pruning network distance $ND_{pruning}$ can be represented as $ND_{q,KNN}([0,5]) + d_{add} = 37$.

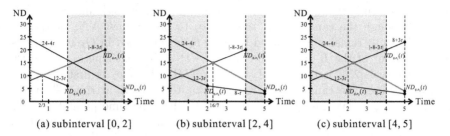

(a) subinterval $[0, 2]$ (b) subinterval $[2, 4]$ (c) subinterval $[4, 5]$

Fig. 5. Example of the refinement phase

Therefore, the objects o_4 to o_7 outside the region starting at n_1 and ending at the cross marks with distance 37 are pruned. Note that these pruned objects' $ND_{q,o}(t)$ need not be computed and the KNN result within $[0, 5]$ would not be affected even though they reach the network nodes. Finally, only three objects o_1 to o_3 are the candidates for the KNNs and will be further verified in the refinement phase.

5.2 Refinement Phase

In the refinement phase, the interval $[t_i, t_j]$ is further divided into n disjoint subintervals $[t_{s_1}, t_{e_1}]$, $[t_{s_2}, t_{e_2}]$, ..., $[t_{s_n}, t_{e_n}]$ by the time points when the candidates (obtained from the filtering phase) reach the network nodes because their $ND_{q,o}(t)$ would be changed at these time points. The goal of the refinement phase is to determine the KNNs-changing time points (t_{NNCs}) within each subinterval $[t_{s_m}, t_{e_m}]$ (for $1 \leq m \leq n$) such that the query object has the same KNNs within two consecutive t_{NNCs}, and find the corresponding KNNs at each t_{NNC}.

To determine the t_{NNCs} with KNNs within a subinterval $[t_{s_m}, t_{e_m}]$, the K-th NN NN_k is continuously monitored from t_{s_m} to t_{e_m} because a t_{NNC} occurs only when NN_k is replaced by another candidate that does not belong to the KNNs. More specifically, the set O_{KNN} of KNNs and the K-th NN NN_k at time t_{s_m} are first determined. Then, for each candidate o (except for NN_k), the time point t_o at which $ND_{q,NN_k}(t) = ND_{q,o}(t)$ is estimated because o is closer to q than NN_k after t_o. By finding t_o of each candidate, the candidate o whose t_o is closest to t_{s_m} is found to be the new NN_k. If o is not contained in O_{KNN}, the KNN result must be changed (i.e., a t_{NNC} occurs) at t_o. In other words, the KNN result remains unchanged from t_{s_m} to t_o so that a tuple $< [t_{s_m}, t_o], O_{KNN} >$ is produced. The interval $[t_o, t_{e_m}]$ is then the remaining period that is considered to produce the other tuples. The above process repeats until the whole interval $[t_{s_m}, t_{e_m}]$ is examined.

Figure 5 continues the previous example of finding 2NNs within $[0, 5]$ in Figure 4. In this example, $ND_{q,o}(t)$ of the candidates o_1 to o_3 are computed and depicted as the line segments in Figure 5(a). As the candidates o_2 and o_1 reach the nodes at time 2 and 4 respectively, the time interval $[0, 5]$ is divided into three subintervals $[0, 2]$, $[2, 4]$, and $[4, 5]$. For the first interval $[0, 2]$, O_{KNN}

and NN_k at time 0 are first determined and set as $\{o_1, o_2\}$ and o_2, respectively. Although the candidate o_1 replaces o_2 to be the new NN_k at time $\frac{2}{3}$, the KNN result remains unchanged because o_1 is contained in O_{KNN} (that is, only the order of the K-th NN and the $(K-1)$-th NN exchanges). When the second interval $[2, 4]$ is considered, the new $ND_{q,o}(t)$ of o_2 is first computed at time 2. Since the candidate o_3 replacing o_1 to be NN_k at time $\frac{16}{7}$ and o_3 is not in O_{KNN}, the time point $\frac{16}{7}$ is a KNNs-changing time point t_{NNC}. As such, a tuple $< [0, \frac{16}{7}], \{o_1, o_2\} >$ is produced, and then o_1 (o_3) is removed from (added into) O_{KNN}. Having examined the third interval $[4, 5]$, the second tuple $< [\frac{16}{7}, 5], \{o_2, o_3\} >$ can be obtained.

6   Performance Evaluation

All the experiments are performed on a PC with AMD 1.61 GHz CPU and 2GB RAM. The algorithm is implemented in JAVA 2 (j2sdk-1.4.0.01). We use a road map of Oldenburg (a city in Germany) from Tiger/Line [12] and generate 100K objects using the generator proposed in [13]. The moving speed of each object is uniformly distributed in between 0 and 20 m/sec. When an object o reaches node n in a network, the next edge on which o moves is randomly chosen from the edges connecting n. In the experimental space, we also generate 30 query objects whose speeds are in the same range as the moving objects mentioned above. Similarly, the next edge that the query object moves on is randomly selected once it reaches a network node. The default values of the number of nearest neighbors requested by each query object (i.e., K) and the length of query time interval are 20 and 100 time units, respectively. In our experiments, we compare the $CKNN$ algorithm to the IMA algorithm proposed by Mouratidis et al. [8]. The IMA algorithm re-evaluates the snapshot KNN query when objects' location updates occur, whose update interval (UI) is set to 5 and 10 time units in this experiment. The IMA algorithm with update interval 5 and 10 time units are termed $IMA(UI = 5)$ and $IMA(UI = 10)$, respectively.

Figure 6 evaluates the CPU time and the precision for the $CKNN$ algorithm and the IMA algorithm under various query interval lengths, where the precision refers to the percentage of time units at which the KNN result retrieved by executing the $CKNN$ and the IMA algorithms is correct. Figure 6(a) shows that for both algorithms, the CPU cost increases with the increase of query length. This is because a large query length results in more objects reaching the network nodes (i.e., more subintervals need to be considered for the $CKNN$ algorithm), and makes more location updates of objects so that more snapshot KNN queries are required for the IMA algorithm. Clearly, the $CKNN$ algorithm has a significantly better performance compared to the IMA algorithm (i.e., $IMA(UI = 5)$ and $IMA(UI = 10)$). Figure 6(b) investigates the effect of the query length on the precision of all algorithms. As the KNNs at each timestamp can be completely determined by executing the $CKNN$ algorithm, the precision of the $CKNN$ algorithm is always equal to 100% under different query lengths. However, if the IMA algorithm is adopted to answer a $CKNN$ query, some of the

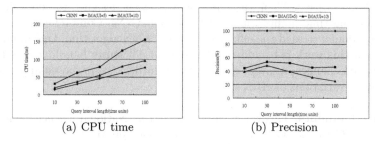

(a) CPU time   (b) Precision

Fig. 6. Effect of query interval length

KNN results are unknown due to the nature of discrete location updates. As we can see, the precision of IMA($UI = 5$) can only reach 40% to 60%. Even worse, the precision of IMA($UI = 10$) is below 30% for a larger query length, which means that most KNN results are incorrect.

Figure 7 illustrates the performance of the $CKNN$ and the IMA algorithms as a function of K (ranging from 10 to 30). As shown in Figure 7(a), when K increases, the CPU overhead for both algorithms grows. The reason is that as K becomes greater, the number of qualifying objects increases so that more distance comparisons between these qualifying objects are required. The experimental result shows that the $CKNN$ algorithm outperforms its competitors in all cases. Figure 7(b) studies how the value of K affects the precision of the $CKNN$ and the IMA algorithms. Similar to Figure 6(b), the precision of the $CKNN$ algorithm remains 100% regardless of the value of K. However, the precisions of IMA($UI = 5$) and IMA($UI = 10$) are both below 60% under various K.

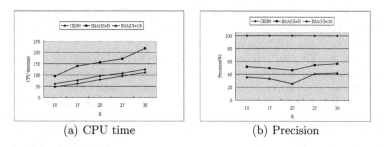

(a) CPU time   (b) Precision

Fig. 7. Effect of K

7   Conclusions

In this paper, we studied the problem of processing the $CKNN$ query over moving objects in road networks. We proposed a technique to represent the network distance between objects as a linear function of time in the Time-ND space. Based on the Time-ND space, we developed the $CKNN$ algorithm to efficiently process a $CKNN$ query. The filtering phase of the $CKNN$ algorithm is used to prune the non-qualifying objects. Then, the refinement phase is designed to determine the KNNs-changing time points with corresponding KNNs. Comprehensive experiments have demonstrated the efficiency of the CKNN algorithm. One important extension of this work is to address the issue of efficiently handling updates of objects' speed changes. Another extension is to utilize the proposed approach to answer other variations of KNN queries.

Acknowledgements. This work was supported by National Science Council of Taiwan (R.O.C.) under Grants NSC95-2221-E-006-206-MY3.

References

1. Huang, Y.-K., Chen, C.-C., Lee, C.: Continuous k-nearest neighbor query for moving objects with uncertain velocity. GeoInformatica (accepted) (to appear)
2. Raptopoulou, K., Papadopoulos, A., Manolopoulos, Y.: Fast nearest-neighbor query processing in moving-object databases. GeoInformatica 7(2), 113–137 (2003)
3. Yu, X., Pu, K.Q., Koudas, N.: Monitoring k-nearest neighbor queries over moving objects. In: Proceedings of the International Conference on Data Engineering (2005)
4. Cho, H.-J., Chung, C.-W.: An efficient and scalable approach to cnn queries in a road network. In: Proceedings of the International Conference on Very Large Data Bases, Trondheim, Norway (2005)
5. Jensen, C.S., Kolar, J., Pedersen, T.B., Timko, I.: Nearest neighbor queries in road networks. In: Proceedings of the ACM GIS, New Orleans, Louisiana, USA, November 7-8 (2003)
6. Kolahdouzan, M., Shahabi, C.: Continuous k nearest neighbor queries in spatial network databases. In: Proceedings of the Spatio-Temporal Databases Management (STDBM), Toronto, Canada, August 30 (2004)
7. Papadias, D., Zhang, J., Mamoulis, N., Tao, Y.: Query processing in spatial network databases. In: Proceedings of the International Conference on Very Large Data Bases, Berlin, Germany, September 9-12 (2003)
8. Mouratidis, K., Yiu, M.L., Papadias, D., Mamoulis, N.: Continuous nearest neighbor monitoring in road networks. In: Proceedings of the International Conference on Very Large Data Bases, Seoul, Korea, September 12-15 (2006)
9. Dijkstra, E.W.: A note on two problems in connection with graphs. Numerische Mathematik 1, 269–271 (1959)
10. Kung, R., Hanson, E., Ioannidis, Y., Sellis, T., Shapiro, L., Stonebraker, M.: Heuristic search in data base system. In: Proceedings of the International Workshop on Expert Database Systems (1986)
11. Kolahdouzan, M., Shahabi, C.: Voronoi-based k nearest neighbor search for spatial network databases. In: Proceedings of the International Conference on Very Large Data Bases, Toronto, Canada (2004)
12. http://www.census.gov/geo/www/tiger/
13. Brinkhoff, T.: A framework for generating network-based moving objects. GeoInformatica 6(2), 153–180 (2002)

Data Quality Aware Queries in Collaborative Information Systems

N.K. Yeganeh, S. Sadiq, K. Deng, and X. Zhou

School of Information Technology and Electrical Engineering
The University of Queensland, St Lucia. QLD 4074 , Australia
{naiem,shazia,dengke,zxf}@itee.uq.edu.au

Abstract. The issue of data quality is gaining importance as individuals as well as corporations are increasingly relying on multiple, often external sources of data to make decisions. Traditional query systems do not factor in data quality considerations in their response. Studies into the diverse interpretations of data quality indicate that fitness for use is a fundamental criteria in the evaluation of data quality. In this paper, we present a 4 step methodology that includes user preferences for data quality in the response of queries from multiple sources. User preferences are modelled using the notion of preference hierarchies. We have developed an SQL extension to facilitate the specification of preference hierarchies. Further, we will demonstrate through experimentation how our approach produces an improved result in query response.

1 Introduction

User satisfaction from a query response is a complex problem encompassing various dimensions including both the efficiency as well as the quality of the response. Quality in turn includes several dimensions such as completeness, currency, accuracy, relevance and many more [28]. In current information environments where individuals as well as corporations are routinely relying on multiple, external data sources for their information needs, absolute metrics for data quality are no longer valid. Thus the same data set may be valuable for a particular usage, but useless for another.

Consider for example a virtual store that is compiling a comparative price list for a given product (such as Google products, previously known as froogle) through a meta search (a search that queries results of other search engines and selects best possible results amongst them). It obviously does not read all the millions of results for a search and does not return millions of records to the user. It normally selects top k results (where k is a constant value) from each search engine and finally returns top n results after the merge.

In the above scenario, when a user queries for a product, the virtual store searches through a variety of data sources for that item and ranks and returns the results. For example the user may query for "Canon PowerShot". In turn the virtual store may query camera vendor sites and return the results. The value that the user associates with the query result is clearly subjective and

Q. Li et al. (Eds.): APWeb/WAIM 2009, LNCS 5446, pp. 39–50, 2009.

related to the user's intended requirements, which go beyond the entered query term, namely "Canon PowerShot" (currently returns 91,345 results from Google products). For example the user may be interested in comparing product prices, or the user may be interested in information on latest models.

More precisely, suppose that the various data sources can be accessed through a view consisting of columns ("Item Title", "Item Description", "Numbers Available", "Price", "Tax", "User Comments"). A user searching for "Canon PowerShot" may actually be interested to:

– Learn about different items (products) - such a user may not care about the "Numbers Available" and "Tax" columns. "Price" is somewhat important to the user although obsoleteness and inaccuracy in price values can be tolerated. However, consistent of "Item Title" and completeness within the populations of "User Comments" in the query results, is of highest importance.
– Compare prices - thus user is sure about the item to purchase but is searching for the best price. Obviously "Price" and "Tax" fields have the greatest importance in this case. They should be current and accurate. "Numbers Available" is also important although slight inaccuracies in this column are acceptable as any number more than 1 will be sufficient.

Above examples indicate that **selection of a good source for data is subjected to what does the term "good" mean to the user**. In this paper, we propose to include user specific quality considerations into query formulations, in order to address user specific requirements. We term this as quality-aware queries. Quality aware queries are a multi-faceted problem. Aggregations across multiple large data sets are infeasible due to the scale of data. Further, ranking approaches based on generic user feedbacks gives a constant rank to the quality of a data source and does not factor in user/application specific quality ranking.

In this paper, we propose to address this problem through a 4-step methodology for quality aware queries. These consist of addressing the following questions:

1. How to obtain information about the generic data quality of data sources?
2. How to model and capture user specific data quality requirements?
3. How to conduct the quality aware query?
4. How to rank results based on the quality aware query?

The rest of the paper is organized as follows to present our approach to address the above questions: In Section 2 related works are reviewed, in order to demonstrate the position and originality of this work. In Section 3 an architecture is presented to provide practical feasibility of the proposed approach. We then present a model for representing generic as well as user specific data quality. Section 4 presents the methods to process the Quality-aware Queries. Finally, in Section 5 effectiveness of the proposed query processing methods is evaluated, and conclusions drawn from the work are presented.

2 Related Works

Consequents of poor quality of data have been experienced in almost all domains. From the research perspective, data quality has been addressed in different contexts, including statistics, management science, and computer science [23] [19] [20] [26]. To understand the concept, various research works have defined a number of quality dimensions [23] [27] [28].

Data quality dimensions characterize data properties e.g. accuracy, currency, completeness etc. Many dimensions are defined for assessment of quality of data that give us the means to measure the quality of data. Data Quality attributes can be very subjective (e.g. ease of use, expandability, objectivity, etc.).

To address the problems that stem from the various data quality dimensions, the approaches can be broadly classified into investigative, preventative and corrective. Investigative approaches essentialy provide the ability to assess the level of data quality and is generally provided through data profiling tools (See e.g. IBM Quality Stage). Many sophisticated commercial profiling tools exist [8], and hence we do not provide further details in this regard. Suffice to say that in this paper we focus on a selected set of data quality dimensions which can be profiled using typical profiling tools. These include completeness (determined through the number of column nulls), consistency (determined through the number of consistency constraint violations), accuracy (determined through format differences), and currency (determined through data relating to time of update). Note that there are several interpretations of these dimensions which may vary from the interpretation above, e.g. completeness may represent missing tuples (open world assumption) [1]. Similarly, accuracy may represent the distance from truth in the real world. Such interpretations are difficult if not impossible to measure through computational means and hence in the subsequent discussion, the interpretation of these dimensions as presented above will be assumed.

A variety of solutions have also been proposed for preventative and corrective aspects of data quality management. These solutions can be categorized into following broad groups : **Semantic integrity constraints** [6] [3]. **Record linkage** solutions. Record linkage has been addressed through approximate matching [10], de-duplicating [11] and entity resolution techniques [2]. **Data lineage or provenance** solutions are classified as annotation and non-annotation based approaches where back tracing is suggested to address auditory or reliability problems [24]. **Data uncertainty** and probabilistic databases are another important consideration in data quality [15].

The data quality problem addressed in this paper, also has relevance to preference theory [25]. The research literature on preferences is extensive. It encompasses preference logics [16], preference reasoning [29], decision theory [7], and multi criteria decision making [22] (the list is by no means exhaustive). For example: multi-criteria decision making approaches can be taken to select from different alternatives based on multiple complex subjective criterias provided through user preferences[22]. We studied and compared different multi-criteria decision making methods and identified suitable techniques to be used for implementation (more details in Section 4).

The preference problem addressed in this paper is intended to select the sources for querying data based on the user's preference for data quality dimensions. The issue of user preferences in database queries dates back to 1987 [14]. Preference queries in deductive databases are studied in [9]. A new operator called Skyline is proposed in [4] that extends generic SQL statements with some powerful preference definitions. In [13] and [5] a logical framework for formulating preferences and its embedding into relational query languages is proposed. However, previous research does not consider the preferences on data quality which is the main focus and contribution of this work.

3   Quality-Aware Queries

Our approach is based on a Collaborative Information Systems (CIS) model as in Fig. 1. Each data source (could be an organization or department within an organization) consists of a base (relational) database and certain meta data. The meta data consists of two main components: i) Schema and ii) Data Quality profiles, which are results obtained through data profiling processes, or what we can term quality-aware metadata (more details in Sec. 3.1). Each source is responsible for providing data quality profiles at attribute level for each table of the database. Profiling can be a periodic task or it can be incrementally updated each time data is modified.

3.1   Quality-Aware Metadata

As previously mentioned in Sec.2, we consider the following data quality metrics (also referred to as dimensions): completeness, consistency, accuracy, and currency. Consider an example data set as below relating to the virtual store example presented previously.

Completeness of an attribute is calculated by counting the nulls over the number of all records. Consistency of the attribute is calculated by counting the number of items conforming to consistency constraints over the number of all records. For example, a consistency constraint on "Item Title" can be defined to ensure that all values have an exact match with a given master data list of item titles. Accuracy of the attribute is calculated by counting the number of values which are not in the correct format over the number of all records. Currency

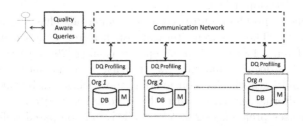

Fig. 1. Quality aware query processing architecture

Item Title	Item Desc	Num Available	Price	Tax	User Comments
S2IS	Canon PShot	8	310	31	
	Camera		1000		Test
S3IS	Canon PShot		375		
XL-2		4	184	18	
DMC-TZ5K	Panasonic Lmx		340		
DSC-W55	S ony Cshot	2	260	26	

(a)

Object	Metric	Value
Shop.ItemTitle	Completeness	0.83
Shop.ItemDesc	Completeness	0.83
Shop.NumAvailable	Completeness	0.50
Shop.Price	Completeness	1.00
Shop.Tax	Completeness	0.50
Shop.User Comments	Completeness	0.16

(b)

Fig. 2. An example data set and data profiling metadata

of a value is calculated by checking the timestamp that DBMS records each time a value is changed (e.g. dividing the last modified timestamp by a constant normalizing factor).

Definition 1. *Let relation $R(A)$ be a relation of attributes $a \in A$. $M = \{m_1 ..m_k\}$ is the set of k data quality metrics and $a.m := \sigma_m(a)$ where σ_m returns the calculated data quality metric value for metric m over attribute a and $0 \leq \sigma_m(a) \leq 1$*

Figure 2(b) shows the profiling data for 2(a) which is, the quality metric m, and stores the result of the function σ_m. Since data values are changed over the time, quality-aware metadata should be updated either periodically or every time the data is modified.

3.2   Quality-Aware SQL

Consider a collaborative information system as given in Fig. 1, wherein an instance of information can be selected from each source, and the quality aware metadata (data profile) as given in Fig. 2, is available for each source. User queries relation $R(A)$ that can be accessed from any source $S_1..S_n$ where n is the number of sources. The data quality metrics allow the user to query for any attribute $a_i \in A_i$ from any source S_i plus the value of any data quality metric $a_i.m$.

For purpose of simple illustration we allow using quality-aware metadata, to query any metric as [Column Name.Metric Name] as part of the SQL query formulation. For example the query 1 below queries sources only if the completeness of the column Title from table ShopItem is more than 80%. Or query 2 sorts the results based on the accuracy of column price in data source. It also returns the accuracy of the column title from the source as a result.

```
1 SELECT Title, Price FROM ShopItem WHERE Title.Completeness>0.8
2 SELECT Title, Title.Accuracy, Price FROM ShopItem ORDER BY Price.Accuracy
```

Generic SQL is not capable of modeling complicated user preferences of higher complexity. Though clause ORDER BY Price.Accuracy models a one dimensional preference that indicates sources with higher price accuracy are preferred, a two dimensional preference can not be intuitively achieved. Further, users may prioritize the various dimensions differently. Therefore, in Sec. 3.3 we propose an extension to the generic SQL (called Hierarchy Clause) to model complicated user preferences relating to data quality.

3.3   Modelling User Preferences

Preference modelling is in general a difficult problem due to the inherent subjective nature. Typically, preferences are stated in relative terms, e.g. "I like A better than B", which can be mapped to strict partial orders [13]. In quality-aware queries, a mixture of preferences could be defined, e.g. the accuracy of the price and completeness of the user comments should be maximized.

The notion of **Hierarchy** in preferences is defined in the literature [5] as prioritized composition of preferences. For example; completeness of user comments may have priority over completeness of prices. We use the term Hierarchy to define prioritised composition of preferences which can form several levels of priority i.e. a hierarchy. The hierarchy over the preference relations is quantifiable such as: a is strongly more important than b, or a is moderately more important than b.

Definition 2. *Let relation $R(A)$ be a relation of attributes $a \in A$. Let $M = m_1..m_k$ be the set of k data quality metrics. Let $S = S_1...S_n$ be a set of possible sources for relation $R(A)$. A preference formula (pf) $C(S_1, S_2)$ is a first order formula defining a preference relation denoted as $\succ$, namely*

$$S_1 \succ S_2 \; iff \; C(S_1, S2).$$

Consider the Relation $ShopItem(Title, Price, UserComments)$ from source S denoted as $R_S(t, p, u)$, quality metric completeness is denoted as m_1 and completeness of the price is denoted as $p.m_1$. A preference relation can thus be defined as C:

$$(t, p, u) \in R_s \succ (t', p', u') \in R_{S'} \equiv p.m_1 > p'.m_1$$

Where $p.m_1$ is from the source S and $p'.m_1$ is from the source S'.

A hierarchy (prioritized order) of preferences is defined as below:[5].

Definition 3. *Consider two preference relations $\succ_1$ and $\succ_2$ defined over the same schema. The prioritized composition $\succ_{1,2}:=\succ_1 \rhd \succ_2$ of $\succ_1$ and $\succ_2$ is defined as:*

$$S_1 \succ_{1,2} S_2 \equiv S_1 \succ_1 S_2 \vee (\neg S_2 \succ_1 S_1 \wedge S_1 \succ_2 S_2).$$

The formal notions above are translated below into pseudo-SQL in order to illustrate the quality aware query. A specialized HIERARCHY clause is defined to identify the hierarchy of preferences. It is assumed that sources with maximial value of a data quality metric are preferred (i.e. source S_k where $\sigma_m(a)$ is greater than or equal to $\sigma_m(a)$ of all $S_i \in S|i : 1..n$), thus, there is no need to explicitly define this preference in the query. Only hierarchy (prioritized order) of those preferences are written in the query.

```
SELECT Title AS t, Price AS p, [User Comments] AS u
FROM ShopItem WHERE ...
HIERARCHY(ShopItem) p OVER (t,u) 7, u OVER (t) 3
HIERARCHY(ShopItem.p) p.Currency OVER (p.Completeness) 3
```

In the `HIERARCHY(a) a.x OVER (a.x',...)` clause, a denotes the object that
the hierarchy is applied for (e.g. ShopItem or column price). `a.x` denotes the
preferred object, `a.x',...` is list of objects on which `a.x` is preferred and the
number 1..9 denotes the intensity of the importance of the hierarchy. 1 denotes
weak or slight importance and 9 denotes strong importance. In the above example two hierarchies are defined which indicate that i) for the ShopItem, price
is strongly more important than title, and user comments attributes is slightly
more important that title. ii) For the column price, Currency is slightly more
important than completeness.

The example hierarchy clause `HIERARCHY(ShopItem.p) p.Currency OVER`
`(p.Completeness)` is defined as below, let p be the "Price" attribute, m_1 and
m_2 be currency and completeness metrics and $p.m_1$ and $p'.m_1$ be the currency
of the column "Price", from quality-aware meta data related to source S and S':

$$(t, p, u) \succ_0 \rhd \succ_1 (t', p', u'),$$
$$(t, p, u) \in S \succ_0 (t', p', u') \in S' \equiv p.m_1 > p'.m_1,$$
$$(t, p, u) \in S \succ_1 (t', p', u') \in S' \equiv p.m_2 > p'.m_2$$

4   Query Processing

In this paper, the method to answer the Quality-aware queries is proposed based
on a multi-criteria decision making technique. The general idea of having a
hierarchy in decision making and definition of hierarchy as strict partial orders
is widely studied. In [21] a decision making approach is proposed which is called
Analytical Hierarchy Process (AHP). Processing the problem of source selection
for Quality-aware queries can be delineated as a decision making problem in
which a source of information should be selected based on a hierarchy of user
preferences that defines what a good source is. In our approach, query processing
consists of two phases; i) each data quality metric from the query is assigned
a weight which is calculated using AHP technique and ii) sources are ranked
against the metric weights and the quality aware metadata of the source.

In AHP, decision hierarchy is structured as a tree on objectives from the
top with the goal (query in our case), then the objectives from a higher level
perspective (attributes), and lower level objectives (DQ metrics) Fig. 3(a) shows
a sample decision tree.

This is then followed by a process of prioritization. Prioritization involves
eliciting judgements in response to question about dominance of one element
over another when compared with respect to a property. Prioritizations form a
set of pair-wise comparison matrices for each level of objective.

Within each objective level (query or attribute) Elements of a pair-wise comparision matrix represent intensity of the priority of each item (e.g. DQ metrics)

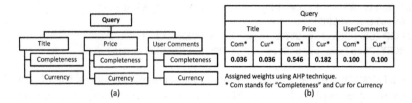

	Query					
	Title		Price		UserComments	
	Com*	Cur*	Com*	Cur*	Com*	Cur*
	0.036	0.036	0.546	0.182	0.100	0.100

Assigned weights using AHP technique.
* Com stands for "Completeness" and Cur for Currency
(a)   (b)

Fig. 3. (a)Decision hierarchy and (b)AHP weightings for querying a virtual shop

over another in a scale of 1 to 9. The Hierarchy Clauses (Sec.3.2) can be directly mapped to pair-wise comparision matrices, thus avoids the tedious task of eliciting judgements.

Algorithm 1 presents the approach. It i) maps a quality-aware query to an AHP decision tree ii) maps Hierarchy clauses to pair-wise comparison matrices and iii) returns weights assigned to each data quality metric using AHP technique over the generated decision tree and comparison matrices.

Data: SELECT $\{a|a \in A\}$ FROM R
HIERARCHY (A) $\{a\,OVER\,a'|a, a' \in A \;\; \land a \neq a'\}$,
$\{Hierarchy(a)\{a.m\,OVER\,a.m'|m, m' \in M \land m \neq m'\}|a \in A\}$
where $A \subset attributes(R)$ and M is a set of DQ metrics.
Result: $W := \{w_{a.m}|a \in A \land m \in M\}$
Define Q as AHP Tree;
r = Q.RootElement; **foreach** $a \in A$ **do**
    r.AddChild(a);
    r.PairwiseComparision$(a, a'|a, a' \in A \land a \neq a')$ =t from Hierarchy a OVER a'
    foreach $m \in M$ **do**
        r.a.AddChild(m);
        m.PairwiseComparision$(m, m'|m, m' \in M \land m \neq m')$ =t from Hierarchy
        a.m OVER a'.m
    end
end
return $AHP(Q)//$ *Which is $w_{a.m}$ for each a and m using AHP technique*

Algorithm 1. AHP Tree from Quality-aware SQL

Assume R is an arbitrary view and M is an arbitrary set of DQ metrics. If a hierarchy a over a' is not defined in the user query, it can be estimated as follows: If there is no hierarchy a' over a with weight t, use a hierarchy of a over a' with weight 1, otherwise use a hierarchy of a over a' with weight $1/t$. AHP method automatically resolves inconsistency in hierarchies like: Hierarchy p over m 1, p over n 5, m over n 1. In this case m over n should have weight of 5 but the provided value is inappropriate. For the sample query in Sec. 3.2, Alg. 1 initially generates a tree root node Q. It then adds all attributes t, p, u (title, price, user comments) as children nodes. The pair-wise comparison matrix is generated for the root node by substituting each matrix pair by hierarchy clause weights e.g. $(p,t) = 7, (p,u) = 7, (u,t) = 3$ etc. Respective tree nodes for each DQ metric in the query are added to each attribute node and pair-wise comparison matrices are generated correspondingly. When all required information for AHP technique (i.e. decision tree and pair-wise comparison matrices) are ready, AHP technique assigns a weight to each DQ metric for each attribute. Figure 3(a) shows the AHP decision hierarchy generated for the example query in Sec. 3.2 and Fig. 3(b) shows resulting weights returned by the algorithm. These weights are representation of user preferences on data quality and can be used against profiling data of each source (from the quality-aware metadata) for the metric of the same attribute to easily rank sources.

Source Selection. Having a fixed weight assigned to each decision criterion, which in our case is each column and metric, and a fixed respective profiling value for each source (from quality-aware meta data), sources can be ranked using a number of ranking methods [17]: Simple Additive Weighting (SAW), Technique for Order Preference by Similarity to Ideal Solution (TOPSIS), Elimination et Choice Translating Reality (ELECTRE) , and Data Envelopment Analysis (DEA). In [17] a comparative analysis of the mentioned ranking methods is provided which shows that the effectiveness of all above methods is not considerably different in regards to the source selection problem. The major difference of these methods is their computational complexity. Hence, we incorporate the SAW method which is easy and fast [12].

The SAW method involves three basic steps: Scale the scores to make them comparable, apply the weighting, and sum up the scores for each source. Data profiles of the sources, represent each column's quality metric in a value in [0,1] (where 1 is the complete quality).

$$dq(S_i) := \sum w_j v_{ij}$$

Where $dq(S_i)$ is the final weight of source S_i, w_j is the weight of (column.metric) j which has been calculated through the AHP process and $v_{ij} := \sigma_{(a.m)_j}$ is the profiling value for source i and (column.metric) j. Because $\sum_{j=1}^{m} w_j = 1$, the final scores are in [0,1].

5   Evaluation and Conclusion

We developed a tool called Quality-aware Query Studio (QQS) to support the idea of quality-aware queries. In QQS, Quality-aware metadata including database schema and data profiling results as well as configuration of data sources are defined as Xml objects. QQS, which is implemented by .Net technology, receives the user quality-aware query and ranks data sources against the query using a plugin that implements the methods described above. Standard query which is formed by removing quality-aware SQL clauses from the quality-aware query can then be sent to the k-top sources as usual.

To evaluate the results of our contributions we generated 120 copies of a table containing product information. We then randomly generated different

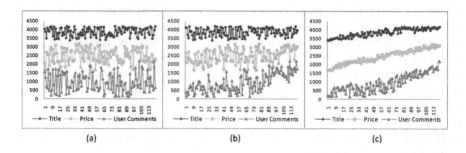

Fig. 4. Comparison of source ranking: a) Random b) Source quality c) Quality-aware

completeness for each column of each table by putting nulls in random places. We then generated profiling results for the quality metric "Completeness" by counting the number of nulls in each column of each table. For simplicity, only completeness of sources is considered in this experiment, but it does not affect generality of the contributions.

Below, three different Quality-aware Queries are defined:

– Making a list of products: price is not important.

```
SELECT Title AS t, Price AS p, [User Comments] AS u
FROM ShopItem AS s HIERARCHY(s) t OVER (p) 9, u OVER (p) 3
```

– Selecting a product to buy: price is very important.

```
SELECT Title AS t, Price AS p, [User Comments] AS u
FROM ShopItem AS s HIERARCHY(s) p OVER (t,y) 9
```

– Reviewing user comments: User comments are important.

```
SELECT Title AS t, Price AS p, [User Comments] AS u
FROM ShopItem AS s HIERARCHY(s) u OVER (p) 9, t OVER (p) 3
```

Figure 4 compares effectiveness of 3 ranking algorithms for above three applications. The x-axis of the graphs represent ranking of sources and the y-axis of the graphs represent the number of useless records which is the number of nulls in the important column for the selected source. In the ideal situation, a higher ranked source should return the least number of useless records, for example if the query application is to find best prices, sources with less null prices should be ranked higher. But if the application is to survey user comments, the sources with less null user comments should be ranked higher. Sources are ranked using three following algorithms:

– No information on quality: Randomly select a source. Figure 4(a) shows number of nulls returned by the query using this method.
– Ranking based on source quality: previous similar source ranking approaches [18] based on the source's data quality. Figure 4(b) represents result of this method.

– Quality-aware source ranking: Referring to our contributions in this paper, ranking sources based on their quality profiles and user requirement gives the desired result for each application. Figure 4(c) represents result of this method.

A comparative study of the graphs in Fig. 4 shows that source ranking method that does not consider information quality has a complete random behaviour. Ranking method based on overall source quality has some, although negligible advantage over the random ranking method, but the quality-aware query processing approach meaningfully ranks sources for less useless records. Higher ranked sources are clearly having minimum useless records and lowest ranked sources have maximum useless records.

In this paper we presented a methodology that includes user preferences for data quality in response of queries from multiple sources. The developed methodology firstly provides a means of obtaining information about the quality of data sources, it then allows user preferences on data quality to be modelled (called Quality-aware Queries), and it finally provideds a ranking of various sources based on user preferences and source quality.

Preference processing in database systems is studied for decades, but to the best of our knowledge attention to information quality as user preferences is under-studied. There are open challenges to adopt existing works on preference such as Skyline Operator [4] to quality-aware ranking. Implication of methods to apply quality-aware queries to web information systems where obtaining profiling information is not easy is also a challenging issue.

References

[1] Batini, C., Scannapieco, M.: Data Quality: Concepts, Methodologies and Techniques (Data-Centric Systems and Applications). Springer, New York (2006)

[2] Benjelloun, O., Garcia-Molina, H., Su, Q., Widom, J.: Swoosh: A generic approach to entity resolution. VLDB Journal (2008)

[3] Bohannon, P., Wenfei, F., Geerts, F., Xibei, J., Kementsietsidis, A.: Conditional functional dependencies for data cleaning. In: ICDE (2007)

[4] Borzsonyi, S., Kossmann, D., Stocker, K.: The skyline operator. In: Proc. of ICDE, pp. 421–430 (2001)

[5] Chomicki, J.: Querying with intrinsic preferences. In: Jensen, C.S., Jeffery, K., Pokorný, J., Šaltenis, S., Bertino, E., Böhm, K., Jarke, M. (eds.) EDBT 2002. LNCS, vol. 2287, pp. 34–51. Springer, Heidelberg (2002)

[6] Cong, G., Fan, W., Geerts, F., Jia, X., Ma, S.: Improving data quality: consistency and accuracy. In: Proceedings of the 33rd international conference on Very large data bases, pp. 315–326 (2007)

[7] Fishburn, P.: Preference structures and their numerical representations. Theoretical Computer Science 217(2), 359–383 (1999)

[8] Friedman, T., Bitterer, A.: Magic Quadrant for Data Quality Tools. Gartner Group (2006)

[9] Govindarajan, K., Jayaraman, B., Mantha, S.: Preference Queries in Deductive Databases. New Generation Computing 19(1), 57–86 (2000)

[10] Gravano, L., Ipeirotis, P.G., Jagadish, H.V., Koudas, N., Muthukrishnan, S., Srivastava, D.: Approximate String Joins in a Database (Almost) for Free. In: Proceedings of the international conference on very large data bases, pp. 491–500 (2001)

[11] Gravano, L., Ipeirotis, P.G., Koudas, N., Srivastava, D.: Text Joins for Data Cleansing and Integration in an RDBMS. In: Proc. of Int. Conf. on Data Engineering (ICDE) (2003)

[12] Hwang, C.L., Yoon, K.: Multiple Attribute Decision Making: Methods and Apllication. Lecture Notes in Economics and Mathematical Systems. Springer, Heidelberg (1981)

[13] Kießling, W.: Foundations of preferences in database systems. In: Proceedings of the 28th international conference on Very Large Data Bases, pp. 311–322. VLDB Endowment (2002)

[14] Lacroix, M., Lavency, P.: Preferences: Putting More Knowledge into Queries. In: Proceedings of the 13th International Conference on Very Large Data Bases, pp. 217–225. Morgan Kaufmann Publishers Inc., San Francisco (1987)

[15] Lakshmanan, L.V.S., Leone, N., Ross, R., Subrahmanian, V.S.: ProbView: a flexible probabilistic database system. ACM Transactions on Database Systems (TODS) 22(3), 419–469 (1997)

[16] Mantha, S.M.: First-order preference theories and their applications. PhD thesis, Mathematics, Salt Lake City, UT, USA (1992)

[17] Naumann, F.: Quality-Driven Query Answering for Integrated Information Systems. LNCS, vol. 2261. Springer, Heidelberg (2002)

[18] Naumann, F., Freytag, J.C., Spiliopoulou, M.: Qualitydriven source selection using Data Envelopment Analysis. In: Proc. of the 3rd Conference on Information Quality (IQ), Cambridge, MA (1998)

[19] Redman, T.C.: Data Quality for the Information Age. Artech House, Inc., Norwood (1997)

[20] Redman, T.C.: The impact of poor data quality on the typical enterprise. Communications of the ACM 41(2), 79–82 (1998)

[21] Saaty, T.L.: How to Make a Decision: The Analytic Hierarchy Process. European Journal of Operational Research 48(1), 9–26 (1990)

[22] Saaty, T.L.: Multicriteria Decision Making: The Analytic Hierarchy Process: Planning, Priority Setting, Resource Allocation. RWS Publications (1996)

[23] Scannapieco, M., Missier, P., Batini, C.: Data quality at a glance. Datenbank-Spektrum 14, 6–14 (2005)

[24] Simmhan, Y.L., Plale, B., Gannon, D.: A Survey of Data Provenance in e-Science. SIGMOD RECORD 34(3), 31 (2005)

[25] von Wright, G.H.: The Logic of Preference. Edinburgh University Press (1963)

[26] Wang, R.Y., Kon, H.B.: Toward total data quality management (TDQM). Prentice-Hall, Inc., Upper Saddle River (1993)

[27] Wang, R.Y., Storey, V.C., Firth, C.P.: A framework for analysis of data quality research. IEEE Transactions on Knowledge and Data Engineering 7(4), 623–640 (1995)

[28] Wang, R.Y., Strong, D.M.: Beyond accuracy: what data quality means to data consumers. Journal of Management Information Systems 12(4), 5–33 (1996)

[29] Wellman, M.P., Doyle, J.: Preferential semantics for goals. In: Proceedings of the National Conference on Artificial Intelligence, pp. 698–703 (1991)

Probabilistic Threshold Range Aggregate Query Processing over Uncertain Data

Shuxiang Yang[1], Wenjie Zhang[2], Ying Zhang[1], and Xuemin Lin[3]

[1] The University of New South Wales, Australia
[2] The University of New South Wales / NICTA, Australia
[3] Dalian Maritime University, China & UNSW / NICTA, Australia
{syang, zhangw, yingz, lxue}@cse.unsw.edu.au

Abstract. Large amount of uncertain data is inherent in many novel and important applications such as sensor data analysis and mobile data management. A probabilistic threshold range aggregate (PTRA) query retrieves summarized information about the uncertain objects satisfying a range query, with respect to a given probability threshold. This paper is the first one to address this important type of query. We develop a new index structure aU-tree and propose an exact querying algorithm based on aU-tree. For the pursue of efficiency, two techniques *SingleSample* and *DoubleSample* are developed. Both techniques provide approximate answers to a PTRA query with accuracy guarantee. Experimental study demonstrates the efficiency and effectiveness of our proposed methods.

1 Introduction

Many emerging important applications involve dealing with uncertain data, such as data integration, sensor data analysis, market surveillance, trends prediction, mobile data management, etc. Uncertainty is inherent in such data due to various factors like randomness or incompleteness of data, limitations of equipment and delay or loss in data transfer. Extensive research effort has been given to model and query uncertain data recently. Research directions include modeling uncertainty [16], query evaluation [2], indexing [18], top-k queries [8], skyline queries [15], clustering and Mining [10], etc. However, though range aggregate query on uncertain data is very important in practice, this problem remains unexplored.

A range aggregate query (RA query) on certain data returns summarized information about objects satisfying a given query range, such as the total number of qualified objects [19]. This type of query is important since users may be interested only in aggregate information instead of specific IDs. For instance, to monitor traffic volume of a crossroad A in rush hours, query "how many vehicles pass A from 8AM to 9AM today" is of more interest than "which vehicles pass A from 8AM to 9AM today". aR-tree [13] is the most popular index structure to answer RA query on spatial space.

While many sophisticated techniques have been developed to answer RA query over certain data [19], the counter problem of RA query over uncertain data

Q. Li et al. (Eds.): APWeb/WAIM 2009, LNCS 5446, pp. 51–62, 2009.

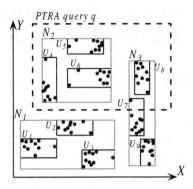

Fig. 1. Uncertain Objects

has not attracted much research attention. Modeling and answering RA query over uncertain data require comprehensive analysis of probabilities, as shown in Figure 1. Assume we still use aR-tree to index the uncertain objects. N_i ($1 \leq i \leq 3$) represents nodes in aR-tree and U_j ($1 \leq j \leq 9$) represents uncertain objects. Each uncertain object contains a set of uncertain instances. A probabilistic threshold range query q fully covers node N_2 and intersects with node N_3. After accessing node N_3, only part of the instances in uncertain object U_7 is in the query range of q. Clearly in this case we have to retrieve the instances of U_7 to compute the probability for it to satisfy q. If this probability is no less than a probability threshold, then U_7 will be counted in the reulst otherwise it will be excluded from final result. Computing such probability is time consuming when the number of uncertain instances from uncertain objects is large.

Challenges and Contributions. Aggregate information retrieval on uncertain objects requires detailed analysis of appearance probabilities from uncertain instances (*discrete case*) or probability density function (PDF) (*continuous case*). Naively computing the probability for every uncertain object to satisfy a *PTRA* query can be very time consuming. Our contributions in this paper are:

- We formally define *PTRA* query over uncertain objects.
- A novel index structure, aU-tree is developed to support exactly execut *PTRA* queries.
- Two techniques, *SingleSample* and *DoubleSample* are proposed to approximately answer *PTRA* queries with accuracy guarantee.
- An extensive experimental study over real and synthetic datasets shows the efficiency and effectiveness of our proposed techniques.

Organization of the paper. The rest of the paper is organized as follows. Section 2 models the problem and introduces preliminaries. Exact and approximate query processing techniques are presented in Section 3 and Section 4, respectively. In Section 5, the efficiency and effectiveness of our proposed techniques are experimentally studied. This is followed by related work in Section 6. We conclude our paper in section 7.

2   Background Information

In this section, we give a formal definition of PTRA query for *discrete* cases. The extension of problem definition and techniques to *continuous* cases will be discussed in the end of this paper in Section 7.

2.1   Problem Statement

A PTRA query q is related with a d-dimensional query range r_q and a probability threshold p_q. In discrete cases, an uncertain object U is represented by a set of uncertain instances $\{u_1, ..., u_l\}$. Each instance u_i is associated with a membership probability $P(u_i)$ and $\sum_{i=1}^{l} P(u_i) = 1$. Let $P_{app}(U, q)$ be the appearance probability that object U satisfies a query with range r_q, then

$$P_{app}(U, q) = \sum_{u \in U, u \vdash r_q} P(u) \qquad (1)$$

where $u \vdash r_q$ denotes uncertain instance u is inside r_q.

Given a set of uncertain objects $\mathcal{U}$, a PTRA query q returns the number of uncertain objects in $\mathcal{U}$ with appearance probability no less than p_q:

$$|\{U \in \mathcal{U} | P_{app}(U, q) \geq p_q\}| \qquad (2)$$

Example 1. As the example illustrated in Figure 1, to process PTRA query q, we calculate the appearance probability of $P(U_7, q)$ using the formulate 1. If $P(U_7, q) \geq p_q$, then the result for q is 5; otherwise, the result is 4.

2.2   Preliminaries

Possible World Semantics.   Given a set of uncertain objects $\mathcal{U} = \{U_1, \cdots, U_n\}$, a *possible world* $W = \{u_1, \cdots, u_n\}$ is a set of n instances - one instance per uncertain object. The probability of W to appear is $P(W) = \prod_{i=1}^{n} P(u_i)$. Let Ω be the set of all possible worlds; that is, $\Omega = U_1 \times U_2 \cdots \times U_n$. Then, $\sum_{W \in \Omega} P(W) = 1$. Namely, Ω enumerates all the possibilities of $\mathcal{U}$.

aR-Tree Based Range Query Processing.   aR-tree is a modification of R-tree [7] by storing the number of objects in each entry. Figure 2 illustrates

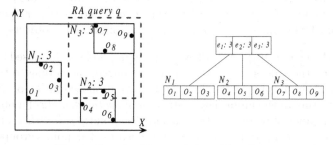

Fig. 2. Certain Objects Indexed by aR-tree

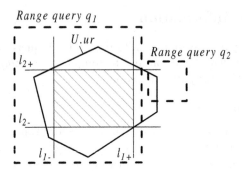

Fig. 3. Pruning/Validating in U-tree

the structure of 2-level aR-tree. Besides R-tree structure information, entries in root node also keep the number of objects contained, such as entry $e_1 \in root$ contains 3 objects. The dashed rectangle is the range of a RA query q. As shown in Figure 2, leaf node N_3 is fully covered by q, so N_3 will not be accessed but adding 3 to the final result. Node N_1 does not need to be accessed either since it does not intersect with q. The only accessed node is N_2 and object O_5 is detected to satisfy q. So the final result is 4.

U-tree. One of the most popular index structure for multi-dimensional uncertain data with arbitrary PDFs is U-tree [18], which is built based on R*-tree with a set of pruning and validating rules to support range queries over uncertain data. In Figure 3, polygon $U.ur$ is the uncertain region of 2-dimensional object U. For a given probability $p = 0.2$, in the horizontal dimension, two lines are calculated. U has probability p to occur on the left side of line l_{1-}, also probability p to occur on the right side of line l_{1+}. In the vertical dimension, two such lines are also computed according to the PDF of U. The intersection of these four lines is called probabilistically constrained regions (PCRs). Suppose probability threshold of range query q_1 is 0.8, U can be validated without accessing the instances of U since it fully contains PCR(p = 0.2). On the other hand, suppose probability threshold of range query q_2 is 0.2, U will be pruned from the result of q_2 because q_2 does not intersect with PCR(p = 0.2). To trade-off between pruning/validating power and space cost, only a set of m probability values are chosen as representatives to compute their PCRs. These m PCRs are further bounded from the "outside" and from "inside", called $o.cfb_{out}$ and $o.cfb_{in}$, respectively. A U-tree is built by organizing the cfb_{out} and cfb_{in} of uncertain objects.

Min-Skew Partitioning. Min-Skew partitioning skill was proposed by [1] aiming at dividing the data space into a number of buckets according to the spatial distribution of input data points. Two metrics are proposed in [1] to capture the underlying feature of the input data distribution: *spatial density* of a point representing the number of rectangles that include the point; *spatial-skew* of a bucket which is the statistical variance of the spatial densities of all points grouped in that bucket. The partitioning procedure is: use a uniform grid of regions with the spatial density in each grid to represent the spatial density of the input data. The process starts from a single bucket including the whole space. Split is processed along the boundary of the grid which will lead to the largest reduce of spatial skewness. The iteration stops when the number of buckets meets users' specification.

3   Exact Query Processing

In this section, we firstly present aU-tree which is modified based on U-tree by integrating aggregate information; this is followed by the exact query processing algorithm based on aU-tree.

3.1   aU-Tree

Similar with the adjustment of aR-tree to R-tree for the RA query over certain data, aU-tree is modified on U-tree by embedding aggregate information in every entry. Updating the aggregate information for aU-tree is similar as for aR-tree: whenever inserting or deleting an object, the aggregate information on entries along the corresponding insertion/deletion path is updated as well.

Specifically, each intermediate entry in an aU-tree keeps the following information: a pointer referencing its child node; two d-dimensional rectangles which are used for pruning as introduced in Section 2; agg which is the number of uncertain objects indexed under the subtree rooted at this intermediate entry. Each leaf entry records the following: conservative functional boxes for uncertain object U for both pruning and validating as introduced in Section 2; MBR of the uncertain region of U; uncertain region $U.ur$; and a set of instances to describe the probability distribution. For a leaf entry, its agg value is assigned to 1.

3.2   Querying Algorithm

With aU-tree, if the intermediate entry is totally covered by the query range, the aggregate result for the subtree underneath can be retrieved immediately without accessing every uncertain object in it. However, given a probability threshold, U-tree can only be used to prune a subtree in the intermediate level but not validate a subtree unless the query range fully covers it. In such cases, we still have to access the children of the intermediate entry. Algorithm 1 illustrates the steps of exact results retrieval based on aU-tree.

Till here, to process a PTRA query, we need to reach the leaf level if an intermediate entry can not be either pruned or fully covered by the query range. Then on the leaf level, the pruning/validating rules of U-tree are applied on individual uncertain object. If an object can not be pruned or validated, its exact appearance probability in r_q will be computed. Two aspects impede the efficiency of the exact algorithm. Firstly, validating on higher levels in aU-tree is not possible and the pruning power is not powerful enough especially when index space is limited; Secondly, computing the appearance probability of an uncertain object is time expensive when the number of instances is large. In the next section, we develop approximate query processing algorithms which are both efficient in time and effective in accuracy.

Algorithm 1. Exact Query Processing

INPUT: root node N of aU-tree;
probabilistic query q with probability threshold p_q; query range r_q;
OUTPUT: $result$:number of objects inside r_q with probability $\geq p_q$;

1. **if** r_q fully covers N **then**
2. $result$ += $N.agg$;
3. **if** r_q partially overlaps with N **then**
4.   **if** N is an intermediate node **then**
5.   **if** N can not be pruned w.r.t p_q **then**
6.   **for** each child $child$ of N **do**
7.   call Algorithm 1 with $child$ and q as input;
8.   **if** N is a leaf node **then**
9.   apply PCR technique of U-tree on N
10.   **if** N is validated **then**
11. $result$ += 1;
12.   **if** N is neither pruned nor validated **then**
13.   compute the exact probability of $P_{app}(N, q)$;
14.   **if** $P_{app}(N, q) \geq p_q$ **then**
15. $result$ += 1;
16. **return** r esult;

4   Approximate Query Processing

For a set of uncertain objects $\mathcal{U} = \{U_1, ..., U_n\}$, a possible world consists of n sampled instances – one instance from one uncertain object. Suppose the number of uncertain instance in an uncertain object U_i is $|U_i|$ ($1 \leq i \leq n$), the total number of possible worlds is $\Pi_{i=1}^{n}|U_i|$. This number can be huge when n is large and each uncertain object is represented by a large number of instances. The basic idea of our approximate algorithm is to sample all possible worlds using a small number m of possible worlds S_i ($1 \leq i \leq m$) where each S_i also contains n instances – one per object. Intuitively, for an uncertain object U, if its sampled instance is inside r_q in m' out of m corresponding sampled possible worlds, $P_{app}(U, q)$ can be approximated by m'/m. If $m'/m \geq p_q$, then U is considered a result contributed to q.

We consider two scenarios in this section. The first one is for the case where the number of uncertain objects is relatively small, we sample the possible worlds only. Techniques developed are called *SingleSample*. The second one is deployed when the number of uncertain objects is also large. It will become too time-consuming to get an approximate answer with decent accuracy performance. In this case, in the first step we sample a small number of uncertain objects $\mathcal{U}_s$, in the second step sampling possible worlds is applied on $\mathcal{U}_s$. The technique is thus named *DoubleSample*. Query processing algorithms and accuracy guarantee are presented for both techniques.

4.1   SingleSample

In this section, an aU-tree is built on the MBR of sampling instances from all uncertain objects, as illustrated in Figure 4. In Figure 4, each uncertain object is

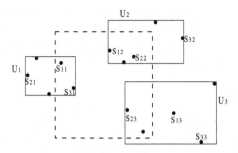

Fig. 4. aU-tree indexing uncertain data

represented using 5 sampling instances. $S_{i,j}$ represents the sampling instance from the i-th sample worlds ($1 \leq i \leq m$) and the j-th uncertain object ($1 \leq j \leq n$).

Approximate Query Algorithm. Theoretically, for each sampled possible world S_i ($1 \leq i \leq m$), we process a PTRA query and record for every uncertain object U_j ($1 \leq j \leq n$) whose sample instance $S_{i,j}$ is inside r_q. Noticing the fact that $S_{i,j}$ is inside r_q is not affected by the sample instances from the other possible worlds or other uncertain objects, a more efficient query algorithm is developed in this section. As shown in Figure 4, two sampled instances from U_1 are inside the query region, so $P_{app}(U_1, q)$ is approximated by 2/3; similarly for U_2 and U_3, the approximated probability is 2/3 and 1/3, respectively. If the given probability threshold is 1/2, then result for this PTRA is 2 (U_1 and U_2). Algorithm detail is described in Algorithm 2.

Algorithm 2. Query Processing for SingleSample

INPUT: aU-tree indexing the MBR of all uncertain objects;
probabilistic query q with probability threshold p_q; query range r_q;
OUTPUT: *result*:number of objects inside r_q with probability $\geq p_q$;

1. sample m possible worlds from all possible worlds;
2. apply pruning/validating techniques on aU-tree as in Algorithm 1;
3. **if** a node N is validated **then**
4.   result += $N.agg$;
5. **for** every uncertain object U_i that can not be pruned/validated **do**
6.   record m' as number of sampled instance inside r_q;
7.   **if** $m'/m \geq p_q$ **then**
8.   result += 1;
9. **return** r esult;

Accuracy Guarantee. Theorem 1 presents the accuracy guarantee for Algorithm 2.

Theorem 1. *Suppose $s_1 = \frac{1}{\epsilon_1^2} log \delta_1^{-1}$ sampling instances are drawn for each uncertain object, then the following inequation holds with probability at least 1 - δ_1,*

$$|P^E - P| \leq \epsilon_1 \tag{3}$$

where P is the actual appearance probability of the uncertain object and P^E is the probability computed using Algorithm 2.

The theorem can be proved using the Chernoff Hoeffding bound. Proof details are omitted due to space limitation.

4.2 DoubleSample

When the number of uncertain objects is huge, the aU-tree in Algorithm 2 may take a large space which prevents the range query from efficiently processing. In this case, we propose a solution to get a sample set of uncertain objects $\mathcal{U}_s$ first, and then sample the possible worlds based on $\mathcal{U}_s$.

Sample Uncertain Objects. A naive way to select uncertain objects is to use uniform sampling; however, this may lead to lose of spatial distribution of uncertain objects. Instead, we utilize Min-Skew partitioning technique to select K nodes from the aU-tree indexing the MBR of uncertain objects. K can be a user-specific parameter to meet with the space requirements. We call the selected K nodes best K nodes ($BKNs$).

The criteria to select $BKNs$ is that the sum of the spatial skewness of the K subtrees is as small as possible meanwhile cover all uncertain objects. To do this, we propose an efficient heuristic. In the first step, identify from aU-tree a level L which has intermediate nodes less than K and the number of intermediate nodes on its child level $L-1$ is larger than K. (Note that if there exists a level L with exactly K nodes then we simply choose them as the K nodes.) Otherwise, each node N_i on level L is split into G_i buckets, making $\sum_{i=1}^r G_i = K$, where r is the number of nodes on level L. G_i $(1 \leq i \leq r)$ is computed according to the spatial skewness of each node. For two nodes N_i and N_j, $G_i : G_j = Sk_i : Sk_j$, where Sk represents the skewness of a node. After getting the $BKNs$, we perform uniform sampling on each selected node and obtain the sampled objects $\mathcal{U}_s$.

Sampling possible worlds is processed the same as in *SingleSample* technique. The difference is we form sample worlds based on $\mathcal{U}_s$ instead of $\mathcal{U}$.

Approximate Query Algorithm. Based on $\mathcal{U}_s$, we process Algorithm 2 which returns *result*. Since we apply sampling twice in DoubleSample, the final result for *DoubleSample* is therefore:

$$result(DoubleSample) = result * \frac{|\mathcal{U}|}{|\mathcal{U}_s|} \tag{4}$$

Accuracy Guarantee. The following theorem states the accuracy guarantee of DoubleSample technique.

Theorem 2. *Let A be number of objects with appearance probability $P_x \geq p_q$. Suppose $s_2 = \frac{1-\delta_1}{\epsilon_2^2} \log \delta_2^{-1}$ sampling objects are drawn. s_1 sampled instances are generated for each sampled uncertain object. Assuming the appearance probability of each uncertain object follows uniform distribution regarding the query, then the following inequality holds with probability at least 1 - δ_2:*

$$|A^E - A| \leq (\epsilon_1 + \epsilon_2) * N \qquad (5)$$

where A^E is the estimated value of A and N is the total number of uncertain objects. s_1, ϵ_1 and δ_1 are the same as in Theorem 1.

This theorem can also be proved using Chernoff Hoeffding bound. Limited by space, proof details are omitted.

5   Experimental Analysis

All algorithms are implemented in C++. Experiments are run on *PCs* with Intel P4 2.8GHz CPU and 2G memory under Debian Linux. The page size is fixed to 8192 bytes.

Two real spatial data sets are used in this section. LB with $53K$ points and CA with $62K$ points, presenting locations in the Long Beach country and California. Data domain along each dimension is $[0, 10000]$. An data U is generated with uncertainty region as a circle with radius rad_U 250. For each uncertain object U, 10000 instances are generated and the spatial distribution follows either *uniform* or *Constrained-Gaussian (Con-Gau)* distribution. A synthetic Aircraft data set consisting of $53K$ points is also generated to investigate the performance in $3D$ space with instances in *Con-Gau* distribution. All data sets are downloaded from *http://www.cse.cuhk.edu.hk/ taoyf/paper/tods07-utree.html.*

The query region r_q is a square/cube with radius rad_q ranging from 500 to 1500. The probabilistic threshold p_q is ranged from 0.3 to 0.9.

For approximate querying algorithms, the number of sampling instances varies among $[10, 1000, 1000]$. We denote $BKNs$ as the number of buckets in *Double-Sample* technique. $BKNs$ ranges from 5 to 30 and each bucket keeps 250 sampled objects. The default value for sampling instances is 1000 and for $BKNs$ is 30.

5.1   Efficiency Evaluation

Figure 5 illustrates the query efficiency of aU-tree and *DoubleSample* technique in terms of CPU cost, I/O cost and total cost. *DoubleSample* techniques outperforms aU-tree significantly. As $BKNs$ increases, the query time increases too since more uncertain objects are sampled to form $\mathcal{U}_s$.

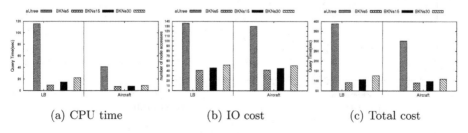

(a) CPU time   (b) IO cost   (c) Total cost

Fig. 5. Efficiency Evaluation

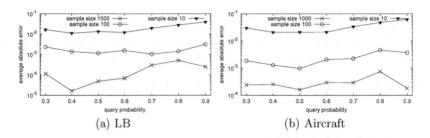

Fig. 6. Accuracy Evaluation for singleSample

5.2   Accuracy Evaluation

Accuracy is defined as relative error of approximate answer w.r.t. exact answer by the formula: $accuracy = |\frac{R_{ex} - R_{ap}}{R_{ex}}|$, where R_{ex} and R_{ap} present the exact result and approximate result, respectively. We use 0.4 as a default value for p_q.

We evaluate the accuracy for *SingleSample* in Figure 6. As expected, accuracy increases as the sample size gets larger.

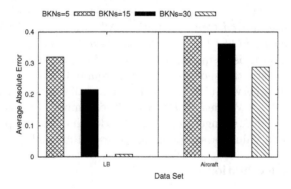

Fig. 7. Accuracy Evaluation for doubleSample

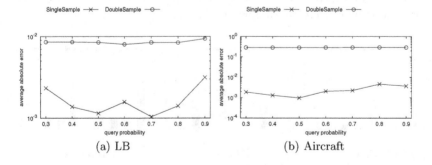

Fig. 8. Accuracy Comparison

In Figure 7, we evaluate the accuracy of *DoubleSample*. We fix the sample instance size at 1000 and vary $BKNs$. Clearly, if more $BKNs$ are deployed, higher accuracy will be obtained. This trend is obvious for 2D dataset LB and less obvious for 3D datasets. We compare the accuracy of *SingleSample* and *DoubleSample* in Figure 8. As shown, in both 2D and 3D datasets, *SingleSample* is more accurate than *DoubleSample*. Both techniques are not affected significantly by p_q values.

6 Related Work

Considerable research effort has been put into modeling and managing uncertain data in recent years due to many emerging applications. Sarma *et al* [16] models uncertain data using *possible world semantics* and a prototype of uncertain data management system, *Trio*, is developed by the Stanford Info Lab [12]. General issues in modelling and managing uncertain data are addressed by Dey and Sarkar in [4], Lee in [11], and Antova, Koch, and Olteanu in [6]. Querying uncertain data by the probabilistic paradigm has been investigated by Dalvi and Suciu in [2] and Sen and Deshpande in [17]. Very recently Dalvi and Suciu [3] have shown that the problem of query evaluation over probabilistic databases is either $PTIME$ or $\#P\text{-}complete$.

A number of problems in querying uncertain data have also been studied, such as indexing [18], similarity join [5], nearest neighbor query [9], skyline query [15], clustering [10], etc. Relatively complete and detailed study of existing techniques on managing uncertainty can be found in [14] and [20].

Range aggregate query over certain data is thoroughly studied in [19]. Range query over uncertain data [18] is the one that is the mostly related with our work, as we introduced in Section 2. To the best of our knowledge, this paper is the first one to address range aggregate query over uncertain data.

7 Conclusion

An important problem, probabilistic threshold range aggregate query over uncertain data is investigated in this paper. After formally defining this problem, we propose a novel index structure aU-tree to retrieve exact answers to $PTRA$ queries. To trade-off between efficiency and accuracy, *SingleSample* and *DoubleSample* methods are developed to approximately answer $PTRA$ queries. Our experimental study confirms the efficiency and effectiveness of the techniques we proposed.

Our techniques proposed can be extended to *continuous cases* directly. Based on the continuous probability density functions of uncertain objects, Monte Carlo sampling technique can be utilized and obtain a set of uncertain instances for each object. Thus both exact and approximate querying algorithms can be applied.

Acknowledgement. The work of the authors was partially supported by ARC Grant (DP0881035, DP0666428 and DP0987557) and a Google Research Award.

References

1. Acharya, S., Poosala, V., Ramaswamy, S.: Selectivity estimation in spatial databases. In: SIGMOD 1999 (1999)
2. Dalvi, N., Suciu, D.: Efficient query evaluation on probabilistic databases. In: VLDB 2004 (2004)
3. Dalvi, N., Suciu, D.: Management of probabilistic data: foundations and challenges. In: PODS 2007 (2007)
4. Dey, D., Sarkar, S.: A probabilistic relational model and algebra. In: TODS 1996 (1996)
5. Kriegel, H.P., et al.: Probabilistic similarity join on uncertain data. In: Li Lee, M., Tan, K.-L., Wuwongse, V. (eds.) DASFAA 2006. LNCS, vol. 3882, pp. 295–309. Springer, Heidelberg (2006)
6. Antova, L., et al.: 10^{10^6} worlds and beyond: Efficient representation and processing of incomplete information. In: ICDE 2007 (2007)
7. Guttman, A.: R-trees: A dynamic index structure for spatial searching. In: SIGMOD 1984 (1984)
8. Hua, M., Pei, J., Lin, X., Zhang, W.: Efficiently answering probabilistic threshold top-k queries on uncertain data. In: ICDE 2008 (2008)
9. Kriegel, H.-P., Kunath, P., Renz, M.: Probabilistic nearest-neighbor query on uncertain objects. In: Kotagiri, R., Radha Krishna, P., Mohania, M., Nantajeewarawat, E. (eds.) DASFAA 2007. LNCS, vol. 4443, pp. 337–348. Springer, Heidelberg (2007)
10. Kriegel, H.P., Pfeifle, M.: Density-based clustering of uncertain data. In: KDD 2005 (2005)
11. Lee, S.K.: Imprecise and uncertain information in databases: an evidential approach. In: ICDE 1992 (1992)
12. Agrawal, P., Benjelloun, O., Das Sarma, A., Hayworth, C., Nabar, S., Sugihara, T., Widom, J.: Trio: A system for data, uncertainty, and lineage. In: VLDB 2006 (2006)
13. Papadias, D., Kalnis, P., Zhang, J., Tao, Y.: Efficient OLAP operations in spatial data warehouses. In: Jensen, C.S., Schneider, M., Seeger, B., Tsotras, V.J. (eds.) SSTD 2001. LNCS, vol. 2121, p. 443. Springer, Heidelberg (2001)
14. Pei, J., Hua, M., Tao, Y., Lin, X.: Query answering techniques on uncertain and probabilistic data. In: SIGMOD 2008 (2008)
15. Pei, J., Jiang, B., Lin, X., Yuan, Y.: Probabilistic skyline on uncertain data. In: VLDB (2007)
16. Sarma, A.D., Benjelloun, O., Halevy, A., Widom, J.: Working models for uncertain data. In: ICDE 2005 (2005)
17. Sen, P., Deshpande, A.: Representing and querying correlated tuples in probabilistic databases. In: ICDE 2007 (2007)
18. Tao, Y., Cheng, R., Xiao, X., Ngai, W.K., Kao, B., Prabhakar, S.: Indexing multidimensional uncertain data with arbitrary probability density functions. In: VLDB, pp. 922–933 (2005)
19. Tao, Y., Papadias, D.: Range aggregate processing in spatial databases. ACM TODS 16(12), 1555–1570 (2004)
20. Zhang, W., Lin, X., Pei, J., Zhang, Y.: Managing uncertain data: Probabilistic approaches. In: WAIM 2008 (2008)

Dynamic Data Migration Policies for Query-Intensive Distributed Data Environments[*]

Tengjiao Wang[1], Bishan Yang[1], Allen Huang[2], Qi Zhang[1], Jun Gao[1],
Dongqing Yang[1], Shiwei Tang[1], and Jinzhong Niu[3]

[1] Key Laboratory of High Confidence Software Technologies, Peking University,
Ministry of Education, China
School of Electronics Engineering and Computer Science, Peking University
Beijing, 100871 China
{tjwang, bishan_yang, rainer, dqyang, gaojun, swtang}@pku.edu.cn
[2] Microsoft SQL China R&D Center
allen.huang@microsoft.com
[3] Computer Science, Graduate School and University Center,
City University of New York
365, 5th Avenue, New York, NY 10016, USA
jniu@gc.cuny.edu

Abstract. Modern large distributed applications, such as telecommunication and banking services, need to respond instantly to a huge number of queries within a short period of time. The data-intensive, query-intensive nature makes it necessary to build these applications in a distributed data environment that involves a number of data servers sharing service load. How data is distributed among the servers has a crucial impact on the system response time. This paper introduces two policies that dynamically migrate data in such an environment as the pattern of queries on data changes, and achieve query load balance. One policy is based on a central controller that periodically collects the query load information on all data servers and regulates data migration across the whole system. The other policy lets individual server dynamically selects a partner to migrate data and balance query load in between. Experimental results show that both policies significantly improve system performance in terms of average query response time and fairness, and communication overhead incurred is marginal.

1 Introduction

As more and more query-intensive applications, such as telecommunication and banking services, are running in large distributed data environments, it is a

---

[*] This work is supported by the Cultivation Fund of the Key Scientific and Technical Innovation ProjectMinistry of Education of China(No.708001), the National '863' High-Tech Program of China(No.2007AA01Z191,2006AA01Z230), and the NSFC(Grants 60873062).

Q. Li et al. (Eds.): APWeb/WAIM 2009, LNCS 5446, pp. 63–75, 2009.

major concern for these service providers how to serve millions of requests with short response time. These environments typically involve a number of data servers sharing service load and the distribution of data among these servers has a crucial impact on how fast queries are fulfilled in average.

A typical example of these applications is telephony service. When a user dials a number to make a call, a request is made to query the callee's location information so as to establish a connection between the two parties. Typically each user's location information is stored as a record in a database that spans across a set of distributed database servers, as the overall data volume is massive. Assuming that each record is unique system-wide, a query on a particular record needs to be routed to the server on which the record is stored. It is usually the case that the number of queries on records are not evenly distributed. A high access frequency of a small set of data records on a single server may easily overload the server; while some other servers may be basically idle if their data is rarely accessed. When load imbalance is severe, the average query response time may easily surpass the maximal time human users can stand. Therefore, how the records are distributed across the system is an important issue for system designers.

Generally, these systems share the following features:

- A massive amount of data is distributed among multiple database server nodes without duplicates. That is a data record is available on exactly one node in the system and queries on the data have to be routed to the node.
- Fast response and high efficiency in serving queries are guaranteed even when the overall query load is high system-wide.
- Data distribution has a significant impact on query load distribution. When data is unevenly distributed among nodes, load imbalance occurs and queries on some records may have to wait a long time to get served.
- The pattern of queries may change over time, and as a result, data records need to be migrated between server nodes dynamically to balance workload.

1.1 Related Work

Load balancing involving data distribution has been addressed in [1,2]. Their approaches are however used to allocate web documents among a cluster of web servers in order to achieve load balance, which is different from the problem we would like to tackle.

A relating problem domain is data distribution in parallel storage systems. In these systems, *horizontal data partitioning* has been commonly used to distribute data among system nodes. There are further three different strategies for horizontal partitioning: *round-robin, hash,* and *value-range partitioning*. The value-range partitioning may create skewed data distributions. The round-robin partitioning is capable of allocating data approximately evenly between partitions, but it requires brute-force searches and is ineffective for queries. The hash partitioning can not only obtain even data distribution, but also perform well for exact matching queries. Although these partitioning strategies each have both advantages and disadvantages, they do not take into consideration the access pattern on data, which would be a problem when the pattern is not a uniform distribution. Dynamic data reallocation is necessary to effectively adapt for changes of access patterns. When access frequencies of data items lead to unbalanced workload on system nodes, data migration should be performed to balance workload. The simple data skew handling method in [3] balances the storage space for data, but does not guarantee balanced data access load across system nodes. Similarly most access load skew handling methods do not consider data balancing [4,5]. In [6], a data-placement method was proposed to balance both access load and storage space of data, but it is applied in parallel storage system, and focuses on achieving availability and scalability for parallel storage configuration.

In this paper, we introduce a series of effective policies for dynamic data migration in query-intensive distributed data environments. Our work differs from previous work in that our approach models patterns of end-user queries and performs dynamic data migration to speed up query processing. To adapt for the ever-changing load distribution in the system, we design dynamic data migration policies to do query load balancing. The idea is basically to collect query load information at system nodes periodically and perform data migration automatically to obtain load balance in the system. In particular, our work has the following contributions:

- We propose a dynamic data migration policy with centralized control. The data migration process is controlled by the central controller that makes decisions from the system-wide viewpoint.
- We propose a dynamic data migration policy with decentralized control. Each individual node has its local viewpoint and makes decisions concurrently along others.
- We develop a platform to evaluate the effectiveness of our proposed policies. Experimental results show that these policies perform almost optimally, and are able to significantly reduce query response time even with the varying query load across the system.

The remainder of this paper is organized as follows: Section 2 presents a model for the query-intensive distributed data environments; Section 3 describes the two proposed dynamic data migration policies in details, one with centralized control and the other with distributed control; Section 4 describes the experimental environment and presents an experimental comparison between the three strategies; and Section 5 concludes this article.

2   Model Description and Formulation

This section presents a query-intensive distributed model, which is commonly used in mobile communications[7]. As Fig. 1 depicts, the model can be divided into three parts: back-end (BE) servers, front-end (FE) servers and application clients (ACs).

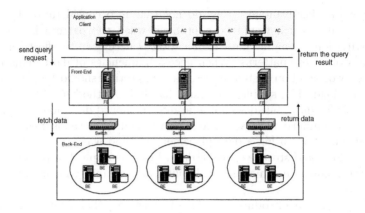

Fig. 1. A model for query-intensive distributed data environment

The BE servers hold data that would be queried about and together form a huge database system. Here we assume that data is distributed between BE servers without overlap and can be migrated between them.[1] The databases on BE servers have a homogeneous structure, but the processing powers of these BE servers may differ. When a user sends a query through an application client, it first arrives at one of the FE servers and then is routed to the BE server where the queried data is located. The BE server then executes the query on its local database, and returns the requested information.

We now make several definitions that would be needed in subsequent sections.

Definition 1. *The query load associated with a data unit is the frequency of read operation[2] on it in a unit of time, denote as l.[3]*

Definition 2. *Given a data group, G, which has k data units with their corresponding loads being $l_1, l_2, ..., l_k$ respectively, the query load associated with G is defined as $L = \sum\limits_{j=1}^{k} l_j$.*

Definition 3. *The process power of a BE sever, μ, is the highest query frequency it can deal with within a unit of time. If there are k data groups on the BE server and $L_1, L_2, ..., L_k$ are their query loads respectively, the load level of the BE server is defined as $\beta = \dfrac{\sum\limits_{i=1}^{k} L_i}{\mu}$.*

---

[1] Multiple BE servers could possibly be used to hold a same set of data units for higher reliability, with one of them being the main sever and the others being backups.

[2] We only consider read operations here because in the environment read operations are the main operation.

[3] In our experiment one unit of time is 10 seconds, which is common in most applications in our application domain.

3   Dynamic Data Migration Policies

To adapt the fluctuation of query load in the system, dynamic data migration policies are needed. We propose two dynamic data migration policies that can adapt for the changing pattern of query load, one with centralized control and the other with distributed control.

Either migration policy consists of three components: (1) information rule, which defines how the query load information is collected and maintained; (2) selection rule, which regulates the selection of source server nodes, with high load, and destination server nodes, with low load, for data migration; (3) migration rule, which determines when and how to migrate data.

3.1   DDMC: Dynamic Data Migration with Centralized Control

We now present the dynamic data migration policy with a central controller (CC). With centralized control, the global information of the query load is collected by asking each BE server to inform the central center of its local load. Data migration decisions are made by the central controller itself. The processing power of the central controller is assumed high enough to fulfill the task. It can be implemented by a low cost processor or even run as a single process on an existent node in the system.

Information Rule. Suppose the query load of data group G_i on FE_j is denoted as $L_{i,j}$, and there are totally N_{fe} FE servers in the system, the total query load of data group G_i is then the sum of query loads recorded on all the FE servers, expressed as $L_i = \sum_{j=1}^{N_{fe}} L_{i,j}$. Since the data groups have no overlap across BE servers, we can calculate the load level of BE_k as the ratio of the overall load of the data groups on BE_k and BE_k's processing power $\beta_k = \sum L_j/\mu_i$.

To avoid high communication overhead, CC periodically notifies all the FE servers to collect load information of each data group in their data group tables. The gathered information is also stored in a table on CC, which has the same structure as the FE data group tables though it records the global information of data groups. CC also maintains a table, called T_{BE}, as a whole picture of the BE servers. Each entry of T_{BE} contains information about a BE server, including its *BE's ID*, its *processing power*, and its current *load level*.

Selection Rule. The objective is to balance the query load in the system through data reallocation. Since BE servers may have various processing capabilities, we can not simply redistribute data to make an uniform load distribution. Intuitively, if all the nodes are at the same level of utilization, the resources are considered to have been made best use of and the servers have achieved system-wide balance. We classify the BE servers into three categories: HL (Heavily Loaded), NL (Normally Loaded), and LL (Lightly Loaded). The classification is based on the average load level of the system: $AvgL = (\sum_{i=1}^{N} L_i)/(\sum_{i=1}^{N} \mu_i)$.

The system is viewed as in an optimal state when the load levels of each BE server converged to to $AvgL$. With a toleration constant ϕ, a node is considered a HL one if its load level $\beta > AvgL + \phi$, a NL one if $AvgL - \phi \le \beta \le AvgL + \phi$, or a LL one if $\beta < AvgL - \phi$. All the HL nodes are chosen as source node candidates and LL nodes are chosen as destination node candidates.

The basic idea is to balance loads between HL nodes and LL nodes. First a HL node is selected, then a LL node is selected for data migration between the two parties. Due to the communication overhead in a large distributed network, only a small set of LL nodes should be selected for a particular HL node. We, in particular, choose the LL node from the k nearest neighbors to the source node. To further minimize the migration overhead, we assign each destination node candidates a *priority*, which represents the priority for it to be chosen. For a destination node candidate BE_i, the priority, p_i, is defined as: $p_i = \left(\frac{\mu_i * (AvgL+\phi) - L_i}{\max_j \{\mu_j * (AvgL+\phi) - L_j\}} \right)$ The priority reflects how much potential a LL node has to take more query load. We select the node with the highest priority as the destination node.

Migration Rule. To speed up data migration, the algorithm is divided into two stages: *decision making* and *data migrating*.

First, the source node candidate with the highest load level is chosen as a source node, called BE_i, and the destination node candidate with the highest priority is chosen as a destination node, called BE_j. In order to move less data, the data groups on BE_i are considered in descending order of their corresponding load values. A data group that is eligible for migration if it satisfy the following two conditions:

Condition 1. $\omega^c \delta_i > \Delta$, where c is the number of times the data group has been migrated, ω ($0 < \omega \le 1$) is a weighting factor used to prevent the data group from being migrated too frequently, δ_i describes how heavier the load level of the source node is relatively, and Δ (> 0) is a constant used to protect the system from potential instability. In particular, δ_i is defined as: $\delta_i = \beta_i / \max_k \beta_k$ where $\max_k \beta_k$ is the highest load level in the system. When $\omega = 1$, c will not effect the data migration decisions, and only the load level of the node matters; and when $\omega = 0$, no data groups would be migrated.

Condition 2. After the migration, the load level of the source node should not be lower than $AvgL - \phi$ and the load level of the destination node should not be higher than $AvgL + \phi$, which can be expressed as $\beta_i - l/\mu_i \ge AvgL - \phi$ and $\beta_j + l/\mu_j \le AvgL + \phi$, where l is the query load of the data group to be migrated from the source node to the destination node.

It is possible that there are no data groups on the source node satisfying the above two conditions. Under this case, data groups on the destination node will also be taken into consideration and we allow data exchanging between the source node and the destination node. A data group with load l on the source node BE_i and a data group with load l' on the destination node BE_j would be exchanged if $\beta_i - (l-l')/\mu_i \geq AvgL - \phi$ and $\beta_j + (l-l')/\mu_j \leq AvgL + \phi$destination node should be searched in the order of increasing load. The first one that meets the above conditions is selected for exchanging.

Once the migration decision is made, the central controller will control the migration process. The priority of the destination node will be updated, and the new destination node will be chosen after migration. This procedure is repeated until all the data groups on the source node are explored for the migration possibilities. Then the next source node candidate is chosen. Until the iterative loop on the source node candidates is finished, the decision process is ended.

Data Migration Mechanism. We provide a data migration mechanism for ensuring data consistency during the migration process. The exchanging process can be divided into two migration processes, by exchanging source and destination.

Step 1. At the beginning, CC sends a command to source node BE_s for migrating data D to destination node BE_d.
Step 2. BE_s performs the migration.
Step 3. If BE_d successfully receives and stores the data, it sends a success signal to CC.
Step 4. CC broadcasts messages to all FE servers to ask for location update in their data group tables.
Step 5. Each FE server performs location update, and then sends a success signal to CC.
Step 6. After receiving the success signals from all FE servers, CC sends a command to BE_s for deleting the migrated data D.
Step 7. BE_s performs the deletion.

3.2 DDMD: Dynamic Data Migration with Distributed Control

This section provides an alternative approach without any centralized control. Individual BE server makes the data migration decision based on their own situation. With this policy, each BE server calculates $AvgL$ by broadcasting its own information to other BE servers in the system. Similar to the policy with centralized control, the BE servers are classified into three categories: HL, NL and LL.

The policy is a HL-initiating one. The objective is to lighten the query load in HL nodes by migrating their frequently queried data to the LL nodes. Only the HL nodes need to store information about the LL nodes. This will enable the HL nodes to make migration decisions, i.e., choosing a LL node that is the best for migration. A LL node may receive migration requests for different HL nodes at the same time. In this case, the LL node will choose the one it thinks the most deserved for migration. The decision of the HL node is valid only if the decision is accepted by the chosen LL node. In this way, a HL node that is too desperate in reducing its load is regulated to some extent.

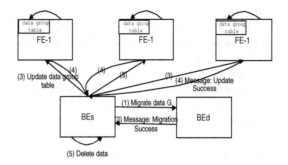

Fig. 2. Data migration process with distributed control

Information Rule. To obtain the overall load, each BE server requests all the FE servers to collect the corresponding data groups information. Then it calculates and stores its own load value L and the load level β, and classifies itself to HL, NL or LL in the light of the threshold. After that, each BE server broadcasts their own state to other BE servers in the system. In this way, every BE server can know the average load level, $AvgL$, of the system. If a server finds itself as a LL node, it will build up a HL-list to record all the HL nodes's IDs, for knowing where its information has been kept. If a server is a HL node, it will build up a LL-List and store the load values of the LL nodes from their broadcasting. The aim of this process is to build up the necessary information for migration decisions.

Selection Rule. The basic idea is that each HL node chooses the destination node candidate that most appropriate for migration regardless of other HL nodes's decisions. On the other hand, the LL node tries to let the most needed HL node to migrate data. This is a concurrent decision making process. Individual HL node can making decisions simultaneously based on their own viewpoints of the system. Specifically, they have different description about the LL nodes in the system, owing to the distance factor. The method used to measure the priority of the LL node is the same way as that used in the centralized policy. It can be computed locally since the HL node knows the current load and the processing power.

Migration Rule. Based on the selection rule, the HL node chooses the destination node based on the priority, from high to low. Since the selection process in various HL nodes can be taken place at the same time, it is likely that a LL node is selected as a destination node by more than one HL nodes. To maintain a coherent view of the system, the load copies of the LL nodes must be consistent. To address this coherency problem, the selected LL node should choose only one source node, and make other HL nodes in the HL-List to see its new load.

The migration mechanism for DDMD is introduced in the follow. It only deals with the information update for the migrated data, and the overall load value of individual node is assumed to get consistent through the above communication process. Fig. 2 shows the mechanism.

Step 1. the data group migrates from the source node to the destination node.
Step 2. if the destination node successfully receives the data, it return a success signal to the source node.
Step 3. the source node notifies all the FE servers to update the location information of the migrated data.
Step 4. All the FE servers performs the update and then sends a signal to the source node.
Step 5. When the source node receives signals from all the FE servers, it deletes the data and its corresponding information in the local table.

This mechanism can also be extended to apply in data exchanging by replacing the source node by a LL node and the destination node by a HL node.

4 Experiments

For the purpose of this study, we prepare real customer data from a mobile communication company. It contains 6,000,000 users covering attributes like the basic ID of the user, the current location of the user and IMSI (International Mobile Subscriber Identification Number). All the data is stored in database, and is queried and processed through SQL sentence. Our platform consists of six BE servers and one FE server. Each server runs a DB2 V8.0 database. The FE server is responsible for dispatching an incoming query request to one of the BE servers. The hardware configurations are presented in Table 1. The processing rate is measured by the number of queries being dealt with by 600 threads in a second, and the processing rate of the slowest computer is set to 1. All the system nodes are assumed to be connected by a communication network with 100Mbps bandwidth. A query was created by generating an id number, which was used as the key for searching the users' location information in the database. They are generated according to a Zipf distribution, the Zipf factor is varied from 0.7 to 0.9. The reason is that in real case, only a few data is queried with high frequency in a period, a medium amount of data with middle query frequency, and a large amount of data is queried not so continually.

Table 1. System configuration

CPU	Memory	Number	Process power(relative)
2.0GHz P4	512M	1	1
4*1.8GHz AMD Opteron 865	18GB	1	5
AMD Athlon 64 3200+	1G	4	2

4.1   Description of Experiments

To simulate the queries in the real applications, we generate 600 threads to send queries to the system concurrently, each thread carrying several SQL queries. We found from experiments that if the number of overall queries reach 6000, the system resources are in the largest extent of use. Hence we generate queries from 3000 to 6000 to vary the system load level from 0.5 to 1.0.

For comparison purpose, an algorithm NoLB was used. It distributes data to databases based on the commonly used horizontal fragmentation method and does not consider any query load information. A static data distribution algorithm DAH in [9] is also used for comparison.

Initially, the data is distributed by NoLB. The data group size is set to be 100 records. The system load level is defined as the ratio of total query frequency to the overall processing power of the system: $AvgL = (\sum_{i=1}^{N} L_i)/(\sum_{i=1}^{N} \mu_i)$. At initial state, we set the system load level at 0.5. Then we varied it from 0.5 to 1.0.

Comparisons between DAH, DDMC, and DDMD. In the second experiment, the effectiveness of the dynamic data migration policies are evaluated, including DDMC and DDMD. After the system query load is moderated by DAH, we vary the query distribution a little by tuning the Zipf factor. At this time, the previous distribution is not adaptable, and DDMC and DDMD policies are considered. As for DDMC, the central controller periodically collected the query load information from the FE server at a time interval of 10 minutes. Similarly for DDMD, BE servers gather information from each other at each 10 minutes. A large number of runs for DDMC and DDMD were conducted with different values of adjustable parameters, and the best combination of those parameter values were used. The tolerable deviation of the average query load ϕ is set at 10% of $AvgL$.

In Fig. 3, we present the average response time of DDMC, DDMD and DAH with the increasing load level of the system. From the result in Fig. 3, DDMC and DDMD both provide substantial speedup of the average response time of the

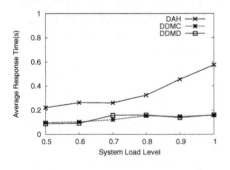

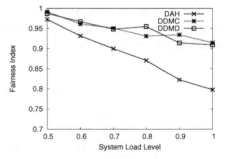

Fig. 3. Average response time

Fig. 4. Fairness indices with DAH, DDMC, and DDMD respectively

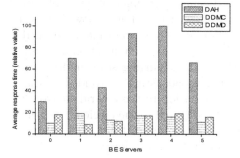

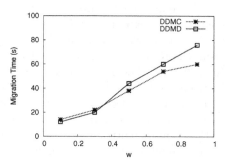

Fig. 5. Average response time at individual servers

Fig. 6. Migration time with different migration weights

system over the situation where there are no dynamic migration decisions. This is because dynamically data migration can contribute to achieve load balancing and adapt the changing query patterns, and the overall performance of the system can be expected to be improved. The comparison results are more obvious with higher load level of the system. The performance provided by DDMD is close to that provided by DDMC. This confirms the fact that making decisions based on individual server's viewpoint will not be inferior to the system-wide viewpoint.

The average response time for individual BE server before and after dynamic migration is also given. Fig. 5 shows the average response time when the system's load level is at 0.7. Both the two policies moderate the response time and make much better utilization of the system resources. Each BE server's load level is inside the expected zone of the average load level.

Fairness index I proposed in [8] is used as a measure of balance: $I = \dfrac{[\sum\limits_{i=1}^{n} F_i]^2}{n \sum\limits_{i=1}^{n} F_i^2}$

where F_i is the expected response time of node i. If the expected response time is the same for all nodes, then $I = 1$, which means the system achieves fairness distribution. Otherwise, if the expected response time differs significantly from one node to another, then I decreases, which means the system suffers load imbalance to some extent. With respect to fair index, Fig. 4 shows that DDMC and DDMD has a close fairness index of about 0.91 to 1.0. The fairness index of DAH varies from 0.97 at high load to 0.79 at low load, while DDMC and DDMD shows stable fair index value with the increasing load level of the system.

Comparisons between DDMC and DDMD on Communication Overhead. In the dynamic policies, the communication overhead costs caused by gathering and exchanging information may negate the benefits of them. In Fig. 6, we show cost of the dynamic migration algorithms DDMC and DDMD in terms of their communication time.

To examine how communication overhead changes at different levels of data migration, we experimented with ω varying from 0.1 to 1.0. Fig. 6 show that with the increasing value of ω, more communication and migration time are needed. From the result, we can see that DDMD incurs more communication overhead than DDMC. The reason is that DDMD needs exchanging state information between individual servers, while in DDMC, broadcasting is avoided in that each server only needs to report its state information to the central controller. However, the cost difference is not large, and compared to the efficiency benefit, DDMD is also an attractive alternative, especially for highly distributed settings, where communication cost is high and reliability is a top priority.

5   Conclusion

In this paper, we propose a series of policies to address the response efficiency challenge in the distributed query-intensive data environment. The main goal is to reduce the query response time by dynamically adjusting data distribution and achieve balanced load. To effectively adapt the system changes, two dynamic policies DDMC and DDMD are designed to adapt for the changing query load and obtain load balance through data migration. DDMC uses a central controller to make migration decisions based on the global load information of the system, while in DDMD, individual server makes decisions based on their own viewpoints of the system without centralized control. Our experiments show that DDMC and DDMD exhibit similar performance in term of average query response time, and as expected, DDMD involves higher communication overhead than DDMC, but the difference is not significant. The experimental results also show that the proposed policies offer favorable response time with increasing query load system-wide.

References

1. Narendran, B., Rangarajan, S., Yajnik, S.: Data distribution algorithms for load balanced fault-tolerant Webaccess. In: Proceedings of The Sixteenth Symposium on Reliable Distributed Systems, pp. 97–106 (1997)
2. Savio, S.: Approximate Algorithms for Document Placement in Distributed Web Servers. IEEE Transactions on Software Engineering, 100–106 (2004)
3. Yokota, H., Kanemasa, Y., Miyazaki, J.: Fat-Btree: An Update-Conscious Parallel Directory Structure. In: International Conference on Data Engineering (ICDE), pp. 448–457 (1999)
4. Lee, M.-L., Kitsuregawa, M., Ooi, B.-C., Tan, K.-L., Mondal, A.: Towards Self-Tuning Data Placement in Parallel Database Systems. In: International Conference on Management of Data (SIGMOD), pp. 225–236 (2000)
5. Feelifl, H., Kitsuregawa, M., Ooi, B.-C.: A fast convergence technique for online heat-balancing of btree indexed database over shared-nothing parallel systems. In: Ibrahim, M., Küng, J., Revell, N. (eds.) DEXA 2000. LNCS, vol. 1873, pp. 846–858. Springer, Heidelberg (2000)

6. Watanabe, A., Yokota, H.: Adaptive Lapped Declustering: A Highly Available Data-Placement Method Balancing Access Load and Space Utilization. In: International Conference on Data Engineering (ICDE), pp. 828–839 (2005)

7. Feldmann, M., Rissen, J.P.: GSM Network Systems and Overall System Integration. Electrical Communication (1993)

8. Jain, R.: The Art of Computer Systems Performance Analysis: Techniques for Experimental Design, Measurement, Simulation, and Modeling. Wiley Interscience, Hoboken (1991)

9. Wang, T., Yang, B., Gao, J., Yang, D.: Effective data distribution and reallocation strategies for fast query response in distributed query-intensive data environments. In: Zhang, Y., Yu, G., Bertino, E., Xu, G. (eds.) APWeb 2008. LNCS, vol. 4976, pp. 548–559. Springer, Heidelberg (2008)

C-Tree Indexing for Holistic Twig Joins

Bo Ning[1,2], Chengfei Liu[1], and Guoren Wang[2]

[1] Swinburne University of Technology, Melbourne, Australia
{BNing,CLiu}@groupwise.swin.edu.au
[2] Northeastern University, Shenyang, China
wanggr@mail.neu.edu.cn

Abstract. With the growing importance of semi-structure data in information exchange, effort has been put in providing an effective mechanism to match a twig query in an XML database. Bruno et al. have proposed a novel algorithm *TwigStack* to deal with the twig query pattern by scanning the tag streams only once. In this paper, we propose a new index called C-Tree and two algorithms named *NestTwigStack* and *ADTwigStack* to speed up the processing of twig pattern queries by omitting some elements that can be processed without scanning. Using *C-Tree*, our algorithms can accelerate both the ancestor-descendant and parent-child edges by skipping the elements with their context in documents. We complement our research with experiments on a set of real and synthetic data, which are intended to show the significant superiority of our algorithms over the previous algorithms.

1 Introduction

The problem of managing and querying XML documents efficiently poses interesting challenges for database researchers. XML documents have a rather complex internal structure; in fact, XML documents can be modeled as ordered trees, where nodes represent elements, attributes and text data.

Twig pattern becomes an essential part of XML queries. A twig pattern can be represented as a node labeled tree whose edges are either parent-child(P-C) or ancestor-descendant(A-D) relationship. For example, $//S[//C]//A//B$. This expression can be represented as a node-labeled twig pattern (shown in Fig.1(b)) with elements and string values as node labels.

Finding all occurrences of a twig pattern in a database is a core operation in XML query processing[2,4,6]. Before the Holistic twig join algorithm[1] was proposed, some typical algorithms have been used to break the twig query into a set of binary relations between pairs of nodes. The Holistic Twig Join develops a new eyeshot to a twig query, which can reduce the size of intermediate result, especially to the queries with only A-D relationships. There are many works that focus on the indexing XML data to accelerate the processing or boost holism in XML twig pattern matching[7,8,9]. In practice, XML data may have many deep nested elements, and sometimes have large fan-outs. Therefore, it will greatly improve the performance, if an effective index can be designed to skip large number of elements which do not participating in a final twig match.

Q. Li et al. (Eds.): APWeb/WAIM 2009, LNCS 5446, pp. 76–87, 2009.
© Springer-Verlag Berlin Heidelberg 2009

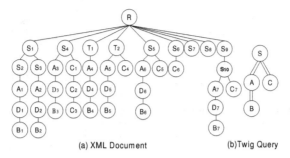

Fig. 1. A sample XML document (a) and a twig query pattern(b)

Bearing this in mind, we propose a new index called C-Tree that captures the context of the indexed elements and develop a series of new algorithms to process holistic twig joins efficiently based on C-Tree. By C-Tree, we can get the context of elements from same tag stream or from different tag streams. The two proposed algorithms, *NestTwigStack* and *ADTwigStack*, can speed up the processing of holistic twig joins, because enormous useless elements can be skipped with the context in C-Tree.

The main contributions of this paper are shown below:

- We propose a new index named *C-Tree* that contains context relationships of elements. We design the streams model named *Context Streaming*, which include the *SelfTagContext streaming* and the *ASDTagContxt streaming*.
- We develop two holistic twig join algorithms which extend the original holistic twig join algorithm and they work correctly on the streaming scheme based on the *C-Tree* index. Those algorithms benefit from skipping lots of elements without matching.
- From experiments we study the performance of our algorithms on both real data and synthetic date. We can see that our algorithms not only accelerate the queries of twig pattern with only A-D edges, but also accelerate the queries of twig pattern with P-C edges.

2 The Context in Tag Streams

In this section, we discuss the context in tag streams which can improve the performance of the twig query. Firstly, we introduce some definitions with respect to a context by using region coding [3,5]. Then, discussions of context information in tag streams are followed.

2.1 Preliminary Definitions

Definition 1 (Transorder). *Given two elements in tag stream T with region E_i ($StartPos_i, EndPos_i$) and E_j ($StartPos_j, EndPos_j$), E_i is said to be **transmigrated** by E_j, or E_j **transmigrates** E_i, if (1)$EndPos_i < StartPos_j$ and (2)$\neg \exists E_k \in T : EndPos_i < StartPos_k < StartPos_j$ The relation between the E_i and E_j is a kind of Partial Ordering Relation, which we call it **Transorder**.*

Fig. 2. An Illustrative Example of Transorder

Fig. 3. An Illustrative Example of SelfTagContext Streaming

It is worth noting that, *transorder* is defined exclusively on a set of elements E of the same tag. The *transorder* can be illustrated in Fig.2. There are seven elements with regions $[StartPos_i, EndPos_i](1 \leq i \leq 7)$ in the tag stream S_E, in which elements are ordered by $StartPos$. There are many *transorder* relations, so we can get a set of Partial Ordering Relations $(e_1,e_5),(e_2,e_3),(e_3,e_5),(e_4,e_5)$ from the stream S_E. It's easy to see that the elements between each Partial Ordering Relation pair is the descendants of the first element in the pair.

Definition 2 (FarthestAnc). *Given an element E_i of tag stream T_D with region $(StartPos_i, EndPos_i)$, and an element E_j from the tag stream T_A with region $(StartPos_j, EndPos_j)$, where tag A is the ancestor of the tag D in DTD. if (1) $StartPos_j < StartPos_i \wedge EndPos_i < EndPos_j$ and (2)$\neg\exists E_k \in T_A : StartPos_k < StartPos_j \wedge EndPos_j < EndPos_k$, E_j is the **FarthestAnc** of the E_i.*

Definition 3 (NextDesc). *Given an element E_m of tag stream T_A with region $(StartPos_m, EndPos_m)$, and an element E_n from the tag stream T_D with region $(StartPos_n, EndPos_n)$, where tag A is the ancestor of the tag D in DTD. If (1) $StartPos_m < StartPos_n \wedge EndPos_n < EndPos_m$ and (2)$\neg\exists E_l \in T_D : StartPos_m < StartPos_l \wedge StartPos_l < StartPos_n$ E_n is the **NextDesc** of the E_m.*

The *FarthestAnc* and *NextDesc* are relations that are exclusively defined on two elements whose tags are A-D or P-C relations in DTD. If element E_A is E_D's *FarthestAnc*, there may be some other elements which are from the stream where E_A comes. These elements are the E_D's ancestors too, however no element whose StartPos is smaller than E_A exists. For example, in Fig.1(a), S_9 is C_7's *FarthestAnc*, while S_{10} is not. A_6 is the *NextDesc* of S_5, and A_7 is the *NextDesc* of S_9.

2.2   SelfTagContext Streaming

The *transorder* context in the tag streams is revealed by the *SelfTagContext* Streaming scheme, where the element in streams focuses on the context of itself's tag. The element of *SelfTagContext* Stream is aware of the position of the element which has *Transorder* relation with it. Therefore the *SelfTagPointer* streaming contains the context of elements which are of the same tag name.

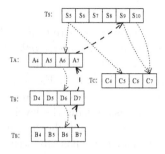 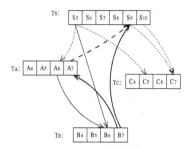

Fig. 4. Example of ADTagContext stream (partial relations displayed)

Fig. 5. Example of Complete Context of *FarthestAnc* and *NextDesc*

The *SelfTagContext* Streaming scheme is a model in concept, and we can define *nextP* operations on its streams. The *nextP* returns the position of the element which transmigrates current XML element in the stream. We can say that the context of the *transorder* in tag streams can be revealed by the operation *nextP*. For example, in Fig.3, $nextP(S_1)$ is S_4, and $nextP(C_1)$ is C_4. We can see that the elements between S_1 and S_4 are the descendants of S_1.

2.3 ADTagContext Streaming

The *ADTagContext* streaming contains more context of the elements. The *ADTagContext* streaming extends *SelfTagContext* streaming by appending the context of *FarthestAnc* in stream of parent tag in DTD, and *NextDesc* in streams of children tag in DTD. Fig.4 shows the *ADTagContext* Streaming of the document in Fig.1, and doesn't show all the relations for simplification.

The element of *ADTagContext* Stream contains the positions of the elements that are in the parent tag streams and the children tag streams. Because a tag in DTD may have multiple children, the element contains multiple positions to elements from streams of children tags in DTD. We can define some operations on the *ADTagContext* Stream: $getA(T, T_P)$ and $getD(T, T_C)$. Given an element E from stream of tag T, we can get E's *FarthestAnc* in the corresponding stream of T_P which is the parent tag of T, and E's *NextDesc* in the corresponding child stream of tag T_C which is one of T's child in DTD. If tag T has n children in the DTD, the element in T's tag stream can get n *NextDesc* elements. In this way, all the possible queries can be evaluated by the *ADTagContext* Streaming. For example in Fig.4, S_5 has two pointers because tag S in DTD has two descendants(A and C).

3   C-Tree Index

In this section, we present the structure of C-Tree, an index specially designed to index XML documents for efficient twig joins. Then we discuss how to get the complete context of nodes in any twig query pattern by using the C-Tree index.

3.1 Data Structure of C-Tree

With the above introduction, we define C-Tree, an index structure that enables elements in tag streams aware of its context (including *transorder,FarthestAnc* and *NextDesc*), and facilitates twig join processing of XML documents.

For simplicity we consider a single, typically large document. Extension to multi-document databases is trivial. Similarly, we concentrate on the ancestor-descendant join; the parent-child join is a simple extension using the level numbers. (Hence in the rest, the *DocID, Level* attributes are not shown but are implicitly assumed.) C-Tree is a B+Tree, and its leaf node has the structure (*currentTag, StartPos, desTag, RType, SPos*). so we can store the relations defined above. The *desTag* means destination tag, and *RType* is the relation type between element of *currentTag* and element of *desTag* including *transorder,FarthestAnc* and *NextDesc*, the *SPos* is destination element's position in *desTag* stream, and it is the position in the respective tag stream of *desTag* in the *ADTagContext* Streaming or *SelfTagContext* Streaming. We assume that a separate index is used to cluster elements from the *currentTag*. This index organization has been shown to be very efficient for some queries. In practice these multiple indices can be combined into a single index, by adding the *StartPos, desTag, RType* in the search key. So the index is built on the (*currentTag,StartPos, desTag, RType*) combination. We define the operation *getSPos* on the C-Tree. Its parameters are (*currentTag, StartPos, desTag, RType*) and it returns the *SPos*. So that, given a element e from stream *currentTag*, we can get its next element in stream of *desTag* of any kind of the relation (*Transorder, FarthestAnc,* or *NextDesc*). When the *RType* is *Transorder* relation, the *desTag* is the same as the *currentTag*.

3.2 Complete Context of *FarthestAnc* and *NextDesc*

For answering all the possible queries, given a twig query pattern Q and a node N in Q, we need to get the context of *FarthestAnc* from the tag stream of N' parent node in Q and the context of *NextDesc* from the tag streams of N' children nodes in Q. While in *ADTagContext* Streaming, we can only get those context from the tag streams of N's parent and children in DTD. For example, in Fig.4, element B_7 can get D_7 the *FarthestAnc* in tag stream D from C-Tree index directly, because node D is the parent of node B in DTD, but it can not achieve the *FarthestAnc* contexts of tag A and S directly.

The complete context of *FarthestAnc* and *NextDesc* contains all the context of *FarthestAnc* and *NextDesc* between any two nodes in query which have ancestor-descendant relationships. By using the contexts between any two parent-child nodes in DTD which are contained in C-Tree, we can deduce the complete context of *FarthestAnc* and *NextDesc*. For example, in Fig.5, element B_7 can get element A_7 the *FarthestAnc* in tag stream A and S_9 the *FarthestAnc* in tag stream S, moreover the element A_6 can get element B_6 the *NextDesc* in tag stream B by searching the C-Tree index twice. In this way, all the possible queries can be evaluated by the C-Tree which is based on the model of *ADTagContext* Streaming.

4   NestTwigStack

Algorithm *NestTwigStack*, which processes the twig join, is presented in Algorithm 1. In our algorithms, we encode the XML element by using the region coding [3,5]. If the streams contain nodes from multiple XML documents, the algorithm is easily extended to test equality of *DocId* before manipulating the nodes in the streams and on the stacks. $IsPCedge(tagP, tagC)$ returns true when $tagP$ is $tagC$' parent in DTD or Schema, and $isADedge(tagA, TagD)$ returns true when $tagA$ is $tagD$'s parent in DTD. $advanceNest(T_q, num))$ skips num elements in stream T_q.

Algorithm 1. $NestTwigStack(q)$

1: **while** not $end(root)$ **do**
2:      $q = getNextNest(root)$;
3:      **if** not $isRoot$(q) **then**
4:          $cleanStack(parent(q), nextL(q))$;
5:          **if** empty($S_{parent}(q)$) **then**
6:              $advanceNested(T_q, q.NestedNum)$;
7:          **end if**
8:      **end if**
9:      **if** $(isRoot(q)$ && $empty(S_{parent}(q)))$ **then**
10:         $cleanStack(q, next(q))$;
11:         **if** $(isPCedge(q, parent(q))$ && $parent(q).Nest == 0)$ **then**
12:             $\{moveStreamToStack(T_q; S_q; pointertotop(S_{parent}(q)))$;
13:             $advanceNested(T_q; q.Nest)\}$
14:         **end if**
15:         **if** $isADedge(q, parent(q))$ **then**
16:             $moveStreamToStack(T_q; S_q; pointertotop(S_{parent}(q)))$;
17:         **end if**
18:         **if** $isLeaf(q)$ **then**
19:             **if** $(isADedge(q, parent(q)$ && $parent(q).Nest==0)$ **then**
20:                 $\{ showSolutionsWithBlocking(S_q; q.Nest + 1)$;
21:                 $advanceNested(T_q; q.Nest);\}$
22:             **else**
23:                 $showSolutionsWithBlocking(S_q; 1)$;
24:             **end if**
25:             $pop(S_q)$;
26:         **end if**
27:         $advance(T_q)$;
28:     **end if**
29: **end while**
30: $mergeAllPathSolutions()$;

In many cases, *NestTwigStack* can skip many elements which don't participate in the final result by the nested relations between the elements of same tag. (1)Given any structural join $A//D$ in the twig query pattern, tag A is nested in the DTD or Schema, and current element e in T_A has many descendants in T_A. If element e in T_A can't participate in the final result, the elements which are the descendants of e can't contribute themselves to the answers too. (2)Given the structural join $A//D$ where D is the leaf node in the twig query pattern, tag D is nested in the DTD or Schema, and the current element e in T_D has many descendants in T_D. If element e in T_D contribute itself to the answers, the elements which are descendants of e contribute themselves to the answers obviously. (3)Given any P-C edge in the twig query pattern, such as the P/C, if the tag C is nested in the DTD or Schema, and the current element e in T_C has many descendants in T_C, If element e in T_C can match the twig query pattern, the elements which are the descendants of e can't be the children of the element in the top of T_P.

Algorithm 2. $getNextNest(q)$

1: **if** $isLeaf(q)$ **then**
2:     return q;
3: **end if**
4: **for** q_i in $children(q)$ **do**
5:     $n_i = getNextNest(q_i)$;
6:     **if** $n_{Si} \neq q_i$ **then**
7:         return n_i;
8:     **end if**
9: **end for**
10: $n_{min} = minarg_{n_i} nextL(T_{n_i})$;
11: $n_{max} = maxarg_{n_i} nextL(T_{n_i})$;
12: $advanceNested(T_q , q.Nest))$;
13: **if** $(nextL(T_q) < nextL(T_{n_{min}}))$ **then**
14:     return q;
15: **else**
16:     return n_{min};
17: **end if**

Example 1. In Fig.1, when S_1 in T_S isn't found in any answer of the query, the algorithm skips S_2, and S_3, because S_2, and S_3 are the descendants of S_1. when C_1 is found having an answer, the algorithm output $(S_4,C_1),(S_4,C_2),(S_4,C_3)$ directly. If the query pattern changes to S[/C]//A//B, when C_1 is output as the intermediate result, C_2, C_3 are skipped because the level of them do not equal to $(S_4.level+1)$.

5   ADTwigStack

In this subsection, we briefly introduce algorithm *ADTwigStack* , which is partly inspired by *TwigStack* algorithm proposed in [1].

The $getDes(e,q)$ in Algorithm 3 returns the position of the element which is the descendant of e in T_q, and $getFarthestAnc(e,q)$ in Algorithm 4 returns the position of the element which is the ancestor of e in T_q. By the function $advanceJumpTo(T_q, num)$, we can find the next element to deal with in T_q, and num is its position.

For each axe in twig query pattern, $ADTwigStack$ can skip both the ancestor and the descendant. Using $getDes(e,q)$, we can achieve the descendant context of the element from the $ADTagContext$ Streams. So we can jump to the first descendant of the element e in e's descendant stream. In order to avoid the loss of the useful results, we have to determinate the present stack of e whether it is empty or not before jumping to the descendant stream. When e has ancestor in stack, the elements whose start positions between the start positions of e and e_a in descendant stream may be in the result, therefore we can't jump the element to the one who has the smallest start position which is larger than e. We also can get the context of ancestor from the streams, so we can jump the element e's ancestor stream T_A to the farthest ancestor of e_{max} whose start position is the maximum of those elements in the child streams of e's tag, to skip the ancestors in T_A that are surely not in the result of the query. In this part we must determinate whether the element which is going to jump to is after the current element in tag stream T_A or not. When e has an ancestor in document e_a, and the elements d with maximum end position in the child streams is e's descendant, the position storied in element d is the highest ancestor of d ,that is e_a. But e_a is before e in stream. The algorithm may run into endless loop, without judgement.

Algorithm 3. $ADTwigStack(q)$

1: **while** not end(root) **do**
2:    $q = getNextC(root)$;
3:    **if** not $isRoot(q)$ **then**
4:       $cleanStack(parent(q), nextL(q))$;
5:    **end if**
6:    **if** $empty(S_{parent}(q))$ **then**
7:       $advanceJumpTo(T_q, getDes(next(T_{parent}(q)) ,q))$;
8:    **else**
9:       $advance(T_q)$;
10:    **end if**
11:    **if** $(isRoot(q)\&\&$not $empty(S_{parent}(q)))$ **then**
12:       $cleanStack(q, next(q))$;
13:       $moveStreamToStack(T_q, S_q, pointertotop(S_{parent}(q)))$;
14:       **if** $isLeaf(q)$ **then**
15:          $showSolutionsWithBlocking(S_q; 1)$;
16:          $pop(S_q)$;
17:       **end if**
18:       $advance(T_q)$;
19:    **end if**
20: **end while**
21: $mergeAllPathSolutions()$;

Example 2. In Fig.1, when the $getNextC$ function returns tag A, the top element in tag stream T_A is A_4, whose end position is smaller than S_5's start position. S_5's descendant context contains the position of S_5's next descendant of tag A, and A_6 is after A_4 in tag stream T_A, so element A_5 is skipped, and the next element to be scanned in stream T_A is A_6; With the algorithm running, the top elements in the respective streams are S_6,A_7,B_6 and C_5. The children of S in twig query Q are A and C, so the element with maximum position is A_7. The ancestor context of A_7 to tag S is S_9, which is A_7's farthest ancestor, and S_9 is after S_6 in T_S, therefore the algorithm scans S_9 in T_S on the next step, and many elements in stream T_S are skipped.

Algorithm 4. $getNextC(q)$

1: **if** $isLeaf(q)$ **then**
2:     return q;
3: **end if**
4: **for** q_i in $children(q)$ **do**
5:     $n_i = getNextC(q_i)$;
6:     **if** $n_i \neq q_i$ **then**
7:         return n_i;
8:     **end if**
9: **end for**
10: $n_{min} = minarg_{n_i} nextL(T_{n_i})$;
11: $n_{max} = maxarg_{n_i} nextL(T_{n_i})$;
12: $advanceNested(T_q, q.Nest)$);
13: **if** $getFarthestAnc(next(T_{n_{max}}), q) > nextL(q)$ **then**
14:     $advanceJumpTo(T_q, getFarthestAnc(next(T_{n_{max}}), q))$;
15: **else**
16:     **while** $nextR(T_q) < nextL(T_{n_{max}})$ **do**
17:         $advance(T_q)$;
18:     **end while**
19: **end if**
20: **if** $(nextL(T_q) < nextL(T_{n_{min}}))$ **then**
21:     return q;
22: **else**
23:     return n_{min};
24: **end if**

---

6   Experiments

In this section, we present experimental results on the performance of the twig pattern matching algorithms. We focus on three kinds of algorithms: *TwigStack*, *NestTwigStack* and *ADTwigStack*. We evaluate the performance of those join algorithms using the following two metrics, the number of elements scanned and the running time of first phase in those algorithms.

6.1   Experiment Setup

We implemented all algorithms in JDK 1.4. All our experiments were performed on 3.2GHz Pentium 4 processor with 512MB RAM running on windows XP system. We used the following real-world and synthetic data sets for our experiments:

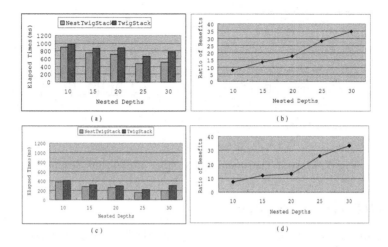

Fig. 6. Elapsed Times and the Ratio of Benefits

- **DBLP.** We obtained the *DBLP* data set from the University of Washington XML repository[10]. And we generate a series of child data set by breaking the *DBLP* document which have different size.
- **Synthetic data.** The definition of the tag *section* is nested in book DTD. We used this DTD to create a set of nested XML documents which are about 50M bytes. The maximal depth of each XML document varies from 10-30.

6.2 *NestTwigStack* Vs. *TwigStack*

We contrasted algorithm *TwigStack* with *NestTwigStack* on synthetic data sets of different nested levels. We evaluated two queries :

Q_1: *section[/title]/figure/image* and Q_2: *section[//title]//figure*

Q_1 only has P-C edges, and Q_2 only has A-D edges. As such, we can evaluate the performance of our algorithm under different kinds of queries. The results of Q_1 are shown in Fig.6(a) and (b), and the results of Q_2 are shown in the Fig.6(c) and (d). The **ratio of the Time benefits** is the ratio of the time benefits from the skipping of the useless elements to the time which original algorithm needs.

From Fig.6(a) and Fig.6(c), we can easily find out *NestTwigStack* outperforms *TwigStack*, because *NestTwigStack* can skip many useless elements. When there is no nested element in the data sets, such as the *DBLP* data set, there is no benefits that we can get to speed up the processing of query. From Fig.6(b) and (d), we can see that the ratio of the benefits increases as the depth of document increases. The deeper the nested level of data set is , the better the *NestTwigStack* performs.

Considering the A-D or P-C edges, we can see that our algorithm has better performance when it evaluates queries with A-D edge only. But we also see that, no matter what query it is, the ratio of the time benefits increase as the depth increases, and the trend is similar. In summary, algorithm *NestTwigStack* performs better than TwigStack on the nested data set.

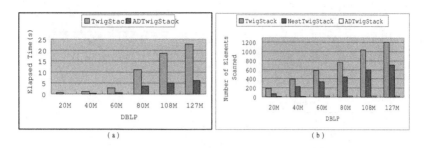

Fig. 7. Elapsed Times and the Number of Elements Scanned

6.3   *ADTwigStack* Vs. *TwigStack*

To compare the performances of *TwigStack* and *ADTwigStack*, we choose *DBLP* data sets of different size, and evaluate the following twig query.

$Q_3 : //inproceedings[//author]//title//i$

It can be observed, from experiment results on Fig.7, that *ADTwigStack* algorithm is usually much more efficient than *TwigStack*. *ADTwigStack* is nearly three times faster than the original *TwigStack*, and *TwigStack*'s numbers of elements scanned are ten times more than the *ADTwigStack*'s. We also can see that the *NestTwigStack* on the *DBLP* documents have about half the elements scanned.

6.4   Performance Analysis

The *ADTwigStack* performs very well against the *TwigStack* while evaluating queries on *DBLP* data set. That is because the *ADTwigStack* algorithm makes use of the context of both ancestor and descendant. Elements in descendant stream can advance the ancestor stream by its ancestor context, and then the current top of the ancestor stream can advance the descendant stream by its descendant context. So *ADTwigStack* can skip most of the useless elements. *NestTwigStack* does not outperform *TwigStack* too much, however, when a document is deeply nested, *NestTwigStack* performs much better than *TwigStack*.

From the experiments we can conclude that the *ADTwigStack* is commonly suitable for all kinds of XML documents, especially for the documents with large fan-outs or wide breadth, while the *NestTwigStack* suits for the documents with nested elements deeply.

7   Conclusions

In this paper, we addressed the problem of efficiently evaluation of holistic twig joins by using C-Tree index. Particularly, we proposed a new series of streaming schemes that are aware of elements' context in the streams. And then we proposed two algorithms to speed up the processing of twig pattern queries by skipping the elements which do not participate in any twig join. The algorithm *NestTwigStack* which is based on *SelfTagContext* Streaming model has better performance on those XML document with deeply nested elements, and the algorithm ADTwigStack which is based on *ADTagContext* Streaming model can significantly improve the original holistic twig joins algorithm, especially on those document with large fan-outs or wide breadth. These algorithms not only accelerate queries of twig pattern with only ancestor-descendant edges, but also accelerate queries of twig pattern with parent-child edges.

Acknowledgement. This research is supported by the National Natural Science Foundation of China (Grant No. 60873011, 60773221, 60773219 and 60803026), National Basic Research Program of China (Grant No. 2006CB303103), and the 863 High Technique Program (Grant No. 2007AA01Z192).

References

1. Bruno, N., Koudas, N., Srivastava, D.: Holistic twig joins: Optimal XML pattern matching. In: SIGMOD, pp. 310–321 (2002)
2. Florescu, D., Kossmann, D.: Storing and querying xml data using an rdmbs. IEEE Data Engineering Bulletin 22(3), 27–34 (1999)
3. Wu, Y., Patel, J.M., Jagadish, H.V.: Structural join order selection for XML query optimization. In: ICDE, pp. 443–454 (2003)
4. McHugh, J., Widom, J.: Query optimization for XML. In: VLDB, pp. 315–326 (1999)
5. Al-Khalifa, S., Jagadish, H.V., Patel, J.M., Wu, Y., Koudas, N., Srivastava, D.: Structural joins: A primitive for efficient XML query pattern matching. In: ICDE, pp. 141–152 (2002)
6. Tatarinov, I., Viglas, S., Beyer, K.S., Shanmugasundaram, J., Shekita, E.J., Zhang, C.: Storing and querying ordered XML using a relational database system. In: SIGMOD, pp. 204–215 (2002)
7. Jiang, H., Lu, H., Wang, W., Ooi, B.C.: XR-tree: Indexing XML data for efficient structural joins. In: ICDE, pp. 253–263 (2003)
8. Chen, T., Ling, T.W., Chan, C.Y.: Prefix path streaming: A new clustering method for optimal holistic XML twig pattern matching. In: Galindo, F., Takizawa, M., Traunmüller, R. (eds.) DEXA 2004. LNCS, vol. 3180, pp. 801–810. Springer, Heidelberg (2004)
9. Lu, J., Chen, T., Ling, T.W.: TJFast: effective processing of XML twig pattern matching. In: WWW (Special interest tracks and posters), pp. 1118–1119 (2005)
10. University of Washington XML Repository,
    http://www.cs.washington.edu/research/xmldatasets/

Processing XML Keyword Search by Constructing Effective Structured Queries

Jianxin Li, Chengfei Liu, Rui Zhou, and Bo Ning

Swinburne University of Technology
Melbourne, Australia
{jianxinli, cliu, rzhou, bning}@groupwise.swin.edu.au

Abstract. Recently, keyword search has attracted a great deal of attention in XML database. It is hard to directly improve the relevancy of XML keyword search because lots of keyword-matched nodes may not contribute to the results. To address this challenge, in this paper we design an adaptive XML keyword search approach, called *XBridge*, that can derive the semantics of a keyword query and generate a set of effective structured queries by analyzing the given keyword query and the schemas of XML data sources. To efficiently answer keyword query, we only need to evaluate the generated structured queries over the XML data sources with any existing XQuery search engine. In addition, we extend our approach to process top-k keyword search based on the execution plan to be proposed. The quality of the returned answers can be measured using the context of the keyword-matched nodes and the contents of the nodes together. The effectiveness and efficiency of *XBridge* is demonstrated with an experimental performance study on real XML data.

1 Introduction

Keyword search is a proven user-friendly way of querying XML data in the World Wide Web [1,2,3,4,5]. It allows users to find the information they are interested in without learning a complex query language or knowing the structure of the underlying data. However, the number of results for a keyword query may become very large due to the lack of clear semantic relationships among keywords. There are two main shortcomings: (1) it may become impossible for users to manually choose the interesting information from the retrieved results, and (2) computing the huge number of results with less meaning may lead to time-consuming and inefficient query evaluation. As we know, users are able to issue a structured query, such as XPath and XQuery, if they already know a lot about the query languages and the structure of the XML data to be retrieved. The desired results can be effectively and efficiently retrieved because the structured query can convey complex and precise semantic meanings. Recently, the study of query relaxation [6,7] can also support structured queries when users cannot specify their queries precisely. Nevertheless, there are many situations where structured queries may not be applicable, such as a user may not know the data schema, or the schema is very complex so that a query cannot be easily

Q. Li et al. (Eds.): APWeb/WAIM 2009, LNCS 5446, pp. 88–99, 2009.
© Springer-Verlag Berlin Heidelberg 2009

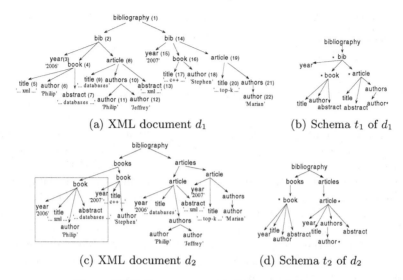

(a) XML document d_1 (b) Schema t_1 of d_1

(c) XML document d_2 (d) Schema t_2 of d_2

Fig. 1. XML Documents with Different Schemas

formulated, or a user prefers to search relevant information from different XML documents via one query.

Consider the example in Figure 1 showing the same bibliography data arranged in two different formats. In this case, a user can quickly issue a keyword query "*Philip, 2006, xml*" to obtain a list of answers. Node #4 *book* and node #8 *article* will be returned as relevant answers. From Figure 1(a), we can see that only node #4 *book* satisfies the searching requirement. For node #8 *article*, only the abstract contains the term xml, as such it does not meet the users' original intention. It would be impossible for an IR-style keyword query to differentiate the semantics, e.g. one term can represent different meanings in different positions. In addition, when the size of XML documents becomes larger, it is difficult to choose the meaningful answers from the large number of returned results. As an alternative, users can construct an XQuery to represent this simple query and specify the precise context. But there are two challenges: first, they have to know that "publication" in the schema is actually presented as *book* and *article* in both schemas; second, they have to know that *title* and *author* are the child elements of "publication", while *year* could be either a child or a sibling. Writing an accurate XQuery is non-trivial even for this simple example due to the complex structure of XML schemas. Therefore, it is highly desirable to design a new keyword search system that not only permits users specify more expressive queries, but also implement keyword search as efficiently as structured queries.

To address this problem, a formalized keyword query consisting of a set of label-term pairs is deployed in [4] and [8]. In [4], labels in the given keyword query are used to filter the node lists. In [8], labels are used to construct answer templates that includes all combinations together according to the schema of XML data stream. When the data stream is coming, all matched nodes will need

```
For $b in bibliography/bib              For $b in bibliography/bib
For $b2 in $b/book                      For $a in $b/article
Where $b/year = '2006'                  Where $b/year = '2006'
     and contains($b2/title, 'xml')          and contains($a/title, 'xml')
     and contains($b2/author, 'Philip')      and contains($a/authors/author, 'Philip')
Return $b                               Return $b
```

Fig. 2. Structured Queries w.r.t. XML Schema t_1

to be maintained until the template-matched results are generated or the end of the stream is reached. Different from them, in this work we develop a keyword search system called *XBridge* that first infers the context of the set of labels and the required information to be returned according to XML data schema. And then it may generate a set of precise structured queries and evaluate them by using existing XML search engines. To evaluate the quality of the results, in *XBridge* we propose a scoring function that takes into account the structure and the content of the results together. In addition, we also design an execution plan to retrieve the more qualified results as soon as possible, which is suitable to process top-k keyword search.

Consider the same example again, the user may change to issue "*author:Philip, year:2006, title:xml*" as a keyword query to search the relevant publications. For this keyword query, *XBridge* is able to automatically construct different structured queries for XML documents conforming to different XML schemas. For example, for the source schemas of the two XML documents shown in Figure 1, we can construct two sets of structured queries as shown in Figure 2 and Figure 3, respectively. After that, we can evaluate the structured queries to answer the original keyword query. The book node #4 will be returned as answers. We do not need to identify whether or not the title node #5 and the author node #11 belong to the same publications. As such, the processing performance would be improved greatly due to the specific context in structured queries.

Contributions of this paper: (1) For different data sources, *XBridge* can infer different semantic contexts from a given keyword query with label-term, which can be used to construct adaptive structured queries. (2) A scoring function is proposed to evaluate the quality of the answers by considering the context of the keyword-matched nodes and the contents of the nodes in the answers. (3) An execution plan, adapting to the proposed scoring function, is designed to efficiently process top-k keyword search. (4) Experiments show that *XBridge* can obtain improved performance over previous keyword search approaches.

The rest of the paper is organized as follows: Section 2 provides the definition of XML schema and presents how to identify the context of terms and derive the

```
For $b in bibliography/books/book       For $a in bibliography/articles/article
Where $b/year = '2006'                  Where $a/year = '2006'
     and contains($b/title, 'xml')           and contains($a/title, 'xml')
     and contains($b/author, 'Philip')       and contains($a/authors/author, 'Philip')
Return $b                               Return $a
```

Fig. 3. Structured Queries w.r.t. XML Schema t_2

returned nodes. Section 3 proposes a scoring function to evaluate the quality of returned answers. Section 4 describes our *XBridge* system and the algorithms for constructing and evaluating the generated structured queries. The experimental results are reported in Section 5. Finally, we discuss the related work and conclude the study of this work in Section 6 and Section 7, respectively.

2   Identifying Context and Returned Nodes

In this section, we show how to identify the context and the types of return nodes for a keyword query w.r.t. XML schema. Here, we use XML schema tree to represent the structural summary of XML documents. Formally, a keyword query consists of the forms $l : k$, $l :$ or $: k$ where l is a label and k is a term in [4]. The query model in our work *XBridge* permits users to distinguish the semantics of predicates from the returned nodes, such as we extend $l :$ to $l : *$ and $l :?$. The former means that one node should exist in the returned answers where the node's tagname is same to the label l and the node's values may be anything. The latter shows that the information of the nodes will be returned as answers if the nodes' tagname is same to the label l.

Definition 1. *(XML Schema Tree) An XML Schema Tree is defined as $T = (V, E, r, Card)$ where V is a finite set of nodes, representing elements and attributes of the schema T; E is set of directed edges where each edge $e(v_1,v_2)$ represents the parent-child (containment) relationship between the two nodes v_1, $v_2 \in V$, denoted by $P(v_2) = v_1$ or $v_2 \in Ch(v_1)$; r is the root node of the tree T; $Card$ is a set of mappings that maps each $v \in V$ to $\{1, *\}$ where "1" means that v can occur once under its parent $P(v)$ in a document conforming to T while "*" means that v may appear many times.*

2.1   Identifying Context of Keywords

From the set of labels given in a keyword query, we can infer the contexts of the terms for a data source based on its conformed XML schema. Each node in an XML document, along with its entire subtree, typically represents a real-world entity. Similarly, given a list of labels $l_1, ..., l_n$ and an input XML schema tree T, an entity of these labels can be represented with a subtree of T such that it contains at least one node labeled as $l_1, ..., l_n$. We define the root node of the subtree as a master entity.

Definition 2. *(Master Entity) Given a set of labels $\{l_i|1 \leq i \leq n\}$ and an XML schema tree T, the master entity is defined as the root node of the subtree T_{sub} of T such that T_{sub} contains at least one schema node labeled as $l_1, ..., l_n$.*

Based on Definition 2, a master entity may contain one or more than one schema nodes taking a label as their tagnames. If one master entity node only contains one schema node for each label, we can directly generate FOR and WHERE clauses. For example, let $q(year:2006, title:xml, author:Philip)$ be a keyword query over the XML document d_2 in Figure 1(c). Based on the schema t_2 in Figure 1(d), we can obtain two master entities *book* and *article*. Since the master entity *book* only contains one node labeled as *year*, *title* and *author* respectively, we can directly construct "For \$b in bibliography/bib/book" and "Where \$b/year='2006' and contains(\$b/title, 'xml') and contains(\$b/author, 'Philip')". Similarly, we can process another master entity *article*. The constructed queries are shown in Figure 3.

If one master entity node contains more than one nodes taking the same label, to construct FOR and WHERE clauses precisely, we need to identify and cluster the nodes based on the semantic relevance of schema nodes within the master entity. To do this, we may deploy the ontology knowledge to precisely estimate the semantic relevance between schema nodes. However, the computation adding additional measurement may be expensive. Therefore, in this paper we would like to infer the semantic relevance of two schema nodes by comparing their descendant attributes or subelements. For example, given any two schema nodes v_1 and $v_2 \in$ a master entity T_{sub}, we can infer that the two schema nodes v_1 and v_2 are semantic-relevant nodes if they hold: $semi(v_1, v_2) \geq \sigma$. Here σ is the similarity threshold. If σ is set to 0.8, then it means that v_1 and v_2 contain 80% similar attributes or subelements.

Consider the same query q and the document d_1 in Figure 1(a). Based on the schema t_1 in Figure 1(b), we know there exists one master entity *bib* that contains one node with the label *year* and two nodes with the same labels *title* and *author* respectively. In this case, we first cluster the five nodes as C: { $year$, {$title$, $author$}$_{book}$, {$title$, $author$}$_{article}$ }$_{bib}$, and then identify whether or not the subclusters in C are semantic-relevant schema nodes. For instance, although {}$_{book}$ and {}$_{article}$ have different labels, both of them contain the same nodes *title* and *author*, i.e., the two nodes *book* and *article* contain 100% similar attributes. Therefore, the cluster C is partitioned into two clusters: C_1: { $year$, {$title$, $author$}$_{book}$ }$_{bib}$ and C_2: { $year$, {$title$, $author$}$_{article}$ }$_{bib}$. When all subclusters cannot be partitioned again, we can generate different sets of FOR and WHERE clauses. For C_1, we have "For \$b in bibliography/bib" for {}$_{bib}$ and "For \$b2 in \$b/book" for {}$_{book}$ together. And its WHERE clause can be represented as "Where \$b/year='2006' and contains(\$b2/title, 'xml') and contains(\$b2/author, 'Philip')" according to the labels in the subclusters of C_1. Similarly, we can process C_2 to generate its FOR and WHERE clauses. Figure 2 shows the expanded structured queries.

The contexts of the terms can be identified by computing all the possible master entities from the source schemas first, and then specifying the precise paths from each master entity to its labels by checking the semantic-relevant schema nodes. Once the contexts are obtained, we can generate the FOR and WHERE clauses of the structured queries for a keyword query. By specifying the detailed context in FOR clauses, we can limit the range of evaluating the structured queries over the XML data sources, which can improve the efficiency of processing the keyword queries.

2.2   Identifying Returned Nodes

Given a keyword query $q = \{l_i : k_i | 1 \leq i \leq n\}$ and an XML schema tree $T = (V, E, r, Card)$, we may retrieve a set of master entities $V_m \subseteq V$ for q w.r.t. T based on the above discussion. In this section, we will derive the returned nodes only by identifying the types of the master entities V_m. For any master entity $v_m \in V_m$, if $Card(v_m) = $ "*", we can determine that the node v_m can be taken as return nodes in the corresponding RETURN clauses because the node represents the real entity at the conceptual level. However, if $Card(v_m) = $ "1", the node v_m may not represent an entity. In this case, we probe its ancestor nodes until we find its nearest ancestor v_a, such that $Card(v_a) = $ "*".

Consider another keyword query $q(title:xml, author:Philip)$ over the XML document d_1 conforming to t_1. We are able to obtain two master entities *book* and *article*. Since $Card(book) = $ "*" and $Card(article) = $ "*", we can take them as the return nodes in the corresponding RETURN clauses. However, if users issue a simple query $q(title:xml)$ over d_1, the master entity of this query is the *title* that is only an attribute of the book or article nodes. In this case, we can trust that users would like to see the information of the whole entity (book or article), rather than one single attribute. Therefore, to generate meaningful RETURN clauses, we have to extend the title node to its parent book or article nodes as the return nodes because book or article nodes belong to *-node type.

If there are some label-term pairs in the form of $\{l_{j_k} :? | 1 \leq k \leq m \leq n \wedge 1 \leq j_k \leq n\}$ in $q = \{l_i : k_i | 1 \leq i \leq n\}$, instead of returning the master entity of q, we will compute the master entity of $\{l_{j_k}\}$ and use it to wrap all return values of l_{j_k} in the RETURN clause. This is because users prefer to see those nodes with the labels $\{l_{j_k}\}$ as the tagnames.

3   Scoring Function

Given a keyword query q and an XML schema tree T, a set of structured queries Q may be constructed and evaluated over the data source conforming to T for answering q. The answer to the XML keyword query q may be a big number of relevant XML fragments. In contrast, the answer to the top-k keyword query is an ordered set of fragments, where the ordering reflects how closely each fragment matches the given keyword query. Therefore, only the top k results with the highest relevance w.r.t. q need to be returned to users. In this section, our scoring function consists of the context of the given terms and the weight of each term.

Let a fragment A be an answer of keyword query q. It is true that we can determine a structured query q_i that matches the fragment A. This is because we first construct the structured query q_i from the keyword query q and then obtain the fragment A by evaluating q_i over XML data. Therefore, we can compute the context score of the fragment A by considering the structure of the query q_i.

Definition 3. *(Context Score) Assume the structured query q_i matching the answer A consists of the labels $\{l_i | 1 \leq i \leq n\}$. Let its master entity be v_m and an XML schema be $T = (V, E, r, Card)$, we can obtain a list of nodes $V' \subseteq V$ that match each label respectively.*

$$ContextScore(A, q_i) = \frac{1}{n} \times \sum_{i=1}^{n} K_{context}(v_i, v_m) \tag{1}$$

where $K_{context}(v_i, v_m)$ is computed based on the distance from the node $v_i \in V'$ matching the label l_i to its master entity v_m, i.e., $LengthOfPath^{-1}(v_i, v_m)$.

In order to effectively capture the weight of each individual node, we are motivated by the $tf * idf$ weight model. Different from IR research, we extend the granularity of the model from document level to element level.

$$tf(v_i, t_i) = |\{v \in V_d | tag(v) = tag(v_i) \& t_i \in v\}| \tag{2}$$

$$idf(v_i, t_i) = log(\frac{|\{v \in V_d : tag(v) = tag(v_i)\}|}{|\{v \in V_d : tag(v) = tag(v_i) \& t_i \in v\}|}) \tag{3}$$

Intuitively, the tf of a term t_i in a node v_i represents the number of distinct occurrences within the content of v_i while the idf quantifies the extent to which the nodes v with the same tagname as v_i in XML document node set V_d contain the term t_i. The fewer v_i nodes whose contents include the term t_i, the larger the idf of a term t_i and a node v_i. Without loss of generality, we will also assume that the weight value of each node is normalized to be real numbers between 0 and 1.

Definition 4. *(Weight of Individual Keyword) The answer A contains a set of leaf nodes as tagnames with the given labels $\{l_i | 1 \leq i \leq n\}$ where each leaf node should contain the corresponding term at least once a time. For each node v_i with label l_i, we have:*

$$\omega(v_i, t_i) = \frac{tf(v_i, t_i) \times idf(v_i, t_i)}{max\{tf(v_i, t_i) \times idf(v_i, t_i) | 1 \leq i \leq n\}} \tag{4}$$

Definition 5. *(Overall Score of Answer) Given a generated structured query q_i and its answer A, the overall score of the answer can be computed as:*

$$Score(A, q_i) = \frac{1}{n} \times \sum_{i=1}^{n} \frac{\omega(v_i, t_i)}{LengthOfPath(v_i, v_m)} \tag{5}$$

Assume q consists of three pairs "*author:Philip, year:2006, title:xml*". We evaluate the keyword query q over the XML document in Figure 1(c). After that, the fragment *book* in the box will be returned as an answer. Based on the pre-computation, we can get the weight of each keyword-matched node in the fragment where we assume each term only occurs once a time in the corresponding nodes.

According to Equation 1, Equation 4, Equation 5, we find that the overall score of an answer is equal to its context score when the weight of each keyword is set to 1 (the maximal value). Therefore, the context score can be taken as the upper bound of the answer.

4   Implementation of XML Keyword Query

In our *XBridge* system, for a given keyword query, we first construct a set of structured queries Q based on the labels in the keyword query and the XML schemas, i.e., query structuring. Then Q will be sorted according to their context score based on Definition 3. After that, we get a query $q_i \in Q$ with the highest context score from the set of structured queries and send it to XQuery engine. We will evaluate the query q_i over XML data and retrieve all the results that match q_i. At last, we will process all the current results, i.e., computing the overall score for each result and caching the k results with the higher overall scores. After we complete the evaluation of the query q_i, we need start a new loop to process another structured query that comes from the current set $\{Q - q_i\}$ until the top k qualified results have been found. Now let us discuss the detailed procedures of query structuring and execution plan.

4.1   Query Structuring

Query structuring is the core part of *XBridge* where we compute the master entities, infer the contexts and derive the return nodes. To search the master entities quickly, Dewey number is used to encode the XML schema. Given a keyword query $q = \{l_i : k_i | 1 \leq i \leq n\}$ and an XML schema tree $T = (V, E, r, Card)$, we first retrieve a list of nodes $V_i \subseteq V$ for each l_i in T and sort them in a descending order. To compute master entities, we propose two approaches: (1) **Pipeline:** We take each node v_i from the list V_i where $|V_i| = \min\{|V_1|, |V_2|, ..., |V_n|\}$ and compute the NCA (nearest common ancestor) of v_i and the nodes in the other lists V_j ($1 \leq j \leq n \wedge j \neq i$) in a sequence. At last, we preserve the NCA nodes as the master entity candidates in V_m; and (2) **Pipeline+σ:** When we get any node v_{ix} from V_i, we may take its next node v_{ix+1} as a threshold σ if the node v_{ix+1} exists in V_i. During the computation of NCA, we only probe the subset of nodes V in the other lists such that for $v \in V$ we have $v \prec v_{ix+1}$. As an optimization approach, the second one can reduce the negative computations of NCA while it can obtain the same structured queries as the first one does.

Algorithm 1. Constructing structured queries

input: a query $q = \{l_i : k_i | 1 \leq i \leq n\}$ and an schema $T = (V, E, r, Card)$
output: a set of structured queries Q

1: Retrieve a list of nodes $V_i \subseteq V$ for each l_i in T;
2: Compute the master entities V_m by calling for **Pipeline+σ** approach;
3: **for all** each master entity $v_m \in V_m$ **do**
4:    Generate **FOR clause** with v_m, i.e. "For \$x in $r/.../v_m$";
5:    Cluster the nodes matching with labels l_i in the subtree of v_m by calling for Cluster_Domain($\{l_i\}$, v_m);
6:    **for all** each cluster **do**
7:      Construct a set of **FOR clauses** for the entity nodes representing different semantic ranges in the cluster;
8:      Generate **WHERE clause** with n paths from v_m to each node l_i;
9:      Generate **RETURN clause** by identifying the types of v_m ($Card(v_m)=$* or 1) and k_i ("?" symbol exists or not) and put the structured query into Q;
10: **return** the set of structured queries Q;

Since the subtree of a master entity v_m may cover one or more than one schema nodes labeled with the same label and the nodes may have different semantic relevances, we develop a function Cluster_Domain() to identify and cluster the nodes matching with the labels l_i $(1 \leq i \leq n)$ in the subtree according to the labels and the data types of their attributes or subelements. For each cluster c, we do not generate a structured query if the cluster c only contains a part of the labels in the given keyword query. Otherwise, we construct a structured query for the cluster c. In this case, we first generate a set of FOR clauses according to the classified clusters in the cluster c and then construct a WHERE clause for c. Finally, a RETURN clause is derived by identifying the type of the master entity node v_m. After all the clusters are processed, we may generate a set of structured queries Q. The detailed procedure has been shown in Algorithm 1.

4.2   Execution Plan for Processing Top-k Query

Given a keyword query q and XML documents D conforming to XML schemas, we may generate a set of structured queries Q. To obtain top-k results, a simple method is to evaluate all the structured queries in Q and compute the overall score for each answer. And then we select and return the top k answers with the k highest scores to users. However, the execution is expensive when the number of structured queries or retrieved results is large.

To improve the performance, we design an efficient and dynamic execution plan w.r.t. our proposed scoring function, which can stop query evaluation as early as possible by detecting the intermediate results. Our basic idea is to first sort the generated structured queries according to their context scores, and then evaluate the query with the highest score where we take the context score of the next query as the current threshold. This is because the weight of each keyword is

---

Algorithm 2. Dynamic Execution Plan

input: A set of ranked structured queries Q and XML data D
output: Top k qualified answers

1: Initialize the answer set SA=null;
2: Initialize a boolean symbol $flag = $ false;
3: **while** $Q \neq null$ and $flag \neq true$ **do**
4:      Get a query $q = $ getAQuery(Q) where q has the maximal context score;
5:      **if** $|SA| = k$ and $(minScore\{A \in SA\} \geq $ ContextScore(q)) **then**
6:          $flag = $ true;
7:      **else**
8:          Issue q to any XQuery search engine and search the matched fragments F_m;
9:          **for all** $A \in F_m$ **do**
10:            Compute overall score $Score(A, q)$;
11:            **if** $|SA| < k$ **then**
12:                Input the result A into SA;
13:            **else if** $\exists A' \in SA$ and $Score(A', q') < Score(A, q)$ **then**
14:                Update the intermediate results $updateSA(A, SA)$;
15: **return** Top k qualified answers SA;

---

assumed to be set as 1 (the maximal value). In this case, the overall score would be equal to the context score. Therefore, if there are k or more than k results, we will compute their overall scores based on Equation 5 and cache k results with the k highest scores. At the same time, we will compare the scores of the k results and the threshold. If all the scores are larger than the threshold, we will stop query evaluation and return the current k results, which means no more relevant results exist. Algorithm 2 shows the detailed procedure of our execution plan.

5   Experiments

We implemented $XBridge$ in Java using the Apache Xerces XML parser and Berkeley DB.To illustrate the effectiveness and efficiency of $XBridge$, we also implemented the Stack-based algorithm[9] because the other related approaches in [10,11,12] are biased to the distribution of the terms in the data sources. All the experiments were carried out on a Pentium 4, with a CPU of 3GHz and 1GB of RAM, running the Windows XP operating system. We selected the Sigmod Record XML document (500k) and generated three DBLP XML documents as the dataset. In this paper, we evaluated the following keyword queries: q_1 ($author : David, title : XML$) and q_2 ($year : 2002, title : XML$).

As we know, the Stack-based algorithm may avoid some unnecessary computations by encoding the documents with the Dewey scheme. But it has to preserve all possible intermediate candidates during query evaluation and lots of them may not produce the results. However, $XBridge$ can infer the precise contexts of the possible results based on the source schemas before query evaluation. Therefore, $XBridge$ can outperform the Stack-based algorithm. For example, the Stack-based algorithm spent 188ms to evaluate q_1 on SigmodRecord.xml while $XBridge$ only used 78ms to process the same keyword search. In addition, if the size of an XML document increases, most of the time it may contain more nodes that match a single keyword. But the number of return nodes that match all the keywords may not be increased significantly. In this case, the performance of query evaluation may be relatively decreased.

Figure 4(a) and Figure 4(b) illustrate the time cost of both methods when we evaluate q_1 and q_2 on the given three DBLP datasets respectively. From the experimental results, we find that in $XBridge$ the change in time is not obvious when the size of dataset is less than 50M. But when the size of document is nearly 70M, the processing speed was decreased by 80%. This is because compared with dblp_01 or dblp_02, dblp_03.xml contains a huge number of nodes that match each single keyword but fail to contribute to return nodes. The figures also shows that XBridge would outperform over Stack-based algorithm in our experiments.

Figure 4(c) and Figure 4(d) compare the performance of the methods when k value is set as 10 or 20 respectively. Since Stack-based algorithm has to retrieve all the results and then select the k qualified answers, its response time for top-k query is nearly same to process general query. However, $XBridge$ depends on the dynamic execution plan, which can stop query evaluation as early as possible and guarantee no more qualified results exist in the data source. Generally, $XBridge$ is suitable to process top-k keyword search and most of the time, it only needs to evaluate parts

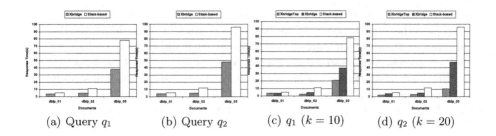

| (a) Query q_1 | (b) Query q_2 | (c) q_1 $(k = 10)$ | (d) q_2 $(k = 20)$ |

of the generated structured queries. But when the number of relevant results is less than k, $XBridge$ also needs to evaluate all the generated structured queries. For example, as we know the document dblp_01 only contains 2 qualified answers for the keyword query q_1. When the specified value of k is 10, there is no change of the response time. Therefore, in this special case, its efficiency cannot be improved.

6   Related Work

Recently, keyword search has been investigated extensively in XML databases. Given a keyword query and an XML data source, most of related work [9,5,10,11,12] first retrieve the relevant nodes matching with every single keyword from the data source and then compute LCAs or SLCAs of the nodes as the results to be returned. XRANK [9] and Schema-Free XQuery [5] develop stack-based algorithms to compute LCAs as the results. [10] proposes the Indexed Lookup Eager algorithm when the keywords appear with significantly different frequencies and the Scan Eager algorithm when the keywords have similar frequencies. [13] takes the valuable LCA as results by avoiding the false positive and false negative of LCA and SLCA. [11,14] takes the similar approaches as [10]. But it focuses on the discussions how to infer RETURN clauses for keyword queries w.r.t. XML data. [12] designs a MS approach to compute SLCAs for keyword queries in multiple ways. In addition, there are several other related work that process keyword search by integrating keywords into structured queries. [15] proposes a new query language XML-QL in which the structure of the query and keywords are separated. [5] embeds keywords into XQuery to process keyword search. Differently, $XBridge$ first constructs the structured queries for the given keyword query based on the source schemas and then evaluates the generated structured queries in a sequence. For a top-k keyword query, $XBridge$ can return the k qualified answers as early as possible without processing all the generated queries.

Ranking schemes have been studied for XML keyword search. XKeyword [16] ranks query results according to the distance between different keywords in the document. The ranking scheme in XSEarch [4] takes the summary of the weights of all keywords within a result to evaluate its relevance. The ranking scheme in XRank [9] takes into account result specificity, keyword proximity and hyperlink awareness together. However,our scoring function considers the context of the matched keywords and the weight of each keyword-matched node.

7 Conclusions

In this paper, we have proposed *XBridge* - an adaptive XML keyword search approach to process keyword search by constructing effective structured queries, which can improve the performance of keyword search greatly by specifying the precise contexts of the constructed structured queries. In addition, we have also provided a scoring function that considers the context of the keywords and the weight of each keyword in the data source together. Especially, an execution plan for processing top-k keyword search has been designed to adapt to our proposed scoring function.

References

1. Schmidt, A., Kersten, M.L., Windhouwer, M.: Querying XML Documents Made Easy: Nearest Concept Queries. In: ICDE, pp. 321–329 (2001)
2. Theobald, A., Weikum, G.: The index-based XXL search engine for querying XML data with relevance ranking. In: Jensen, C.S., Jeffery, K., Pokorný, J., Šaltenis, S., Bertino, E., Böhm, K., Jarke, M. (eds.) EDBT 2002. LNCS, vol. 2287, pp. 477–495. Springer, Heidelberg (2002)
3. Hristidis, V., Koudas, N., Papakonstantinou, Y., Srivastava, D.: Keyword Proximity Search in XML Trees. IEEE Trans. Knowl. Data Eng. 18(4), 525–539 (2006)
4. Cohen, S., Mamou, J., Kanza, Y., Sagiv, Y.: XSEarch: A Semantic Search Engine for XML. In: VLDB, pp. 45–56 (2003)
5. Li, Y., Yu, C., Jagadish, H.V.: Schema-Free XQuery. In: VLDB, pp. 72–83 (2004)
6. Amer-Yahia, S., Cho, S., Srivastava, D.: Tree pattern relaxation. In: Jensen, C.S., Jeffery, K., Pokorný, J., Šaltenis, S., Bertino, E., Böhm, K., Jarke, M. (eds.) EDBT 2002. LNCS, vol. 2287, pp. 496–513. Springer, Heidelberg (2002)
7. Li, J., Liu, C., Yu, J.X., Zhou, R.: Efficient top-k search across heterogeneous XML data sources. In: Haritsa, J.R., Kotagiri, R., Pudi, V. (eds.) DASFAA 2008. LNCS, vol. 4947, pp. 314–329. Springer, Heidelberg (2008)
8. Yang, W., Shi, B.: Schema-aware keyword search over xml streams. In: CIT, pp. 29–34 (2007)
9. Guo, L., Shao, F., Botev, C., Shanmugasundaram, J.: XRANK: Ranked Keyword Search over XML Documents. In: SIGMOD Conference, pp. 16–27 (2003)
10. Xu, Y., Papakonstantinou, Y.: Efficient Keyword Search for Smallest LCAs in XML Databases. In: SIGMOD Conference, pp. 537–538 (2005)
11. Liu, Z., Chen, Y.: Identifying meaningful return information for XML keyword search. In: SIGMOD Conference, pp. 329–340 (2007)
12. Sun, C., Chan, C.Y., Goenka, A.K.: Multiway slca-based keyword search in xml data. In: WWW, pp. 1043–1052 (2007)
13. Li, G., Feng, J., Wang, J., Zhou, L.: Effective keyword search for valuable lcas over xml documents. In: CIKM, pp. 31–40 (2007)
14. Liu, Z., Chen, Y.: Reasoning and identifying relevant matches for xml keyword search. PVLDB 1(1), 921–932 (2008)
15. Florescu, D., Kossmann, D., Manolescu, I.: Integrating keyword search into XML query processing. Computer Networks 33(1-6), 119–135 (2000)
16. Hristidis, V., Papakonstantinou, Y., Balmin, A.: Keyword Proximity Search on XML Graphs. In: ICDE, pp. 367–378 (2003)

Representing Multiple Mappings between XML and Relational Schemas for Bi-directional Query Translation[*]

Ya-Hui Chang and Chia-Zhen Lee

Department of Computer Science, National Taiwan Ocean University
yahui@mail.ntou.edu.tw

Abstract. There is an increasing need to translate between SQL and XQuery. This task could be successfully performed only when the differences between the underlying schemas are properly resolved. In this paper, we propose to represent each schema by value, collection, and structure constructs, and design a set of mappings to represent their correspondence. The cases of multiple mappings are particularly considered. Based on the mapping information, queries could be transformed to each other and information could be therefore shared. We give a formal proof to validate the proposed approach, and also show some experimental results to demonstrate its performance.

1 Introduction

Relational databases have been widely used in enterprises to support critical business operations, while XML has emerged as the de facto standard for data representation and exchange on the World-Wide-Web. There is a need to translate an SQL against a relational database to an XQuery for XML data, and vice versa. For example, suppose originally data are stored in the relational database. An SQL which is embedded in existing applications, should be translated into an XQuery to get the XML data from the Web. On the other hand, an XQuery posed against an XML view, might need to be translated to SQL if the underlying database engine is relational.

In this paper, we propose a general framework, where relational queries and XML queries can be easily transformed to each other. The main issue to support this task is to properly represent the mapping between the relational schema and the XML schema, where representational conflicts might exist. Among all the conflicts, we mainly focus on those related to the characteristics of the XML representation. Take the *structure* constructs as an example. A relational schema is usually considered as *flat*, since no explicit structures exist between relations and the relationship is constructed by joining attribute values. On the contrary, the relationship between XML data could be directly represented through the *nested* structure. The correspondence between a join and a nested structure will need to be presented. Moreover, we also consider the possibility of *multiple mappings* between the different constructs in two schemas, which is usually neglected to simplify the transformation task.

---

[*] This work was partially supported by the National Science Council under Contract No. NSC 96-2221-E-019-048-.

Q. Li et al. (Eds.): APWeb/WAIM 2009, LNCS 5446, pp. 100–112, 2009.

The contributions of this paper could be summarized as follows:

- Classification of the representational conflicts:
  We discuss the possible representational conflicts between the relational and XML schemas. We focus on the characteristics of the XML format, and discuss the possible sources of multiple mappings.
- Specification of the mapping of schemas:
  We design a set of mappings to represent the correspondence between different constructs of the relational schema and the XML schema. The type of multiple mappings is represented to facilitate the selection among multiple choices.
- Design of the transformation algorithms:
  We have designed a set of algorithms which could perform bi-directional query transformation. It utilizes the schema mapping information, and adopts a priority-based strategy to handle multiple mappings. A formal proof is given to show the correctness of the proposed approach.
- Implementation of the system:
  A prototype is built to perform query transformation between the most common type of SPJ (selection-projection-join) SQL and those XQuery statements with equivalent expressive capabilities. Experimental results show the efficiency of the proposed approach.

The remaining of this paper is organized as follows. In Section 2, we describe the representation of relational and XML schemas, and formulate the problem to solve. In Section 3, we define a set of mappings between different schema constructs, and the transformation algorithms along with examples are presented in Section 4. We evaluate our proposed approach in Section 5, and a brief comparison with other works is provided in Section 6. Finally, we conclude this paper in Section 7.

2   Problem Definition

In this section, we discuss how to represent the relational schema and the XML schema, and show the differences of the corresponding query languages. We also formalize the problem to solve in this paper.

2.1   Schema Representations

The relational schema considered in this paper follows the traditional definition and satisfies the First Normal Form. A sample relational schema, which is illustrated as a graph, is shown in Figure 1(a). In the graph, each box corresponds to a relation, and the attributes associated with the relation are represented within the box, with the primary key on the top. The foreign key is depicted as an arrow pointing to the corresponding primary key. We further classify the relations into two types. The *E-relation* corresponds to a relation which describes the information of an entity, *e.g.*, *supplier*, *part*, and *order*. The *R-relation* corresponds to a relation which describes the relationship among other entities and has a composite primary key, *e.g.*, *partsupp* and *lineitem*.

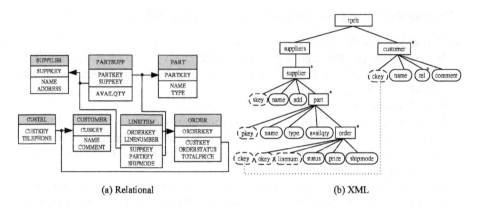

(a) Relational   (b) XML

Fig. 1. The sample schema

The XML schema proposed by W3C consists of many complex constructs. We only process those conforming to the definition of DTD (Document Type Definition). It can be represented as a quadruple (E, A, r, P), where E is a set of elements, A is a set of attributes, r is the root element, and P is a production rule defined as follows:

$$e ::= \epsilon \mid l, l \in E \mid e|e \mid ee \mid e^*$$

where ϵ is the empty string; $e|e$, ee, and e^* represent the union, concatenation, and Kleene star, respectivley.

In this paper, the XML schema is represented as a graph, where elements and attributes are represented as nodes, and the nesting relationship between elements is represented by the relationship of parent/child nodes in the graph. Note that we do not distinguish the representation of union and concatenation. The sample XML schema, which represents similar information as in Figure 1(a), is illustrated in Figure 1(b). We also define several special kinds of nodes. The *repeatable element* refers to an internal node which is allowed to have multiple occurrences under the same parent element, and is annotated by the symbol "*", *e.g.*, *part*. On the other hand, the *dummy element, e.g.,* *suppliers*, is an internal node which is usually introduced just to group elements that appear beneath it. They are both represented as square nodes. The *leaf nodes*, which are represented by rounded rectangles, are nodes associated with values. They might be elements, which are named as *value elements*, or they might be *attributes*, which are represented using dotted lines. We also use dashed lines to connect two leaf nodes which are semantically equivalent.

To uniformly refer to the construct represented in different schemas but with the same functionality, we define the following terms. A *value construct* is the construct which directly represents data, *e.g.*, *attribute* or *leaf node*. The *collection construct* is a construct which represents a set (multi-set) of data with homogeneous structures, eg., *relation* or *repeatable element*. The *structure construct* is used to connect two collections. In XML, it can be directly represented by the *nested construct*, or indirectly represented by a *join statement* in the relational database. This will be explained further in the next subsection.

Scope of SQL	Scope of XQuery
SQL ::= SelClause FromClause WhereClause?	XQuery ::= ForClause WhereClause? RetClause
SelClause ::= SELECT {Rel.Attr}	RetClause ::= RETURN {P_i}
FromClause ::= FROM {Rel}	ForClause ::= For {$Var IN P_i}
WhereClause ::= WHERE W_i {AND W_i}*	WhereClause ::= WHERE W_i {AND W_i}*
W_i ::= Rel.Attr CompOP Value	W_i ::= P_i CompOP Value | P_i CompOP P_i
| Rel.Attr CompOP Rel.Attr	P_i ::= PathExpr | $Var / PathExpr
CompOP ::= > | = | <	PathExpr ::= NameTest | NameTest/PathExpr
	NameTest ::= Element | @Attr

Fig. 2. Scope of Queries

2.2 Query Representations

The XQuery and SQL statements considered in this paper, are formally specified in Figure 2. We use two sample queries to explain. Suppose the user intends to identify the type of all parts with the name "dvd", and retrieve the name of its suppliers. The SQL query posed against the relational schema in Figure 1(a) will be as follows:

SQ1:
SELECT supplier.name, part.type
FROM   supplier, part, partsupp
WHERE part.name = "dvd" AND .... (1)
        supplier.suppkey = partsupp.suppkey AND .... (2)
        part.partkey = partsupp.partkey .... (2)

To briefly explain, the FROM clause is used to enumerate all the relations consulted, and the SELECT clause lists the attributes for output. The conditional statements listed in the WHERE clause could be classified into two types: the one marked with (1) is called the *selection statement*, which restricts the values of certain attributes; the one marked with (2) is called the *join statement*, which constructs the relationship between two relations. Note that the two join statements construct the relationship between the two E-relations *supplier* and *part* through the R-relation *partsupp*.

    The XQuery statement which performs the same function as *SQ1* does, but is appropriate for the XML schema in Figure 1(b), will be as follows:

XQ1:
FOR $t0 in /tpch/suppliers/supplier , $t1 in $t0/part
WHERE $t1/name = "dvd"
RETURN $t0/name, $t1/type

An XQuery statement uses the FOR clause to list a sequence of variable bindings. In this query, the variable $t0$ considers all supplier elements, and the variable $t1$ examines all the parts supported by a supplier. The WHERE clause is used to specify the selection condition, and the RETURN clause specifies what to output. Note that XQuery uses *path expressions* such as */tpch/suppliers/supplier* to navigate the nested structure of the XML schema.

Similarly, we define some terms which have the common functionality in different query languages. First, *value literals* refer to those statements represented in the SELECT/RETURN clause, such as *supplier.name*, or the selection statement in the WHERE clause. The *collection literals* refer to the relations or elements extracted from the FROM/FOR clause, such as */tpch/suppliers/supplier*. Finally, the structure literal will be the *join statement* in the WHERE clause, such as *supplier.suppkey = partsupp.suppkey*, or the *nested statement* in the FOR clause, such as *$t1 in $t0/part*. Note that *value literals* operate on *value constructs*, and the correspondence is same for the *collection* and the *structure literals*.

In this paper, we intend to translate an input XQuery into an equivalent SQL query, or vice versa. Traditionally, two equivalent queries in the same database mean that they can retrieve the same set of data. This definition is not applied in our heterogenous environment. Therefore, we define the equivalency based on the query statements themselves. An intuitive definition is first given as follows:

Definition 1. *Given the input query q_i and the output query q_o, q_i and q_o will be strongly equivalent, denoted $q_i \equiv q_o$, if they have the same numbers of value literals, collection literals, and structure literals, and the corresponding literals are equivalent.*

However, based on this definition, the sample query SQ1 is not equivalent to XQ1, since the numbers of collection literals are different. The weaker definition is then specified as follows:

Definition 2. *Given the input query q_i and the output query q_o, q_i and q_o will be weakly equivalent, denoted $q_i \simeq q_o$, if they have the same number of equivalent value literals, but with different numbers of equivalent collection literals. However, the collections in q_o are connected by proper structure literals.*

We can see SQ1 $\simeq$ XQ1. In this paper, we will consider the translated query as *correct*, if it is strongly or weakly equivalent to the input query. The problem to solve in this paper could be therefore formally stated as follows: *"Given two relational or XML schemas, where there might exist multiple mappings between the value, collection, or structure constructs, and an input query, find the correct translated query."*

3   Representations of Schema Mappings

We discuss how to represent schema mappings between the relational schema and the XML schema in this section. The two sample schemas described in Section 2 will be used as examples, and they will be called rdb and xdb, respectively. We also use the suffix $s2x$ to represent the mapping from rdb to xdb, and $x2s$ for the reverse direction.

3.1   Representing Value and Collection Mappings

There might exist multiple mappings between values in two databases, due to redundant representations, or keys which are represented in two relations to construct joins. We define the following Value Mapping (VM) to represent the correspondence between values in two schemas:

Definition 1. *Given a value construct v_i from schema$_1$, $VM(v_i)$ will return the set of tuples (v_o, type), where v_o represents the equivalent value construct represented in schema$_2$, and the value of types could be PK (standing for primary keys), FK (standing for foreign keys), or ANY.*

In the case of multiple mappings, types are used to define the priority, and PK $>$ FK $>$ ANY, since primary keys have the important identifying characteristics. For example, VM_{x2s}(supplier@skey) = {(supplier.suppkey, PK), (partsupp.suppkey, FK), (lineitem.suppkey, FK)}, and we will use the attribute associated with the relation *supplier* in the translated query.

The mappings between collections are more complicated. We first discuss the case of $1 : n$ mappings from xml schemas to relational schemas. This refers to the situation where the value elements or attributes under a repeatable element are scattered in different relations. This is usually caused by the *normalization* process in relational databases. A special case concerns the R-relation, which consists of information from several E-relations. In the XML representation, such relationship could be represented by the nested construct. For example, *partsupp* is a relationship between the two E-relations *part* and *supplier*. In the sample XML schema, there is no such explicit element, and *part* is directly represented as a child element of *supplier* instead. Therefore, we will let *partsupp* corresponds to the more specific element *part*, since the path associated with *part* is *//suplier/part*, which represents the relationship between the repeatable elements *supplier* and *part*.

As to the case of $1 : n$ mappings from relational schemas to xml schemas, it refers to the situation where the attributes in one relation are represented under several repeatable elements. It might be similarly caused by different partitions of an entity, *e.g.*, using two repeatable elements to represent normal customers and VIP customers, respectively. It is also possibly caused by the *dummy element*. For example, since the functionality of the dummy element *suppliers* is to group all the *supplier* elements, both *suppliers* and *supplier* will map to the same relation *supplier* in the sample relational schema.

The Collection Mapping (CM) summarizes the discussion above and is defined as follows:

Definition 2. *Given a collection construct c_i from schema$_1$, $CM(c_i)$ will return the set of tuples (c_o, type), where c_o represents the corresponding collection construct represented in schema$_2$. For an XML schema, the type could be REP (standing for repeatable elements) or DUM (standing for dummy elements). For a relational schema, the type could be E (standing for E-relations) or R (standing for R-relations).*

For example, CM_{s2x}(supplier) = {(suppliers, DUM), (supplier, REP)}, and CM_{s2x} (partsupp) returns the single tuple (part, REP). As to the case of multiple mappings, the priority of REP will be higher than DUM. However, we let E and R have the same priority, since which relation to output depends on the required attributes. For example, CM_{x2s}(part) = {(part, E), (partsupp, R)}, where the two relations have the same priority.

(a) Join statements in *rdb*

R_ID	Condition1	Condition2
RE1	supplier.suppkey	partsupp.suppkey
RE2	supplier.suppkey	lineitem.suppkey
RE3	part.partkey	lineitem.partkey
RE4	order.custkey	customer.custkey
RE5	order.custkey	custel.custkey
RR1-1	supplier.suppkey	lineitem.suppkey
RR1-2	supplier.partkey	lineitem.partkey
IJ1	part.partkey	partsupp.partkey
IJ2	order.orderkey	lineitem.orderkey
IJ3	customer.custkey	custel.custkey

(b) Mapping

R_ID	X_ID
RE1	XD1
RE1	XN1
RE2	XD2
RE2	XN2
RE3	XN3
RE1	XN3
RE4	XF1
RE5	XN1

(c) Path statements in *xdb*

X_ID	Xpath1	Xpath2
XD1	/tpch/suppliers	part
XN1	/tpch/suppliers/supplier	part
XD2	/tpch/suppliers	order
XN2	/tpch/suppliers/supplier	order
XN3	/tpch/suppliers/supplier/part	order
XF1	customer@ckey	order@ckey

Fig. 3. Structure mappings between rdb and xdb

3.2 Representing Structure Mapping

We define the structure mapping in this section. Recall that the structure construct might be represented as join statements or nested paths. For easy explanation, we will represent them using tables and give each construct an identifier, which is used to identify the type of the construct.

Consider the relational schema. If the second letter of the identifier is the letter E, it will represent a join between two E-relations or one E-relation and one R-relation. If the second letter of the identifier is the letter R, it will represent a join between two R-relations. The join statements for the sample relational schema are represented in Table 3(a). For example, RE1 is a join between the E-relation *supplier* and the R-relation *partsupp*, and RR1 is a join between two R-relations *partsupp* and *lineitem*. Note that RR1 consists of the two joins RR1-1 and RR1-2. The reason is that the R-relation *partsupp* has a composite primary key which consists of two attributes *partkey* and *suppkey*. To make the query uniquely identify the correct tuples, we need to let the join statements involve all component keys.

Similarly, the structure constructs in the sample XML database are represented in Table 3(c), and the identifiers are used to denote their types. If the second letter is "F", which stands for "flat", the two paths represented by the fields *Xpath1* and *Xpath2*, will be used to construct a join expression. If the second letter is "N", it will represent two *nested* elements. Here *Xpath1* represents the ancestor, and we specify the complete path from the root, while *Xpath2* represents the descendent, and we only specify the element name. If a *dummy* element is involved, the second letter will be "D".

Table 3(b) represents the structure mappings between the two sample schemas. Recall that two structure constructs are equivalent if they connect equivalent collection constructs. As specified in the second row of Table 3(b), XN1 is equivalent to RE1, since XN1 connects the repeatable elements *supplier* and *part*, and RE1 connects the equivalent counterparts, the E-relation *supplier* and the R-relation *partsupp*. Observe that there exist multiple mappings between structural expressions, as seen in collection constructs. If the case happens in the XML schema, the priority will be XN > XD, *i.e.*, the construct without involving dummy elements will have the higher priority. As to the relational schema, we let RE > RR, since the construct involving the E-relations is preferred.

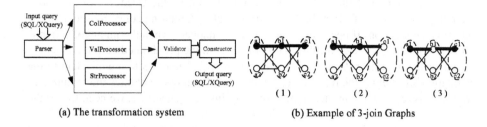

(a) The transformation system   (b) Example of 3-join Graphs

Fig. 4. The transformation process

In conclusion, we define the Structure Mapping (SM) as follows:

Definition 3. *Given a structure construct s_i from schema$_1$, SM(s_i) will return the set of tuples (s_o, type), where s_o represents the corresponding structure construct in schema$_2$, and the type could be RE or RR in the relational schema, or XF, XN, or XD in the XML schema.*

Note that in a special case of multiple mappings between collection constructs, where several equivalent collections are selected for output at the same time, we will need to connect those collections to identify that they are mapped from the same source. For example, the two relations *PART* and *PARTSUP* both map to the repeatable element *part*, so we need to provide an *internal join* between the two relations. Several examples are listed in Table 3(a), whose identifiers are started with IJ. Also note that internal joins map to nothing, as seen in Table 3(b).

4   Transformations of Queries

We discuss how to perform query transformation using the mapping information in this section, and the sample queries in Section 2 will be used to illustrate the whole translation process.

4.1   Processing Each Literal

The transformation system is depicted in Figure 4(a). The input query, either in SQL or XQuery, will be parsed into the internal representation, where the value literals, collection literals, and structure literals are extracted, and are then sent to the corresponding processor for transformation in sequence.

First, Algorithm ColProcessor processes each collection literal, and gets all the collection constructs with the highest priority through CM. If there are several of them, each of which will be associated with a flag called UsedFlag, with the initial value FALSE. It will be set TRUE later if the data represented in the collection is "accessed", either for projection, selection, or join. All the identified output collections will be represented in the structure ColSet.

Algorithm ValProcessor then identifies equivalent value literals through VM. In contrast, only one with the highest priority will be obtained, since it alone can get the most relevant information. Note that this algorithm will update the UsedFlag of the corresponding collection in ColSet. It also needs to deal with missing collections and inserts additional collections if necessary. Such situation happens when the input collection is an R-relation, but the used attributes do not occur in the pre-defined output collection. Consider the query: *"select suppkey from partsupp"*. The collection equivalent to the R-relation *partsupp* is the repeatable element *part*, but it does not have the required attribute. It is represented in the element *supplier* instead. Therefore, we need to add the element *supplier* into the structure ColSet for output.

In the next step, Algorithm StrProcessor will examine and identify all the equivalent structure literals. To perform this task, we first represent all the relevant information into graphs as defined in the following:

Definition 4. *A 2-join graph for two input collections c_1 and c_2 is a bipartite graph, where the nodes of partition$_i$ represent the output collections equivalent to c_i. A node will be marked black if its UsedFlag has the value TRUE. The edge connecting two nodes represents the structure literal between the associated output collections, and annotated with the structure identifier.*

Definition 5. *A n-join graph consists of a sequence of n partitions, and i_{th} and $i+1_{th}$ partitions and the edges between them form a 2-join graph, where $1 \leq i \leq n-1$.*

A 3-join graph is illustrated in Figure 4(b), where there are three input collections, and each input collection has two equivalent output collections. We omit the identifier of the structure construct. The idea is to select the structure construct if the associated collection is marked black. Otherwise, we will choose the one with the highest priority. In cases (1)-(3), the structure literals corresponding to edges (a_1, b_1) and (b_1, c_1) will be output. The complete algorithm is omitted due to space limitation. Note that an original unused collection may be marked black if the corresponding structure construct is chosen.

4.2 Formulating the Query

After processing the collection, value, and structure literals, all the intermediate structure will be processed by Algorithm Validator. There are mainly three parts in this Algorithm. The first part forces each input collection to have an output collection. The second part of the algorithm removes unused collections. The last part identifies required internal joins, as discussed in Section 3.

Finally, Algorithm Constructor will insert the proper keywords, and combine those intermediate structures as a syntactically correct statement and produce the transformed query statement. Note that when processing the structure literals for XML, if the two identified path statements are nested, as denoted in the identifier, a nested statement in the FOR clause will be produced. Otherwise, the paths will be transformed into a proper join statement in the WHERE clause.

We use the two sample queries in Section 2 to illustrate the process of translation. If the input query is SQ1, the two repeatable elements *supplier* and *part* will be first identified, and their UsedFlags will be set TRUE by Algorithm ValProcessor. The first join statement, which is RE1, will map to XN1, based on Tables 3(a)-3(b). XN1 will be selected since both the associated collections are marked black. Therefore, a nested path between *supplier* and *part* will be output, as seen in XQ1. Note that the second join statement is an internal join and has no corresponding output. In the reverse direction, the two repeatable elements will identify three relations, *supplier*, *part*, and *partsupp*, where only the first two relations are marked by Algorithm ValProcessor. The nested path in the FOR clause will then identify the join statement *supplier.suppkey = part-supp.suppkey*. Note that here StrProcessor will mark the relation *partsupp*. Algorithm Validator will find out that relations *part* and *partsupp* are both introduced by the element *part*, and the internal join *part.partkey = partsupp.partkey* will be identified, as seen in SQ1.

5   Evaluation

We evaluate the correctness and the efficiency of the proposed approach in this section. A formal proof of the correction is shown first, followed by the presentation of the experimental results.

5.1   Correctness

To prove the correctness of the proposed approach, we suppose all the schema constructs have mapping counterparts, since the output query will not be correct if the transformation system cannot get all the required information. Processing missing attributes are outside of the scope of this paper.

Lemma 1. *Given the input query q_i, the output query q_o transformed by the proposed system will have the same number of value literals as that of q_i, and they are equivalent.*

Proof. Since Algorithm ValProcessor will process each value literal, and output the one with the highest priority, this statement is trivially proved.   □

Theorem 1. *Given the input query q_i, the output query q_o transformed by the proposed system will be correct.*

Proof. The proof is provided by induction on the number of collection literals and value literals. Suppose the statement is true when the input query consists of collection literals $CI_1 \cdots CI_m$ and value literals $VI_1 \cdots VI_l$, and the output query consists of collection literals $CO_1 \cdots CO_n$. We consider the case when the input query has an additional CI_{m+1}, and an additional value literal VI_{l+1}. Note that to make the input query correct, there must exist a collection literal CI_i, where $1 \le i \le m$, which is connected with CI_{m+1}. Without loss of generality, we will assume only one structural construct between any two collection constructs. We call the structural literal connecting CI_i and CI_{m+1} as SI_i.

Suppose all the collection constructs corresponding to CI_{m+1} are represented by $\{CO_{n+1}\}$, and CO_j corresponds to CI_i, where $1 \le j \le n$. The structural constructs connecting $\{CO_{n+1}\}$ and CO_j are represented as $\{SO_j\}$. Based on Lemma 1, there will be an VO_{l+1} corresponding to VI_{l+1}, so we only need to analyze the collection and

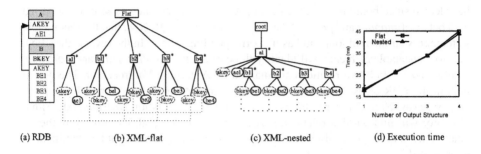

(a) RDB   (b) XML-flat   (c) XML-nested   (d) Execution time

Fig. 5. Analysis of the effect caused by the number of output literals

structural literals in the output query. In the following, we discuss how many CO_{n+1} instances appear in the output query.

(1) zero: This case happens when no SO_j and VO_{l+1} access any CO_{n+1} instances, but it is only possible when all SO_j instances are empty mapping, and all CO_{n+1} instances are equal to CO_j. This is a case of $n : 1$ mappings, and each SI_i is an internal join and maps to nothing. Therefore, the collection and structure parts of the output query do not change. It is still a correct output query.

(2) one: One CO_{n+1} instance appears in the final output query if VO_{l+1} accesses it. Algorithm StrProcessor will choose the proper SO_j instance to output. Therefore, the output query will have $n + 1$ collection literals and all the collection literals are connected. If $m = n$, the output query will satisfy the definition of strong equivalency. Otherwise, it will satisfy the definition of weak equivalency.

(3) more than one: This case happens when VO_{l+1} and SO_j access different CO_{n+1} instances. This is a case of $1 : n$ mappings, and Algorithm StrProcessor will output additional internal joins to make the query correct.   □

5.2   Efficiency

We have designed several experiments to evaluate the efficiency of the proposed approach. Due to space limitation, we will only show the result of one of them. All experiments are performed on a P4-2.4GHz machine, with 512 MB of DDR-RAM.

The schemas used for the experiment are shown in Figure 5(a)-(c). In the relational database, the attributes BE1-BE4 are all represented within the relation B, but they are scattered under different repeatable elements in the XML schema. The difference between the two XML schemas, is that the repeatable elements in (b) are in a flat structure, but in a nested structure in (c). We test four SQL query statements, with increasing BE attributes. Note that the numbers of output collection and structure literals will increase along. The transformation time is shown in Figure 5(d). We can see that the numbers of output literals possess a linear effect on the transformation time, which is acceptable. Also note that the effect of the XML schema structure is quite minor.

6   Related Work

There has been many researches on representing XML documents in relational databases systems. In this type of researches, users pose XQuery to manipulate XML data, but the query needs to be translated into SQL for being executed in the underlying relational databases. The issues regarding this translation process have been addressed in [10]. Particularly, the authors in [3] consider the efficient ways to handle wild cards in the path expression, the authors in [6,5] tackle the problem of recursive path expressions or paths, and the authors in [7] discuss how to produce optimal queries. Note that the platform of our research is more general than this type of work, since we consider any kind of mappings between relational databases and XML databases, and can perform bi-directional query translation. However, We could include their approach in our system if we want to consider more complex XQuery.

Some researchers intend to present complex mappings between schemas as well. The mapping proposed in [4] could handle "composite" mappings, such as the object *address* consisting of many components like *street*, *city* and *state*. The authors in [2] consider the case of multiple elements mapping to the same relation, and they use the input query to uniquely identify one. We have not seen other works which tackle the same problem of multiple mappings as we do.

On the other hand, many researchers have investigated the technique on automatic schema matching [9,8,1]. This type of work can be incorporated in our work to help identify equivalent constructs. We can also use such information in the assignment of priorities.

7   Conclusions

In this paper, we represent the schema mapping between relational databases and XML databases. A prototype utilizing these mapping information is built to translate the core expressions between SQL and XQuery. We have shown that our translator could produce transformation correctly and efficiently. In the future, we plan to extend our prototype, so that the full-fledge syntax of XQuery and SQL could be handled.

References

1. An, Y., Borgida, A., Miller, R.J., Mylopoulos, J.: A semantic approach to discovering schema mapping expressions. In: Proceedings of the ICDE Conference (2007)
2. Atay, M., Chebotko, A., Lu, S., Fotouhi, F.: XML-to-SQL query mapping in the presence of multi-valued schema mappings and recursive XML schemas. In: Wagner, R., Revell, N., Pernul, G. (eds.) DEXA 2007. LNCS, vol. 4653, pp. 603–616. Springer, Heidelberg (2007)
3. DeHaan, D., Toman, D., Consens, M.P., Ozsu, M.T.: A comprehensive xquery to sql translation using dynamic interval encoding. In: Proceedings of the SIGMOD International Conference on Management of Data (2003)
4. Embley, D.W., Xu, L., Ding, Y.: Automatic direct and indirect schema mapping: experiences and lessons learned. SIGMOD Record 33(4) (2004)
5. Fan, W., Yu, J.X., Lu, H., Lu, J., Rastogi, R.: Query translation from xpath to sql in the presence of recursive dtds. In: Proceedings of the 31th VLDB Conference (2005)

6. Krishnamurthy, R., Chakaravarthy, V.T., Kaushik, R., Naughton, J.F.: Recursive XML schemas, recursive xml queries, and relational storage: Xml-to-sql query translation. In: Proceedings of the 20th International Conference on Data Engineering (2004)
7. Krishnamurthy, R., Kaushik, R., Naughton, J.F.: Efficient xml-to-sql query translation: Where to add the intelligence? In: Proceedings of the 30th VLDB Conference (2004)
8. Lee, M.L., Yang, L.H., Hsu, W., Yang, X.: Xclust: Clustering xml schemas for effective integration. In: Proceedings of the ACM CIKM conference (2002)
9. Rahm, E., Bernstein, P.A.: A survey of approaches to automatic schema matching. The VLDB Journal 10, 334–350 (2001)
10. Krishnamurthy, R.K.R., Naughton, J.F.: Xml-to-sql query translation literature: The state of the art and open problems. In: Proceedings of the XML Database Symposium (2003)

Finding Irredundant Contained Rewritings of Tree Pattern Queries Using Views

Junhu Wang[1], Kewen Wang[1], and Jiuyong Li[2]

[1] Griffith University, Australia
{J.Wang, k.Wang}@griffith.edu.au
[2] University of South Australia, Adelaide, Australia
Jiuyong.Li@unisa.edu.au

Abstract. Contained rewriting and maximal contained rewriting of tree pattern queries using views have been studied recently for the class of tree patterns involving /, //, and []. Given query Q and view V, it has been shown that a contained rewriting of Q using V can be obtained by finding a *useful embedding* of Q in V. However, for the same Q and V, there may be many useful embeddings and thus many contained rewritings. Some of the useful embeddings may be redundant in that the rewritings obtained from them are contained in those obtained from other useful embeddings. Redundant useful embeddings are useless and they waste computational resource. Thus it becomes important to identify and remove them. In this paper, we show that the criteria for identifying redundant useful embeddings given in a previous work are neither sufficient nor necessary. We then present some useful observations on the containment relationship of rewritings, and based on which, a heuristic algorithm for removing redundant useful embeddings. We demonstrate the efficiency of our algorithm using examples.

1 Introduction

Query rewriting using views has many important applications such as in data integration and query caching [1]. A *view* is a pre-defined query which may or may not be materialized. A *contained rewriting (CR)*, or simply *rewriting*, of query Q using view V is a new query Q' such that, when evaluated over the answers to the view, Q' produces part or all of the correct answers to the original query Q, for any instance of the database.

The problem of finding contained rewritings has been well studied in relational databases [4,5,1]. Recently, contained rewritings of tree pattern queries using views were studied in [2] for the class $P^{\{/,//,[]\}}$. A tree pattern in $P^{\{/,//,[]\}}$ represents an `XPath` expression. It is a tree with every node labeled with an XML tag, and every edge labeled with either / or // (see Section 2 for more details). It was shown in [2] that, given query Q and view V in $P^{\{/,//,[]\}}$, a contained rewriting of Q using V can be found by finding a *useful embedding (UE)* of Q in V. However, there can be many UEs, and some of them may be redundant in that the rewritings obtained from them are contained in those obtained from

Q. Li et al. (Eds.): APWeb/WAIM 2009, LNCS 5446, pp. 113–125, 2009.

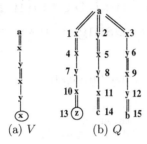

13 //	14 //	15 //
13 //	14 //	12 /
13 //	14 //	9 //
13 //	5 //	15 //
⋮	⋮	⋮
1 //	2 //	3 //

(a) V (b) Q (c) UE matrix

Fig. 1. (a) and (b): Example tree patterns; (c): part of a UE matrix

other UEs. For example, for the view V and query Q shown in Fig.1 (a) and (b), there are 80 UEs, but 79 of them are redundant. Redundant UEs are useless and they waste computational resource because all results produced by them can be produced by other rewritings. Therefore it is important for the redundant UEs to be efficiently identified and removed. As observed in [2], identifying and removing all redundant UEs is a challenging problem.

In this paper, we first show that the criterion given in [2] for identifying irredundant UEs is wrong, and in fact, it is neither sufficient nor necessary. This is done in Section 3.1. We then present some useful observations on the relationships between UEs (Section 3.1). Based on these observations, we give a heuristic algorithm for computing all irredundant UEs (in Section 4). We demonstrate the efficiency of our algorithm by examples.

We will introduce some background knowledge and notations in the next section, before proceeding to present our major results in Sections 3 and 4.

2 Preliminaries

XTree and Tree Patterns. Let Σ be an infinite set of tags. An XML *tree* (XTree) is a tree with every node labeled with a tag in Σ. A *tree pattern* (TP) in $P^{\{/,//,[]\}}$ is a tree with a unique *distinguished node*, with every node labeled with a tag in Σ, and every edge labeled with either / or //. The path from the root to the distinguished node is called the *distinguished path*[1]. Fig.2 shows some example TPs, where single and double lines are used to represent /-edges and //-edges respectively, and a circle is used to indicate the distinguished node. A TP corresponds to an XPath expression. The TPs in Fig.2 (a) and (b) correspond to the XPath expressions $a[c]//b[d]$ and $a[c]//b[x]//y$ respectively.

Let P be a TP. We will use DN(P), and DP(P) to denote the distinguished node and the distinguished path of P respectively. For any tree T, we will use $N(T)$, $E(T)$ and $rt(T)$ to denote the node set, the edge set, and the root of T

---

[1] Note the 1:1 correspondence of our tree patterns to those in [2]: the root of a TP in [2] is marked / or //. These tree patterns can be changed to ours by adding an artificial root and connect it to the original root with a / or //-edge, respectively.

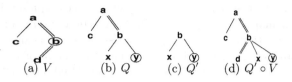

Fig. 2. View (a), Query (b), CP (c), and CR (d)

respectively. We will also use $label(v)$ to denote the label of node v, and call a node labeled a an a-*node*. In addition, if (u,v) is a $/$-edge (resp. $//$-edge), we say v is a $/$-child (resp. $//$-child) of u.

A *matching* of TP P in an XTree t is a mapping δ from $N(P)$ to $N(t)$ that is (1) *root-preserving*: $\delta(rt(P)) = rt(t)$, (2) *label-preserving*: $\forall v \in N(P)$, $label(v) = label(\delta(v))$, and (3) *structure-preserving*: for every edge (x,y) in P, if it is a $/$-edge, then $\delta(y)$ is a child of $\delta(x)$; if it is a $//$-edge, then $\delta(y)$ is a descendant of $\delta(x)$. Each matching δ produces a subtree of t rooted at $\delta(\text{DN}(P))$, denoted $t^{\delta(\text{DN}(P))}$, which is known as an *answer* to the TP. We use $P(t)$ to denote the set of all answers of P over t. For a set T of XTrees, we define $P(T) = \cup_{t \in T} P(t)$.

Containment and Containment Mapping. Let P and Q be TPs. P is said to be *contained* in Q, denoted $P \subseteq Q$, if for every XTree t, $P(t) \subseteq Q(t)$. For P and Q in $P^{\{/,//,[]\}}$, $P \subseteq Q$ iff there is a containment mapping from Q to P. Recall [3]: A *containment mapping* (CM) from Q to P is a mapping δ from $N(Q)$ to $N(P)$ that is label-preserving, root-preserving as discussed in the last section, structure-preserving (which now means for any $/$-edge (x,y) in Q, $(\delta(x),\delta(y))$ is a $/$-edge in P, and for any $//$-edge (x,y), there is a path from $\delta(x)$ to $\delta(y)$ in P) and is output-preserving, which means $\delta(\text{DN}(V)) = \text{DN}(P)$.

Rewriting TP Using Views. A *view* is a pre-defined TP. Let V be a view and Q be a TP. In this paper we implicitly assume that $rt(Q)$ and $rt(V)$ have the same label. A *compensation pattern* (CP) of Q using V is a TP Q' such that (1) for any XTree t, $Q'(V(t)) \subseteq Q(t)$, and (2) there exists an XTree t such that $Q'(V(t))$ is non-empty. It is clear that $label(rt(Q')) = label(\text{DN}(V))$. Let Q' be a CP of Q using V. We use $Q' \circ V$ to represent the TP obtained by merging $rt(Q')$ and $\text{DN}(V)$. The distinguished node of Q' becomes that of $Q' \circ V$. We call $Q' \circ V$ a *contained rewriting* of Q using V. Fig.2 shows a view V, a TP Q, a CP Q' of Q using V, and the corresponding CR $Q' \circ V$. Note that condition (1) in the definition of contained rewriting is equivalent to $Q' \circ V \subseteq Q$. The *maximal contained rewriting (MCR)* of Q using V is defined to be the union of all CRs of Q using V [2].

Given TP Q and view V in $P^{\{/,//,[]\}}$, [2] shows that the existence of a CR of Q using V can be characterized by the existence of a *useful embedding (UE)* of Q in V. In brief, a useful embedding of Q in V is a partial mapping f from $N(Q)$

to $N(V)$ that is root-preserving, label-preserving, structure-preserving as defined in a containment mapping (except that they are required only for the nodes of Q on which the function f is defined), and satisfies the following conditions. (1) if $f(x)$ is defined, then f is defined on $parent(x)$ (the parent of x). (2) for every node x on the distinguished path $DP(Q)$, if $f(x)$ is defined, then $f(x) \in DP_V$, and if $f(DN(Q))$ is defined, then $f(DN(Q)) = DN(V)$; (3) for every path $p \in Q$, *either* p is fully embedded, i.e., f is defined on every node in p, *or* if x is the last node in p such that $f(x)$ is defined and $f(x) \in DP(P)$ (x is called the *anchor node*), and y the node immediately following x (y is call the *successor* of x), then either $f(x) = DN(V)$, or the edge (x, y) is a $//$-edge.

Given a UE f from Q to V, a *clip-away tree* (CAT) CAT_f can be constructed as follows. The root of CAT_f is labeled $label(DN(V))$. For each path $p \in Q$ that is not fully embedded in V, find the anchor node x and its successor y, then connect the root of Q^y (Q^y is subtree of Q rooted at node y) to $rt(CAT_f)$ with the same type of edge as (x, y). The distinguished node of CAT_f is the node corresponding to $DN(Q)$. The MCR of Q using V is the union of all CRs found this way.

3   Redundant UEs Re-examined

We fix query Q and view V. For UE f, we define $CR_f = CAT_f \circ V$, namely CR_f is the CR obtained from f. Let f and g be UEs. We say f is contained in g if $CR_f \subseteq CR_g$.

Definition 1. *A UE f and the rewriting CR_f are said to be* redundant *if CR_f is contained in the union of the CRs generated by other UEs. Otherwise they are said to be* irredundant.

Using the independence of containing patterns property proved in [6], we know that f is redundant if and only if there is another UE g such that $CR_f \subseteq CR_g$. Thus the above definition of redundancy coincides with that in [2].

Intuitively redundant UEs and their corresponding CRs may be ignored because all answers produced by them can be produced by CRs obtained from other UEs.

3.1   Re-examining the Conditions Given in [2]

For TPs in $P^{\{/,//,[]\}}$, Lemma 1 of [2] gave a condition for irredundant UEs, and it claims the condition is sufficient and necessary. For easy reference we copy the lemma below.

A node $v \in Q$ is said to be *special* wrt useful embedding f, if $f(v) = DN(V)$, and v has a $/$-child u which is undefined by f. Two nodes are said to be *incomparable* if neither is the ancestor of the other. Given useful embeddings f and g, if g is defined on every node on which f is defined, then g is called an *extension* of f. A $/$-path is a path consisting of $/$-edges only.

Lemma 1 of [2]: Let $Q, V \in P^{\{/,//,[]\}}$. Suppose f is a useful embedding and CAT_f is the CAT induced by f. Then CR_f is irredundant iff, for every node $x \in Q$ on which f is undefined, one of the following conditions holds:

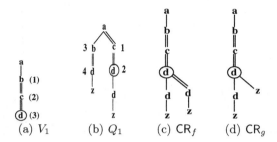

Fig. 3. Counter example to Lemma 1 of [2]

1. there is no extension g of f such that $g(x)$ is defined;
2. $\exists z \in Q$: Q has a /-path from x to z or $x = z$ (note: $x = z$ is the case when the /-path is of length 0, as allowed in [2]), and z is special wrt every extension g of f for which $g(x)$ is defined;
3. x is the distinguished node of Q, and every extension g of f for which $g(x)$ is defined, maps x to DN(V), and there is a node $y \in Q$, incomparable with x, such that $g(y)$ is undefined.

The following examples show that the above lemma is wrong, and in fact, the condition is neither sufficient nor necessary for f to be irredundant. Since the root of the query is always mapped to the root of the view, we omit such explicit description in the examples.

Example 1. Consider the view V_1 and query Q_1 in Fig.3. Let f be the UE which only maps the nodes 1, 2, 3 to (2), (3), and (1) respectively. Let g extend f by mapping the node 4 to the node (3). CR_f and CR_g are shown in Fig.3 (c), (d) respectively. Clearly, $CR_g \subseteq CR_f$ (and $CR_f \nsubseteq CR_g$). Therefore g is redundant. But for every node in Q_1 that is not embedded by g, there is no extension of g which embeds that node. In other words, for every node on which g is undefined, condition 1 of the lemma is true, but g is not irredundant.

The example above also demonstrates that (1) $CR_g \subseteq CR_f$ does not imply that f must be an extension of g, and (2) g is redundant does not imply there is an extension h of g such that $CR_g \subseteq CR_h$.

Example 2. Consider V_2 and Q_2 in Fig.4. Let f map the nodes 1, 2, 3 and 5 to the nodes (2), (3), (4) and (1) respectively. Let g extend f by mapping the nodes 6 and 7 to the nodes (3) and (4) respectively as well. g is the only extension of f, and g does not embed the two u-nodes, and the node (6) and (7) are special wrt g. In other words, for every node on which f is not defined, either condition 1 or condition 2 is true, but f is redundant because $CR_f \subseteq CR_g$ (CR_f and CR_g are as shown in Fig.4).

Example 3. Consider V_1 and Q_3 in Fig.4. Let f' map the nodes 1, 3, 4 to the nodes (2), (1), (3) respectively. Let g' extend f' by mapping node 2 to node (3)

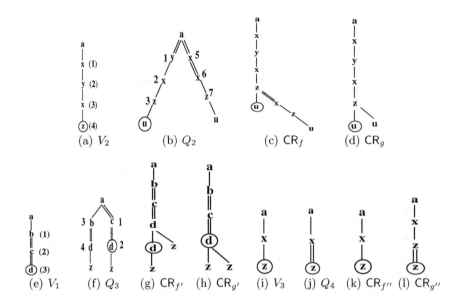

Fig. 4. Counter examples to Lemma 1 of [2]

as well. g' is the only extension of f', and every node that is not embedded by f' is either not embedded by g' (the z-nodes), or it (node 2) is the distinguished node of Q_3 and it is mapped to $\mathtt{DN}(\mathsf{V_1})$ by g', and there is a node (the left z-node) incomparable to node 2, which is not defined by g'. In other words, for every node on which f' is not defined, either condition 1 or condition 3 is true. But $\mathsf{CR}_{f'}$ is redundant because $\mathsf{CR}_{f'} \subseteq \mathsf{CR}_{g'}$.

The above examples show that the condition in Lemma 1 of [2] is insufficient. The next example will show it is not necessary either.

Example 4. Consider V_3 and Q_4 in Fig.4. Let f'' map the x-node (in Q_4) to the x-node in V_3. The z-node is the only node not embedded by f'', and it is the distinguished node of Q. The only extension g'' of f'' maps the z-node in Q_4 to the z-node in V_3 (the distinguished node of V). Clearly f'' is irredundant ($\mathsf{CR}_{f''}$ and $\mathsf{CR}_{g''}$ are as show in the figure), but there is no y, incomparable with z, such that $g''(y)$ is undefined, z is not special with respect to g'', and z is embedded by g''. That is, for the z-node, none of the conditions in the lemma holds.

3.2   Sufficient Conditions for Redundant UEs and Other Observations

Finding a simple, exact condition for irredundant UEs turns out to be quite challenging. From practice point of view, it is more important to identify UEs that are redundant, because such UE can be ignored when computing the MCR. In this section we provide some sufficient conditions and some related observations. They will be used in our heuristic algorithm (to be presented in the next

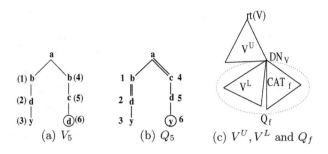

Fig. 5. (a) and (b) Example view and query used in proof of Lemma 1. (c) V^U, V^L and Q_f.

section) for finding irredundant UEs. By definition, a UE f is redundant if there is another UE g such that $\mathsf{CR}_f \subseteq \mathsf{CR}_g$. Therefore, we will first look at conditions under which $\mathsf{CR}_f \subseteq \mathsf{CR}_g$.

In the following, if UE f is defined on node $x \in N(Q)$, we say f *embeds* x. If f embeds all nodes on path $p \in Q$, we say f *fully embeds* p.

Lemma 1. *Let f and g be two useful embeddings. If f and g embed the same set of nodes in Q, then $\mathsf{CAT}_f = \mathsf{CAT}_g$ and $\mathsf{CR}_f = \mathsf{CR}_g$. However, $\mathsf{CAT}_f = \mathsf{CAT}_g$ does not imply f and g embed the same set of nodes.*

Proof. By definition, if f and g embed the same set of nodes in Q, then $\mathsf{CAT}_f = \mathsf{CAT}_g$ and $\mathsf{CR}_f = \mathsf{CR}_g$. The example below shows that $\mathsf{CAT}_f = \mathsf{CAT}_g$ does not imply f and g embed the same set of nodes. For V_5 and Q_5 in Fig.5 (a) and (b), let f and g be as follows:

$$f : 1 \to (1), 2 \to (2), 3 \to (3), 4 \to (5), 5 \to (6).$$
$$g : 1 \to (4), 2 \to (6), 4 \to (5), 5 \to (6).$$

Then $\mathsf{CAT}_f = \mathsf{CAT}_g$ although g and f do not embed the same set of nodes.

Lemma 2. *Let Q be the query, and V be the view in $P^{\{/,//,[]\}}$. $\mathsf{CAT}_f \circ V \subseteq \mathsf{CAT}_g \circ V$ if, but not only if, $\mathsf{CAT}_f \subseteq \mathsf{CAT}_g$.*

The proof of 'if' is straightforward. As an example to show $\mathsf{CAT}_f \circ V \subseteq \mathsf{CAT}_g \circ V$ does not imply $\mathsf{CAT}_f \subseteq \mathsf{CAT}_g$, consider $V = a//x$, and $Q = a//x/y$. Let f map $rt(V)$ to $rt(Q)$, let g extend f by mapping the x-node in Q to the x-node in V. Then $\mathsf{CAT}_f = x//x/y$ and $\mathsf{CAT}_g = x/y$. Thus $\mathsf{CR}_f \circ V = a//x//x/y$ and $\mathsf{CAT}_g \circ V = a//x/y$. Clearly, $\mathsf{CAT}_f \not\subseteq \mathsf{CAT}_g$, but $\mathsf{CAT}_f \circ V \subseteq \mathsf{CAT}_g \circ V$.

The theorem below provides a sufficient condition for $\mathsf{CR}_f \subseteq \mathsf{CR}_g$, which is the main idea behind our algorithm in the next section. In the theorem, $x(p, f)$ denotes the anchor node of path p with respect to UE f.

Theorem 1. *Let f and g be two useful embeddings. $\mathsf{CR}_f \subseteq \mathsf{CR}_g$ if the following conditions are satisfied.*

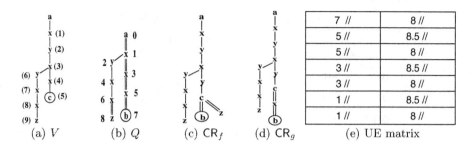

7 //	8 //
5 //	8.5 //
5 //	8 //
3 //	8.5 //
3 //	8 //
1 //	8.5 //
1 //	8 //

(a) V   (b) Q   (c) CR_f (d) CR_g (e) UE matrix

Fig. 6. V, Q and two CATs in Example 5

1. *Every path in Q that are fully embedded by f is also fully embedded by g;*
2. *For every path p in Q that is not fully embedded by g either $x(p, f) = x(p, g)$, or $x(p, f)$ is an ancestor of $x(p, g)$, and $x(p, g)$ and its successor are connected by a //-edge.*
3. *Either both $g(DN(Q))$ and $f(DN(Q))$ are defined, or both are undefined.*

Proof. By definition, in Q only paths that are not fully embedded by g will contribute to CAT_g. For each path p that is not fully embedded by g, it is not fully embedded by f either (condition (1)). Furthermore, by condition (2), either (A) $x(p, f) = x(p, g)$, or (B) $x(p, f)$ is an ancestor of $x(p, g)$, and $x(p, g)$ and its successor are connected by a //-edge. In case (A), the subtree contributed by p to CAT_g is identical to that contributed by p to CAT_f; in case (B), the subtree contributed by p to CAT_g is a subtree of the subtree contributed by p to CAT_f. Since $x(p, g)$ and its successor are connected by a //-edge, the edge between $rt(CAT_g)$ and the successor of $x(p, g)$ is also a //-edge. Therefore, there is a homomorphism from CAT_g to CAT_f. Furthermore, condition (3) ensures that either both $g(DN(Q))$ and $f(DN(Q))$ are defined, or both are undefined. In the former case, $g(DN(Q))$ and $f(DN(Q))$ are both $DN(V)$; in the later case, $g(DN(Q))$ can be mapped to $f(DN(Q))$ by the homomorphism. Thus in both cases, there is a containment mapping from CAT_g to CAT_f. That is, $CAT_f \subseteq CAT_g$. Hence $CR_f \subseteq CR_g$.

The above theorem provides a sufficient condition for f to be redundant: f is redundant if there is another UE g that satisfies the conditions (1) to (3) in the theorem. Note that the conditions (1) and (2) are equivalent to say g is an extension of f, and for every path on which g embeds more nodes than f, the anchor node (wrt g) is connected to its successor via a //-edge. Note also that the condition (3) in Theorem 1 is necessary. As an example, consider $V = a/x$ and $Q = a//x$. Let g map x in Q to x in V, but f do not. Then g is an extension of f, and CAT_g has no /-child. But $CR_f \not\subseteq CR_g$.

Using the above theorem, we can verify that, for the view V and query Q in Fig.1, the UE that embeds all nodes except 13, 14, and 15 contains all other UEs.

One might wonder whether it is possible for there to be two UEs f and g such that f embeds more nodes on path p_1 than g, but less nodes on path p_2 than g, $rt(\mathsf{CAT}_f)$ and $rt(\mathsf{CAT}_g)$ have no $/$-child, and both f and g are irredundant. The next example gives a positive answer to this question.

Example 5. Consider the view V and query Q in Fig.6. Let f and g be as follows:

$$f\colon 1 \to (1),\, 3 \to (3),\, (5) \to (4),\, 2 \to (2),\, 4 \to (3),\, 6 \to (4)$$
$$g\colon 1 \to (3),\, 3 \to (4),\, 2 \to (6),\, 4 \to (7),\, 6 \to (8),\, 8 \to 9$$

Compared with g, f embeds more nodes on the distinguished path, but less nodes on the non-distinguished path. CR_f and CR_g are as shown in Fig.6 (c) and (d). Both f and g are irredundant. Note there is no UE that embeds both node 5 and node 8.

In what follows, we will use P^y to denote the subtree of P rooted at node y, and use V^L as a shorthand for $V^{\mathsf{DN}(V)}$. We will also use Q_f to denote the pattern obtained by merging the roots of V^L and CAT_f (see Fig.5 (c)). For a UE f and a path p (in Q) that is not fully embedded by f, we will use $y(p, f)$ to denote the successor of $x(p, f)$ (recall: $x(p, f)$ is the anchor node of p with respect to f). As a shorthand we will use x_f to denote the last node on $\mathsf{DP}(Q)$ that is embedded by f.

Proposition 1. *Let f and g be useful embeddings.*

(1) If x_f is a descendant of x_g, then $\mathsf{CR}_f \not\subseteq \mathsf{CR}_g$.
(2) If $x_f = x_g$, then $\mathsf{CR}_f \subseteq \mathsf{CR}_g$ iff $Q_f \subseteq \mathsf{CAT}_g$,
(3) If x_f is an ancestor of x_g, and there are no $//$-edges on $\mathsf{DP}(V)$, then $\mathsf{CR}_f \subseteq \mathsf{CR}_g$ iff $Q_f \subseteq \mathsf{CAT}_g$.

Proof. We use the fact that $\mathsf{CR}_f \subseteq \mathsf{CR}_g$ iff there is a containment mapping from CR_g to CR_f.

(1) If x_f is an descendant of x_g then $|\mathsf{DP}(\mathsf{CR_g})| > |\mathsf{DP}(\mathsf{CR_f})|$, and thus cannot be a containment mapping from CR_g to CR_f.
(2) If $x_f = x_g$, there is an obvious one-to-one correspondence between the nodes on $\mathsf{DP}(\mathsf{CR_f})$ and those on $\mathsf{DP}(\mathsf{CR_g})$, and $\mathsf{DN}(V)$ corresponds to $\mathsf{DN}(V)$ on the two paths. Therefore, $\mathsf{CR}_f \subseteq \mathsf{CR}_g$ iff $Q_f \subseteq \mathsf{CAT}_g$.
(3) If x_f is an ancestor of x_g, and there are no $//$-edges on $\mathsf{DP}(V)$, then $\mathsf{DN}(V)$ on $\mathsf{DP}(\mathsf{CR_g})$ corresponds to $\mathsf{DN}(V)$ on $\mathsf{DP}(\mathsf{CR_f})$. Hence $\mathsf{CR}_f \subseteq \mathsf{CR}_g$ iff $Q_f \subseteq \mathsf{CAT}_g$.

Proposition 1 implies that (1) UEs that embed more nodes on $\mathsf{DP}(Q)$ can not be contained in those that embed less nodes on $\mathsf{DP}(Q)$. (2) if f and g embed the same nodes on $\mathsf{DP}(Q)$, or they satisfy condition (3), then testing for $\mathsf{CR}_f \subseteq \mathsf{CR}_g$ can be done by testing $Q_f \subseteq \mathsf{CAT}_g$, which is easier because Q_f and CAT_g are usually smaller than CR_f and CR_g.

The next proposition can be used to determine non-containment of UEs quickly.

Proposition 2. *If $g(\mathtt{DN(Q)})$ is defined, but $f(\mathtt{DN(Q)})$ is not, then $\mathsf{CR}_f \subseteq \mathsf{CR}_g$ implies there is a homomorphism from V_L to Q^L (Q^L is the subtree of Q rooted at $\mathtt{DN(Q)}$), and there is a homomorphism from CAT_g to Q^L, and there is at least one //-edge on $\mathtt{DP(V)}$.*

The above proposition is true because $\mathsf{CR}_f \subseteq \mathsf{CR}_g$ implies there is a containment mapping from CR_g to CR_f, which maps $g(\mathtt{DN(Q)})$ to $f(\mathtt{DN(Q)})$. Thus there must be a //-edge on $\mathtt{DP(V)}$. If we restrict the containment mapping to nodes in Q_g, it becomes a homomorphism from Q_g to Q^L. Thus there is a homomorphism from V_L to Q^L, and from CAT_g to Q^L.

4   A Heuristic Algorithm for Finding Irredundant UEs

We are interested in a set of UEs that has the following property: (1) the union of CRs generated by these UEs is equivalent to the MCR; and (2) no UE in the set is contained in any other. We call such a set of UEs a *minimal cover set*.

A naive approach to finding a minimal cover set is to, first, find all of the UEs using the algorithm in [2], and then, for each UE f, test whether it is contained in another g using the method of containment mapping, and if yes, discard it. This method is very inefficient in general. For example, for the query Q and view V in Fig.1, there are 80 UEs, but only one of them is irredundant. If we compare every pair of UEs randomly we would have to make at least 79 containment tests (e.g., in the lucky case we have chosen to compare the irredundant one with every other)[2]. Recall that finding the existence of a containment mapping from tree pattern P_1 to P_2 takes $O(|P_1||P_2|)$ where $|P_1|$ represents the number of edges in P_1 [3].

We now present a heuristic algorithm for finding a minimal cover set of UEs. Our algorithm uses Theorem 1 to filter out some redundant UEs first (see Algorithm 1). Then it compares the UEs that embed the same nodes on $\mathtt{DP(Q)}$ to see whether further redundancy can be removed. This can be done by comparing their CATs (Proposition 1 (2)). Finally it deals with the case where f embeds less nodes than g on $\mathtt{DP(Q)}$. In such case we know $\mathsf{CR}_g \not\subseteq \mathsf{CR}_f$ (Proposition 1 (1)). So we only need to test whether $\mathsf{CR}_f \subseteq \mathsf{CR}_g$. We do this using Proposition 1 (3) and Proposition 2 first, before resorting to the method of containment mapping. The algorithm is summarized into the following steps.

1 Filter redundant UEs using Algorithm 1.
2 If there are two or more UEs left after the above step, divide them into sets $S_1, \ldots, S_N$, such that each UE in S_i embeds k_i nodes on $\mathtt{DP(Q)}$, and $k_1 > k_2 > \ldots > k_N$. For each pair of UEs f and g within the same group, test whether $Q_f \subseteq \mathsf{CAT}_g$. If yes, remove f.
3 For $i = 1$ to $N - 1$, and $j = i + 1$ to N, if there are $g \in S_i$ and $f \in S_j$, then test whether $\mathsf{CR}_f \subseteq \mathsf{CR}_g$ (if yes, remove f) as follows. (1) If the are no //-edges on $\mathtt{DP(V)}$, simply test $Q_f \not\subseteq \mathsf{CAT}_g$; (2) if $g(\mathtt{DN(Q)})$ is defined, and

---
[2] We may have to make many more containment tests.

(there is no //-edge on $\text{DP}(\text{Q})$ *or* there is no homomorphism from CAT_g to Q^L, then $\text{CR}_f \not\subseteq \text{CR}_g$. In other cases test whether $\text{CR}_f \subseteq \text{CR}_g$ using containment mapping.

Steps 2 and 3 are easy to understand, they are applications of Propositions 1 and 2. We now describe Step 1 in more detail. Given query Q, we give each node v a unique identifying integer N_v such that for any two nodes u and v in any path $p \in Q$, u is a descendant of v iff $N_u > N_v$. This number is assigned using *breadth-first traversal*, starting from $rt(Q)$ and the number 0, with increment 1. Fig.6 (b) shows such a Q. Note this process takes time linear in the size of Q.

Suppose $p_1 = \text{DP}(\text{Q})$ and $p_2, \ldots, p_k$ are all other paths in Q. These paths can be identified by finding the leaf nodes. Paths corresponding to leaf nodes which are descendants of $\text{DN}(\text{Q})$ can be ignored. Furthermore, paths that have a common ancestor which can never be embedded into V by any UE (such nodes are easily found after generating all UEs using the algorithm in [2]) can be treated as a single path.

Suppose we have n UEs $f_1, \ldots, f_n$. We represent each UE f_i by a vector $\langle a_{i,1}, a_{i,2}, \ldots, a_{i,k} \rangle$, where $a_{i,j} = N_{y(p_j, f_i)}$ if p_j is not fully embedded by f_i, and $y_i(f) = M + 0.5$ otherwise, where M is the maximal value among the identifiers of all possible successor nodes across all UEs [3]. For example, for the query Q and view V in Fig.6, $M = 8.5$. All UEs together form a $n \times k$ matrix, denoted $A = (a_{i,j})$. We mark each value in the matrix with either / or //, representing the fact the corresponding node is connected to its parent with a /-edge or a //-edge. The value $M + 0.5$ is marked with //.

In Algorithm 1, we first sort the rows in descending order of the sum of values in the row, and then in descending order of the number of values marked //. The aim of this is to arrange those UEs which are more likely to contain other UEs at the top, so as to remove redundant UEs early, thus reducing the number of subsequent comparisons. Intuitively, the more //-children the CAT has, and the more nodes an UE embeds, the more likely it contains more other UEs. Sorting using the sum of values in a row will also help to arrange UEs in such a way that if $i \leq j$, then row j cannot be an extension of row i, because for any UE f and g, g is an extension of f implies the sum of successor nodes of g is greater than the sum of successor nodes of f. This is partly due to our numbering scheme in which descendants have a larger number than ancestors on each path. This will also explain why we have chosen the breadth-first (rather than depth-first) traversal when numbering the nodes in Q: we want to treat the paths in Q as evenly as possible, but slightly favor paths that have more nodes embedded by some UE. The reason we have chosen $M + 0.5$ rather than ∞ for fully embedded paths can also be explained here: if two or more UEs both fully embed some paths of Q, we are still able to arrange them by the number of nodes they embed on other paths. We chose $M + 0.5$ rather than $M + 1$ in order to distinguish this

---

[3] It is obvious that the CR generated by a UE is determined by the anchor nodes (or their successors) of those paths that are not fully embedded. Also by Lemma 1, UEs that have embed the same set of nodes into V will generate equivalent CRs, so they can be regarded as the same UE.

Algorithm 1. Filtering matrix $A = (a_{i,j})$ using Theorem 1

1: sort the rows in A in descending order of the sum of all values in the row, and then in descending order of the number of values marked //
2: **for** (for $i = 1$ to n) **do**
3: **if** (row i is not removed) **then**
4:   **for** $j = i + 1$ to n and $j \neq i$ **do**
5:    **if** row i is not removed **then**
6:     **if** for all $s = 1$ to k, either $(a_{i,s} = a_{j,s})$ or $(a_{i,s} > a_{j,s}$ and $a_{i,s}$ is marked //) **then**
7:      **if** $\neg(s = 1 \wedge a_{i,1} = M \wedge a_{j,1} \neq M)$ **then**
8:       remove row j

---

special value from the *real* successor nodes (which are all integers) and avoid the confusion in the case where there is a node in Q which is numbered $M + 1$. Fig.6 (e) shows the sorted matrix representing all UEs for the V and Q shown in Fig.6 (a),(b). The algorithm then proceeds to check the rows one by one from top to bottom, to see if any other row is contained in the current row (thus can be removed) using Theorem 1.

Example 6. Consider the query Q and the view V in Fig.1. Using the algorithm in [2] we can find 80 UEs. The possible set of successor nodes for the three paths in Q are $\{13, 7, 4, 1\}$, $\{14, 8, 5, 2\}$, and $\{15, 12, 9, 6, 3\}$ respectively (Thus $M = 15.5$). For all of these UEs we will have an UE matrix A which contains 80 rows and three columns. Part of the matrix is shown in Fig.1 (c). The first row of the sorted matrix contains the vector $(13, 14, 15)$, where each element in the vector is associated with //. Using Algorithm 1, we will compare this row with all other rows. Since the values in this row are greater than or equal to the corresponding values in other rows, and each value in this row is marked //, and $a_{1,1} = 13 \neq M$, we can remove all other rows. We are left with only one row, which represents a single UE. The minimal cover set contains only this UE.

Example 7. Consider the view V and Q in Fig.6. There are 7 UEs which make up the matrix shown in Fig.6 (e). All values in the matrix are marked //. Using Algorithm 1, we can remove the 3rd, 5th, and the last rows when they are compared with the first row; we can then remove the 4th and 6th rows by comparing them with the 2nd row. The remaining two rows turned out to be not contained in each other. So they form the minimal cover set.

5 Conclusion

We showed that the conditions for irredundant UE given in [2] is incorrect, and it is neither sufficient nor necessary. We then provided some sufficient conditions and separate necessary conditions for UE containment. Using these results, we designed a heuristic algorithm for removing redundant UEs from a given set. We demonstrated, using examples, that our algorithms can be significantly faster than the brute-force method of testing UE containment using containment mappings pair by pair.

References

1. Halevy, A.Y.: Answering queries using views: A survey. VLDB J. 10(4), 270–294 (2001)
2. Lakshmanan, L.V.S., Wang, H., Zhao, Z.J.: Answering tree pattern queries using views. In: VLDB (2006)
3. Miklau, G., Suciu, D.: Containment and equivalence for a fragment of XPath. J. ACM 51(1) (2004)
4. Nash, A., Segoufin, L., Vianu, V.: Determinacy and rewriting of conjunctive queries using views: A progress report. In: Schwentick, T., Suciu, D. (eds.) ICDT 2007. LNCS, vol. 4353, pp. 59–73. Springer, Heidelberg (2006)
5. Wang, J., Topor, R.W., Maher, M.J.: Rewriting union queries using views. Constraints 10(3), 219–251 (2005)
6. Wang, J., Yu, J.X., Liu, C.: Independence of containing patterns property and its application in tree pattern query rewriting using views. WWWJ (to appear)

IRank: A Term-Based Innovation Ranking System for Conferences and Scholars

Zhixu Li[1,2], Xiaoyong Du[1,2], Hongyan Liu[3], Jun He[1,2], and Xiaofang Zhou[4]

[1] Key Labs of Data Engineering and Knowledge Engineering, Ministry of Education, China
[2] School of Information, Renmin University of China, Beijing, China
{lizx,lp,hejun,duyong}@ruc.edu.cn
[3] Department of Management Science and Engineering, Tsinghua University, Beijing, China
hyliu@tsinghua.edu.cn
[4] University of Queensland, Australia
zxf@itee.uq.edu.au

Abstract. Since the proposition of Journal Impact Factor [1] in 1963, the classical citation-based ranking scheme has been a standard criterion to rank journals and conferences. However, the reference of a paper cannot list all relevant publications and the citation relationships are not always available especially when related to copyright problem. Besides, we cannot evaluate a newly published paper before it is cited by others. Therefore, we propose an alternative method, term-based evaluation scheme which can evaluate publications by terms they use. Then we can rank conferences, journals and scholars accordingly. We think this term-based ranking scheme can be used to evaluate innovation quality for conferences and scholars. To evaluate our scheme and to facilitate its application, we develop an innovation ranking system called *IRank* to rank conferences and authors in the field of *Database Systems*. The performance of *IRank* demonstrates the effectiveness of our scheme.

Keywords: Ranking System, Term-based, Innovation Quality, IRank.

1 Introduction

Innovation spirit is not only an important quality for scholars, but also a key factor to push a research field to go forward. Scholars like to discuss with creative people in order to stimulate their inspirations, and prefer to attend highly innovative conferences to touch cutting edge research topics. The establishment of some innovation awards such as *"SIGMOD Edgar F. Codd Innovations Award"* in *Database Systems domain* also proves that people are becoming more and more concerned about innovation quality. With expecting to propose an innovation ranking criterion, there are several questions need to be discussed beforehand:

(1) What is the Definition of Innovation?
An official government definition of innovation is that it is "the successful exploitation of new ideas". This implies that it is not only about the invention of a new idea that we are interested in, but also about its potential usage, which means this idea is actually "brought to market", used, put into practice, exploited in some way,

Q. Li et al. (Eds.): APWeb/WAIM 2009, LNCS 5446, pp. 126–137, 2009.
© Springer-Verlag Berlin Heidelberg 2009

e.g., leading to new products, processes, systems, attitudes or services that improve something or add some value.

(2) Why do We Evaluate Innovation Quality Based on Terms?

Generally speaking, the only possible way to judge innovation quality of conferences and scholars automatically is by evaluating their research products - publications. All the publications are innovative working achievement, but they are different in terms of their innovation quality and contribution to respective research field. The classical citation-based statistical evaluation criterion [1, 2] which has been widely accepted tries to evaluate the value of a publication according to its quotation rate. The higher of the publication's quotation rate, the higher of its value.

Different publications have different innovation points. Some may propose new theories or algorithms; others may point out new directions or develop new systems. Apparently, it is impossible to set up a perfectly fair criterion to evaluate the innovation quality of publications. However, we can use some reasonable ways to approach fair judgment or at least provide some references to an expert judgment. After conducting a manual observation to thousands of publications in *Database Systems*, we found out that most of the time a new-born innovative idea (including new theory or technology, new algorithm, fresh topic, rising direction and so forth) has always been proposed together with a new domain term which is used to denote it. This implies that every domain term represents a new innovative idea; therefore, it is a reasonable way to evaluate the innovation ability of a research individual or institution by mining new domain terms proposed by its publications.

(3) How to Evaluate Innovation of publication by Terms?

According to classical citation-based criterion, a publication cited by lots of others can get a high mark for its great influence to others. We use the same strategy in our term-based criterion: We introduce a new factor into our ranking scheme, which we call **In**novation **F**actor (or *InF* for simplicity). If a new term proposed by a publication which becomes a popular topic or direction later, then this publication deserves a high *InF* score since it was not only once a fresh idea but also a successful leader to a new research direction of the domain. The details for *InF* scoring will be discussed in section 3.

(4) Where to Get Authentic Domain Terms?

A good domain term set which should contain nearly all the authentic terms in a certain domain is a necessary precondition for our work. In order to have an authentic term set of a certain domain, we not only use statistical method [3] to select terms from keywords and general terms of publications in a domain, but also extract terms from session names of some famous conferences using a hybrid method of statistical and rule-based methods devised by ourselves. Finally, it is necessary to have some domain experts to do a manual judgment and selection to these terms. The detail of this part will be given in section 4.

After we know how to do *InF* scoring, an innovation ranking system which we call **IRank** (Term-based **I**nnovation **Rank**ing System for Conferences and Scholars) is developed to rank conferences and scholars in the field of *Database System*.

The rest of this paper is organized as follows: Section 2 reviews some related works in history. In Section 3 we will explain how to do *InF* scoring. Our method to get domain term set, especially algorithm for term extraction from session names is described in Section 4. Then we give the system architecture of *IRank* in Section 5. Section 6 evaluates the results of *IRank*. Finally we make a conclusion in Section 7.

2   Related Works

Despite the oversimplification of using just a few numbers to quantify the scientific merit of a body of research, the entire science and technology community is relying more and more on citation-based statistics as a tool for evaluating the research quality of individuals and institutions [4]. Since the Institute of Scientific Information (ISI) Journal Impact Factor (JIF) [5] was proposed firstly to rate scientific journals in 1963, it has been widely applied in various fields including CS (Computer Science). There have been some computer science conference ranking reports such as the NUS 1999 report on CS conference rankings (We will call it *NUSRank* in the rest of this paper) [6] and CS conference ranking based on Estimated Impact of Conference (EIC) [7] as shown in Fig. 1 below:

(a) NUS 1999   (b) EIC 2008

Fig. 1. The snapshots of NUS 1999 report on CS conference rankings and EIC2008

However, JIF can be misleading [8]. Someone has proposed that we should judge a paper by the number of citations the paper itself receives [9], but it may take decades to accumulate citations. For a recently published article, it is an adaptive heuristic method to use the journal in which the paper is published as a good proxy for the ultimate impact of a paper [10]. Lately, in [11] its authors have developed a model to quantify both the typical impact and the range of impacts of papers published in a journal, and then proposed a journal-ranking scheme that maximizes the efficiency of locating high impact research.

How to get domain terms is another problem we have to face. There have been lots of works on term extraction from texts in Ontology Learning research field. Most of them are based on information retrieval methods for term indexing [12], but many others [13, 14□and 15] also take inspiration from terminology and NLP research. Term extraction implies advanced levels of linguistic processing, such as identify complex noun phrases that may express terms by phrase analysis. The state-of-the-art is mostly to run a part-of-speech tagger over the domain corpus used for the ontology learning task and then to identify possible terms by manually constructing ad-hoc patterns [16, 17], whereas more advanced approaches to term extraction for ontology learning build on deeper linguistic analysis as discussed above [18, 19]. Additionally, in order to identify only relevant term candidates, a statistical processing step may be included that compares the distribution of terms between corpora. [20] There are also some hybrid methods, which use linguistic rules to extra term candidates combining with statistical (pre- or post-) filtering. In fact, the algorithm we propose in this paper to extract terms from session names in proceedings is also a hybrid method. We not only combine linguistic analysis and statistical methods, but also apply association rules in our algorithm. Our algorithm for term extraction from them is given in section 4.

3   Innovation Evaluation

We propose a method to calculate *InF* score for publications according to the terms they used; accordingly we can calculate *InF* score for scholars and conferences which have published these publications. For the facility of better explanation, assume we have already had a qualified domain term set T for a domain, which contains nearly all the necessary domain terms appeared in history. For a publication, there might be several terms used in it, each of which will denote a value to its innovation score.

3.1   Weight and Value of Term

Different terms should have different weights according to their influence to the development of certain domain. In the classical citation-based ranking scheme, paper with a higher citation rate deserves a higher value, and vice versa. Here we use the same principle in valuing terms. A term with a higher appearance frequency in publications deserves a higher weight than others since it has attracted lots of attentions just like a high-citation paper. Besides, a term can have an additional weight when appear in a session name since session name is given by experts of a certain domain which represents important research directions in this domain. Therefore terms appeared in session names are more valuable than other terms.

Term Weight: If we assume a term t has appeared f_{tp} times in publications and f_{ts} times in session names in history, then the weight of t can be represented by:

$$w_t = C \cdot (log f_{tp} + f_{ts}) \tag{1}$$

In this equation, C is a constant factor which can normalize the value of w_t to be between 0 and 1. Since f_{ts} is always much less than f_{tp} but term can be much more valuable when it appears in session names, so we use $log f_{tp}$ to reduce the weight of f_{tp}, which enhances the weight of f_{ts} in the meantime.

Term Value: As the time passes by, the innovation value of a term t is reducing at an exponential rate. Since different terms have different attenuation speed, so we consider something similar to "half-life period" of each term. In *IRank*, we use a simple function shown in Equation (2) to simulate the attenuation value of a term t when it has been appearing for y years after it was born. Fig. 2 depicts the attenuation trend of the function below.

$$v_t(t, y) = e^{-\alpha(t) \cdot y} \cdot w_t \tag{2}$$

In this function, $\alpha(t)$ can be calculated by $(t) = A_{domain}/A_t$, in which A_{domain} denotes the age of the domain while A_t denotes the year span from the year t was born until the latest year in which t appears more than a threshold (We use 2 in our experiments).

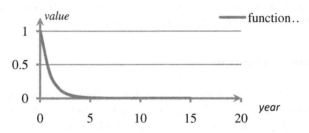

Fig. 2. The trend of function $v_t(t, y) = e^{-\alpha(t) \cdot y} \cdot w_t$

3.2 *InF* Scoring for Publications

As we know, a publication may mention dozens of terms at a time, each of which donates a value to the *InF* score of the publication. After we define how to get term values, it is convenient to evaluate the innovation value of publications.

For a publication p published in the year y_p, if we use $T(p)$ to represent the set of terms appeared in p, then its innovation value can be calculated as:

$$v_p(p, y_p) = \Sigma_{t \in T(p)} v_t(t, y_p - y_t) \tag{3}$$

In this function, t denotes a term in $T(p)$; y_t represent the year in which the term t was born.

3.3 Evaluation of Conference and Scholars

After we get *InF* score for each publication, it is easy to evaluate *InF* score for conferences and scholars. In fact, we define two *InF* scores for each conference and author.

1.   Tc-InF Score:
Tc-*InF* Score is the standard according to which we rank conferences and scholars for their **T**otal **C**ontribution to the innovation development of a domain. *Tc-InF* score of a conference or scholar can be given by the summation of term values that first proposed by this conference or author, which is given in the formula (4).

$$Score_{Tc-InF}(c\_s) = \Sigma_{t \in T(c\_s)} v_t(t, 0) \tag{4}$$

$T(c\_s)$ is a set of terms first proposed by the conference or scholar c_s.

2. *Aiq-InF Score:*

As we know, different conferences and scholars are different in publication quantity. Maybe someone only published several publications but all of which are high-quality papers. In order to eliminate the scale effect caused by different production, we also define this Aiq-InF score to evaluate the **A**verage **I**nnovation **Q**uality of conference or scholars' publications.

Different from *Tc-InF* score, *Aiq-InF* score of a conference equals to the average value of all its publications, no matter if they have proposed any new terms or not. The *Aiq-Inf* score of a conference or a scholar c_s can be calculated by formula (5):

$$Score_{Aiq-InF}(c\_s) = \frac{\sum_{p \in P(c\_s)} v_p(p, y_p)}{sizeof(P(c\_s))} \tag{5}$$

In this function, $P(c\_s)$ represents the publication set of a conference or scholar c_s.

4 Domain Terms Extraction

In Section 4 we assume that we have already had a good enough term set, however, it is also a big challenge to get a qualified domain term set which should contain most of the domain terms we have ever used in a domain.

4.1 How to Get Domain Terms

In order to get a good domain set and later find out those terms' appearance locations in history, we have to collect information (including paper title, keywords, abstract) of nearly all the publications which have ever been published in famous international conferences of the domain. Our domain terms can be gotten from those keywords as well as session names of famous international conferences. Different from keywords, a session name may constructed by several terms in a special form. Following examples listed some session names from domains "Databases" and "Data Mining":

Example 1: Query Processing
            XML
            Concurrency Control
Example 2: Data Warehousing **and** Data Mining
            Data Warehousing **&** Mining
            Data, Text **and** Web Mining
            Databases **and** information retrieval / Data mining
Example 3: AI and Knowledge-Based Systems – Reasoning
            Data Mining – Association Rules and Decision Trees

In fact, except for some meaningless ones, there are mainly three types of session names as shown above. Session names in example 1 are **Single-term Sessions**, which only contain one domain term. Session names in example 2 and example 3 are **Multi-term Sessions**, which are composed by several terms. Session names in example 2 always have some list separators in it like "and", or "," or "/" or "&", while session names in example 3 must contain a dash ("–").

Term extraction strategy from a Single-term Session is very simple. But it is much harder to extract terms from Multi-term session. Session name in example 2 is the most complicated situation we should dispose since they are composed by several concepts and we have no idea about their structures. Our method to dispose them will be introduced in section 4.2. Session names in example 3 can be divided into two parts. The first one is the phrase before "–", while the second is the one after "–". Therefore, we would dispose the two parts according to their forms respectively.

4.2   Term Extraction from Multi-term Session

A multi-term session name in example 2 can be represented by:

$$(prefix)\ W_1\ (separator)\ W_2\ ...\ (separator)\ W_m\ (suffix) \tag{6}$$

In the formula (6) above, **prefix, suffix** denotes possible prefix and suffix phrase. **Separator** includes various separator signs such as: $',', '\&'\ and\ '/'$, and conjunction words "and". W_i represents other words or phrase which doesn't contain separators. There are mainly three different types of structures:

- No prefix, no suffix: "Data Warehousing & Data Mining"
- No prefix, but a suffix: "Data, Text and Web Mining"
- A prefix, no suffix: "Data Warehousing and Mining"

It is impossible to know the structure of each session name without knowing its semantic meaning. Therefore, we prefer to split a session name in all possible ways and then use *statistical language model* to judge which way is the most reasonable one. Major steps of this algorithm are outlined below:

Step 1: First of all, we put all single-term sessions names in the same domain into a table with their frequencies. This table can be called **candidate-term table**, which is used to store candidate terms' names and their frequencies appeared in all session names.

Step 2: In the second step, we split each multi-term session of this domain in various ways. Assume we could get a phrase set $C = \{c_i | i = 1,2, ... n\}$ each time we split it in a way. In order to get credit of each split method, we match $c_i (i = 1,2, ... n)$ to term in the candidate-term table. In this way we can get its frequency $f(c_i)$ in candidate-term table.

It is reasonable to believe that the bigger $f(c_i)$ is, the more possible that c_i is a term. So we calculate the credit of this split method by the following formula (7):

$$Prob(C) = F_1 * \prod_{i=1}^{n}(f(c_i) + F_2) \tag{7}$$

F_1 is a scaling constant factor which can normalize the value of Prob(C) to (0,1); F_2 is a smoothing constant factor which is used to prevent zero probability).

Most of the time, we regard the most credible split method as the right one and split the session accordingly. All terms we get will be put into our candidate-term table until all multi-term session names have been disposed. But there are also some multi-term session names that we cannot determine how to split them since terms they contain may not be in the candidate-term table. In this situation, we will meet two kinds of problems: **Zero-score Problem** and **Equal-score Problem**.

Definition 3 (Zero-score Problem): When we are trying to split a multi-term session name in all possible ways, if all those splitting ways including the right one can only get zero score, we say a zero-score problem happens.

Definition 4 (Equal-score Problem): When we are trying to split a multi-term session name in all possible ways, if there are more than one splitting ways have the same largest scores, we say an equal-score problem happens.

Step 3: For those multi-term session names which meet either of the two problems above, we use text classification method to help splitting them properly in this step.

As we discussed above, there are three structure types for each composition phrase, we can find out whether there are relationships between words used in a composition phrase and its structure type. Although we don't have training set in the beginning, but we can use those split composition phrases to create a training set, then learn relationships (association rules) between words and structure types of a composition phrase from it. Major steps are given below:

1) Firstly, each time we split a phrase which doesn't have zero-score or equal-score problems in step 2, a record which records all words and their frequencies in the phrase as well as its splitting way will be put into a data set. Therefore, after we split all composition phrases that we can split, the data set contains all the records, each of which records all words of a phrase as well as its right structure type (splitting way).

2) Secondly, we can learn association rules from the date set. Each association rule represents relationships between several words and a split type. Our algorithm to learn association rules between words and structure type is similar to the classification based on association classifier CBA [21].

3) Finally, with all the association rules, we can establish our forecasting model which can be used to split those composition phrases with any of the two problems.

Step 4: After we split all session multi-term names, we have all candidate terms in the candidate-term table. Except for some stop words, we regard those candidate term whose frequency is larger than a threshold (we set 3) as formal domain terms.

Of course, after we get all those terms, it is necessary to invite domain experts to have a judgment and selection to those terms. Our experiment in Database System tells us that more than 70% of these terms (about 9,100 terms) we get from session names and keywords are qualified to use. For the left 30%, some of them are dropped for meaningless while others can become qualified after being modified.

5 Introduction to *IRank*

IRank has mainly three modules as shown in Fig. 3.

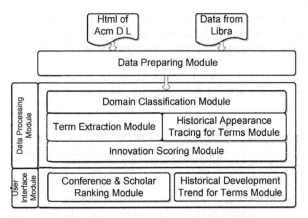

Fig. 3. The system architecture of *IRank*

Module 1: Data Preparing Module

Our experimental data mainly comes from ACM Digital Library[1] and Microsoft Libra[2]. We choose ACM Digital Library corpus since it contains information of publications published in famous conferences and journals in Computer Science. In order to separate conferences in Database domain from others, we have to use domain classification of CS conferences from Microsoft Libra.

Module 2: Data Processing Module

This is the major function module of *IRank* which is constituted by four sub-modules. The **Domain Classification Module** classifies data into 23 different domains. The **Term Extraction Module** serves to get domain terms. The **Historical Appearance Tracing for Terms Module is** responsible for finding out all the appearance locations (in session names, paper titles, keywords, abstracts) of each term in history. The **Innovation Scoring Module** is used to calculate *InF* score for publications, then for conferences and scholars.

Module 3: User Interface Module

1)   Conference & Scholar Innovation Ranking List:

This is the principal demonstration. There is a ranking list to present our ranking results for conferences and scholars in a certain domain. Besides, we list all fresh terms proposed firstly by each conference or scholars, with a graph to visualize the term-proposing trend as shown in Fig. 4(a).

2)   Historical Development Trend for Terms:

In *IRank*, we also help users to trace the historical development trend of each term. Our interface can clearly show the tracing for each term in history including when and where a term appeared each time, proposed by whom; there is also a graph to visualize its trend as it is shown in Fig. 4(b).

---

[1] http://portal.acm.org/portal.cfm

[2] http://libra.msra.cn/

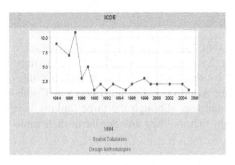

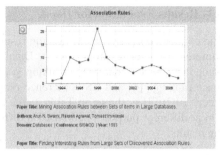

(a) New terms proposed by ICDE

(b) The historical trend of the term "Association Rules"

Fig. 4. Two screen capture images of *IRank*'s User Interface Module

6   Results Evaluation

6.1   Comparison between Term-Based Ranking and Citation-Based Ranking

Although the classical citation-based ranking scheme has been accepted as an authority rank for journals and conferences, it may not be proper to judge the innovation quality. Let's take a look at an interesting example in *Database System* domain. According to the NUSRank, SIGMOD, VLDB, PODS and ICDE are **rank one** authoritative conferences which also proposed lots of new terms in history. But surprisingly, ER, DEXA which are only **rank two** conferences also have proposed lots of new terms, even more than PODS. Therefore, the citation-based ranking scheme can't tell us everything about a conference such as innovation quality. A relative low authority conference or author can also be very creative and filled with innovation power.

We think our term-based ranking scheme is better than citation-based ranking scheme in the following several aspects:

1)   Term-based ranking scheme considers the innovation quality from the term-level granularity, which contains much more information than the paper-level granularity.

2)   Citation-based ranking scheme cannot evaluate a newly published paper before it is cited by others, but our term-based ranking scheme can give an evaluation to it according to terms it used.

3)   The reference of a paper can only lists some most relevant or most recently published publications it cited. There are still lots of other publications which have a long-time influence to the domain cannot be listed. But we can find the relationship by tracing the history development trend of domain terms.

4)   The citation relationships are not always available especially when related to copyright problem. But the paper title, keywords and abstracts are easier to get from DBLP or ACMDL and so forth.

6.2 Results Evaluation for *IRank*

The comparison between our ranking results (only part of our results, we select some conferences in different levels) in Database Systems and the other two rankings: computer science conference ranking given by the NUS 1999 report on CS conference rankings [6] and CS conference ranking based on Estimated Impact of Conference (EIC) [7] is shown in Table 1. As we can see, although the three ranking lists aim at different purposes and base on different factors, they are similar in their ranking results.

Table 1. A comparison of the conferences ranked by NUS 1999, EIC 2008 and IRank 2008

NUS1999	EIC2008	IRank 2008
SIGMOD (level 1)	SIGMOD 0.99	SIGMOD
PODS (level 1)	VLDB 0.99	VLDB
VLDB (level 1)	ICDE 0.97	CIKM
ICDE (level 1)	PODS 0.94	ICDE
DEXA (level 2)	CIKM 0.90	DEXA
DASFAA (level 2)	DEXA 0.90	WIDM
CIKM (level 2)	DASFAA 0.79	PODS
MobiDE (level 4)	MobiDE 0.71	DASFAA
WIDM (level 4)	WIDM 0.70	MobiDE

In order to further evaluate our results, we try to find out similarity between our results and four SIGMOD awards. Those scholars who have ever received "SIGMOD Edgar F. Codd Innovations Award" or "SIGMOD Contributions Award" have excellent ranks (most of the 34 people are top 10%, even top 5%) in our ranking list, which proves that our method may provide a credible reference to the evaluation of research individuals in the future. Besides, although there are hundreds of papers being published annually, those papers which have received "SIGMOD Best Paper Award" and "SIGMOD Test of Time Award" are always in the top 5% (sometimes even in top 1%) in our annual publication ranking. All these results demonstrate the effectiveness of our system.

7   Conclusion

In this paper, we try to propose an automatic term-based innovation ranking scheme which can be used to rank publications, conferences and scholars based on their innovative contributions. We make the observation that an innovative idea is mostly presented by introducing a new domain term, hence we can use such terms to identify innovative ideas. The quality of an idea is gauged by looking at the time-varying popularity of its associated domain term. According to our ranking scheme, we also develop a demonstration system named *IRank*. The results of *IRank* demonstrate the effectiveness of our criterion. We think our term-based ranking scheme can be a good alternative scheme for the classical citation-based ranking scheme.

Acknowledgments

This work was supported in part by the National Natural Science Foundation of China under Grant No. 70871068, 70621061, 70890083, 60873017, 60573092 and 60496325.

References

1. Garfield, E., Sher, I.H.: Genetics Citation Index. Institute for Scientific Information, Philadelphia (July 1963)
2. Weingart, P.: Impact of bibliometrics upon the science system: Inadvertent consequences? Scientometrics 62, 117–131 (2005)
3. Church, K.: A stochastic parts program and noun phrase parser for unrestricted text. In: Proceedings of the Second Conference on Applied Natural Language Processing (1988)
4. Weingart, P.: Impact of bibliometrics upon the science system: Inadvertent consequences? Scientometrics 62, 117–131 (2005)
5. Garfield, E., Sher, I.H.: New factors in the evaluation of scientific literature through citation indexing. American Documentation 14(3), 195–201 (1963)
6. http://www.cc.gatech.edu/~guofei/CS_ConfRank.htm
7. http://www.cs-conference-ranking.org/home.htm
8. Moed, H., van Leeuwen, T.: Impact factors can mislead. Nature 381, 186 (1996)
9. Zhang, S.D.: Judge a paper on its own merits, not its journal's. Nature 442, 26 (2006)
10. Gigerenzer, G., Todd, P.M.: The ABC Research Group, Simple Heuristics That Make Us Smart. Oxford University Press, Oxford (1999)
11. Stringer, M.J., Sales-Pardo, M., Nunes Amaral, L.A.: Effectiveness of Journal Ranking Schemes as a Tool for Locating Information 3(2) (February 2008)
12. Salton, G., Buckley, C.: Term-weighting approaches in automatic text retrieval. Information Processing & Management 24(5), 515–523 (1988)
13. Borigault, D., Jacquemin, C., L'Homme, M.-C. (eds.): Recent Advances in Computational Terminology. John Benjamins Publishing Company, Amsterdam (2001)
14. Frantzi, K., Ananiadou, S.: The c-value / nc-value domain independent method for multiword term extraction. Journal of Natural Language Processing 6(3), 145–179 (1999)
15. Pantel, P., Lin, D.: A statistical corpus-based term extractor. In: Stroulia, E., Matwin, S. (eds.) Canadian AI 2001. LNCS, vol. 2056, pp. 36–46. Springer, Heidelberg (2001)
16. Ciramita, M., Gangemi, A., Ratsch, E., Saric, J., Rojas, I.: Unsupervised Learning of Semantic Relations between Concepts of a Molecular Biology Ontology. In: Proceedings of the 19th International Joint Conference on Artificial Intelligence (IJCAI) (2005)
17. Buitelaar, P.: Learning Web Service Ontologies: an Automatic Extraction Method and its Evaluation. Ontology Learning and Population. IOS Press, Amsterdam (2005)
18. Reinberger, M.-L., Spyns, P., Pretorius, A.J., Daelemans, W.: Automatic initiation of an ontology. In: Meersman, R., Tari, Z. (eds.) OTM 2004. LNCS, vol. 3290, pp. 600–617. Springer, Heidelberg (2004)
19. Buitelaar, P., Olejnik, D., Sintek, M.: A protégé plug-in for ontology extraction from text based on linguistic analysis. In: Bussler, C.J., Davies, J., Fensel, D., Studer, R. (eds.) ESWS 2004. LNCS, vol. 3053, pp. 31–44. Springer, Heidelberg (2004)
20. Buitelaar, P., Cimiano, P., Magnini, B.: Ontology Learning from Text: An Overview (2003)
21. Liu, B., Ma, Y., Wong, C.-K.: Classification Using Association Rules: Weaknesses and Enhancements. Data mining for scientific applications (2001)

A Generalized Topic Modeling Approach for Maven Search

Ali Daud[1], Juanzi Li[1], Lizhu Zhou[1], and Faqir Muhammad[2]

[1] Department of Computer Science & Technology, 1-308, FIT Building, Tsinghua University, Beijing, China, 100084
[2] Department of Mathematics & Statistics, Allama Iqbal Open University, Sector H-8, Islamabad, Pakistan
ali_msdb@hotmail.com, ljz@keg.cs.tsinghua.edu.cn, dcszlz@tsinghua.edu.cn, aioufsd@yahoo.com

Abstract. This paper addresses the problem of semantics-based maven search in research community, which means identifying a person with some given expertise. Traditional approaches either ignored semantic knowledge or temporal information, resulting in some right mavens that cannot be effectively identified because of non-occurrence of keywords and un-exploitation of time effects. In this paper, we propose a novel semantics and temporal information based maven search (STMS) approach to discover latent topics (semantically related soft clusters of words) between the authors, venues (conferences or journals) and time simultaneously. In the proposed approach, each author in a venue is represented as a probability distribution over topics, and each topic is represented as a probability distribution over words and year of the venue for that topic. Through discovered latent topics we can search mavens by implicitly modeling word-author, author-author and author-venue correlations with continuous time effects. Inference making procedure for topics and authors of new venues is explained. We also show how authors' correlations can be discovered and the bad effect of topics sparseness on the retrieval performance. Experimental results on the corpus downloaded from DBLP show that proposed approach significantly outperformed the baseline approach, due to its ability to produce less sparse topics.

Keywords: Maven Search, Research Community, Topic Modeling, Unsupervised Learning.

1 Introduction

Web is unanimously the biggest source of structured, semi-structured and unstructured data and automatic acquirement of useful information from the web is interesting and challenging from academic recommendation point of view. With the advancement of information retrieval technologies from traditional document-level to object-level [14], expert search problem has gained a lot of attention in the web-based research communities. The motivation is to find a person with topic relevant expertise to automatically fulfill different recommendation tasks. Such as, to find appropriate

Q. Li et al. (Eds.): APWeb/WAIM 2009, LNCS 5446, pp. 138–149, 2009.
© Springer-Verlag Berlin Heidelberg 2009

collaborators for the project, choose mavens for consultation about research topics, reviewers matching for research papers, and to invite program committee members and distinguished speakers for the conferences.

TREC has provided a common platform for researchers to empirically assess approaches for maven search. Several approaches have been proposed to handle this problem. In particular, Cao et al. [6] proposed a two-stage language model which combines a co-occurrence model to retrieve documents related to a given query, and a relevance model for maven search in those documents. Balog et al. [3] proposed a model which models candidates using its support documents by using language modeling framework. He also proposed several advanced models for maven search specific to sparse data environments [4].

Based on the methods employed in previous language models, they can be classified into two categories: *composite* and *hybrid*. In composite approach, $D_m = \{d_j\}$ denotes the support documents of candidate author r. Each support document d_j is viewed as a unit and the estimation of all the documents of a candidate r are combined. While hybrid approach is very much similar to the composite model, except that it describes each term t_i by using a combination of support documents models and then used a language model to integrate them. Composite approaches suffers from the limitation that all the query terms should occur in one support document and hybrid approaches suffers from the limitation that all the query terms should occur in the support documents.

Generally speaking, language models are lexical-level and ignores semantics-based information present in the documents. Latent semantic structure can be used to capture semantics-based text information for improved knowledge discovery. The idea of using latent semantic structure traces back to Latent Semantic Analysis (LSA) [8], which can map the data using a Singular Value Decomposition (SVD) from high dimensional vector space to reduced lower representation, which is a so-called latent semantic space.

LSA lacks strong statistical foundation; consequently Hofmann [11] proposed probabilistic latent semantic analysis (PLSA). It was based on likelihood principal and defines a proper generative model for the data. By using PLSA, Zhang et al. [20] proposed a mixture model for maven search. They used latent topics between query and support documents to model the mavens by considering all documents of one author as one virtual document. So far all approaches only used authors and supported documents information. While, Tang et al. argued the importance of venues by saying that the documents, authors and venues all are dependent on each other and should be modeled together to obtain the combined influence [17]. Based on it they proposed a unified topic modeling approach by extending Latent Dirichlet Allocation (LDA) [5]. However to the limitation of their approach, they considered the venues information just as a stamp and did not utilize semantics-based text and author's correlations present between the venues.

Previous approaches ignored venues internal semantic structure, author's correlations and time effects. Firstly in real world, venues internal semantic structure and authors correlations are very important, as authors of the renowned venues are more likely to be mavens than authors of not renowned venues. Additionally, renowned venues are more dedicated to specific research areas than not renowned venues, e.g. in famous conferences submission are carefully judged for relevance to the conference areas, while in not renowned venues it is usually ignored by saying that the topics are not limited to above mentioned research areas on the call for papers page. Some people may think that one has to use impact factors of venues to influence the ranking of mavens, but unluckily there is no standard dataset available to the best of our knowledge. Secondly, continuous-time effects can be very handy in a case if author changes his research interests. For example, an author A was focusing on biological data networks until 2004 and published a lot of papers about this topic; afterwards he switched to academics social network mining and not published many papers. He still can be found as a biological data networks expert in 2008 if we ignore time effects, while it is not an appropriate choice now. The reason for this is occurrence of biological word many times in 2004 and preceding years. However, by attaching time stamp one can minimize the high rate of occurrence effect of one word (e.g. biological) for all years.

In this paper, we investigate the problem of maven search by modeling venues internal semantic structure, author's correlations and time all together. We generalized previous topic modeling approach [17] from a single document to all publications of the venues and added continuous-time considerations, which can provide ranking of mavens in different groups on the basis of semantics. We empirically showed that our approach can clearly produce better results than baseline approach due to topics denseness effect on retrieval performance. We can say that the solution provided by us is well justified and produced quite promising and functional results.

The novelty of work described in this paper lies in the; formalization of the maven search problem, generalization of previous topic modeling approach from document level to venue level (STMS approach) with embedded time effects, and experimental verification of the effectiveness of our approach on the real world corpus. To the best of our knowledge, we are the first to deal with the maven search problem by proposing a generalized topic modeling approach, which can capture word-author, author-author and author-venue correlations with non-discretized time effects.

The rest of the paper is organized as follows. In Section 2, we formalize maven search problem. Section 3 provides maven search modeling related models and illustrates our proposed approach for modeling mavens with its parameter estimation details. In Section 4, corpus, experimental setup, performance measures with empirical studies and discussions about the results are given. Section 5 brings this paper to the conclusions.

2   Maven Search in Research Community

Maven search addresses the task of finding the right person related to a specific knowledge domain. It is becoming one of the biggest challenges for information management in research communities [10]. The question can be like "Who are the mavens on topic Z?" A submitted query by user is denoted by q and a maven is denoted by m. In general semantics-based maven finding process, main task is to probabilistically rank discovered mavens for a given query, i.e. $p\ (m/q)$ where a query is usually comprised of several words or terms.

Our work is focused on finding mavens by using a generalized topic modeling approach. Each conference accepts many papers every year written by different authors.

To our interest, each publication contains some title words and names, which usually covers most of the highly related sub research areas of conferences and authors, respectively. Conferences (or journals) with their accepted papers on the basis of latent topics based correlations can help us to discover mavens. We think that latent topics based correlations between the authors publishing papers in the specific venues by considering time effects is an appropriate way to find mavens. Our thinking is supported by the facts that 1) in highly ranked venues usually papers of mavens or potential mavens of different fields are accepted so venues internal topic-based author correlations are highly influential, 2) highly ranked venues are the best source for analyzing the topical trends due to the reason of mostly accepting and highlighting relatively new ideas, and 3) all accepted papers are very carefully judged for relevance to the venue research areas, so papers are more typical (strongly semantically related).

We denote a venue c as a vector of N_c words based on the paper accepted by the venue for a specific year y, an author r on the basis of his accepted paper (s), and formulize maven search problem as: Given a venue c with N_c words having a stamp of year y, and $\mathbf{a}_c$ authors of a venue c, discover most skilled persons of a specific domain. Formally for finding specific area mavens, we need to calculate the probability $P(z|m)$ and $P(w|z)$ where z is a latent topic, m is maven and w is the words of a venue.

3 Maven Search Modeling

In this section, before describing our STMS approach, we will first describe how mavens can be modeled by using Language Model (LM) [19], Probabilistic Latent Semantic Analysis (PLSA) [11], and Author-Conference-Topic (ACT) Model [17] and why our approach is indispensable.

3.1 Related Models

LM is one of the state-of-the-art modeling approaches for information retrieval. The basic idea is to relate a document given to a query by using the probability of generating a query from given document. In eq. 1, w is a query word token, d is a document and $P(q|d)$ is the probability of the document model generating a query. $P(w|d)$ is the probability of the document model generating a word by using a bag of words assumption.

$$P(q|d) = \prod_{w=q} P(w|d) \tag{1}$$

In eq. 1 one can simply merge all support documents of one author and treat it as a virtual document r representing that author [3,15]. Mavens discovered and ranked with respect to a specific query can be retrieved by using the following equation.

$$P(q|m) = \prod_{w=q} P(w|r) \tag{2}$$

Document and object retrieval with the help of LM has gained a lot of success. But LM faces the inability to exactly match query with the support documents. Hofmann proposed PLSA [11] with a latent layer between query and documents to overcome LM inability by semantically retrieving documents related to a query. The core of

PLSA is a statistical model which is called aspect model [12]. The aspect model is a latent variable model for co-occurrence data which associates an unobserved class variable $z \in Z = \{z_1, z_2, ..., z_T\}$ with each observation. A joint probability model over d x w is defined by the mixture, where, each pair (d,w) is assumed to be generated independently, corresponding to bag of words assumption words w are generated independently for the specific document d conditioned on topic z.

$$P(d,w) = P(d)P(w|d),$$
$$where\ P(w|d) = \sum_{z=z} P(w|z)P(z|d) \tag{3}$$

In eq. 3 by changing word w with query q and document d with author r, where r can be seen as a composition of all documents of one author as one virtual document [20]. Here, $P\ (m)$ can be calculated by a variety of techniques [4,21] to obtain the joint probability of maven and query which can be used to discover and rank mavens.

$$P(m,q) = P(r)P(q|r),$$
$$where\ P(q|r) = \sum_{z=z} P(q|z)P(z|r) \tag{4}$$

Recently, Author-Conference-Topic (ACT) model was proposed for expertise search [17]. In ACT model, each author is represented by the probability distribution over topics and each topic is represented as a probability distribution over words and venues for each word of a document for that topic. Here venue is viewed as a stamp associated with each word with same value. Therefore, the modeling is just based on semantics-based text information and co-authorship of documents, while semantics-based intrinsic structure of words and authors' correlations with embedding time effects present in the venues on the basis of writing for the same venue are ignored, which became the reason of topic sparseness that resulted in poor retrieval performance. The generative probability of the word w with venue c for author r of a document d is given as:

$$P(w,c|r,d,\emptyset,\Psi,\theta) = \sum_{z=1}^{T} P(w|z,\emptyset_z)P(c|z,\Psi_z)P(z|r,\theta_r) \tag{5}$$

3.2 Semantics and Temporal Information Based Maven Search (STMS) Approach

We think it is necessary to model venues internal semantic structure and author correlations than only considering venues as a stamp [17] and time factor for maven search. The basic idea presented in Author-Topic model [16], that words and authors of the documents can be modeled by considering latent topics became the intuition of modeling words, authors, venues and years, simultaneously. We generalized the idea presented in [17] from documents level (DL) to venues level (VL) by considering research papers as sub-entities of the venues to model the influence of renowned and not renowned venues on the basis of participation in same venues. Additionally, we considered continuous-time factor to deal with the topic drift in different years. In the proposed approach, we viewed a venue as a composition of documents words and the authors of its accepted publications with year as a stamp. Symbolically, for a venue c (a virtual document) we can write it as: $C = [\{(\mathbf{d}_1,\mathbf{a}_{d1}) + (\mathbf{d}_2,\mathbf{a}_{d2}) + (\mathbf{d}_3,\mathbf{a}_{d3}) + ... + (\mathbf{d}_i,\mathbf{a}_{di})\} + y_c]$ where $\mathbf{d}_i$ is a word vector of document published in a venue, $\mathbf{a}_{di}$ is author vector of d_i and y_c is paper publishing year.

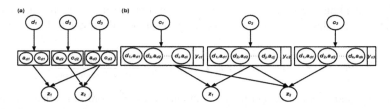

Fig. 1. Semantics-Based Maven search a) ACT and b) STMS approaches

DL approach considers that an author is responsible for generating some latent topics of the documents on the basis of semantics-based text information and co-authorship. While, VL approach considers that an author is responsible for generating some latent topics of the venues on the basis of semantics-based text information and authors correlations with time is not-discretized (please see fig. 1). In STMS approach, each author (from set of K authors) of a venue c is associated with a multi-nomial distribution θ_r over topics and each topic is associated with a multinomial distribution Φ_z over words and multinomial distribution Ψ_z with a year stamp for each word of a venue for that topic. So, θ_r, Φ_z and Ψ_z have a symmetric Dirichlet prior with hyper parameters α, β and γ, respectively. The generating probability of the word w with year y for author r of venue c is given as:

$$P(w, y | r, c, \emptyset, \Psi, \theta) = \sum_{z=1}^{T} P(w|z, \emptyset_z) P(y|z, \Psi_z) P(z|r, \theta_r) \tag{6}$$

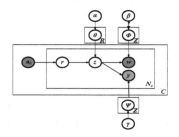

Fig. 2. STMS approach

The generative process of STMS is as follows:

1. For each author $r = 1, \ldots, K$ of venue c
    Choose θ_r from Dirichlet (α)
2. For each topic $z = 1, \ldots, T$
    Choose Φ_z from Dirichlet (β)
    Choose Ψ_z from Dirichlet (γ)
3. For each word $w = 1, \ldots, N_c$ of venue c
    Choose an author r uniformly from all authors $\mathbf{a}_c$
    Choose a topic z from multinomial (θ_r) conditioned on r
    Choose a word w from multinomial (Φ_z) conditioned on z
    Choose a year y associated with word w from multinomial (Ψ_z) conditioned on z

Gibbs sampling is utilized [1,9] for parameter estimation in our approach, which has two latent variables z and r; the conditional posterior distribution for z and r is given by:

$$P(z_i = j, \, r_i = k | w_i = m, y_i = n, \, \mathbf{z}_{-i}, \mathbf{r}_{-i}, \mathbf{a}_c) \propto \frac{n_{-i,j}^{(wi)} + \beta}{n_{-i,j}^{()} + W\beta} \frac{n_{-i,j}^{(yi)} + \gamma}{n_{-i,j}^{()} + Y\gamma} \frac{n_{-i,j}^{(ri)} + \alpha}{n_{-i,}^{(ri)} + R\alpha} \tag{7}$$

where $z_i = j$ and $r_i = k$ represent the assignments of the ith word in a venue to a topic j and author k respectively, $w_i = m$ represents the observation that i^{th} word is the mth word in the lexicon, $y_i = n$ represents i^{th} year of paper publishing attached with the nth word in the lexicon and z_{-i} and r_{-i} represents all topic and author assignments not including the ith word. Furthermore, $n_{-i,j}^{(wi)}$ is the total number of words associated with topic j, excluding the current instance, $n_{-i,j}^{(yi)}$ is the total number of years associated with topic j, excluding the current instance and $n_{-i,j}^{(ri)}$ is the number of times author k is assigned to topic j, excluding the current instance, W is the size of the lexicon, Y is the number of years and R is the number of authors. "." Indicates summing over the column where it occurs and $n_{-i,j}^{()}$ stands for number of all words and years that are assigned to topic z respectively, excluding the current instance.

During parameter estimation, the algorithm needs to keep track of W x Z (word by topic), Y x Z (year by topic) and Z x R (topic by author) count matrices. From these count matrices, topic-word distribution Φ, topic-year distribution Ψ and author-topic distribution θ can be calculated as:

$$\emptyset_{zw} = \frac{n_{-i,j}^{(wi)} + \beta}{n_{-i,j}^{()} + W\beta} \tag{8}$$

$$\Psi_{zy} = \frac{n_{-i,j}^{(yi)} + \gamma}{n_{-i,j}^{()} + Y\gamma} \tag{9}$$

$$\theta_{rz} = \frac{n_{-i,j}^{(ri)} + \alpha}{n_{-i,}^{(ri)} + R\alpha} \tag{10}$$

Where, $\emptyset_{zw}$ is the probability of word w in topic z, Ψ_{zy} is the probability of year y for topic z and θ_{rz} is the probability of topic z for author r. These values correspond to the predictive distributions over new words w, new years' y and new topics z conditioned on w, y and z. The mavens related to a query can be found and ranked with respect to their probabilities as:

$$P(m|q) \propto P(q|m) = \prod_{w \in q} \sum_{z \in Z} P(w|z) P(z|r) \tag{11}$$

Where, w is words contained in a query q and m denotes a maven.

4   Experiments

4.1   Corpus

We downloaded five years publication Corpus of venues from DBLP [7], by only considering conferences for which data was available for years 2003-2007. In total, we extracted 112,317 authors, 62,563 publications, and combined them into a virtual document separately for 261 conferences each year. We then processed corpus by a) removing stop-words, punctuations and numbers b) down-casing the obtained words of publications, and c) removing words and authors that appear less than three times in the corpus. This led to a vocabulary size of V=10,872, a total of 572,592 words and 26,078 authors in the corpus.

4.2 Parameter Setting

In our experiments, for 150 topics Z the hyper-parameters α, β and γ were set at 50/Z, .01, and 0.1. Topics are set at 150 at a minimized perplexity [5], a standard measure for estimating the performance of probabilistic models with the lower the best, for the estimated topic models. Teh et al. proposed a solution for automatic selection of number of topics, which can also be used for topic optimization [18].

4.3 Performance Measures

Perplexity is usually used to measure the performance of latent-topic based approaches; however it cannot be a statistically significant measure when they are used for information retrieval [Please see [2] for details]. In our experiments, firstly we use average entropy to measure the quality of discovered topics, which reveals the purity of topics. Entropy is a measure of the disorder of system, less intra-topic entropy is better. Secondly, we used average Symmetric KL (sKL) divergence [17,20] to measure the quality of topics, in terms of inter-topic distance. sKL divergence is used here to measure the relationship between two topics, more inter-topic sKL divergence (distance) is better.

$$Entropy\ of\ (Topic) = -\sum_z P(z)log_2[P(z)] \tag{12}$$

$$sKL(i,j) = \sum_{z=1}^{T}\left[\theta_{iz}log\frac{\theta_{iz}}{\theta_{jz}} + \theta_{jz}log\frac{\theta_{jz}}{\theta_{iz}}\right] \tag{13}$$

4.4 Baseline Approach

We compared proposed STMS approach with ACT approach and used same number of topics for comparability. The number of Gibbs sampler iterations used for ACT is 1000 and parameter values same as the values used in [17].

4.5 Results and Discussions

4.5.1 Topically Related Mavens
We extracted and probabilistically ranked mavens related to a specific area of research on the basis of latent topics. Tab. 1 illustrates 5 different topics out of 150, discovered from the 100th iteration of the particular Gibbs sampler run.

Table 1. An illustration of 5 topics with related mavens. The titles are our interpretation of the topics.

Topic 27 "XML Databases"		Topic 123 "Software Engineering"		Topic 98 "Robotics"		Topic 119 "Data Mining"		Topic 35 "Bayesian Learning"	
Word	Prob.	Word	Prob.	Word	Prob.	Word	Prob.	Word	Prob.
Data	0.110127	Software	0.083778	Robot	0.093050	Mining	0.164206	Learning	0.146662
XML	0.068910	Development	0.035603	Control	0.041873	Data	0.10328	Bayesian	0.039022
Query	0.047482	Oriented	0.031238	Robots	0.039397	Clustering	0.064461	Models	0.030610
Databases	0.038257	Engineering	0.028553	Motion	0.032675	Patterns	0.039122	Classification	0.027550
Database	0.037364	Systems	0.022006	Robotic	0.021119	Frequent	0.030041	Markov	0.018373
Queries	0.034239	Model	0.018649	Planning	0.018761	Series	0.023889	Kernel	0.015123
Processing	0.027543	Component	0.017306	Force	0.016638	Streams	0.023010	Semi	0.013785
Relational	0.020401	Tool	0.016467	Tracking	0.015695	Dimensional	0.012611	Regression	0.013785
Mavens	Prob.	Mavens	Prob.	Mavens	Prob.	Mavens	Prob.	Mavens	Prob.
Divesh Srivastava	0.011326	Gerardo Canfora	0.012056	Gerd Hirzinger	0.014581	Philip S. Yu	0.021991	Zoubin Ghahramani	0.012506
Elke A. Rundensteiner	0.010205	Tao Xie	0.009842	Paolo Dario	0.013586	Wei Wang	0.01785	Andrew Y. Ng	0.011882
Kian-Lee Tan	0.009747	Jun Han	0.005930	Ning Xi	0.011657	Jiawei Han	0.017792	John Langford	0.006266
Tok Wang Ling	0.008881	Nenad Medvidovic	0.005192	Toshio Fukuda	0.011470	Hans-Peter Kriegel	0.014686	Michael H. Bowling	0.006032
Surajit Chaudhuri	0.008779	Johannes Mayer	0.004970	Atsuo Takanishi	0.010475	Christos Faloutsos	0.01066	Sanjay Jain	0.005798
Rakesh Agrawal	0.008269	Jason O. Hallstrom	0.004601	Vijay Kumar	0.008857	Reda Alhajj	0.010373	Harry Zhang	0.005486
Sharma Chakravarthy	0.008218	S. C. Cheung	0.004158	Yoshihiko Naka.	0.007737	Eamonn J. Keogh	0.009567	Doina Precup	0.005018
Jeffrey F. Naughton	0.007913	Lu Zhang	0.003715	Joel W. Burdick	0.007426	Jian Pei	0.008992	Kai Yu	0.004706

The words associated with each topic are quite intuitive and precise in the sense of conveying a semantic summary of a specific area of research. For example, topic # 27 "XML Databases" shows quite specific and precise words when a person is searching for databases experts with move from simple databases to XML databases. Other topics shown in the tab. 1 are also quite descriptive that shows the ability of STMS approach to discover compact topics. The mavens associated with each topic are quite representative, as we have analyzed that, top ranked mavens for different topics are typically mavens of that area of research. For example, in case of topic 35 "Bayesian Learning" and topic 119 "Data Mining" top ranked mavens are well-known in their respective fields.

Proposed approach discovered several other topics related to data mining such as neural networks, multi-agent systems and pattern recognition, also other topics that span the full range of areas encompassed in the corpus.

In addition, by doing analysis of mavens home pages and DBLP [7], we found that 1) all highly ranked mavens have evenly published papers on their relevant topics for all years, no matter they are old or new researchers and 2) all of their papers are usually published in the well-known venues. Both findings provide qualitative supporting evidence for the effectiveness of the proposed approach.

Fig. 3 provides a quantitative comparison between STMS and ACT models. Fig. 3 (a) shows the average entropy of topic-word distribution for all topics calculated by

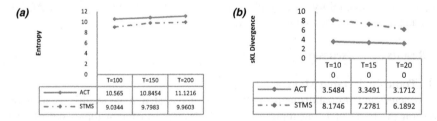

Fig. 3. a) Average Entropy curve as a function of different number of topics and **b)** Average sKL divergence curve as a function of different number of topics

using eq. 12. Lower entropy for different number of topics T= 100, 150, 200 proves the effectiveness of proposed approach for obtaining dense (less sparse, clearer) topics. Fig. 3 (b) shows the average distance of topic-word distribution between all pairs of the topics calculated by using eq. 13. Higher sKL divergence for different number of topics T= 100, 150, 200 confirms the effectiveness of proposed approach for obtaining dense topics.

One would like to quickly acquire the topics and mavens for new venues that are not contained in the training corpus. Provided parameter estimation Gibbs sampling algorithm requires significant processing time for large number of dataset. It is computationally inefficient to rerun the Gibbs sampling algorithm for every new venue added to the corpus. For this purpose, we can apply eq. 7 only on the word tokens and authors of the new venue each time temporarily updating the count matrices of (word by topic) and (topic by author). The resulting assignments of words to topics can be saved after a few iterations (10 in our simulations) and then eq. 11 can be used to search query related mavens.

4.5.2 Effect of Topic Sparseness on Retrieval Performance

Topic by author matrix (influenced by venues and time information) can also be used for automatic correlation discovery between authors more appropriately, than previously used topic by author matrix (not influenced by venues and time information) [16]. To illustrate how it can be used in this respect, distance between authors i and j is calculated by using eq. 13 for topic-author distribution.

We calculated the dissimilarity between the authors; smaller dissimilarity value means higher correlation between the authors. Tab. 2 shows top 7 semantics-based mavens ids discovered related to the first maven of each topic for STMS and ACT approaches. For example, in case of "XML Databases" topic 4808, 337, 5194, 4457, 4870, 4775, 640 are top 7 mavens correlated with "Divesh Srivastava" for STMS approach in terms of sKL divergence and so on.

Table 2. An illustration of 5 topics sparseness effect on retrieval performance in terms of error rate (ER)

STMS Approach					ACT Approach				
XML Databases	Software Engineering	Robotics	Data Mining	Bayesian Learning	XML Databases	Software Engineering	Robotics	Data Mining	Bayesian Learning
4808	6871	12723	4477	11094	9398	12823	24645	14131	3627
337	14588	12508	5119	3289	14221	6700	24952	14409	19973
5194	2531	12887	2644	924	14401	3403	24828	1467	3655
4457	19610	1898	4743	3250	13696	7786	24808	1499	19988
4870	832	12496	10282	1877	6275	7637	24699	14589	23912
4775	13304	12486	10326	9637	13620	2525	9202	4410	5922
640	25680	4915	323	1682	14248	18352	24643	815	10974
ER=57.14	ER=57.14	ER=71.43	ER=42.85	ER=28.57	ER=71.43	ER=71.43	ER=71.43	ER=71.43	ER=57.14
Average Error Rate = 51.43					Average Error Rate = 68.57				

The highlighted blocks in tab. 2 shows that similar results are found for discovered topics and sKL divergence. For example, in case of STMS approach top eight mavens shown in tab. 1 for "XML Databases" topic has three mavens in common, which are 337 "Sharma Chakravarthy", 4457 "Tok Wang Ling", and 4775 "Surajit Chaudhuri". From top 7 related mavens for five selected topics (same is the case with non selected topics) shown in the tab. 1 the error rate (ER) for STMS is less than ACT and STMS has 17.14 % less average error rate than ACT. It shows the bad effect of topics sparseness on maven retrieval performance for ACT, and inability of ACT to discover better results in comparison with STMS.

5   Conclusions and Future Works

This study deals with the problem of maven search through latent topics. Initially we generalized this problem to VL with embedding continuous time effects and discussed the motivation for it. We then introduced STMS approach, which can discover and probabilistically rank experts related to specific knowledge domains (or queries) by modeling semantics-based correlations and temporal information simultaneously. We demonstrated how it can be used to rank experts for unseen data and to find mavens correlations. We studied the effect of generalization on topics denseness when modeling entities and concluded that more dense topics will results in better performance of the approach. Empirical results show better performance on the basis of compact topics as compared to the baseline approach. As a future work, we plan to investigate how to use STMS approach for ranking mavens related to a topic for different years and discover changing trends in their expertness, as we think for different years the mavens are usually different and the expertness of an author can be dynamic over different time span.

Acknowledgements. The work is supported by the National Natural Science Foundation of China under Grant (90604025, 60703059) and Chinese National Key Foundation Research and Development Plan under Grant (2007CB310803). We are thankful to Jie Tang and Jing Zhang for sharing their codes, valuable discussions and suggestions.

References

1.   Andrieu, C., Freitas, N.D., Doucet, A., Jordan, M.: An Introduction to MCMC for Machine Learning. Journal of Machine Learning 50, 5–43 (2003)
2.   Azzopardi, L., Girolami, M., Risjbergen, K.V.: Investigating the Relationship between Language Model Perplexity and IR Precision-Recall Measures. In: Proc. of the 26th ACM SIGIR, Toronto, Canada, July 28-August 1 (2003)
3.   Balog, K., Azzopardi, L., de Rijke, M.: Formal Models for Expert Finding in Enterprise Corpora. In: Proc. of SIGIR, pp. 43–55 (2006)
4.   Balog, K., Bogers, T., Azzopardi, L., Rijke, M., Bosch, A.: Broad Expertise Retrieval in Sparse Data Environments. In: Proc. of SIGIR, pp. 551–558 (2007)
5.   Blei, D.M., Ng, A.Y., Jordan, M.I.: Latent Dirichlet Allocation. Journal of Machine Learning Research 3, 993–1022 (2003)
6.   Cao, Y., Liu, J., Bao, S., Li, H.: Research on Expert Search at Enterprise Track of TREC (2005)
7.   DBLP Bibliography Database,
     http://www.informatik.uni-trier.de/~ley/db/
8.   Deerwester, S., Dumais, S.T., Furnas, G.W., Landauer, T.K., Harshman, R.: Indexing by Latent Semantic Analysis. Journal of the American Society for Information Science 41(6), 391–407 (1990)

9. Griffiths, T.L., Steyvers, M.: Finding Scientific Topics. In: Proc. of the National Academy of Sciences, USA, pp. 5228–5235 (2004)
10. Hawking, D.: Challenges in Enterprise Search. In: Proc. of the 15th Conference on Australasian Database, vol. 27, pp. 15–24 (2004)
11. Hofmann, T.: Probabilistic Latent Semantic Analysis. In: Proc. of the 15th Annual Conference on UAI, Stockholm, Sweden, July 30-August 1 (1999)
12. Hofmann, T., Puzicha, J., Jordan, M.I.: Learning from Dyadic Data. In: Advances in Neural Information Processing Systems (NIPS), vol. 11. MIT Press, Cambridge (1999)
13. Mimno, D., McCallum, A.: Expertise Modeling for Matching Papers with Reviewers. In: Proc. of the 13th ACM SIGKDD, pp. 500–509 (2007)
14. Nie, Z., Ma, Y., Shi, S., Wen, J., Ma, W.: Web Object Retrieval. In: Proc. of World Wide Web (WWW), pp. 81–90 (2007)
15. Petkova, D., Croft, W.B.: Generalizing the Language Modeling Framework for Named Entity Retrieval. In: Proc. of SIGIR (2007)
16. Rosen-Zvi, M., Griffiths, T., Steyvers, M., Smyth, P.: The Author-Topic Model for Authors and Documents. In: Proc. of the 20th International Conference on UAI, Canada (2004)
17. Tang, J., Zhang, J., Yao, L., Li, J., Zhang, L., Su, Z.: ArnetMiner: Extraction and Mining of Academic Social Networks. In: Proc. of the 14th ACM SIGKDD (2008)
18. Teh, Y.W., Jordan, M.I., Beal, M.J., Blei, D.M.: Hierarchical Dirichlet Processes. Technical Report 653, Department of Statistics, UC Berkeley (2004)
19. Zhai, C., Lafferty, J.: A Study of Smoothing Methods for Language Models Applied to Ad-hoc Information Retrieval. In: Proc. of the 24th ACM SIGIR, pp. 334–342 (2001)
20. Zhang, J., Tang, J., Liu, L., Li, J.: A Mixture Model for Expert Finding. In: Washio, T., Suzuki, E., Ting, K.M., Inokuchi, A. (eds.) PAKDD 2008. LNCS (LNAI), vol. 5012, pp. 466–478. Springer, Heidelberg (2008)
21. Zhang, J., Tang, J., Li, J.: Expert Finding in a Social Network. In: Kotagiri, R., Radha Krishna, P., Mohania, M., Nantajeewarawat, E. (eds.) DASFAA 2007. LNCS, vol. 4443, pp. 1066–1069. Springer, Heidelberg (2007)

Topic and Viewpoint Extraction for Diversity and Bias Analysis of News Contents

Qiang Ma and Masatoshi Yoshikawa

Graduate School of Informatics, Kyoto University
{qiang,yoshikawa}@i.kyoto-u.ac.jp
http://www.db.soc.i.kyoto-u.ac.jp

Abstract. News content is one kind of popular and valuable information on the Web. Since news agencies have different viewpoints and collect different news materials, their perspectives on news contents may be diverse (biased). In such cases, it is important to indicate this bias and diversity to newsreaders. In this paper, we propose a system called TVBanc (Topic and Viewpoint based Bias Analysis of News Content) to analyze diversity and bias in Web-news content based on comparisons of topics and viewpoints. The topic and viewpoint of a news item are represented by using a novel notion called a content structure consisting touple of subject, aspect and state terms. Given a news item, TVBanc facilitates bias analysis in three steps: first, TVBanc extracts the topic and viewpoint of that news item based on its content structure. Second, TVBanc searches for related news items from multi-sources such as TV-news programs, video news clips, and articles on the Web. Finally, TVBanc groups the related news items into different clusters, and analyzes their distribution to estimate the diversity and bias of the news contents. The details of clustering results are also presented to help users understand the different viewpoints of the news contents. This paper also presents some experimental results we obtained to validate the methods we propose.

1 Introduction

Constant advances in information technologies and the spread of these technologies have changed our everyday lives. A major change is the ready availability of cross-media news content. News is now available from TV, Web sites, and newspapers, and people can access these information sources as they choose. Video news clips (TV news program and video news on the Web) offer immediacy and realism, but they suffer from time restrictions and often need to maintain audience ratings, which limits the details and scope of the information they provide. In addition, because the news agencies have different viewpoints and they have collected different news materials, the perspective the news content has may be biased. For example, news material, especially which used in a news item, is a kind of evidence that supports the viewpoint of news agencies in a certain organization. That is, news material stands for a viewpoint on a news event and the news professionals select news materials based on what their intentions are and what they want to convey to the audience. As a result, the perspective of news content may be diverse and biased. Hence, detecting bias and intent are very important and necessary

Q. Li et al. (Eds.): APWeb/WAIM 2009, LNCS 5446, pp. 150–161, 2009.

to help users understand the news event and acquire balanced information. If the news contents are diverse, we need to read more articles to acquire balanced information. On the other hand, when the diversity of news contents is small, we do not need to check other articles because there may be little additional information on that news event. However, in this case, the credibility of news contents may be questionable.

We call an event or an activity a "topic". To represent such "topics" described in a news item (e.g., a video news clip or a news article) and analyze the bias (diversity) in news, we propose a novel notion of content structure to represent the topic and viewpoint described in a news item. The content structure is made up of three-tuple of subject, aspect, and state terms. Intuitively, a subject term denotes the most dominant term. An aspect term means a term that has strong co-occurrence relationships with the subject terms and denotes the aspect described in that news item. State terms are keywords that denote the current state or status of a *subject*'s *aspect*. The topic of a news item is represented by using the content structure extracted from the whole text. However, because the last paragraph of a news article usually describes the conclusion or scoop material that clarifies the authors' focus, we can represent the viewpoint of that news item by using the content structure extracted from the last paragraph. We currently use a video news clip's closed caption to analyze its topic and viewpoint.

Based on the extraction of topics and viewpoints, we propose a system called TVBanc (Topic and Viewpoint based Bias Analysis of News Content) to analyze and present the bias (diversity) in news content to help users understand news and acquire balanced information. Given a news item, TVBanc facilitates analysis of bias and diversity in three steps: First, TVBanc extracts the topic and viewpoint from the text data of a news item by using the content structure. Second, TVBanc searches for related news items from multi-sources such as TV news programs, video news clips, and articles on the Web. Finally, TVBanc groups the related news items into different clusters, and analyzes their distribution to estimate the diversity and bias of news content. The details on clustering results are also presented to help users understand the different viewpoints in the news contents.

The remainder of this paper is organized as follows. Section 2 introduces related work. Section 3 describes the method used to extract content structure from news content. Section 4 introduces a mechanism for analyzing bias and an application system TVBanc. The experimental results are described in Section 5 and we conclude this paper in Section 6.

2 Related Work

Topic detection and tracking (TDT) research has led to algorithms for discovering and weaving together topically related material in data streams such as newswires and broadcast news [6, 9, 11]. In contrast, as one of the next steps in TDT, we have focused on how to extract the topic and viewpoint to analyze the bias and diversity in news content.

The CORPORUM [3] is a tool designed to extract content-representation models from natural language texts and use them for information retrieval. It consists of two basic concepts: an extraction concept that focuses on the semantics of a text, and a resonance algorithm that enables the content of two texts to be compared as a way of analyzing their conceptual structure. WebMaster [5] is a software tool for knowledge-based verification of Web pages designed to help maintain the content of Web sites. It is able to reveal hidden cluster relationships between different conceptual information by analyzing the content of weakly or semi-structured information sources. These tools are based on an ontological and knowledge-based approach to representing content. In contrast, by identifying the roles of various keywords, our keyword-based content structure enables us to analyze the difference in news items from different aspects.

Newsjunkie [4] generates personalized news for users by identifying the novelty of topics within the context of topics they have already reviewed. This approach focuses on the novelty of news topics using a $tf \cdot idf$-based method and personalized presentation, while our research uses the content structure to analyze the bias and diversity of viewpoints in news content.

Numerous studies have also been done on news browsers. For example, Columbia's Newsblaster is a tracking and summarization system that clusters news into events, categorizes the events into broad topics, and summarizes multiple articles on each event [2, 7]. The Comparative Web Browser (CWB), which is a system that allows users to concurrently browser different news articles for comparison, has also been proposed [8]. Unlike our system, the CWB did not aim at analyzing bias.

3   Extraction of Content Structure

3.1   Content Structure

As previously mentioned, we have considered a content structure to be a three-tuple of subject, aspect, and state terms. To reiterate, the subject terms denote the most dominant terms of a news item, while the aspect terms are those that have a strong co-occurrence relationships with the subject terms. The state terms, which have strong co-occurrence relationships with subject and content terms, denote the state or status of a certain topic's certain aspect. In other words, the subject terms are the centric keywords that play the title role in a news item. The aspect terms indicate which part (i.e., subtopic or viewpoint) of that topic has been focused on, while state terms describe the state and status of that aspect. Content structure CS is defined as

$$
\begin{aligned}
CS &:= (Subject, Aspect, State) \\
Subject &:= subject-term+ \\
Aspect &:= aspect-term+ \\
State &:= state-term+
\end{aligned}
\tag{1}
$$

where "+" means that the element(s) should appear one or more times. In addition, a keyword should not occur more than once in a content structure, i.e., the subject, aspect, and state terms should be different. For example, content structure,

$$(\{Beijing\ Olympic\ Games\}, \{Opening\ Ceremony\}, \{Spectacular\})$$

denotes that the news item is reporting that the opening ceremony (*aspect*) of the Beijing Olympics (*topic*) was spectacular (*state*).

To present a content structure to a user, we used an edge-labeled directed graph. In this graph, the source vertex denotes the subject terms and the destination vertex denotes the aspect terms. The label of the directed edge stands for the state terms.

3.2   Content-Structure Extraction

We respectively define the notions of *subject degree*, *aspect degree*, and *state degree* to determine whether a keyword has a high probability of being a subject, aspect, or state term.

First, we extract the news item's keywords (nouns and unknown-words) as a set of K. For each keyword $w \in K$, we compute its subject degree by using the formula below. If w appears frequently, and is close to the leading part and has a strong co-occurrence relationship with other terms, then w has a high probability of being a subject term. Moreover, a proper noun is assigned higher probability.

$$subject(w, K) = tf(w) * weight(w) * dis(w) * \Sigma_{w_i \in K, w_i \neq w} cooc(w, w_i) \qquad (2)$$

where $tf(w)$ stands for the term frequency of w. The $weight(w)$ is a weight function depending on the word class of w; the proper noun has been assigned a higher weight than the others. The $dis(w)$ denotes the distance between w and the first term of that news item. The $cooc(w, w_i)$ denotes the co-occurrence relationship between w and w_i.

The $dis(w)$ is currently computed as the number of words appearing between the first term and w. Because w may appear more than once, we only use the first location of w to compute $dis(w)$.

The co-occurrence relationship between two words w and w_i is defined as the ratio of sentences containing both w and w_i in all sentences in that news item.

$$cooc(w, w_i) = \frac{sf(w, w_i)}{N} \qquad (3)$$

where $sf(w, w_i)$ means the number of sentences containing both w and w_i. N is the total number of sentences.

The keywords with the top n subject degrees will be selected as the subject terms. After this, we compute the aspect degrees of the remaining keywords as

$$aspect(w, S) = tf(w) * weight(w) * dis(w) * \Sigma_{s \in S} cooc(w, s), \qquad (4)$$

where S denotes subject terms. $w \in K - S$

Keywords with the top m aspect degrees will be selected as the aspect terms. After that, we compute the state degrees of remaining keywords as

$$state(w, S, A) = \frac{tf(w) * weight'(w) * (\Sigma_{s \in S} cooc(w, s) + \Sigma_{a \in A} cooc(w, a))}{dis(w) + 1}, \qquad (5)$$

where S and A are subject and aspect terms, respectively. $w \in K - S - A$. In contrast to the formula for subject and aspect degrees, a larger $dis(w)$ leads to a smaller state degree. This is because we have assumed that the state and viewpoint will be described in the latter parts of a news article. $weight'(w)$ is a weight function depending on the word class of w; we assign the adjective noun the highest weight, and assign proper noun higher weight than the other. Keywords having the top l state degrees will be selected as state terms.

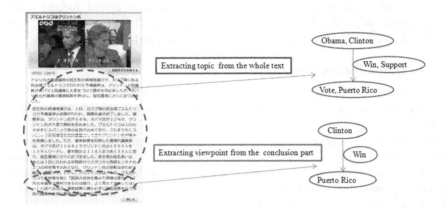

Fig. 1. Analysis of Topic and Viewpoint

3.3   Topic and Viewpoint Analysis

We used the content structure to represent the topic and viewpoint of a news item. Generally, a news article consists of three parts: the lead is used to give a summary, the nut graf is used to show the details, and the conclusion is used to describe additional information and original material to indicate the viewpoint of the authors. Based on such structure of a news article, as shown in Fig. 1, we extract the content structure from the whole text as the topic of that news item. The viewpoint is the content structure extracted from the conclusion of the news article.

To analyze the topic and viewpoint of a group of similar news items, we merge paragraphs corresponding to these news items together to construct one virtual news item. For example, as shown in Fig. 2, to construct the virtual article from articles A and B, we first merge the titles of A and B to that of the virtual article. Similarly, the lead, nut graf, and conclusion of articles A and B are merged into those of the virtual article. We then extract the topic and viewpoint from that virtual news item as those of that group.

4   Diversity Analysis

By using the method of extracting the topic and viewpoint, we propose a mechanism that can be used to analyze the diversity and bias of news content and an application system TVBanc.

As we can see from Fig. 3, we first extract the topic and viewpoint for a given news item (e.g., video news clip or news article) based on the content structure by using its text data or closed caption data. We then use the keywords in the topic and viewpoint to search for related news items from the Internet and a local news database. After the search results are obtained, we group these news items into clusters. We also extract the topic and viewpoint of each cluster. Finally, we analyze the distribution of clusters to estimate the diversity and bias of news content and present users with the results to help them understand the news.

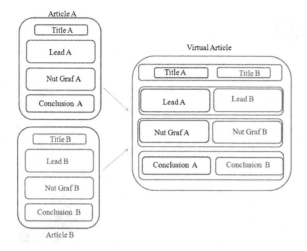

Fig. 2. Construction of Virtual Article

4.1   Topic and Viewpoint Analysis

We extract the topic and viewpoint for a given news item. Hereafter, the topic will be considered to be based on the content structure consisting of two-subject, two-aspect, and two-state terms, while the viewpoint consists of one-subject, one-aspect, and one-state terms.

4.2   Related News-Item Retrieval

We search for related news items of that given one from the Internet and a local news database.

To search for related news items from the Internet, we generate keyword queries by using the subject and aspect terms of the topic extracted from the given news item. Such keyword queries will be issued to a search engine via a Web search API. The strategy to generate keyword queries is that at least one subject term should appear in the title and at least one aspect term should appear in the body of a searched news page. For example, suppose the subject terms are s_1, s_2, and the aspect terms are a_1, a_2, the query should be

$$q = (intitle : s_1 \lor s_2) \land (intext : a_1 \lor a_2), \tag{6}$$

where "intitle" means the following term should appear in the title and "intext" means it should appear in the body of the searched news item. "$\land$" stands for logical AND and "$\lor$" means logical OR.

The local news database implemented by using MySQL currently consists of three kinds of news contents (in Japanese). The topic and viewpoint based on the content structure are also extracted and stored in the database.

- TV news program with closed caption data. Eight news TV programs from digital terrestrial broadcasting with closed captions are recorded daily. Because a news

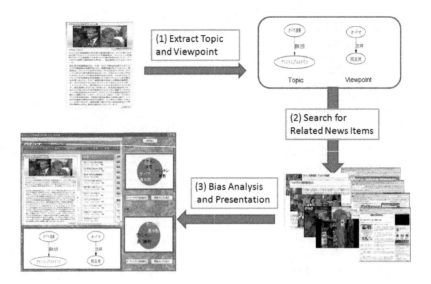

Fig. 3. Overview of Diversity Analysis

TV program includes more than one news item, we used a method based on text (closed caption data) analysis [10] to segment news programs. Each segment was considered as a news item.

- News article on the Web. Four well-known news sites in Japan were the targets for collecting the news. These Web sites included Asahi.com[1], Yomiuri Online[2], MSN Sankei[3], and Mainichi JP[4]. We use pre-defined patterns to extract pure text (the title and the body of the news article) from news articles on these Web sites.
- Video news clip and its surrounding text. In Japan, major TV stations also publish video news clips on their Web sites. We collected these video news clips and their surrounding text (title and body) from the Web sites of TV stations including NHK News Online[5], FNN[6], ANN News[7], News-i[8] and News24[9].

To find related news items from the news database, we issued an SQL query to the database. The subject and aspect terms (of the topic and viewpoint based on the content structure) of a searched news item should contain at least two of those extracted from the given news item. In the above example, the SQL query is generated as

---

[1] http://www.asahi.com

[2] http://www.yomiuri.co.jp

[3] http://sankei.jp.msn.com

[4] http://mainichi.jp

[5] http://www3.nhk.or.jp/news/

[6] http://www.fnn-news.com

[7] http://www.tv-asahi.co.jp/ann/

[8] http://news.tbs.co.jp

[9] http://www.news24.jp

$$select\ item\ from\ newsitem\ where\ subject\ =\ s_1\ or\ s_2\ or\ a_1\ or\ a_2\ and$$
$$aspect\ =\ s_1\ or\ s_2\ or\ a_1\ or\ a_2 \qquad (7)$$

4.3 Diversity Analysis and Presentation

After the related news items were searched, we grouped them including the given one into clusters by using the complete linkage method. The similarity between news items, i.e., the measure used in the complete-linkage method, is computed using the vector space model as

$$sim(s_1, s_2) = \frac{k_{11} \cdot k_{21} + k_{12} \cdot k_{22} + ... + k_{1n} \cdot k_{2n}}{\sqrt{k_{11}^2 + ... + k_{1n}^2} \cdot \sqrt{k_{21}^2 + ... + k_{2n}^2}}, \qquad (8)$$

where s_1 and s_2 stand for the news items. The $(k_{11}, k_{12}, ..., k_{1n})$ and $(k_{21}, k_{22}, ..., k_{2n})$ correspond to the $tf \cdot idf$ keyword vectors of s_1 and s_2. Here, keywords denote the nouns and unknown-words appearing in that news item and idf value is computed within the related news items.

We compute the entropy of the related news items as

$$E = -\Sigma_{i=1}^{i=m} \frac{c_i}{N} * log \frac{c_i}{N}, \qquad (9)$$

where c_i denotes the cluster size, i.e., the number of news items in cluster i. Here, m is the number of clusters and N is the number of news items. Obviously, $N = \Sigma_{i=1}^{i=m} c_i$.

Intuitively, E is a score that indicates the diversity of news on a certain topic or event. We hence called E the *diversity degree*. A large E means the news report on that topic is diverse. A small E, on the other hand, denotes the news reports may be biased. That is, the smaller the E, the greater the bias.

We presented the results to users to help them understand the news and bias. Additionaly, we also extract the topic and viewpoint of each cluster (a group of similar news items) and use the viewpoint as the label for each cluster. If the news content is biased, we do not need to check other news because there is little new information. However, the credibility may be questionable. However, if the news content is diverse, we need to check other items to acquire news that has balanced viewpoints.

4.4 Prototype System

We developed a prototype system of TVBanc based on the methods above. Fig. 4a is a screenshot of TVBanc. The top left window is a viewer for the news item that can be used to browse news pages and to watch video news clips. The bottom left window displays the topic and viewpoint of the current news item. The right window is used to present clustering and diversity-analysis results. In our prototype system, we grouped related news items per media (news sites, and videos (TV)). The top right pie chart shows the results of clustering news items on a news site. The bottom right pie chart shows the results of clustering news on the TV and Web sites of a TV station.

The window presenting details of each cluster will pop up if a user clicks the clustering results pie chart. As shown in Fig. 4b, each tab window stands for a respective

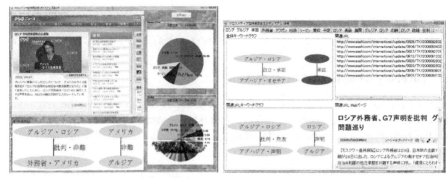

(a) *Main Screenshot of TVBanc*   (b) *Details of Clustering Results*

Fig. 4. Running Examples of TVBanc

cluster. The top left window is the viewer for the topic and the viewpoint of the cluster. The top right window is the list of related news items. When a news item in the list has been selected, the bottom left window will display its topic and viewpoint, and its text data will be presented in the bottom right window.

The prototype system was developed on a Windows Server 2003 machine by using Visual Studio 2008 Professional Edition Service Pack 1. The target news items are currently in Japanese. We used ChaSen[1] to analyze the Japanese morphology and only nouns and unknown-words as the keywords for further processing. To exclude stop words, we built a stop-word dictionary that contained 593 terms in English and 347 terms in Japanese. Google Web API and Yahoo! Web API were used to search for related news items from the Web. Top 50 results returned by Google and Yahoo! were selected as the related news items on the Web.

5   Experimental Results

This section describes the experimental results. We carried out two kinds of experiments: 1) an experiment to evaluate methods of extracting the topic and viewpoint based on the content structure, and 2) an experiment to evaluate the method of analyzing diversity and bias.

We selected 10 experimental news items as the target data for evaluation from video-news Web sites, such as NHK News, FNN, News-i and News24. The local news database contained 26111 news items at the time of the experiment. Google Web API and Yahoo! Web API were used to search for related news items from the Web.

5.1   Experiment 1: Topic and Viewpoint Extraction

We asked 10 students to evaluate the methods of extracting the topic and viewpoints. First, each student was asked to read each given news article and then extract six keywords to represent the topic and three to represent the viewpoint of that news item. We asked them to compare the topic and viewpoint given by TVBanc with those of their

Table 1. Topic and Viewpoint Scores Given by Evaluators and Common Keywords Ratio

	News Item	News Site			TV Station			Average
		cluster 1	cluster 2	cluster 3	cluster 1	cluster 2	cluster 3	
topic score (average)	6.06	6.03	5.17	4.91	6.34	5.51	5.47	5.64
viewpoint socre (average)	3.84	4.81	3.92	3.75	5.13	5.24	5.2	4.56
r_t(average)	0.53	0.45	0.34	0.31	0.52	0.32	0.37	0.406
r_v(average)	0.26	0.31	0.203	0.17	0.37	0.29	0.25	0.265

own. Then we asked them give a score ranging from 0 to 10 to show how well the topic and viewpoint given by TVBanc.

Similarly, we asked these students to check the news items in a cluster (of related news items) and extract six keywords to represent the topic and three to represent the viewpoint of that cluster. These students were then asked to compare the topic and viewpoint given by TVBanc with their own topic and viewpoint. Then these students were asked to score the topic and viewpoint given by TVBanc, respectively. In experiment 1, we only asked these students to evaluate the largest top-three clusters. We also computed the common keywords ratios of topic (r_t) and viewpoint (r_v) given by the 10 students and TVBanc as follows.

$$r_t = \frac{1}{10} * \sum_{i=1}^{i=10} \frac{|T_{u_i} \cap T_s|}{6} \tag{10}$$

$$r_v = \frac{1}{10} * \sum_{i=1}^{i=10} \frac{|V_{u_i} \cap V_s|}{3} \tag{11}$$

where T_{u_i} and V_{u_i} are the topic and viewpoint keywords given by student i. T_s and V_s are the respective keywords of the topic and viewpoint given by TVBanc.

The experimental results listed in Table 1 revealed that our method can extract keywords that are similar to those given by human (the 10 students). They also revealed that our method can extract better topics than viewpoints based on the content structure. The topic extracted from a news item is better than that from a cluster, while the viewpoint extracted from the cluster is better. One significant reason is that the conclusion of a news item is too short to extract its viewpoint, while the cluster's conclusion is merged from more than one news item and has long text data for extracting the viewpoint. Another reason is that although the conclusion may describe original material to reveal the authors' focal points, the reader may focus on other areas from his/her own viewpoint. The content structure of larger clusters is better than that of smaller ones. This is to say, when we extract the content structure from a virtual news item, a small cluster is easily influenced by the noise of news items, which is a false result in the clustering method. We intend to discuss these issues in the near future.

5.2 Experiment 2: Diversity Analysis

We carried out a user study to evaluate our mechanism for analyzing diversity and bias. Experiment 2 was actually carried out before Experiment 1. We asked the same 10

Table 2. Precison and Recall Ratios

θ	2.5	2.8	3.2
Recall Ratio	0.556	0.444	0.444
Precision Ratio	0.71	0.80	1.0

students to access the given news items carefully and then do a survey on the Internet to estimate the news diversity and bias. If there were many news items describing the same topic (event) from different aspects and viewpoints, then the news on that topic (event) was diverse and the bias was small. From this viewpoint, we asked these students to score the diversity (bias) on a five-point Likert scale ranging from one to five. If there were six students who scored a news topic greater than three, we considered the news topic to be diverse and regarded it as a relevant result.

However, we also computed the degree of diversity in the given 10 news items based on our mechanism for analyzing diversity and bias. When the degree of diversity was greater than threshold θ, we regarded it as a positive result (i.e., the news content on that topic is diverse.) returned by our method. We computed the precision and recall ratios (Table 2). The average degree of diversity of the given 10 news items was 2.92. Although Experiment 2 was a preliminary evaluation and further experiments are necessary, at least the results denoted that the proposed method based on the extraction of topics and viewpoints is potentially an efficient way of analyzing diversity and bias in news.

6   Conclusion

We proposed a system of analyzing diversity and bias in Web news called TVBanc based on the extraction of topics and viewpoints. We also proposed a novel notion called the content structure to extract topics and viewpoints. First, for a given news item, we extracted its topic and viewpoint based on the content structure. We then searched for its related news items from the Internet and a local news database. We computed the degree of diversity in the news content to estimate the bias in news by analyzing related news items. We intend to present the analysis results to users to help them understand news items. The experimental results revealed that the methods of extracting topics and viewpoints could extract keywords that were similar to those given by human. A preliminary study of users validated our mechanism for analyzing diversity and bias based on the extraction of topics and viewpoints.

Further studies on the mechanism for analyzing diversity and bias, and evaluations are necessary. For instance, other core technical areas, i.e., retrieval of related news items and clustering, need to be studied in the near future. The notion of content structure and methods of extracting topics and viewpoints should be improved.

Acknowledgment

This work is supported in part by the National Institute of Information and Communications Technology. This research is partly supported by the research for the grant of Scientific Research (No.20700084 and 20300042) made available by MEXT, Japan.

References

[1] ChaSen (2009), `http://chasen.aist-nara.ac.jp/index.html.en`

[2] NewsBlaster (2005), `http://www1.cs.columbia.edu/nlp/newsblaster/`

[3] Engels, R., Bremdal, B.A.: CORPORUM: a workbench for the semantic Web. In: Proceedings of Semantic Web Mining Workshop, PKDD/ECML 2001, pp. 1–10 (2001)

[4] Gabrilovich, S.E., Horvitz, E.: Newsjunkie: providing personalized newsfeeds via analysis of information novelty. In: Proceedings of the 13th International World Wide Web Conference, pp. 482–490 (2004)

[5] Harmelen, F.V., Meer, J.V.D.: WebMaster: knowledge-based verification of web pages. In: Proceedings of the 12th International Conference on Industrial and Engineering Applications of Artificial Intelligence and Expert Systems, pp. 256–265 (1999)

[6] Allan, J., Papka, R., Lavrenko, V.: On-line new event dectection and tracking. In: Proceedings of the 21st International ACM SIGIR Conference on Research and Development in Information Retrieval, pp. 37–45 (1998)

[7] McKeown, K., Barzilay, R., Evans, D., Hatzivassiloglou, V., Klavans, J., Sable, C., Schiffman, B., Sigelman, S.: Tracking and summarizing news on a daily basis with Columbia's Newsblaster. In: Proceedings of the 2002 Human Language Technology Conference (HLT) (2002)

[8] Nadamoto, A., Tanaka, K.: A comparative web browser (CWB) for browsing and comparing web pages. In: Proceedings of WWW 2003, pp. 727–735 (2003)

[9] Wayne, C.L.: Multilingual topic detection and tracking: Successful research enabled by corpora and evaluation. In: Proceedings of the Language Resources and Evaluation Conference (LREC) 2000, pp. 487–1494 (2000)

[10] Uchiyama, M., Isahara, H.: A statistical model for domain -independent text segmentation. In: Proceedings of the 9th Conference of the European Chapter of the Association for Computational Linguistics, pp. 491–498 (2001)

[11] Yang, Y., Pierce, T., Carbonell, J.: A study on retrospective and on-line event detection. In: Proceedings of the 21st International ACM SIGIR Conference on Research and Development in Information Retrieval, pp. 28–36 (1998)

Topic-Level Random Walk through Probabilistic Model*

Zi Yang, Jie Tang, Jing Zhang, Juanzi Li, and Bo Gao

Department of Computer Science & Technology, Tsinghua University, China

Abstract. In this paper, we study the problem of topic-level random walk, which concerns the random walk at the topic level. Previously, several related works such as topic sensitive page rank have been conducted. However, topics in these methods were predefined, which makes the methods inapplicable to different domains. In this paper, we propose a four-step approach for topic-level random walk. We employ a probabilistic topic model to automatically extract topics from documents. Then we perform the random walk at the topic level. We also propose an approach to model topics of the query and then combine the random walk ranking score with the relevance score based on the modeling results. Experimental results on a real-world data set show that our proposed approach can significantly outperform the baseline methods of using language model and that of using traditional PageRank.

1 Introduction

Link-based analysis has become one of the most important research topics for Web search. Random walk, a mathematical formalization of a trajectory that consists of taking successive random steps, has been widely used for link analysis. For example, the PageRank algorithm [11] uses the random walk techniques to capture the relative "importance" of Web pages from the link structure between the pages. Intuitively, a Web page with links from other "important" pages is highly possible to be an "important" page. Other random walk based methods have been also proposed, e.g. HITS [7].

Unfortunately, traditional random walk methods only use one single score to measure a page's importance without considering what topics are talked about in the page content. As a result, pages with a highly popular topic may dominate pages of the other topics. For example, a product page may be pointed by many other advertising pages and thus has a high PageRank score. This makes the search system susceptible to retrieve such pages in a top position. An ideal solution might be that the system considers the topics talked about in each page and ranks the pages according to different topics. With such a topic-based ranking score, for queries with different topics (intentions), the system can return different topic-based ranking lists.

---

* The work is supported by NSFC (60703059), Chinese National Key Foundation Research and Development Plan (2007CB310803), and Chinese Young Faculty Research Funding (20070003093).

Q. Li et al. (Eds.): APWeb/WAIM 2009, LNCS 5446, pp. 162–173, 2009.

Recently, a little effort has been made along with this research line. For example, Topic-sensitive PageRank [5] tries to break this limitation by introducing a vector of scores for each page. Specifically, the method assumes that there are multiple topics associated with each page and uses a bias factor to capture the notion of importance with respect to a particular topic. Nie et al. [10] investigate the topical link analysis problem for Web search and propose Topical PageRank and Topical HITS models. However, these proposed methods have a critical limitation: All topics are predefined, which is not applicable to new domains.

Another problem for Web search is how to "understand" the query submitted by the user. With a query, the user typically wants to know multiple perspectives related to the query. For instance, when a user asks for information about a product, e.g., "iPod touch", she/he does not typically mean to find pages containing these two words. Her/his intention is to find documents describing different features (e.g., price, color, size, and battery) of the product. However, existing methods usually ignore such latent information in the query.

Therefore, several interesting research questions are: (1) How to automatically discover topics from Web pages and how to conduct the random walk at the topic-level? (2) How to discover topics related to a query and how to take advantage of the modeling results for search? To the best of our knowledge, this problem has not been formally addressed, although some related tasks have been studied, such as topical PageRank and query reformulation, which we will further discuss in Section 2.

In this paper, we aim at conducting a thorough investigation for the problem of topic-level random walk. We identify the major tasks of the problem and propose a four-step approach to solve the tasks. The approach can discover topics from documents and calculate a vector of topic-based ranking scores for every page. We apply the approach to academic search. Experiments on a real-world data set show that our method outperforms the baseline methods of using language model and those of using traditional PageRank algorithm.

2   Prior Work

We begin with a brief description of several related work, including: language model [20], LDA [1], random walk [11], and topical PageRank [5] [10].

Without the loss of generality, we represent pages on the Web as a collection of linked documents, i.e. $G = (D, E)$, where D represents all pages (documents), and the directed edge $d_1 \rightarrow d_2 \in E$ suggests the page (document) d_1 has a hyperlink pointing to the page (document) d_2. Table 1 summarizes the notations.

2.1   Language Model

Language model is a classical approach for information retrieval. It interprets the relevance between a document and a query word as a generative probability:

$$P(w|d) = \frac{N_d}{N_d + \lambda} \cdot \frac{tf(w, d)}{N_d} + (1 - \frac{N_d}{N_d + \lambda}) \cdot \frac{tf(w, D)}{N_D} \qquad (1)$$

Table 1. Notations

SYMBOL	DESCRIPTION
T	set of topics
D	document collection
E	hyperlinks
V	vocabulary of unique words
N_d	number of word tokens in document d
$L(d)$	out-degree of document d
$r[d, z]$	the ranking score of document d on the topic z
w_{di}	the ith word token in document d
z_{di}	the topic assigned to word token w_{di}
θ_d	multinomial distribution over topics specific to document d
ϕ_z	multinomial distribution over words specific to document z
α, β	Dirichlet priors to multinomial distributions θ and ϕ
γ	preference to the topic-intra transition or the topic-inter transition

where $tf(w, d)$ is the word frequency (i.e., occurring number) of word w in d, N_D is the number of word tokens in the entire collection, and $tf(w, D)$ is the word frequency of word w in the collection D. λ is the Dirichlet smoothing factor and is commonly set according to the average document length in the collection [20]. Further, the probability of the document d generating a query q can be defined as $P_{\text{LM}}(q|d) = \prod_{w \in q} P(w|d)$.

2.2 Latent Dirichlet Allocation

Latent Dirichlet Allocation (LDA) [1] models documents using a latent topic layer. In LDA, for each document d, a multinomial distribution θ_d over topics is first sampled from a Dirichlet distribution with parameter α. Second, for each word w_{di}, a topic z_{di} is chosen from this topic distribution. Finally, the word w_{di} is generated from a topic-specific multinomial distribution $\phi_{z_{di}}$. Accordingly, the generating probability of word w from document d is:

$$P(w|d, \theta, \phi) = \sum_{z \in T} P(w|z, \phi_z) P(z|d, \theta_d) \tag{2}$$

Wei and Croft have applied the LDA model to information retrieval and obtained improvements [17].

2.3 Random Walk

Considerable research has been conducted on analyzing link structures of the Web, for example PageRank [11] and HITS [7].

Many extensions of the PageRank model have been proposed. For example, Haveliwala [5] introduces a vector of scores for each page, with each score representing the importance of the page with respect to a topic. Nie et al. [10] propose a topical link analysis method for Web search. They first use a classification method to categorize each document and then conduct a category-based PageRank on the network of the documents. However, these methods need to pre-define a set of categories, which is not applicable to a new domain.

Richardson and Domingos [12] propose a method for improving PageRank by considering the query terms in the transition probability of the random walk. However, the method does not consider the topical aspects of documents.

Tang et al. [15] propose a method by integrating topic models into the random walk framework. However, they do not consider the topic-level random walk.

2.4   Query Reformulation

A widely used method for query reformulation is to re-weight the original query vector using user click-through data or pseudo-feedback techniques, e.g. [14].

One type of approach is to expand the query terms with synonyms of words, various morphological forms of words, or spelling errors, e.g. Zhai and Lafferty's model-based feedback [19], Lavrenko and Croft's relevance models [8]. Both approaches use relevant documents to expand queries.

Other approaches consider not solely the input text, but also the behavior of the user or the related community, e.g. the click-through data in a search system log of the user, the entire user group or a specific related users in a community, including discovering semantically similar queries [18]. However, all of these methods do not consider the topical semantics of the query.

3   Our Approach

At a high level, our approach primarily consists of four steps:

1. We employ a statistical topic model to automatically extract topics from the document collection.
2. We propose a topic-level random walk, which propagates the importance score of each document with respect to the extracted topics.
3. We propose a method to discover topics related to the query.
4. Given a query, we calculate the relevance score of a document given the query, based on the discovered topics. We further combine the topic-level random walk ranking score with the relevance score.

3.1   Step 1: Topic Modeling

The purpose of the first step is to use a statistical topic model to discover topics from the document collection. Statistical topic modelings [1] [6] [9] [16] are quite effective for mining topics in a text collection. In this kind of approaches, a document is often assumed to be generated from a mixture of $|T|$ topic models. LDA is a widely used topic model. In this model, The likelihood of a document collection D is defined as:

$$P(\mathbf{z}, \mathbf{w} | \Theta, \Phi) = \prod_{d \in D} \prod_{z \in T} \theta_{dz}^{n_{dz}} \times \prod_{z \in T} \prod_{v \in V} \phi_{zv}^{n_{zv}} \tag{3}$$

where n_{dz} is the number of times that topic z has been used associated with document d and n_{zv} is the number of times that word w_v is generated by topic z.

Intuitively, we assume that there are $|T|$ topics discussed in the document collection D. Each document has a probability $P(z|d)$ to discuss the topic z. The topics with the highest probabilities would suggest a semantic representation for the document d. According to the topic model, each document is generated by following a stochastic process: first the author would decide what topic z to write according to $P(z|d)$, which is the topic distribution of the document. Then a word w_{di} is sampled from the topic z according to the word distribution of the topic $P(w|z)$.

For inference, the task is to estimate the unknown parameters in the LDA model: (1) the distribution θ of $|D|$ document-topics and the distribution ϕ of $|T|$ topic-words; (2) the corresponding topic z_{di} for each word w_{di} in the document d. We use Gibbs sampling [4] for parameter estimation. Specifically, instead of estimating the model parameters directly, we evaluate the posterior distribution on just z and then use the results to infer θ and ϕ. The posterior probability is defined as: (Details can be referred to [1] [4].)

$$p(z_{di}|\mathbf{z}_{-di}, \mathbf{w}, \alpha, \beta) = \frac{n_{dz_{di}}^{-di} + \alpha}{\sum_z (n_{dz}^{-di} + \alpha)} \cdot \frac{n_{z_{di}w_{di}}^{-di} + \beta}{\sum_v (n_{z_{di}v}^{-di} + \beta)} \tag{4}$$

where the number n^{-di} with the superscript $-di$ denotes a quantity, excluding the current instance (the i-th word token in the query q).

3.2   Step 2: Topic-Level Random Walk

After applying the topic model to the document collection, we obtain a topic distribution for each document. Formally, for each document we use a multinomial distribution of topics $\{p(z|d)\}$ (equally written as θ_{dz}) to represent it. We then define a random walk at the topic level.

Specifically, for every document d, we associate it with a vector of ranking scores $\{r[d, z]\}$, each of which is specific to topic z. Random walk is performed along with the hyperlink between documents within the same topic and across different topics.

For document d_k having a hyperlink pointing to document d_l, we define two types of transition probabilities between documents: topic-intra transition probability and topic-inter transition probability, i.e.,

$$P(d_l|d_k, z_i) = \frac{1}{|L(d_k)|} \tag{5}$$

$$P(d_l, z_j|d_k, z_i) = P(z_j|d_l)P(z_i|d_k) \tag{6}$$

where $P(d_l|d_k, z_i)$ is the transition probability from document d_k to d_l on the same topic z_i; $P(d_l, z_j|d_k, z_i)$ is the transition probability from document d_k on topic z_i to page d_l on topic z_j.

Further we introduce a parameter γ to represent preference to the topic-intra transition or the topic-inter transition. Thus, this transition graph formalizes a random surfer's behavior as follows. The random surfer will have a γ probability to access the same topical content on document d_l and will have a $(1 - \gamma)$ probability to find different topics on d_l.

Given this, similar to PageRank, we can define a general form of the random walk ranking score for each document d as:

$$r[d, z_i] = \lambda \frac{1}{|D|} P(z_i|d) + (1 - \lambda) \sum_{d':d' \to d} \left[\gamma P(d|d', z_i) + (1 - \gamma)\frac{1}{T} \sum_{j \neq i} P(d, z_i|d', z_j) \right] \tag{7}$$

where $P(z|d)$ is the probability of topic z generated by document d; similar to the PageRank algorithm, we also introduce a random jump parameter λ, which allows a surfer to randomly jump to different pages in the network:

3.3   Step 3: Modeling Query

The third step is to find topics related to the query. This is not a necessary step, but it can help find semantic information of the query. Modeling query is not an easy task, as the query is usually short. To ensure the coverage of topics, we perform query expansion, a commonly used method in information retrieval. Specifically, for each word w in the query q, we extract its frequent words in the document collection and add them into the query. We consider words appearing in a window-size of the word w as its co-occurring words, i.e. words before and after the word w. We set the window size as 1. We then apply the topic model on the expanded query and discover the query-related topics. For each word w_{qi} in the expanded query q, we sample its topic z according to the probability:

$$P(z_{qi}|\mathbf{z}_{-qi}, \mathbf{w}_q, \alpha_q, \beta) = \frac{n_{qz_{qi}}^{-qi} + \alpha_q}{\sum_z (n_{qz}^{-qi} + \alpha_q)} \frac{n_{z_{qi}w_{qi}}^{-qi} + n_{z_{qi}w_{di}}^{d} + \beta}{\sum_v (n_{z_{qi}v}^{-qi} + n_{z_{qi}v}^{d} + \beta)} \tag{8}$$

where n_{qz} is the number of times that topic z has been sampled from the multinomial distribution specific to query q; α_q is a Dirichlet prior for the query-specific multinomial; the number n^d with the superscript d denotes that we count the numbers in all documents after inference in Step 1. For example, n_{zw}^{d} denotes the number of times that word w assigned to topic z in all documents.

Essentially, we perform another generative process particularly for the query q. The generative process is analogous to that in Step 1, except that we combine the document modeling results for modeling the query. Specifically, we sample a multinomial θ_q for each query q from a Dirichlet prior α_q; we then, for each word w_{qi} in the query, sample a topic z_{qi} from the multinomial; finally, we generate the word w_{qi} from a multinomial specific to the topic z_{qi}. The generative process accounts for dependencies between query and documents. After the process, we obtain a query-specific topic distribution $\{p(z|q)\}$, which suggests a semantic representation of the query.

3.4   Step 4: Search with Topic-Level Random Walk

The last step is to employ the topic-level random walk for search. Specifically, for each document d we combine the relevance score between the query q and the document, with the ranking score of document d from the topic-level random walk. We calculate the relevance score by

$$P_{\text{LDA}}(q|d, \theta, \phi) = \prod_{w \in q} P(w|d, \theta, \phi) = \prod_{w \in q} \sum_{z \in T} P(w|z, \phi_z) P(z|d, \theta_d) \tag{9}$$

However, the learned topics by the topic model are usually *general* and not *specific* to a given query. Therefore, only using topic model itself is too coarse for search [17]. Our preliminary experiments [15] also show that employing only topic model to information retrieval hurts the retrieval performance. In general, we would like to have a balance between *generality* and *specificity*. Thus, we derive a combination form of the LDA model and the word-based language model:

$$P(q|d) = P_{\text{LM}}(q|d) \times P_{\text{LDA}}(q|d) \tag{10}$$

where $P_{\text{LM}}(q|d)$ is the generating probability of query q from document d by the language model and $p_{\text{LDA}}(q|d)$ is the generating probability by the topic model.

Then, we first consider two ways to combine the relevance score with the random walk ranking score, i.e.,

$$S_{\text{TPR}*}(d, q) = P_{\text{LM}}(q|d) \cdot \left[\prod_{w \in Q} \sum_{z \in T} P(w|z) \cdot P(z|d) \right] \cdot \left[\prod_{w \in Q} \sum_{z \in T} r[d, z] \cdot P(w|z) \right] \tag{11}$$

and

$$S_{\text{TPR}+}(d, q) = \left[(1 - t) P_{\text{LM}}(q|d) + t \prod_{w \in Q} \sum_{z \in T} P(w|z) \cdot P(z|d) \right] \cdot \prod_{w \in Q} \sum_{z \in T} r[d, z] \cdot P(w|z) \tag{12}$$

where $P(z|d)$ denotes the probability of document d generating topic z and $r[d, z]$ denotes the importance of the document d on topic z.

The above methods sum up the generative probabilities on all topics. We also consider another combination by taking the query modeling results into consideration. The combination score is defined as:

$$S_{\text{TPRq}}(d, q) = P_{\text{LM}}(q|d) \cdot \sum_{z \in T} r[d, z] \cdot P(z|q) \tag{13}$$

where z_w is the topic selected for the word w in query q in the sampling process (cf. Section 3.3).

3.5   Computational Complexity

We analyze the complexity of the proposed topic models. The topic model has a complexity of $O(I_s \cdot (|D| \cdot \bar{N}_d) \cdot |T|)$, where I_s is the number of sampling iterations and $\bar{N}_d$ is the average number of word tokens in a document. The topic-level random walk has a complexity of $O(I_p \cdot |E| \cdot |T|)$, where I_p is the number of propagating iterations, $|E|$ is the number of hyperlinks in graph G, and $|T|$ is the number of topics.

4   Experimental Results

4.1   Experimental Settings

Data sets. We evaluated the proposed methods in the context of ArnetMiner (http://arnetminer.org) [16]. 200 example topics with their representative persons/papers are shown at http://arnetminer.org/topicBrowser.do.

We conducted experiments on a sub data set (including 14,134 authors and 10,716 papers) from ArnetMiner. As there is not a standard data set with ground truth and also it is difficult to create such one, we collected a list of the most frequent queries from the log of ArnetMiner. For evaluation, we used the method of pooled relevance judgments [2] together with human judgments. Specifically, for each query, we first pooled the top 30 results from three similar systems (Libra, Rexa, and ArnetMiner). Then, two faculties and five graduate students from

CS provided human judgments. Four-grade scores (3, 2, 1, and 0) were assigned respectively representing definite expertise, expertise, marginal expertise, and no expertise. We will conduct two types of searches (paper search and author search) on this selected collection. The data set was also used in [16] and [15].

Evaluation measures. In all experiments, we conducted evaluation in terms of P@5, P@10, P@20, R-pre, and mean average precision (MAP) [2] [3].

Baseline methods. We use language model (LM), BM25, LDA, PageRank (PR) as the baseline methods. For language model, we used Eq.1 to calculate the relevance between a query term and a paper and similar equations for an author, who is represented by her/his published papers). For LDA, we used Eq.2 to calculate the relevance between a term and a paper/author. In BM25, we used the method in [13] to calculate the relevance of a query and a paper, denoted by $S_{BM25}(d,q)$. For PageRank $S_{PR}(d)$, we use equation described in [11]. For Topical PageRank S_{TPR}, we use equation described in [10].

We also consider several forms of combination of the baseline methods, including LM+LDA, LM*LDA, LM*PR, LM+PR, LM*LDA*PR, and BM25*PR.

$$S_{LM+LDA}(d,q) = (1-t)P_{LM}(d|q) + tP_{LDA}(d|q) \tag{14}$$

$$S_{LM*LDA}(d,q) = P_{LM}(d|q) \cdot P_{LDA}(d|q) \tag{15}$$

$$S_{LM*PR}(d,q) = P_{LM}(d|q) \cdot PR(d) \tag{16}$$

$$S_{LM+PR}(d,q) = (1-t)P_{LM}(d|q) + tPR(d) \tag{17}$$

$$S_{LM*LDA*PR}(d,q) = P_{LM}(d|q) \cdot P_{LDA}(d|q) \cdot PR(d) \tag{18}$$

$$S_{BM25*PR}(d,q) = S_{BM25}(d,q) \cdot PR(d) \tag{19}$$

4.2 Experimental Results

In the experiments, parameters were set as follows: for the LDA model, we set the hyperparameters as $\alpha = 0.1$ and $\beta = 0.1$. The number of topics was set different values (5, 15, and 80). In our topic-level random walk, we set the random jump factor as $\lambda = 0.15$ and ranged the factor γ from 0 to 1.0, with interval 0.1. The combination weight t of LM and LDA (Eq.12) was tested with 0, 0.1, 0.2, 0.4, 0.6, 0.8, 1.0. We tuned the parameters and report the best performance.

Table 2 and Table 3 illustrate the experimental results of our approaches (TPR+, TPR*, and TPRq) and the baseline methods. We see that our proposed methods outperform the baseline methods in terms of most measures. The best performance is achieved by TPR+. We also note that although TPRq combines a query modeling process, it does not perform well as we expected. The problem might be the combination methods used in Eq.13. How to find a better combination is also one of our ongoing work.

4.3 Parameter Tuning

Tuning Parameter γ. Figure 1(a) shows the performance (in terms of MAP) of retrieving papers using TPR+ and TPR* with fixed values of the parameter

Table 2. Performance of retrieving papers

Model	p@5	p@10	p@20	R-pre	MAP	MRR	B-pre
LM	0.4286	0.4000	0.4000	0.1094	0.4876	0.6786	0.2732
LDA	0.0286	0.0429	0.0500	0.0221	0.1272	0.1180	0.0238
LM+LDA	0.4000	0.4714	0.4000	0.1561	0.5325	0.7659	0.2808
LM*LDA	0.4000	0.4714	0.4714	0.1605	0.5009	0.4728	0.3236
TPR	0.7143	0.5143	0.4857	0.2201	0.6311	0.8333	0.3687
BM25	0.4286	0.4571	0.4143	0.1196	0.4712	0.6000	0.2734
LM*PR	0.6571	**0.6714**	0.5214	0.1985	0.6541	0.7976	0.4055
LM+PR	0.4286	0.4286	0.4000	0.2100	0.4918	0.6905	0.3107
BM25*PR	0.6571	0.5429	0.4857	0.2094	0.6115	0.8571	0.3780
LM*LDA*PR	0.6571	0.6000	0.5071	0.2146	0.6404	0.7262	0.4091
TPR+	0.7143	0.5714	0.5286	0.2065	**0.7139**	**1.0000**	0.4193
TPR*	**0.7429**	0.6000	**0.5429**	**0.2255**	0.6961	0.8333	**0.4251**
TPRq	0.6286	0.6286	0.4929	0.1838	0.6235	0.7262	0.3874

Table 3. Performance of retrieving authors

Model	p@5	p@10	p@20	R-pre	MAP	MRR	B-pre
LM	0.4000	0.2714	0.1500	0.3274	0.5744	0.8095	0.3118
LDA	0.0857	0.0571	0.0286	0.0595	0.2183	0.2063	0.0506
LM+LDA	0.6571	0.4429	0.2500	0.5881	0.7347	0.8929	0.6012
LM*LDA	0.5714	0.3714	0.2429	0.5107	0.6593	0.8929	0.4926
TPR	0.3714	0.2286	0.1500	0.2964	0.3991	0.4976	0.2267
BM25	0.4857	0.2857	0.1500	0.4214	0.6486	0.7857	0.3775
LM*PR	0.6857	**0.5286**	**0.2786**	0.6357	0.7864	0.9286	**0.7183**
LM+PR	0.6571	0.4429	0.2500	0.5881	0.7338	0.8929	0.6012
BM25*PR	0.6286	0.4857	0.2714	0.6179	0.7392	**1.0000**	0.6526
LM*LDA*PR	0.6571	0.4857	0.2571	0.6655	0.7661	0.9048	0.6831
TPR+	**0.7143**	0.4857	0.2714	**0.7179**	**0.8029**	**1.0000**	0.7151
TPR*	**0.7143**	0.4857	0.2500	0.6512	0.7424	0.8571	0.6650
TPRq	0.6571	0.4571	0.2714	0.6179	0.7084	0.7857	0.6261

γ (ranging from 0 to 1.0 with interval 0.1), denoted as constant γ. We see that
for TPR+, the best performance was obtained at $\gamma = 1$ for retrieving papers.
We also tried to set γ as $P(z|d')$, denoted as variable γ. For retrieving authors
(as shown in Fig. 1(b)), we tuned the parameter with the setting as that for
retrieving papers. We see that the performance of retrieving authors is more
stable with different values for γ.

Tuning Parameter $|T|$. For tuning the parameter of topic number $|T|$, we
varied $|T|$ with 5, 15, 80. The results for retrieving papers are shown in Fig. 2(a),

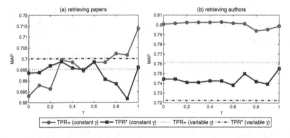

Fig. 1. MAP values of retrieving papers and authors (varying γ)

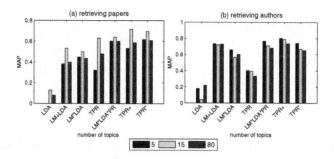

Fig. 2. MAP values of retrieving papers and authors (varying $|T|$)

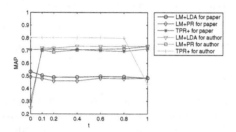

Fig. 3. MAP values of retrieving papers and authors (varying t)

and that for retrieving authors are shown in Fig. 2(b). For papers, we obtained the best performance when setting $|T|$ as 15, for almost all the methods. For authors, the best performance was obtained by setting the topic number as a smaller value 5. It seems that it is more difficult to distinguish the interest of authors than papers.

Tuning Parameter t. Shown in Fig. 3, we tuned the parameter t from 0 to 1, with interval 0.2. We see that the performances keep stable when setting t between 0.2 and 0.8.

4.4   Example Analysis

We give a case study to show the motivation of our work. We selected two queries ("natural language processing" and "intelligent agents") to analyze their semantics. For each word in the two queries, we selected the most representative topics for them, i.e. #4, #7, #10, #13 in Table 4. We can see from Table 4 that the query "natural language processing" prefers to topic #10 and "intelligent agents" prefers to topic #4 and #7.

Table 5 shows the importance scores calculated by PageRank and our topic-level random walk (TPR+). When using TPR+ to search "natural language processing", the first document *Verifiable Semantics for Agent Communication Languages* is not retrieved, since the importance score of Topic #10 is small, while the second document *Probabilistic Parsing Using Left Corner Language*

Table 4. Topic distribution for different query words

Query Word	Topic #4	Topic #7	Topic #10	Topic #13
natural	0.000018	0.000018	0.018966	0.000022
language	0.000018	0.002946	0.043322	0.000022
processing	0.000018	0.000018	0.012652	0.000022
intelligent	0.002363	0.022158	0.000023	0.000022
agents	0.037541	0.034784	0.000023	0.000022

Table 5. Importance scores of 4 documents by TPR+ and PageRank

Paper	TPR+				PageRank
	Topic #4	Topic #7	Topic #10	Topic #13	
Verifiable Semantics for Agent Commu- nication Languages	0.000113	0.000026	0.000007	0.000005	0.000612
Probabilistic Parsing Using Left Corner Language Models	0.000002	0.000002	0.000055	0.000014	0.000306
The GRAIL concept modelling language for medical terminology	0.000062	0.000052	0.000050	0.000037	0.003042
Agent-Based Business Process Manage- ment	0.000236	0.000179	0.000027	0.000029	0.002279

Models was retrieved. However, when using PagePank to search "natural language processing", the first document was retrieved due to its high score of PageRank, while the second document was not. This result indicates our approach is more reasonable than PageRank. When searching for papers using query "intelligent agents", the fourth document *Agent-Based Business Process Management* can be successfully extracted by our TPR+, while PageRank fails. PageRank chooses the third document *The GRAIL concept modelling language for medical terminology*, which is not retrieved by TPR+, though it has large number of out-links.

5   Conclusion

In this paper we investigate the problem of topic-level random walk. We propose a four-step approach for solving this problem. Specifically, we employ a probabilistic topic model to automatically extract topics from documents. We perform the random walk at the topic level. We also propose an approach to model topics of the query and combine the random walk ranking score with the relevance score based on the modeling results. Experimental results on a real-world data set show that our proposed approach can significantly outperform the baseline methods using language model and those of using traditional PageRank.

There are many potential directions of this work. It would be interesting to investigate how to extend our approach to a heterogeneous network, e.g., a social network consisting of users, documents, and communities. It would also be interesting to investigate how to integrate the user click-through data into the topic model.

References

1. Blei, D.M., Ng, A.Y., Jordan, M.I.: Latent dirichlet allocation. Journal of Machine Learning Research 3, 993–1022 (2003)
2. Buckley, C., Voorhees, E.M.: Retrieval evaluation with incomplete information. In: SIGIR 2004, pp. 25–32 (2004)
3. Craswell, N., de Vries, A.P., Soboroff, I.: Overview of the trec-2005 enterprise track. In: TREC 2005 Conference Notebook, pp. 199–205 (2005)
4. Griffiths, T.L., Steyvers, M.: Finding scientific topics. In: Proceedings of the National Academy of Sciences, pp. 5228–5235 (2004)
5. Haveliwala, T.H.: Topic-sensitive pagerank. In: Proceedings of the 11th international conference on World Wide Web (WWW 2002), pp. 517–526 (2002)
6. Hofmann, T.: Probabilistic latent semantic indexing. In: Proceedings of SIGIR 1999, pp. 50–57 (1999)
7. Kleinberg, J.M.: Authoritative sources in a hyperlinked environment. Journal of the ACM 46(5), 604–632 (1999)
8. Lavrenko, V., Croft, W.B.: Relevance-based language models. In: Proceedings of 24th Ann. Intl. ACM SIGIR Conf. on Research and Development in Information Retrieval, pp. 120–127 (2001)
9. Mei, Q., Cai, D., Zhang, D., Zhai, C.: Topic modeling with network regularization. In: Proceedings of WWW 2008, pp. 101–110 (2008)
10. Nie, L., Davison, B.D., Qi, X.: Topical link analysis for web search. In: SIGIR 2006, pp. 91–98 (2006)
11. Page, L., Brin, S., Motwani, R., Winograd, T.: The pagerank citation ranking: Bringing order to the web. Technical Report SIDL-WP-1999-0120, Stanford University (1999)
12. Richardson, M., Domingos, P.: The intelligent surfer: Probabilistic combination of link and content information in pagerank. In: NIPS 2002 (2002)
13. Robertson, S.E., Walker, S., Hancock-Beaulieu, M., Gatford, M., Payne, A.: Okapi at trec-4. In: Text REtrieval Conference (1996)
14. Rocchio, J.J.: Relevance feedback in information retrieval, pp. 313–323. Prentice Hall, Englewood Cliffs (1971)
15. Tang, J., Jin, R., Zhang, J.: A topic modeling approach and its integration into the random walk framework for academic search. In: ICDM 2008 (2008)
16. Tang, J., Zhang, J., Yao, L., Li, J., Zhang, L., Su, Z.: Arnetminer: Extraction and mining of academic social networks. In: KDD 2008, pp. 990–998 (2008)
17. Wei, X., Croft, W.B.: Lda-based document models for ad-hoc retrieval. In: SIGIR 2006, pp. 178–185 (2006)
18. Xue, G.-R., Zeng, H.-J., Chen, Z., Yu, Y., Ma, W.-Y., Xi, W., Fan, W.: Optimizing web search using web click-through data. In: CIKM 2004, pp. 118–126 (2004)
19. Zhai, C., Lafferty, J.: Model-based feedback in the language modeling approach to information retrieval. In: CIKM 2001, pp. 403–410 (2001)
20. Zhai, C., Lafferty, J.: A study of smoothing methods for language models applied to ad hoc information retrieval. In: Proceedings of SIGIR 2001, pp. 334–342 (2001)

Query-Focused Summarization by Combining Topic Model and Affinity Propagation[*]

Dewei Chen[1], Jie Tang[1], Limin Yao[2], Juanzi Li[1], and Lizhu Zhou[1]

[1] Department of Computer Science and Technology, Tsinghua University, China
{chendw,ljz}@keg.cs.tisnghua.edu.cn,
{jietang,ndcszlz}@tsinghua.edu.cn
[2] Department of Computer Science, University of Massachusetts Amherst, USA
lmyao@cs.umass.edu

Abstract. The goal of query-focused summarization is to extract a summary for a given query from the document collection. Although much work has been done for this problem, there are still many challenging issues: (1) The length of the summary is predefined by, for example, the number of word tokens or the number of sentences. (2) A query usually asks for information of several perspectives (topics); however existing methods cannot capture topical aspects with respect to the query. In this paper, we propose a novel approach by combining statistical topic model and affinity propagation. Specifically, the topic model, called qLDA, can simultaneously model documents and the query. Moreover, the affinity propagation can automatically discover key sentences from the document collection without predefining the length of the summary. Experimental results on DUC05 and DUC06 data sets show that our approach is effective and the summarization performance is better than baseline methods.

1 Introduction

Summarization is a common task in many applications. For example, in a Web search system, the summarization can return a concise and informative summary of the result documents for each query. A precise summary can greatly help users to digest the large number of returned documents in the traditional search engine.

Unlike the *generic* summarization task, which aims at extracting a summary about general ideas of document(s), *query-focused* summarization tries to extract a summary related to a given query. Therefore, the summarization results of different queries might be different, for the same document collection.

Previously, much effort has been made for dealing with this problem. For example, methods based on word frequency (e.g., TF) have been proposed to score each sentence and sentences with the highest scores have been extracted as the summary. However, the word-based methods tend to be overly specific in matching words of the query and documents. For example, it cannot discover semantic similarity between different terms such as "data mining" and "knowledge discovery". In addition, with a

---

[*] The work is supported by NSFC (60703059), Chinese National Key Foundation Research and Development Plan (2007CB310803), and Chinese Young Faculty Funding (20070003093).

Q. Li et al. (Eds.): APWeb/WAIM 2009, LNCS 5446, pp. 174–185, 2009.

query, the user may want to know information about different perspectives of the query, not just documents containing words in the query.

Therefore, how to discover topics, more accurately, how to discover the query-related topics from the document collection has become a challenging issue. Recently, probabilistic topic models have been successfully applied to multiple text mining tasks [18] [21] [22] [23] [24]. In a topic model, documents are represented in a latent topic space. Some methods also consider using the estimated latent topical aspects for each document to help extract the generic summary [2] [5] [12]. However, such methods are inapplicable to query-focused summarization, because the existing topic models do not consider the query information, thus the learned topic distribution is general on all documents.

From another perspective, how to determine the length of the summary is an open issue in all document(s) summarization tasks. It would be more serious in the query-focus summarization. This is because, with a query, the number of returned documents may vary largely (from a few documents to thousands of documents). Previously, the length of summary is predefined with a fix value (e.g., a fixed number of sentences/words). However, it is obviously not reasonable to specify a same length value to the summary of several document and that of thousands of documents.

In this paper, we aim at breaking these limitations by combining statistical topic models and affinity propagation for query-focused summarization. We first identify the major tasks of query-focused summarization and propose a probabilistic topic model called query Latent Dirichlet Allocation (qLDA). The qLDA represents each document as a combination of (1) a mixture distribution over general topics and (2) a distribution over topics that are specific to the query. Specifically, in the qLDA model, we treat the query as a kind of prior knowledge of documents. Thus the learned topic distribution of each document depends on the query. Based on the modeling results, we propose using an affinity propagation to automatically identify key sentences from documents.

Experimental results on the DUC data show that our proposed method outperforms the baseline methods of using term frequency and LDA. Experimental results also indicate that the length of the extracted summary by our approach is close to that of the human labeled summary.

The remainder of the paper is organized as follows. In Section 2, we formalize the major tasks in query-focused summarization. In Section 3, we present our approach and in Section 4 we give experimental results. We discuss related work in Section 5 and conclude the paper in Section 6.

2   Query-Focused Summarization

In this section, we first present several necessary definitions and then define the tasks of query-focused summarization.

We define notations used throughout this paper. Assuming that a query q consists of a sequence of words, denoted as a vector $\mathbf{w}_q$ and a document d consists of a sequence of N_d words, denoted as $\mathbf{w}_d$, where each word is chosen from a vocabulary of size V. Then a cluster of M documents can be represented as $C=\{\mathbf{w}_d | 1 \leq d \leq M\}$. Table 1 summarizes the notations.

Table 1. Notations (Notations w.r.t. the query is denoted with subscript q.)

Notation	Description
a, a_q	Dirichlet priors of the topic mixtures for document and query
β	Dirichlet priors of the multinomial word distribution for each topic
γ, γ_q	Beta parameters for generating λ
λ	Parameters for sampling switch x
x	Indicate whether a word inherits the topic from the query (x=0) or from the document (x=1)
θ, θ_q	The topic mixture of document and query
Φ	Word distribution for each topic
z, t	Topic assignment of document word and query word respectively
w, w_q	Document word and query word respectively
T	Number of topics
N_d, N_q	Number of words of a document and a query

Definition 1 (Summary and Key Sentence). *A **summary** is comprised of multiple **key sentences** extracted from the document cluster. The key sentence is also called center sentence in affinity propagation, which will be described in Section 3.4.*

Definition 2 (Query-focused Summarization). *Given a document cluster C and a query q, the task of query-focused summarization is to identify the most representative sentences that are relevant to the query, from the document cluster.*

The query represents the information need of the user. With different queries, the user asks for summaries of the document cluster from different perspectives. Following [3] [11] [18], we can define topical aspects of a document cluster as:

Definition 3 (Topical Aspects of Document Cluster). *A document cluster contains information of different aspects, each of which represents one topic. Hence, a document cluster can be viewed as a combination of different topics.*

For example, in people comments of "global climate change", the topical aspects may include "pollution", "disaster", "extinction of bio-species", etc. Discovery of the topical aspects from the document cluster can be helpful to obtain an overview understanding of the document cluster.

Definition 4 (Query-focused Topical Aspects of Document Cluster). *Different from general topical aspects, query-focused topical aspects mainly include topics related to the query.*

For the example of "global climate change", suppose the query is about the "impacts" of the global climate change, we may only want to highlight relevant topics such as "the sea level rise" and "the extinction of bio-species", and treat the other topics in the second place. Discovery of the query-focused topical aspects thus can be greatly helpful for extracting the query-focused summary.

Based on these definitions, the major tasks of query-focused summarization can be defined as:

1) Modeling query-focused topical aspects. Given a document cluster and a query, model the latent topic distribution for each document.
2) Identifying key sentences for the query from the document cluster. Based on the learned topic model, estimate the similarity between sentences and then apply affinity propagation to extract key sentences from the document cluster.

It is challenging to perform these tasks. First, existing topic models only consider the general topical aspects in the document cluster, but cannot capture the query-focused topical aspects. It is not clear how to balance the generality and specificity in a principled way. Second, it is unclear how to determine how many sentences should be included in the summary.

3 Our Approach

3.1 Overview

At a high level, our approach primarily consists of the following three steps:

1) We propose a unified probabilistic topic model to uncover query-focused topical aspects for each query-focused summarization task (i.e., each pair of query and document-cluster).
2) After building the topic model, we calculate similarities between sentences based on the modeling results. Then, affinity propagation has been applied to discover "centers" from these sentences.
3) We take the "center" sentences as the summary. The number of sentences in the summary is automatically determined while in the affinity propagation.

We see from above discussions that our main technical contributions lie in the first and second steps. In the first step, we use qLDA to capture the dependencies between document contents and the query, in other words, to obtain the *query-focused topical aspects* of the document cluster. In the second step, we propose using an affinity propagation to *identify the key sentences*. In the remainder of this section, we will first introduce the baseline method based on an existing topic model: LDA [3]. We will then describe our proposed topic model in detail and explain how we make use of the affinity propagation to identify the key sentences.

3.2 Baseline Model

Latent Dirichlet Allocation (LDA). LDA [3] is a three-level Bayesian network, which models documents by using a latent topic layer. In LDA, for each document d, a multinomial distribution θ_d over topics is first sampled from a Dirichlet distribution with parameter α. Second, for each word w_{di}, a topic z_{di} is chosen from this topic distribution. Finally, the word w_{di} is generated from a topic-specific multinomial $\phi_{z_{di}}$.

Accordingly, the generating probability of word w from document d is:

$$P(w \mid d, \theta, \phi) = \sum_{z=1}^{T} P(w \mid z, \phi_z) P(z \mid d, \theta_d) \tag{1}$$

For scoring a sentence, one can sum up the probability $P(w|d)$ of all words in a sentence.

3.3 Modeling Documents and Query

Our topic model is called query Latent Dirichlet Allocation (qLDA). The basic idea in this model is to use two correlated generative processes to model the documents and the query. The first process is to model the topic distribution of the query. The process is similar to the traditional topic model LDA [3]. The second process is to model the documents. It uses a Bernoulli distribution to determine whether to generate a word from a document-specific distribution or from a query-specific distribution.

Formally, the generative process can be described as follows:

1. For each topic t, draw the word distribution Φ_t from Dirichlet(β)
2. For each query q
   a) Draw a topic distribution θ_q from Dirichlet(α_q)
   b) For each word w_{qi} in query,
      i) Draw a topic t_{qi} from θ_q
      ii) Draw a word w_{qi} from Φ_{tqi}
3. For each document d
   a) Draw a topic mixture θ_d from Dirichlet(α)
   b) Draw the proportion between words associated with a specific query and those associated with its general topic
      $\lambda = p(x=0|d) \sim beta(\gamma_q, \gamma)$.
   c) For each word w_{di},
      i) Toss a coin x_{di} from λ with $x=bernoulli(\lambda)$
      ii) if $x=0$,
         a) Draw a topic z_{di} from θ_q
         b) Draw a word w_{di} from Φ_{zdi}
      iii) else
         a) Draw a topic z_{di} from θ_d
         b) Draw a word w_{di} from Φ_{zdi}

For inference in the qLDA model, the task is to estimate the two sets of unknown parameters: (1) the distribution θ of M document-topics, the distribution θ_q of query-topics, the distribution λ of M document-Bernoulli, and the distribution Φ of T topic-words; (2) the corresponding coin x_{di} and topic z_{di} for each word w_{di} in the document d, and the corresponding topic t_{qi} for each word w_{qi} in the query. We choose Gibbs sampling [9] for parameter estimation due to its ease of implementation. Additionally, instead of estimating the model parameters directly, we evaluate (a) the posterior distribution on just t and then use the results to infer θ_q and Φ for the first generative process; (b) the posterior distribution on x and z, and then use the sampling results to infer θ, λ and Φ for the second generative process. (Due to space limitation, we omitted the derivation of the posterior probability. Details can be referred to [22].)

During the Gibbs sampling process, the algorithm keeps track of an MxT (document by topic) count matrix, a TxV (topic by word) count matrix, an Mx2 (document by coin) count matrix, and a $|q|$xT (query by topic) count matrix. Given these four count matrices, we can easily estimate the probability of a topic given a query θ_{qt}, a topic given a document θ_{dt}, the probability of a word given a topic Φ_{zv} by:

$$\theta_{dz} = \frac{n_{dz} + \alpha_z}{\sum_{z'}(n_{dz'} + \alpha_{z'})}, \quad \phi_{zv} = \frac{n_{zv}^d + n_{zv}^q + \beta_v}{\sum_{v'}(n_{zv'}^d + n_{zv'}^q + \beta_{v'})}, \quad \theta_{qz} = \frac{n_{qz} + \alpha_{qz}}{\sum_{z'}(n_{qz'} + \alpha_{qz'})} \quad (2)$$

where n_{dz} (or n_{qz}) is the number of times that topic z has been sampled from the multinomial specific to document d (or query q); n_{zv} is the number of times that word w_v has been sampled from the multinomial specific to topic z; the superscript "q" denotes that we count the numbers in all the queries.

Query Expansion. Sometimes the query may be so short that the qLDA model cannot yield good results. We employ a method to expand the query. Specifically, we preprocess a query by word stemming and stop-words filtering. After that, we obtain a sequence of query words w_1, w_2, ..., w_n. For each query word w_i, we extract a set of co-occurring words from the document cluster and add them into the query. We consider words appearing in a window-size of the word w_i as its co-occurrence words, i.e. words before and after the word w_i. We set the window size as 1.

3.4 Key Sentences Extraction

We use affinity propagation [8] to discover key sentences from the document cluster. Specifically, we first employ the learned topic model from the above step to calculate an asymmetric similarity between sentences; we then use all sentences and similarities between them to generate a graph; next, we perform propagations for links on the graph. The basic idea is that if a sentence has a high "representative" score for other sentences, then the sentence is likely to be selected as a key sentence in the summary. In the rest of this section, we will introduce how we generate the sentence graph and describe how we conduct the propagation on the graph.

Sentence Graph Generation. From the topic modeling step (cf. Section 3.3), we obtain a topic distribution for each sentence. Based on the distribution, we calculate a similarity matrix between sentences. Specifically, we define an asymmetric similarity between two sentences according to the negative *KL*-divergence [14]:

$$sim(S_1, S_2) = -KL(S_1, S_2) = -\sum_{z=1}^{T} p(z \mid S_1) \cdot \log_2 \left(\frac{p(z \mid S_1)}{p(z \mid S_2)} \right) \tag{3}$$

where $p(z|S_1)$ is the probability of topic z sampled from sentence S_1.

Another optional input of the affinity propagation is a "preference" score for each sentence, which can be taken as the priori information of how likely a sentence can be selected as the key sentence. Note that the input is optional, which provides a flexible way to integrate the prior information for summarization, thus the method has the flavor of semi-supervised learning. In this work, we do not combine the human provided prior information. We initialize the "preference" score of each sentence by its similarity with the query, i.e.,

$$pref_i = \lambda \cdot sim(S_i, q) + (1 - \lambda) \cdot \frac{1}{N-1} \sum_{k \neq i} sim(S_i, S_k) \tag{4}$$

where N is the number of sentences, $sim(S_i, q)$ is calculated in a similar form with Equation (3) and it represents the similarity between sentence S_i and the query q according to the topic model; λ is a parameter to balance the two terms and it is empirically set as 0.5.

We combine the preference scores into the similarity matrix. They are put in the diagonal of the matrix. The combined matrix is the input of propagation, denoted as $s(i,k)$.

Following [8], we define two types of links between sentences "responsibility" and "availability". The former, denoted as $r(i, k)$, pointed from sentence i to k, represents how well-suited sentence k is to serve as the center of sentence i, by considering the potential centers for sentence i. The latter ("availability"), denoted as $a(i, k)$, pointed from sentence k to i, represents how appropriate the sentence i to choose sentence k as the center, by considering for what sentences that sentence k can be the center. Fig. 1. shows an example of the generated sentence graph.

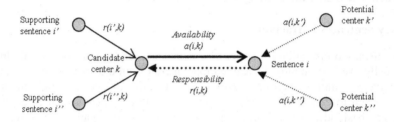

Fig. 1. The generated sentence graph with two types of links

Affinity Propagation. Different from other propagation algorithms (e.g. PageRank [20]), which usually propagate scores of nodes along with links on a graph; the affinity propagation used in this paper propagates score of links. Two scores, i.e. $r(i, k)$ and $a(i, k)$, are propagated on the graph. The propagation runs in an iterative process (as shown in algorithm 1).

Algorithm 1. affinity propagation

Input: Similarities and preferences: $s(i,k)$

Algorithm:

1. Initialize availabilities to zero: $a(i,k) = 0$

2. do{

   a. update responsibilities:
   $$r(i,k) = \lambda \cdot r(i,k) + (1-\lambda) \cdot \begin{cases} s(i,k) - \max_{k' \neq k}\{a(i,k') + s(i,k')\} & (i \neq k) \\ s(i,k) - \max_{k' \neq k}\{s(i,k')\} & (i = k) \end{cases}$$

   (λ is a damping factor used to avoid numerical oscillations.)

   b. update availabilities:
   $$a(i,k) = \lambda \cdot a(i,k) + (1-\lambda) \cdot \begin{cases} \min\{0, r(k,k) + \sum_{i' \notin \{i,k\}} \max\{0, r(i',k)\}\} & (i \neq k) \\ \sum_{i' \neq k} \max\{0, r(i',k)\} & (i = k) \end{cases}$$

   } while some terminal condition is satisfied.

* **Terminal condition:** the maximum number of iterations reached, or the relationship of data points kept stable for some iterations.

After affinity propagation, we identify key sentences by combining the "availability" and the "responsibility" scores. For example, for sentence i, the sentence k with $max_k(a(i,k)+r(i,k))$ will be identified as its "center", thus key sentence. In this way, key sentences are selected without a predefined length of the summary.

4 Experiments

4.1 Experimental Setup

Data Sets. Document Understanding Conference (DUC), now moved to Text Analysis Conference (TAC), provides a series of benchmarks for evaluating approaches for document summarization. We conducted experiments on data sets of DUC2005 and DUC2006. Each of them contains 50 document clusters, accordingly 50 query-focused summarization tasks. Documents are from Financial Times of London and Los Angeles Times. For each summarization task, given a query and a document cluster containing 25-50 documents, the objective is to generate a summary for answering the query. (In DUC, the query is called "narrative" or "topic").

Evaluation measures. We conducted evaluations in terms of ROUGE [17]. The measure evaluates the quality of the summarization by counting the numbers of overlapping units, such as n-grams, between the automatic summary and a set of reference summaries (or human labeled summaries in DUC). In our experiment, we calculated the macro-average score of ROUGE-N on all document sets. We utilized the tool ROUGE 1.5.5 with the parameter setting "-n 4 -w 1.2 -m -2 4 -u -c 95 -r 1000 -f A -p 0.5 -t 0 -a".

Summarization methods. We define the following summarization methods:

TF: it uses only term frequency for scoring words and sentences. The basic idea is to use terms in the query to retrieve the most relevant sentences. The method is similar to [17], except that [17] also adjusts the scoring measure to reduce redundancy.

    LDA + TF: it combines the LDA based scores with the TF based scores.
    qLDA + TF: it combines the qLDA based scores with the TF based scores..
    LDA+AP: it combines the LDA model with affinity propagation.
    qLDA+AP: it combines the qLDA model with affinity propagation.

For both LDA and qLDA, we performed model estimation with the same setting. The topic number is set as $T=60$. It is determined by empirical experiments (more accurately, by minimizing the perplexity). One can also use some solutions like [4] to automatically estimate the number of topics.

4.2 Results on DUC

All experiments were carried out on a server running Windows 2003 with two Dual-Core Intel Xeon processors (3.0 GHz) and 4GB memory. For each dataset, it needs about 55 minutes for estimating the qLDA model. For each query, it needs about 40 seconds to identify the key sentences (including the I/O time). Because the qLDA model estimating results will be shared for all summarization tasks, and only the time spent on identifying the key sentences is concerned for online applications, then if we do qLDA modeling offline and cache the model parameters to save the I/O time, our solution will be efficient enough for online summarizations.

Table 2 shows the performance of summarization using our proposed methods (*T=60*) and the baseline methods on the DUC 2006 data set. We see from Table 2 that, in terms of Rouge1, qLDA+AP results in the best performance; while in terms of Rouge2 and RougeSU4, qLDA+TF obtains the highest score.

Table 2. Results of the five methods on DUC 2006

System	TF	LDA+TF	qLDA+TF	LDA+AP	qLDA+AP
Rouge1	.36956	.37043	.37571	.39795	**.40433**
Rouge2	.07017	.06862	**.07144**	.06540	.06468
RougeSU4	.12721	.12710	**.13015**	.12793	.12893

We also compared our results with the three best participant systems in DUC2005 and DUC2006. Table 3 shows the evaluation results.

Table 3. Comparison with participant systems on DUC2005 (left) and DUC2006 (right)

System	Rouge1	Rouge2	RougeSU4	System	Rouge1	Rouge2	RougeSU4
System15	.37469	07020	**.13133**	System24	.41062	**.09514**	.15489
System4	.37436	.06831	.12746	System15	.40197	.09030	.14677
System17	.36900	.07132	.12933	System12	.40398	.08904	.14686
qLDA +TF	.37571	**.07144**	.13015	qLDA +TF	.40410	.08967	**.15800**
LDA+AP	.38104	.05079	.11798	LDA+AP	.39795	.06540	.12793
qLDA+AP	**.39089**	.05196	.12075	qLDA+AP	**.41433**	.06468	.12893

We see that our proposed approaches are comparable with the three best systems. In terms of Rouge1, qLDA+AP achieves the best results (0.39089 and 0.41433) on both data sets. In terms of Rouge2, qLDA+TF achieves the best performance (0.07144) in DUC2005. In terms of RougeSU4, qLDA+TF achieves the best perform-ance (0.15800) in DUC2006. We need to note that our approach does not make use of any external information; while the systems usually (heavily) depend on some exter-nal knowledge, for example System 15 [25] and System 17 [15]. We can also see that qLDA+AP performs better than LDA+AP on almost all measures, which confirms us that qLDA for the query-focused summarization task is necessary.

We also evaluated the length of summary by comparing the sentence numbers ex-tracted by different methods with that in the referred (human labeled) summary. Table 4 shows the comparison result. We see that the sentence number extracted by our proposed approach is closer to that of the referred summary.

We conducted a detailed analysis on the summarization task D307 (DUC2005), which talks about "new hydroelectric projects". The query is "what hydroelectric

Table 4. Comparison of the length (number of sentences) of summary (DUC2006)

Tasks	System12	System15	System24	qLDA+AP	Average of manual results
D0601	9	12	11	17	17
D0602	8	10	10	17	13
D0603	11	11	9	14	14.5
D0604	9	14	14	15	15.5
D0638	7	12	11	14	15.75
D0646	10	12	10	16	15.5

projects are planned or in progress and what problems are associated with them?". Four human summaries (A, B, C, and D) are given for this task. They mainly focus on problems of projects, like environmental problems, social problems, and so on. The summary obtained by TF contains many sentences about the financial problem and the power of the projects. It does not cover so many problems. Our summary covers more aspects. Table 5 shows some sentence samples.

Table 5. Some sentences from summaries of D307 (DUC2005)

Summary obtained by TF	The Pangue dam, already under construction, is likely to cost Dollars 450m.
	However, it needs foreign capital to complete even the first dam.
	If built, it would be by far the largest electrical project in the world financed by the private sector.
	The first dam alone will produce almost as much power as Nepal's existing generating stations which have a capacity of just 230MW.
	The dam will cost 1.13m people their homes by the time the project nears completion in the year 2008.
Summary obtained by qLDA+AP	A new 50,000-line telephone exchange is being installed.
	Tenaga Nasional says the 600MW dam is a vital component in its plan to double Malaysia's generating capacity over the next decade.
	The fear was that carcinogenic sediment will build up in the reservoir and pollute the falling water table.
	Ankara's control of the Euphrates and the Tigris rivers, both of which rise in its central highlands and flow into the Gulf, has long soured relations with its Arab neighbours.
	This impasse was sorely tested towards the end of last year when the Qeshm island Free Trade Authority, flagship of Iran's free market experiment, signed a letter of agreement with Siemens for the finance and construction of the first of four 250MW CCGT units.

Fig. 2. shows a snippet of the sentence graph after affinity propagation on the summarization task D0641 (DUC2006). Circles with the gray color are selected as key sentences. The sentences are listed in the right table.

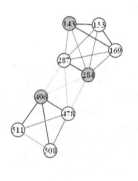

ID	Sentence
153	Except that there was no global cooling.
143	The declaration, plus a wealth of information on every aspect of the global warming controversy, is posted at the Web site of the Science
169	What we're saying is that you can understand much of the observed climate change in terms of that picture," said Dr. Tim Palmer
287	In the next decade, we might see the same amount of warming that we saw over the last century.
284	Researchers are increasingly convinced that carbon dioxide gas from the burning of fossil fuels is at least partly responsible for the marked warming of recent years.
496	Higher temperatures and severe water shortages will bring about a rise in winter wheat production, but a drop in spring wheat and corn.
478	Global warming is happening all the time,"
511	Countries attending the meeting, hosted by South Africa, include Angola, Lesotho, Malawi, Mauritius, Mozambique,...
508	It is more and more a matter which is impacting on all people.

Fig. 2. A snippet of the sentence graph. The dark nodes denote 3 centers.

5 Related Work

Most of extraction-based document summarization methods aim to rank sentences by different scoring schemes and select ones with the highest scores as summaries. Features such as term frequency [19] [26], cue words, stigma words, topic signature [16] and topic theme [10] are used to calculate the importance score of words. A composition function is then utilized to score the sentences.

Nenkova et al. [19] uses term frequency as the important factor for choosing words to be included in summaries and utilize an average sum of the word score to compute the score of a sentence.

Conroy et al. [6] propose an oracle score based on the probabilistic distribution of unigrams of human summaries. They utilize the query words and the topic signature terms to approximate the probabilistic distribution. Our qLDA model is relevant to this model to some sense, except that this model does not consider the topical aspects in the document cluster.

Over the last several years, several methods have been proposed for documents summarization using latent topics in the document cluster. For example, Barzilay and Lee [1] use the Hidden Markov Model to learn a V-topic. Their method requires a labeled training data to learn the probability. Topic models like pLSI has been applied to spoken document summarization [12] and text document summarization [2]. The word topical mixture model [5] in spoken document summarization resembles LDA. However, all of the aforementioned topic-based methods do not consider the query information or integrate the query information in a heuristic way.

More recently, some research effort has been made for incorporating the query information into the topic model. For example, Daumé and Marcu [7] propose a hierarchical Bayesian model to compute the relevance of a sentence to a query. However, the method needs a predefined length for the generated summary. Kummar et al.[13] proposed a method to uncover storylines of multi-documents based on graph theoretic. But the results they get are bags of words instead of natural sentences.

Frey and Dueck [8] propose the notion of affinity propagation, which can cluster data points without setting the number of clusters. Inspired by this idea, we propose the method for query-focused summarization by combining a query-focused topic model and the affinity propagation. Different from existing methods, our method can discover multiple topics related to the query. Moreover, our method can automatically discover the number of sentences for each summarization task.

6 Conclusion

In this paper we investigate the problem of query-focused summarization on multiple documents. We propose a novel method to deal with the problem by combining a statistical topic model (i.e. qLDA) with the affinity propagation. The method can automatically identify key sentences from the document cluster without a predefined summary length. Experiments on DUC data show that our proposed method can outperform the general topic-based methods as well as a word-based method (TF) for query-focused summarization tasks. Experiment results also indicate that the length of the summary extracted by our approach is close to that of the referred summary.

The proposed solution is quite general. It can be applied to many other applications, such as question answering, clustering information retrieval results. There are many potential future directions of this work. One potential issue is how to obtain more accurate prior information, in particular, query expansion when the query is short. In addition, the similarity between the query and each sentence can be further considered during the affinity propagation process.

References

1. Barzilay, R., Lee, L.: Catching the drift: probabilistic content models, with applications to generation and summarization. In: Proceedings of HLT-NAACL 2004 (2004)
2. Bhandari, H., Shimbo, M., Ito, T., Matsumoto, Y.: Generic text summarization using probabilistic latent semantic indexing. In: Proceedings of IJCNLP 2008 (2008)

3. Blei, D., Ng, A.Y., Jordan, M.I.: Latent Dirichlet Allocation. JMLR 3, 993–1022 (2003)
4. Blei, D., Griffiths, T., Jordan, M., Tenenbaum, J.: Hierarchical topic models and the nested Chinese restaurant process. In: Proceedings of NIPS 2004 (2004)
5. Chen, B., Chen, Y.: Word Topical Mixture Models for Extractive Spoken Document Summarization. In: Proceedings of ICME 2007 (2007)
6. Conroy, J., Schlesinger, J., O'Leary, D.: Topic Focused Multi-document Summarization Using an Approximate Oracle Score. In: Proceedings of ACL 2006 (2006)
7. Daumé III, H., Marcu, D.: Bayesian Query-Focused Summarization. In: Proceedings of ACL 2006 (2006)
8. Frey, B., Dueck, D.: Clustering by passing messages between data points. Science 315(5814), 972–976 (2007)
9. Griffiths, T., Steyvers, M.: Finding scientific topics. In: Proceedings of NAS, pp. 5228–5235 (2004)
10. Harabagiu, S., Lacatusu, F.: Topic Themes for Multi-Document Summarization. In: Proceedings of SIGIR 2005 (2005)
11. Hofmann, T.: Probabilistic latent semantic indexing. In: Proceedings of SIGIR 1999 (1999)
12. Kong, S., Lee, L.: Improved Spoken Document Summarization Using Probabilistic Latent Semantic Analysis (PLSA). In: Proceedings of ICASS 2006 (2006)
13. Kumar, R., Mahadevan, U., Sivakumar, D.: A graph-theoretic approach to extract story-lines from search results. In: Proceedings of KDD 2004, pp. 216–225 (2004)
14. Kullback, S., Leibler, R.A.: On information and sufficiency. Annals of Mathematical Statistics, vol. 22, pp. 79–86 (1951)
15. Li, W., Li, W., Li, B., Chen, Q., Wu, M.: The Hong Kong Polytechnic University at DUC2005. In: Proceedings of DUC 2005 (2005)
16. Lin, C., Hovy, E.: The Automatic Acquisition of Topic Signatures for Text Summarization. In: Proceedings of COLING 2000 (2000)
17. Lin, C., Hovy, E.: Automatic evaluation of summaries using N-gram co-occurrence statistics. In: Proceedings of HLT-NAACL 2003 (2003)
18. Mei, Q., Ling, X., Wondra, M., Su, H., Zhai, C.: Topic Sentiment Mixture: Modeling Facets and Opinions in Weblogs. In: Proceedings of WWW 2007 (2007)
19. Nenkova, A., Vanderwende, L., McKeown, K.: A compositional context sensitive multi-document summarizer: exploring the factors that influence summarization. In: Proceedings of SIGIR 2006 (2006)
20. Page, L., Brin, S., Motwani, R., Winograd, T.: PageRank Bringing Order to the Web. Stanford University (1999)
21. Steyvers, M., Smyth, P., Griffiths, T.: Probabilistic author topic models for information discovery. In: Proceedings of SIGKDD 2004, pp. 306–315 (2004)
22. Tang, J., Yao, L., Chen, D.: Multi-topic based query-oriented summarization. In: Proceedings of SDM 2009 (2009)
23. Tang, J., Zhang, J., Yao, L., Li, J., Zhang, L., Su, Z.: ArnetMiner: extraction and mining of academic social networks. In: Proceedings of SIGKDD 2008, pp. 990–998 (2008)
24. Wei, X., Bruce Croft, W.: LDA-based document models for Ad-hoc retrieval. In: Proceedings of SIGIR 2006 (2006)
25. Ye, S., Qiu, L., Chua, T., Kan, M.: NUS at DUC2005: Understanding documents via concept links. In: Proceedings of DUC 2005 (2005)
26. Yih, W., Goodman, J., Vanderwende, L., Suzuki, H.: Multi-document summarization by maximizing informative content-words. In: Proceedings of IJCAI 2007 (2007)

Rank Aggregation to Combine QoS in Web Search

Xiangyi Chen and Chen Ding

Department of Computer Science, Ryerson University, Canada
x26chen@ryerson.ca, cding@scs.ryerson.ca

Abstract. In this paper, we consider the quality of the page delivery as an important factor in the page ranking process. So if two pages have the same level of relevancy to a query, the one with a higher delivery quality (e.g. faster response) should be ranked higher. We define several important quality attributes and explain how we rank the web page based on these attributes. The experiment result shows that our proposed algorithm can promote the pages with a higher delivery quality to higher positions in the result list. We also compare several rank aggregation algorithms trying to find out the best performing one.

1 Introduction

Issues related to Quality of Service (QoS) have been addressed in telecommunication, networking and multimedia delivery areas, and recently in web service area. Some of the common QoS attributes include response time, throughput, reliability, availability, reputation, cost, etc. In a web search engine, when a user submits a query, there might be millions of web pages relevant to the query. In most of popular search engines, the results are ranked based on their qualities, and the most commonly used and effective quality attributes are authority, popularity and connectivity, which in a way all reflect the reputation to a certain extent. Usually the quality attributes that are related to page delivery performance (e.g. response time, reliability) are not considered at all. However, they are also very important factors to determine a user's searching experiences. For instance, if the top ranked pages in the result list respond very slowly when users want to access them, or couldn't be displayed correctly, or the hosting web site crashes when too many people try to access at the same time, it will impact users' searching experiences negatively.

If we take response time as an example, when two pages have similar content, page A has a faster response time than page B, definitely users would prefer A than B if the difference is obvious. When the network is fast and the hosting web site is fast, response time is normally not a concern at all. But if the hosting web site is very slow or the Internet user has a very slow connection, then it could be an issue. It is the purpose of this paper to take the delivery-related quality attributes into consideration to rank the search results, and we believe that it will be very beneficial to end users to improve their overall Internet experiences. When we say a page has a high quality, it means it has the content highly relevant to queries, it has a high level of reputation, and it has a good quality of delivery.

In the paper, we define how we can measure and calculate the delivery QoS, and how we can rank web pages based on QoS. The basic idea is that if two pages are

Q. Li et al. (Eds.): APWeb/WAIM 2009, LNCS 5446, pp. 186–197, 2009.
© Springer-Verlag Berlin Heidelberg 2009

both relevant to a query with the same degree of relevancy, since usually the user prefers the page with a low response time, the page with a lower response time will be ranked higher than the other page. The same principle can be applied to other QoS attributes.

Since the content, reputation and QoS are all important to decide the page ranking, it is necessary to find a way to combine them together. In this study, both content and reputation based ranking are obtained from the meta search engine, in which the numeric relevance score for each page is not available, and QoS based score is calculated by our proposed QoS-based ranking algorithm. Usually there are two ways to do the combination, one is to combine the scores, and the other is to combine the ranks. We choose the rank aggregation as our way to combine them for several reasons: firstly, score is not available when we get the result from an existing search engine; secondly, if we convert a rank to a score, it would be a skewed value different from the original score; thirdly, the converted score would not be in line with the QoS values because the ways to calculate them are different.

In the experiment, we compare several popular rank aggregation algorithms to see which one has the best performance to combine QoS. In order to measure the performance, we consider both the improvement of QoS of the top ranked pages and the loss of precision as compared to the original top ranked pages from the meta search engine. We find that the median rank aggregation algorithm [7] is the best to improve the QoS, and its loss of precision is the second least, and overall, it is the best performing algorithm.

The rest of the paper is organized as follows. Section 2 reviews the related works in several areas. Section 3 defines the delivery-related quality attributes and how to measure them, and it also describes the QoS-based ranking algorithm. Then in section 4, we explain the rank aggregation process and the system architecture. In section 5, we explain our experiment design and analyze the results. Finally in section 6, we conclude the paper.

2 Related Works

As indicated in a number of research works [12], every search engine crawls different part of the web and thus has a different web page collection, and also indexes and ranks differently. As a consequence, meta search engines become popular because they are believed to be able to have a wider coverage of the web and provide a more accurate result than each of the individual search engines. When a user submits a query to a meta search engine, it will issue the same query to n search engines and get back n ranked lists, and then generate a single ranked list as the output, which is expected to outperform any of the n lists. There are two types of meta search algorithms. One is called the rank aggregation algorithm [1] [2] [6] [7] [10] which only requires the ranks from each search engine, and the other is the score combination algorithm [4] which requires the knowledge of relevance scores for each returned result. For the rank aggregation, there are works on unsupervised learning approach requiring no training data and supervised learning approach with the training data [9]. Since the score information is usually unavailable when building meta search engines, and it is also not easy to get the training data when doing aggregation, we choose to use the unsupervised rank aggregation method in this study. In this section, we only review those algorithms that we have tested and compared in the experiment.

Borda-fuse [1] [6] and Condorcet-fuse [10] are two commonly used rank aggregation algorithms. Borda-fuse (or Borda Count) was proposed to solve the problem of voting. It works as follows. Suppose there are n search engines, given a query, each of them will return a set of relevant pages, and suppose altogether there are c pages in the result set. For each search engine, the top ranked page is given c points, the second ranked page is given $c-1$ points, and so on. If one page is not ranked by a search engine, the remaining points are divided evenly among all the unranked pages. Then we could calculate the total points earned by each page, and rank all the c pages in the descending order. It is a very simple procedure, but it has been proven to be both efficient and effective. Weighted Borda-fuse was also proposed in [1], which assigns a weight to each search engine based on its average precision, and its performance is better than Borda-fuse.

Condorcet-fuse is another major approach to solve the voting problem. The basic idea is that the Condorcet winner should win (or ties in) every possible pair-wise majority contest. If we need to compare two pages A and B, the initial value of a counter is set to zero, then for each search engine, if it ranks A higher than B, counter will be incremented, otherwise, it will be decremented, after the iteration over all search engines, if the counter is a positive value, A will be ranked better than B, otherwise B better than A. By going through all pages in the result set, a ranked order is obtained. In the weighted Condorcet-fuse, a weight is assigned to each search engine.

Dwork et al. [6] explains that extended Condorcet criterion (ECC) has the excellent anti-spamming properties when used in the context of the meta search, and this target can be achieved by their proposed Kendall-optimal aggregation method with several heuristic algorithms to improve the efficiency. Since Kendall-optimal aggregation itself is a NP-complete problem, footrule-optimal aggregation is a good approximation and is more efficient. Fagin et al. [7] uses a heuristic called median rank aggregation for the footrule-optimal aggregation. The heuristic is to sort all objects based on the median of the ranks they receive from the voters. They also propose a very efficient algorithm MEDRANK to median rank aggregation, in which all the ranked lists are accessed one page at a time, the winner is the candidate that is seen more than half the lists, and the whole process is continued to find all top pages. By applying the genetic algorithm, Beg and Ahmad [2] try to optimize the footrule distance when only partial lists are available. It achieves the better performance although it takes longer time, and the progressive result presentation can compensate it. They also test the fuzzy rank ordering and several other soft computing techniques.

QoS has been an active research topic in several areas. Our review focuses on the web service area because how they use QoS in web services is similar to how we can use QoS for web search. In [11], Ran described a list of QoS attributes and how they can be used in web service discovery. The QoS attributes are categorized into four types: runtime related, transaction support related, configuration management and cost related, and security related. It is quite a complete list, and we define the search-related QoS attributes based on this list. We mainly focus on the first type because other types are not direct concerns of search engine users. Usually, QoS is a matter of personal preference or restrictions from the user device. Different users have different expectations on QoS and their ratings on the service will be closely related to their expectations [9]. So in this work, we only consider QoS attributes that users prefer. In a number of other works such as [13], several common QoS attributes they choose include reliability, availability, response time, cost, etc.

3 QoS-Based Ranking

3.1 Delivery-Related QoS Attributes

There are many QoS attributes that have been proposed in the communication area or web service area, but not all of them are applicable to the web search. Based on our observation on the average user behavior in the web searching process, we identify a list of QoS attributes that we believe are quite relevant to the web search. In this study, we only focus on the quality attributes that are used to measure the delivery performance.

The delivery-related QoS attribute measures how well a page is delivered to the end user. Users normally want a page with a good delivery speed, a high reliability and availability, and from a web site that can handle a large volume of concurrent requests. Sometimes users need to balance between the speed and the rich multimedia experience. If a page has lots of embedded multimedia or flash objects, definitely the speed is slower compared with a page with only text information. In this study, we divide the QoS attributes into two types, one is the standard attribute, on which all users have the same requirements, and the other is the personalized attribute, which are based on user's personal preference or the capacity of the user device.

We only choose several most important and representative delivery-related quality attributes in the paper. For the standard QoS attributes, we choose reliability and response time. The ability to handle a big volume of requests is not considered because it would be hard to measure it without the cooperation from the web site owner. For the personalized QoS attributes, we choose the file size and the number of embedded multimedia objects. The file size and multimedia objects can be reflected in the response time to a certain extent. The reasons we consider them as separate QoS attributes are twofold: firstly, response time is determined not only by the file size, but also by the network speed, the volume of traffic, the stability of the web server, etc; secondly, the file size and multimedia objects can contribute to other quality attributes too, such as the cost of the Internet access. Also compared with the response time, they are more of a personal preference.

We would not give an exhaustive list of all QoS attributes, but our system framework would be flexible and extensible enough so as to easily include new QoS attributes. The granularity of these attributes can be adjusted if necessary. For example, if a user does not like any flash file in a web page, the multimedia objects can be divided into several types depending on the media types. The data type of these QoS attributes can also be adjusted. Currently, all of them are continuous numeric values, but they could also be a discrete or categorical value. For instance, instead of using a numeric value, the reliability can have three ranges: low, medium, and high. Below we explain the four QoS attributes we choose, and how we measure and calculate them.

Reliability. It measures how reliable a web site is and whether it could deliver a page without error. It is defined as the percentage of successful delivery among all the requests to this page. In order to measure it, we make a number of requests to the same web page during different time of a day and different days of a week. Each time, if the HTTP response code is 2** [8], it is considered as a successful delivery, or otherwise, if it is a server error, a client error, or a page move, it means a failure. Whether a delivery is successful is also dependent on the location, a request from one location might be successful, whereas the request for the same page from another location, even at the same time, might be unsuccessful. So ideally, we should also send requests from different locations in order to get a more accurate value. Currently, because of the restriction of the experiment, we only measure from one location.

Response time. It measures how soon a user can see the web page after a request is submitted to the web server. It is defined as the time difference between the time when a user sends a request and the time when the user receives the response. Since the network speed is not a constant number and the web server is not in the same state all the time because of the change of the incoming traffic and the internal status of the server itself, the response time may vary for the same page when it is accessed at different times. In order to more accurately measure the response time, we submit the request to the same page several times, and get the average value. Same as the reliability, currently, we only measure it from one location.

File size. It measures the size of the web page itself excluding all the embedded objects. It can be obtained from the HTTP header or by downloading the page.

Media richness. It measures the number of embedded multimedia objects within a web page.

Our selected QoS attributes are especially important for mobile users or users with slow Internet connections. For this type of users, the first two directly affect the satisfaction level of their Internet experiences, and the latter two determine the cost of their Internet access. Although file size and media richness are both based on personal preferences, for this particular type of users, we assume that they prefer a smaller file size and a smaller number of multimedia objects, which we believe is reasonable assumption.

3.2 QoS-Based Ranking Algorithm

For any given page, we calculate all four QoS attributes based on the previous definitions, and normalize the values to (0, 1) range using a simple normalization method. The reliability value is in (0, 1) range already, and thus we don't do the normalization for this attribute. For the response time, because the page with a lower response time should be ranked higher, the normalization formula is as below:

$$NQoS_{ij} = \frac{Max_i(QoS_{ij}) - QoS_{ij}}{Max_i(QoS_{ij}) - Min_i(QoS_{ij})} \tag{1}$$

Where QoS_{ij} is the value of the j-th QoS attribute for page i, $NQoS_{ij}$ is the normalized value of QoS_{ij}, Min and Max are operations to get the minimum and maximum values among all pages. In the current system, j is a value from 1 to 4 because we have only 4 QoS attributes, and it could be expanded later.

For the other two attributes (file size and media richness), because they are preference related, the page with a lower value could be ranked higher or lower depending on user's preference. If we take media richness as an example, when a user prefers a lower value on media richness, the normalization formula is the same as the previous one, when a user prefers a higher value, we would use a different formula as shown below:

$$NQoS_{ij} = \frac{QoS_{ij} - Min_i(QoS_{ij})}{Max_i(QoS_{ij}) - Min_i(QoS_{ij})} \tag{2}$$

In the current experiment, we assume a lower value is preferred on these two attributes. After the normalization step, we do the linear combination of the four normalized values, and each page would have an overall QoS value.

$$QoS_i = \sum_j w_j \cdot NQoS_{ij} \tag{3}$$

Where QoS_i is the overall QoS value of page i, and w_j is the weight of the j-th QoS attribute. The value of this weight is set between 0 to 1, 0 means that a user does not care about this QoS attribute, 1 means the QoS attribute is important to the user, any value in between could indicate the relative importance of the QoS attribute, and currently we only consider two borderline cases – 0 and 1.

4 Rank Aggregation

After we define the search-related QoS attributes and QoS-based ranking, we need to decide how we can integrate them into the final ranking score. The content of a page, whether it is relevant to a query, and whether it has a high quality based on its linkage connectivity or its click rate, are still the main sources in the page ranking process. Since most of search engines consider all of them, by building a meta search engine based on those existing search engines, we can deploy them directly, and then QoS-based ranking can be combined with the ranking from the meta search engine using the rank aggregation methods.

The first step is to build the meta search engine. We choose the three most popular search engines – Google, Yahoo and MSN Search as the underlying search engines. Since only the rank information is available from each of these search engines, we decide to use the rank aggregation method to combine the three ranked lists. We choose Condorcet-fuse as the rank aggregation method for the meta search engine in order to get a single ranked list.

Usually users won't go beyond the first one or two pages when browsing the search results. Therefore, we only download the top-ranked n pages, and calculate their normalized QoS values. Then we rank these n pages based on each QoS attribute and hence get four rankings. Whether a QoS attribute is included in the final ranking is determined by the user preference. Finally we apply the rank aggregation on the original ranking from the meta search engine and the user preferred QoS rankings to get the final ranked list. Figure 1 shows the system architecture.

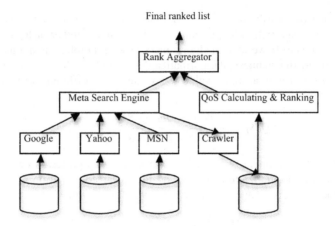

Fig. 1. System architecture for rank aggregation with QoS information

First, a query is submitted to three underlying search engines: Google, Yahoo and MSN. The top ranked pages from each search engine are combined and duplications are removed. Secondly, for each page in this combined list, we need to calculate the values of its QoS attributes. We submit the request on each page a few times and get the reliability and the average response time. Then we download the actual page in order to calculate the file size and the number of embedded objects. Using equation (3), we can calculate the QoS value for each page and get the QoS-based ranking. Finally different rank aggregation algorithms are tested to combine the QoS-based ranking with the ranking from the meta search engine.

5 Experiments

5.1 Experiment Design

We collected 36 queries from a group of users (most of them are graduate students), some example queries are *Disney, schizophrenia research, image processing, Buenos Aires travel, sea monster video, Object Oriented programming language, Taiwan president election 2008,* and a complete list could refer to [3]. They are on a wide variety of subjects such as computer science, biology, tourism, etc. We specially chose some topics that are country related because the web sites hosted in different countries may have a different response time and reliability. We also did the same experiment on another query set [2] [6] to test whether we could get the same conclusion for different queries.

Since users normally look at the first one or two pages from the search engine, and by default, there are 10 results in one page, we downloaded the top 20 results from the meta search engine and calculated their QoS values, and then re-ranked them trying to promote the pages with a good delivery quality to top 10 positions. Our assumption here is that by only re-ranking the top 20 pages, the result relevancy won't be affected much. In order to calculate the QoS value more accurately, we downloaded each page ten times and then we used the average value for each QoS attribute.

In the experiment, we tested four rank aggregation methods – Condorcet-fuse [10], Borda-fuse [1], MEDRANK [7], and GA-based [2] rank aggregation. We also compared them with the baseline – QoS-based re-ranking algorithm, in which the top 20 results are simply re-ranked based on their overall QoS values. The weight of the QoS attribute could be either 0 or 1 as we explained before. We tested all the possible combinations of four weights in the experiment, but we only showed the case when all of the weights are 1 in the figures below.

5.2 Result Analysis

Since the main purpose of our proposed algorithms is to improve QoS of the search results, we use the top 10 QoS improvement to measure the effectiveness of the algorithm. It is defined as the improvement on the average QoS value (one attribute or overall) of the top 10 results,

$$\Delta QoS = \frac{Avg_{i \in new-top10}(QoS_i) - Avg_{i \in old-top10}(QoS_i)}{Avg_{i \in old-top10}(QoS_i)} \tag{4}$$

In the mean time, we don't want to sacrifice the accuracy of the results. Usually, the accuracy is measured by the precision value, due to the lack of the user evaluated relevancy data, we assume the precision from the meta search engine is our baseline, and then take it as the reference, to see how many original top 10 can be retained by the new algorithm, the higher the value, the higher the precision of the top 10 results.

$$\Delta Top10 = \frac{new\_top10 \cap old-top10}{10} \tag{5}$$

We believe that improving QoS and keeping precision are equally important to our system. In our experiment, we use the following formula to measure the overall effectiveness of the tested algorithm.

$$E = \Delta QoS \cdot \Delta Top10 \tag{6}$$

In the following figures, CF refers to Condorcet-fuse, MR refers to MEDRANK, BF refers to Borda-fuse, GA refers to GA-based rank aggregation, and RR refers to the QoS based re-ranking. We only look at the results that are averaged on all queries.

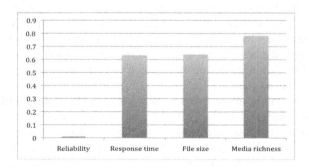

Fig. 2. Average ΔQoS when only one attribute weight is 1

Figure 2 shows ΔQoS on each QoS attribute for the average query when the weight of that particular QoS is set to 1 and all other weights are set to 0. RR is the ranking algorithm used here.

From this figure, we can see that the improvement on reliability is very little. The reason is that the original reliability is very high already (i.e. above 90%), and the room for improvement is little. The improvements on other QoS attributes are very good, on average, both response time and file size have over 60% increases on QoS values, and media richness could achieve almost 80% increase.

Since the improvement on reliability is very low, almost negligible compared with other QoS attributes, when we do the comparison on different rank aggregation algorithms in the following analyses, we only consider three attributes – response time, file size and media richness. Table 1 shows the improvement on the values of these three QoS attributes from the four aggregation algorithms and QoS-based re-ranking algorithm. We could see that QoS-based re-ranking algorithm is consistently the best for all three attributes, and among the four aggregation algorithms, MEDRANK is the best performing one. It is quite understandable because in QoS-based re-ranking, only QoS values are used to re-rank the results without considering the original ranks from the meta search engine.

Table 1. Comparison of Condorcet-fuse, MEDRANK, Borda-fuse, GA-based rank aggregation with the QoS-based re-ranking algorithm on their improvements on 3 QoS attributes

	Response time	File size	Media richness
CF	0.4609	0.5331	0.6108
MR	0.5198	0.5717	0.6385
BF	0.4536	0.5273	0.6315
GA	0.4922	0.5190	0.5538
RR	0.5576	0.5805	0.6622

Figure 3 shows three overall measurements as we defined above – ΔQoS, ΔTop10 and the overall effectiveness (E). We could see that overall speaking, MEDRANK is the best performing method, it achieves the highest improvement on E, is the second best on ΔQoS and ΔTop10 and only slightly worse than the best one on these two measurements. Borda-fuse is the second best one on E and the best on ΔTop10. GA-based algorithm is the worst on all three measurements. Regarding ΔQoS, Condorcet-fuse is worse than QoS-based re-ranking, but is better in overall effectiveness. QoS-based re-ranking is the best on individual QoS improvement and also on overall QoS improvement, but it changes the top 10 results too much and thus its performance on ΔTop10 and E is not so good.

Figure 4 shows the result on another query set [2] [6] and the distribution of the data is consistent with that in Figure 3. We can get the exact same conclusion from this figure. Among the four aggregation algorithms, MEDRANK performs consistently the best when combining the QoS-based ranking, Borda-fuse is the second best one, the next is Condorcet-fuse, and the worst performing one is GA-based algorithm. When we compare their efficiencies, the first three are more or less the same, whereas the GA-based algorithm is the slowest one.

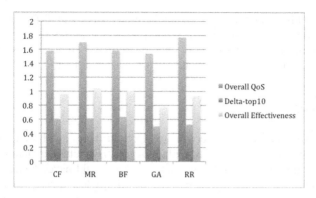

Fig. 3. Comparison of Condorcet-fuse, MEDRANK, Borda-fuse, GA-based rank aggregation with the QoS-based re-ranking algorithm

5.3 Discussions

Our current way of collecting QoS data may not work in a real system, because it will affect the response time to get the retrieved results due to the extra page downloading and QoS calculation time. There are two possible solutions. The first one is to implement QoS based ranking in a search engine instead of a meta search engine. It wouldn't affect the query response time because the QoS data can be collected when the document is crawled and downloaded. The second solution is to apply QoS based ranking only for queries which have been issued before. Whenever a query is submitted, QoS data could be collected for all the retrieved results and saved in a database for the future use, and it could also be updated regularly. The query response time wouldn't be affected because QoS processing is done offline.

The current implementation of measuring reliability and response time is not very accurate. We send the request on the same page during different time period. However, we didn't consider the effect of different locations. If the request is from different locations, most likely, the reliability and response time will be different. One

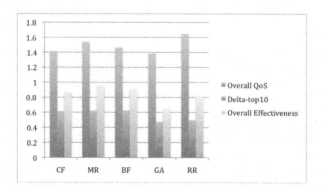

Fig. 4. Comparison of Condorcet-fuse, MEDRANK, Borda-fuse, GA-based rank aggregation with the QoS-based re-ranking algorithm on a new query set

possible solution is to send requests via public proxy servers in different locations and then get the average response time or reliability, which makes result more accurate. Another solution is to implement the QoS ranking in personalized search, so that reliability or response time is only related to one particular user.

6 Conclusions

In this paper, we define four delivery-related QoS attributes and how we could measure them and calculate the QoS-based ranking scores. By including them in the ranking procedure to re-rank the top positioned pages, users could have chance to see pages with a high delivery quality earlier than pages with a low quality, and as a consequence, it could improve users' searching experiences. The experiment shows the promising results. Generally speaking, the top ranked pages have a better delivery performance than before, and in the mean time also keep a high precision on relevancy. And by comparing four rank aggregation algorithms, MEDRANK achieves the best performance regarding the improvement on QoS values and the overall effectiveness.

There are several directions of works we would like to work in the future such as the ones we listed in the previous section. We would also like to conduct some experiments to collect the user relevancy data and thus have a more accurate measurement on precision. Another direction we would like to work on is that we could test our algorithm on a wireless device such as a cell phone or PDA, and on Internet connections with different speed, to see whether it could improve the searching experiences for this type of users. We could also implement a user interface so that users can choose their own preferences on QoS attributes, and a user study on their satisfaction level can be done afterwards.

Acknowledgments. This work is sponsored by Natural Science and Engineering Research Council of Canada (grant 299021-07).

References

1. Aslam, J.A., Montague, M.: Models for Metasearch. In: Proceedings of the 24th Annual International ACM SIGIR Conference on Research and Development in Information Retrieval, pp. 276–284 (2001)
2. Beg, M.M.S., Ahmad, N.: Soft Computing Techniques for Rank Aggregation on the World Wide Web. Journal of the World Wide Web: Internet and Web Information Systems 6(1), 5–22 (2003)
3. Chen, X.: QoS-based Ranking for Web Search. Master Thesis, Ryerson University, Toronto, Canada (2008)
4. Croft, W.B.: Combining Approaches to Information Retrieval. In: Croft, W.B. (ed.) Advances in Information Retrieval: Recent Research from the Center for Intelligent Information Retrieval, pp. 1–36. Kluwer Academic Publishers, Boston (2000)
5. Deora, V., Shao, J., Gray, W.A., Fiddian, N.J.: A quality of service management framework based on user expectations. In: Orlowska, M.E., Weerawarana, S., Papazoglou, M.P., Yang, J. (eds.) ICSOC 2003. LNCS, vol. 2910, pp. 104–114. Springer, Heidelberg (2003)

6. Dwork, C., Kumar, R., Naor, M., Sivakumar, D.: Rank Aggregation Methods for the Web. In: Proceedings of the 10th World Wide Web Conference, pp. 613–622 (2001)
7. Fagin, R., Kumar, R., Sivakumar, D.: Efficient Similarity Search and Classification via Rank Aggregation. In: Proceedings of the 22nd ACM International Conference on Management of Data, pp. 301–312 (2003)
8. HTTP – Hypertext Transfer Protocol Overview, http://www.w3.org/Protocols
9. Liu, Y.T., Liu, T.Y., Qin, T., Ma, Z.M., Li, H.: Supervised Rank Aggregation. In: Proceedings of the 16th International World Wide Web Conference, pp. 481–489 (2007)
10. Montague, M., Aslam, J.A.: Condorcet Fusion for Improved Retrieval. In: Proceedings of the 11th International Conference on Information and knowledge management, pp. 538–548 (2002)
11. Ran, S.P.: A Model for Web Services Discovery with QoS. ACM SIGecom Exchanges 4(1), 1–10 (2003)
12. Selberg, E.W., Etzioni, O.: On the Instability of Web Search Engines. In: Proceedings of the 6th International Conference on Content-Based Multimedia Information Access, pp. 223–235 (2000)
13. Zeng, L., Benatallah, B., Dumas, M., Kalagnanam, J., Sheng, Q.Z.: Quality Driven Web Services Composition. In: Proceedings of the 12th International World Wide Web Conference, pp. 411–421 (2003)

Personalized Delivery of On–Line Search Advertisement Based on User Interests

Guangyi Xiao and Zhiguo Gong

Faculty of Science and Technology
University of Macau
Macao, PRC
{ma66568,fstzgg}@umac.mo

Abstract. Search engine advertising has become the main stream of online advertising system. In current search engine advertising systems, all users will get the same advertisement rank if they use the same query. However, different users may have different degree of interest to each advertisement even though they query the same word. In other words, users prefer to click the interested ad by themselves. For this reason, it is important to be able to accurately estimate the interests of individual users and schedule the advertisements with respect to individual users' favorites. For users that have rich history queries, their interests can be evaluated using their query logs. For new users, interests are calculated by summarizing the interests of other users who use similar queries. In this paper, we provide a model to automatically learn individual user's interests based on features of user history queries, user history views of advertisements, user history clicks of advertisements. Then, advertisement schedule is performed according to individual user's interests in order to raise the clickthrough rate of search engine advertisements in response to each user's query. We simulate user's interests of ads and clicks in our experiments. As a result, our personalized ranking scheme of delivering online ads can increase both search engine revenues and users' satisfactions.

1 Introduction

Search engines can earn millions of dollars each day through their sponsored search advertisement systems such as Adwords in Google, Overture in Yahoo. They sell the advertisement slots in the right side of the search results which are responded to the query words. There are millions of queries each day, however, frequencies of queries are quite different. For example, the query word "apple nano" may occur 10000 times each day while the query word "mathematic book" may only happen 10 times. So a lot of advertisers may bid the keyword "apple nano" for advertisements. Only limited number of slots, as a result, can be scheduled in response to the query word "apple nano". Therefore, sponsored systems often make use of auction principles. Though it may not be always true, most systems suppose that high positions in the advertisement space get more attention and more click-through rate (CTR), and the bidder may get more chances to draw consumer to its product page or homepage. With this idea, the bidders who bid the bigger values can get the higher slot in the advertisement space.

Q. Li et al. (Eds.): APWeb/WAIM 2009, LNCS 5446, pp. 198–210, 2009.
© Springer-Verlag Berlin Heidelberg 2009

The key task for a search engine advertising system is to determine which ads should be displayed, and in what order for each query. The ranking of advertisements has a strong impact on the revenue of the search engine from clicks of the ads. At the same time we should also show the user those ads that they prefer to click in order to improve user satisfactions. For this sake, most search engines rank the ads by taking into account of both bid values per click and the ad's click-through rate. Some other systems may also consider conversion rate of the advertisement, however it is difficult for a search engine system to get detail information regarding advertisement conversion.

Most of past algorithms for delivery of online advertisements are based on common users. In other words, all users get the same advertisement scheduling if they use the same word to query. As a matter of fact, different users show quite different interests to the same advertisements even if they query the same word. For example, user1 searches 'computer' only for this time while user 2 has queried this word and other semantically related words for many times. It is clear that they have different interests to the same word. In this paper, we tackle the problems of personalized delivering advertisements based on individual user's interests. In our approach, individual user's interests are evaluated based on their query histories, advertisement click histories, as well as advertisement showing histories to the user. Our algorithm is also aimed to deliver online advertisements with best revenue of search engines. Both of those objectives are complementary, since the first target is to satisfy user's searches, and the second target is to increase the revenues of search engines. Our algorithm combines these two targets together to raise the overall performances of the advertisement systems.

Research group from Yahoo has built a static linear programming model [1] and a dynamic model [7] to maximize the utility slate for advertisement scheduling. Both of them make use of common search behaviors of users. In this study, we construct our model with reference to their basic ideas, but take into account of individual users' interests.

In remainder of this paper, we introduce some related work in section 2; advertisement schedule algorithm is addressed in section 3; user interest model is investigated in section 4 and 5; section 6 discusses experiment results; and in section7, we conclude the paper.

2 Related Work

Mehta et al. use a trade-off function with the fraction of the bidder's budget that has been spent so far, and it get the competitive rate $1 - 1/e$ [5]. Karp et al. indicate that no randomized online algorithm can have a competitive ratio better than $1 - 1/e$ for the bi-matching problem. Abrams, Mendelevitch, and Tomlin in yahoo research make use of stochastic information and model the allocating problem with column generation linear programming model [1]. Mahdian, Nazerzadeh, and Saberi do not only consider the stochastic information but also consider whether the estimate of the query frequency is accurate [2]. Recently, Ghose and Yang use a dataset of several hundred keywords collected from a large nationwide retailer that advertises on Google, and they find that the presence of retailer-specific information in the keyword increase click-through rates, and the presence of brand-specific information in the keyword increases conversion rates [4].

There is a very important constraint for the advertisement bid with each keyword. Each advertisement has a budget each day. When the budget of the advertisement balance becomes 0, the advertisement will not be displayed when the query received by the search engine, even though all the other advertisements have not bidden this query word. The daily budget constrains the number of times their ads appear each day. In the generalized second price auction scheme, this constraint produces impact to the revenue of the search engine. If one of the advertisements uses up its budget completely, the corresponding advertisement will be kept out from the auction system. Such situation will influence other advertisements cost per click with the same keywords, thus reduce the revenue of the search engine. Therefore, it is very important to protect all the advertisements from using up their daily budgets completely. Another constraint is the distribution of the query frequency. The distribution of the query frequency can be predicted from history query log, but the data may have big differences with real values.

With these two constraints, Abrams et al. propose an approach based on linear programming to optimize ads delivery [1]. In their approach, linear programming model with column generation is employed to find the optimal scheduling of advertisements [8]. For each linear programming, there is a dual problem for the prime problem (for more detail, see [8]). Then we can get a important parameter ρ_j from solving the dual problem with column generation model. Where parameter ρ_j is called a "utility factor" associated with the appearance of bidder's ad in response to query [7]. Keerthi et al. make use of ρ_j to generate an algorithm for constructing an optimal slate of online advertisements [7]. They model the problems with an optimal path approach searching (minimum path algorithm) to construct an optimal slate. Their work obtains a good result from their simulation as long as the predicted clickthrough rate really reflects the real situation. This assumption is obviously too subjective.

Feng Qiu and JunghooCho find that users are reluctant to provide any explicit input on their personal preference [10]. They study how a search engine can learn a user's preference automatically based on users' past click history and how it can use the user preference to personalize search results. Benjamin Piwowarski and Hugo Zaragoza in yahoo research find that query logs have the potential to partially alleviate the search engines from thousand of searches by providing a way to predict answers for a subset of queries and users without knowing the content of a document [11]. They present three different models for predicting user click, ranging from most specific ones (using only past user history for the query) to very general ones (aggregating data over all users for a given query). The former model has a very high precision at low recall values, while the latter can achieve high recalls. Their method is based on recall which reflects the probabilities that users will re-use the same query in search engine, then the probability to click the same document is very large.

3 Personalized Delivery On-Line Ads Algorithm

Our goal is to deliver a set of ads (slate) in response to a query of an individual user. Similar to [7], we denote the n bidders on one query by $j = 0, \ldots, n$, and suppose that Generalized Second Price auction (GSP) is used to rank the advertisements [13]. Work [1] extends GSP to what is known as revenue ranking, where the ads are ranked not by bid values, but by general expected revenue of the search engine which is modeled as the product of the advertiser's bid value A_j and Q_j which is called as quality score, or clickability, for bidder j's ad. Let the number of indexes also indicate the revenue ranking. This quantity is supposed to be better to represent the value of bidden ad than the raw bid value. Thus, the ads are ranked as:

$$A_1 Q_1 \geq A_2 Q_2 \geq .. \geq A_n Q_n \qquad (1)$$

In this paper, we propose a new scheme called "personalized revenue ranking", where the ads are not ranked by general expected revenue, but ranked by personalized expected revenue which is taken into account of both bid values, and individual user's interest score. This rank scheme makes use of the advertiser's bid A_j, and user u's interest $I_{u,j}$ score to adj in response to the query. For personalized delivery of online ads, we claim that this quantity is much better to catch interests of individual users when scheduling advertisements. Thus, in our proposed method, personalized revenue scheme ranks the advertisements as:

$$A_1 I_{u,1} \geq A_2 I_{u,2} \geq ... \geq A_n I_{u,n} . \qquad (2)$$

With the principles of Generalized Second Price scheme [13]. In our personalized revenue ranking, we require that the expected cost per click (CPC) of bidder j_p must be at least

$$A_{j_{p+1}} \frac{I_{u,j_{p+1}}}{I_{u,j_p}} . \qquad (3)$$

Because the limited space for displaying advertisements, only a subset of ads can be displayed for each query. We set m positions for displaying advertisements in the search result page. In [7], the authors define a slate as an ordered subset $S = \{j_1,..., j_k\}$, where $k<m$, of the ordered set of ads $\{1,..., n\}$. Since we assume a second-price auction is used for the slot action, bidder j_p in position p pays the bid $A_{j_{p+1}}$ of the bidder occupying position $p+1$. In addition, there is a minimum bid ε, which is paid by the last bidder in the slate if and only if there are no lower bidders on the query. Under these assumptions we wish to solve maximum problem:

$$\text{Maximize } \mu = \sum_{p=1}^{m} \rho_{j_p} A_{j_{p+1}} \frac{I_{u,j_{p+1}}}{I_{u,j_p}} \qquad (4)$$

subject to the requirement that $j_1,...,j_k$ is an increasing set of indices, where $k<m$, and ρ_{j_p} is a parameter used to reduce the desirability of showing an ad when it is at or near to its budget.

We make use of optimal path approach as in [7] to solve the optimal scheduling problems for the personalized delivery of the online ads, which successfully transforms dynamic programming to the optimal path problem.

4 User Interest Model

It is obvious that, if one gets more interest in an ad, he may have more probability to click the ad when it is displayed. Hence, we couple click-through rate of individual users with their interests to ads. For example, suppose that two ads ad_1 and ad_2 are responded to one query. If a user feels that ad_1 is getting more relevant to his interest, he has more probability of clicking ad_1 when seeing these two ads. Therefore, how to model the user's interests is a key point in our approach. We assume that interest score of a user to an ad is between 0 and 1. We cast it as a regression problem—that is, to predict the Interest Score, given a set of features. Similar to work [3], we define user's interest to an advertisement as:

$$I_{u,j} = \frac{1}{1+e^{-Z}} \quad Z = \sum_i w_i f_i(u,j) \tag{5}$$

where $f_i(u,j)$ is the value of user u's i^{th} feature for the ad j, and w_i is a weight for the corresponding feature. Features may be anything, such as a session of queries performed by user u, overall query history of user u, or overall click history for ads. (We will discuss it in detail in the next section).

Although our model for user interests to each ad seems like click through rate, but it is difficult to collect the value of click through rate. It can not be calculated as the number of clicks divided by the number of views (number of showing to users). Firstly, the number of views is limited when the same ad is viewed by the same user. Secondly, if one user has clicked one ad, he may not want to click the same ad again. It seems that it is difficult to change the weight of features because it is difficult to collect the true value corresponded to predicting user interest. However, if user u has clicked one ad (ad_{click}), then it indicate that the user prefer this ad than others. We suppose the weight of feature i can be automatically adjusted based on user's click actions to ads. In our implementation, if the interest score of user u to the clicked ad (ad_{click}) is larger than other $N-1$ ads, the weight of the feature is increased for this feature. Otherwise, it is decreased. The automatically adjusting formula is defined as:

$$w_i \leftarrow w_i + \eta \frac{\sum_{j=0}^{N} (f_i(u,ad_{click}) - f_i(u,j))}{N-1} \tag{6}$$

where i is the ith feature, and N is the number of total ads responded to user's query, and η is a small constant controlling the learning rate.

5 Estimating User Interest

As we known, there two types of history queries. One is session oriented, in which only the queries within the same access session of the user are taken into account to predict the user's interests. Another one involves all the historical queries of the same user since he firstly registered to the system. In this paper, we define two measures for estimating user interest with respect to these two types of history queries.

5.1 User Interest with Queries in Session

Many personalized search methods make use of history queries in the same session to expand the current search in order to derive the user's search intention. In our work, we do not only use history queries to expand the search, but also to find more relevant ads to schedule for the user. For example, there are three ads in the figure 1. Ad_1 bids three keywords, 'thinkpad', 'ideapad' and 'notebook'; ad_2 bids two keywords, 'thinkpad' and 'notebook'; and ad_3 bids five keywords, 'lenove', 'notebook', 'IBM thinkpad', 'X200' and 'T400'.

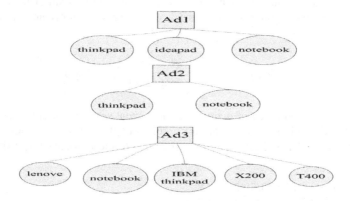

Fig. 1. Three ads bid different keywords

We further suppose there are four queries in the same session from a user with a search engine. The history queries and current query are as following:

q1: notebook
q2: thinkpad notebook
q3: lenove notebook
q4: thinkpad X200
q5: T400 (Current query)

To simplify the model, we just calculate the ratio of ad's keywords over the total words in the user's query session. So the user's interest to adj is defined as:

$$f_0(u, j) = \frac{N_{u,j}}{M_j} * \sum_{s=1}^{|Q_{S_U}|} \frac{\theta(k_j \in q_s)}{|Q_{S_U}|} \tag{7}$$

$$\theta(k_j \in q_s) = \begin{cases} 1 & \text{if any keyword of } ad_j \text{ contained in } q_s \\ 0 & \text{otherwise} \end{cases} \tag{8}$$

where $|Q_s|$ is the number of user u's queries in the current session s, and M_j is the number of keywords bidding for adj, and $N_{u,j}$ is the number of those bidding keywords which also occur in any queries of user u, and θ is Boolean function to indicate whether query q_s contains some bidding keyword of adj. In this example, the degrees of user's interest to ad_1, ad_2 and ad_3 are 4/5, 4/5, and 5/5 respectively. So the user will get more interest in ad_3 in current query session.

However, we have two objects for allocating these ads to current query. The first object is to expand current query in order to alleviate ambiguous of the query (short queries may not clearly indicate the user's search intention). After expanding keywords, all three ads in the example may respond to the current query. The second object is to find user's preference degree to those ads. In general, history queries which are more close in time to current query are more relevant and important to derive the user's current intention for search, so the query position in the session should be taken into account. For such sake, we define session oriented user interest to adj as:

$$f_0(u,j) = \frac{N_{u,j}}{M_j} * \sum_{s=1}^{|Q_{S_U}|} \frac{s}{|Q_{S_U}|} \theta(k_j \in q_s) \tag{9}$$

where s is the position index of query q_s in the current session, and all other parameters are defined similarly as in equation(8). For example, keywords of ad_1 contained in q_1, q_2 and q_3, and 1/5, 2/5, 3/5 are corresponding to these queries respectively. For each ad, we get:

1. $f_0(u,1)=(1/5+2/5+3/5)*2/3 =0.8$;
2. $f_0(u,2)=(1/5+2/5+3/5)*2/2 =1.2$;
3. $f_0(u,3)=(1/5+2/5+3/5+4/5+5/5)*5/5 =3$;

Therefore, with respect to session feature (history queries of session), ad_3 is more important than ad_1 and ad_2 to user u. This formula is based on both timestamp of keywords and the frequency of keywords in the query session. Our experiments show this consideration can significantly improve the performance of the advertisement system.

5.2 User Interest with Overall History Queries

The basic idea of user interest model with respect to overall history queries is discussed in this subsection. Firstly, there are some of keywords bidden for each ad, and each keyword in the queries indicates some intention of the user. However, there are different degrees of interests for each keyword individually. In this work, the second feature for the user interest model is defined as:

$$f_1(u,j) = \frac{N_{u,j}}{M_j} * \sum_{i=1}^{M_j} \frac{|k_{i,j} \in Q_u|}{|Q_u|} \tag{10}$$

where M_j is the total number of keywords associated to ad_j, and $N_{u,j}$ is the number of keywords contained in queries of user u, and Q_u denotes the set of user u's history queries in logs, and $|Q_u|$ denotes the number of user u's history queries in logs, and $k_{i,j}$ denotes the ith keyword in the adj. So $|k_{i,j}$ in $Q_u|$ represents the number of quires contained in the overall query history of user u, thus, $|k_{i,j}$ in $Q_u|/|Q_u|$ denotes degree of user interest to the adj with respect to keyword $k_{i,j}$. So the user's overall interest to adj is the sum of individual keyword interests to adj.

For example, if an ad bid some keywords, such as 'tobacco', 'retail', 'marijuana', and 'get high'. There are two users querying 'tobacco and pipes', how to know their different interest degrees to the same ad? From history queries within three months, u_1 performed 218 queries and u_2 did 121 queries respectively. And the frequencies of query words of u_1 and u_2 are as in the following table.

Table 1. Frequency of keywords contained in queries of u1 and u2

	Query Frequency of u_1	Query Frequency of u_2
tobacco	9	0
retail	5	0
marijuana	13	0
get high	1	4

From table 1, u_1's degree of interest to keyword "tobacco" is 9/218, and u_2 is 0. u_1's degree of interest to keyword "retail" is 5/218, and u_2 is 0. u_1's degree of interest to keyword "marijuana" is 13/218, and u_2 is 0. u_1's degree of interest to keyword "get high" is 1/218, and u_2's is 4/121. So the interest scores with the ad are 28/218 and (4/121)*(1/4) respectively to u_1 and u_2, and u_1's interest is bigger than u_2.

As another example, there are two ads, such as ad_1 and ad_2. Ad_1 bids keywords 'map', 'guide', 'hotel' and 'Florida'. Ad_2 bids keywords 'travel', 'map', 'Georgina' and 'university'. A user searches with query "hotel" in search engine and the search engine delivers these two ads, so how different about scores of interests of these two ads to the same user. For history queries in three months, this user with id 220 got 332 queries. The frequencies of query words of user 220 are as following table:

Table 2. Frequncy of keywords in queries and degree of interest with keywords

	Frequency in user u	Degree of interest with keyword
travel	0	0
guide	5	5//332
hotel	4	4/332
Florida	116	116/332
map	11	11/332
Georgina	18	18/332
university	10	10/332

Therefore, the interest scores of this user are (129/332)*(4/4) and (39/332)*(3/4) respectively to ad_1 and ad_2, and u get more interest in ad_1 than ad_2.

5.3 User Interest with History Views and Clicks

If user u has clicked the ad_j, then u may not want to click the ad_j again. In other words, u loses the interest to ad_j. So the third feature is based on clicks of the ad_j

performed by u, and this feature produces negative impact to interest. Then we define the third feature as:

$$f_2(u, j) = -\frac{\lambda \mid Clicks_{u,j} \mid}{1 + \lambda \mid Clicks_{u,j} \mid}. \tag{11}$$

where $Clicks_{u,j}$ is the number of clicking ad_j of u.

Similarly, if user u has viewed the ad_j, u may not prefer to click the ad_j either. In other word, u loses the interest to ad_j. So the forth feature is the number of ad_j showing to u, and this feature also has negative impact to interest. Then, we similarly define the forth feature as:

$$f_3(u, j) = -\frac{\pi \mid Views_{u,j} \mid}{1 + \pi \mid Views_{u,j} \mid}. \tag{12}$$

where $Views_{u,j}$ is the number of clicks when user u clicked on the ad j.

We find that it is promising with our experiments when we set the two parameters λ, π as 1/3, 1 respectively. Though advertisement visual features, such as visual or text description quality [12], may also influence the user's interest to the advertisement, we ignore them in the current study.

Fig. 2. Our Advertisement Scheduling System

6   Experiments

Figure 2 shows an example of our search advertisement scheduling. The left column shows the scheduled advertisements of our algorithm. In order to test our approach, we do not only evaluate the performance of predicting user's interest to ads responded to each query, but also evaluate the performance of the personalized schedule algorithm. For the first objective, we assume the click through rate is independent of the position of ads in the same search result page. If a user clicks the ad with largest predicted degree of interest, we denote the click as an **accurate click**, and otherwise we denote this click as an error click. The **accurate click ratio** is defined as:

$$ACR = \frac{|\,Clicks_{accurate}\,|}{|\,Clicks_{all}\,|} \tag{13}$$

where ACR is the accurate click ratio, $|Clicks_{accurate}|$ is the number of the accurate clicks, and $|Clicks_{all}|$ is the number of all clicks. The bigger ACR is, the better performance the user interest model works.

For the second objective, we evaluate revenue of our proposed approach against the baseline algorithm which schedule ads based on general click-through rate.

We collect some ads bidden with multiple keywords. For each query, ads are scheduled and displayed on the pages of query results. We capture each ad's degree of interests to different users, views and clicks of ads from each user. Then we compare the degree of interest of the clicked ad with other ads'. If the degree of interest of the clicked ad is the largest one, then we increase the number of the accurate clicks. If the degree of interest of the ad is not the largest one, we will improve the weight of each feature using equation (6).

To evaluate our algorithm, we perform the following steps for each query:

1. Get all the ads bidden with the keyword contained in this query
2. Calculate all the features and the degree of interest as discussed in the section 5 according to the user interest model
3. Log those interest degrees, feature values, the number of views of ads showing to users.

We further perform the following actions when any click happens:

1. Record the cost of this click, and increase the number of clicks of this ad.
2. Check if this clicked ad's degree of interest is the largest one. If yes, we increase the number of accurate clicks. Otherwise, we will change the weight of features depend on equation (6).

For baseline algorithm, we perform the following steps: for each query we get all the ads bidden with the keyword contained in this query, ordered by the bid per click and the click through rate; for each click, we record the cost per click and clicking users.

In figure 3, there are nine bidders in response to query "fabric". For different users, the user degree of interests to these bidders is different. In our experiment, we set each page with 5 slots for ads, and we record user's degree of interests to these displayed ads. So d1 is the largest one when u1 queries "fabric". If u1 click one ad from these 5 ads, we check if it is accurate click or not. If u1 click the ad with the largest degree of interest, we get more confidence for our prediction. Then, we increase the number of accurate clicks in order to increase the accurate click ratio. Otherwise, we should change the weight of features instead.

Figure 4 shows that our simulation results are quite promising. Even through our accurate click ratio is very low in the beginning, it increases with consequently adjusting the weight of features through equation (6). It means that the system can adjust the weights of features automatically.

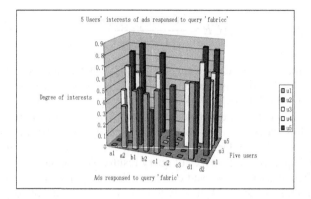

Fig. 3. Users' degree of interests of ads responded to query 'fabric'

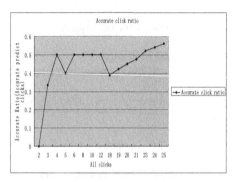

Fig. 4. The change of the accurate click ratio

Fig. 5. The change of weights of each feature

From figure 5, we find that the feature weights of session history, global query history, views of ads, and clicks of ads become more and more stable along increasing clicks. It tells us those features are very important for scheduling advertisements. But we also find the weight of quality score of ads has no fixed pattern in change.

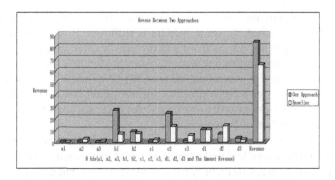

Fig. 6. Payment for 8 ads and the total revenue

Figure 6 shows the revenue improvement of our proposed scheduling mechanism over general click-through model. The revenue improvement of our proposed model is obvious, comparing with baseline algorithm, with about 28% increase.

7 Conclusion

In this paper we proposed a framework for personalized delivery of online advertisements based on users' past search histories. In particular, we use optimal network path approach to generate maximum utility slate. The proposed approach can generate maximum utility for different users based on their individual interests to ads. In our implementation, we construct the user interest model with taking into account of both words associated with ads and words in users' query history.

We have conducted experiments to evaluate our approach. In our simulation experiments, we found that our model can predict user interest accurately. It shows that our method is both effective and easily applicable to real-word search engines. If the history queries are very limited, we should take the combining predictions. We are currently working on probabilistic schedule algorithm that can combine dynamic programming algorithm as in yahoo-research and our approach. In the future, we plan to expand our framework to take more user-specific information into consideration.

Acknowledgement. This Work was supported in part by the University Research Committee under Grant No. RG069/05-06S/07R/GZG/FST and by the Science and Technology Development Fund of Macao Government under Grant No. 044/2006/A.

References

[1] Abrams, Z., Mendelevitch, O., Tomlin, J.A.: Optimal Delivery of Sponsored Search Advertisements Subject to Budget Constraints. In: Proceedings 8th ACM Conference on Electronic Commerce (EC 2007), San Diego, California, USA, June 11-15, 2007, pp. 272–278 (2007)

[2] Mahdian, M., Nazerzadeh, H., Saberi, A.: Allocating Online Advertisement Space with Unreliable Estimates. In: Proceedings 8th ACM Conference on Electronic Commerce (EC 2007), San Diego, California, USA, June 11-15, pp. 288–294 (2007)

[3] Richardson, M., Dominowska, E., Ragno, R.: Predicting Clicks: Estimating the Click-Through Rate for New Ads. In: Proceedings of the 16th International Conference on World Wide Web, WWW 2007, Banff, Alberta, Canada, May 8-12, pp. 521–530 (2007)

[4] Ghose, A., Yang, S.: An Empirical Analysis of Sponsored Search Performance in Search Engine Advertising. In: Proceedings of the International Conference on Web Search and Web Data Mining, WSDM 2008, Palo Alto, California, USA, February 11-12, pp. 241–250 (2008)

[5] Mehta, A., Saberi, A., Vazirani, U., Vazirani, V.: Adwords and Generalized Online Matching. Journal of ACM 54(5) (October 2007)

[6] Aggarwal, G., Goel, A., Motwani, R.: Truthful Auctions for Pricing Search Keywords. In: Proceedings of the 7th ACM Conference on Electronic Commerce (EC 2006), Ann Arbor, Michigan, USA, June 11-15, pp. 1–7 (2006)

[7] Keerhi, S.S., Tomlin, J.A.: Constructing a maximum utility slate of on-line advertisements. Yahoo! Research, Technical Report, YR-2007-001

[8] Dietrich, B., Forrest, J.J.: A column generation approach for combinatorial auctions. In: Workshop on Mathematics of the Internet: E-Auction and Markets Institute for mathematics and its Applications (2001)

[9] Pass, G., Chowdhury, A., Torgeson, C.: A Picture of Search. In: Proceedings of the first International Conference on Scalable Information Systems, Hong Kong (June 2006)

[10] Qiu, F., Cho, J.: Automatic Identification of User Interest for Personalized Search. In: Proceedings of ACM WWW 2006, Edinburgh, Scotland, MAY 23-26 (2006)

[11] Piwowarski, B., Zaragoza, H.: Predictive User Click Models Based on Click-through History. In: Proceedings of ACM CIKM 2007, Lisboa, Portugal, November 6-8 (2007)

[12] http://adwords.google.com/support/bin/answer.py?answer=10215

[13] Edelman, B., Ostrovsky, M., Schwarz, M.: Internet advertising and the generalized second price auction: Selling billions of dollars worth of keywords. In: Proceedings of the Second Workshop on Sponsored Search Auctions, Ann Arbor, MI (June 2006)

Automatic Web Image Annotation via Web-Scale Image Semantic Space Learning

Hongtao Xu[1], Xiangdong Zhou[1], Lan Lin[2], Yu Xiang[1], and Baile Shi[1]

[1] Fudan University, Shanghai, China
{061021054,xdzhou,072021109,bshi}@fudan.edu.cn
[2] Tongji University, Shanghai, China
linlan@mail.tongji.edu.cn

Abstract. The correlation between keywords has been exploited to improve Automatic Image Annotation(AIA). Differing from the traditional lexicon or training data based keyword correlation estimation, we propose using Web-scale image semantic space learning to explore the keyword correlation for automatic Web image annotation. Specifically, we use the Social Media Web site: Flickr as Web scale image semantic space to determine the annotation keyword correlation graph to smooth the annotation probability estimation. To further improve Web image annotation performance, we present a novel constraint piecewise penalty weighted regression model to estimate the semantics of the Web image from the corresponding associated text. We integrate the proposed approaches into our Web image annotation framework and conduct experiments on a real Web image data set. The experimental results show that both of our approaches can improve the annotation performance significantly.

1 Introduction

Automatic Image Annotation(AIA) has attracted a great deal of research interests [11,7,6,10,13], due to its critical role in keyword based image retrieval and browsing. However, the long lasting *Semantic Gap* problem still challenges the effectiveness of AIA. It is urgent to improve the annotation performance to meet the increasing requirement of practical applications.

Recently, the correlation between annotated keywords was explored to improve the performance of image annotation. For instance, keyword set {sky, grass} usually has a larger probability to be an image caption than {ocean, grass}. Only a few work had been done to investigate the keyword correlation on AIA, such as CLM [8] and WordNet-based approaches [9,16]. The former employed the co-occurrence of keywords indirectly by using EM algorithm to fit a language model for generating annotations, while the latter made use of Word-Net to exploit the hierarchy of the keywords. Zhou [22] proposed an iterative image annotation approach which learn the keywords correlation by "Automatic Local Analysis". Rui et al. [13] presented a bipartite graph reinforcement model (BGRM) for image annotation, which exploit the keywords semantic correlation

Q. Li et al. (Eds.): APWeb/WAIM 2009, LNCS 5446, pp. 211–222, 2009.

based on a large-scale image database maintained by themselves. In general, most of the previous work infer the correlations between keywords according to the co-occurrence of keywords in the training set or the hierarchy of lexicon. However, for the scenario of annotating Web images, the keywords correlation estimation is more subtle and complicated in some extent, due to the problems of the unlimited number of keywords and the intrinsic diversity of Web data space.

To improve the performance of the Web image annotation, we exploit Web Social Media, that is, we use the popular Web Photo Community site Flickr [1] as a Web-Scale Image Semantic Space to learn the keywords correlation graph. Then we propose a novel Web image annotation approach, which incorporates the keywords correlations and the semantic contributions of visual features and associated texts of Web image. Our method conducts the probability estimation using not only Web image textual and visual features, but also the semantic correlations between the keywords and annotated keyword subset assigned previously. In particular, for the Web scale semantic space learning, we submit semantic keywords of the annotation vocabulary to Flickr to obtain the Relative Tag (RT)[1] set as the neighborhood for keyword graph generation. We estimate the contribution of the textual features in deriving the semantics of Web image by a new constraint piecewise penalty weighted regression model. The keyword which brings the maximum annotation conditional probability is selected to be added into the annotation set. Experiments on 4,000 real Web images data set demonstrate the effectiveness of the proposed Web AIA approach.

Our contributions are as follows:

1. We exploit the popular Web Photo Community site Flickr as a Web-Scale Image Semantic Space to analyze the correlations between keywords and incorporate it into our annotation framework.
2. We propose a new constraint piecewise penalty weighted regression model to combine the adaptive estimation of the weight distribution of associated texts and the prior knowledge together for estimating the semantic contributions of the textual features of Web image.

The rest of the paper is organized as follows. Section 2 introduces the related work. Section 3 presents our Web image annotation framework. We discuss the experiment results in Section 4. Section 5 concludes this paper.

2   Related Work

In recent years, much attention has been paid to the research on AIA. Various of machine learning techniques or statistical models have been employed to develop a variety of AIA models, which can mainly be divided into two

---

[1] RT can be obtained by using Flickr's APIs: flickr.tags.getRelated. It returns "a list of tags 'related' to the given tag, based on clustered usage analysis "–refer to: http://www.flickr.net/services/api/flickr.tags.getRelated.html

categories–probabilistic model based methods and classification based methods. The first category focuses on inferring the correlations or joint probabilities between images and annotation keywords. The representative work include Translation Model(TM) [4], CMRM [7], CRM [10], MBRM [6], etc. The classification based methods try to associate keywords or concepts with images by learning classifiers. Methods like SVM-based approach [3] and Multi-instanced learning [21] fall into this category. However, these approaches do not focus on annotating Web images and neglect the available textual information of Web images, so they cannot be applied directly to annotate Web images.

Sanderson and Dunlop [14] were among the first to model image contents using a combination of texts from associated Web pages, however, they modeled the contents as a bag of keywords without any structure information. Wang et al. [20] proposed a search-based annotation system–AnnoSearch. This system requires an initial keyword as a seed to speed up the search by leveraging on text-based search technologies. Li et al. [11] proposed a search and mining framework to tackle the AIA problem. Given an unlabeled image, content-based image retrieval(CBIR) was firstly performed to find a set of visually similar images from a large-scale image database. Then clustering was performed to find the most representative keywords from the annotations of the retrieved image subset. These keywords, after saliency ranking, were used to annotate the unlabeled images eventually. Its annotation performance was highly dependent on the result of CBIR. Feng et al. [5] described a bootstrapping framework by adopting a co-training approach involving classifiers based on two orthogonal set of features–visual and textual. Tseng et al. [18] built two models based on image visual and textual features, and weighted them to annotate the unlabeled Web images. Xu et al. [19] presents a Web Image Semantic Analysis (WISA) system to adaptively model the distributions of the semantic labels of the web image on its surrounding text.

Some previous work demonstrated that keyword correlations can be utilized to improve the performance of image annotation. Jin et al. [8] address the problem by using EM algorithm to fit a language model to generate an annotation keyword subset. However, the annotation speed is lower due to the EM algorithm. Munirathnam et al. [16] propose a hierarchical classification approach for image annotation. They use a hierarchy induced on the annotation words derived from WordNet. Jin et al. [9] make use of the knowledge-based WordNet and multiple evidence combination to prune irrelevant keywords. Zhou et al. [22] proposed an iterative image annotation approach which learn the keywords correlation by "Automatic Local Analysis". Wang et al. [2] annotate image in the progressive way which explore the keywords correlation by the co-occurrence of keywords in the training images. Tang et al. [17] propose a graph-based learning approach SSMR to measure the pairwise concept similarity. However, these approaches usually learn the keywords correlations according to the appearance of keywords in the training set or lexicon, and the correlation may not reflect the real correlation for annotating Web images. Recent years, with the rapid development of Web social knowledge network, the applications which exploit the manually tagging resources of Web image, such as Flickr, have attracted researcher's great interests [15]. In this paper, we propose to learn the keywords correlation graph by exploiting Web social knowledge network Flickr.

3   The Web Image Annotation Framework

3.1   The Overview of Web Image Annotation Framework

For a given training set L_{train}, each labeled image $J \in L_{train}$ is demoted by $J = \{W, V, T\}$, where the annotation keywords W is a binary annotation keyword vector indicating whether a keyword is the annotation of J; V is a set of region-based visual features of J; and $T = \{T_1, T_2, \ldots, T_n\}$ is a set of the types of associated texts.

Our annotation framework annotates Web images in an iterative way [22,2]. That is, given a new image I, the annotation keywords set after $i-1$ iterative is denoted as $AN_{i-1}(i = 1, \ldots, k, AN_0 = null$, where k is the size of the annotation keywords set). In the ith iterative, the probability of keyword w to be annotated for I is:

$$P(w|I, AN_{i-1}) = \frac{P(w|I)P(w|AN_{i-1})}{P(w)} = \frac{P(w|I_V, I_T)P(w|AN_{i-1})}{P(w)}, \quad (1)$$

where I_V and I_T is the visual and textual feature of image I respectively. Assuming that $P(w)$ is uniformly distributed, and I_V and I_T are independent, we have:

$$w_i^* = argmax_w P(w|I)P(w|AN_{i-1}) \quad (2)$$
$$= argmax_w P(w|I_V)P(w|I_T)P(w|AN_{i-1}).$$

Then

$$AN_i = AN_{i-1} \cup w_i^* \quad (3)$$

Note that the maximum likelihood estimation for $P(w|AN_i)$ is:

$$P_M(w|AN_i) = \frac{\#\{J|w, AN_i \in J\}}{\#\{J|AN_i \in J\}}, \quad (4)$$

where $\#\{J|w, AN_i \in J\}$ denotes the number of images in which keyword w and keywords subset AN_i appear together. For a limited training set, when $|AN_i|$ is large, the co-occurrence of w and AN_i is rare, which means there will be many zero values in the probability estimation. However, a zero probability event in the limited training set does not mean it never happen in the future, thus smoothing is necessary.

In the text information retrieval, smoothing is usually performed by making use of a large background collection to assign a non-zero probability to the unhappened event in current model. For instance, we can choose a larger training image set for smoothing. However, it is hard to obtain sufficient training images for the Web-scale image annotation task. Therefore, we propose to explore the Web Social Multimedia to infer semantic correlation, rather than maintaining a large scale image database by ourselves, or using the limited training image set to generate the keywords correlation graph. It is expected that Web-scale semantic space learning is more flexible to deal with the scalability problem of Web images annotation.

Denoting the keywords correlation graph as Sim, then we can smooth the maximum likelihood estimation for $P(w|AN_i)$ by the keyword semantic similarity graph [12] as follows:

$$P(w|AN_i) = (1 - \gamma)P_M(w|AN_i) + \gamma \sum_{v \in V} \frac{Sim(w, v)}{Degree_+(v)} P(v|AN_i), \qquad (5)$$

where γ is the smoothing factor. V is the vertex set of the graph Sim. $Degree_+(v)$ is the outside degree of vertex v in Sim, that is:

$$Degree_+(v) = \sum_{u \in V} Sim(v, u) \qquad (6)$$

Different keywords have different importance for smoothing, here $Degree_+(v)$ captures the importance of keyword v, that is, if keyword v only associates to few keywords, then v is more important for smoothing than those associating to more keywords. Eqn.5 shows that the more similar between keyword v and w, the more important of keyword v for smoothing the probability of w.

The visual generation probability $P(w|I_V)$ is computed as the expectation over the images in the training set, that is:

$$P(w|I_V) \propto P(w, I_V) = \sum_{i=1}^{|T|} P(w, I_V|J_i)P(J_i) = \sum_{i=1}^{|T|} P_V(I|J_i)P(w|J_i)P(J_i), (7)$$

where $P_V(I|J_i)$ is the probability of I being generated from J_i based on their visual features. $P(w|J_i)$ denotes the probability of word w generated from J_i, which can be estimated by maximum likelihood estimation. And we assume $P(J)$ is uniformly distributed.

Based on the assumption that the regions of image are independent each other, $P_V(I|J_i)$ equals to the product of the regional generation probabilities. The regional generation probability $P_V(f_j|J_i)$ can be estimated by non-parameter kernel-based density estimation [10].

3.2 The Keyword Correlation Graph Generation by Web Semantic Space Learning

The directed keyword correlation graph is denoted by $Sim = <V, E>$, where the vertex set V consists of all the annotation keywords, and a directed edge from keyword w to keyword w' is denoted by $e_{ww'} \in E$ which is established if and only if $Sim(w, w') > 0$, where $Sim(w, w')$ is the similarity between w and

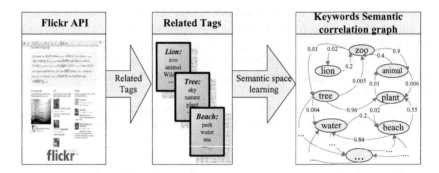

Fig. 1. The generation process of keywords correlation graph. Only part of the vertices and edges are given.

w'. Note that in our keyword correlation graph, $Sim(w, w')$ may not equal to $Sim(w', w)$.

Figure 1 gives the generation process of keywords correlation graph. Firstly, we submit the annotation vocabulary to Filckr to retrieve a neighborhood Image Semantic Subspace (ISS), which is composed of the returned Related Tags (RT)(or concepts). In order to further explore the correlations between the pairs of concepts(keywords) that cannot be directly obtained from Flickr, we build a directed graph $G = < V', E' >$ to represent the semantic correlations obtained from Flickr directly, where the vertex set V' consists of the keywords in ISS, and a directed edge from keyword w to keyword w' is denoted by $e_{ww'} \in E'$ which is established if and only if $w' \in RT(w)$, where $RT(w)$ is the set of Related Tags (RT) of w.

The definition of the directed graph G shows that, if concept (keyword) w' is similar to w, then there exists a path from vertex w to w'. If concept(keyword) w is accessible from w_1 and w_2 in G, and the number of the accessible concept(keyword) of w_1 is larger than w_2, then we can conclude that the similarity between w_1 and w is smaller than the one between w_2 and w. Therefore we can estimate the semantic similarity between w and w' as follows:

$$Sim(w, w') = e^{-dis(w,w') \times \frac{|acess(w)|}{|ISS|}}, \qquad (8)$$

where $dis(w, w')$ refers to the distance from w to w' in graph G, which is measured by the length of the shortest path from w to w' in graph G, $acess(w)$ is the set of the accessible concepts(keywords) from w, and $|acess(w)|$ and $|ISS|$ denotes the number of keywords in $acess(w)$ and ISS respectively.

3.3 The Estimation of Textual Generation Probability $P(w|I_T)$

The Object Function. We adopt the linear basic expansion model for estimating the textual generation probability $P(w|I_T)$. Let $H(T)$ denote the set of expansion functions, which represents the associated texts T and their interaction structures; $\omega = \{\omega_1, \ldots, \omega_N\}$ represents the weights of semantic contributions of

$H(T)$ to I. Then the probability $P(w|I_T)$ can be estimated by a linear model as follows:

$$P(w|I_T) = \sum_{j=1}^{N} \omega_j(w)p(w|h_j(T)),\qquad(9)$$

where N is the number of original and extended parts of the associated texts.

The Estimation of Probability $p(w|h_j(T))$. We extend the structure of the associated texts to further explore the relationship between image semantic labels and the different types of associated texts. That is, we consider the higher order pairwise structures of the different types of the associated texts by estimating their pairwise joint generation probability $p(w|T_kT_l) = p(w|T_k)p(w|T_l)$, where $(k \neq l) \leq n$ and $p(w|T_i)$ can be estimated by the textual multinomial distribution estimation [22]. Here we define the expansion function set $H(T)$ to represent the associated texts and their higher-order interaction structures identically. For simplicity, we just consider the semantic contributions of the textual data T and their order 2 interaction structures. The probability of keyword w being generated by $h_j(T) \in H(T)$ is estimated as follows:

$$\hat{p}(w|h_j(T)) = \begin{cases} p(w|T_j) & j = 1,\ldots,n \\ p(w|T_i)p(w|T_l) & (i \neq l) \leq n, n < j \leq N \end{cases}\qquad(10)$$

The Estimation of Weights $\omega(w, I)$. According to Eqn.9, the weights distribution are crucial to estimate the textual generation probability $P(w|I_T)$. Thus we propose the following constraint piecewise penalized weighted regression model to learn the weight distribution $\omega(w, I)$:

1. For a given unlabeled image I and the corresponding associated texts T, a neighborhood in the training Web image set (denoted as $neighbor(I)$) is first generated under the visual and textual features similarity measurement.

2. Since the textual structures have higher order, we impose different penalty to the associated texts and their pairwise higher order structure. We partition the weight coefficients into k subsets corresponding to T and their $i^{th}(i = 2,\ldots,k)$ order interaction structures. Our aim is to shrinkage the regression coefficients by imposing a L_2 penalty to each part, where the penalty parameters are $\gamma = \{\gamma_1,\ldots,\gamma_k\}(\gamma_1 \geq \ldots \geq \gamma_k)$. Especially, to rectify the statistic error brought by the limitation of the training data, we add the additional prior knowledge into our regression model. Denote the prior knowledge as D_{pre}, which is the set of important dimensions, and the corresponding penalty parameter is γ_{pre}. Then the constraint piecewise penalty weighted regression estimation is defined as follows:

$$\hat{\omega}(w) = arg\min_{\omega(w)}\{\sum_{i=1}^{K} \mu_i(y_i - \omega_0 - \sum_{j=1}^{N} X_{ij}\omega_j(w))^2\}$$

$$subject by : \sum_{\omega_j \in D_s} \omega_j(w)^2 \leq t_s, (s = 1,\ldots,k), \sum_{\omega_j \in D_{pre}} \omega_j(w)^2 \geq t_{pre}. \quad(11)$$

Eqn.11 equals to the following constraint piecewise penalty weighted residual sum of squares:

$$\hat{\omega}(w) = arg \min_{\omega(w)} \{ \sum_{i=1}^{K} \mu_i (y_i - \omega_0 - \sum_{j=1}^{N} X_{ij} \omega_j(w))^2 \qquad (12)$$

$$+ \sum_{s=1}^{k} \gamma_s \sum_{\omega_j \in D_s} \omega_j(w)^2 - \sum_{\omega_j \in D_{pre}} \gamma_{pre} \omega_j(w)^2 \},$$

where K is the number of images in $neighbor(I)$, $X_{ij} = p(w|h_j(T))$, y_i refer to the likelihood of the semantic concept w as the label of image J_i (i^{th} Web image in $neighbor(I)$), and T_i denotes the textual features of image J_i, μ_i denotes the similarity between image I and J_i.

4   Experiments

All the data used in our experiments are crawled from Internet. The image data set is obtained by HTML parsing and small icons are filtered out. The size of the image data set L is 4000. We use a heuristic method to generate training set automatically from the download Web pages. The idea is similar to tf/idf heuristic, here we consider two kinds of term frequency, that is, the frequency of the keyword w appears in one type associated text, and the frequency that accounts for the number of the types of associated texts that w appears in. The heuristic rule is that the keyword with higher frequency is more important for the corresponding Web image. At last, we obtain 640 training images, and the rest is used as test set. Each test image is manually labeled with 1-7 keywords as ground truth. The vocabulary of manual annotations consists of about 137 keywords. Each image of L is segmented into 36 blobs based on fixed size grid, and 528 dimensional visual feature for each blob is extracted according to $MPEG7$ standards. Each image associates 5 types of associated texts: image file name, ALT text (ALT tag), caption text (Heading tag), associated text and page title.

We partition half of the training set as validation set to determine the model parameters, such as the smoothing parameter λ, the regularization parameter γ_1, γ_2 and γ_{pre}. Their values are set to 0.6, 0.7, 0.4 and 0.25 respectively in our experiments. The *recall, precision* and *F1* measures are adopted to evaluate the annotation performance. That is, given a query keyword w, let $|W_G|$ denote the number of human annotated images with label w in the test set, $|W_M|$ denote the number of images annotated with the same label by our algorithm. The *recall, precision* and *F1* are respectively defined as: $Recall = \frac{|W_M \cap W_G|}{|W_G|}$, $Precision = \frac{|W_M \cap W_G|}{|W_M|}$, $F1 = \frac{2(Precision \times Recall)}{Precision + Recall}$. The number of annotation keywords is set to 5, and the average *recall, precision* and *F1* over all keywords are calculated as the evaluations of the overall performance.

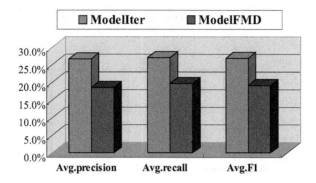

Fig. 2. The overall performance of our Web image annotation approach

4.1   The Overall Performance of Our Annotation Approach

Figure 2 compares the ModelIter approach proposed in this paper and the ModelFMD approach [18]. The ModelFMD model don't use the keywords correlation and learn a fix textual model for estimating the semantic from the associated texts of Web image in the training stage. According to the figure, the performance of our approach is superior to the ModelFMD method significantly. Our annotation framework incorporates two approaches: the keywords correlation and the constraint piecewise penalty weighted regression, we need to test their effectiveness for improving the performance of Web image annotation respectively.

4.2   The Effectiveness of the Web Semantic Space Learning Based Keywords Correlation

To test the effectiveness of our Web semantic space learning based keywords correlation, we compare the performance of our annotation approach which uses the Web concept space learning based keywords correlation(Flickr) and the WordNet-based keywords correlation(WordNet), against the annotation approach without using keywords semantic correlation(noKWCor), that is the probability estimation(Eqn.1) doesn't consider the keywords correlation. Both methods use the constraint piecewise penalty weighted regression model to estimation the semantic contribution of the associated texts, and consider the contribution of visual features. Figure 3 gives the comparison result.

According to Figure 3, the performance of Flickr and WordNet both be superior to noKWCor, which show that the keywords correlation could be an effective way to smooth the maximum likelihood estimation to exploit the keywords correlation in the process of Web image annotation. Meanwhile, we found that the performance of Flickr is better than WordNet significantly, this demonstrates our Web semantic space learning method is more effective than WordNet for measuring the keyword correlation to improve the performance of Web image annotation.

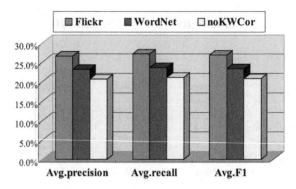

Fig. 3. The effectiveness of keywords semantic correlation

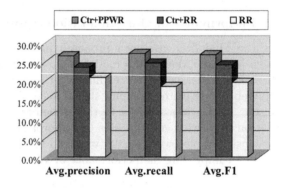

Fig. 4. The effectiveness of prior constraint and Piecewise Penalty Weighted Regression

4.3 The Effectiveness of Prior Constraint and Piecewise Penalty Weighted Regression

To test the effectiveness of prior knowledge constraint(Ctr) and Piecewise Penalty Weighted Regression(PPWR) in the process of associated texts based generation probability estimation, we compare the performance of our approach(Ctr+PPWR) and the approaches which only consider the contribution of Ctr or PPWR. The baseline approach does not consider the prior knowledge constraint and applies Ridge Regression(RR) to learn the weights distribution. All approaches consider the contributions of the keywords correlation and the visual features. Figure 4 gives the comparison result.

The results in Fig.4 show that: (a) The "Ctr+RR" approach is superior to RR approach. This demonstrates that it is effective to impose the prior knowledge constraints in the regress model for Web image semantic annotation. (b) The "Ctr+PPWR" approach is superior to "Ctr+RR" approach. It demonstrates that PPWR algorithm is more effective than ridge regression in learning the weight distribution of the associated texts and their higher order structures when annotating Web images.

5   Conclusions

Ubiquitous image resources on the Web have long been attractive to research community. Web-based AIA is a promising way to manage and retrieve the fast growing Web images. However, its effectiveness still needs to be improved. In this paper, we developed and evaluated a novel automatic Web image annotation approach, which incorporates the image semantic keywords correlations and the semantic contributions of visual features and associated texts of Web image. In particular, we estimate the keywords semantic correlation by using Web image semantic space learning, as well as adaptively model the distribution of semantic labels of Web images on their associated texts by using the proposed constraint piecewise penalty weighted regression. The experimental results demonstrate that both the Web semantic space learning based keywords correlation and the constraint piecewise penalty weighted regression model improve the performance of Web image annotation significantly.

Acknowledgment

This work was partially supported by the Natural Science Foundation of China under Grant No.60403018 and No.60773077.

References

1. http://www.flickr.com
2. Wang, B., Li, Z., Yu, N., Li, M.: Image annotation in a progressive way. In: ICME, pp. 1483–1490 (2007)
3. Chang, E., et al.: Cbsa: Content-based soft annotation for multimodal image retrieval using bayes point machines. CirSysVideo 13(1), 26–38 (2003)
4. Duygulu, P., Barnard, K., de Freitas, J., Forsyth, D.: Object recognition as machine translation: Learning a lexicon for a fixed image vocabulary. In: Heyden, A., Sparr, G., Nielsen, M., Johansen, P. (eds.) ECCV 2002. LNCS, vol. 2353, pp. 97–112. Springer, Heidelberg (2002)
5. Feng, H., Shi, R., Chua, T.: A bootstrapping framework for annotating and retrieving www images. In: ACM Multimedia, pp. 960–967 (2004)
6. Feng, S., Manmatha, R., Lavrenko, V.: Multiple bernoulli relevance models for image and video annotation. In: CVPR, pp. 1002–1009 (2004)
7. Jeon, J., Lavrenko, V., Manmatha, R.: Automatic image annotation and retrieval using cross-media relevance models. In: SIGIR, pp. 119–126 (2003)
8. Jin, R., Chai, J., Si, L.: Effective automatic image annotation via a coherent language model and active learning. In: ACM Multimedia (2004)
9. Jin, Y., Khan, L., Wang, L., Awad, M.: Image annotations by combining multiple evidence & wordnet. In: ACM Multimedia, pp. 706–715 (2005)
10. Lavrenko, V., Manmatha, R., Jeon, J.: A model for learning the semantics of pictures. In: NIPS (2003)
11. Li, X., Chen, L., Zhang, L., Lin, F., Ma, W.: Image annotation by large-scale content-based image retrieval. In: ACM Multimedia, pp. 607–610 (2006)

12. Mei, Q., Zhang, D., Zhai, C.: A general optimization framework for smoothing language models on graph structures. In: SIGIR, pp. 611–618 (2008)
13. Rui, X., Li, M., Li, Z., Ma, W., Yu, N.: Bipartite graph reinforcement model for web image annotation. In: ACM Multimedia, pp. 585–594 (2007)
14. Sanderson, H., Dunlop, M.: Image retrieval by hypertext links. In: SIGIR (1997)
15. Song, Y., Zhuang, Z., Li, H., Zhao, Q.: Real-time automatic tag recommendation. In: SIGIR, pp. 515–522 (2008)
16. Srikanth, M., Varner, J., Bowden, M., Moldovan, D.: Exploiting ontologies for automatic image annotation. In: SIGIR, pp. 552–558 (2005)
17. Tang, J., Hua, X.-S., Qi, G.-J., Wang, M., Mei, T., Wu, X.: Structure-sensitive manifold ranking for video concept detection. In: ACM Multimedia (2007)
18. Tseng, V., Su, J., Wang, B., Lin, Y.: Web image annotation by fusing visual features and textual information. In: SAC, pp. 1056–1060 (2007)
19. Xu, H., Zhou, X., Lin, L.: Wisa: A novel web image semantic analysis system. In: SIGIR (2008)
20. Wang, X., Zhang, L., et al.: Annosearch: Image auto-annotation by search. In: CVPR, pp. 1483–1490 (2006)
21. Yang, C., Dong, M.: Region-based image annotation using asymmetrical support vector machine-based multiple-instance learning. In: CVPR, pp. 2057–2063 (2006)
22. Zhou, X., Wang, M., Zhang, Q., Zhang, J., Shi, B.: Automatic image annotation by an iterative approach:incorporating keyword correlations and region matching. In: CIVR, pp. 25–32 (2007)

Representing and Inferring Causalities among Classes of Multidimensional Data[*]

Kun Yue[1], Mu-Jin Wei[1], Kai-Lin Tian[2], and Wei-Yi Liu[1]

[1] Department of Computer Science and Engineering,
School of Information Science and Engineering, Yunnan University,
Kunming, 650091, P.R. China
[2] Library of Southwest Forestry University,
Kunming, 650224, P.R. China
yue.kun@gmail.com

Abstract. When adopting Bayesian network (BN) to represent and infer probabilistic causalities among multidimensional variables, the size of the conditional probability table (CPT) associated with each variable is doomed to be large, and the causality inferences cannot be done for arbitrary evidences. In this paper, we first extend the general BN by augmenting parameters for describing causalities among classes instead of specific instances of multidimensional variables. In the extended BN, called CBN, the CPT of a variable includes the probability of each class given parent classes, while a classifier of each variable is associated to determine the class that the given evidence belongs to. Further, we give the method for approximate inferences of the CBN for arbitrary evidences. Preliminary experiments verify the feasibility of our methods.

Keywords: Multidimensional random variable, Class, Bayesian network, Approximate inference, Classification.

1 Introduction

Multidimensional random variables are widely used to describe complex phenomena in real-world applications. For example, *Customer* is multidimensional and described by *age, income, student* and *credit-rating*. *Sold Product* is described by *price, unit, private-use* and *after-sale service*. Multidimensional data on multidimensional variables are inherent in many applications such as e-commerce, telecommunication, retail, financial markets, scientific data, bioinformatics data. Mining multidimensional data plays a critical role in both theoretical data mining research and data mining development [1, 2].

It is well known that Bayesian network is the effective framework of representing uncertainties among random variables. The representation and inference

---

[*] This work was supported by the National Natural Science Foundation of China (No. 60763007), the Natural Science Foundation of Yunnan Province (No. 2008CD083) and the Research Foundation of the Educational Department of Yunnan Province (No. 08Y0023).

Q. Li et al. (Eds.): APWeb/WAIM 2009, LNCS 5446, pp. 223–234, 2009.
© Springer-Verlag Berlin Heidelberg 2009

of probabilistic causalities based on Bayesian network are widely used in realistic diagnosis, decision support, prediction, etc [3, 4]. A BN is a directed acyclic graph (DAGs) where nodes represent random variables, and edges represent conditional dependencies between these random variables. Each variable in a BN is associated with a CPT to give the probability of each state given parent states. A BN can be constructed by means of statistical learning from sample data [5, 6, 7]. By inferences [12], the posterior probability distribution of query variables can be obtained. It is also natural to adopt Bayesian network to represent and infer probabilistic causalities implied in multidimensional data.

Example 1. Suppose a simple BN of *Customer* and *Sold Product* is learned and shown in Fig. 1, in which we assume *Customer*=(*age, income, student, credit-rating*) and *Sold Product*=(*price, unit, private-use, after-sale service*).

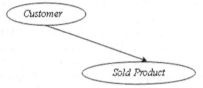

Customer	$P(C)$
$cust_1$=(≤30, high, no, fair)	0.4
$cust_2$=(31...40, low, yes, excellent)	0.3
$cust_3$=(>40, medium, no, excellent)	0.1
$cust_4$=(>40, low, yes, fair)	0.2

P (Sold Product|Customer)	$cust_1$	$cust_2$	$cust_3$	$cust_4$
$prod_1$=(medium, S, yes, no)	0.1	0.4	0.5	0.2
$prod_2$=(high, Z, no, yes)	0.2	0.4	0.3	0.8
$prod_3$=(low, T, yes, yes)	0.7	0.2	0.2	0

Fig. 1. A simple Bayesian network

Unfortunately, current research findings of BN inferences mainly focus on the efficiency-related issues [8, 9, 10, 11]. The need for BN inferences under arbitrary evidences is pervasive, but the inference of a BN cannot be made when the given evidence is not appearing in corresponding CPTs. For example, by ordinary inference method on the BN in Fig. 1, we cannot obtain the probability of *Sold Product* under the evidence of $c = (\leq 30, medium, no, excellent)$, since there is no data about such *customers* in sample data and accordingly no corresponding entries in the CPT associated to *Customer*. This frequently happens in BN inferences. Especially for multidimensional situations, the CPT size of each variable is doomed to be large because of the large amount of combinations of values on various dimensions. The more the dimensions of a variable, the less probably the CPT includes all possible values. Furthermore, although the BN could be revised and improved gradually based on inferences and domain experts, it is still not realistic to enumerate all possible values for a multidimensional variable.

Actually, in real applications, higher level knowledge on classes may be paid more attention when too precise details are not necessary. Above observations motivate our idea of extending the general BN by considering inherent relationships among classes of multidimensional data. To make inferences on a BN given arbitrary evidences, we will have to consider the following 2 aspects of extensions to the general BN: (1) The CPT of each variable should represent the probabilities of each class given parent classes instead of those for specific multidimensional values. (2) The class that the given evidence belongs to should be determined with respect to the corresponding variable.

For the first aspect, we obtain the CPT of classes easily by statistics on the sample data with class labels. For the second aspect, the ultimate class may not be distinguished even though the classification rules for every dimension have been given. Moreover, it is brute force to give the classification rules by considering all possible combinations of the values on respective dimensions.

Decision tree induction is the learning of decision trees from class-labeled training sample [2]. Based on the algorithms such as ID3, C4.5, etc. [13, 14], we can learn the decision tree, which can be easily converted to classification rules. Decision tree classifiers [2, 15, 16] are appropriate for exploratory knowledge discovery and can handle high dimensional data [2]. In this paper, taking decision tree classifier as the representative, we adopt the classical method of classification analysis in data mining and machine learning to obtain the classifier for each multidimensional variable respectively. Therefore from the multidimensional perspective, we obtain the class-parameter-augmented BN, called CBN, as the extension to a general BN. Consequently, we propose the method for approximate inferences of the CBN based on the Gibbs sampling algorithm [8].

Generally speaking, the main advantages of the CBN we propose in this paper can be summarized as follows: by incorporating the classical classification method into Bayesian network modeling, probabilistic causalities among classes instead of those among specific values of multidimensional variables are represented and inferred. The inference can be made approximately no matter whether the given evidence is appearing in the sample data from which the network is learned. Based on the CBN and its inference method, causally relevant concepts can be discovered automatically.

The remainder of this paper is organized as follows: In Section 2, we define and construct the CBN, the BN with parameters of describing causalities among classes of multidimensional data. In Section 3, we give the method for approximate inferences of the CBN. In Section 4, we show experimental results and performance analysis. In Section 5, we conclude and discuss future work.

2   Bayesian Network with Class Parameters

Definition 1. *A Bayesian network is a directed acyclic graph, in which the following holds [3, 4]:*

(1) A set of random variables makes up the nodes of the network.

(2) A set of directed links connects pairs of nodes. An arrow from node X to node Y means that X has a direct influence on Y. Each node X is independent of its non-descendants given its parents.

(3) Each node has a conditional probability table that quantifies the effects that the parents have on the node. The parents of node X are all those that have arrows pointing to X.

A BN represents the joint probability distribution (JPD) in products, and every entry in the JPD can be computed from the information in the BN by the chain rule $P(x_1, \cdots, x_n) = \prod_{i=1}^{n} P(x_i | Parents(x_i))$.

Based on the definition of a general BN, following we give the definition of the BN with class parameters on multidimensional random variables.

Definition 2. *Let* $\mathcal{V} = \{\overrightarrow{V_1}, \overrightarrow{V_2}, \cdots, \overrightarrow{V_n}\}$ *be the set of multidimensional random variables, where* $\overrightarrow{V_i} = (A_{i1}, A_{i2}, \cdots, A_{im_i})$ *is a multidimensional variable and* A_{ij} *is the j-th dimension of the i-th variable* $(1 \leq i \leq n, 1 \leq j \leq m_i)$. *Let* $C_i = \{c_{i1}, c_{i2}, \cdots, c_{il_i}\}$ *be the classes on* $\overrightarrow{V_i}$.

Example 2. $\overrightarrow{V_1} = Customer = (A_{11} = age, A_{12} = income, A_{13} = student, A_{14} = credit-rating)$ *and* $\overrightarrow{V_2} = Sold\ Product = (A_{21} = price, A_{22} = unit, A_{23} = private-use, A_{24} = after-sale\ service)$ *are 2 multidimensional variables, which constitute the set* $\mathcal{V} = \{Customer, Sold\ Product\}$. *Suppose buys-computer =yes and buys-computer=no are 2 classes of Customer, while is-software=yes and is-software=no be 2 classes of Sold Product. Then, we have* $C_1 = \{c_{11} : buys\text{-}computer=yes, c_{12} : buys\text{-}computer=no\}$ *and* $C_2 = \{c_{21} : is\text{-}software=yes, c_{22} : is\text{-}software=no\}$.

Definition 3. *A class-parameter-augmented BN, abbreviated as CBN, is a DAG* $G = (\mathcal{V}, E)$, *where* $\mathcal{V}$ *is the set of multidimensional variables and* E *is the set of directed edges in G. The random variables in* $\mathcal{V}$ *make up the nodes of G, and for each node* $\overrightarrow{V_i}(1 \leq i \leq n)$ *in G:*

(1) There is a CPT to represent the probability of each class of $\overrightarrow{V_i}$ *given all classes of parents of* $\overrightarrow{V_i}$, *denoted* $P(c_i | Parents(c_i))$, *where* $c_i \in C_i$ *and* $Parents(c_i)$ *is the set of classes of parents of* $\overrightarrow{V_i}$. *Thus every entry of the JDP on classes can be computed by* $P(c_1, \cdots, c_n) = \prod_{i=1}^{n} P(c_i | Parents(c_i))$.

(2) There is a classifier f for mapping each value e_i *of* $\overrightarrow{V_i}$ *to the corresponding class* c_i *in* C_i, *and* $f : \overrightarrow{V_i} \rightarrow C_i$.

We assume that sample data are given with class labels for every variable, such as those shown in Table 1. In particular, the sample data for constructing the CBN structure can be easily extracted from the original sample set by selecting the class labels for each variable. For example, the sample data on classes of *Customer* and *Sold Product* can be extracted from Table 1 and shown as Table 2. Then the methods for constructing a BN [5, 6, 7] can be executed. Accordingly, we can obtain the CPTs based on the DAG structure and the sample data of classes, as shown in Table 2. At the same time, the classifier for each variable can be learned based on the classification methods in data mining and machine learning [2]. In this paper, we adopt decision tree induction [13, 14, 15, 16] as the basis of classification.

Example 3. Suppose the CBN structure is learned from sample data and shown in Fig. 1. By statistics on the sample data in Table 2, we can obtain

Table 1. Sample data with class labels

A_{11}	A_{12}	A_{13}	A_{14}	C_1	A_{21}	A_{22}	A_{23}	A_{24}	C_2
	$\overrightarrow{V_1}=Customer$				$\overrightarrow{V_2}=Sold\ Product$				
≤ 30	high	no	fair	no	high	Z	no	yes	no
≤ 30	high	no	excellent	no	medium	S	yes	no	yes
$31\cdot\cdot40$	high	no	fair	yes	low	T	yes	yes	no
> 40	low	yes	fair	yes	high	S	no	no	yes
> 40	low	yes	excellent	no	high	T	yes	yes	no
$31\cdot\cdot40$	medium	no	excellent	yes	medium	T	no	yes	no
≤ 30	medium	yes	excellent	yes	low	S	yes	no	yes
≤ 30	low	yes	fair	yes	high	Z	no	yes	yes
> 40	medium	yes	fair	yes	medium	S	no	yes	yes
$31\cdot\cdot40$	low	yes	excellent	yes	low	T	yes	no	no

Table 2. Sample data on classes of *Customer* and *Sold Product*

$\overrightarrow{V_1}=Customer$	$\overrightarrow{V_2}=Sold\ Product$	Count
yes	yes	4
yes	no	3
no	yes	1
no	no	2

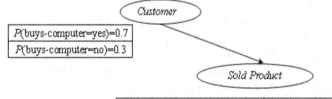

Fig. 2. The Bayesian network with class CPTs

$P(buys-computer = yes) = 0.7, P(buys-computer = no) = 0.3$. Then, we have $P(is-software = yes|buys-computer = yes) = \frac{4}{7}, P(is-software = no|buys-computer = yes) = \frac{3}{7}$. Meanwhile, $P(is-software = yes|buys-computer = no) = \frac{1}{3}$ and $P(is-software = no|buys-computer = no) = \frac{2}{3}$. Therefore, the Bayesian network with class CPTs is shown in Fig. 2.

Based on the sample data as shown in Table 1 and the classical decision tree algorithms [2, 13, 14], the decision trees of *Customer* can be obtained in Fig. 3, while the tree of *Sold Product* will not be given for space limitation.

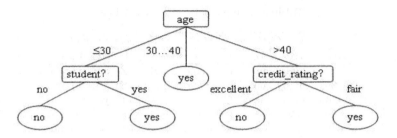

Fig. 3. The decision tree for *buys-computer* of *Customer*

Consequently, for any given evidence, we can determine the class that it belongs to at first. For example, if the value of *Customer* is given as *cust* = (*age* ≤ 30, *income* = *medium*, *student* = *yes*, *credit-rating=fair*), then we can obtain *cust* belongs to the class of *buys-computer=yes* based on the decision tree in Fig. 3.

3   Approximate Inferences of the CBN

It is known that when computing posterior probabilities, the efficiency and tractability of exact BN inferences depends on the network structure [4]. In this paper, we are to discuss the inference on the classes of multidimensional variables, no matter whether the given evidence values are included in the sample data or not. This leads to the approximate result given any specific multidimensional values. We consider adopting the Gibbs sampling algorithm [8] for approximate inferences of CBN on the basis of the property of Markov blankets. Pearl pointed out that in any BN, the union of the following 3 types of neighbors is sufficient for forming a Markov blanket of X: the direct parents of X, the direct successors of X and all direct parents of X's direct successors [2, 10].

The Gibbs sampling algorithm [8] generates each variable by making a random change to the proceeding variable. We look upon the network as being a particular current state specifying a value for every variable. The next state is generated by randomly sampling a value for one of the nonevidence variables, conditioned on the current values of the variables in the Markov blanket of this currently-sampled variable. We extend the Gibbs sampling algorithm to the CBN inference and give the approximate inference method in Algorithm 1, whose convergence and corretness can be guaranteed by the conclusion given by Russel [8].

Algorithm 1. CBN_Inference /* Approximate inferences of the CBN */

Input:

   G: a CBN, where the classifier of multidimensional variable $\vec{V_i}$ is denoted as f_i,

   $\vec{Z}$: the nonevidence nodes in G,
   $\vec{E}$: the evidence nodes in G,
   X: the query variable,

$\overrightarrow{e}$: the set of values of the evidence nodes $\overrightarrow{E}$, $\overrightarrow{e} = (\overrightarrow{e_1}, \overrightarrow{e_2}, \cdots, \overrightarrow{e_m})$,
n: the total number of samples to be generated.

Output: The estimates of $P(X|\overrightarrow{e})$.
Variables:

$\overrightarrow{z}$: the set of values of the nonevidence nodes $\overrightarrow{Z}$,
$\overrightarrow{q} = \overrightarrow{z} \cup \overrightarrow{e}$: the current state of CBN,
$N_X(x_i)(i = 1, \cdots, l)$: a vector of counts over probabilities of x_i, where
$x_i(i = 1, \cdots, l)$ is a class of X.

Step 1. Initialization:
$\overrightarrow{z} \leftarrow$ random values of nonevidence nodes $\overrightarrow{Z}$
$\overrightarrow{e} \leftarrow$ evidence values of $\overrightarrow{E}$
for $i \leftarrow 1$ to m do
$c_i \leftarrow f_i(\overrightarrow{e_i})$ /*get the class of each value in the give evidence*/
end for
$\overrightarrow{e} \leftarrow (c_1, c_2, \cdots, c_m)$
for $j \leftarrow 1$ to l do
$N_X(x_j) \leftarrow 0$
end for

Step 2. for $i \leftarrow 1$ to n do
(1) Compute the probabilities of X in the next state:
$\hat{P}(x_i|MB(X))$, where $MB(X)$ is the set of the current values in the Markov
blanket of X
$B \leftarrow \sum_i \hat{P}(x_i|MB(X))$
(2) Generate a random number $r \in [0, B]$ and we determine the value of X:

$$X = \begin{cases} x_1 & r \leq P(x_1|MB(X)) \\ x_2 & P(x_1|MB(X)) < r \leq P(x_1|MB(X)) + \leq P(x_2|MB(X)) \\ \cdots & \cdots \end{cases}$$

(3) Count: If $X = x_i$ then $N_X(x_i) \leftarrow N_X(x_i) + 1$

Step 3. Estimate $P(X|\overrightarrow{e})$: $P(x_i|\overrightarrow{e}) \leftarrow \frac{N_X(x_i)}{n}$.

Example 4. We illustrate the application of Algorithm 1 based on the CBN of
$\overrightarrow{V_1}$, $\overrightarrow{V_2}$, $\overrightarrow{V_3}$ and $\overrightarrow{V_4}$, shown in Fig. 5. Following we consider the query $P(C_3|C_1 = c_{11}, C_4 = c_{42})$ applied to the CBN. The evidence variables $\overrightarrow{V_1}$ and $\overrightarrow{V_4}$ are fixed to
their observed classes and the nonevidence variables $\overrightarrow{V_2}$ is initialized randomly.
We suppose the initial state is $[C_1 = c_{11}, C_2 = c_{21}, C_3 = c_{31}, C_4 = c_{42}]$. The
following steps are executed repeatedly.
(1) $\overrightarrow{V_2}$ is sampled, given the current classes of its Markov blanket variables.
In this case, we can obtain the probabilities when $\overrightarrow{V_2}$ is sampled as its classes
c_{21} and c_{22} respectively.

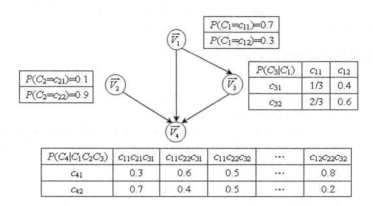

Fig. 4. An example of CBN

$P(C_2 = c_{21}|C_1 = c_{11}, C_3 = c_{31}, C_4 = c_{42})$
$= P(C_1 = c_{11})P(C_2 = c_{21})P(C_3 = c_{31}|C_1 = c_{11})P(C_4 = c_{42}|C_1 = c_{11}, C_2 = c_{21},$
  $C_3 = c_{31}) = 0.3 \times 0.1 \times \frac{1}{3} \times 0.7 = 0.007,$

$P(C_2 = c_{22}|C_1 = c_{11}, C_3 = c_{31}, C_4 = c_{42}) = 0.036.$

Suppose $r_2 = 0.04$, and then the result will be c_{22}. The new state will be $[C_1 = c_{11}, C_2 = c_{22}, C_3 = c_{31}, C_4 = c_{42}]$.

(2) $\vec{V_3}$ is sampled, given the current classes of its Markov blanket variables.
$P(C_3 = c_{31}|C_1 = c_{11}, C_2 = c_{22}, C_4 = c_{42})$
$= P(C_1 = c_{11})P(C_2 = c_{22})P(C_3 = c_{31}|C_1 = c_{11})P(C_4 = c_{42}|C_1 = c_{11}, C_2 = c_{22},$
  $C_3 = c_{31}) = 0.3 \times 0.9 \times \frac{1}{3} \times 0.4 = 0.036,$
$P(C_3 = c_{32}|C_1 = c_{11}, C_2 = c_{22}, C_4 = c_{42}) = 0.09.$

Suppose $r_3 = 0.1$, and then the result will be c_{32}. The new state will be $[C_1 = c_{11}, C_2 = c_{22}, C_3 = c_{32}, C_4 = c_{42}]$.

If the process visits 40 states of $C_3 = c_{31}$ and 160 states of $C_3 = c_{32}$, then $P(c_{31}|\vec{e}) = 0.2$ and $P(c_{32}|\vec{e}) = 0.8$, which are answers to the given query. It should be noted that each state visited during above processes is a sample that contributes to the estimate for the query variable $\vec{V_3}$.

4   Experimental Results

To test the feasibility of the methods proposed in this paper, we implemented those for constructing and inferring CBNs. Centered on CBN inferences given by Algorithm 1, the performance of CBN is tested. We first tested the influence of noise data in the sample set on CBN inference results. We then tested the precision of CBN inferences compared with those of corresponding BNs.

4.1   Experiment Setup

We generated the test data manually according to the *AllElectronics* customer database [13]. All test data were stored in MS SQL Server. Our codes were

Table 3. Multidimensional variables and their classes in test data

Multidimensional variables	Classes
$Branch = (in\_shopping\_center, trained\_for\_sale)$	$backend = \{yes, no\}$
$Customer = (age, income, student)$	$buy\_computer = \{yes, no\}$
$Employee = (full\_of\_frontend\_experience, skill)$	$software\_salesman = \{yes, no, other\}$
$Sold\_Product = (price, after\_sale\_service)$	$is\_software = \{yes, no\}$

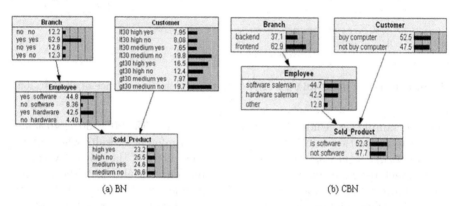

(a) BN   (b) CBN

Fig. 5. BN and CBN learned from test data

written in Java and JDBC-ODBC was used to connect the database. We constructed the CBN based on PowerConstructor [7], a learning tool of BNs, and we implemented the decision tree induction algorithm [13] to learn the classifiers for concerned variables. The multidimensional variables and their classes are shown in Table 3. The domains of each dimension of every variable are ignored here for space limitation while shown in corresponding CPTs of the generated BN.

On the generated sample data without noises, the BN and CBN are learned and shown in Fig. 5 (a) and Fig. 5 (b) respectively.

According to the sample data with class labels, for each variable the specific values and their respective classes are given as follows. For *Branch*, (no, no), (no, yes) and (yes, no) belong to *backend* = *yes*, while (yes, yes) belongs to *backend* = *no* (i.e, *frontend* in Fig. 5 (b)). For *Customer*, (lt30, high, yes), (lt30, high, no), (lt30, medium, yes), (gt30, high, yes) and (gt30, high ,no) belong to *buy_computer* = *yes*, while others belong to *buy_computer* = *no* (i.e, *not_buy_computer* in Fig. 5 (b)). For *Employee*, (yes, software) belongs to *software_saleman* = *yes*, (yes, hardware) belongs to *software_saleman* = *no* (i.e, *hardware_saleman* in Fig. 5 (b)), while others belong to *software_saleman* = *other*. For *Sold_Product*, (high, yes) and (medium, yes) belong to *is_software* = *no* (i.e, *not_software* in Fig. 5 (b)), while others belong to *is_software=yes*.

4.2   Influences of Noise Data on CBN

To explore the influential factors on the CBN and its inferences, we tested the tendency of CBN inferences with the increase of proportions of noise data in the

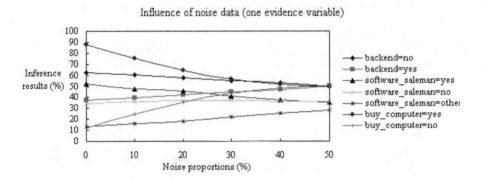

Fig. 6. Influence of noise data on CBN inferences given the evidence of one variable

generated sample set. Under the proportions of 0%, 10%, 20%, 30%, 40% and 50%, we recorded the conditional probabilities of inference results when adopting $is\_software = yes$ as the evidence and every class of the other 3 variables as the target respectively. The tendency of inference results with the increase of noise proportions is shown in Fig. 6. Meanwhile, we also recorded the inference results when adopting $backend = no$, $buy\_computer = yes$ as the evidence and every class of *Employee* and *Sold_Product* as the target respectively. The similar tendency could be obtained, which will not be shown here for space limitation.

In our sample data, for one class of a a certain variable, the noise data are exactly the non-noisy ones for other classes. From Fig. 6, we can see that for some certain classes, the posterior probabilities are decreased with the increase of noise proportions, while the probabilities are increased for other classes of the same variable. Moreover, we can conclude that the CBN inferences are dependent on the proportions of noise data in the sample set, and the sensitivity to noise data is determined by the adopted classification method. Therefore, the applicability of the CBN can be guaranteed to a great extent by choosing appropriate classification methods according to specific realistic situations.

4.3   Precision of CBN Inferences

For all possible specific values of the evidence and target variables in sample data, we obtained the inference results by the BN in Fig. 5 (a). Accordingly, we derive corresponding conditional probabilities of the classes that the target values belong to under the evidence classes. These probabilities were adopted as the criteria to test the precision of CBN inferences.

Meanwhile, we obtained the inference results directly by the CBN in Fig. 5 (b) with respect to the same evidence and target variables as those in above BN inferences. The classes are considered instead of the specific values concerned in the BN inferences. Then, as for some representative cases, we specifically

Table 4. Errors of CBN inferences

	BN(%)	CBN(%)	Error(%)
$P(software\_saleman = yes|is\_software = yes)$	52.8	52.6	0.2
$P(software\_saleman = no|is\_software)$	34.35	34.3	0.05
$P(buy\_computer = yes|is\_software = yes)$	88	88	0
$P(buy\_computer = no|is\_software = yes)$	12	12	0
$P(backend = no|is\_software = yes)$	62.5	62.4	0.1
$P(backend = yes|is\_software = yes)$	37.5	37.6	0.1
$P(software\_saleman = yes|is\_software = no)$	36	36	0
$P(software\_saleman = no|is\_software = no)$	51.4	51.5	0.1
$P(buy\_computer = yes|is\_software = no)$	13.8	13.7	0.1
$P(buy\_computer = no|is\_software = no)$	87.55	86.3	1.25
$P(backend = no|is\_software = yes)$	63.3	63.4	0.1
$P(backend = yes|is\_software = no)$	36.75	36.6	0.15
$P(backend = no|is\_softwarebuy\_computer = no)$	61.45	61.1	0.35
$P(backend = yes|is\_software = yes, buy\_computer = no)$	38.55	38.9	0.35
$P(software\_saleman = yes|is\_software = yes, buy\_computer = no)$	77.42	76.6	0.82
$P(software\_saleman = no|is\_software = yes, buy\_computer = no)$	9.56	9.64	0.08

compared the conditional probabilities obtained by BN inferences with those obtained by CBN inferences, and we gave the errors in Table 4. It can be seen that the maximal, minimal and average errors are 1.25%, 0% and 0.23% respectively, which can be accepted generally and looked upon as precise basically. This means that the CBN inference results are precise and thus well verifies the applicability and feasibility of CBN.

To sum up, above experimental results show the soundness and feasibility of CBN, as well as the corresponding inference method proposed in this paper. These results make CBN be practical in real-world situations for representing and inferring causalities among classes of multidimensional data.

5   Conclusions and Future Work

In this paper, we first extended the general BN by augmenting parameters of describing causalities among classes instead of specific instances of multidimensional data. Thus the class-parameter-augmented BN, called CBN, was obtained. We then proposed the approximate inference method for the CBN no matter whether the given evidence is appearing in the sample data from which the network is learned. Preliminary experiments showed the feasibility of our methods.

The methods proposed in this paper also raise some other interesting research issues. For example, based on the CBN and its inference method, causally relevant concepts can be discovered automatically. Based on the CBN, we can establish the mechanisms that are used to represent the probabilistic causalities under general-specific relationships. The representation and inferences of the causal relationships can be fulfilled by various hierarchies, and thus queries can be answered according to the requirements by an on-demand manner. These are exactly our future work.

References

1. Pedersen, T.B., Jensen, C.S.: Multidimensional data modeling for complex data. In: Proceedings of the 15th International Conference on Data Engineering (ICDE), pp. 336–345. IEEE Computer Society, Los Alamitos (1999)
2. Han, J., Kamber, M.: Data mining: Concepts and techniques, 2nd edn. Morgan Kaufmann, San Francisco (2006)
3. Heckerman, D., Wellman, M.P.: Bayesian networks. Communication of the ACM 38(3), 27–30 (1995)
4. Pearl, J.: Probabilistic reasoning in intelligent systems: network of plausible inference. Morgan Kaufmann, San Mates (1988)
5. Buntine, W.L.: A guide to the literature on learning probabilistic networks from data. IEEE Transactions on Knowledge and Data Engineering 8(2), 195–210 (1996)
6. Cheng, J., Greiner, R., Kelly, J., Bell, D., Liu, W.: Learning Bayesian network from data: An information-theory based approach. Artificial Intelligence 137(2), 619, 43–90 (2002)
7. Cheng, J.: PowerConstructor system (1998), http://www.cs.ualberta.ca/~jcheng/bnpc.htm
8. Russel, S.J., Norvig, P.: Artificial intelligence – a modern approach. Pearson Education, Publishing as Prentice-Hall (2002)
9. Cooper, G.F.: The computational complexity of probabilistic inference using Bayesian belief networks. Artificial Intelligence 42(2-3), 393–405 (1990)
10. Pearl, J.: Evidential reasoning using stochastic simulation of causal models. Artificial Intelligence 32, 245–257 (1987)
11. Guo, H., Hsu, W.: A survey on algorithms for real-time Bayesian network inference. In: Proc. of the joint AAAI-02/KDD-02/UAI-02 workshop on Real-Time Decision Support and Diagnosis Systems (2002)
12. Norsys Software Corp. Netica 3.17 Bayesian network software from Norsys (2007), http://www.norsys.com
13. Quinlan, J.R.: Induction of decision trees. Machine Learning 1, 81–106 (1986)
14. Murthy, S.K.: Automatic construction of decision trees from data: A multi-disciplinary survey. Data Mining and Knowledge Discovery 2, 345–389 (1998)
15. Rastogi, R., Shim, K.: PUBLIC: A decision tree classifier that integrates building and pruning. In: Proceedings of the 14th International Conference on Data Engineering (ICDE), Florida, USA, pp. 404–415. IEEE Computer Society, Los Alamitos (1998)
16. Alsabti, K., Ranka, S., Singh, V.: CLOUDS: A decision tree classifier for large datasets. In: Proceedings of the 4th International Conference on Knowledge Discovery and Data Mining (KDD), New York, USA, pp. 2–8. AAAI Press, Menlo Park (1998)

Efficient Incremental Computation of CUBE in Multiple Versions What-If Analysis[*]

Yanqin Xiao[1,2,3], Yansong Zhang[1,2,4], Shan Wang[1,2], and Hong Chen[1,2]

[1] Key Laboratory of Data Engineering and Knowledge Engineering, Renmin University of China, MOE, Beijing 100872, P.R. China
[2] School of Information, Renmin University of China, Beijing, P.R. China 100872
[3] Computer Center, Hebei University, Baoding Hebei 071002, China
[4] Department of Computer Science, Harbin Finance College, Harbin 150030, China
{xyqwang,zhangys_ruc,swang,chong}@ruc.edu.cn

Abstract. What-if analysis is an important method to analyze the hypothetical scenarios based on the historical data. It provides useful information for the decision-maker. Multiple versions are critical to what-if analysis. In this paper, we analyze the problem of multiple versions data processing and the incremental computation of cubes in what-if analysis, proposing a strategy to process multiple versions of what-if data. Our solution can adopt different data processing methods according to the type of aggregate functions, the efficiency of multiple versions data processing is improved. Furthermore, we proposes an algorithm of incremental computation of CUBE for max(min) function to reduce the access time of fact table, and the experimental results showed that the performance is improved about 10%.

Keywords: what-if analysis, cube, Delta cube, incremental computation, multiple versions.

1 Introduction

What-if analysis is a particularly common and very important decision support process. It has applications in marketing, production planning, and other areas. What-if analysis is used to forecast future performance under a set of assumptions related to past data. It also enables the evaluation of past performance and the estimation of the opportunity cost taken by not following alternative policies in the past. For example, if the sale price of drink was increased by 10%, how would it affect the profit of the company? What-if analysis can provide important predicting information for decision-maker.

The what-if analysis based on the hypothetical schema update was studied by[1][2][3]. [5]analyzed the types of dimension updates, and the method of maintaining cube under dimension updates. Most of what-if analyses are based on the hypothetical

---

[*] Supported by the National Natural Science Foundation of China under Grant No. 60473069,60496325; the joint research of HP Lab China and Information School of Renmin University(Large Scale Data Management);the joint research of Beijing Municipal Commission of education and Information School of Renmin University(Main Memory OLAP Server).

Q. Li et al. (Eds.): APWeb/WAIM 2009, LNCS 5446, pp. 235–247, 2009.

fact table updates[6][7][8]. It analyzes the influence of the change of aggregation value on the base data from top to bottom.

The development of multiple scenarios is critical to what-if analysis. The what-if update data of each scenario is a what-if version. The what-if version and its original data form a what-if view. Each what-if view shows in detail the result of a combination of decisions, and allows the decision maker to observe the global effect of those decisions. The decision maker can compare the results of multiple what-if views to select the optimal plan. Since what-if view differs only on a few items from its original data, we only store the what-if update data into delta tables to reduce storage cost. Because a what-if version can base on another what-if version, how to these what-if versions and original data to get correct what-if view is a problem to be considered. We formulate the problem and propose a novel method to process multiple version what-if update data.

To speedup the processing of query, some cubes are materialized in data warehouse and OLAP system. As assumed changes are made to the schema, dimension or fact table, a new cube, which has same aggregate function, dimension and measure attributes as the materialized cube, must be updated to reflect the changed state of the data sources. The materialized cube can be recomputed from scratch. It also can be incremental computed by calculate the changes to the views due to the source changes. The delta cube[7] was proposed to incremental compute the new cube. By the method, the procedure of incremental computation of a cube view is divided into two stages: propagate and refresh. In the propagate stage, a delta cube ΔC was computed from the changes of the data source. The ΔC has same aggregate function as the cube view. In the refresh stage, the cube view was refreshed by applying the ΔC to it. The method fits the incremental computing self-maintainable[7] aggregate functions, such as SUM, COUNT, etc.. But, the MAX and MIN functions are self-maintainable with respect to insertions, non-self-maintainable with respect to deletions. For deletion operations, when a tuple having the maximum or minimum is deleted, the new maximum or minimum value for the group must be recomputed. The new maximum or minimum value was recomputed from the base table and delta table[7]. Due to the size of base table is very large, the recompute is very costly. This paper proposed a novel incremental computation data cube for MAX and MIN aggregation functions(abbr. MAX and MIN cube): R_NLC. It uses delta cube method incremental computation MAX and MIN cube. If the tuple having the maximum or minimum is deleted, the new maximum or minimum value is recomputed from the next level cuboid. It reduced the access frequency of base table. As a result, the cost of incremental compute for MAX and MIN cube is substantially reduced.

The contributions of this paper are as follows:

- We propose a novel method to process multiple version what-if update data.
- We propose a novel incremental compute method for MAX and MIN cube that can reduce the access time of base table. Consequently, the cost of incremental compute for MAX and MIN cube is reduced.

This paper is organized as following: Section 2 will discuss the related work. Then, in section 3, the method to process multiple-version of what-if update data is presented. In section 4, we will describe the algorithm to incrementally computing MAX and MIN cube in detail. The results of performance experiments are given in section 5. Finally, we conclude our work in section 6.

2 Related Work

In data warehouse and OLAP system, some cuboids are materialized to improve the query performance. When the data of fact table changed, the materialized cuboids must be updated to reflect the changed state of the data source. There are two strategies to re-compute the data cube: (1) discarding the materialized cuboids, and re-computing new cuboids from the fact table. (2) incremental computing the new cuboids based on the original materialized cuboids. The advantage of first strategy is that it can be used in any situations. Its disadvantage is that it can not make full use of existing resource. The re-computation of data cube consumes a lot of system resources. The advantage of second strategy is that it can make full use of existing cuboids, saving system resources.

In what-if analysis, the materialized cube must be updated to reflect the hypothetical update of the data sources. [2][3]describe different approaches to track schema changes and enable simultaneous queries on different schema version. [3]proposes a SQL query rewriting technique to support the OLAP query analysis in multidimensional database systems modeled on multiversion schema. [4] introduces independently-updated views for creating multiple scenarios. But, they do not consider the processing of data of the cascade what-if version. The incremental compute of data cube under hypothetical dimension update are researched by [5][9] [10][11]. They do not consider the incremental maintenance of holistic functions. [7][8] research the incremental maintenance of data cube under hypothetical fact table update. Both of them adopt delta cube method to incremental compute data cube under fact table update. [7] describes an approach for incremental maintenance of MAX and MIN cube. By the approach, when the tuple having the maximum or minimum value is deleted, the new maximum or minimum value of the group is recomputed from the base table and delta table. It does not use the cuboids in the data cube which has been updated. [8] proposes an incremental computation method for data cube that can maintain a data cube by using only $\binom{n}{\lceil n/2 \rceil}$ delta cuboids. Because the amount of delta cuboids is reduced largely, the cost of computing delta cuboids is substantially reduced. The algorithm does not discuss how to update the MAX and MIN cube with respect to deletions.

3 Processing Multiple Versions of What-If Update Data

In what-if applications, what-if operations are temporary. The what-if update data based on what-if operations does not exist. These what-if update data should not destroy the real data in the OLAP system. Delta table is adopted to store what-if update data. This paper uses two types of delta tables to store what-if data. One is insert_delta table (Δ^+), which stores the data of hypothetical insertions. The other is delete_delta table (Δ^-), which stores the data of hypothetical deletions. The update operations in hypothetical conditions are transformed into deletions and insertions.

3.1 Basic Conceptions about Multiple Versions

Definition 1. what-if operation: A what-if operation $OP_{wi} = \{OP_A, OP_A \in \{Ins, Del, Upd\}\}$. Where, OP_A is an atomic hypothetical update operation. It may be insert(Ins), delete(Del) or update(Upd). That is, a what-if operation is a set of atomic hypothetical update operation.

In what-if analysis, there may be many what-if operation based on the same base table. What-if operation may base on tables or what-if views.

Definition 2: what-if update data: The data produced by what-if operation is named what-if update data, denoted as D_{wi}. $D_{wi} \leftarrow OP_{wi}$ means that the what-if operation OP_{wi} produces a set of data D_{wi}.

Definition 3: base table: A base table T_B is the object of what-if operation. In OLAP system, T_B may be fact table, view or dimension table.

Definition 4: what-if view: The merging of what-if update data D_{wi} with its base table T_B forms a what-if view, denoted as VI, then $VI = T_B \odot D_{wi}$.

Definition 5: what-if version: the what-if update data produced by a what-if operation forms a what-if version, denoted as $\mathcal{V}$. $\mathcal{V} \leftarrow OP_{wi}$ means the what-if version $\mathcal{V}$ is produced by OP_{wi}.

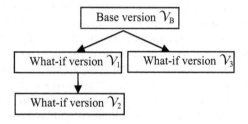

Fig. 1. What-if version

The data of base table T_B forms the base version $\mathcal{V}_B$ of what-if analysis. The data of same what-if version is stored into the same delta table. The relationship of what-if versions may be independently, or relatively. As shown in Figure 1.

Definition 6. cascade what-if version: if $\mathcal{V}_1 \leftarrow OP_{wi1}$, $\mathcal{V}_2 \leftarrow OP_{wi2}$, and OP_{wi2} is based on VI_1, then the what-if version $\mathcal{V}_2$ is the cascade what-if version of $\mathcal{V}_1$, denoted as $\mathcal{V}_2 \bowtie \mathcal{V}_1$. In Figure 1, the $\mathcal{V}_2$ is a cascade what-if version of $\mathcal{V}_1$.

3.2   The Processing of Multiple Versions of What-If Data

Assumed, there is a base table T_B and multiple what-if versions $\mathcal{V}_1, \mathcal{V}_2, \ldots, \mathcal{V}_n$, and $\mathcal{V}_n \bowtie \mathcal{V}_{n-1} \bowtie \cdots \bowtie \mathcal{V}_1$, which corresponding to delta table $\Delta_1, \Delta_2, \ldots, \Delta_n$ respectively, then the what-if view VI_n of what-if version $\mathcal{V}_n$ need to be computed while processing the hypothetical query on VI_n.

There are two methods to compute VI_n:

(1) merging Δ_1 and T_B to get the what-if view of $\mathcal{V}_1$, i.e., VI_1. Then, merging Δ_2 and VI_1 to get VI_2 for $\mathcal{V}_2$, ..., merging Δ_n and VI_{n-1} to get VI_n for $\mathcal{V}_n$.

(2) merging Δ_1, Δ_2, ..., and Δ_n to get a total what-if data table, namely Δ_t. Then, merging Δ_t and T_B to get VI_n for $\mathcal{V}_n$.

Because the number of tuples in base table is much more than the number of tuples in delta table, the second method has higher efficiency than the first method. But, both of them need merge the delta tables and the base table.

When incremental computing cube by delta method, it only need compute the delta_cube based on the delta tables. Therefore, it need not merge the delta tables and the base table. Only the merge of delta tables is needed. Based on different aggregate functions, the method of merge of delta tables is different. The method of merge is discussed as following:

A direct method to merge Δ_1, Δ_2, ..., Δ_n is Δ_1 *Union all* Δ_2 *Union all* ... *Union all* Δ_n. It is assumed that there are two what-if versions, corresponding to four delta tables: Δ_1^+, Δ_1^-, Δ_2^+ and Δ_2^-, respectively. Then, the insert_delta table Δ^+ can be computed from Δ_1^+ *Union all* Δ_2^+. The delete_delta table Δ^- can be computed from Δ_1^- *Union all* Δ_2^-. The processing method is correct to some aggregate functions, for instants, SUM, COUNT, etc.. But, it is not correct to other aggregate functions, for instance, MAX, MIN, etc.. For example, the data in Δ_1^+, Δ_1^-, Δ_2^+ and Δ_2^- is shown in Figure 2:

Δ_1^+:

P	T	C	Unit sales
1	1	1	5
1	1	2	3

Δ_1^-:

P	T	C	Unit sales
3	1	1	6
1	2	2	3

Δ_2^+:

P	T	C	Unit sales
2	1	1	6
1	3	2	3

Δ_2^-:

P	T	C	Unit sales
1	1	1	5
2	2	1	8

Fig. 2. Multiple versions delta data

Δ^+:

P	T	C	Unit sales
1	1	1	5
1	1	2	3
2	1	1	6
1	3	2	3

Δ^-:

P	T	C	Unit sales
3	1	1	6
1	2	2	3
1	1	1	5
2	2	1	8

Fig. 3. The delta data after union all

After delta tables are merged by above method, Δ^+ and Δ^- are shown in Figure 3.

It is assumed that the aggregate function is MAX. Then, the maximum value of group (1,1,1) in Δ^+ is 5. But, the tuple(1,1,1,5) is inserted in Δ_1^+, and then is deleted from Δ_2^-, so it does not exist. The maximum value 5 is not the correct maximum value of group (1,1,1).

For aggregate functions, such as SUM and COUNT, this processing method is correct. For SUM, the total of Unit_sales can be computed by subtracting the subtotal of Unit_sales in Δ^- from the subtotal of Unit_sales in Δ^+. The tuple (1,1,1,5) is added once, and then is subtracted once, so it has no effect on the result.

Definition 7. Invertible aggregate function: For a function F, if existing a function F' satisfies that F'(F(a, b), b) = a, F'(F'(a, b), c) = F'(a, F(b, c)), F and F' are distributive or algebraic functions, then F is an invertible aggregate function, F' is the inverse function of F.

For example, aggregate functions SUM, COUNT are invertible aggregate function. The inverse function of SUM and COUNT is SUBSTRACT.

The aggregate functions MAX, MIN and MEDIAN are not invertible aggregate function.

For different type of aggregate functions, different method can be adopted to process multiple versions of what-if data to improve the efficiency.

Rule 1. For invertible aggregate function, using the following method to process multi-version what-if data:

$$\Delta^+ = \Delta_1^+ + \Delta_2^+ + \ldots\ldots + \Delta_n^+$$
$$\Delta^- = \Delta_1^- + \Delta_2^- + \ldots\ldots + \Delta_n^-$$

Proof. Assumed Δ_p^+ denotes the set of tuples in pure delta_insert table, Δ_p^- denotes the set of tuples in pure delta_delete table, then $\Delta_p^+ = \Delta^+ - \Delta^-$, $\Delta_p^- = \Delta^- - \Delta^+$, $\Delta_p^+ \cap \Delta_p^- = \Phi$, $F(\Delta) = F'(F(\Delta_p^+), F(\Delta_p^-))$, then:

(1) if $\Delta^+ \cap \Delta^- = \Phi$, then $\Delta_p^+ = \Delta^+$, $\Delta_p^- = \Delta^-$, $F(\Delta) = F'(F(\Delta^+), F(\Delta^-)) = F'(F(\Delta_p^+), F(\Delta_p^-))$.

(2) if $\Delta^+ \cap \Delta^- \neq \Phi$, assumed $\Delta^+ \cap \Delta^- = \Delta_\cap$, then
$$\Delta_p^+ = \Delta^+ - \Delta_\cap, \quad \Delta_p^- = \Delta^- - \Delta_\cap.$$
$F(\Delta) = F'(F(\Delta^+), F(\Delta^-)) = F'(F(\Delta_p^+, \Delta_\cap), F(\Delta_p^-, \Delta_\cap))$. Due to F is a distributive aggregate function, therefore:
$$F(\Delta) = F'(F(F(\Delta_p^+), F(\Delta_\cap)), F(F(\Delta_p^-), F(\Delta_\cap)))$$
$$= F'(F'(F(F(\Delta_p^+), F(\Delta_\cap)), F(\Delta_\cap)), F(\Delta_p^-))$$
$$= F'(F(\Delta_p^+), F(\Delta_p^-))$$

End.

For non_invertible aggregate function, Δ^+ and Δ^- should satisfies the follow conditions: for $\forall t \in \Delta^+$, if $t \in \Delta_i^+$, then $t \notin \Delta_j^-$, $i < j <= n$. Similarly, $\forall t \in \Delta^-$, if $t \in \Delta_i^-$, then $t \notin \Delta_j^+$, $i < j <= n$.

Rule 2: For non_invertible aggregate function, Δ^+ and Δ^- can be computed by following formula[12]:

$$\Delta^+ = \Delta_1^+ - \Delta_2^- + \Delta_2^+ - \ldots\ldots - \Delta_n^- + \Delta_n^+$$
$$\Delta^- = \Delta_1^- - \Delta_2^+ + \Delta_2^- - \ldots\ldots - \Delta_n^+ + \Delta_n^-$$

The sequence of above operations is based on the sequence of versions. Otherwise, it may lead to wrong result. For example, if the what-if data in Δ_3^+ is (2,2,1,8), the what-if data in Δ_3^- is (2,2,3,3), then the result of $\Delta^+=\Delta_1^+-\Delta_2^-+\Delta_2^+-\Delta_3^-+\Delta_3^+$ is show as following:

P	T	C	Unit_sales
1	1	2	3
2	1	1	6
1	3	2	3
2	2	1	8

the result of $\Delta^+=\Delta_1^++\Delta_2^++\Delta_3^+-\Delta_2^--\Delta_3^-$ is:

P	T	C	Unit_sales
1	1	2	3
2	1	1	6
1	3	2	3

In this result, the tuple(2,2,1,8) is deleted. But, it is deleted from Δ_2^-, then inserted into Δ_3^+. Because Δ_3^+ is later than Δ_2^-, it should be kept in the result.

Obviously, the set operations of Rule 1 are much less than Rule 2. It has no demand on operation order and can be optimized by parallel or pipeline, etc. to improve the processing performance.

4   Incremental Computation of CUBE

While the number of dimension attributes is n, the CUBE contains 2^n cuboids, and each cuboid is defined by a single SELECT-FROM-WHERE-GROUP BY block, having identical aggregate functions, identical FROM and WHERE clauses, and one of the 2^n subsets of the dimension attributes as the group-by columns. Figure 4 shows a CUBE with three dimension attributes (level i indicates the level of cuboid).

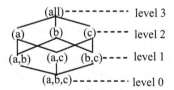

Fig. 4. A cube with three dimension attributes

The what-if analysis is based on what-if view. Generally, the what-if view VI is computed by $T_B \odot D_{wi}$, then what-if analysis is performed on VI. But, when there exists a materialized view on T_B, we can carry out the what-if analysis by incremental computation the new view with the materialized view, T_B and D_{wi}. In this section, we discuss the incremental computation of MAX and MIN cube.

MAX and MIN function are self-maintainable with respect to insertions. The new maximum or minimum value of a group can be computed by comparing the old maximum or minimum value with the maximum or minimum value of the group in delta table. Therefore, for insertions, the incremental computation of MAX or MIN cube is same to the SUM cubes. This paper does not discuss it again. The discussion about it refers to the paper[7][8]. This paper mainly discusses the problem about incremental computation of MAX and MIN cube with respect to deletions.

MAX and MIN functions are not self-maintainable with respect to deletions, so the incremental computation of MAX and MIN cube are more complex than the SUM or COUNT cube. [7]introduces a method for incremental computing MAX and MIN cube. When the tuple holding maximum or minimum value is deleted, the new maximum or minimum value for the group can be recomputed from the base table and the delta table. Because the number of tuples in the base table is very large, the cost of re-computation of maximum(minimum) value is very high.

This paper proposed a novel algorithm for incremental computing MAX(MIN) CUBE: R_NLC(Refreshed by Next_Level Cuboid):

   (1) incrementally computing CUBE from bottom to up;
   (2) when incrementally computing the cuboid of level 0, and the tuple with the maximum(minimum) value has been deleted, re-computing new maximum(minimum) value from the base table and delta table;
   (3) when incrementally computing the cuboids of level $i(i>0)$, and the tuple with the maximum(minimum) value has been deleted, re-computing new maximum(minimum) value from the cuboid of level $i-1$;

The correctness of the algorithm is proofed as following:

Proof : Assumed $S = \{ S_1, S_2, ..., S_n \}$, where S is a set of tuples, $S_1, S_2, ..., S_n$ are the disjoint subsets of S. S_D is a subset of S, denoting the set of tuples which need to be deleted from S. Then $S-S_D = \{ S_1 - S_D, S_2 - S_D, ..., S_n - S_D \}$. Therefore, $MAX(S-S_D) = MAX(\{ S_1 - S_D, S_2 - S_D, ..., S_n - S_D \})$. Because MAX functions is distributive aggregate function, $MAX(S-S_D) = MAX(\{ MAX(S_1 - S_D), MAX(S_2 - S_D), ..., MAX(S_n - S_D) \})$. Simultaneously, $MIN(S-S_D) = MIN(\{ MIN(S_1 - S_D), MIN(S_2 - S_D), ..., MIN(S_n - S_D) \})$. End.

For example, for a cube C with three dimension attributes: a, b and c, it has eight cuboid: abc, ab, ac, bc, a, b, c and Φ. The incremental computing procedure for cuboid new_abc(Figure 5(a)) and new_ab(Figure 5(b)) is shown in Figure 5. In the procedure of incremental compute of cuboid new_abc, when the tuple of cuboid abc is (1,1,2,5,1), there exists a tuple (1,1,2,5,1) in cuboid Δabc, that is the tuple with maximum value 5 is deleted. Therefore, the maximum value of group(1,1,2) needs to be recomputed. Because abc is the cuboid of level 0, the new value is recomputed from the base table and delta table. The result of recomputed is4, then inserting tuple(1,1,2,4,1) into new_abc. When incrementally computing of cuboid new_ab and the tuple of cuboid ab is (1,1,5,1), there exists a tuple(1,1,5,1) in cuboid Δab. Therefore, the maximum value of group(1,1) needs to be recomputed. Because cuboid new_abc has been computed, we can compute the new maximum value from new_abc to get 4, then inserting the tuple (1,1,4,1) into cuboid new_ab.

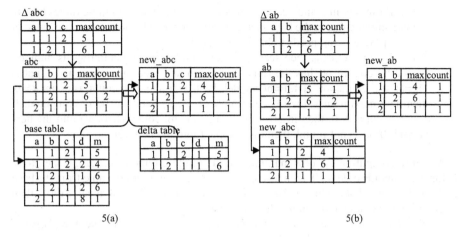

Fig. 5. The procedure of incremental com

4.1 Computing Delta Cube

R_NLC algorithm computes a delta cube from delete_delta table Firstly. The delta cube has the same aggregate function, dimension and measure attributes as base cube. In a group, there may be more than one tuples having the same measure values, the algorithm adopts count to determine if all of tuples with the same measure values have been deleted. Therefore, besides of the measure attributes of the result of MAX or MIN function, the delta cube has another measure attribute: COUNT. The values of COUNT attribute of cuboids in the delta cube is the inverse of the number of tuples with the maximum or minimum value in the delete_delta table. Therefore, during computing delta cube, not only the maximum value of each group need to be computed, but also the number of tuples having the maximum value in the group need to be computed.

4.2 Incremental Compute the MAX and MIN Cube

After the delta cube is computed, we compute the new cube by the base cube and delta cube. The incremental computing procedure for MAX cubes is the following (The algorithm for MIN cubes is similar to it.):

For each tuple $t(a_1,a_2,...,a_i, m, count)$ in base cuboid, searching the tuple $\delta t(a_1,a_2,...,a_i, m\_del, count\_del)$ which has the same dimension attribute values as t in the compounding delta cuboid:

(1) If δt is not found, then insert tuple t into the new cuboid.
(2) If $\delta t(a_1,a_2,...,a_i, m_{del}, count_{del})$ is found, then:

    (i) If $t.m > \delta t.m_{del}$, then the tuples with the maximum value in the base fact table have not been deleted. So, insert tuple t into the new cuboid..
    (ii) If $t.m = \delta t.m_{del}$ and $t.count + \delta t.count_{del} > 0$, then some of these tuples with the maximum value in the base fact table are deleted, others are not deleted. So, insert tuple$(a_1,a_2,...,a_i, m, t.count + \delta t.count_{del})$ into the new cuboid;

(iii) If $t.m = \delta t.m_{del}$ and $t.count + \delta t.count_{del} <= 0$, then all of these tuples having the maximum value in the base fact table are deleted. The new maximum value of the $group(a_1,a_2,...,a_i)$ needs to be re-computed.

(iv) If $t.m < \delta t.m_{del}$, then searching these tuples $\delta t'$ and counting its number, named by count'. $\delta t'$ has the same dimension and measure values as t. That is $t.a_1=\delta t'.a_1$, ..., $t.a_i=\delta t'.a_i$, $t.m=\delta t'.m\_del$. If $\delta t'$ is found, and count'+t.count <= 0, then new maximum value of the $group(a_1,a_2,...,a_i)$ needs to be re-computed. If count'+t.count > 0, then insert $tuple(a_1,a_2,...,a_i, m, count'+t.count)$ into the new cuboid. If $\delta t'$ is not found, then insert tuple t into the new cuboid..

When re-computing the new maximum value of the $group(a_1,a_2,...,a_i)$, if the group belongs to cuboids of level 0, then re-computing the new maximum value from the base table and delete_delta table. If the group belongs to cuboids of level i(i>0), then re-computing the new maximum value from the cuboid of level i-1.

If all of the tuples of delete_delta table come from the base fact table, the (2)(iv) of above algorithm need not to be considered.

5   Performance Evaluation

In this section, we present the results of performance evaluation. In the experiments, our method R_NLC is compared with an incremental CUBE maintenance method that re-computes the maximum(minimum) value from the source data and delta data. We use an Oracle 10g database system running on a PC with 3GHZ Pentium4 CPU and 1G RAM. The methods used in the experiments are implemented using PL/SQL in Oracle 10g.

The FoodMart database of Microsoft Analysis Service is used in the experiments. The table sales_fact_1997 is used as base table. The attributes product_id, time_id, customer_id and store_id in the sales_fact_1997 are used as dimension attributes, while unit_sales is used as a measure attribute.

In this experiments, we firstly compare the performance of the two merge methods(denoted as merge_by_rule1 and merge_by_rule2 in Figure 6 respectively). We use three insert_delta tables and three delete_delta tables in this experiment. The size of each delta table varies from 200000 tuples to 800000 tuples. The result is shown in Figure 6. It shows that the efficiency of merging delta tables by Rule1 is better than merging delta tables by Rule2. Therefore, if an aggregate function is invertible aggregate function, the delta table should be merged using Rule1.

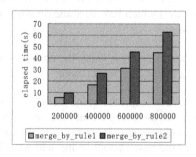

Fig. 6. The performance of merge

Then, we evaluated the performance of R_NLC. Table 1 shows two cuboids, i.e., C1 and C2 used in the experiments. In the experiments, we compared R_NLC with an incremental cube compute algorithm which uses the base table and delta table to re-compute the new maximum or minimum value when the tuple with the maximum or minimum value is deleted(denoted as R_BFT).

Table 1. CUBEs used in the experiments

CUBE	Dimension attributes	Measure attribute
C1	product_id, time_id, customer_id	Unit_sales
C2	product_id, time_id, customer_id, store_id	Unit_sales

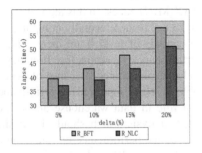
Fig. 7(a). C1

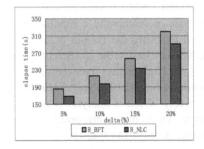
Fig. 7(b). C2

Fig. 7. The performance evaluation

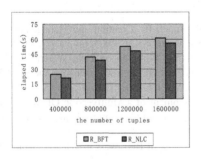
Fig. 8(a). C1

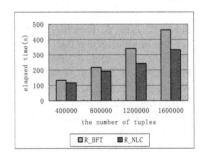
Fig. 8(b). C2

Fig. 8. Scalability evaluation

Figure 7 shows the results of performance experiment when we varied update ratio for a fixed size of the base table. In the experiments, the size of base table is 800000 tuples. We varied the update ratio of the fact table from 5% to 20%. In Figure 7, the time of the R_NLC algorithm is less than another algorithm. The performance is improved about 10% by R_NLC. The results of experiments show that reducing the access times to base fact table can improve the performance effectively. The savings in time mainly come from the fact that R_NLC uses the result of previous computation.

We next vary the size of base table from 400000 to 1600000 tuples for a fixed update ratio 10% to evaluate the scalability performance. The result (shown in Figure 8) shows that R_NLC has better performance than R_BFT. The benefit to R_NLC increases when the size of base table increases.

6 Conclusion

This paper proposed a strategy for multiple versions what-if update data: improving the merging efficiency by using different methods to merge multiple versions what-if update data according to different types of aggregate functions. We also proposed an algorithm R_NLC to incrementally compute MAX(MIN) cubes. In this algorithm, the cubes are incrementally computed from bottom to up. It incrementally computes the cuboid by using the cuboid of next level. The cuboids of level 0 are incrementally computed by base table and delta table. In the performance evaluation, the efficiency of the R_NLC algorithm over the previous method is demonstrated. Because of benefit from the reduction of the access times of base table, R_NLC improved the performance effectively. What-if analysis plays an important role in data warehouse and OLAP system. There are many problems about what-if analysis which should be further studied, such as, the framework of what-if analysis, the performance optimization of what-if analysis, the storage strategy of what-if data, etc.. We will investigate these problems in the future work.

References

1. Bebel, B., Eder, J., Koncilia, C., Morzy, T., Wrembel, R.: Creation and Management of versions in Multiversion Data Warehouse. In: The Proc. of SAC 2004, pp. 717–723 (2004)
2. Morzy, T., Wrembel, R.: On Querying Versions Of Multiversion Data Warehouse. In: Proceedings of the 7th ACM international workshop on Data warehousing and OLAP, Washington, DC, USA, vol. 2004, pp. 92–101. ACM, New York (2004)
3. Mitrpanont, J.L., Fugkeaw, S.: Design and Development of a Multiversion OLAP Application. In: The 21st Annual ACM Symposium on Applied Computing, Dijon, France, pp. 493–497. ACM, New York (2006)
4. Ramirez, R.G., Kulkarni, U.R., Moser, K.A.: The Cost of Retrievals in What-if Database, pp. 136–145. IEEE, Los Alamitos (1991)
5. Hurtado, C.A., Mendelzon, A.O., Vaisman, A.A.: Maintaining data cubes under dimension updates. In: ICDE, pp. 346–355 (1999)
6. Balmin, A., Papadimitriou, T., Papakonstantinou, Y.: Hypothetical Queries in an OLAP environment. In: Proc. of the 26th VLDB conference, pp. 220–231 (2000)
7. Mumick, I.S., Quass, D., Mumick, B.S.: Maintenance of Data Cubes and summary Tables in a Warehouse. In: Proc. of the ACM SIGMOD Conference, pp. 100–111 (1997)
8. Lee, K.Y., Kim, M.H.: Efficient Incremental Maintenance of Data Cubes. In: Proc. of the VLDB Conference, pp. 823–833 (2006)

9. Lakshmanan, L.V.S., Russakovsky, A., Sashikanth, V.: What-if OLAP Queries with Changing Dimensions. In: Proceedings of the 24th International Conference on Data Engineering, pp. 1334–1336. IEEE Computer Society, Los Alamitos (2008)
10. Vaisman, A.A., Mendelzon, A.O., Ruaro, W., Cymerman, S.G.: Supporting Dimension Updates In An OLAP Server. Information Systems 29(2), 67–82 (2004)
11. Mendelzon, A.O., Vaisman, A.A.: Temporal queries in OLAP. In: Proceedings of the 26th International Conference on Very Large Data Bases, Cairo Egypt, pp. 242–253. Morgan Kaufmann, San Francisco (2000)
12. Ghandeharizadeh, S., Hull, F., Jacobs, D.: Heraclitus: Elevating Deltas to be First-Class Citizens in a Database Programming Language. ACM Transactions on Database Systems 22(3), 370–426 (1996)

Incremental Computation for MEDIAN Cubes in What-If Analysis[*]

Yanqin Xiao[1,2,3], Yansong Zhang[1,2,4], Shan Wang[1,2], and Hong Chen[1,2]

[1] Key Laboratory of Data Engineering and Knowledge Engineering, Renmin University of China, MOE, Beijing 100872, P.R. China
[2] School of Information, Renmin University of China, Beijing, P.R. China 100872
[3] Computer Center, Hebei University, Baoding Hebei 071002, China
[4] Department of Computer Science, Harbin Finance College, Harbin 150030, China
{xyqwang,zhangys_ruc,swang,chong}@ruc.edu.cn

Abstract. What-if analysis is an important type of DSS analysis processing procedure. It analyzes hypothetical scenarios based on historical data. The data cube view must be updated when the what-if condition is changed. Since source data must be kept in order to compute the new aggregate value when new tuples are inserted or deleted, in what-if analysis, incrementally computing a data cube for holistic aggregation functions is a difficult problem. In this paper, we adopt delta cube strategy and work area technique to incrementally compute data cube for MEDIAN function. The size of work area has important influence on the efficiency of the incremental computing. This paper optimizes the size of work area based on the number and the cardinality of dimension attributes of the cuboid. Performance study shows that our algorithms are effective over large databases.

Keywords: what-if analysis, cube, delta cube, incremental computation.

1 Introduction

What-if analysis is a very important decision support process. It is used to forecast future information under a set of hypothetical scenarios related to historical data. It can provide important predicting information for DSS users. What-if analysis also enables the evaluation of past performance and the estimation of the opportunity cost taken by not following alternative policies in the past. For example:

Q1: if the sale price of drink was increased by 10%, how would it affect the profit of the company?

Q2: if introducing production A into CA area, how would it affect the profit of the enterprise?

Q3: if the total profit need to be increased 10%, how to adjust the product structure of enterprise?

---

[*] Supported by the National Natural Science Foundation of China under Grant No. 60473069,60496325; the joint research of HP Lab China and Information School of Renmin University(Large Scale Data Management);the joint research of Beijing Municipal Commission of education and Information School of Renmin University(Main Memory OLAP Server).

Q. Li et al. (Eds.): APWeb/WAIM 2009, LNCS 5446, pp. 248–259, 2009.

According to the object of hypothetical update, the what-if analysis in OLAP can be classified into three categories:

A. the what-if analysis based on hypothetical schema update;
B. the what-if analysis based on hypothetical dimension update;
C. the what-if analysis based on hypothetical fact table or view update.

The what-if analysis based on the hypothetical schema update was studied by [1][2][3]. [4]analyzed the types of dimension updates, and the method of maintaining cube under dimension updates. The Q2 belongs to the what-if analysis based on hypothetical dimension update. Most of what-if analyses are based on the hypothetical fact table updates[5][6][7]. The Q1 belongs to the what-if analysis based on hypothetical fact table update. Q3 belongs to the what-if analysis based on hypothetical view update[7]. It analyzes the influence of the change of aggregation value on the base data from top to bottom.

The implementation approaches of what-if analysis(what-if query) are divided into two classes: "lazy" and "eager". In the "lazy" approach, the hypothetical update expression is transformed into an equivalent explicit substitution, and then applying these substitutions to what-if query, to obtain a pure relational algebra query. The "lazy" approach stores the hypothetical update expression, but not hypothetical update data. The what-if query implemented by "lazy" approach is transformed into the queries on the fact data[5]. In the "eager" approach, the hypothetical update expression is materialized and stored into the system to form an assumed scenario, and then the what-if query is processed on the assumed scenario. The "lazy" approach can save massive memory space due to not store hypothetical update data. But, some complex hypothetical update can not be presented by update expression, so it can not be adopted in some situations.

The data cube is a convenient way to exhibit multi-dimension aggregating views in data warehouse and OLAP system. To speedup the processing of query, some cubes are materialized in data warehouse and OLAP system. As assumed changes are made to the schema, dimension or fact table, a new cube, which has same aggregate function, dimension and measure attributes as the materialized cube, must be updated to reflect the changed state of the data sources. The materialized cube can be recomputed from scratch. It also can be incrementally computed by calculate the changes to the views due to the source changes. The delta cube[6] was proposed to incrementally compute the new cube. By the method, the procedure of incremental computation of a cube view is divided into two stages: propagate and refresh. In the propagate stage, a delta cube ΔC was computed from the changes of the data source. The ΔC has same aggregate function as the cube view. In the refresh stage, the cube view was refreshed by applying the ΔC to it. The method fits the distributive and algebraic aggregate functions for incremental computing, such as SUM, COUNT, etc.. The delta cube method belongs to the "eager" method.

The aggregate functions are divided into two classes: self-maintainable and non-self-maintainable[6]. A set of aggregate functions are self-maintainable if the new value of the functions can be computed solely from the old values of the aggregate functions and from the changes to the base data. According to the definition, a self-maintainable aggregate function must be a distributive or algebraic aggregate function. It can not be holistic function. Therefore, the delta cube method can not be used directly to incrementally compute a cube with holistic aggregation functions such as MEDIAN and QUANTILE. In this paper, we propose a solution for the incremental computation of a cube with MEDIAN aggregation function.

The contributions of this paper are as follows:

- We adopt delta cube method to incrementally compute MEDIAN cube. In our algorithm, the delta cube has different content from the source cube.
- We propose a novel work area technique to efficiently incrementally compute a cube with MEDIAN aggregation function.
- We vary the size of work area based on the number and the cardinality of dimension attributes of cuboid to reduce the cost of management of work area.

This paper is organized as following: Section 2 will discuss the related work. Section 3 introduces the background information of our study. Then, in section 4, we will describe the algorithm to incrementally compute MEDIAN cube in detail. The results of performance experiments are given in section 5. Finally, we conclude our work in section 6.

2   Related Work

An d-dimension data cube contains 2^d cuboids, which come from the same fact table, with the same aggregate function and different group-bys. Because the computation of data cube is a time-consuming work, many research works have been done to improve the efficiency of data cube computation. In data warehouse and OLAP system, some cuboids are materialized to improve the query performance. When the data of fact table changed, the materialized cuboids must be updated to reflect the changed state of the data source.

There are two strategies to re-compute the data cube: (1) discarding the materialized cuboids, and re-computing new cuboids from the fact table. (2) incrementally computing the new cuboids based on the original materialized cuboids. The advantage of first strategy is that it can be used in any situations. Its disadvantage is that it can not make full use of existing resources. The re-computation of data cube consumes a lot of system resources. The advantage of second strategy is that it can make full use of existing cuboids, saving system resources. However, it is fit to distributive and algebraic aggregate functions only.

In what-if analysis, the materialized cube must be updated to reflect the hypothetical update of the data sources. The incremental computation of data cube under hypothetical dimension update is under research by [4][9][10][11]. They do not consider the incremental computation under fact table update. [6][8] research the incremental maintenance of data cube under hypothetical fact table update. Both of them adopt delta cube method to incrementally compute data cube under fact table update. [8]proposes an incremental computation method for data cube that can maintain a data cube by using only delta cuboids. Because the $\binom{n}{\lfloor n/2 \rfloor}$unt of delta cuboids is much smaller, the cost of computing delta cuboids is substantially reduced. The algorithm does not fit to incrementally maintain a cube with holistic function. The work area technique is used in [12] and [13]. But, they do not discuss the incremental maintenance of cube. [15] introduces the methods of efficient computation of iceberg cubes with complex measures. [14] proposes a methods of incremental maintenance of quotient cube for MEDIAN. But, it does not discuss the deletion operations.

3 Backgroud

This section provides the necessary background for discussion in the rest of this paper. In section 3.1, we define some notations about what-if analysis, and discuss the storage of what-if data in section 3.2. Section 3.3 introduces the conception about data cube.

3.1 Notation Definitions

In this section, we introduce some general notations about what-if analysis.

Definition 1. base table: A base table T_B is the object of what-if operation. In OLAP system, T_B may be fact table, view or dimension table.

Definition 2. what-if operation: A what-if operation $OP_{wi} = \{OP_A, OP_A \in \{Ins, Del, Upd\}\}$. Where, OP_A is an atomic hypothetical update operation. It may be insert(Ins), delete(Del) or update(Upd). That is, a what-if operation is a set of atomic hypothetical update operations.

In what-if analysis, there may be many what-if operations based on the same base table. What-if operation may base on tables or what-if views.

Definition 3. what-if data: The data produced by what-if operation is named what-if data, denoted as D_{wi}. $D_{wi} \leftarrow OP_{wi}$ means that the what-if operation OP_{wi} produces a set of data D_{wi}.

Definition 4. what-if view: The merging of what-if update data D_{wi} with its base table T_B forms a what-if view, denoted as VI, then $VI = T_B \odot D_{wi}$.

3.2 Storage of What-If Data

In what-if applications, what-if operations are temporary. The what-if update data based on what-if operations does not exist. Therefore, these what-if update data should not change the real data in the data warehouse and OLAP system. Delta table is adopted to store what-if update data. That is, what-if update data are stored into one or more delta tables which are independent to real fact table. The what-if queries are processed on the union of the base table and delta table.

This paper uses two types of delta tables to store what-if data. One is insert_delta table (Δ^+)which stores the data of hypothetical insertions. The other is delete_delta table (Δ^-), which stores the data of hypothetical deletions. The update operations in hypothetical conditions are transformed into deletions and insertions: deleting the old data firstly, inserting the new data secondly, then storing the data of hypothetical deletions into delete_delta table and the data of hypothetical insertions into insert_delta table.

3.3 Data Cube

In data warehouse and OLAP applications, it is often necessary to analyze the data in fact table based on various combinations of dimension attributes. The DATA CUBE operator supports such analysis in data warehouse and OLAP. In a CUBE, attributes are categorized into dimension attributes, on which grouping operation may be performed, and measure attributes are the results of aggregate functions. While the number of dimension attributes is d, the CUBE contains 2^d cuboids, and each cuboid is defined by a single SELECT-FROM-WHERE-GROUP BY block, having identical aggregate functions, identical FROM and WHERE clauses, and one of the 2^d subsets of the dimension attributes as the group-by columns. Figure 1 shows a CUBE with three dimension attributes (level i indicates the level of cuboid).

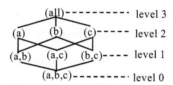

Fig. 1. A cube with three dimension attributes

4   Incremental Computing Data Cube for MEDIAN

The what-if analysis is based on what-if view. Generally, the what-if view *VI* is computed by $T_B \odot D_{wi}$, then what-if analysis is performed on *VI*. But, when there exists a materialized view on T_B, we can carry out the what-if analysis by incremental computation the new view with the materialized view, T_B and D_{wi}. In this section, we discuss the incremental computation of MEDIAN cube.

4.1 Work Area

For a set having n tuple, the number of tuples that are larger or smaller than the median values of the n values is same. When inserting new tuple into the set or deleting tuple from the set, if storing m ($1 \leq m \leq n$) tuples around the median value in the work area, then the new median value can be got from the m tuple in the work area when the number of tuples of inserting or deleting is smaller than m.

Assumed a sorted set $S = \{t_0, t_1, \ldots, t_{n-1}\}$, the median value of the set can be computed as following when the set is updated:

1. Let t_{med} represent the position of the median of the n tuples of S. We maintain a work area to store m tuples around t_{med}, and let t_{low} and t_{high} represent the position of the smallest tuple and the biggest tuple in the work area, respectively.
2. When deleting a tuple t from the set S, we compute the new t_{med} of S according the following rules, and adjust the t_{low} and t_{high} correspondently.

   (1) If $t < t_{low}$, then t_{med} moves 1/2 position to the right.
   (2) If $t > t_{high}$, then t_{med} moves 1/2 position to the left.

(3) If $t_{low} \leq t \leq t_{high}$, then t_{med} moves 1/2 position to the left, and the count of tuple in the work area reduces 1, t_{high} moves 1 position to the left.

3. When inserting a tuple t into the set S, we compute the new t_{med} of S according the following rules, and adjust the t_{low} and t_{high} correspondingly.

(1) If $t < t_{low}$, then t_{med} moves 1/2 position to the left.
(2) If $t > t_{high}$, then t_{med} moves 1/2 position to the right.
(3) If $t_{low} \leq t \leq t_{med}$, then t_{med} moves 1/2 position to the left, and the value of t_{low} is updated as the value of its right neighnor.
(4) If $t_{med} \leq t \leq t_{high}$, then t_{med} moves 1/2 position to the right, and the value of t_{high} is updated as the value of its left neighnor.

4.2 Computing Delta Cube

The paper[6] proposed delta cube method to incrementally maintain data cube. Delta cube has the same aggregation function, measure attributes and dimension attributes as the source data cube. For example, if the aggregation function of source data cube is SUM, then the aggregation function of the delta cube is SUM also. The new data cube can be computed by merging the source data cube and the delta cube.

But, the MEDIAN function is not a distributive aggregation function. It is not a self-maintainable function with respect to insertions and deletions. If the MEDIAN is used to compute the delta cube, then merging the delta cube and the source data cube can not get the correct median value of the new cube. Therefore, the MEDIAN can not be used to compute the delta cube. According to the analysis of work area in 4.1, for a group(tuple or member) in a cuboid, if we have known the number of tuples which is larger than the median of the group in the source cube, and the number of tuples which is smaller than the median of the group in the source cube, then we can get the new median value of the group from the tuples in the work area. The tuples that is between the t_{low} and t_{high} can affect the content of the work area. To maintain the effectiveness of the work area, we build a work area for each group of cuboids in the cube to store the tuples between the t_{low} and t_{high}.

Therefore, the following values need to be computed when computing the delta cube:

(1) NUMOFSMALL: recording the number of tuples which is smaller than the median of the group in the source cube.
(2) NUMOFLARGE: recording the number of tuples which is larger than the median of the group in the source cube.
(3) NUMOFEQUAL: recording the number of tuples which is equal to the median of the group in the source cube.
(4) WORKAREA: storing the tuples between t_{low} and t_{high} of delta data.

For each cuboid of the delta cube, we can compute it by the algorithm 1(as shown in Figure 2).

```
Algorithm 1: delta cube(setofgroup, cuboid)
Input: setofgroup: the grouped delta data according to some
dimension attributes;
        cuboid: a cuboid of the source data cube, which has
the same grouping attributes as the datagroup.
Output: delta_cuboid
1. Begin
2. For each group G in the setofgroup:
3.     Find the t_med, t_low, t_high of the corresponding group in
cuboid.
4.     i = 0;
5.     while(G[i].value != t_med && i < G.count) i++;
6.     G.numofsmall = i;
7.     j = i;
8.     while(j < G.count && G[j].value == t_med) j++;
9.     G.numofequal = j - i;
10.    G.numoflarge = G.count - j;
11.    i = 0;
12.    while(G[i].value < t_low && i < G.count) i++;
13.    j = i;
14.    while(j < G.count && G[j].value < t_high) j++;
15.    store the tuples between G[i] and G[j] into the work
area for the group G.
16. end for;
17. end.
```

Fig. 2. The algorithm for computing delta_cube

4.3 Incremental Computing the MEDIAN Cube

In this section, we illustrate how to compute the new MEDIAN cube using the source data cube and the delta cube. In the propagate stage, we have computed two delta cubes, one for delta insertion tuples, and the other for delta deletion tuples. Now, we compute the new MEDIAN cube using the source data cube and the two delta cubes. The new MEDIAN cube is computed from top to down. For each group in each cuboid in the source cube, we compute the new median value according to the following steps.

1. Merging the work areas:

There are three work areas for each group G of cuboid, for the source data cube, delta insertion cube and delta deletion cube respectively, denoted as W_s, W_i, and W_d. The new work area W is equal to $W_s - W_d + W_i$ (the subscript s, d and i represent the variable if for the source data cube, delta deletion cube and delta insertion cube respectively.).

(1) If the number of tuples of which the value is equal to the median in W_d is larger than 0, then recording two variables: Equal$_{left}$ and Equal$_{right}$ when deleting them from W_s. The two variables represent the number of tuples which are located on the left of t_{med} and the number of tuples which are located on the right of t_{med} respectively, among the deleted tuples that are equal to the median.

(2) If there are some tuples of which the value is equal to the median in the W_i, then inserting them into those positions which are located on the left of t_{med}.

2. computing the new median

We compute the new median according to the following criteria:

(1) If the old t_{med} is deleted, then the t_{med} moves 1/2 position to the left.
(2) If $NUMOFLARGE_i + NUMOFSMALL_d + Equal_{left} = NUMOFLARGE_d + NUMOFSMALL_i + NUMOFEQUAL_i + Equal_{right}$, then t_{med} keeps invariant.
(2) If $NUMOFLARGE_i + NUMOFSMALL_d + Equal_{left} < NUMOFLARGE_d + NUMOFSMALL_i + NUMOFEQUAL_i + Equal_{right}$, then t_{med} moves $(NUMOFLARGE_d + NUMOFSMALL_i + NUMOFEQUAL_i + Equal_{right} - NUMOFLARGE_i - NUMOFSMALL_d - Equal_{left})/2$ position to the left.
(3) If $NUMOFLARGE_i + NUMOFSMALL_d + Equal_{left} > NUMOFLARGE_d + NUMOFSMALL_i + NUMOFEQUAL_i + Equal_{right}$, then t_{med} moves $(NUMOFLARGE_i + NUMOFSMALL_d + Equal_{left} - NUMOFLARGE_d - NUMOFSMALL_i - NUMOFEQUAL_i - Equal_{right})/2$ position to the right.
(4) If t_{med} moves out of the range of work area, then the new t_{med} needs to be computed from the base table and the delta table.

The algorithm for incremental computing a cuboid is shown in figure 3.

```
Algorithm  2  Newcuboid(delta D cuboid,  delta I cuboid,
s_cuboid)
Input: delta_D_cuboid: a cuboid computed from the delta
delete data.
       delta_I_cuboid: a cuboid computed from the delta insert
data.
       s_cuboid: a cuboid of the source data cube and its
grouping attributes are same as the grouping attributes of
delta_D_cuboid and delta_I_cuboid.
Output: new_cuboid
1. Begin
2. For each group G of delta_D_cuboid and delta_I_cuboid
3. Get work area W_d and W_i from G_d and G_i respectively;
4. Get work area W_s, t_med, t_low and t_high from the corresponding
        group G_s of s_cuboid;
5. W_s = W_s - W_d + W_i; compute variables Equal_left and Equal_right;
6. if t_med has been deleted, then t_med moves 1/2 position to
        the left;
7. if t_med is still in the work area and has not been deleted,
        then moving t_med to new position.
8. if (t_med < t_low || t_med > t_high)
9.      compute the new t_med from base table and the delta table.
10. end for;
11. end.
```

Fig. 3. The algorithm for computing a new cuboid

In the algorithm, we only re-compute the median of the affected groups. It can reduce the amount of calculation. When there are three work area W_s, W_d and W_i need to be merged, we delete the tuples of W_d from W_s firstly, then insert the tuple of W_i into W_s. It can keep the tuples in the W_s as many as possible. If the t_{med} need to be moved to a new position which is out of the range of W_s, then it needs to be computed from the base table and the delta table. Because we incrementally compute the new cube from top to down, the sorting and grouping results of the merging result of base table and the delta table, which is completed when computing the higher level cuboid, can be used to compute the lower level cuboid. It avoids the repeated merging and sort-group operations of the base table and the delta table.

4.4 Optimizing the Size of Work Area

To a great extent, the size of work area affects the performance of the algorithm. If the size of work area is very small, then the probability of the new median occurring in the work area is decreased which increases the probability for re-computing the new median. If the size of work area is very large, it can keep the median in the work area effectively. But, it increases the cost for management of the work area. When the increased cost of management is larger, although the cost of re-computation is reduced, the performance of algorithm may decrease.

The cuboids of a data cube implement aggregate operations based on different dimension attributes. Focusing on the number of tuples, there is very great difference among the groups of different level cuboids. Generally, the number of tuples of the group in the cuboids of higher level is not less than the number of tuples of the group in the cuboids of lower level. Therefore, they have different size requirements of the work area. In the advanced algorithm, we optimize the size of the work area based on the number d of the dimension attributes of the cuboid and the cardinality C of the dimension attributes. The size of work area can be defined by the formula (1):

$$\text{SizeofWA} = \begin{cases} A/C^d & \text{if } A/C^d > B \\ B & \text{if } A/C^d \le B \end{cases} \tag{1}$$

Where, A is the max value of the size of the work area, B is the min value of the size of the work area. The value of A and B can be set flexibly by the user. They can be a fixed value or a function.

5 Performance Evaluation

In this section, we present the results of performance experiments. In the experiments, we implement three algorithms: the complete re-computation algorithm for the MEDIAN cube(denoted as Re-com), the incremental computation algorithm for the MEDIAN cube with fixed size of the work area(denoted as In-com-FS), and the incremental computation algorithm for the MEDIAN cube with variable size of the work area(denoted as In-com-VS). In the experiments, the number of dimension attributes of each tuple is 4, and the cardinality of each dimension is 20. The delta data includes what-if insertion data and what-if deletion data, account for 50% and 50%, respectively. Our experiments are carried out on a PC with an Intel Pentium 3GHz CPU and 1GB main memory.

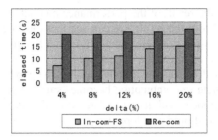

Fig. 4. The performance evaluation of Re-com and In-com-FS

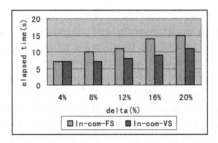

Fig. 5. The performance evaluation of In-com-FS and In-com-VS

We compare the performance of Re-com and In-com-FS firstly. The result is shown in figure 4. In this experiment, the number of tuples is 2,000,000, and the update ratio varies from 4% to 20%. Figure 4 shows that the time taken by the In-com-FS algorithm is less than the time taken by the Re-com algorithm. But, with the update ratio increasing, the gap between then In-com-FS and Re-com algorithm is becoming small.

Next, we compare the In-com-FS and In-com-VS algorithm, In this experiment, the number of tuples is 2,000,000. We vary the update ratio from 4% to 20%. Figure 5 shows the result. It shows that the In-com-VS algorithm has better performance than the In-com-FS algorithm.

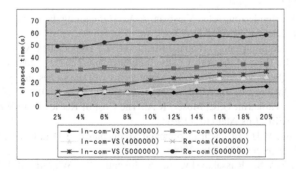

Fig. 6. The performance evaluation of In-com-VS

Figure 6 shows that the result of the performance evaluation when we increase the size of the base table. From the result, we can see that the In-com-VS algorithm can save about 50% in processing time. Hence, the In-com-VS algorithm can be effectively used for fast incremental maintenance of cube views.

In all experiments, with the update ratio increasing, the gap between the incremental computation and complete re-computation becomes small. Therefore, the incremental computation fit to be adopted with lower update ratio. If update ratio is very high, the complete re-computation should be adopted.

6 Conlusion

Incremental computation of data cube is an extremely important aspect of data warehouse and OLAP system under hypothetical scenarios. However, the incremental computation of data cube for holistic functions is a hard work(or big problem). This paper addresses the problem of incrementally maintaining the existing MEDIAN cube. We adopt the delta cube strategy and work area technique to incrementally compute the MEDIAN cube under the what-if operations, and we optimize the algorithm by varing the size of the work area with the number and the cardinality of dimension attributes of the cuboid. Through performance evaluation, we can draw a conclusion that the efficiency of the proposed algorithm is superior to the complete re-computation method, and the efficiency of the In-com-VS algorithm is superior to the In-com-FS algorithm.

References

1. Bebel, B., Eder, J., Koncilia, C., Morzy, T., Wrembel, R.: Creation and Management of versions in Multiversion Data Warehouse. In: The Proc. of SAC 2004, Nicosia, Cyprus, pp. 717–723 (2004)
2. Morzy, T., Wrembel, R.: On Querying Versions Of Multiversion Data Warehouse. In: Proceedings of the 7th ACM international workshop on Data warehousing and OLAP, Washington, DC, USA, pp. 92–101. ACM, New York (2004)
3. Mitrpanont, J.L., Fugkeaw, S.: Design and Development of a Multiversion OLAP Application. In: The 21st Annual ACM Symposium on Applied Computing. Dijon, France, pp. 493–497. ACM, New York (2006)
4. Hurtado, C.A., Mendelzon, A.O., Vaisman, A.A.: Maintaining data cubes under dimension updates. In: ICDE, pp. 346–355 (1999)
5. Balmin, A., Papadimitriou, T., Papakonstantinou, Y.: Hypothetical Queries in an OLAP environment. In: Proc. of the 26th VLDB conference, pp. 220–231 (2000)
6. Mumick, I.S., Quass, D., Mumick, B.S.: Maintenance of Data Cubes and summary Tables in a Warehouse. In: Proc. of the ACM SIGMOD Conference, pp. 100–111 (1997)
7. Ramirez, R.G., Kulkarni, U.R., Moser, K.A.: The Cost of Retrievals In What-if Database, pp. 136–145. IEEE, Los Alamitos (1991)
8. Lee, K.Y., Kim, M.H.: Efficient Incremental Maintenance of Data Cubes. In: Proc. of the VLDB Conference, pp. 823–833 (2006)
9. Lakshmanan, L.V.S., Russakovsky, A., Sashikanth, V.: What-if OLAP Queries with Changing Dimensions. In: Proceedings of the 24th International Conference on Data Engineering, pp. 1334–1336. IEEE Computer Society, Los Alamitos (2008)

10. Vaisman, A.A., Mendelzon, A.O., Ruaro, W., Cymerman, S.G.: Supporting Dimension Updates In An OLAP Server. Information Systems 29(2), 67–82 (2004)
11. Mendelzon, A.O., Vaisman, A.A.: Temporal queries in OLAP. In: Proceedings of the 26th International Conference on Very Large Data Bases, Cairo, Egypt, 2000, pp. 242–253. Morgan Kaufmann, San Francisco (2000)
12. Yi, K., Yu, H., Yang, J., Xia, G., Chen, Y.: Efficient Maintenance of Materialized Top-k Views. In: Proceedings of 19th International Conference on Data engineering, pp. 189–200 (2003)
13. Palpanas, T., Sidle, R., Cochrane, R., Pirahesh, H.: Incremental Maintenance for Non-Distributive Aggregate Functions. In: Proceedings of the 28th VLDB Conference, Hong Kong, China, pp. 802–813 (2002)
14. Cuiping, L., Gao, C., Anthony, K.H.T., Shan, W.: Incremental Maintenance of Quotient Cube for median. In: Proceedings of the Tenth ACM SIGKDD International Conference on Knowledge Discovery and Data Mining, Seattle, Washington, USA, pp. 226–235 (2004)
15. Jiawei, H., Jian, P., Guozhu, D., Ke, W.: Efficient Computation of Iceberg Cubes with Complex Measures. In: Proc. of the ACM SIGMOD Conference, pp. 1–12 (2001)

The Tradeoff of Delta Table Merging and Re-writing Algorithms in What-If Analysis Application[*]

Yansong Zhang[1,2,3], Yu Zhang[1,2,3], Yanqin Xiao[1,2,4],
Shan Wang[1,2], and Hong Chen[1,2]

[1] Key Laboratory of the Ministry of Education for Data Engineering and Knowledge
Engineering, Renmin University of China, Beijing 100872, China
[2] School of Information, Renmin University of China, Beijing 100872, China
[3] Department of Computer Science, Harbin Finance College, Harbin 150030, China
[4] Computer Center, Hebei University, Baoding Hebei 071002, China
{zhangys_ruc,zhangyu,xyqwang,swang,chong}@ruc.edu.cn

Abstract. What-if analysis can provide more meaningful information than classical OLAP. Multi-scenario hypothesis based on historical data needs efficient what-if data view support. In general, delta table for what-if analysis is more general than other solutions such as query re-writing prototype of Sesame. Delta table is independent of base table and is more suitable to represent complex hypothetical updates and multi-version hypothetical updates. Due to low efficiency of traditional delta table merging algorithm which is based on set operation, there are few researches focus on delta table merging algorithm but on query re-writing algorithm. By analyzing the difference between what-if query and what-if analysis and improving the delta table merging algorithm, we propose novel algorithms without set operation of difference. Considering the feature of aggregate operations in OLAP analysis, pre-merge algorithm is presented without the generation of what-if data view before group-bys in the scenario of SUM, AVERAGE and COUNT function OLAP queries. In our experiments, the pre-merge algorithm greatly improved the efficiency of delta table merging procedure which is close to base group-by statement and superior to the re-writing algorithm. A complete comparison between all candidate delta table merging algorithms and re-writing algorithm with different what-if update conditions of update, deletion, insertion and mixed what-if updates is exhibited in our experiments, the policy of what-if analysis among different types is also discussed.

Keywords: delta table, delta table merging, what-if analysis, re-writing query.

1 Introduction

What-if queries based on hypothetical scenarios can provide adequate information for decision support applications. What-if analysis is based on hypothetical OLAP DB,

---

[*] Supported by the National Natural Science Foundation of China under Grant No. 60473069,60496325; the joint research of HP Lab China and Information School of Renmin University(Large Scale Data Management);the joint research of Beijing Municipal Commission of education and Information School of Renmin University(Main Memory OLAP Server).

Q. Li et al. (Eds.): APWeb/WAIM 2009, LNCS 5446, pp. 260–272, 2009.

including ad-hoc what-if queries, roll-up or drill-down functions etc. The challenges are how to represent hypothetical updates in a read-only historical dataset and how to process multi-dimension queries in a hypothetical data warehouse just like in an ordinary data warehouse.

What-if analysis includes two types. One is based on hypothetical updates targets, the other is based on the expression of hypothetical update which can be seen as the realization for all types of what-if analysis. For hypothetical updates target, hypothetical update can occur on schema, dimension and fact table. For the expression of hypothetical update, there are three typical sub-types of delta table merging, re-writing query and delta cube.

What-if analysis tools based on spreadsheet are also employed under research[4,5]. But this kind of what-if analysis usually focuses on key data view and key enterprise parameters; it is not general enough for the purpose of OLAP analysis.

Among the realization techniques of what-if analysis, delta cube is based on incremental cube maintenance [1,3,6,7] and materialized views are needed. In our project of "Main Memory OLAP Server", few materialized views are created because of limited memory capacity and higher query processing performance, so we limit our topic on ad-hoc what-if analysis queries. We also research on delta cube algorithm under main-memory OLAP server in other research papers. [8,9] mainly focus on multi-version schema hypothetical updates, and can provide what-if OLAP analysis functions on multi-version scenarios. A general multi-version what-if analysis still need further research work. [10] discusses the framework of what-if analysis and the main research works in the future.

Hypothetical update can be presented either as delta tuples or as update rules. Re-writing query algorithm [2] proposes a set of query algebra operators to perform what-if queries and hypothetical update can be presented with some query algebra operators. Delta table can be used as materialized results of re-writing hypothetical update queries.

The performance of what-if query is important for what-if applications. More researchers focus on re-writing query algorithm than delta table merging algorithm because of the low performance with set operations of the latter and there is no comprehensive comparison between the two solutions. We extend the delta table merging algorithm under what-if OLAP scenario with novel algorithms which can improve the performance and we carefully design experiments to measure the two types in different what-if data scenarios. Our work is the base research work for algorithm selection module in our prototype system.

Our contributions are as follows:

1. We improve traditional delta table merging algorithm based on set operation.
2. We propose a novel pre-merging algorithm based on delta table and analyze the boundary of this algorithm.
3. We carefully analyze the performance of re-writing algorithm under uniform what-if update delta tuples, uniform what-if deletion delta tuples, uniform what-if insertion delta tuples and mixed what-if delta tuples. We also compare the performance of each type with relative delta table merging algorithm.
4. We propose the tradeoff policy of delta table merging and re-writing algorithms in what-if analysis application.

The background notions and conventions are presented in Sec.2. Algorithms are discussed in Sec. 3. Sec.4 shows the result of experiments. In Sec.5 we conclude the tradeoff policy of algorithm selection. Finally, Sec.6 summarizes the paper and discusses future work.

2 Preliminaries

Traditional delta table merging algorithm is based on set operation, where each base table maintains two delta tables, one for insertion delta tuples, and the other for deletion delta tuples. Update operation is divided into a deletion operation and an insertion operation.

Definition 1 (delta table). In this paper, delta table is a 4-atom structure <D,F,V,C>, where D denotes updated tuple values; F denotes three what-if update flags with values of "U", "I", "D"; V denotes version information and C denotes count value of delta tuple with 0 for updated delta tuple, 1 for inserted delta tuple and -1for deleted delta tuple.

Hypothetical query or what-if query is a general concept in hypothetical scenarios, hypothetical data view must be obtained firstly and then queries are performed on hypothetical data view. The following queries on hypothetical data view are arbitrary.

Definition 2 (what-if analysis). What-if analysis can be seen as a kind of specific what-if queries limited in OLAP areas. What-if analysis={OLAP query//what-if updates}, "//" means under the condition of given what-if updates. What-if updates are same as ordinary what-if queries but the following queries are only queries with aggregate computation, i.e. what-if analysis $\subset$ {what-if queries}.

Traditional delta table merging algorithm based on set operation is described as: W=R $\cup$ S$-$T, where W denotes what-if data view, R denotes fact table, S denotes set of insertion delta tuples, and T denotes set of deletion delta tuples. $\bowtie$ is defined as the operation of delta table merging with base table. With query re-writing algorithm, there are no delta tuples, what-if data view can be obtained by re-writing query. The costly set difference operation can be re-written with proper where clauses, for example, R-{t|t[A]>1500$\wedge$t$\in$R} can be represented with $\delta_{A \leqslant 1500}(R)$ which will improve the performance remarkably. Because of the advantages in space cost and performance, more researchers focus on query re-writing algorithm than delta table merging algorithm. In practical applications, some what-if updates are difficult to represent with re-writing query, some what-if updates contain complex computing but small amount of delta tuples while re-writing algorithm has to re-compute those costly what-if update operations whenever queries are called. There are more and more requirements in co-analysis under multi-user and multi-version what-if scenarios, the performance of query re-writing algorithm will decline as cascade re-writing statements increase.

For what-if updated and deleted delta tuples, the process is: factTuple$_i$ $\bowtie$ deltaUorI$_j$ $\rightarrow$SUM accumulator (or other aggregate operations). For sum function, factTuple$_i$ and deltaUorI$_j$ independently entering SUM accumulator results in the same result, we can regard this feature as associative law in aggregating computation. If we pre-merge the fact table and delta table with union operation and then perform the unique group-by operation, the merging cost can be saved. The process is described as:

Fact table $\cup$ delta table $\xrightarrow{pre-merge}$ pre-merge set of what-if data view $\xrightarrow{group-by}$ OLAP query results.

But with COUNT function, merged tuple get the result of 1, but fact tuple and delta tuple get the result of 2. The additional field of count in delta table can solve this problem. We perform "sum(count) " instead of count(*), for what-if updated delta tuple, 1 (fact tuple count value)+0(updated delta tuple count value)=1; for deleted delta tuple, 1 (fact tuple count value)+(-1)(deleted delta tuple count value)=0; for inserted delta tuple, 0 (fact tuple count value)+1(deleted delta tuple count value)=1. We can get AVERGE value by SUM/COUNT. MAX and MIN functions need to be further studied; other aggregate functions such as MEDIAN do not suit the pre-merge algorithm.

We compare the key procedure of join and set difference operations, in main-memory database MonetDB, the result shows that time cost of set difference operation is about 100 times longer than join operation, so the bottleneck of delta table merging algorithm is set difference operation. In general delta table merging algorithm, we improve the algorithm with join operation instead of set difference operation, the results of experiments show that the merge performance can be improved more than 10 times. We will discuss detailed algorithms in next section.

3   What-If Analysis with Delta Table

In our research work, we classify delta table merging algorithm by thee types, one is based on set operation, another is based on join operation and the last type is based on pre-merge algorithm. We realize those algorithms with FoodMart dataset and SQL statement in order to perform experiment in a real ROLAP system. We select a typical MDX query for performance testing and transform it into SQL statement.

```
SELECT
   store.store_state, time_by_day.the_year,
   time_by_day.quarter, product_class.product_family,
   SUM(sales_fact_1997.store_sales)
FROM
   store,time_by_day,product_class,product,sales_fact_1997
WHERE
   sales_fact_1997.store_id=store.store_id AND
   sales_fact_1997.time_id=time_by_day.time_id AND
   sales_fact_1997.product_id=product.product_id AND
   product.product_class_id=product_class.product_class_id
GROUP BY
   store.store_state,time_by_day.the_year,
   time_by_day.quarter,product_class.product_family;
```

3.1 Set Operation Based Delta Table Merging Algorithm

A. Simple set operation based delta table merging algorithm

In traditional set operation based delta table merging algorithm, what-if update operation generates deletion and insertion delta tuples stored in T and S tables. The additional fields carry on necessary information, the value fields are same structure as fact table, and changed values are stored in relative fields. We assumed that in the phase of what-if update, new delta tuples satisfy the entity integrity check, we can use UNION ALL instead of UNION which is much more efficient. So we can simply describe set operation based delta table merging algorithm(set BDM) as:

Algorithm1. set BDM

    R UNION ALL S EXCEPT T

B. Improved set operation based delta table merging algorithm

In practical applications, single delta table is usually used to store the what-if updates of fact table. Hypothetical update operation generates delta tuples with flag of "D(deleted tuples) ", "I(inserted tuples) " and "U(updated tuples) ". We improve set operation based delta table merging algorithm(improvedSetBDM) as:

Algorithm2. improvedSetBDM

$R \bowtie (\pi_T(R)$ EXCEPT $(\delta_{flag='U' \ or \ flag='D'} (\text{deltaTable})))$
UNION ALL
$(\delta_{flag='U' \ or \ flag='D'} (\text{deltaTable}))$

Algorithm 1 and 2 in this section are the same time complexity, unique delta table simplify the maintenance cost of applications and there no need to transform hypothetical update operation.

3.2 Join Based Delta Table Algorithm

Set operation is simple but low efficient. In our experiments, join operation is faster than set difference operation, so we propose join based delta table merging algorithms.

As table 1 shows, we generate the whole tuple map of fact tuple and delta tuple. We can see four types of maps: No.1 denotes current fact tuple has a U type delta

Table 1. Relation map of fact tuple and delta tuple

	Base table	Delta table
1:	1 2 1 3.5 5.6 7	1 2 1 3.5 5.6 2 ... U 0
2:	2 3 7 5.8 7.6 9	
3:	3 2 4 5.1 2.8. 6	3 2 4 5.1 2.8 6 ... D -1
4:		4 2 6 1.4 5.7 3 ... I 1
5:	4 5 9 7.4 5.1 9	

tuple and updated delta tuple is visible; No.3 denotes current fact tuple has a D type delta tuple and current fact tuple is invisible; No.4 denotes that there is a new inserted delta tuple and the delta tuple is visible; No.2 and No.5 denote that current fact tuple has no what-if update, current fact tuple is visible. We can get this map with SQL statement of "full join ", and output merged tuples through table scan operation on the full join table.

In DW, if what-if update only change measure fields, we can store final measure value m_i' or delta value of $m_i\text{-}m_i'$. If we use final values, then merging operation is to decide whether fact tuple or delta tuple is visible. If we use delta value, the merging operation is just to compute the sum value of fact tuple field and delta tuple field, if the result equals to 0 then use a flag field to mark the delete type. We select delta value in delta table in the following algorithms.

A. Full join based delta table merging algorithm

We use"case...when...then..." clause to process the merging procedure.

Algorithm3. `fullJoinBDM`

```
SELECT

(CASE WHEN b.product_id IS NULL THEN d.product_id
            ELSE b.product_id END) AS product_id,
(CASE WHEN b.time_id IS NULL THEN d.time_id ELSE
            b.time_id END) AS time_id,
(CASE WHEN b.customer_id IS NULL THEN d.customer_id
            ELSE b.customer_id END) AS customer_id,
(CASE WHEN b.promotion_id IS NULL THEN d.promotion_id
            ELSE b.promotion_id END) AS promotion_id,
(CASE WHEN b.store_id IS NULL THEN d.store_id ELSE
            b.store_id END) AS store_id,
(CASE WHEN b.store_cost IS NULL THEN d.store_cost
            ELSE b.store_cost END) AS store_cost,
(CASE WHEN b.unit_sales IS NULL THEN d.unit_sales
            ELSE b.unit_sales END) AS unit_sales,
(CASE WHEN b.store_sales IS NOT NULL AND
            d.store_sales IS NULL THEN b.store_sales
        WHEN b.store_sales IS NOT NULL AND d.store_sales
            IS NOT NULL THEN
            (b.store_sales+d.store_sales)
        WHEN b.store_sales IS NULL AND d.store_sales IS
            NOT NULL THEN d.store_sales
        END) AS store_sales, d.flag AS what_if_flag
FROM
            sales_fact_1997 b FULL JOIN delta_table d
ON
            b.product_id=d.product_id AND
            b.time_id=d.time_id AND
            b.customer_id=d.customer_id AND
            b.promotion_id=d.promotion_id AND
            b.store_id=d.store_id;
```

We use "CASE" clause to decide which tuple field should be visible on the full join table, all the delta tuple type of U, I and D can be process in the single SQL statement. We can't skip the what-if deleted delta tuple in the SQL statement, so we add flag field of output tuple to distinguish different tuple types for the following group-by operation.

With fullJoinBDM algorithm, set difference operation is replaced with full join operation which is much faster. What-if inserted delta tuples can be directly merged with fact table but updated and deleted delta tuples can't, so we improve this algorithm with different delta tuple type policy.

B. left join based delta table merging algorithm

What-if inserted delta tuples can be merged with fact table with union all command, so we separate the inserted delta tuples from delta table and directly merge them with the results. We can see in table 1 that the other three types of tuple mapping don't have NULL values in the left part, so we can perform left outer join operation instead of full join, the time cost can be lower and only measure fields need to be re-written with case clause, which will greatly reduced the comparison process time .

```
Algorithm4.  leftJoinBDM
SELECT
            b.product_id,b.time_id,b.customer_id,
            b.promotion_id,b.store_id,
            b.store_cost,b.unit_sales,
CASE   WHEN d.product_id IS NULL THEN
            b.store_sales
        WHEN d.product_id IS NOT NULL THEN
            (b.store_sales+d.store_sales)
      END as store_sales
FROM
            sales_fact_1997 b LEFT OUTER JOIN
            delta_table d
ON
            b.product_id=d.product_id AND
            b.time_id=d.time_id AND
            b.customer_id=d.customer_id AND
            b.promotion_id=d.promotion_id AND
            b.store_id=d.store_id
UNION ALL
SELECT
            d.product_id,d.time_id,d.customer_id,
            d.promotion_id, d.store_id, d.store_cost,
            d.unit_sales, d.store_sales
FROM        delta_table
WHERE       d.flag= 'I';
```

3.3 Pre-merge Based Delta Table Algorithm

As we discussed in section 2, we can perform SUM based OLAP queries with pre-merge algorithm, we also develop algorithms for COUNT and AVERAGE functions.

Algorithm5. `pre-mergeBDM(SUM)`

```
SELECT
            store.store_state , time_by_day.the_year,
            time_by_day.quarter ,
            product_class.product_family ,
            SUM(sales_fact_view.store_sales)
FROM
            store, time_by_day , product_class,
            product ,
(SELECT
            product_id,time_id,customer_id,promotion_id,
            store_id,store_sales,store_cost,unit_sales
FROM        sales_fact_1997
UNION ALL
SELECT
            product_id,time_id,customer_id,promotion_id,
            store_id,store_sales,store_cost,unit_sales
FROM        delta_table ) sales_fact_view
WHERE
            sales_fact_view.store_id = store.store_id
            AND   sales_fact_view.time_id =
            time_by_day.time_id AND
            sales_fact_view.product_id =
            product.product_id
            AND product.product_class_id =
            product_class.product_class_id
GROUP BY store.store_state, time_by_day.the_year,
            time_by_day.quarter,
            product_class.product_family;
```

If aggregate function is COUNT, we use the SUM value of count field as COUNT value. This method can be applied in multi-version scenario too. Multi-version what-if update only occur on updated delta tuple such as $factTuple_i(1)$, $UdeltaTuple_{j1}(0)$, $UdeltaTuple_{j2}(0)$,..., the SUM value of all the count field of all delta tuples and fact tuple is still 1, what-if delete on one fact tuple can only occur once, the re-insertion of deleted fact tuple is just stored as a new what-if insertion delta tuple.

To COUNT and AVG aggregate operation, we use "SUM(sales_fact_view.count) " instead of "COUNT (measure$_i$) " , and "SUM(measure$_i$)/COUNT(count) " as AVERAGE(measure$_i$).

4 Experiments

In order to get a credible experiment results, we perform our test with typical dataset FoodMart and design all our algorithms with SQL statements. We perform our test on a HP Intel Itanium-64 server with two 1.6G CPUs running Red Hat Enterprise Linux ES release 4 (kernel version 2.6.9) with 2GB RAM, 2GB Swap and 80GB hard disk. We develop a tool for test data generation, which can generate the give amount of fact tuples and ten delta tuple datasets with the delta rate to fact tuples varies from 1% to 10%. In our experiments, we use a dataset of 8,000,000 fact tuples.

We also construct query re-writing algorithms for comparison, what-if updates are represented as the union of multiple sub selection statements. We design test SQL statements for pure what-if update, what-if deletion, what-if insertion and mixed type of 40% update delta tuples, 30% deletion delta tuples and 30% insertion delta tuple. The granularity of delta tuple is decided by adjusting where clauses on dimension fields. We develop two groups of re-writing query with simple algebra update and complex aggregate based update with join and aggregate operations.

We select the open source main-memory MonetDB database as our testing database which is column sorage(DSM) MMDB and has better performance in OLAP queries than many databases. The Power@Size、Throughput@Size 、QphH@Size measures in TPC-H test of MonetDB are much more better than typical DRDBs, and also better than typical row-storage(NSM) MMDBs.

A. Performance analysis of different delta table merging algorithms

We select algorithm 1-5 as our test targets and the original MDX query as a benchmark test. We can find that two set operation based algorithms are much costly than other algorithms, in both 1% and 10% delta rate conditions, more than 90% percents of time spends on the merging procedure between fact table and delta table, as figure 1(a) and (b) show.

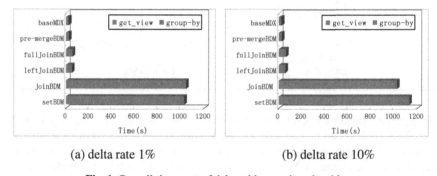

(a) delta rate 1%   (b) delta rate 10%

Fig. 1. Overall time cost of delta table merging algorithms

We do further test on the great gap between set operation based algorithms and join based algorithms, we design test SQL statements for set operation and join operation between fact table and delta table. The average rate of set difference operation to join operation is 97.23, the rate remains stable as fact table or delta table increase.We can draw a conclusion that in MonetDB, the set difference operation is the bottleneck of delta table merging algorithm.

DSM model is less efficient than NSM model in set operations, so we perform the same test in other databases, the results show that set operations are usually less efficient than join operation, but the difference of set operation and join operation in NSM database is smaller than DSM database. We will do further experiments with other databases in our future work.

B. Performance of different delta table merging algorithms

Now we focus on the performance curve of different delta table merging algorithms as delta rate increases.

In figure 2, time cost grows larger with the increase of delta rate at average ratio of about 1% in setBDN algorithm. In improvedSetBDM algorithm, the amount of set difference operation tuples grow larger and the join operation tuples grow smaller, the whole time cost curve slowly decline as delta rate increases at average ratio of about 0.2%. Time cost of both join based algorithms increases with nearly the same ratio of about 0.5%, the average time cost of fullJoinBDM algorithm is about 10% percents more than leftJoinBDM algorithm, it owes to the smaller delta tuple set without what-if insertion delta tuples in left-JoinBDM algorithm and the benefit of left join substitutes full join.

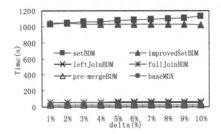

Fig. 2. Time cost of different delta table merging algorithms

We can see the time cost curve rises as delta rate increases because the amount of group-by tuples grows. The average ratio of time cost increase is about 1%. The average time cost of join based algorithms is about 2.25 times of the average time cost of pre-merge algorithm. The time cost of group-by and join operations based on delta table merging is almost equal, it is important to improve the performance of delta `table merging algorithm in massive OLAP applications. Among all the delta table based algorithms, pre-merge algorithm has the best performance.

C. Comparison between delta table merging and query re-writing algorithm

We measure the performance of simple query re-writing, complex query re-writing with aggregate function and delta table merging algorithm under four what-if update scenarios of pure update, pure deletion, pure insertion and mixed what-if updates. We have the following several observations in our experiments as figure 3 shows:

(1) In re-writing query algorithm, mixed hypothetical updates take more time than single types because of more clauses involved.

(2) Time cost of complex query re-writing is about 28% more than that of simple query re-writing.

(3) Time costs of three types of hypothetical updates are close to each other because each tuple should be scanned to determine whether it should be output or output with re-computing values.

(4) To delta table merging algorithm, inserted delta tuples can be directly merged with base table with UNION ALL operation while deleted or updated delta tuples have to perform join operation to generate target tuples in left join merging algorithm and deleted delta tuples spend more time on checking whether current merged tuple should be output.

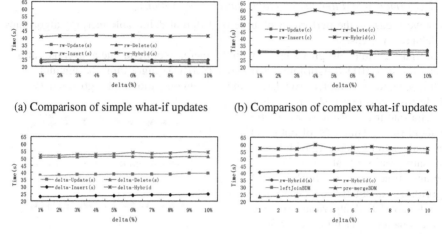

(a) Comparison of simple what-if updates

(b) Comparison of complex what-if updates

(c) Comparison of delta table what-if updates

(d) Comparison of mixed query re-writing and delta table merging algorithms

Fig. 3. Time cost of different algorithms between query re-writing and delta table merging algorithms

(5)   The comparison of four algorithm shows that complex query re-writing algorithm is more costly than delta table merging algorithm, we can easily construct more complex rules for what-if updates and the time cost of query re-writing would be even longer; pre-merge algorithm has better performance than others but has limitation with aggregate functions which don't follow the associative law in aggregate computations.

5   Policy of Tradeoff in What-If Analysis

It is difficult to tell which type of algorithm is better in different application scenarios; the policy of tradeoff in what-if analysis applications should consider four factors:

(1)   Usability: delta table can represent either regular what-if updates based on other tables or arbitrary delta tuples without relation to other tables; re-writing query can only represent regular what-if updates based on SQL statements. Query re-writing and join based delta table merging algorithms can satisfy any SQL statements by the generation of what-if data view while pre-merge algorithm only supports OLAP queries with aggregate function of SUM, COUNT and AVERAGE which are most popular used in practical OLAP applications.

(2)   Space cost: query re-writing algorithms are superior to delta table merging algorithms by zero space cost. The width of delta table is a bit more than the size of fact table. If the amount of delta tuples exceed a threshold (for example 10%), the space cost should be considered.

(3) Time cost: OLAP application on massive dataset provides poor performance, and what-if analysis decreases the performance further. The performance of what-if analysis is the key problem in many research works. SUM aggregate function based what-if analysis should employ pre-merge algorithm to obtain high performance. Whether query re-writing algorithm is better than delta table merging algorithm depends on the complexity of re-written queries, it is necessary to develop a cost model for what-if updates evaluation. If what-if updates involve much costly aggregate computation and only generate small amount of delta tuples, then delta table merging algorithm can provide better performance with less time cost in merging procedure, but the costly computation repeats whenever query is called in query re-writing algorithm.

(4) Application scenario feature: most what-if analysis applications are under multi-version, multi-user or co-analysis among users of different levels, cascade what-if update is also required. Delta table can store different types of delta tuples together with flags to distinguish and easily perform merging procedure with given version delta tuples. Multi-version cascade re-writing would lead to complex sub query nesting problem and result in poor performance.

6 Conclusions and Future Work

We analyze the performance bottleneck of delta table merging algorithm and propose improved algorithms based on join operation and pre-merge policy, by experiments we obtain the performance feature of each algorithm. We also design complete experiments for query re-writing algorithms under all kinds of conditions to compare with delta table merging algorithm. Finally, we discuss the policy of how to select proper algorithm for practical what-if analysis application. We will continue our research on delta cube algorithm, version merging, cascade version accessing and version management in our future work.

References

1. Hurtado, C.A., Mendelzon, A.O., Vaisman, A.A.: Maintaining Data Cubes under Dimension Updates. In: Proc. workshop on DOLAP, pp. 60–66 (1999)
2. Balmin, A., et al.: Hypothetical Queries in an OLAP Environment. In: Proc. VLDB 2000, pp. 220–231 (2000)
3. Lee, K.Y., Kim, M.: Efficient Incremental Maintenance Of Data Cubes. In: Proc. VLDB 2006, pp. 823–833 (2006)
4. Witkowski, A., Bellamkonda, S., Bozkayz, T., Folkert, N., Gupta, A., Sheng, L., Subramanian, S.: Business Modeling Using SQL Spreadsheets. In: Proc. VLDB 2003, pp. 1117–1120 (2003)
5. Witkowski, A., Bellamkonda, S., Bozkayz, T., Naimat, A., Sheng, L., Subramanian, S., Waingold, A.: Query By Excel. In: Proc. VLDB 2005, pp. 1204–1215 (2005)
6. Lehner, W., Sidle, R., Pirahesh, H., Cochrane, R.: Maintenance Of Cube Automatic Summary Tables. In: Proc. SIGMOD, pp. 512–513 (2000)
7. Lakshmanan, L.V.S., Russakovsky, A., Sashikanth, V.: What-if OLAP Queries with Changing Dimensions. In: Proc. ICDE 2008 (2008)

8. Bebel, B., Eder, J., Koncilia, C., Morzy, T., Wrembel, R.: Creation And Management Of Versions In Multiversion Data Warehouse. In: Proceedings of the 2004 ACM symposium on Applied computing, Nicosia, Cyprus, pp. 717–723. ACM, NY (2004)
9. Morzy, T., Wrembel, R.: On Querying Versions Of Multiversion Data Warehouse. In: Proceedings of the 7th ACM international workshop on Data warehousing and OLAP, Washington, DC, USA, pp. 92–101. ACM, NY (2004)
10. Wang, S., Xiao, Y.-Q., Zhang, Y.-S., Chen, H.: Research on OLAP System Supporting What-if Analysis. Chinese Journal of Computers 31(9), 1573–1587 (2008)

Efficiently Clustering Probabilistic Data Streams*

Chen Zhang[1], Cheqing Jin[2,3], and Aoying Zhou[2,3]

[1] Department of Computer Science and Engineering, Fudan University, P.R.C
zhangchen@fudan.edu.cn
[2] Software Engineering Institute of East China Normal University, P.R.C
[3] Shanghai Key Laboratory of Trustworthy Computering, P.R.C
{cqjin,ayzhou}@sei.ecnu.edu.cn

Abstract. Data mining on uncertain data stream has attracted a lot of attentions because of the widely existed imprecise data generated from a variety of streaming applications in recent years. The main challenge of mining uncertain data streams stems from the strict space and time requirements of processing arriving tuples in high-speed. When new tuples arrive, the number of the possible world instances will increase exponentially related to the volume of the data stream. As one of the most important mining task, how to devise clustering algorithms has been studied intensively on deterministic data streams, whereas the work on the uncertain data streams still remains rare. This paper proposes a novel solution for clustering on uncertain data streams in point probability model, where the existence of each tuple is uncertain. Detailed analysis and the thorough experimental reports both on synthetic and real data sets illustrate the advantages of our new method in terms of effectiveness and efficiency.

Keywords: Uncertain data stream, Clustering.

1 Introduction

Recently, data mining on uncertain data stream has attracted a lot of attentions because of the widely existed imprecise data generated from a variety of streaming applications, such as the data from sensor network and RFID application. In such applications, the volume of data is very huge, and tuple's arriving rate is quite fast so that it is infeasible to reserve all tuples in memory to be visited for multiple times. A key task for streaming applications is to devise one-pass algorithms to calculate the high-quality approximate result efficiently with each tuple visited at most once. Undoubtedly, mining uncertain data streams is much more difficult than mining deterministic data streams because of the usage of uncertain data model. As the most widely used uncertain data model, the *possible world* model generates a huge number of the possible world instances from an uncertain data set with the sum of probabilities equal to 1. However, the number of the possible world instances blowups exponentially when new tuples arrive, making it impossible to combine medial results generated from all of possible world instances for the final query results.

---

* This work is supported by Shanghai Leading Academic Discipline Project (Project Number: B412) and National Natural Science Foundation of China (NSFC) under grant No. 60803020.

Q. Li et al. (Eds.): APWeb/WAIM 2009, LNCS 5446, pp. 273–284, 2009.

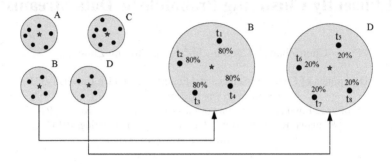

Fig. 1. Cluster Quality Problem

As one of the most important tasks in data mining field, clustering algorithm aims at finding some clusters on one data set, with the inner similarity of clusters maximized and the outer similarity among clusters minimized. Currently, most traditional clustering methods on deterministic data sets treat the distances between tuples as the unique factor to construct clusters. For example, the k-means algorithm[1] aims at finding k clusters with smallest value of the sum of squared errors (SSE). Let C_j be the j-th cluster, and c_j the central point of C_j, the SSE of C_j is calculated as follows.

$$SSE(C_j) = \sum_{j=1}^{k} \sum_{x_i \in C_j} \| c_j - x_i \|^2$$

However, the distance between tuples may not be the only critical metric when clustering uncertain data streams. Consider an example that all tuples arriving from an uncertain data stream at some time point are deployed in Figure 1. The value affiliated to each tuple represents the corresponding occurrence probability. Clearly, all tuples can be easily categorized into four clusters according to distance metric, denoted as A, B, C and D. The distributions of B and D are same, but the probabilities of them differ a lot. In D, the probability of tuples are small (equal to 20%), but in B, the probability of tuples are large (equal to 80%). According to the semantic of the probability value, we can learn that more tuples will occur in B than in D in a randomly selected possible world instance, which means that the quality of cluster B is better than D (We will depict it in detail in next section).

Currently, the uncertainty of a tuple can be described in several ways, such as continuous (/discrete) probability density function[13] and distribution parameter[2]. Aggarwal and Yu proposed one method to handle the distribution parameter case[2]. Their method can be extended to cope with probability density function case. Above two cases belong to attribute-level uncertainty, which implies that each tuple must occur but the values of some attributes are uncertain. Another important case is the *point probability model*, in which each tuple t is affiliated with a probability $p(t)$, meaning the occurring probability of t. The *point probability case* is a modest and realistic model in the uncertain stream field, and can be treated as an existential uncertainty [12,13]. The work by Aggarwal and Yu can not be extended to cope with our model directly [2].

We made following contributions in this paper. First, we define a new metric to measure the quality of a cluster on uncertain data streams, where the distance between

Table 1. Possible world instances for B

No.	PW	prob.	No	PW	prob.	No.	PW	prob.	No.	PW	prob.
1	$\{t_1\}$	.0064	2	$\{t_2\}$	.0064	3	$\{t_3\}$	.0064	4	$\{t_4\}$	.0064
5	$\{t_1, t_2\}$	.0256	6	$\{t_1, t_3\}$	.0256	7	$\{t_1, t_4\}$	.0256	8	$\{t_2, t_3\}$	.0256
9	$\{t_2, t_4\}$	.0256	10	$\{t_3, t_4\}$	.0256	11	(t_1, t_2, t_3)	.1024	12	(t_1, t_2, t_4)	.1024
13	$\{t_1, t_3, t_4\}$	.1024	14	$\{t_2, t_3, t_4\}$	.1024	15	$\{t_1, t_2, t_3, t_4\}$	.4096	16	empty	.0016

Table 2. Possible world instances for D

No.	PW	prob.	No	PW	prob.	No.	PW	prob.	No.	PW	prob.
1	$\{t_5\}$	.1024	2	$\{t_6\}$	.1024	3	$\{t_7\}$	.1024	4	$\{t_8\}$	.1024
5	$\{t_5, t_6\}$	.0256	6	$\{t_5, t_7\}$	.0256	7	$\{t_5, t_8\}$	.0256	8	$\{t_6, t_7\}$	.0256
9	$\{t_6, t_8\}$	.0256	10	$\{t_7, t_8\}$	.0256	11	(t_5, t_6, t_7)	.0064	12	(t_5, t_6, t_8)	.0064
13	$\{t_5, t_7, t_8\}$	.0064	14	$\{t_6, t_7, t_8\}$	.0064	15	$\{t_5, t_6, t_7, t_8\}$	.0016	16	empty	.4096

tuples and the probabilities of tuples are considered together. Second, we propose one novel method to cluster uncertain data streams with detailed analysis and extend a hybrid decay mechanism. Finally, experimental results on synthetic and real data sets can evaluate the effectiveness and efficiency of our proposed method.

This paper is organized as follows. After defining the semantics of clusters' quality in Section 2, the detailed algorithms are shown in Section 3 and 4. In Section 5, we describe the confident decay mechanism as an extension. Section 6 reports experimental results. Section 7 reviews the related work. Finally, we give a brief conclusion and point out the future work in Section 8.

2   Semantics of Clusters

Traditionally, we construct clusters over a deterministic database by using the distance metric, such as Manhattan distance, Euclidean distance, etc. Tuples close to each other are treated to create one cluster because these tuples share some similar characteristics in general. However, the distance metric is not enough to judge the quality of a cluster on uncertain database. Actually, the uncertainty information must be considered together.

As the most widely used model in managing uncertain data, the *possible world model* consists of numerous *possible world instances* where parts of tuples will occur and the rest will not occur. The probability of each possible world instance is calculated as the product of the probability of tuples in the instance and the product of the probability of tuples not existed in the instance. It is important that the sum of the probability of all possible world instances is equal to 1.

For example, the possible world instances constructed based on B and D in Figure 1 are shown in Table 1 and 2 respectively. Because each tuple is independent to others, there are 2^4 possible world instances for each data set. For the possible world instance only containing t_1, t_2 and t_3, the probability is calculated as: $0.8 \times 0.8 \times 0.8 \times (1 - 0.8) = 0.1024$ (Table 1). For the possible world instance only containing t_5, t_6 and t_7, the probability is calculated as: $0.2 \times 0.2 \times 0.2 \times (1 - 0.2) = 0.064$

(Table 2). Let E denote the expected number of tuples in a randomly selected possible world instance based on an uncertain data set. Then,

$$E = \sum_{w \in W} |w| \cdot p(w)$$

where W is the whole possible world space, $p(w)$ is the probability of a possible world instance w, and $|w|$ is the number of tuples in w. In this example, $E(B) = 3.2, E(D) = 0.8$. In fact, the value of E for an uncertain data base C can be calculated in an easier way, as shown below.

$$E(C) = \sum_{x \in C} p(t)$$

According to above analysis, we can learn that the quality of an uncertain cluster C is deeply influenced by two factors, $E(C)$ and $r(C)$, where $r(C)$ is the radius of C. Either larger $E(C)$ or smaller $r(C)$ means better quality. We define the quality of an uncertain cluster in Definition 1.

Definition 1. *The Quality of a cluster $Q(C)$ is defined as $Q(C) = \frac{E(C)}{r(C)} = \frac{\sum_{x \in C} p(x)}{r(C)}$.*

The goal of this paper is to find some groups of tuples with high quality in streaming model.

3   Algorithm

In this paper, we propose a novel method, named PMicro, to cluster uncertain data stream with the use of a micro-clustering framework. Firstly proposed in [5] for large data sets, the micro-clustering model was also adapted in [4] for the case of deterministic data streams and in [2] for the case of uncertain data streams. In [2], Aggarwal and Yu proposed Error based Cluster Feature (ECF) to handle uncertain data streams. However, their work cannot be applied into the point probability model directly, which is the core task in our paper.

Obviously, the main challenge is how to integrate the uncertainty into the micro-clustering statistics and algorithms. Assume a data stream consists of a set of multi-dimensional records $x_1, \cdots, x_k, \cdots$ arriving at time stamps $t_1, \cdots, t_k, \cdots$. Each record x_i contains d dimensions denoted by $x_i = (x_i^1, \cdots, x_i^d)$. Different from the deterministic data streams, each record x_k is also affiliated with a probability value $p(x_i)$. In order to cope with the point probability model, we define a variety of micro-clusters, named *Probability Cluster Feature* (PCF) (Definition 2).

Definition 2 (PCF). *A Probability Cluster Feature (PCF) for a set of d-dimensional points $x_{i_1}, \cdots, x_{i_n}$ with time stamps $t_{i_1}, \cdots, t_{i_n}$ and probabilities $p(x_{i_1}), \cdots, p(x_{i_n})$ is defined as a $(2 \cdot d + 3)$-dimension vector $(CF2^x(C), CF1^x(C), E(C), t(C), n(C))$, wherein $CF2^x(C)$ and $CF1^x(C)$ each correspond to a vector of d entries. The definition of each entry is as follows.*

- *$CF2^x(C)$ maintains the sum of squares of the data values for each dimension, i.e, the value of p-th entry is $\sum_{j=1}^{n} (x_{i_j}^p)^2$.*

- $CF1^x(C)$ *maintains the sum of data values for each dimension, i.e, the value of p-th entry is* $\sum_{j=1}^{n} x_{i_j}^p$.
- $E(C)$ *maintains the expected number of tuples in a randomly selected possible world instance for dataset C, i.e.,* $E(C) = \sum_{j=1}^{n} p(x_{i_j})$.
- $n(C)$ *maintains the number of points in C, i.e,* $n(C) = n$.
- $t(C)$ *is the time point of the most recent point, i.e.,* $t(C) = t_{i_n}$.

Property 1 (PCF additive). Let C_1, C_2 denote two sets of points. $PCF(C_1 \bigcup C_2)$ can be calculated based on $PCF(C_1)$ and $PCF(C_2)$.

The correctness of the additive property is straightforward. The values of entries $CF2^x$, $CF1^x$, E and n in $PCF(C_1 \bigcup C_2)$ are the sum of the corresponding entries in $PCF(C_1)$ and $PCF(C_2)$. The value of $t(C_1 \cup C_2)$ is equal to $\max(t(C_1), t(C_2))$. Besides the simplicity, the additive property is also very useful to cluster a huge dataset.

Algorithm 1. *PMicro*(c_{micro}, α)

1: Create a buffer S to reserve at most c_{micro} PCF, $S = \emptyset$;
2: **repeat**
3:     receive current stream point x_t;
4:     pcf=SelectPCF(S, x_t, α)
5:     **if** (pcf is not NULL)
6:         update pcf by using property 1 to absorb point x_t.
7:     **else**
8:         Create a new PCF for x_t in S, because x_t cannot be absorbed by any present PCF.
9:     **if** ($|S| > c_{micro}$)
10:         remove the least recently updated PCF entry in S.
11: **until** data stream ends;

Based on PCF, we propose Algorithm *PMicro* (Algorithm 1) to cluster uncertain data streams. It uses two input parameters, c_{micro} and α. The parameter c_{micro} limits the maximum number of PCF in the buffer S. The elastic coefficient α is used in Algorithm *SelectPCF* (Algorithm 2) to select one cluster for merging. We will explain the usage of parameter α in detail in the next section.

Initially, a buffer S is created to reserve some PCFs during the running time. As mentioned earlier, the maximum size of S is c_{micro} (Line 1). For each newly arrival point x_t at time t, we invoke subroutine *SelectPCF* (Algorithm 2) to find one cluster for maintenance. The goal of subroutine *SelectPCF* is to return NULL if x_t is an outlier, or to return a PCF if any cluster in S is near to x_t (see details in next section). If x_t happens to be near to a PCF (denoted as pcf), we update pcf by using property 1. Otherwise, we also generate a new PCF for x_t in S. Finally, if the size of S has exceeded the predefined parameter c_{micro}, the least recently updated PCF must be removed from S.

4   Select One Candidate Cluster

As mentioned in Algorithm 1, the goal of subroutine *SelectPCF* is to select one cluster in a set S "near" to x_t. One main issue left is how to judge the priority of the cluster? In traditional methods, researchers often treat the cluster nearest the point as the optimal cluster by using the distance metric. However, the distance metric is not enough in the uncertain database management field. The Quality of a cluster defined in Definition 1 integrates both the distance metric and the uncertainty. According to this semantic, the quality of the candidate cluster in S should benefit a lot after absorbing the new point. In other words, the goal of Algorithm *SelectPCF* is to select one cluster with maximum value of $\triangle Q(C) = \frac{E(C)+p(x)}{r(C\cup\{x\})} - \frac{E(C)}{r(C)}$, where C is any cluster in S and the radius can be calculated by $r = \sqrt{CF2^x(C) - (CF1^x(C))^2}$.

A direct solution is to calculate $\triangle Q(C)$ for all $C \in S$. However, this method will result in lots of computation cost. Instead, we propose a heuristic method to handle this problem. This heuristic method consists of two phases. First, it will select only a small part of clusters, close to the point. Second, it will select the candidate cluster from such a small dataset. Clearly, although this method cannot get the optimal cluster in theory, it is very efficient in practice.

Algorithm 2. SelectPCF(S, x, α)

1: **if** ($\forall C \in S, d(C, x) > \tau \cdot r(C)$)
2:   **return** NULL;   //x is an outlier;
3: **else**
4:   Create an empty set S'.
5:   **forall** ($C \in S$ satisfying $d(C, x) \leq \alpha \cdot \min_{C\in S}(d(C, x))$)
6:     add C into S'.
7:   **return** a PCF C in S' with maximum value of $\frac{E(C)+p(x)}{r(C\cup\{x\})} - \frac{E(C)}{r(C)}$.

Algorithm 2 describes the detailed steps of how to select a candidate cluster out of S. At first, the algorithm will try to check whether the new tuple is an outlier or not by checking the distance between the point and the center of a cluster. Parameter τ is predefined to restrict the range. A NULL will be returned if an outlier is detected. Otherwise, we will try to construct S', containing only a small part of tuples in S. In this step, only part of near clusters from the new point are selected. The parameter α will be suggested by the experiment. Finally, we should calculate the change of the cluster quality. It returns the cluster with maximum quality change in S' (at Lines 4-7).

Example 1. We will take an example as shown in Figure 2. Suppose a new stream point $x(p = 50\%)$ arrives, firstly the distance between x and all clusters will be calculated , then the nearest distance (between x and cluster C) will be selected as the minimal distance $d_{min}(\min_{C\in S}(d(C, x)))$, then according to the parameter elastic coefficient α (suppose α=1.2), except B, C, D, all clusters whose distance is more than $1.2 \times d_{min}$ will be discarded. Next, our approach will run the second round selection to calculate the quality change of cluster $\triangle Q(C)$, the results are respectively $Q(B) = -0.27, Q(C) = -0.07, Q(D) = +0.13$, the calculation detail is elaborated in Figure 2 (B and B' respectively represent the cluster before and after x insert into it, and the radius is supposed to be calculated already). As a result, the cluster D will be selected as the final cluster to absorb the new arrival point.

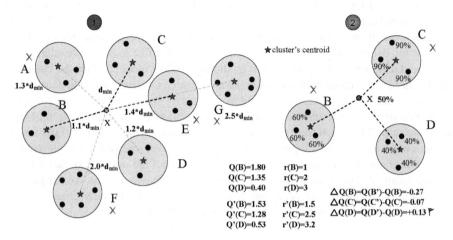

Fig. 2. SelectPCF Example

5 Extension: Confident Weighted Decay

In static database, each record has same weight on the clustering result, while in the data stream model, user would like to pay more attention to the recent data comparing to the old data, thus stream algorithms adopt a time decay mechanism to reduce the affection of old data. Such as UMicro [2] improved the CluStream [4], the latter adopted the land mark window, which means points in each time stamp are equal weight. UMicro used an exponential decay function in order to define the weights of the different data points for constructing the micro-clusters. The weight $W_t(x)$ of a data point x, at the end of time interval between $[t_{begin}, t_{end}]$ is defined as

$$W_t(x) = 2^{-\lambda(t_{end} - t_{begin})} \qquad (1)$$

The limitation of this method is that in the decay process, it does not consider the data quality (uncertainty) factor, for instance the 90% confident point and 10% confident point arrive at the same time, then they will suffer same decay ratio and be eliminated at the same time. However, in practice, we suppose, the high confident point should have more weight not only on clustering but also decaying, since the high confident points are more valuable than the lower ones. We should use this high confident information during a longer period of time. Based on this intuition, we design a confident weighted decay mechanism, the decay function is defined as follows:

$$\widetilde{W_t(x)} = 2^{-\frac{1}{Q(C)} \times (t_{end} - t_{begin})} \qquad (2)$$

We note that all the data structure and algorithms discussed in this paper can be easily extended to the case of the confident weighted discussed above. As in UMicro [2], the lazy approach is employed to reduce the update time for the continuously decaying statistics in the micro-clusters. In the lazy approach, the time decay factor for a micro-cluster is updated only when it is modified by the addition of a new data point to the micro-cluster. This approach can maintain the PCF at a modestly accurate, while without significantly affect on the complexity of the algorithm.

Now we will interpret the decay function from another perspective, for one thing, when static database progress to data stream, user would like to pay more attention to the recent data, so the damped window or sliding window model replace the landmark model. In fact, the inherent different among above three models is the time decay function, for example, in the landmark window, older data and recent data have the same weight on the mining result, while, in the sliding window the older points (out of window) are overall removed from the result, and the damped window model (Equation 1) balance between above two supreme situations. For another, when the deterministic data stream develops to uncertain data stream, the time is not the only decay factor under consideration, due to the data quality also should be emphasized. Thus, we design such probability and time hybrid decay mechanism comparing to the only time-weighted decay function which is widely used in deterministic stream.

6   Experiment

In this section, we demonstrate the effectiveness of PMicro on improving the quality of uncertain data stream clustering. Specifically, we show: i) PMicro can get high quality clustering result to the pure distance-based method; ii) Our method can achieve superior performances; iii) We show how sensitive the clustering quality is in relevance to the parameters. We implemented PMicro as well as comparative method with Matlab v7R14. All experiments were conducted on a PentiumIV 3.0 GHz PC with 1GB memory, running Microsoft Windows XP.

Experimental Competitor. In the experiments, we compare our technique against *UMicro*[2] method. *UMicro* is an uncertainty stream clustering method focused on the effect of uncertainty on the computations of distance. We extend it to include probability calculation component. Note that although *UMicro*[2] is also designed to tackle the problem of clustering on uncertain stream, but the uncertain model is different, direct comparison is unreasonable. Thus, before comparison we change our point probability dataset to the standard error model data. The transform method is described as: larger deviation or error in *UMicro*, means the worse of the data quality, as the same, in *PMicro*, the more smaller existence probability of the data, the worse of the data quality. So we use $x_t \times p_t$ to simulate the standard error used in *UMicro*.

Note that the parameters of implementations for the two stream clustering methods are set to match each other for the purpose of the comparison, and the micro-cluster number c_{micro} is set to match the class number of each data set.

Validation Measure. We use Average of Quality (AQ) to measure the effectiveness of the cluster result. The AQ is defined as the mean of clusters' quality. Given a set of clusters $C_i(1 < i < k)$, then we have $AQ = \frac{1}{k} \sum_{i=1}^{k} Q(C_i)$. AQ is a probability number that allows comparison of the variation of clusters that have significantly different points. In general, the larger of AQ value is, the greater quality of the clustering result.

Experimental Data Sets. For our competition, we use a number of classic benchmark data sets which are widely employed in [4], [7] and [2].

- *Synthetic* data set is generated using continuously drifting clusters. The probability of each tuple is given by a random variable which is drawn from the uniform distribution on $(0, 1)$. As [2], a 20-dimensional data stream containing 600,000 points are generated by using its methodology. The data set is referred as SynDrift corresponding to the fact that it is a synthetic data set with drifting clusters.
- *Network Intrusion Detection* data set consists of raw TCP connection records from a local area network. Each record in the data set corresponds to either a normal connection or a kind of attack types. Also as [4,2], all 34 out of the total 42 continuous attributes available are used for clustering. The probability part is created from the uniform distribution on $(0, 1)$.
- *Forest Covertype* data set is obtained from the UCI machine learning repository [14]. This data set composes of 581012 points and we choose the all 10 quantitative attributes from 54 variables. It is converted into an uncertain data stream by adding each data input with a probability obtained from the gaussian distribution on $(0.5, 3)$ (eliminate the data outrage [0,1]) and flow-in with an even speed.

6.1   Clustering Quality Evaluation

First, the clustering quality of *PMicro* is compared with that of *UMicro*. In Figure 3, the effectiveness of the approach is illustrated with progression of the stream. On the X-axis, we have illustrated the progression of the data stream in terms of the number of points. It is clear that in each case, the PMicro method provides superior quality to UMicro under the AQ criterion. This advantage in cluster quality remains over the progression of the entire data stream. In the case of the Forest Cover data set, the advantage is a little less, since most of the probability centralized, thus clusters are relatively less improved in the cluster quality. In the cases of synthetic data sets, the probability quality improvement of the our PMicro method is quite high, and could often be greater than 8%. It is because PMicro uses the probability information in order to decide assignment of points to clusters.

6.2   Performance Results

Second, we test the efficiency of our stream clustering method. In Figure 4, we have compared the efficiency of the clustering method on different data sets. On the X-axis,

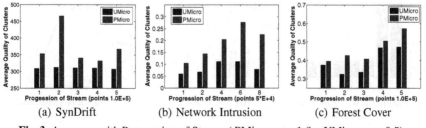

(a) SynDrift   (b) Network Intrusion   (c) Forest Cover

Fig. 3. Accuracy with Progression of Stream (PMicro $\alpha = 1.2$; UMicro $\eta = 0.5$)

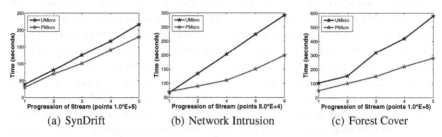

(a) SynDrift   (b) Network Intrusion   (c) Forest Cover

Fig. 4. Performance of Probability Stream Clustering

we also illustrate the progression of the data stream in terms of the number of points, whereas on the Y-axis, we illustrate the time cost with the data stream progression.

The result shows the execution time for the three data sets. We can see that both the execution time of PMicro and UMicro grow linearly as the stream proceeds, and PMicro is more efficient than UMicro. It is because the processing bottleneck mainly lies in high dimensions calculation and UMicro usually has double size dimensions (due to error dimensions) comparing to PMicro. We note that this is quite modest considering the fact that even the processing step of the PMciro method is twice as much as UMciro because of the addition of probability information calculation. Thus, PMicro is not only effective, but is also a very efficient clustering method for data streams.

6.3 Sensitivity Analysis

Finally, we show how sensitive the clustering quality is in relevance to the elastic factor α, the decay rate λ, and the radius threshold τ.

The elastic factor plays an important role in choosing a proper set of clusters that is used for the second phase selection. The experiment about the sensitivity of elastic factor is shown in Figures 5a. It demonstrates that as long as we choose this parameter not too variant from the standard deviation, PMicro could generate very fairly clustering result. Another important parameter which may affect the clustering quality is the different decay mechanism elaborated in Section 5. Figures 5b shows that our probability and time hybrid decay mechanism has significantly improved the quality of clusters result which is in accordance to the intuition. The sensitivity of the radius threshold τ

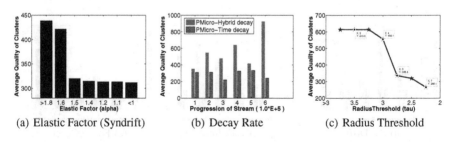

(a) Elastic Factor (Syndrift)   (b) Decay Rate   (c) Radius Threshold

Fig. 5. Sensitivity of parameter Selection

(in Algorithm 2) is described in Figures 5c. Further analysis on this parameter provides a deep understanding of skewness of different data set.

7   Related Work

Clustering a data stream is one of the most challenging tasks in data stream field. The rate of stream is quite fast and the volume of a stream is huge, so that we can only devise a space-efficient one-pass algorithm to answer the query in an online style. As the first piece of work on clustering data streams, **STREAM** algorithm [3] extended the well known k-means method for data stream computation. Subsequently, a **CluStream** named framework was shown for more flexible clustering analysis by using the micro-cluster methodology. Zhou et al. also proposed **CluWin**[6], a solution for sliding-window model, to cluster data streams by combining Exponential Histogram with Cluster Feature.

Most recently, clustering a probabilistic data stream became more and more important because of the uncertain nature in many applications. However, most of present work focused on processing a *static* uncertain database, which required to visit tuples multiple times, thus were incapable of processing an uncertain stream. The goals of **FDBSCAN**[9] and **FOPTICS**[10] algorithms were to find density based cluster from uncertain data. The **UK-means**[11] algorithm also extended the K-means method.

The first work on clustering a probabilistic data stream was named as **UMicro**[2], proposed by Aggarwal and Yu. UMicro used a very general model of the uncertainty in which it assumed that only standard error of individual entries was available. Authors showed that the use of even such modest uncertainty information during the mining process was sufficient to greatly improved the quality of the results than a purely deterministic method. However, this work can not be simply extended to handle the probabilistic point model, which is the goal of this paper.

8   Conclusion

Clustering uncertain data streams is one of the most critical tasks in data mining field. In this paper, we propose a novel algorithm to handle this problem, named *PMicro*. Traditional methods often use the distance as the only metric to construct clusters on deterministic data streams. Contrarily, PMicro method also uses the uncertainty information besides the distance metric. According to the experimental results, both the time consumption and the per-tuple processing cost are very low. One piece of our ongoing work is to devise novel solutions on other uncertain data models.

References

1. MacQueen, J.B.: Some Methods for classification and Analysis of Multivariate Observations. In: Proceedings of 5th Berkeley Symposium on Mathematical Statistics and Probability, Berkeley, vol. 1, pp. 281–297. University of California Press
2. Aggarwal, C.C., Yu, P.S.: A Framework for Clustering Uncertain Data Streams. In: Proc. of ICDE (2008)

3. OCallaghan, L., Meyerson, A., Motwani, R., Mishra, N., Guha, S.: Streaming-Data Algorithms for High-Quality Clustering. In: Proc. of ICDE (2002)
4. Aggarwal, C.C., Han, J., Wang, J., Yu, P.S.: A Framework for Clustering Evolving Data Streams. In: Proc. of VLDB (2003)
5. Zhang, T., Ramakrishnan, R., Livny, M.: BIRCH: An Efficient Data Clustering Method for Very Large Databases. In: Proc. of SIGMOD (1996)
6. Zhou, A., Cao, F., Qian, W., Jin, C.: Tracking clusters in evolving data streams over sliding windows. Knowledge and Information System Journal (KAIS) (2007)
7. Aggarwal, C.C., Han, J., Wang, J., Yu, P.S.: A Framework for Projected Clustering of High Dimensional Data Streams. In: Proc. of VLDB (2004)
8. Tasoulis, D.K., Adams, N.M., Hand, D.J.: Unsupervised Clustering In Streaming Data. In: Proc. of ICDM (2006)
9. Kriegel, H.-P., Pfeifle, M.: Density-Based Clustering of Uncertain Data. In: Proc. of KDD (2005)
10. Kriegel, H.-P., Pfeifle, M.: Hierarchical Density-Based Clustering of Uncertain Data. In: Proc. of ICDM (2005)
11. Ngai, W.K., Kao, B., Chui, C.K., Cheng, R., Chau, M., Yip, K.Y.: Efficient Clustering of Uncertain Data. In: Proc. of ICDM (2006)
12. Cormode, G., Garofalakis, M.N.: Sketching probabilistic data streams. In: Proc. of SIGMOD (2007)
13. Jayram, T.S., McGregor, A., Muthukrishnan, S., Vee, E.: Estimating statistical aggregates on probabilistic data streams. In: Proc. of PODS (2007)
14. Newman, D.J., Hettich, S., Blake, C.L., Merz, C.J.: UCI Repository of machine learning databases, http://www.ics.uci.edu/~mlearn/MLRepository.html

Detecting Abnormal Trend Evolution over Multiple Data Streams[*]

Chen Zhang[1], Nianlong Weng[1,2], Jianlong Chang[1,3], and Aoying Zhou[4,5]

[1] Department of Computer Science and Engineering, Fudan University, P.R.C
{zhangchen,ayzhou}@fudan.edu.cn
[2] ShangHai Stock Exchange, P.R.C
nlweng@sei.ecnu.edu.cn
[3] ShangHai Telecom Company, P.R.C
jlchang@online.sh.cn
[4] Software Engineering Institute of East China Normal University, P.R.C
[5] Shanghai Key Laboratory of Trustworthy Computering, P.R.C
ayzhou@sei.ecnu.edu.cn

Abstract. In this paper, we present a method to trace evolution of trend over multiple data streams and detect the abnormal ones. First of all, a definition of trend for single data stream is provided, the advantage of our definition lies in its low time and space cost. Second, we improve a SVD-based method in order to select a pair of optimal initial parameters, then a novel *chessboard* named sketch is also illustrated aim at adjusting the parameters dynamically. Then, utilizing the skewness of trend distribution, an anomaly detection strategy is briefly introduced. Finally, we implement experiment on a variety of real data sets to illustrate effectiveness and efficiency of our approach.

Keywords: Data Stream Trend Analysis, Anomaly Detection.

1   Introduction

Data stream mining has recently received much attention in several communities including theory, data mining, financial and telecom network due to several important applications. Recently, some efforts concentrate on abnormal stream detection over co-evolving multiple streams [2]. In practice, the inherent evolution of trend over multiple streams are often correlated (e.g., temperatures in the same building, traces in the same road network, stock prices in the same market, etc). Thus, it is possible to detect the anomaly from hundreds of streams by using such inherent correlation. First of all, we illustrate our motivation with two real examples.

Stock Market. Figure 1(a) shows a part of constituent stock streams of Heng Seng Index (Hong Kong) in one year. It exhibits a clear trend (upward movement) in a long-term period. Although in a short-term time scale, there are lots of fluctuations, the prices of mass of stocks generally follow a similar moving direction such as bull market in a

---

[*] This work is supported by Shanghai Leading Academic Discipline Project (Project Number: B412).

long time scale (e.g., quarters or years). However, because of the mechanism of insider trading, it changes the nature balance of random oscillatory, such factor changes the price trend of one stock which will present a significant different characteristic of evolution. As shown in Figure 1(a), the *Northeast Electrical* stock stream shows an obvious opposite moving direction compare with others over T_1 time interval. In practice, detecting such abnormal stream will provide the speculators a well opportunity to make a profit, or for supervisors a well monitoring of the market status.

Telecom. Consider a few number of SNMP streams collected from telecom backbone network (Figure 1(b) shows 7 days of 4 streams). Under normal situation, the network traffics (in terms of flows, packets or byte counts) follow a similar evolution of trend periodically, for example, at midnight the amount of flow gets down to the valley and in the morning the traffic volume begins to rise and gets its peak around 3:00 pm, then trend begins to move downward again. Although there are some short-term fluctuations, the main evolution of trend exhibits this regular pattern day by day. Assume during an abnormal period T_2, some malfunctions occur in link 3, the pattern of evolution will be broken up and exhibit a totally different feature of trend. During such kind of periods, our mission is to detect the link whose evolution of trend is significant different from others as soon as possible.

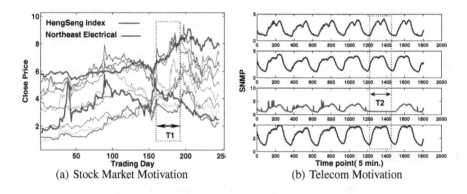

(a) Stock Market Motivation   (b) Telecom Motivation

Fig. 1. Motivation of Abnormal Trend Detection

The main challenge of above problems lies in three aspects: (i) How to define the trend, it is easy to inspect visually, but quantify trend is non-trivial task, moreover, the definition must easy to be calculated in stream setting with restricted resource. (ii) How to select the optimal time scale, actually, it's a balance between sensitivity and accuracy. (iii) How to monitor the status over multiple data streams efficiently, not only focus on volume but also distribution.

In this paper, we present a method to resolve above challenges. Our main contributions are:

- We provide a **SWAD** named definition of trend, it will be calculated under constant space and time cost, meanwhile an accuracy evaluation criterion among different trend definitions is also proposed.

- We improve a SVD (singular value decomposition) based offline period initialization method and illustrate a novel sketch **chessboard** for online, automatically
  adjustment.
- A strategy to label the abnormal periods based on the skewness is also briefly
  elaborated.

The paper is organized as follows: after introducing the related work in Section 2.
We give our definition of trend and accuracy criterion in section 3. Then we illustrate
our approach to choose optimal parameters in Section 4. Section 5 briefly introduce the
strategy to label the abnormal period. Finally, we evaluate our algorithms on a variety
of real datasets in section 6 and conclude in section 7 .

2 Related Work

Trend analysis has a long history in time series analysis field. There is extensive literature in the statistics community regarding trend analysis [1], as well as in the database
community [4]. Generally, a time series can be decomposed into trend, seasonal, periodic, and irregular fluctuation components. The traditional solution for this problem is
to fitting a trendline [8] which is widely used by regression analysis. However, it's not
fit for the case of data stream, for one thing, due to there are various of fitting functions
such as linear, logarithmic, power, exponential or polynomial, without prior knowledge,
the selection of appropriate function is a nontrivial task, for another, the drawback of
trendline-fitting technique is its high complexity, even in the simplest case, the complexity of linear regression is also $O(n \log n)$ [4] which is also unacceptable by the
stream setting.

The work of anomaly detection on data stream has received much attention in recent
years and can be loosely classified in two categories: single-stream and multiple-stream.
Zhou et al. [5] presented a fractal based method to monitor aggregates and detect burst
with multi-granularity windows over a single data stream. The authors of [6] provide
a method to detect approximate top-k subspace anomalies in multi-dimensional time-
series data. However, these methods were too sensitive to the fluctuations and focused
on the outliers in single stream or burst over multiple data streams. Spiros et al. [2] introduced *SPIRIT* which was a streaming pattern discovery method in multiple time series
data streams. It can incrementally detect changes and trends based on correlations.

However, none of above methods simultaneously satisfy the requirement in our introduction: "streaming operation, scalability on the number of streams, evolution of
trend analysis and abnormal detection". Our method monitors the evolution of trend
over multiple data streams with low cost without any predefined models.

3 Definition of Trend

In order to trace trend in data stream, we have to define *what is a trend*.

Problem 1. Given a time series data stream, how to find an appropriate *trend indicator* that can trace the evolution of trend and an accuracy evaluation criterion among
different trend indicators.

In order to describe our definition of trend smoothly, we take the stock data stream of Boeing Inc for example. Figure 2a shows the close price, it is plotted as the stair line. The periods *A* and *C* and periods *B* and *D* respectively represent the upward and downward trend cursorily. A careful study reveals the following observations:

Observation 1. *In the sliding window model, the average of long sliding window changes smoothly compared to the average of short sliding window. Generally speaking, the shorter of sliding window's length, the faster moves of the window's average.*

Through calculating the divergence between two different sliding windows' averages, we can trace the evolution of trend. A positive difference indicates the short sliding window average is larger than the long window average, it means most recent points are larger than the older ones, the evolution of trend is in an upward movement and vice verse. If the difference enlarges, the gap between the two averages is widening, it indicates the upward trend is being strengthened. According to above analysis, we propose our SWAD trend in Definition 1.

Definition 1 (SWAD Trend). *Suppose there is a data stream and two different length sliding windows α, β, respectively represent the short and long period, and averages of such windows at time t is $swa_{\alpha,t}$ and $swa_{\beta,t}$, then our SWAD (sliding window average difference) trend is defined as $SWAD_t = swa_{\alpha,t} - swa_{\beta,t}$;*

Example. Figure 2 gives an example to illustrate our SWAD trend definition. For convenience, the parameters 12/26 (short/long) are used for explanation. In Figure 2a, 12-unit time sliding window average ($swa12$) with the 26-unit time sliding window average ($swa26$) overlaid the stream plot. The short sliding window average ($swa12$) moves faster and fluctuates larger than the long sliding window average ($swa26$), the $swa26$ is relatively more smooth. Generally speaking, an upward trend occurs when $swa12$

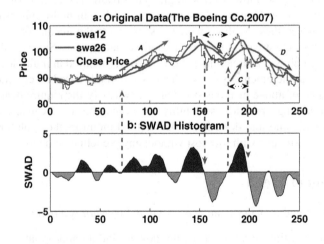

Fig. 2. SWAD Example

moves above the line $swa26$ (A time interval) and a downward trend occurs when $swa12$ moves below the $swa26$ (B time interval). The divergence of two sliding windows' averages make up the SWAD histogram in Figure 2b. The histogram is positive when $swa12$ is above $swa26$, that means an upward trend and vice versa.

Ground Truth. As mentioned in Section 2, there are various trend definition in time series analysis, even if for our SWAD trend, through tuning the parameters, we can get different judgement of the trend status, thus some ground truth as criterion is needed to evaluate those accuracy. Our design idea is based on the intuition that if a trader in a bull (upward trend) market, he will make a profit, the more he earns, the more accurate of his trend tracing method and vice versa. Although this assumption is derived from finance domain, it is easy extended to other field. Thus, we provide our *Trend Accuracy Score* named criterion in Definition 2.

Definition 2 (Trend Accuracy Score). *Suppose there is a data stream $< x_1, x_2..., x_n >$ and related trend indicator is T_t $(t = 1, 2..., n)$. Initialize $m_0 = 1, s_0 = 0$ and $T_0 < 0$, with the stream advances, we will update m and s at each time stamp t according to following equation:*

$$\begin{cases} s_t = s_{t-1}, m_t = m_{t-1}; \ if \ (T_{t-1} \geq 0 \ and \ T_t \geq 0) \ or \ (T_{t-1} < 0 \ and \ T_t < 0) \\ s_t = 0, m_t = s_{t-1} \cdot x_t; \ if \ T_{t-1} \geq 0 \ and \ T_t < 0 \\ m_t = 0, s_t = m_{t-1}/x_t; \ if \ T_{t-1} < 0 \ and \ T_t \geq 0 \end{cases}$$

(1)

The Trend Accuracy Score G at time t is calculated as $G_t = m_t + s_t \cdot x_t$;

The example of *Trend Gain Score* is that, suppose in the stock stream scenario, an investor is assigned 1\$$(m_0)$ initial fund. His operation strategy is that he will use all the money (m) to buy the stock (s) when the trend (T)presents upward beginning or sell all his stocks when the trend changes to downward. If the trend does not change direction, he will hold the current stock and money without operation. Eventually, we use the final gain (sell all stocks plus money) as the score of his trend tracing method.

4 Period Selection

With the optimal short and long sliding windows length (parameters α, β in Definition 1), we could calculate the SWAD trend easily. However, in practice, we have no idea of the proper length of sliding windows beforehand. Thus, we have to resolve following problem:

Problem 2. Given a data stream, and a set of windows $w_i = w_1, w_2, ..w_n$. How to find the optimal short and long windows w_s and w_l from those in order to make sure they can describe two representative time periods and get high Trend Accuracy Score when calculating SWAD.

4.1 Parameter Initialization

Note that from the view of signal processing, the short and long sliding windows respectively represent the main low-frequency and high-frequency periods. They can be used to separate low and high frequency components from the signal. Then, the need of choosing optimal window size transforms to the requirement of selecting proper frequency. Our solution is based on following observations.

Observation 2. *If there is a trend pattern and it repeats in a period of T, then different non-overlapping subsequences of the stream should be highly correlated when the length of time interval is $w \approx T$.*

According to Observation 2, an SVD-based algorithm is designed for optimal time period exploration. The basic step in Algorithm 1 is that firstly according to each window length w, the stream is cut into non-overlapping windows to build a matrix M_w (lines 2), then use the standard SVD algorithm [3] to obtain the top k singular values $\sigma_i, i = 1, 2, ...k$ (lines 3) of M_w. The sum of squares of discarded singular values $\sigma_j, j = k + 1, ...w$ consist of $err^{(w)}$ sequence corresponding to every window w (lines 4).

Algorithm 1. InitialPara()

Input:
sample stream is S, length is n.
top k singular value is remained.

1. **for** $w = 1$ **to** $n/2$ **do**
2.   cut the stream S into non-overlapping windows to build athe matrix M_w;
3. σ=SVD (M_w);
4. $err^{(w)} = \sum_{i=k+1}^{w} \sigma_i^2/w = (\sum_{t=1}^{m} x_t^2 - \sum_{i=1}^{k} \sigma_i^2)/w$;
5. **end for**

Then, we choose two appropriate windows that best capture the main periods according to $err^{(w)}$. First of all, select the first window w_s that exhibits a sharp drop in $err^{(w)}$ and ignore all other drops occurring at windows $c \times w_s, c = 2, 3...$ that are approximately multiples of w_s. In the same way, choose the second window w_l that exhibits a sharp drop to be the long sliding window length. If there are several sharp drops in $err^{(w)}$. We simply choose two smallest ones as the SWAD parameters.

Example. Although the time series stream can be arbitrary, we illustrate our idea with a signal $x = \sin(2\pi t/50) + \sin(2\pi t/40)$ for example in Figure 3a. It composes of two sine waves whose frequency components are at 20Hz and 25Hz respectively.

In Figure 3b, we plot the $err^{(w)}$ line, as expected, the line drops sharply at 20 and 25 (unit window length) which essentially corresponds to the potential frequencies of the original signal. In addition, the 40 and 50 unit window length sharp drops are also from the 20Hz and 25Hz series respectively.

4.2   Parameter Adjusting

In practice, the stream does not always exhibit stable feature (eg. constant period), such as the financial stream, it is one kind of typical unstable signal, the periods change with time lapse. Thus, how to adjust the SWAD calculation parameters dynamically is

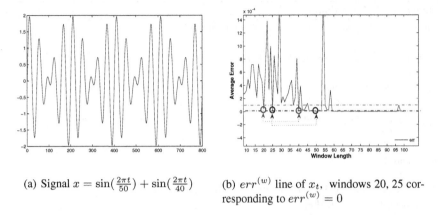

(a) Signal $x = \sin(\frac{2\pi t}{50}) + \sin(\frac{2\pi t}{40})$

(b) $err^{(w)}$ line of x_t, windows 20, 25 corresponding to $err^{(w)} = 0$

Fig. 3. Illumination of Period Analysis

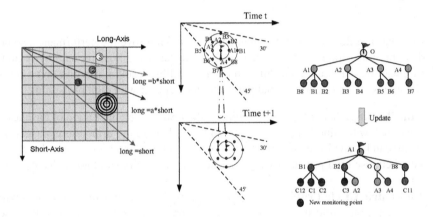

Fig. 4. Chessboard Sketch and Update Example

a new challenge. For this purpose, we propose a ***chessboard*** named sketch for adjusting parameters dynamically. As shown in the left of Figure 4, in our sketch, the short-axis and long-axis respectively denote the length of short and long sliding windows. Every positive integer point in this coordinate system represents a pair of input parameters for SWAD calculation.

The background data structure to support our sketch is a quadtree which can be used to efficiently index regions in a grid. It has been used successfully in image coding and has the benefit of small overhead and very efficient construction [7]. The main idea has been illustrated in Algorithm 2. It maintains Trend Accuracy Score (introduced in Section 3) of candidate points and updates the quadtree if a more superior point is detected in order to make sure the root has the largest accuracy score. When calculating the trend at t, the SWAD procedure will pick up the root as the iuput. A simple update example is illustrated in Figure 4, at time t_0, point O is labelled as the optimal parameter, while at t_1, it is replaced by $A1$, due to the trend accuracy score of $A1$ is larger than O.

Algorithm 2. AdjustPara(x_t, C, S)

Input: new arrival point x_t, chessboard C, trend accuracy score S,

1. Initialize $adjust = FALSE$ and $header = C \rightarrow root$;
2. **repeat**
3.   inorder traversing C to get a point p_{ij}.
4.   calculate $S_{ij,t}$ according to x_t;
5.   **if** $S_{ij,t} > S_{header,t}$ **then**
6. $header = p_{ij}$;
7. $adjust = TRUE$;
8.   **end if**
9. **until**
10. all the points in the high probability region has been checked.
11. **if** $adjust = TRUE$ **then**
12.   update C to ensure $C \rightarrow root = p_{ij}$;
13. **end if**;

Note that in order to reduce the maintain cost and improve the performance, some heuristic pruning strategies have been designed to label the high probability region. For instance, we can use the information from the offline result, it can give some commendatory region where the optimal parameter may appear with high probability.

- **Line Pruning:** According to Definition 1, the lower triangular region of the chessboard below the line $long = short$ could be easily cutout. Furthermore, the chessboard can be divided by the line $long = a \times short$ and line $long = b \times short$. We could choose high probability region according to the domain knowledge case by case.
- **Concentric Circle Pruning:** The points nearby the current using parameters will be given a high priority, from the center of concentric circle we inorder traverse every point by each circle until the radius is beyond a threshold.

5   Anomaly Detection

In this section, we have a shift of focus from single stream and volume-based analysis to multiple streams and distribution-based diagnosis. Based on the preparation of previous sections, the next problem to be resolved is:

Problem 3. When and How to find the abnormal stream whose character of trend is significantly different from others?

The problem of trend anomaly detection is to find the stream that is not similar to the majority in the collection. The two important issues of this problem is when to detect the anomaly and which one is abnormal. Through the observation of stream evolution, we discover the abnormal grade is correlated with the asymmetry degree of the distribution. In a lopsided distribution (such as obviously bull trend in stock market or upward trend in telecom). The mass of the streams are concentrated on a coherent trend direction, there are a few streams in a downward trend movement, thus it is an appropriate period to pick up the abnormal ones. In statistics theory, skewness is used as a measure of the asymmetry of variable distribution. It denotes as γ and has been defined as $\gamma = \frac{\mu_3}{\sigma^3}$, where μ_3 is the third moment about the mean and σ is the standard deviation.

Our approach uses skewness to label the abnormal period, meanwhile returns the minority according to the degree of skewness. The symmetry of trend distribution means the streams are approximate to a random distribution, the count of anomaly is small, while asymmetry of trend distribution indicates the trend evolve to one direction, then the number of abnormal ones is larger than the symmetry stage, hence it is an appropriate period to detect the anomaly.

6 Experiment

In this section, we implement experiments on real dataset to demonstrate the effectiveness and efficiency of our approach in trend tracing and abnormal detection. We compare our method with two widely used trend chasing methods. They are piecewise polynomials trendline (TrendLine) [8] and stream linear regression (LRTrend) [9]. TrendLine is a software package for fitting trendline. LRTrend is a multi-dimensional linear regression analysis method for time-series data stream [9].

We employ three groups of real datasets in different domains for evaluating our algorithms. A brief description is provided in Table 1. The experiments are conducted on an Intel Pentium IV 3.0 GHZ CPU with 1 GB RAM and the experiment environment is windows XP Professional and matlab v7R14.

6.1 SWAD Accuracy and Efficiency

In Figure 5(a), we illustrate four representative stock streams on the X-axis and use the Trend Accuracy Score (Definition 2) as the criterion illustrated on the Y-axis. As the same, in Figure 5(b), we sample four telecom SNMP streams from the backbone network and the length of each stream is respectively week, month, quarter and year. It is clear that in each case, the accuracy of our method is better than other competitors. Furthermore, the superiority of our method is relatively more pronounced in the case of the telecom stream than financial stream. This is because the financial stream has insignificant period, moreover the characteristic of period is also unstable.

Figure 5(c) presents the wall-clock time and space cost of our Matlab implementation for each algorithm on the sensor network data set. The result shows that the efficiency

Table 1. Experimental Data Sets

Domain	Name	#Stream	Size (MB)	#Point	Source
Finance	Heng Seng	29	4.66	83668	HengSeng H&A Stock.
	Dow Jones	30	11.6	193423	US. Dow Jones Stock
	Shang Hai	50	23	154230	Shang Hai 50 ETF
Telecom	SHTelecom07	144	66.8	252126	ShangHai Telecom
Sensor	Sensor Network	200	20	≈ 30000	MERL Detector Dataset [10]

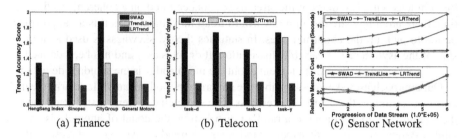

Fig. 5. SWAD Accuracy and Performance

of our SWAD method is obviously superior to others. It is because, SWAD only needs to examine the new incoming point, thus it has almost scalable time cost and constant memory need with the streams advance. While, the TrendLine and LRTrend require scanning the points sequentially and buffering points in order to adjust the parameters, thus the performance of two methods decrease rapidly when more stream data come.

6.2   Period Analysis Offline and Online

Figure 6(a) shows the result of SVD-based period analysis method in telecom data stream. The result shows that it has significant multiple periods such as 3 hours (36 points) , 6 hours (72 points) , 8 hours (96 points) , 12 hours (144 points). We will use 36 and 96 as the short and long sliding windows in the SWAD calculation. The 72 and 144 are integer multiples of 36, thus they can be interpreted as from the same period.

As the same, Figure 6(b) shows the finance stream result, we can see that the error line of Dow Jones Index shows decline at windows 12 and 20. The result of part of sensor network analysis is shown as 32 and 91 in Figure 6(c).

Then, we compare our adaptable, streaming and *chessboard*-based approach against the constant parameter method in Figure 7(a). From the result, we can see our dynamic method achieves near 5% -10% accuracy improvement. In order to test the effectiveness of two pruning method, we illustrate the result in Figure 7(b)-7(c) through tuning the radius and angle of pruning lines. The result is consistent with the intuition, the larger angel and radius, the more accuracy of trend tracing.

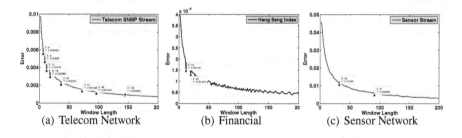

Fig. 6. Offline Period Analysis

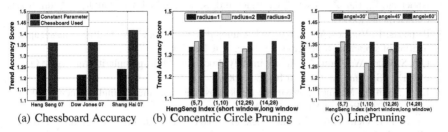

Fig. 7. Online Parameter Adjustment

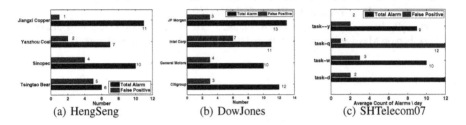

Fig. 8. Skewness Effectiveness

6.3 Anomaly Detection

Our approach find out anomaly streams in different time intervals . Professional domain experts examine the result and judge whether they are meaningful alarms or just false alarms. The result shows that our skewness-based method performs accurate and stable. Figure 8(a)-8(b) illustrate parts of result over Heng Seng and Dow Jones data sets and Figure 8(c) shows the average count of anomalies per day over telecom data set.

7   Conclusion

In this paper, we focus on the problem of trend tracing in single stream and anomaly detection over multiple data streams. We define trend as the difference between two sliding windows averages, then extend the SVD approach and provide a *chessboard* named sketch in order to support the parameter initialization and dynamic adjustment in the calculating process. We also briefly discussed how to incorporate skewness into the anomaly detection. Finally, we conduct extensive experiments on a variety of real world data sets to show the efficiency and effectiveness of our approach.

References

1. Chatfield, C.: The analysis of time series: Theory and Practice, p. 1975. Chapman and Hall, London (1998)
2. Papadimitriou, S., Sun, J., Faloutsos, C.: Streaming Pattern Discovery in Multiple Time-Series. In: Proc. of VLDB (2005)

3. Anderson, E., Bai, Z., Bischof, C., Blackford, S.: LAPACK User's Guide, 3rd edn. SIAM, Philadelphia (1999)

4. Palpanas, T., Vlachos, M., Keogh, E.J., Gunopulos, D., Truppel, W.: Online amnesic approximation of streaming time series. In: Proc. of ICDE (2004)

5. Qin, S., Qian, W., Zhou, A.: Approximately Processing Multi-granularity Aggregate Queries over Data Streams. In: Proc. of ICDE (2006)

6. Li, X., Han, J.: Mining approximate top-k subspace anomalies in multi-dimensional time-series data. In: Proc. of VLDB (2007)

7. Vaisey, D.J., Gersho, A.: Variable block-size image coding. In: Proc. of ICASSP (1987)

8. Microsoft Website (2005),
   `http://office.microsoft.com/en-us/excel/HP100074611033.aspx`

9. Chen, Y., Dong, G., Han, J., Wah, B.W., Wang, J.: Multi-Dimensional Regression Analysis of Time-Series Data Streams. In: Proc. of VLDB (2002)

10. Wren, C.R., Ivanov, Y.A., Leigh, D., Westhues, J.: The MERL Motion Detector Dataset. In: Workshop on Massive Datasets (MD), pp. 10–14. ACM Press, New York (2007)

Mining Interventions from Parallel Event Sequences[*]

Ning Yang[1], Changjie Tang[1], Yue Wang[1], Rong Tang[1], Chuan Li[1],
Jiaoling Zheng[1], and Jun Zhu[2]

[1] School of Computer, Sichuan University
[2] China Birth Defect Monitoring Centre, Sichuan University
610065 Chengdu, China
yneversky@gmail.com,
chjtang@vip.sina.com,
{wangyue,tangrong,lichuan,zhengjiaoling}@cs.scu.edu.cn

Abstract. Discovering temporal patterns from sequence data has been an important task of data mining in recent years. In this paper a novel temporal pattern, *Intervention*, is proposed to capture the partial ordering relations in parallel event sequences. It is demonstrated that *Intervention* is essentially a deviation of generalized Markov property holding in parallel event sequences. A measure to evaluate the degree of such deviation, *Intervention Intensity*, is suggested, which has an important mathematical property, non-symmetry. As a result, an algorithm called MIPES for mining interventions is proposed. The time complexity of MIPES is of $O(m^2)$ and is independent of data size, where m is the number of event types and is far smaller than the data size in practice. The experimental results show MIPES is applicable and scalable.

Keywords: Parallel Event Sequence, Intervention.

1 Introduction

Discovering temporal patterns from event sequences has received considerable attention recently[1,2,3,4,5]. The goal of temporal patterns mining is to discover hidden relations between sequences and subsequences of events widely appearing in a variety of knowledge discovery applications, such as biology, medicine, business, finance, etc [5]. The existing researches, however, mostly focus on mining frequent subsequences from serial event sequences with two major defects[6,7,8,11]. (1) Only serial sequences are considered, though the parallel sequences are more common in practice. In serial sequences, only one event happens at an instant, whereas several events of different types could happen at an instant in parallel sequences. (2) Little attention is paid to the intervening phenomenon between different event types. Frequent event subsequences, such as episodes[7,8], can only reveal the correlations between the events of different types. However, in many cases we are more concerned about the

---

[*] Supported by the National Natural Science Foundation of China under Grant No.600773169 and the 11th Five Years Key Programs for Sci. &Tech. Development of China under grant No. 2006BAI05A01.

Q. Li et al. (Eds.): APWeb/WAIM 2009, LNCS 5446, pp. 297–307, 2009.

intervention imposed on one type of events by another. For example, in Birth Defect monitoring, we are interested in which kind of event types are more effective to intervene the occurrences of the birth defects.

In this paper, we propose a new class of temporal patterns, *Intervention*, existing widely in parallel event sequences. Our idea is illustrated in the following example.

Example 1. The parallel sequence in Fig.1 is composed by two single type event sequences $X(t)$, $Y(t)$, where X and Y are event types and t is the occurrence time of events, $1 \le t \le 12$. Let the length of a period is 2, then the whole lasting time can be split into 6 periods. The number of events of type X in i-th period is denoted by x_i, $1 \le i \le 6$. We say there exists an intervention $X \rightarrow Y$ if the distribution of y_{i+1} is influenced by x_i, i.e., the occurrences of events of type X in current period can influence the occurrences of events of type Y in next period, and not the other way inverse. X is referred to as an intervention source of Y. The precise definition of intervention is given in Section 4.

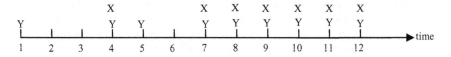

Fig. 1. A piece of parallel 2-type event sequence

The next problem is how to evaluate the effect of intervention. This study assumes that the event sequences can be essentially modeled by Markov Chains, hence the intervention can be regarded as the deviation from generalized Markov property which states that the frequency distribution of $Y(t)$ in the current period is independent on that of $X(t)$ in the previous period. Consequently, we suggest a measure, *Intervention Intensity*, to evaluate the intervention effect. Intervention intensity is based on Kullback-Leibler divergence[9] which is rooted in information theory and is used to measure the distance between two distributions. Kullback-Leibler divergence is asymmetric and is suitable to resist noise. The intervention intensity could be computed by appropriately conditioned transition probabilities. These probabilities can be approximated by kernel estimation from data. More details are described in Section 5.

Integrating above ideas, we implement an algorithm called MIPES to Mine the Interventions from a given Parallel Event Sequence. MIPES is with time complexity of $O(m^2)$ where m is the number of event types and is far smaller than the amount of events and will not degrade the scalability of MIPES.

The rest of this paper is organized as follows. Section 2 discusses the related works. Section 3 gives concise description of Markov Chain and Kullback-leibler divergence. Section 4 formulates the problem and describes the intervention in details. Section 5 presents the definition of intervention intensity and the estimation method. Section 6 presents the comprehension of intervention from the perspective of the fluctuation of entropy rate. Section 7 presents the outline and time complexity of MIPES. We demonstrate the effectiveness and efficiency of MIPES through experiments in section 8 and present the conclusions in section 9.

2   Related Work

The first popular task of mining from temporal data is mining associate rule [6,10,11], whose common approach is based on the apriority method [10]. When a transactional database is represented as a set of sequences of transactions performed by one entity is used, the manipulation of temporal sequences requires that some adaptations be made to the apriori algorithm. To mining frequent patterns from temporal data, the most important modification is on the notion of support: support is now the fraction of entities, which had consumed the itemsets in any of their possible transactions [11], i.e. an entity could only contribute one time to increment the support of each itemset, beside it could had consumed that itemset several times.

Another important temporal pattern is episode, which is essentially small, partially ordered collections of event types [7]. An episode is considered interesting if it occurs often enough in the data. The frame work of discovering frequent episodes in event sequences was introduced by Mannila, et al [7]. They define the frequency of an episode as the number of time windows and propose a counting algorithm using finite state automata for obtaining the frequencies of a set of candidate episodes. Some extensions to this windows-based frequency have also been proposed[12,13]. S. Laxman, et al, have proposed a new notion for episode frequency based on the non-overlapped occurrences of an episode in the given event sequences[8].

A few of work has been done with classification and clustering in temporal data mining. One classification method to deal with temporal sequences is based on the merge operator[14]. This operator receives two sequences and returns "a sequence whose shape is a compromise between the two original sequences"[14]. The basic idea is to iteratively merge a typical example of a class with each positive example, building a more general model for the class. Since traditional classification algorithms are difficult to apply to sequential examples, mostly because there is a vast number of potentially useful features for describing each example, an interesting improvement consists on applying a preprocessing mechanism to extract relevant features. One approach to implement this idea consists on discovering frequent subsequences, and then using them, as the relevant features to classify sequences with traditional methods, like Naive Bayes[15]. Besides classification, Ketterlin proposed an approach to use hierarchical clustering method to cluster temporal sequences data[16].

The existing work did nothing with the intervention phenomenons in parallel event sequences, which are common in real world. We focuses on the mining of interventions, which is discussed in detail in the rest of this paper.

3   Preliminaries

For the sake of completeness, we begin with an overview of Markov chain and Kullback-Leibler divergence.

A Markov chain is a sequence of random variables X_1, X_2, X_3, ... with the Markov property,

$$P(X_{n+1}=x|X_n=x_n,\ldots,X_1=x_1)=P(X_{n+1}=x\mid X_n=x_n).$$

The possible values of X_i form a countable set S called the state space of the chain. Define the probability of going from state i to state j in n time steps as

$$p^{(n)}_{ij} = P(X_n = j \mid X_0 = i),$$

and the single step transition as

$$p_{ij} = P(X_1 = j \mid X_0 = i).$$

The n-step transition satisfies the Chapman-Kolmogorov equation, that for any k such that $0 < k < n$,

$$p^{(n)}_{ij} = \sum_r \in_S p^{(k)}_{ir} p^{(n-k)}_{rj}.$$

According to probability theory and information theory, the Kullback-Leibler divergence is a non-commutative measure of the difference between two probability distributions P and Q. Kullback-Leibler divergence measures the expected difference in the number of bits required to code samples from P when using a code based on P, and when using a code based on Q. Typically P represents the "true" distribution of data, observations, or a precise calculated theoretical distribution. The measure Q typically represents a theory, model, description, or approximation of P. For probability distributions P and Q of a discrete random variable, the Kullback-Leibler divergence of P from Q is defined as

$$D_{KL}(P\|Q) = \sum_i P(i) \log(P(i) / Q(i)).$$

Although it is often intuited as a distance metric, the Kullback-Leibler divergence is not a true metric since it is not symmetric (hence 'divergence' rather than 'distance'), i.e., $D_{KL}(P\|Q) \neq D_{KL}(Q\|P)$. This property is favorable since it is suitable to resist the noise.

4 Intervention

Before defining the intervention, we first formalize the framework of the event sequences. Given a set E of event types, a single type event sequence is $X(t)$ where $X \in E$ is an event type and t is the occurrence time of event. A parallel m-type event sequence is a tuple $(X_1(t), \ldots, X_m(t), T_s, T_e)$, where $T_s \leq t \leq T_e$. Let the length of sampling period is w, then the whole lasting time is split into $k = (T_s - T_e) / w$ periods. The number of events of type X in i-th period is denoted by x_i, where $1 \leq i \leq k$. We consider only the discrete timelines in which occurrence times of events are positive integers.

Assuming that the event sequences under study can be approximated by Markov Chains, the evolutions of the event sequences are described by the transition probabilities. If two sequences $X(t)$ and $Y(t)$ are independent, then the generalized Markov property

$$P(y_{i+1} \mid y_i) = P(y_{i+1} \mid y_i, x_i) \tag{1}$$

holds for all i. Based on this idea, the intervention can be defined as the following.

Definition 1. *Let $X(t)$ and $Y(t)$ be two single type event sequences,* Intervention *from $X(t)$ to $Y(t)$, denoted by $X \rightarrow Y$, exists if $P(y_{i+1} | y_i) \neq P(y_{i+1} | y_i, x_i)$, where $P(\cdot)$ is the probability function. X is referred as to the* intervention source *of Y, and Y is referred as to the* intervention target *of X.*

By definition 1, $X \rightarrow Y$ holds if the frequency distribution of Y in next period depends on not only the frequency of Y in current period, but also the frequency of X in current period. In other words, the occurrences of X in current period can influence the occurrences of Y in next period.

Proposition 1. *The relation of $X \rightarrow Y$ is reflexive, antisymmetrical, and transitive.*

Proof. The reflexivity and transitivity are obvious, hence we only prove the antisymmetry by contradiction. Suppose $Y \rightarrow X$ holds, by Definition 1

$$P(x_{i+1} | x_i) \neq P(x_{i+1} | x_i, y_i)$$

holds, which implies that x_{i+1} depends on y_i. Since

$$P(y_i | y_{i-1}) \neq P(y_i | y_{i-1}, x_{i-1})$$

implies y_i depends on x_{i-1}, x_{i+1} depends on x_{i-1}, i.e.,

$$P(x_{i+1} | x_i) \neq P(x_{i+1} | x_i, x_{i-1}),$$

which contradicts the Markov property of $X(t)$.   □

Thus interventions are partial ordering relations on different types of events. Based on the concept of intervention, the intervention chain is defined as follows.

Definition 2. *A N-order intervention chain is a partially ordered set V of event types and has form: $V = X_1 \rightarrow X_2 \rightarrow \ldots \rightarrow X_N$.*

As the consequence of Proposition 1, it is easy to prove the Proposition 2.

Proposition 2. *Two $(N-1)$-order intervention chains: $X_1 \rightarrow X_2 \rightarrow \ldots \rightarrow X_{N-1}$, $X_2 \rightarrow X_3 \rightarrow \ldots \rightarrow X_N$ can be composed to a N-order intervention chain $X_1 \rightarrow X_2 \rightarrow \ldots \rightarrow X_n$.*

By emploiting the Proposition 2, the intervention chains with higher order can be easily constructed after all the 2-order intervention chains are discovered. Therefore this study only focuses the algorithm on the discovery of 2-order intervention chains.

5 Intervention Intensity

We refer to the effect of intervention as *intervention intensity* which reflects the impact imposed by one single type event sequence on another. In this section we described the measure schema of intervention intensity. We first discuss the noise-free scenarios, then the noisy ones.

5.1   The Noise-Free Scenarios

As per Definition 1, the intervention between two single type event sequences is essentially deviation from the generalized Markov property equation (1). The quantity of this deviation can be measured by Kullback-Leibler divergence[9], which results to definition of intervention intensity.

Definition 3. *Let $X(t)$ and $Y(t)$ be two single type event sequences, the* intervention intensity *of $X \rightarrow Y$ is defined as*

$$I(X \rightarrow Y) = \sum P(y_{i+1}, y_i, x_i) \log \frac{P(y_{i+1} \mid y_i, x_i)}{P(y_{i+1} \mid y_i)}. \tag{2}$$

Intuitively, the intervention intensity indicates the excess amount of bits that must be used to encode the information of the state of the process by erroneously assuming the actual transition probability distribution function is $p(y_{i+1} \mid y_i)$, instead of $p(y_{i+1} \mid y_i, x_i)$.

The joint probability $P(y_{i+1}, y_i, x_i)$ can be approximated by kernel estimation. Given a resolution r, $P(y_{i+1}, y_i, x_i)$ can be obtained by

$$P(y_{i+1}, y_i, x_i) = \frac{1}{N} \sum_j K \left(\max \left(\left| y_{i+1} - y_{j+1} \right|, \left| y_i - y_j \right|, \left| x_i - x_j \right| \right) - r \right), \tag{3}$$

where $K(\cdot)$ is step kernel such that $K(x>0)=1$, $K(x \leq 0)=0$, N is the total number of ij pairs available. Similarly, $P(y_{i+1}, y_i)$, $P(y_i, x_i)$ and $P(y_i)$ can be estimated respectively by

$$P(y_{i+1}, y_i) = \frac{1}{N} \sum_j K \left(\max \left(\left| y_{i+1} - y_{j+1} \right|, \left| y_i - y_j \right| \right) - r \right), \tag{4}$$

$$P(y_i, x_i) = \frac{1}{N} \sum_j K \left(\max \left(\left| y_i - y_j \right|, \left| x_i - x_j \right| \right) - r \right), \tag{5}$$

and

$$P(y_i) = \frac{1}{N} \sum_j K \left(\left| y_i - y_j \right| - r \right). \tag{6}$$

Then the conditional probabilities $P(y_{i+1} \mid y_i, x_i)$ and $P(y_{i+1} \mid y_i)$ are computed respectively, i.e., $P(y_{i+1} \mid y_i, x_i) = P(y_{i+1}, y_i, x_i) / P(y_i, x_i)$ and $P(y_{i+1} \mid y_i) = P(y_{i+1}, y_i) / P(y_i)$.

It follows from the nonsymmetry of Kullback-Leibler divergence that the nonsymmetry of $I(X \rightarrow Y)$, which reflects the fact that information flow between two event sequences is directional.

It is obvious that if $P(y_{i+1} \mid y_i, x_i) = P(y_{i+1} \mid y_i)$ then $I(X \rightarrow Y) = 0$, which means $X(t)$ does not influence $Y(t)$. It is consistent with the Markov property and shows the definition of intervention intensity is reasonable.

5.2 The Noisy Scenarios

So far, only the noise-free cases are considered in which if $X{\rightarrow}Y$ holds then $I(Y{\rightarrow}X)$ will be zero because the Markov property $P(x_{i+1} \mid x_i) = P(x_{i+1} \mid x_i, y_i)$. In noisy cases, however, we can not assure $I(Y{\rightarrow}X)$ is zero even if $X{\rightarrow}Y$ holds.

Fortunately, it is trivial to prove the directional property of intervention intensity, as it follows from the nonsymmetry of Kullback-Leibler divergence[9].

Proposition 3. $I(X{\rightarrow}Y){\neq}I(Y{\rightarrow}X)$.

By Proposition 3, $X{\rightarrow}Y$ not $Y{\rightarrow}X$ can be confirmed if $I(X{\rightarrow}Y) > I(Y{\rightarrow}X)$. It is easy to show that this tradeoff will not violate the antisymmetry of intervention which is stated in proposition 1.

6 Understanding Intervention from Perspective of Entropy Rate

We can understand the intervention from the perspective of entropy rate. As stated in section 4, the single type event sequences can be approximated by Markov chains. The entropy rate of stochastic process $Y(t)$ is defined as

$$h_y = \lim_{n \to \infty} \frac{1}{n} H\left(y_1, y_2, \cdots, y_n\right),$$

and if $Y(t)$ is a Markov chain, then

$$h_y = H\left(y_{i+1} \mid y_i\right) \text{ [17]},$$

where $H(\cdot)$ is entropy.

Proposition 4. *Let $X(t)$ and $Y(t)$ be two single type event sequences, $I(X{\rightarrow}Y) = h_y - h_{y|x}$, where $h_x = H(y_{i+1}|y_i)$, $h_{y|x} = H(y_{i+1}|y_i, x_i)$.*

Proof. By definition 1 and the relation between the joint and conditional probabilities, we can obtain:

$$
\begin{aligned}
I(X \to Y) &= \sum P(y_{i+1}, y_i, x_i) \log \frac{P(y_{i+1} \mid y_i, x_i)}{P(y_{i+1} \mid y_i)} \\
&= \sum P(y_{i+1}, y_i, x_i) \log P(y_{i+1} \mid y_i, x_i) - \sum P(y_{i+1}, y_i, x_i) \log P(y_{i+1} \mid y_i) \\
&= -\sum P(y_{i+1}, y_i) \log P(y_{i+1} \mid y_i) - \left(-\sum P(y_{i+1}, y_i, x_i) \log P(y_{i+1} \mid y_i, x_i)\right) \\
&= H\left(y_{i+1} \mid y_i\right) - H\left(y_{i+1} \mid y_i, x_i\right) \\
&= h_y - h_{y|x}.
\end{aligned}
$$

$\square$

Note the following two remarks on proposition 4:

(1)   As a Markov chain, the uncertainty, namely the entropy, of one event sequence $Y(t)$ will increase at the rate of h_y without any intervention.

(2)   With intervention imposed on $Y(t)$ by $X(t)$, the increasing rate of the uncertainty of $Y(t)$ will be reduced, i.e., $H(y_{i+1}|y_i) \geq H(y_{i+1}|y_i,x_i)$, which is proved strictly in literature [17].

Consequently, Propostion 4 confirms that intervention from $X(t)$ to $Y(t)$ is equivalent to the change of $Y(t)$'s entropy rate resulted by $X(t)$, more explicitly, the intervention imposed on $Y(t)$ by $X(t)$ leads to the reduction of entropy rate of $Y(t)$.

7   Algorithm to Mine Interventions

By integrating the ideas stated by previous sections, this section proposes an algorithm called MIPES to Mine Interventions from Parallel Event Sequences. The outline of MIPES is shown as follows.

> **Algorithm** MIPES $(X_1(t), ..., X_m(t), T_s, T_e, w, r, R)$
> **input**: $X_1(t), ..., X_m(t)$ the event sequences; T_s the start time; T_e the end time;
> m the number of event types; w length of sampling period;
> r the resolution
> **output**: R the set of 2-order intervention patterns with intervention intensity;
> **begin**
> 1. $k = (T_e - T_e) / w$;
> 2.   **foreach** $X(t)$ in $\{ X_1(t), ..., X_m(t) \}$ **do begin**
> 3.   count the x_i, $1 \leq i \leq k$;
> 4.   **end for**;
> 5.   Generate candidates set $C = \{(X_p(t), X_q(t)): 1 \leq p \neq q \leq m \}$;
> 6.   **foreach** $(X_p(t), X_q(t)) \in C$ **do begin**
> 7.   Compute $I(X_p \rightarrow X_q)$ and $I(X_q \rightarrow X_p)$ by equation (2) with resolution r.
> 8.   **if** $I(X_p \rightarrow X_q) > I(X_q \rightarrow X_p)$ **then do begin**
> 9.   add $\{I(X_p \rightarrow X_q)\}$ to R;
> 10.   **else**
> 11.   add $\{I(X_q \rightarrow X_p)\}$ to R;
> 12.   **end if**;
> 13.   **end for**;
> 14. **end**

The algorithm is straightforward. The major work is computing the intervention intensity between every two single type event sequences (Line 7). Since the number of single type sequence pairs is not more than $|C| = m! / (2!(m-2)!) = (m^2 - m) / 2$, the temporal complexity of MIPES is $O(m^2)$, which will not lower the performance in practice when the number of event types is small.

8   Experimental Results

In this section, we apply MIPES to dataset CalIT2 downloaded from UCI[18] to validate the effectiveness and performance of MIPES. CalIT2 consists of 3 months of count data of two event types, In-event and Out-event, recorded every 30 minutes at the front door of the Calit2 institute building on the UC Irvine campus.

The first experiment investigates how the resolution r in equation (3), (4), (5), (6) impacts the evaluation of intervention intensity. Without loss of generality, we use the records on July 24 of CalIT2. The frequency distribution of In-event and Out-event are shown in Fig.2, and the results are depicted in Fig.3.

A few observations on Fig.3 are as follows.

(1) The data have noise, and consequently both I(In-event→Out-event) and I(Out-event →Out-event) are not zero.

(2) In-event→Out-event can be confirmed because I(In-event→Out-event) > I(Out-event→In-event), which is reasonable and consistent with common sense that more people into building leads to more people out building.

(3) Along with r increasing, I(In-event→Out-event) and I(Out-event→In-event) tends to close together, which can be explained as the smaller r favours the isolation of noise.

(4) Nevertheless, excessively small r will limit the amount of available data for the estimation of the probabilities and causes small intervention intensity, which is observed from the part of curve with $r < 3$.

The second experiment examines the performance of MIPES over different scales of data, i.e. 100K, 500K, 1000K, 2000K, 5000K. The result is depicted in Fig.4, from which we can observe that the time cost increases almost linearly along with the increasing of data size, which implies nice scalability of MIPES. The reason is that although the time complexity of MIPES is $O(m^2)$ where m is the number of event types, the value of m (m is 2 in Calit2) is far smaller than the size of data in practice, so the time complexity can be regarded as independent of the size of data. Consequently, MIPES will not sharp decline in performance in practice even though the size of data increases remarkably.

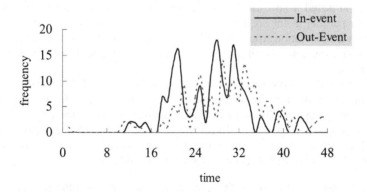

Fig. 2. Frequencies of In-event and Out-event. The data is sampled every 30 minutes.

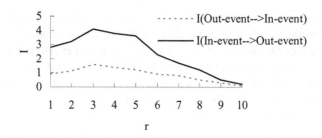

Fig. 3. Intervention intentisties in different resolutions

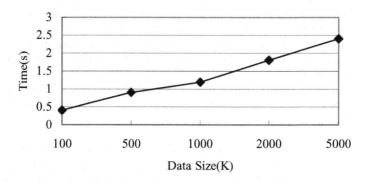

Fig. 4. Performance on scales of data

9 Conclusions

In this paper we have studied the intervening phenomenons appearing widely in parallel event sequences. The main contributions include:

(1) We proposed a new class of temporal patterns called *Intervention*, which is common in parallel event sequences. Furthermore, we revealed that the essence of intervention is a deviation from generalized Markov property and is equivalent to the change of entropy rate. Additionally, we demonstrated that *Intervention* is partial ordering relation.

(2) We defined *Intervention Intensity* based on the Kullback-Leibler divergence to measure the effect of intervention imposed on one type of events by another. We also proved an important property of the intervention intensity, nonsymmetry.

(3) We proposed algorithm MIPES by integrating the above ideas. We have applied it to real data set to examine the effectiveness and efficiency. The examination results shows that MIPES is funtionable and scalable.

In this paper, we concentrate on the intervention mining algorithm that presupposes the whole data of event sequences have been observed. Hence, it is an important future work to research incremental and outline algorithms which are capable of mining interventions in streaming fashion from time evolving event sequences.

References

1. Unnikrishnan, K.P., Uthurusamy, R. (eds): Temproal Data Mining Workshop Notes. In: 8th ACM SIGKDD International Conference on Knowledge Discovery and Data Mining (KDD 2002) (2001)
2. Vautier, A., Cordier, M., Quiniou, R.: An Inductive Database for Mining Temporal Patterns in Event Sequences. In: International Joint Conference on Artificial Intellegence (2005)
3. Roddick, J.F., Hornsby, K.S., Spiliopoulou, M.: An updated bibliography of temporal, spatial, and spatio-temporal data mining research. In: Roddick, J.F., Hornsby, K.S. (eds.) TSDM 2000. LNCS, vol. 2007, pp. 147–163. Springer, Heidelberg (2001)
4. Wu, S., Chen, Y.L.: Mining Nonambiguous Temporal Patterns for Interval-Based Events. IEEE Transactions on Knowledge and Data Engineering 19(6), 742–758 (2007)
5. Antunes, O.: Temporal data mining: An overview. In: 7th ACM SIGKDD International Conference on Knowledge Discovery and Data Mining (KDD 2001) (2001)
6. Han, J., Cheng, H., Xin, D., Yan, X.: Frequent Pattern Mining: Current Status and Future Directions. Data Mining and Knowledge Discovery 15(1), 55–86 (2007)
7. Mannila, H., Toivonen, H., Verkamo, A.I.: Discovery of frequent episodes in event sequences. Data Mining and Knowledge Discovery 1(3), 259–289 (1997)
8. Laxman, S., Sastry, P.S., Unnikrishnan, K.P.: Discovering Frequent Episodes and Learning Hidden Markov Models: A Formal Connection. IEEE Transactions on Knowledge and Data Engineering 19(11), 742–758 (2005)
9. Kullback, S., Leibler, R.A.: On Information and Sufficiency. Ann. Math. Stat. 22, 79–86 (1951)
10. Agrawal, R., Srikant, R.: Fast Algoritms for Mining Association Rules. In: VLDB, pp. 487–499 (1994)
11. Agrawal, R., Srikant, R.: Mining sequential patterns. In: ICDE, pp. 3–14 (1995)
12. Casas-Garriga, G.: Discovering unbounded episodes in sequential data. In: Lavrač, N., Gamberger, D., Todorovski, L., Blockeel, H. (eds.) PKDD 2003. LNCS, vol. 2838, pp. 83–94. Springer, Heidelberg (2003)
13. Meger, N., Rigotti, C.: Constraint-based mining of episode rules and optimal window sizes. In: Boulicaut, J.-F., Esposito, F., Giannotti, F., Pedreschi, D. (eds.) PKDD 2004. LNCS, vol. 3202, pp. 313–324. Springer, Heidelberg (2004)
14. Keogh, E., Pazzani, M.: An Enhanced Representation of Time Series Which Allows Fast and Accurate Classification, Clustering and Relevance Feedback. In: Knowledge Discovery in Databases and Data Mining, pp. 239–241 (1998)
15. Lesh, N., Zaki, M., Ogihara, M.: Mining Features for Sequence Classification. In: The 5th ACM SIGKDD International Conference on Knowledge Discovery and Data Mining (KDD 1999) (1999)
16. Ketterlin, A.: Clustering Sequences of Complex Objects. In: The 3th ACM SIGKDD International Conference on Knowledge Discovery and Data Mining (KDD 1997) (1997)
17. Cover, T.: Elements of Information Theory. John Wiley & Sons, Inc., San Francisco (2006)
18. http://archive.ics.uci.edu/ml/datasets/Dodgers+Loop+Sensor

Effective Similarity Analysis over Event Streams Based on Sharing Extent

Yanqiu Wang, Ge Yu, Tiancheng Zhang, Dejun Yue, Yu Gu, and Xiaolong Hu

School of Information Science and Engineering, Northeastern University, China
wangyanqiu86@sina.com

Abstract. With the development of event-driven applications, event stream processing has received more and more attentions in database community. However, little work has focused on the problem of data mining and similarity analysis among event streams. As the foundation for the data mining such as frequent or abnormal event pattern detection, efficient similarity search is desired to be first executed. In this paper, we attempt to take the first step into the similarity search in the context of vast event streams. We propose a simple but effective model to improve the efficiency of the similarity search. To avoid redundant pair-wise comparison, we adopt the definition of sharing extent to dramatically filter dissimilar event streams and speed up the calculation of similarity. Extensive simulated experiments have demonstrated that our model and algorithm can lead to higher efficiency when guaranteeing expected accuracy.

1 Introduction

Huge amount of stream data implicating primitive or composite events can be generated in many scenarios such as RFID (Radio Frequency Identification), sensor network and stock analysis. Consequently, the events will shape into rapid and time-varying event streams [1, 2]. Different from the general data stream, the data of event stream may represent the state change of the system or monitored object instead of numeric values. Furthermore, event stream is more flexible and abundant in semantics and the data in the event stream can be correlated for the advanced pattern match [3]. Just like data stream processing, event stream processing needs incremental online analysis and real-time response. The existing studies mainly focus on the temporal pattern match of event stream [4, 5, 6] rather than the advanced analysis and mining. Specially, similarity analysis, as the fundamental problem of data mining over event stream, has not been tackled. Without effective similarity analysis, further clustering, classification, frequent pattern and novelty detection can not be executed. Also, with the prior knowledge obtained from similarity analysis, we can build some inherent and valuable rules in lots of complicated applications.

Emphatically, the event definition and transformation are closely related to specific application scenarios. For example, the dramatic increase of the stock price measured by some threshold level can be defined as an event, the situation when an object enter a logic area monitored by RFID or video can be modeled as an event, and each access

Q. Li et al. (Eds.): APWeb/WAIM 2009, LNCS 5446, pp. 308–319, 2009.

of websites of interests can be transformed as an event. In this paper, we only propose a general similarity analysis framework for all kinds of event streams. The semantics of the events will be left to applications. Besides, we will divide the original event streams into different logic streams according to some key attributes such as stock transaction code, monitored object EPC or web user ID. And similarity analysis will be conducted among these logic event streams to potentially reflect the relationships of target objects. Furthermore, our proposed similarity measurement model is intuitive and efficient. It is intuitional that the distance between two event streams should reflect the workload to transform one stream to the other one [7], and thus we will utilize edit distance as our similarity measurement guideline, which is commonly used in similar strings pattern matching and comparison of biological sequences.

The basic method for similarity analysis is to compare all pairs of event streams and output the results once the similarity conditions are evaluated true. However, we note only quite a few fraction of the whole has potential similarities in the real-world applications. When the number of event streams or event type is huge and only a few pairs of event streams can satisfy the similarity condition, too much computation cost would be consumed for the pair-wise way. In this paper, we will solve this problem by putting forward the notion of *sharing extent*, which approximately reflects the proportion of shared sub-sequence of event streams. The *sharing extent* is on the assumption that two similar event streams must have "enough" shared sub-sequences. Compared to the pair-wise method, the computation of *sharing extent* is very efficient due to one pass scan.

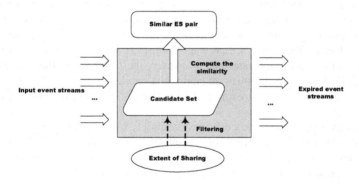

Fig. 1. Process of similarity analysis

Figure 1 represents the processing procedure of similarity analysis. Firstly, we compute the *sharing extent* of each event stream pair, and then discard the dissimilar ones to construct the candidate set according to certain threshold. Secondly, we can compute the final similar pairs in the candidate set. Specifically, our major contributions are listed as follows:

- We utilize the edit distance as the similarity measurement, which is a common and simple formalization of the distance between strings or sequences. Also, the definition of event stream similarity is first proposed.

- A concept of *sharing extent* is introduced, which is easily computed and can approximately reflect the extent of event stream similarity.
- An efficient way is proposed to query the similar pair from massive event streams, and the results can be updated incrementally.

The rest of the paper is organized as follows. Section 2 discusses the related work. Section 3 introduces necessary definitions. Section 4 presents the method for the computation of *sharing extent*. Section 5 proposes the algorithms for similarity analysis. Section 6 evaluates the results of the experiments and section 7 concludes the paper.

2 Related Work

As for similarity analysis over event sequences, H. Mannila et al. [7] propose a similarity model using edit distance as distance measurement. However, the time complexity of this method is very high. DR et al. [8] focus on the probabilistic characteristic of events, and they propose the relation algebra and probabilistic method to describe the distance of events. A.Unal et al. [9] aim at finding the frequencies of events in an event stream at different time granularities by examining the distance distributions of event occurrences. Besides, different event types can be potentially distinguished by different characters, so the approximate string matching technique may be helpful for our model. L. Gravano et al. [10] develop a technique of approximate string joins, which relies on matching q-gram, taking into account of both positions of individual matches and total number of such matches. A. Arasu et al. [11] propose an exact algorithm for set similarity joins, which can give a precise performance guarantee. Furthermore, S. Sarawagi et al. [12] execute set joins on predicates involving various similarity measures. However, the target of all these work differs from ours. Their methods perform on static data, while we aim at efficient similarity analysis on dynamic event streams and consider the problem of cost optimization as well.

Also, many recent event processing systems [1, 6] offer relatively simple event languages and stream-based processing. They detect the complex event from real-time event streams based on known rule, but none of them consider the similarity analysis of event streams. In summary, to the best of our knowledge, what we focus on is an original and valuable topic of event stream processing.

3 Problem Definition

An event stream is composed of continuous arrival events. An event can be classified into primitive event (e.g., a product attached with RFID tag is detected at some reader) and composite event (e.g., a theft event takes place at a store, which is defined as the sequence of product detected at shelf, detected at the exit and un-detected at the counter). The composite event is the composition of primitive events using event operators, such as OR, AND, NOT, SEQ etc.

Let *ES* be an event stream $(e_1, e_2,..., e_k,...)$, where e_i represents the ith event. An event can be described as $e_i = \langle E_i, t_i \rangle$, where E_i and t_i represent the event type and the timestamp respectively. And the event type set is $\Sigma = \{E_1, E_2,..., E_m\}$, where $E_i \in \Sigma$.

The occurrence time of events is not uniform, which may burst occasionally. For each logic event stream, we assume that no more than one event will happen at each time point. In the gap time point without any occurrence of events, we just insert a symbol '#' to represent a special "null" event type. Note that each event type has its own weight, which is defined according to the prior knowledge. The more important event type has higher weight, and the '#' type has the minimun weight. Especially, we need to compare the event streams in the current sliding window to reflect the recent situation. We will illustrate our problem definition as follows.

Problem Definition. Given m event streams ES_1, $ES_2,...,$ ES_m, with the window size n, design efficient model and algorithm to search all pairs of event streams that are similar in the current window and update the results automatically and incrementally.

According to the characteristics of event streams and the demands of applications, the similarity analysis over event streams should satisfy the following requirements:

(1) **Reasonable Similarity Measurement.** The choice of similarity measurement is the key problem which needs to reflect the change trend and temporal feature of event streams.

(2) **Efficient way.** In practice only a few pairs of event streams have similarities, so pair-wise method will have tremendous cost which is unnecessary especially when the amount of event streams is huge. How to search the similar event streams in an efficient way is highly necessary for fast on-line stream processing.

4 Similarity of Event Streams

In this section we will give the definition of similarity measured by distance function between event streams. Firstly, the distance function should support the local time shifting, because the time delay of transformation and detection can not be avoided completely in the distributed system. Secondly, different event types have different impacts on the event stream. We select the *weighted edit distance* (*WED*) as the measurement of similarity.

Definition 1 (Similarity). Given two event streams ES_i and ES_j, where N is the length of current sliding window, then the similarity coefficient of these two event streams is

$$Sim(ES_i, ES_j) = 1 - WED(ES_i, ES_j) / N \ . \tag{1}$$

Where $WED(ES_i, ES_j)$ is the *weighted edit distance* between two event streams. Let the similarity threshold be δ, if $Sim(ES_1, ES_2) > \delta$, we consider those two event streams have similarity.

Next we will introduce the computation of $WED(ES_i, ES_j)$, which is a common distance measurement for the string match and comparison of biological sequences. The value of $WED(ES_i, ES_j)$ is the minimal cost to transform ES_i into ES_j with insertion, deletion and substitution operations. Given two event streams ES_1 and ES_2, denoted as

$(e_1, e_2, ..., e_m)$ and $(f_1, f_2, ..., f_n)$ respectively, the value of *weighted edit distance* can be obtained by the dynamic programming algorithm. We use $r(i, j)$ to denote the minimal cost of operations transforming the first i events of ES_1 into the first j events of ES_2. The basic conditions and the recurrence relation for the value $r(i, j)$ are

$$r(i, j) = \min\{r(i-1, j) + w(e_i), r(i, j-1) + w(f_i), r(i-1, j-1) + k(i, j)\} \ . \tag{2}$$

where $w(e_i)$, $w(f_i)$ are costs of inserting (deleteing) a e_i-type or f_i-type event, respectively, and

$$k(i, j) = \begin{cases} 0, & if \ \ e_i = f_i \\ \max\{w(e_i), w(f_i)\}, & if \ \ e_i \neq f_i \end{cases} . \tag{3}$$

With this definition, the *weighted edit distance* of ES_1 and ES_2, namely $WED(ES_1, ES_2)$, is $r(n, m)$. For those event streams arrive simultaneously, let $m = n = N$(sliding window size), we have

$$Sim(ES_1, ES_2) = 1 - r(N, N) / N \ . \tag{4}$$

The computation of *weighted edit distance* has high time complexity and the pair-wise comparison is not feasible. We will introduce a much more efficient method combined with the features of event stream in the next section.

5 Similarity Analysis Based on Sharing Extent

In many applications, only a few parts of event streams have similarity. There will be huge computation cost by naive method which compares every two event streams. To solve this problem, we propose a *sharing extent* based method which can avoid the pair-wise-way computation. We utilize this technique to quickly discard most of the dissimilar event stream pairs, and then compute the similarity coefficient to get the accurate similar event stream pairs in a smaller candidate set.

5.1 ES-Sequence

The *sharing extent* technique relies on matching ES-sequences (*ESS*) of length q, which are subsequences composed with adjacent events. The intuition behind the use of ES-sequences is that if two event streams ES_1 and ES_2 are similar, they share a large number of ES-sequences in common.

We generate ES-sequence from continuous event streams by sliding the window of size q ($q > 1$). The ES-sequences are generated on the fly and we need to consider the number of each distinct ES-sequence for each event stream. For some event stream ES with length of $|ES|$, then the number of ESS is $|ES| + 1 - q$.

Suppose the current event stream ES_1 in a sliding window is $\{e_1, e_3, e_2, e_2, e_3, e_5\}$ and the length of ES-sequence is defined as 3, then the ES-sequences generated from ES_1 are $e_1 e_3 e_2$, $e_3 e_2 e_2$, $e_2 e_2 e_3$, $e_2 e_3 e_5$. We maintain a set Σ_q to store all the distinct ES-sequences.

The generation of ES-sequence borrows the idea from the q-gram technique in character string matches, but they have totally different semantics. Each event belongs to some event type and has an occurrence time, and the ES-sequence reflects the state change of system or represents some composite events. Meanwhile, since each event has different impact on the system, the ES-sequence has different weights too. We define the weight of ES-sequence as the summation of involed events' weights. That is

$$W(ESS) = [W(e_1) + W(e_2) + ,..., + W(e_q)] . \tag{5}$$

The length q of ES-sequence is selected according to the prior knowledge. The effect of filtration and the semantics of event composition should be taken into account. How to choose the most suitable value of q is our future work.

5.2 Sharing Extent

Definition 2 (Sharing Extent). Given two event streams ES_i and ES_j, the *sharing extent* is defined as follows:

$$SE_{ij} = \frac{\sum_k W(ESS_k) * SN_k * 2}{\sum_k W(ESS_k) * (C_{ik} + C_{jk})} . \tag{6}$$

where
$$SN_k = \begin{cases} C_{ik}, & if \quad C_{ik} < C_{jk} \\ C_{jk}, & if \quad C_{ik} > C_{jk} \end{cases} . \tag{7}$$

We use Σ_q to store all the distinct ES-sequences appearing in the ES_i and ES_j, and the size of Σ_q is m. EES_k ($k=1, 2,..., m$) represents an ES-sequence of Σ_q. C_{ik} and C_{jk} denote the count of ESS_k appearing in the event stream ES_i and ES_j respectively. And SN_k denotes the count of shared ESS_k in ES_i and ES_j, as the equation (7) shows. The $W(ESS_k)$ denotes the weight of ESS_k.

Next we use an example to illustrate the computation process. For n event streams ES_1, ES_2,..., ES_n, all distinct ES-sequences are stored in Σ_q where elements are denoted as ESS_k ($k=1,...,m$). We construct a table called *ES_distribution_table*, which stores the distribution situation of ESS_k for each event stream. The rows represent event streams and the columns represent ES-sequences at the Σ_q set. The value of each cell represents the count of ES-sequence appearing in this event stream. When a new ESS_k is generated, the value of the corresponding cell will be updated by 1. As shown in table 1, the value in row 1 & column 1 is 2, which means that ESS_1 appears twice in ES_1.

Table 1. Example of ES_distribution_Table

	ESS_1	...	ESS_m
ES_1	2		6
...			
ES_n	4		7

As the event streams pass by, the *ES_distribution_table* is established at the same time. We can compute the *sharing extent* utilizing this table by the equation (6) and discard dissimilar event stream pairs. Assuming the *sharing extent* threshold is δ, if $SE_{ij}>\delta$, the event stream pair (ES_i, ES_j) will be put into the candidate set. The reason of filtration by *sharing extent* will be discussed in section 5.4 and the sharing extent based similarity algorithm is given in algorithm 1.

Algorithm 1. Similarity analysis algorithm based on sharing extent

---

Input: n event streams $ES_1,ES_2,...,ES_n$, the size of the sliding window N
Output: Similar set S

Procedure:
1. Compute the Sharing Extent of each event stream pair,
   if $SE(ES_i,ES_j)>\delta$, add (ES_i,ES_j) into candidate set C;
2. Compute the similarity coefficient $Sim(ES_i,ES_j)$ from all pairs of C
   if $Sim(ES_i,ES_j)>\varepsilon$, add (ES_i,ES_j) into the Similar set S.

5.3 Update Mechanism of Sharing Extent

We adopt the sliding window model to process the continuous and infinite event streams. The *sharing extent* is computed based on the current sliding window. Because of the limitation of the sliding scope, compared to the previous sliding window, most events in the current sliding window will not be changed. We will design an incremental maintenance strategy reusing middle results to avoid the repeated computations.

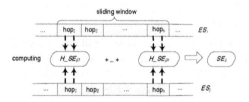

Fig. 2. The update mechanism of SE

We partition the whole window into h hops with each size N/h. When the window is updated, the oldest hop is dropped and a new one enters. In this way, we revise the computation process. The *sharing extent* will be computed at each hop, denoted as H_SE and then be accumulated to obtain the final sharing extent. The process is illustrated in figure 2. When the window slides, we just simply compute the H_SE of the newly arriving hop, and the value of *sharing extent* can be dynamically maintained by the addition of new hop H_SE and deletion of the oldest hop H_SE. The update algorithm is given in algorithm 2.

Algorithm 2. SE-Update Algorithm
Input: n event streams $ES_1, ES_2, \ldots, ES_n$ of the current sliding window, newly arriving event streams ΔES_1, $\Delta ES_2, \ldots,$ ΔES_n with one hop size
Output: Updated Candidate Set C'
Procedure:

1. Delete the expired events and form the new sliding window;
2. Compute the H_SE of newly arriving event streams;
3. For each event stream pair ES_i, ES_j, subtract the H_SE_{ij} of the oldest hop from the SE_{ij}, and add H_SE_{ij} of new hop into it;
4. if $SE_{ij} > \delta$, put (ES_i, ES_j) into Candidate Set C'.

There are several benefits to partition the sliding window into several hops:

(1) Compared with the basic method, this update mechanism considers the position factor. For example, an ES-sequence ABC appears at the beginning of the current window in ES_1 and the same ES-sequence appears at the end of the current window in ES_2. They may have totally different influences on these two event streams. Computing the sharing extent in the small bound of hop can take more account of position factors.

(2) Saving the computation time. When the window slides, we can reuse the H_SE value without re-computing the *sharing extent* for the whole sliding window.

5.4 Threshold of Sharing Extent

In this section we give a theoretical analysis to show the efficiency and accuracy of *sharing extent* based methods.

The value of edit distance is related with the total number of operations and the cost of each operation. The *sharing extent* considers both the number of shared ES-sequences and the weight value of shared ES-sequences. It is based on two facts: similar event streams share more identical ES-sequences, and the weight of shared ES-sequence influences the value of similarity coefficient.

The sharing extent resembles the intuitional idea of edit distance but with simple computations. We can acquire a much smaller candidate set by comparing *sharing extent* with threshold δ. For better correctness and efficiency we need to obtain a compact candidate set under the precision requirement, so how to define the value of δ is the key problem of *sharing extent* method.

Firstly we give a simple observation: suppose each length of two event streams in the current sliding window is N. If there is only one different event, their ES-sequence sets almost have q differences (the length of ES-sequence is q). If they have two different events, then their ES-sequences sets may have $[q+1, 2q]$ differences. We can infer that if the number of different events between two event streams is not larger than k, the same ES-sequences are at least $N+1-(k+1)*q$.

Next we will choose a suitable value of δ for the given similarity threshold ε. From the definition of similarity, the corresponding edit distance value of two event streams is $(1-\varepsilon)*N$. We use k to denote the operation counts of transforming one event stream to another. Assuming the weight of each event is 1, then $k = (1-\varepsilon)*N$, representing the number of different events. If the weight of event is not equal to 1, approximately we have $t = (1-\varepsilon)*N*\lambda/W_{avg}(E)$ to represent the number of different events, where $W_{avg}(E)$ represents the average weight of event, and λ is an adjusting parameter. Based on the observation ahead, the number of the same ES-sequences is at least $N+1-((1-\varepsilon)*N*\lambda/W_{avg}(E)+1)*q$.

According to the definition of *sharing extent* and the factor of different events weights, we can use the number of the same ES-sequences to divide the total number of ES-sequences and acquire the corresponding threshold of sharing extent as follows:

$$\delta = \frac{N+1-((1-\varepsilon)*N*\lambda/W_{avg}(E)+1)*q}{N+1-q} . \tag{8}$$

We can adjust the value of λ to get the most suitable threshold δ to obtain efficient processing time with high accuracy in terms of the intrinsic feature of the original event streams.

5.5 Complexity

Next, we will compare the time complexity of *sharing extent* based similarity analysis method and the exact method.

Assuming m event streams arrive simultaneously and the size of sliding window is N, we analyze the time complexity in the current window. The main cost of exact method comes from the computation of the edit distance. Computing the edit distance of two event streams with length N has time complexity of $O(N^2)$. Due to the pairwise way, the total time complexity of exact method will be $O(m^2N^2)$. There will be tremendous cost if m or N is very large.

The execution process of *sharing extent* method contains two parts: generation of the candidate set and exact computation over the candidate set. The time cost of first part comes from the generation of *ES-distribution-table* and the computation of *sharing extent*. The time complexity is about $O(mN+ m^2)$. The time cost of second part resembles the exact algorithm. Suppose the size of candidate set is l, then the cost is $O(l^2N^2)$. The candidate set is much smaller than the raw set, so the time cost will be reduced dramatically.

6 Experiments

In this section we present some results of experiments on exact similarity analysis method and SE method. All the experiments were conducted on a PC with 2.5G Pentium processor and 512 MB main memory. We design experiments to evaluate the efficiency and precision of these two methods. All these experiments were performed on synthetic datasets.

In the real world, the data implicating the occurrence of events usually conform to the normal distribution. Thus, we first simulate 1000 event streams with attributes following the normal distribution. Then, event values are divided into five ranges: $(-\infty,-2],(-2,-1],(-1,1],(1,2],(2,+\infty)$, and each range represents an event type of E_1,E_2,E_3,E_4,E_5 respectively. Consequently, we can transform every event into corresponding event type according to the range it belongs to. Based on the property of the normal distribution, we can infer the appearance probability of E_3 is 68.24%, and the other event types have smaller probability. For simplicity, we assume the event occurrence time is uniform with the same interval. In practice, '#' event type needs to fill in the blank gap. The weights of each event type are defined as 0.6, 0.8, 0.5, 0.8, 0.6 respectively and the length of event streams can be represented by the number of events.

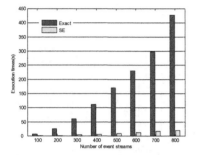

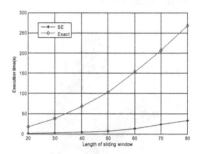

Fig. 3. Comparison of execution time with different number of event streams

Fig. 4. Comparison of execution time with different length of sliding window

The basic parameters are set as follows. The threshold is 0.8, the adjusting parameter λ of SE is 0.35, the length of event stream is 600, the sliding window size is 40 and the hop number in each window is 4. Figure 3 compares the execution time of SE and exact method with different number of event streams. With the increase of the number of streams, the execution time of exact method increases rapidly, while the SE method is much better than the exact pair-wise method in terms of execution time.

Figure 4 shows the effect of sliding window size on the execution time. The execution time of exact method rises in a quadratic way with increase of sliding window size, but the SE method achieves a dramatic reduction in computation time.

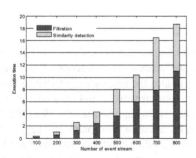

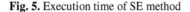

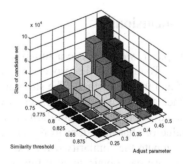

Fig. 5. Execution time of SE method

Fig. 6. Size of candidate set with different thresholds

As shown in Figure 5, the execution time of SE is divided into two parts: generation of the candidate set and similarity detection. Compared to generating candidate set, the time consumed by similarity detection is much less because the candidate set is smaller than the raw event stream set.

From the previous section, we know that the SE threshold δ is very important for the precision and efficiency of the SE method, which is influenced by the adjusting parameter λ. We test the size of candidate set, the execution time and the precision of SE method among 500 event streams with different thresholds and adjusting parameters.

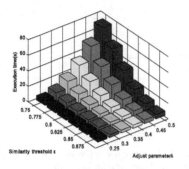
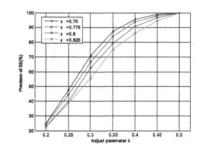

Fig. 7. Execution time with different thresholds   **Fig. 8.** Precision with different thresholds

According to the theoretical analysis, we know that the SE threshold δ will be increased with the decreasing of λ. Consequently, the performance of SE will be varied with different adjusting parameters. From figure 6 and in figure 7, we can observe that the size of candidate set decreases with λ, and the execution time is correspondingly reduced as well. However, the precision decreases from 100% to a smaller value as shown in figure 8. There will be a tradeoff between the precision and the efficiency with different thresholds. We will choose the most suitable threshold range to get efficient processing time with high accuracy.

7   Conclusions

In this paper, we introduce a suitable measurement for similarity analysis over event streams. To improve the processing efficiency, we propose the concept of sharing extent and the SE method. The sharing extent implicates the proportion of shared ES-sequence and the weight influences of compared event streams, which can be simply computed and reflects the similarity extent of event streams approximately. We utilize the value of sharing extent to discard the dissimilar event stream pairs quickly, and then compute the similarity using exact method in a smaller candidate set. By the SE method, we can obtain great efficiency and expected precision by adjusting the parameter. Our future research direction is to choose suitable size of sliding window adaptively to achieve best effectiveness according to different applications.

Acknowledgment

The research was partially supported by NSF of China under Grants Nos. 60773220, 60873009 and 60773222.

References

1. Wu, E., Diao, Y., Rizvi, S.: High-performance complex event processing over streams. In: Proc. of SIGMOD, pp. 407–418. ACM press, New York (2006)
2. Wang, F., Liu, S., Liu, P., et al.: Bridge physical and virtual worlds: complex event processing for RFID data streams. In: Ioannidis, Y., Scholl, M.H., Schmidt, J.W., Matthes, F., Hatzopoulos, M., Böhm, K., Kemper, A., Grust, T., Böhm, C. (eds.) EDBT 2006. LNCS, vol. 3896, pp. 588–607. Springer, Heidelberg (2006)
3. Chakravarthy, S., Adaikkalavan, R.: Events and Streams: Harnessing and Unleashing Their Synergy! In: Proc. of DEBS, pp. 1–12. ACM press, New York (2008)
4. Rozsnyai, S., Schiefer, J., Schatten, A.: Concepts and Models for Typing Events for Event-Based Systems. In: Proc. of DEBS, pp. 62–70. ACM press, New York (2007)
5. Barga, R.S., Goldstein, J., Ali, M., Hong, M.: Consistent streaming through time: A vision for event stream processing. In: Proc. of CIDR, pp. 363–373 (2007)
6. Brenna, L., Demers, A., Gehrke, J., et al.: Cayuga: a high-performance event processing engine. In: Proc. of CIDR, pp. 1100–1102. ACM Press, New York (2007)
7. Mannila, H., Ronkainen, P.: Similarity of Event Sequences. In: Temporal Representation and Reasoning, pp. 136–139. IEEE press, Dayton Beach (1997)
8. Goodman, I.R.: Similarity Measures of Events, Relational Event Algebra, and Extensions to Fuzzy Logic. In: Fuzzy Information Processing Society, pp. 187–191. IEEE press, Berkeley (1997)
9. Ünal, A., Saygin, Y., Ulüsoy, Ö.: Processing count queries over event streams at multiple time granularities. Information Sciences 176, 2066–2096 (2005)
10. Gravano, L., Ipeirotis, P.G., Jagadish, H.V., et al.: Approximate String Joins in a Database (Almost) for Free. In: Proc. of VLDB, pp. 491–500. VLDB Endowment, Italy (2001)
11. Arasu, A., Ganti, V., Kaushik, R.: Efficient Exact Set-Similarity Joins. In: Proc. of VLDB, pp. 918–929. VLDB Endowment, Seoul (2006)
12. Sarawagi, S., Kirpal, A.: Efficient set joins on similarity predicates. In: Proc. of SIGMOD, pp. 743–754. ACM Press, New York (2004)

A Term-Based Driven Clustering Approach for Name Disambiguation

Jia Zhu, Xiaofang Zhou, and Gabriel Pui Cheong Fung

School of ITEE, The University of Queensland, Australia
{jiazhu,zxf,uqgfung}@itee.uq.edu.au

Abstract. Name disambiguation in databases is a non-trivial task because people's names are often not unique and usually only a limited information is associated with each name in the database. For example, in DBLP many authors share the same name, whereas we do not have any unique identifier to distinguish them. To make it worst, we may not always be able to access the full contents of the materials, unless we have joined those organizations (e.g. ACM) who publish them. As such, how to disambiguate different names with a very limited information is a very challenging task. In this paper, we focus ourselves on such situation. We propose a term-based driven clustering approach for solving it. Specifically, we first construct some term-based taxonomies to mimic the expert knowledge of the domain by linking the related terms that appear in there automatically. Each taxonomy is then transformed into a graph, and we group the entries that belong to the same author by using either of the two novel models, namely, graph-based similarity model and graph-based random walk model. The former model aims at computing the similarity among terms, whereas the later model aims at investigating how likely would a set of terms be transformed to another set of terms. Extensive experiments are conducted by using the entries in DBLP. The favorable results indicated that our proposed approach is highly effective.

Keywords: Taxonomy, Clustering, Name Disambiguation, Graph.

1 Introduction

Name disambiguation in digital libraries refers to the task of attributing the publications to the proper authors. It is very common that several authors share the same name in a digital library. For instance, there are at least 50 different authors who call "Wei Wang" in DBLP and there are more than 400 entries under this name. At present, although DBLP provides some mechanisms to solve this problem already but wrong associations can still be found occasionally.

One may argue that we can use some existing techniques such as record linkages [1], duplication detection [2,3] and citation matching [4] for solving the name disambiguation problem. Unfortunately, this may not always be practical for all digital libraries. For example, in DBLP, we cannot access the full contents of most papers, unless we have joined those organizations (e.g. ACM and IEEE) who publish them. In this paper, we are interested in solving such kind of problem – disambiguate the authors in a digital library with only a limited amount of information such as authors' names and papers'

Q. Li et al. (Eds.): APWeb/WAIM 2009, LNCS 5446, pp. 320–331, 2009.

titles. Hence, Our problem is how to cluster the entries that belong to the same author based on the entries' information only.

Broadly speaking, there are four different directions to solve this problem, namely, supervised learning [6], unsupervised learning [7], semi-supervised learning [5] and topic-based modeling [17]. However, supervised learning and topic-based modeling both require expert knowledge to label the data, which is very time consuming. On the other hand, the effectiveness of unsupervised learning method and semi-supervised learning both depend heavily on the data selection and the data preprocessing, which is difficult to guarantee reliable results are obtained.

In order to avoid obtaining unreliable results, such as in the case of unsupervised learning, we believe that having domain knowledge, such as in the case of supervised learning, is very important. In this paper, we try to obtain the domain knowledge automatically by constructing some term-based taxonomies. In each term-based taxonomy, each term is regarded as a node and all terms are linked together according to their relationships with each others. Thus, a graph model is resulted after a term-based taxonomy is built. Based on these graphs, we compute the similarity among entries by not just on how many common terms entries shared, but also on how the terms are related based on the graphs (i.e. domain knowledge). For computing the similarity, two models are proposed in this paper, namely, graph-based similarity model and graph-based random walk model. The former model aims at computing the similarity among terms, whereas the later model aims at investigating how likely would a set of terms be transformed to another set of terms. Eventually, the similar entries will be gradually grouped together according to either of the models, such that each set of entries, ideally, belongs to a unique author. In other words, the authors' names can be disambiguated.

To summarize, the main contributions of this paper are as follows: (1) Up to our knowledge, we are the first ones to apply taxonomy to perform name disambiguation in DBLP, even though taxonomy have long been used in text classification problem [17]; (2) We propose a generic clustering framework which is not only applicable on DBLP, but also can be applied to any other similar domains. This is very different from the existing works which are highly domain specific, therefore, attributes like co-authorship are not considered in this approach; (3) We propose how graph models can be integrated to the taxonomy for calculating whether two entries belong to the same author. This technique is not being reported elsewhere; (4) Our approaches do not need to obtain the domain knowledge manually as supervised approaches.

2   Related Work

In this paper, we propose two models, graph-based similarity model and graph-based random walk model, to be used together with the hierarchical clustering model to solve the name disambiguation problem. Both of the graph models rely on some graphs generated by some term-based taxonomies.

2.1   Name Disambiguation

[7] proposes an unsupervised learning approach using K-way spectral clustering to solve the name disambiguation problem. Although this method is fast but it depends

heavily on the initial partition and the order of processing each data point. Furthermore, this clustering method may not work well when there are many ambiguous authors in the dataset. [6] proposes a supervised learning framework (by using SVM and Naive Bayes) to solve the name disambiguation problem. Although the accuracy of this method is high but it relies heavily on the quality of the training data, which is difficult to obtain. In addition, this method will assemble multiple attributes into a training network without any logical order, which may deteriorate the overall performance of the framework. [5] proposes a semi-supervised learning method which uses SVM to train different linkage weights in order to distinguish objects with identical names. In general, this approach may obtain sound results. Yet, it is quite difficult to generate appropriate linkage weights if the linkages are shared by many different authors. [17] uses a topic-based model to solve the name disambiguation problem. Two hierarchical Bayesian text models, Probabilistic Latent Semantic Analysis (PLSA) and Latent Dirichlet Allocation (LDA), are used. In general, it achieves better experimental results over the other approaches. Unfortunately, this approach is difficult to implement because it requires too much labor intensive preprocessing (e.g. manually extract the first page of each paper) and relies on some special assumptions (e.g. the entries of each paper is clean and complete). In addition, an approach has been proposed in [24] that analyzes not only features but also inter relationships among each object in a graph format to improve the disambiguation quality which provides very good reference to our work, however, the attributes has been used in their approach like affiliation are not available in DBLP.

2.2 Taxonomy Building

Taxonomy is a structure that is used to reflect the domain knowledge. A taxonomy is normally built base on terms. Building a term-based taxonomy generally involves two steps: terms extraction and terms linking. For terms extraction, some of the most popular techniques are: domain terminology extraction based on lexical cohesion measure [8], traditional C/NC-value method combined with statistic measures of term frequency [9] and Yahoo! Term Extraction[1] [10]. Recently, an approach called PAT-Tree-based local maxima algorithm is proposed [11], which provides another efficient way to extract terms in a large text collection. In this paper, we adopt this latest algorithm for extracting terms from DBLP. The implementation details will be given in the next section.

For terms linking, the classical way to solve it is either to create a regular expression list on the lexical level [12] or to use machine learning methods based on the co-occurrence of terms in text [13,14]. Recently, some studies [15,16] propose linking terms together by uncovering their semantic relationships. In this paper, we use the classical methods to solve the term linking problem. The reason is that semantic relationships among terms cannot be easily obtained in DBLP. Certainly, our framework is independent of choosing whichever term linking method, and choosing classical method to solve it is for the shake of convenience.

---

[1] http://developer.yahoo.com

3   Proposed Models

In this section, we will discuss our models in details, which include how the term-based taxonomy is built and how the two proposed graph models, graph-based similarity model and graph-based random walk model, are constructed.

3.1   Overview

Recall that our goal is to cluster the papers which belong to the same author based only on the entries' information (e.g. titles and authors). Given a paper titled "An effective way to performing indexing in multimedia database" and another paper titled "Q-SQL: A quicker SQL query", then even though they do not share any common terminology, we understand that these two entries both belong to the database paradigm because the terms such as "indexing", "SQL", "query" and "database" are highly related to this area. Hence, we claim that simply using some string matching strategies to measure the similarity between two entries is not enough. We should include domain knowledge when we have to compute their similarity. Yet, obtaining domain knowledge is usually very expensive as it requires experts' judgments.

In this paper, we try to obtain the domain knowledge automatically by constructing three different term-based taxonomies. In each term-based taxonomy, each term is regarded as a node and all terms are linked together according to their relationships with each other. Hence, a graph model will be resulted naturally on each constructed term-based taxonomy.

Based on these three graphs, we compute the similarity between two entries by not just how many common terms they shared, but also how the terms are related to each others based on the taxonomy graph. For computing the similarity between two entries, two models are proposed: a graph-based similarity model and a graph-based random walk model. In short, graph-based similarity model calculates how similar two entries are by computing the similarity of the terms exhibit in two entries according to the three taxonomy graphs, whereas the graph-based random walk model calculate the similarity by measuring how many steps that the terms in one entry need to "walk" to the terms in another entry.

Finally, the entries are gradually grouped together based on a traditional hierarchical clustering algorithm. The stopping criteria of the clustering will be discussed in the later sections.

3.2   Term-Based Taxonomy Construction

In our term-based taxonomy, we define term as follows:

Definition 1 (Term). *Each term, t_k consists of at least one word, w_i, e.g. "Data" is a term and "Data mining" also is a term. $t_k = \{w_1, w_2 \ldots w_n\}$, $n \leq k$; k is an input parameter that represents the number of words to make up a term.*

The ordering of words in a term is important because changing their orders may result in different meanings. Hence, the traditional bag-of-words representation, which means the words in a text is unordered, is not appropriated here [23]. In the followings, we will describe how we extract the terms and how they are linked together.

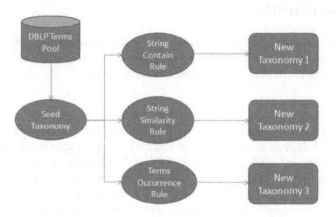

Fig. 1. Overview of Taxonomy Construction

Terms Extraction. All terms are extracted from the paper titles and the session titles (if the paper is from a conference and the conference contains session titles) in DBLP. We implement the extraction method called PAT-Tree-based local maxima algorithm [11][2].

Yet, in order to use this PAT-Tree-based local maxima algorithm, we need to define an association measurement first. In this paper, we define an association measurement called *LSCPCD*, which is a modification of the traditional *SCPCD* [11][3]:

$$LSCPCD(t_1 \dots t_n) = \frac{LC(t_1 \dots t_n)}{\frac{1}{n-1} \sum_{i=1}^{n} f(t_1 \dots t_i)}, \tag{1}$$

where $(t_1 \dots t_n)$ is the set of unique left adjacent terms for the n-gram; $LC(t_1 \dots t_n)$ is the total number of terms in the set, and $f(t_1 \dots t_n)$ is the frequency of unique left adjacent terms in the pool.

Terms Linking. Once we have collected all terms, the next step is to uncover their relationships. In general, terms can be linked together based on their lexical similarity, their co-occurrences in texts, and some semantic rules. In this paper, we take the first two measurements only (i.e. without semantic rules). This is because semantic information can hardly be obtained in our domain (DBLP text). Fig. 1 shows an overview of the taxonomy construction. Initially, we obtain a seed taxonomy from Microsoft Libra Academic Search[4] which can roughly partition DBLP into several distinct computer science disciplines. We iterate each term in the set of terms and identify their relationships against the seed taxonomy by using lexical similarity and term co-occurrence. For lexical similarity, two rules are applied:

---

[2] http://rt.openfoundry.org/Foundry/Project/?Queue=367

[3] *SCPCD* is designed for the domains with both English terms and Chinese terms, whereas our domain contains English terms only. Hence, modification is necessary. Please refer to [11] for details.

[4] http://libra.msra.cn/Default.aspx

Definition 2 (String Contain Rule). *Suppose W_1 is a set of words that belongs to term t_1 and W_2 is another set of the words that belongs to term t_2; if $W_1 \subseteq W_2$, then t_1 and t_2 are very likely to refer to the same thing and they will be linked together.*

Definition 3 (String Similarity Rule). *Suppose s_1 is a string representation of the term t_1, and s_2 is a string representation of the term t_2; if S_1 and S_2 are very similar, then t_1 and t_2 are very likely to refer to the same thing and they will be linked together.*

The string contain rule is based on a hyper relationship in terms level, whereas the string similarity rule is based on a lexical level comparison. Now, the only problem remained here is that how we define "similar" in the string similarity rule. There are many algorithms that aim at computing the similarity between two strings, such as soundex, edit distance, and longest common substring [20]. In this paper, we use Jaccard coefficient to do so because it considers not only the single longest common substring, but also the other common substrings in there, as it uses every character pairs for the string matching. Eventually, if the similarity between two terms is greater than a pre-defined threshold, then a link is established between these two terms.

Apart from the above two rules, from [13], two different terms may both be related to the same area frequently if they appear in the same paper title simultaneously. Hence, in this paper, we propose term co-occurrence rule:

Definition 4 (Term Co-occurrence Rule). *Suppose there are totally N paper titles, and each paper title, P_t, contains a list of terms $P_t = \{t_1, t_2 \ldots t_n\}$. If two terms are in the same paper, $t_1 \in P_t$ and $t_2 \in P_t$, then t_1, t_2 co-occur together, and they will be linked together.*

According to our observations and our preliminary studies, the accuracy to link terms together by using term co-occurrence (Definition 4) is generally lower than that of using lexical similarity (Definition 2 and Definition 3). The reason is that terms that co-occur in the paper title may usually mean different issues. For example, consider the paper titled: "Discovering a Term Taxonomy from Term Similarities Using Principal Component Analysis". In this title, "Term Taxonomy" and "Principal Component Analysis" refer to different issues but they co-occur together. In this paper, we are trying to demonstrate the feasibility of applying these rules to improve the retrieval results. Certainly, some rules may have side-effects. We will study this issue more in-depth in our future work.

Finally, three taxonomies will be generated based on the above three rules (refer to Fig. 1 also). We named them as: (1) String Similarity Rule Taxonomy; (2) String Contain Rule Taxonomy; and (3) Co-occurrence Rule Taxonomy. Note that the relationships among terms are undirectional. Eventually, three undirected graphs will be generated where each vertex denotes a term and each link denotes there is a relationship between two terms (vertex).

3.3 Taxonomy-Based Clustering Approach

Once we have the taxonomies, we can use it as the source to perform hierarchical agglomerative clustering. The clustering process proceeds according to the following steps: (1) Initially, each paper represents a single cluster; (2) Compare clusters pairwisely and decide if the two clusters with the highest similarity should be merged to form a new cluster based on a pre-defined threshold; (3) Repeat Step 2 until there are no more clusters that can be merged. Among these three steps, Step 2 is obviously the key step. In this paper, we try two different measurements to see if two clusters should be merged. The first one is called graph-based similarity model, which computes statistical similarity based on only the linking structure of the taxonomy, and the other one is called graph-based random walk model, which calculates the distance of two clusters with random walk algorithm [21].

Graph-Based Similarity Model. Let $G = (V, E)$ be a graph, where V is a set of vertexes and E is a set of edges. As we discussed in the previous sections, our term-based taxonomy is in fact an undirected graph with $t_k \in V$ and the links among terms are edges. As we mentioned earlier, initially, each cluster contains one entry where an entry is represented by a list of terms. In order to decide whether two clusters should be merged, we compute the overlapping of the two clusters based on the similarity of the terms in two clusters. Certainly, the similarity measurement is based on G. Specifically, we follow the existing approaches which use the well-known Dice Coefficient for modeling the similarity between two clusters, which calculate it by the weights of the edges:

$$Similarity(c, c') = \frac{2 \times W(e_1...e_n)}{W_c(e_1...e_i) + W'_c(e_1...e_j)} \qquad (2)$$

where $(e_1...e_i)$ and $(e_1...e_j)$ are respectively the set of edges linked to the terms in c and c' from other nodes in the taxonomy; $W_c(e_1...e_i)$ and $W'_c(e_1...e_j)$ are the total weights of the corresponding sets of edges. In our model, each edge has a weight of 1; $W(e_1...e_n)$ is the total weight of the set of edges linked by other terms to those terms that are shared by c and c'. The set of terms that are shared by c and c' can simply be identified by: $(v_1...v_i) \cap (v_1...v_j)$. where $(v_1...v_i)$ and $(v_1...v_j)$ are the sets of vertexes c and c', respectively. Finally, if $Similarity(c, c')$ is greater than a pre-defined threshold parameter θ, then c and c' will be merged together.

Graph-Based Random Walk Model. The graph-based random walk model is a mathematical formalization of a trajectory that consists of taking successive random steps [22]. In our term-based taxonomy, each cluster is a DBLP entry, which is represented by a list of terms that can be found in the taxonomy. The basic idea of our graph-based random random walk model is to compute the probability of the terms from one cluster can "walk" to the terms in another clusters within a certain number of steps. We use this probability to denote the similarity between two clusters.

The probability is represented as a score manner such that the higher the score is, the more dissimilar is two clusters because more number of steps is involved for one cluster to walk to another cluster. Specifically, given a term-based taxonomy graph $G = (V, E)$, let $S(v_1)$ be a set of vertexes that connect to the vertex v_1, then the score from v_1 to another vertex v_2 is:

$$Score(v_1, v_2) = 1 + \frac{1}{d(v_1)} \sum_{v \in S(v_1)} Score(v, v_2) . \qquad (3)$$

where $d(v_1)$ is the number of degrees of v_1 (i.e. number of edges in v_1). Eq. (3) is a classical random walk model. Based on this equation, we extend it and apply its

---

Algorithm 1. GraphBasedRandomWalkModel($G(V,E), \theta, \omega, c, c'$)

---

input : A term-based taxonomy graphs $G(V,E)$; v is a node representing a term, $v \in V$; e is an edge linked among terms, $e \in E$; θ is a threshold parameter; ω is a parameter for walking steps; c and c' are clusters that need to be validated

1 **foreach** $G(V,E)$ **do**
2   **foreach** *term v in cluster c* **do**
3   **foreach** *term v' in cluster c'* **do**
4   **if** $Score(v,v') < \omega$ **then** $Score(c,c') \mathrel{+}= Score(c,c')$;
5   **else** $Score(c,c') \mathrel{+}= \omega$;
6   **end**
7   **end**
8 **end**
9 **if** $Score(c,c') < \theta$ **then** merge c and c' to form cluster;

---

extension to measure the similarity between two clusters, c and c' as follows:

$$Score(c,c') = \sum_{v \in c, v' \in c'} Score(v,v') \tag{4}$$

Eventually, we apply Eq. (4) to our model to see whether two clusters can be merged. Algorithm 1 outlines the major steps for this graph-based random walk model.

In Algorithm 1, we need to set up a parameter ω in order to limit the maximum number of walking steps for each iteration. In addition, the value of ω has to be changed in different graphs (taxonomies) because the confidences of linking two terms are different in different graphs, as we mentioned in the previous section. Quite obviously, the lower the confidence is, the higher value ω should be as there are more links existing from each vertex. It is worth noting that a larger taxonomy usually requires more walking steps for a term to walk to another term. Considering the size of taxonomy, the size of String Contain Rule Taxonomy is smaller than that of String Similarity Rule Taxonomy and the size of String Similarity Rule Taxonomy is smaller than that of Co-occurrence Rule Taxonomy, Thus, in this paper, we have the following settings according to our preliminary testing and observations: (1) For the String Contain Rule Taxonomy, $\omega = \delta$; (2) For the String Similarity Rule Taxonomy, $\omega = 1.25\delta$; and (3) For the Co-occurrence Rule Taxonomy, $\omega = 1.5\delta$. In our model, by default, we set the value of δ is 1. These parameters are tuned based on some empirical studies. The sensitivity of the parameters will be left as our future work.

4 Experiment Design and Results

In our experiments, we selected ten common names from the DBLP. These names are shown in Table 1. We then generate ten datasets, each of which contains entries that appeared in DBLP from one of these authors. We manually checked each of the dataset through Google, read the information in the papers, and sent e-mails to the authors to validate whether the datasets are correct. We removed those entries that are ambiguous and some authors who only have one entry are also removed compared to the datasets

Table 1. Test Standard Dataset

Name	Authors	Entries	Name	Authors	Entries
Kai Xu	7	14	Lei Wang	30	108
Qin Zhang	7	20	Jun Zhang	21	91
Wei Wang	16	69	Wei Li	27	118
Ying Liu	12	23	Michael Wagner	5	29
Tao Wang	18	34	Jim Smith	3	19

in [5]. Hence, the datasets that we generated are clean and can be used for evaluating our proposed work.

For the term-based taxonomy, we extracted terms from over 400K paper titles, and we only considered the terms that contained up to three words according to our observations and the general length of a paper title. Punctuation, numbers and stopwords are removed. There were three taxonomies that were generated from seed taxonomy against four linking decision rules. Following the standard evaluation process, we use precision to evaluate our model:

$$Precision = \frac{PC}{PC + PIC} \tag{5}$$

where PC is the number of pairs of entries being clustered correctly and PIC is the number of pairs being clustered incorrectly. In the following sections, we use this measurement to evaluate the performances of our graph-based similarity model and graph-based random walk model. Note that we will use the traditional string match hierarchal agglomerative clustering method as a baseline method for comparison.

4.1 Evaluation of Graph-Based Similarity Model

In this section, we are going to evaluate the graph-based similarity model. As described in the earlier sections, when the similarity between two clusters that is calculated by the intersection of clusters in the graph which means the set of terms in the taxonomy have linking with both clusters. For the case where exact terms match, cluster A and cluster B both have term X in their terms lists, for example. We make a rule that states that two clusters can be merged when at least N% terms in one cluster's term list are matched in the other cluster's term list, and if the percentage of term matches is lower than N%. Thereafter, the merge will be based on the similarity between clusters. For evaluation purpose, we set up three levels, namely, 10%, 20%, and 30%. The criterion also applies to the random walk model and the baseline method.

In this model, there is a threshold to control if the similarity of two clusters is enough to perform clustering. As shown in Fig. 2(a) and Fig. 2(b), there are two thresholds in the evaluation, 0.5 and 0.7. Thus, a higher threshold means that the higher similarity between clusters is required. From the data chart, we know that mean precision values over 70% have been reached in both thresholds when the terms match level is 20%. We also note that in the case of authors 'Jun Zhang' and 'Wei Li', the precision values are very low if the terms match level is 10%, which means that authors often use same terms in their paper titles although they are not the same people. However, the precision values in the

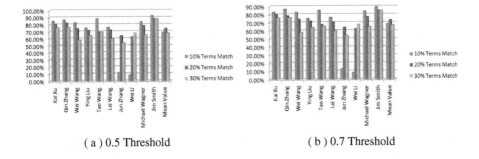

(a) 0.5 Threshold   (b) 0.7 Threshold

Fig. 2. Precision of Graph-based Similarity Model

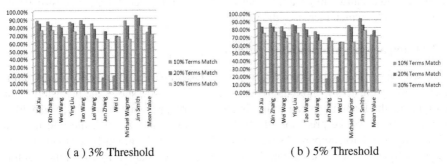

(a) 3% Threshold   (b) 5% Threshold

Fig. 3. Precision of Graph-based Random Walk Model

10% terms match level are generally better than the other two levels for most other authors because a lower level can make the merge process easier. The same scenario exists for two thresholds; the model performs slightly better when the threshold is 0.5.

4.2   Evaluation of Graph-Based Random Walk Model

For this model, we set the process that will run up to ten steps when we try to link the term from one cluster to the other. In addition, there is a need for a total score to check if two clusters can be merged. The total score for each linking process between two clusters will be the number of taxonomy multiple by the size of term list multiple by the max numbers of steps. If the score of one cluster link to the other cluster is less than a certain threshold of the total score in this model, then we merge these two clusters together. In the evaluation, we evaluate two thresholds, 3% and 5%.

From Fig. 3(a) and Fig.3(b), we can see the same problem occurs in the random walk model as well for authors 'Jun Zhang' and 'Wei Li' when the terms match level is 10%. The mean values are higher at about 80% compared to the graph-based similarity model because this model does not only look at the neighboring linking terms of the cluster's term list in the graph, but also has the probability to run through the graph to get a term start from term list up to ten edges. Hence, this model will cover a greater range of the taxonomies. The performance of the 5% threshold is worse than that of the

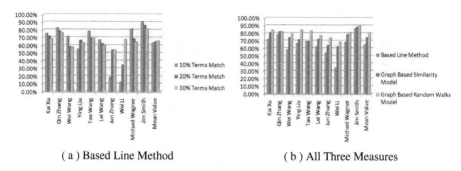

(a) Based Line Method   (b) All Three Measures

Fig. 4. Comparisons of Precision with Baseline Method

3% threshold because more walking steps are allowed for each linking process between the two clusters which increases the probability of errors.

4.3   Comparisons with Baseline Method

This section compares the two models with the baseline method that only checks terms match among clusters without applying taxonomy. Fig. 4(a) shows the precision of the baseline method, and the mean values in all levels of terms match are lower than 70%. Figure 4(b) shows the precision of all three methods, and we pick the mid level of 20% terms match to compare. As we can see from Fig. 9, the graph-based random walk model performs better than the others and the mean value is about 15% over the baseline method. The mean value of the graph-based similarity model is also acceptable and is about 10% over that of the base line method.

5   Conclusions and Future Works

This paper describes a clustering approach for name disambiguation in DBLP based on term-based taxonomies. By utilizing the graph-based similarity model and graph-based random walk model and combining them with general hierarchal agglomerative clustering, the approach efficiently classifies authors and generates sound results than baseline method. The approach is only based on internal references in DBLP, which avoids the problems in other previous clustering approaches, manually label and extract data for example. Last but not least, the taxonomies we built are generic and can be applied to other digital libraries or can be used for other purposes, examples of which include paper ranking and social network discovery.

References

1. Fellegi, I.P., Sunter, A.B.: A theory for record linkage. Journal of the American Statistical Association 21 (1969)
2. Bitton, D., Dewitt, D.J.: Duplicate record elimination in large data files. ACM TODS 8 (1983)
3. Monge, A.E., Elkan, C.P.: An efficient domain-independent algorithm for detecting approximately duplicate database records. In: SIGMOD Wshp. on Research Issues on Data Mining and Knowledge Discovery (1997)

4. McCallum, A., Nigam, K., Ungar, L.H.: Efficient clustering of high-dimensional data sets with application to reference matching. In: 6th ACM SIGKDD (2000)
5. Yin, X.X., Han, J.W.: Object Distinction: Distinguishing Objects with Identical Names. In: IEEE 23rd ICDE. ACM Press, New York (2007)
6. Han, H., Giles, C.L., Hong, Y.Z.: Two supervised learning approaches for name disambiguation in author citations. In: 4th ACM/IEEE Joint Conference on Digital Libraries (2004)
7. Han, H., Zhang, H., Giles, C.L.: Name disambiguation in author citations using a k-way spectral clustering method. In: 5th ACM/IEEE Joint Conference on Digital Libraries (2005)
8. Park, Y., Byrd, R.J., Boguraev, B.K.: Automatic Glossary Extraction:Beyond Terminology Identification. In: 19th International Conference on Computational Linguistics (2002)
9. Hliaoutakis, A., Zervanou, K., Petrakis, E.G., Milios, E.E.: Automatic document indexing in large medical collections. In: International Workshop on Healthcare information and Knowledge Management (2006)
10. Aleman-Meza, B., Decker, S., Cameron, D., Arpinar, I.B.: Association Analytics for Network Connectivity in a Bibliographic and Expertise Dataset. In: Semantic Web Engineering in the Knowledge Society (2008)
11. Wang, H., Teng, J.W., Lu, W.H., Chien, L.F.: Translating Unknown Cross-Lingual Queries in Digital Libraries Using a Web-based Approach. In: 4th ACM/IEEE Joint Conference on Digital Libraries (2004)
12. Rion, S., Daniel, J., Andrew, N.: Learning syntactic patterns for automatic hypernym discovery. In: Advances in Neural Information Processing Systems, vol. 17 (2004)
13. Bast, H., Durpret, G., Piwowarski, B.: Discovering a term taxonomy from term similarities using principal component analysis. In: Ackermann, M., Berendt, B., Grobelnik, M., Hotho, A., Mladenič, D., Semeraro, G., Spiliopoulou, M., Stumme, G., Svátek, V., van Someren, M. (eds.) EWMF 2005 and KDO 2005. LNCS, vol. 4289, pp. 103–120. Springer, Heidelberg (2006)
14. Arpinar, B., Hassell, J., Aleman-Meza, B.: Ontology-driven automatic entity disambiguation in unstructured text. In: Cruz, I., Decker, S., Allemang, D., Preist, C., Schwabe, D., Mika, P., Uschold, M., Aroyo, L.M. (eds.) ISWC 2006. LNCS, vol. 4273, pp. 44–57. Springer, Heidelberg (2006)
15. Rion, S., Daniel, J., Andrew, N.: Semantic taxonomy induction from heterogenous evidence. In: 21st International Conference on Computational Linguistics (2006)
16. Velardi, P., Cucchiarelli, A., Petit, M.: A Taxonomy Learning Method and Its Application to Characterize a Scientific Web Community. In: IEEE TKDE, vol. 19 (2007)
17. Yang, S., Jian, H., Isaac, G.C., Jia, L., Lee, G.: Efficient topic-based unsupervised name disambiguation. In: 7th ACM/IEEE Joint Conference on Digital Libraries (2007)
18. Dunning, T.: Accurate Methods for the Statistics of Surprise and Coincidence. Computational Linguistics 19 (1994)
19. Breaux, T.D., Reed, J.W.: Using Ontology in Hierarchical Information Clustering. In: 38th Annual Hawaii International Conference (2005)
20. Luján-Mora, S., Palomar, M.: Comparing string similarity measures for reducing inconsistency in integrating data from different sources. In: Wang, X.S., Yu, G., Lu, H. (eds.) WAIM 2001. LNCS, vol. 2118, p. 191. Springer, Heidelberg (2001)
21. Aldous, D.J.: Low bounds for covering times for reversible markov chains and random walks on graph. J. Theoretical probability 2 (1989)
22. Coppersmith, D., Feige, U., Shearer, J.: Random walks on regular and irregular graphs. SIAM J. Discret. Math. 9 (1996)
23. Lewis, D.: Naive Bayes at Forty: The Independence Assumption in Information Retrieval. In: Nédellec, C., Rouveirol, C. (eds.) ECML 1998. LNCS, vol. 1398. Springer, Heidelberg (1998)
24. Kalashnikov, D.V., Mehrotra, S.: Domain-independent data cleaning via analysis of entity-relationship graph. ACM Trans. Database Syst. 31 (2006)

Sentiment Clustering: A Novel Method to Explore in the Blogosphere[*]

Shi Feng, Daling Wang, Ge Yu, Chao Yang, and Nan Yang

College of Information Science and Engineering, Northeastern University,
Shenyang 110004, P.R. China
wondertime@gmail.com, dlwang@mail.neu.edu.cn,
yuge@mail.neu.edu.cn,
yang90408@sina.com, nanan1986@126.com

Abstract. In recent years, blogs have become the major platform for people to express their opinions and sentiments in the Web age. The traditional blog search engines usually employ topic-oriented techniques, which are not easy for users to make better understanding of bloggers' feelings and emotions. In this paper, an emotion-oriented clustering approach is proposed according to the sentiment similarities between blog search result titles and snippets. Extensive experiments were conducted based on a real world blog search engine and the experiments show that our approach can cluster blog search result items into sentiment groups to allow for better organization and easy navigation, which provides users a novel method to explore in the blogosphere.

1 Introduction

Most recently, Weblogs (also referred to as blogs) have become a very popular type of media on the Web. Blogs are often online diaries published and maintained by individual users (bloggers), reporting on the bloggers' daily activities and feelings. The contents of the blogs include commentaries or discussions on a particular subject, ranging from mainstream topics (e.g., food, music, products, politics, etc.), to highly personal interests [9]. So blogs have turned to be the major platform for people to express their opinions and sentiments in the Web age.

Currently, there is a great number of blogs on the Internet. By February 2008, the famous blog search engine Technorati [16] had tracked more than 112.8 million blogs. As the number of blogs and bloggers increases dramatically, how to provide an effective way to access the blogs and make full use of the blog contents have become the major concern for both research and industrial communities. There are some previous literatures about blog content based analysis [1, 5, 7, 10]. Different kinds of tools are also being provided to help users retrieve, organize and analyze the blogs. Several commercial blog search engines and blog tagging systems [6, 16] have been published on the Web.

---

[*] This work is supported by National Natural Science Foundation of China (No. 60573090, 60703068, 60673139) and the National High-Tech Development Program (2008AA01Z146).

Q. Li et al. (Eds.): APWeb/WAIM 2009, LNCS 5446, pp. 332–344, 2009.

Most of the blog contents are about bloggers' opinions, feelings and emotions [12]. Some commercial web search engines have launched blog retrieval services. But these blog search tools are just the applications of the traditional web search techniques in the blog domain, that is, given a query, they only search for topic-related information in the blogs. However, the opinions and sentiments, which are very important feature of the blog documents, are not searched by these tools [22]. For example, if a person wants to know other bloggers' opinions and reviews about certain movie, he or she may type the movie's name, such as "*Hancock*", in Google blog search engine, and get thousands of searching results which can be divided into two types: the informative blogs and the affective blogs. Usually, he or she only reads several top blogs in the results, so he or she only gets the opinions about the movie in a relatively small area because the opinion of one person may not represent all the other people's opinions. Moreover, the affective blogs are mixed with the informative blogs, so it is difficult for users to choose the appropriate blogs to read. From the discussions above, we can see that the traditional blog search engines do not provide a convenient way for Internet users to explore in the Weblog space and there are still some obstacles and limitations for people to make better understanding of bloggers' opinions and sentiments when exploring in the blogosphere.

In this paper, we propose a novel method to group blog search results by sentiment clustering. Usually, bloggers express their feelings and emotions in a much complex way, but most previous work on sentiment analysis simply classify blogs as positive and negative, and do not provide a comprehensive understanding of the sentiments reflected in the blogs. Different from binary classification problem, we present a sentiment clustering algorithm to group blog search results. The sentiment similarity between blog search result items is calculated and by this method we can partition blogs into clusters to allow better organization and easy navigation. For business companies and governments, they can quickly collect people's attitudes about their products and services. For individual users, our methods can provide them a brand new way to explore in the blogosphere.

The rest of the paper is organized as follows. Section 2 introduces the related work on blog mining and sentiment analysis. Section 3 analyzes the sentiment characteristics of blog search result items. Section 4 describes sentiment clustering and navigation words extraction algorithms for blog search results. Section 5 provides experimental results on real world blog search engine. Finally we present concluding remarks and future work in Section 6.

2 Related Work

2.1 Blog Mining

Blogs have recently attracted a lot of interest from both computer researchers and sociologists. According to Mei's study, blogs mainly have two unique characteristics [10]: (1) blogs are mainly maintained by individual persons and the contents are generally highly personal opinions, and (2) the link structures between blogs generally form localized communities.

One direction of blog mining focuses on analyzing the contents of blogs. Gruhl et al. [7] studied the dynamics of information propagation in blog space. Glance et al. [5] gave a temporal analysis on blog contents and proposed a method to discover trends across blogs. In [15], the similarity between two blogs was calculated at the topic level, and Shen et al. presented the approach to find the latent friends who shared the similar topic distribution in their blogs. The other direction of blog mining focuses on studying the link structure of blogosphere. Tseng et al. [17] combined blog rankings with their social connections to provide a framework to understand multiple blog communities and a mountain view visualization was provided to explore different communities of interest in blogosphere. Kumar et al. [8] built a time graph for blog space, and developed views of the graph as a function of time.

Blog search is also a hot topic in this area. Mishne et al. [11] studied blog searchers' behavior and the result showed that the users are usually interested in the first few results only. Google recently has launched a new homepage for blog search so that the users can browse and discover the most interesting stories in the blogosphere [6]. It groups the blogs into several predefined categories, such as politics, business and technology, and the users can find a collection of the most interesting and recent posts on the topic. In [20] and [4], the Web search result snippets were clustered into groups according to sub-topics of the given query and highly readable names were extracted from each cluster to facilitate users' quick browsing through search results.

Our work is quite different from the existing studies on blogs. Most of existing work focuses on providing topic-based search interface or trend analysis for blogs. We proposes a novel method to group the blog search results into sentiment clusters, which provides a brand new way for uses to explore in the blogosphere and facilitate public opinion monitoring for governments and business organizations.

2.2 Sentiment Analysis

Sentiment analysis is the main task of opinion mining, and most of existing work focuses on determining the sentiment orientations of documents, sentences and words. In document level sentiment analysis, documents are classified into positive and negative according to the overall sentiment expressed in them. Turney et al. [18] measured the strength of sentiment by the difference of the Pointwise Mutual Information (PMI) between the given phrase and the seed words. In [13], Pang et al. treated the problem as a topic-based text classification problem, i.e. there were only two kinds of topic, namely positive and negative. The authors employed three machine learning approaches (Naive Bayes, Maximum Entropy, and Support Vector Machine) to label the polarity of IMDB movie reviews.

Most of prior studies about sentiment analysis attempt to classify the text by the overall sentiment the author expressed. However, the emotions that bloggers want to express are much more complex. Therefore, it would be too simplistic to classify blogs into just positive and negative categories. Different from the traditional classification approaches for sentiment analysis, in this paper, we propose a lexicon based sentiment clustering method for blog search result items. An interactive clustering method for grouping documents by sentiment has been proposed in [2]. However, the experiments in that paper are based on long text dataset, and the clustering results are restricted to four categories: strongly disliked, somewhat disliked, somewhat liked and strongly liked. Our method can cluster blog search result snippets which are usually very short, and extract navigational sentiment words in each cluster.

3 Blog Search Results Sentiment Analysis

Given a query key word, our goal is to find the topic relevant blogs and cluster them into groups according to their sentiment similarities. For practical usage, the algorithm proposed should take the blog search result titles and snippets instead of the whole blog articles as input, since the downloading of original blogs is time-consuming and the clustering algorithm should be fast enough for online calculation.

3.1 Blog Search Results Analysis

The commercial blog search engines usually use traditional Web search techniques and rank the results by their topic relevance, which does not consider the latent sentiments and opinions expressed in the blogs. Each search result item is composed of title, date, author, snippet and URL. Our work focuses on sentiment analysis for the titles and snippets. We issue the movie name "*Hancock*" in Google Blog Search, and then several results are collected and shown in Table 1.

Table 1. The titles and snippets for the blog search key word "*Hancock*"

Title	Snippet
Hancock	Go see Hancock. Much respect to Will Smith and the directors behind the film, truly inspirational.
It's a comedy, It's a fantasy, Yes, It's 'Hancock'	Hancock was enjoyable, but no without it problems thanks to many unanswered questions…
Hancock Sucks	There are a number of factors in Hancock that fuel promise and expectations to its viewers.
Hancock 2008 DVDRip direct links	Hancock 2008 Language: English Runtime: 92 min Country: USA Release Date: 2 July 2008...
Hancock.DVDRip.XviD-ALLiANCE	Category: Movies-XviD Size: 827.28 MB Files: 65 (12 pars)…

We can see from Table 1 that the blog search result titles and snippets have the following characteristics:

(1) The titles and snippets are highly relevant to the given query key word. That's because the blog search engine employs sophisticated and mature techniques to get the most topic relevant articles and snippets;

(2) Some of the titles and snippets contain the bloggers' sentiments and opinions. As the search results are highly topic-coherent, the sentiment words in search results mainly reflect the bloggers' opinions about the given query key word;

(3) Since not all the blogs contain authors' emotions, there are some informative results mixed up with affective ones. For example, the last two result items in Table 1 tell us the downloading information of Hancock DVDRip.

From the analysis above we can find that the traditional blog search engines provide good topic relevant retrieval interface, however, they do not provide a convenient way for users to get bloggers' opinions and sentiments. In Section 3.2, we employ a lexicon based method to filter out the informative result items and extract the sentiment words in search result items.

3.2 Lexicon Based Sentiment Word Mining

The emotion orientation of words is the foundational and indispensable resource for sentiment analysis. The statistics based and learning based sentiment word acquisition methods are not suitable for our clustering algorithm, because they always need training sets or the calculation is really time-consuming. Therefore, we employ a lexicon based sentiment word acquisition method to find the emotion words in blog search result items.

WordNet [19] is a large lexical database of English. Nouns, verbs, adjectives and adverbs are grouped into sets of cognitive synonyms (synsets), each expressing a distinct concept. Synsets are interlinked by means of conceptual-semantic and lexical relations. The resulting network of meaningfully related words and concepts can be accessed by programming interface. WordNet's structure makes it a useful tool for computational linguistics and natural language processing.

SentiWordNet [3] is a lexical resource for opinion mining. SentiWordNet assigns to each synset of WordNet three sentiment scores: positivity, negativity, objectivity. The basic assumption of SentiWordNet is that terms with similar polarity tend to have "similar" glosses: for instance, that the glosses of *honest* and *intrepid* will both contain appreciative expressions, while the glosses of *disturbing* and *superfluous* will both contain derogative expressions. Therefore SentiWordNet is developed based on the quantitative analysis of the glosses associated to synsets, and on the use of the resulting vector term representations for semi-supervised synset classification. The three scores are derived by combining the results produced by a committee of eight ternary classifiers, all characterized by similar accuracy levels but different classification behavior. SentiWordNet has been widely used in the previous literatures. Words with positive or negative score above a threshold in SentiWordNet are used by some participants of the TREC opinion retrieval task [21]. Extensive experiments show that SentiWordNet is a very effective and efficient lexicon tool for sentiment analysis.

Using SentiWordNet, we propose a lexicon based method to mine the sentiment words in blog search results. The method includes the following steps:

(1) Label the words in the titles and the snippets with Part-of-Speech (POS) tags.

(2) Remove the stop words.

(3) Stemming. Notice that we only convert the plural nouns into singular form and transform the verbs into the present tense. We do not conduct other stemming algorithm to the words because we must keep their original sentiment meanings.

(4) Word sense disambiguation. To get the correct score of a given word in SentiWordNet, we must know its POS tag and the sense in the given context. WordNet

based sense disambiguation algorithm is employed to find the appropriate meaning of the word in result items.

(5) Words with positive or negative score above a threshold in SentiWordNet are picked out as sentiment words in blog search results. Different threshold settings may affect sentiment titles and snippets recognizing results, which can be seen in the experiment section.

After these 5 steps, we can extract the sentiment words from the search result items. If an item contains at least one sentiment word, we classify it into affective category, where we conduct our further sentiment clustering algorithm; otherwise, the result item is regarded as having no sentiment indication, and classified into informative category.

3.3 Sentiment Word Similarity Computing

In Section 3.2 we introduce the lexicon based method to extract sentiment words and classify blog search result items into affective and informative categories. For further clustering steps, we must calculate the similarity between words in the affective category. Several methods have been proposed to compute word similarity based on WordNet.

Extended Gloss Overlaps: Each concept (or word sense) in WordNet is defined by a short gloss. The Extended Gloss Overlaps (*Lesk* for short) measure uses the text of that gloss as a unique representation for the underlying concept [14]. The *lesk* measure assigns similarity by finding and scoring overlaps between the glosses of the two concepts, as well as concepts that are directly linked to them according to WordNet. The bigger number of overlap gloss words is, the bigger similarity value between two concepts is.

Gloss Vector: The Gloss Vector (*Vector* for short) measure creates a co-occurrence matrix from a corpus made up of the WordNet glosses [14]. Each content word used in a WordNet gloss has an associated context vector. Each gloss is represented by a gloss vector that is the average of all the context vectors of the words found in the gloss. Similarity between concepts is measured by finding the cosine between a pair of gloss vectors.

Gloss Vector (pairwise): The Gloss Vector (pairwise) measure (*Vector_Pairs* for short) is very similar to Gloss Vector measure, except in the way it augments the glosses of concepts with adjacent glosses [14]. The regular Gloss Vector measure first combines the adjacent glosses to form one large "super-gloss" and creates a single vector corresponding to each of the two concepts from the two "super-glosses". The pairwise Gloss Vector measure, on the other hand, forms separate vectors corresponding to each of the adjacent glosses (does not form a single super gloss).

These three WordNet based algorithms make use of the glosses to find the similarity between words. As terms with similar emotion meanings tend to have similar glosses, these algorithms also reflect the sentiment similarity between words. We will test the effectiveness of these three algorithms in the experiment section.

4 Sentiment Clustering and Navigation Words Extraction for Blog Search Results

4.1 Sentiment Clustering for Blog Search Results

Different from traditional classifying the blogs into positive and negative categories, the goal of sentiment clustering is to make best use of the complexity of people's emotions, and group the documents by their sentiment similarity. That is to say, if two documents express similar emotions, they should be clustered into the same group. People's feelings and emotions are complex, so there should be more than just positive and negative groups in the clustering results.

We define the affective blog search result items as the items that contains sentiment words and the rest are informative items. After sentiment analysis for the blog search results, we classify the result items set R into affective and informative categories. So we get:

$$R = R_{affective} \cup R_{informative} \tag{1}$$

Suppose the result items r_i, $r_j \subset R_{affective}$, and r_i, r_j contain the sentiment word set SW_i, SW_j respectively, where $SW_i = \{sw_1, sw_2, \dots, sw_m\}$ and $SW_j = \{sw_1, sw_2, \dots, sw_n\}$. So according to the definition of $R_{affective}$, the number of elements in SW_i, SW_j is greater than or equal to 1. We define two kinds of measurement $MaxSim_{ij}$ and $AvgSim_{ij}$ for sentiment similarity between r_i and r_j:

$$MaxSim_{ij} = \max WordSim(SW_i, SW_j) \tag{2}$$

$$AvgSim_{ij} = \operatorname{avg} WordSim(SW_i, SW_j) \tag{3}$$

where the function $WordSim$ calculates the sentiment word similarity based on the algorithms introduced in Section 3. Therefore, the max value and average value of the word sentiment relatedness are used to represent the similarity between blog search results in $R_{affective}$. We will check the effectiveness of these two different measurements in the experiment section.

Based on the defined similarity measurements, we can employ the existing clustering methods to do sentiment clustering tasks. In this paper, we cluster blog search result items based on K-Medoids algorithm, which has already been proved to be an effective Web search snippets clustering method.

4.2 Key Sentiment Words Extraction for Exploring Navigation

It is not enough for our algorithm to create sentiment coherent clusters, so we must also convey the emotion contents of the clusters to the users concisely and accurately. The navigation sentiment words are most useful when the user can decide at a glance whether the contents of a cluster are of interest.

Here we define three properties for the given cluster SC_k as follows:

Sentiment Word Frequency: Since the cluster SC_k has the blog search result items with the coherent emotions, the sentiment words which appear most frequently can represent the emotions expressed in SC_k. The property is defined as:

$$SWF = f(sw) \tag{4}$$

where $f(sw)$ calculates the appearance frequency of sentiment word sw in SC_k.

Sentiment Word Count per Result: The blog search result item which contains the most number of sentiment words can provide a wealth of information about bloggers' emotions. Given a result item r_i, the Sentiment Word Count per Result property is defined as:

$$SWC_i = Count(SW_i) \tag{5}$$

where $Count(SW_i)$ calculates the number of sentiment word in SW_i of the given result item r_i.

Sentiment Word Strength: The word with higher sentiment strength can be a better indicator for bloggers' emotions. We use sentiment score value of given word sw in SentiWordNet to represent sw's emotion strength. The Sentiment Word Strength property is defined as:

$$SWS = SentiWordNet(sw) \tag{6}$$

Given a cluster SC_k, we can extract the sentiment navigation words by the following rules:

(1) Find the word with the top 3 biggest SWF in SC_k;
(2) If more than 3 words are found, choose the top 3 words according to SWS.

By the properties of the clusters and the rules, we can find the key sentiment words which can help users to make decisions when exploring in the blogosphere. Another problem to be considered is how to reorganize the search result items in each sentiment cluster. We employ Sentiment Rank function to sort the result items by their average emotion strength. The formula is defined as:

$$SentiRank(r_i) = \frac{\sum_{p=1}^{m} SWS(sw_p)}{SWC_i} \tag{7}$$

Given a cluster SC_k, we calculate *SentiRank* of each result item r, and rerank the items according to *SentiRank* value in descent order. Together with key sentiment words extraction, we can provide users more readable and useful blog search results.

5 Experiments

Our experiment is conducted on a commodity PC with Windows XP, Pentium D CPU and 1GB RAM. Given a query word, we use Google SOAP Search API to find the topic relevant blog entries. Titles and snippets can be easily accessed by this API. However, this component is not designed for blog search, so the special query term "*inurl:*" is used to restrict the search results in the blogosphere. We restrict the search results in the world's three major blog web sites: MSN Live Spaces, Google Blogspot and WordPress.com, which provide us a wealth of information about the bloggers' opinions all around the blogosphere.

The algorithms of word sentiment similarity computing and word sense disambiguation are all implemented based on WordNet. WordNet::Similarity [14] is a freely available software package that makes it possible to measure the semantic similarity between a pair of concepts in WordNet. We use this package to calculate the *Lesk*, *Vector* and *Vector_Pairs* similarity of the given sentiment words. Another tool WordNet::SenseRelate is used to perform word sense disambiguation. It measures the semantic similarity between a word and its neighbors and a word is assigned to the sense that is most related to its neighbors.

First we evaluate our algorithm for identifying the affective blogs in the search result items. We collect 1258 result items from Google with the query key word "Hancock". Since our goal is to cluster the result items with the similar emotions, we do not pay attention to the sentiment polarity of the words. Three human annotators label the items with Y/N tag, i.e. sentiment result/non-sentiment result. We use the algorithm proposed in Section 3 to mine the sentiment words. Different thresholds of the score in SentiWordNet are set to determine whether the one is a sentiment word. If the result item contains at least one such word, it is classified into $R_{affective}$ category and the results are compared to the human annotators. The result is shown in Figure 1.

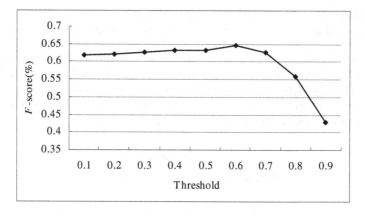

Fig. 1. The F-score of blog search sentiment results identification

Figure 1 describes the F-score of recognizing blog search result sentiment items by using the algorithm in Section 3. The X-axis is the threshold of sentiment score. From this figure, we can see that the highest *F*-score 0.65 (with precision 0.49 and recall 0.93) can be archived when threshold is set to be 0.6. So 0.6 is set to be the final threshold of SentiWordNet when we classify the result items. Since our goal is to cluster titles and snippets by their embedded emotions, this confirms very high blog search sentiment results coverage and an acceptable precision.

We have ranked the result item in each cluster by their average sentiment strength, and we use precision (*P*) at top *N* results to measure the performance:

$$P @ N = \sum_{i=1}^{k} \frac{S_i \cap N_i}{N_i} \tag{8}$$

where given k clusters, N_i denotes the top N result items in the *ith* cluster, and S_i denotes the set of items in the *ith* cluster which have been manually recognized as coherent to the extracted key sentiment words. The result is shown in Figure 2:

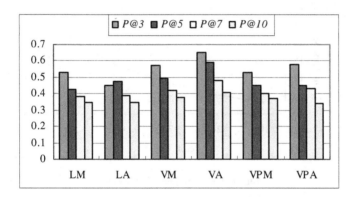

Fig. 2. *P@N* performance for sentiment clustering blog search result items

In Figure 2, LM represents *Lesk* word similarity computing algorithm and *MaxSim* result items sentiment similarity computing algorithm. So LA, VM, VA, VPM, VPA represent *Lesk-AvgSim, Vector-MaxSim, Vector-AvgSim, Vector_pairs-MaxSim, Vector_pairs-AvgSim* respectively. We can see that *Vector-AvgSim* can produce the best *P@N* performance. We employ *Vector-AvgSim* algorithm to cluster 200 blog search result items with the query key word "*Hancock*" and extract the key sentiment words in each cluster, as shown in Figure 3.

We can see from Figure 3 that, our method have clustered the dataset into 6 groups and the largest group has about 57% of all result items and the smallest group has about 2% of all items. According to Figure 2, the key sentiment words in Figure 3 can provide a very brief but effective summarization of the sentiments in each cluster. To verify the time complexity of our algorithm, we select a query as the example and record the time cost of our *Vector-AvgSim* sentiment clustering algorithm as shown in Figure 4, in which the X-axis stands for the number of results returned from blog search engine, and the Y-axis is the time spent in second.

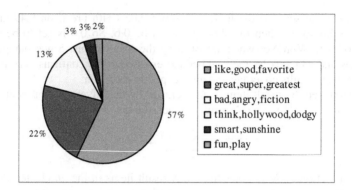

Fig. 3. Covering distribution and key sentiment words for "*Hancock*"

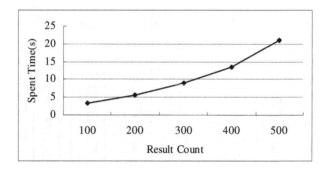

Fig. 4. Time complexity analysis

Google uses traditional Web search engine techniques to retrieval the blogs and ranks the results by topic relevance and authority scores. Previous sentiment analysis systems simply classify the blogs into positive and negative categories. The experiments prove that compared to the other approaches, our method has the following advantage. Firstly, our method can reflect the complexity of emotions and it employs clusters to group bloggers' sentiments. Secondly, our method provides a novel and effective sentiment-oriented blog search approach. Thirdly, the blog search result items can be reorganized by embedded emotions and ranked by sentiment strength. Key sentiment words extracted in each cluster are good navigation guidelines for users to explore in the blogosphere.

6 Conclusion and Future Work

The traditional blog search engines only provide topic-oriented retrieval interface and it is not convenient for users to make better understanding of bloggers' opinions and emotions. In this paper, we propose a lexicon-based sentiment clustering algorithm to group blog search result items according to their emotions. Experimental results demonstrate that we can classify the blog search result items into affective and informative categories and generate correct sentiment clusters by their embedded emotions, thus could improve users' browsing efficiency through blog search results.

The product reviews are useful data source for both individual customers and business companies. Further research directions include using sentiment clustering method to analyze people's opinions about certain product and help latent customers and company leaders to make decisions.

References

1. Bar-Ilan, J.: An Outsider's View on "Topic-oriented" Blogging. In: 13th International Conference on World Wide Web Alternate Papers Track, pp. 28–34 (2004)
2. Bekkerman, R., Raghavan, H., Allan, J., Eguchi, K.: Interactive Clustering of Text Collections According to a User-Specified Criterion. In: 20th International Joint Conference on Artificial Intelligence, pp. 684–689 (2007)
3. Esuli, A., Sebastiani, F.: SentiWordNet: A Publicly Available Lexical Resource for Opinion Mining. In: Fifth International Conference on Language Resources and Evaluation, pp. 417–422 (2006)
4. Ferragina, P., Gulli, A.: A Personalized Search Engine based on Web-snippet Hierarchical Clustering. In: 14th International Conference on World Wide Web, pp. 801–810 (2005)
5. Glance, N., Hurst, M., Tornkiyo, T.: Blogpulse: Automated Trend Discovery for Weblogs. In: WWW 2004 Workshop on the Weblogging Ecosystem (2004)
6. Google Blog Search, http://blogsearch.google.com
7. Gruhl, D., Guha, R., Liben-Nowell, D., Tomkins, A.: Information Diffusion through Blogspace. In: 13th International Conference on World Wide Web, pp. 491–501 (2004)
8. Kumar, R., Novak, J., Raghavan, P., Tomkins, A.: On the Bursty Evolution of Blogspace. In: 14th International Conference on World Wide Web, pp. 159–178 (2005)
9. Kumar, R., Novak, J., Raghavan, P., Tomkins, A.: Structure and Evolution of Blogspace. Commun. ACM 47(12), 35–39 (2004)
10. Mei, Q., Liu, C., Su, H., Zhai, C.: A Probabilistic Approach to Spatiotemporal Theme Pattern Mining on Weblogs. In: 15th International Conference on World Wide Web, pp. 533–542 (2006)
11. Mishne, G., de Rijke, M.: A Study of Blog Search. In: Lalmas, M., MacFarlane, A., Rüger, S.M., Tombros, A., Tsikrika, T., Yavlinsky, A. (eds.) ECIR 2006. LNCS, vol. 3936, pp. 289–301. Springer, Heidelberg (2006)
12. Ni, X., Xue, G., Ling, X., Yu, Y., Yang, Q.: Exploring in the Weblog Space by Detecting Informative and Affective Articles. In: 16th International Conference on World Wide Web, pp. 281–290 (2007)
13. Pang, B., Lee, L., Vaithyanathan, S.: Thumbs up? Sentiment Classification Using Machine Learning Techniques. In: 2002 Conference on Empirical Methods in Natural Language Processing, pp. 79–86 (2002)
14. Pedersen, T., Patwardhan, S., Michelizzi, J.: WordNet:Similarity-Measuring the Relatedness of Concepts. In: 19th National Conference on Artificial Intelligence, pp. 1024–1025 (2004)
15. Shen, D., Sun, J., Yang, Q., Chen, Z.: Latent Friend Mining from Blog Data. In: 6th IEEE International Conference on Data Mining, pp. 552–561 (2006)
16. Technorati, http://technorati.com

17. Tseng, B.L., Tatemura, J., Wu, Y.: Tomographic Clustering to Visualize Blog Communities as Mountain Views. In: 2nd Annual Workshop on the Weblogging Ecosystem (2005)
18. Turney, P.: Thumbs Up or Thumbs Down? Semantic Orientation Applied to Unsupervised Classification of Reviews. In: 40th Annual Meeting of the Association for Computational Linguistics, pp. 417–424 (2002)
19. WordNet, http://wordnet.princeton.edu
20. Zeng, H., He, Q., Chen, Z., Ma, W., Ma, J.: Learning to Cluster Web Search Results. In: 27th Annual International ACM SIGIR Conference on Research and Development in Information Retrieval, pp. 210–217 (2004)
21. Zhang, M., Ye, X.: A Generation Model to Unify Topic Relevance and Lexicon-based Sentiment for Opinion Retrieval. In: 31th Annual International ACM SIGIR Conference on Research and Development in Information Retrieval, pp. 411–418 (2008)
22. Zhang, W., Yu, C.T., Meng, W.: Opinion Retrieval from Blogs. In: 16th ACM Conference on Information and Knowledge Management, pp. 831–840 (2007)

Kernel-Based Transductive Learning with Nearest Neighbors

Liangcai Shu, Jinhui Wu, Lei Yu, and Weiyi Meng

Dept. of Computer Science, SUNY at Binghamton
Binghamton, New York 13902, U.S.A.
{lshu,jwu6,lyu,meng}@cs.binghamton.edu

Abstract. In the k-nearest neighbor (KNN) classifier, nearest neighbors involve only labeled data. That makes it inappropriate for the data set that includes very few labeled data. In this paper, we aim to solve the classification problem by applying transduction to the KNN algorithm. We consider two groups of nearest neighbors for each data point — one from labeled data, and the other from unlabeled data. A kernel function is used to assign weights to neighbors. We derive the recurrence relation of neighboring data points, and then present two solutions to the classification problem. One solution is to solve it by matrix computation for small or medium-size data sets. The other is an iterative algorithm for large data sets, and in the iterative process an energy function is minimized. Experiments show that our solutions achieve high performance and our iterative algorithm converges quickly.

Keywords: KNN, transductive learning, semi-supervised learning, kernel function.

1   Introduction

The k-nearest neighbor (KNN) algorithm [1] is a simple and effective supervised learning algorithm. One of the disadvantages of supervised learning is that it requires significant amount of labeled data for training in order to achieve high performance. But in applications like object identification, text categorization, a large amount of labeled data need a lot of human efforts, whereas unlabeled data is quite cheap and easy to obtain. This is the reason that transductive learning or semi-supervised learning has been developed in recent years [2][3][4][5][6][7][8][9][10].

Transduction or transductive learning was first introduced by Vladimir Vapnik who thinks it is preferable to induction because induction requires solving a more general problem before solving a specific problem, e.g. decision trees, SVMs, and transduction solves the specific problem directly. This is so-called Vapnik's principle [11]. Transductive learning takes advantage of unlabeled data to capture the global structure of the data that is assumed to be helpful to predict class labels. But this approach requires assumptions on the structure of data — first, cluster assumption that data in the same cluster are supposed to be in

Q. Li et al. (Eds.): APWeb/WAIM 2009, LNCS 5446, pp. 345–356, 2009.

Fig. 1. (left) The two interlock rings problem in 3D space. Only two points, one for each class and marked by O and +, are labeled. **(right)** An example in which simple local propagation fails and TKNN succeeds. • and ▲ represent labeled points.

the same class, and second, manifold assumption that the high-dimensional data lie on a low-dimensional manifold [11]. The second assumption is for algorithms using pairwise distance to avoid the curse of dimensionality.

Consider the two interlock rings problem in Fig. 1 (left). We aim to find the class label for each point, given only two labeled points. Mixture models cannot solve it since distribution of points in a cluster is unknown instead of standard distributions like Gaussian. Radial basis functions (RBFs) and KNN cannot solve it because of too few labeled points. The difficulty of applying support vector machines (SVMs) on such data is that it is hard to find the right kernel function to map the data from the original space to a higher dimensional space where the data becomes linearly separable. Also, kernel learning is time-consuming. So we do not consider SVMs in this paper.

Instead, we propose an approach called TKNN, which applies transduction to the KNN algorithm to solve this problem. Traditionally, the nearest neighbors in the KNN algorithm are defined on the labeled data only. In this paper, we re-examine the method and extend KNN by redefining nearest neighbors as from both labeled data and unlabeled data. Two groups of nearest neighbors are simultaneously considered for each data point — one group from labeled data, and the other from unlabeled data. A kernel function or radial basis function is used to increase weights for nearer neighbors. Then global structure of data is captured in a graph. Taking advantage of global structure, labels propagate through the graph, and the two interlock rings problem can be solved completely.

Note that TKNN is more sophisticated than simple local propagation starting from the labeled points. For the example in Fig. 1 (right), simple local propagation fails because, without global structure of data, the label propagates across the sparse area between two classes. Otherwise in TKNN, due to the kernel function, label propagation in sparse areas gives way to that in dense areas and finally the sparsest area is identified as the boundary.

Our Contributions are as follows:

- We apply transductive learning to KNN, with the kernel function adjusting weights for nearest neighbors.
- We derive the recurrence relation between labeled data and unlabeled data, and propose solutions to the classification problem.

- We propose an efficient and flexible iterative algorithm for large data sets. Techniques are applied to achieve fast convergence. And an outlier detection in labeled data is presented too.
- Experimental results show the performance and efficiency of our algorithm.

The remaining part of this paper is organized as follows. In Section 2 we discuss the model formulation, including deriving recurrence relation between labeled data and unlabeled data. In Section 3 we propose solutions and derive the iterative algorithm and its ability to converge, and also discuss outlier detection. In Section 4 we report our experimental results and analysis. Section 5 is related work. Section 6 is the conclusion.

2 Model Formulation

2.1 Problem Description and Notation

In this paper, we aim to solve the problem of classification by applying transductive learning to the KNN algorithm, i.e., taking advantage of unlabeled data as well as labeled data.

Assume we are given the set of points $\mathcal{D} = \mathcal{D}_L \cup \mathcal{D}_U$ in a p-dimensional Euclidean space $\mathbf{R}^p$, where $\mathcal{D}_L = \{\mathbf{x}_i\}_{1 \leq i \leq l}$ is the set of labeled points and $\mathcal{D}_U = \{\mathbf{x}_i\}_{l+1 \leq i \leq l+u}$ is the set of unlabeled points. One class label should be assigned to each member of $\mathcal{D}$. The set of classes is denoted by $\mathcal{C} = \{c_r\}_{1 \leq r \leq |C|}$ and correspondingly, the set of class labels is denoted by $C = \{r\}_{1 \leq r \leq |C|}$. Here, $|C|$ is the cardinality of C and r is an integer.

The class label of point $\mathbf{x}_i \in \mathcal{D}$ is denoted by $y(\mathbf{x}_i)$ or y_i, where $y_i \in C$. Each point in $\mathcal{D}_L$ has been labeled or assigned to one class in C, while the class label of each point in $\mathcal{D}_U$ is unknown and needs to be determined. In this paper, k_l denotes the number of nearest neighbors in labeled data, and k_u denotes the number of nearest neighbors in unlabeled data. The k_l nearest neighbors of a given point $\mathbf{x}_i \in \mathcal{D}$ are denoted by $\{\mathbf{n}_{iq}\}_{1 \leq q \leq k_l}$, where $\mathbf{n}_{iq}$ is $\mathbf{x}_i$'s qth nearest neighbor in $\mathcal{D}_L$. Similarly, the k_u nearest neighbors of $\mathbf{x}_i$ are denoted by $\{\mathbf{m}_{iq}\}_{1 \leq q \leq k_u}$, where $\mathbf{m}_{iq}$ is $\mathbf{x}_i$'s qth nearest neighbor in $\mathcal{D}_U$.

2.2 Extending KNN

In supervised KNN learning [12][1], the nearest neighbors included only labeled points. Therefore, it needs a lot of labeled points for good performance. In this paper, we propose TKNN after extending KNN by including both labeled and unlabeled points in the nearest neighbors of a point.

After applying the KNN to density estimation [1], for a point $\mathbf{x}$ the posterior probability $p(c_r|\mathbf{x})$ can be estimated as

$$\hat{p}(c_r|\mathbf{x}) = \frac{k(r)}{k}, \tag{1}$$

where $k(r)$ is the number of points with class label r among the k nearest neighbors of $\mathbf{x}$. Obviously, $\sum_{r=1}^{|C|} k(r) = k$.

In supervised KNN learning [12][1], $k(r)$ in Eq. (1) is computed based on labeled points and $\hat{p}(c_r|\mathbf{x})$ can be directly obtained. Differently, in our model TKNN, there are two groups of nearest neighbors for each point. One group includes k_l nearest neighbors from labeled data, and the other includes k_u nearest neighbors from unlabeled points. With the help of unlabeled data, the amount of labeled data can be small.

Let $k_l(r)$ be the number of labeled points in class c_r, and $k_u(r)$ be the number of unlabeled points in class c_r. $k_l(r)$ can be easily obtained, but $k_u(r)$ is not available. Then $\hat{p}(c_r|\mathbf{x})$ in Eq. (1) cannot be directly obtained anymore. But we can learn $\hat{p}(c_r|\mathbf{x})$ by deriving recurrence relation among neighboring data points. For convenience of description, we define the class matrix.

Class Matrix. The class matrix for a data set describes probabilities that data are assigned to classes, data as rows and classes as columns. Then the sum of values in each row is 1.

Definition 1. *The* class matrix *of data set* $\mathcal{D} = \{\mathbf{x}_i\}_{1 \leq i \leq l+u}$ *is defined as* $\mathbf{P} = (p_{ij})_{(l+u) \times |C|}$, *where* $p_{ij} = p(c_j|\mathbf{x}_i)$.

The class matrix $\mathbf{P}$ can be written in two blocks:

$$\mathbf{P} = \begin{bmatrix} \mathbf{P}_L \\ \mathbf{P}_U \end{bmatrix}, \tag{2}$$

where $\mathbf{P}_L$'s rows correspond to labeled data and $\mathbf{P}_U$'s rows correspond to unlabeled data.

$\mathbf{P}_L$ is defined according to prior knowledge of labeled data. For any point $\mathbf{x}_i \in \mathcal{D}_L$, assume its class label is $y_i \in C$. Then the element at the ith row and y_ith column is set to 1. Formally, $\mathbf{P}_L = (p_{ij})_{l \times |C|}$, where

$$p_{ij} = p(c_j|\mathbf{x}_i) = \begin{cases} 1, & j = y_i, \\ 0, & j \neq y_i. \end{cases}$$

2.3 Kernel-Based Weights for Neighbors

Every neighbor of $\mathbf{x}_i$ has a weight according to its distance to $\mathbf{x}_i$. Nearer neighbors are given greater weights. We use a kernel function, also called radial basis function, to compute weights.

Kernel Function. In this paper, we use Gaussian function as kernel

$$K(\mathbf{x}_i, \mathbf{x}_j) = \frac{1}{\sqrt{2\pi}h} exp\left(-\frac{\|\mathbf{x}_i - \mathbf{x}_j\|^2}{2h^2}\right), \tag{3}$$

where $\| \cdot \|$ is Euclidean distance and h is the bandwidth.

The bandwidth h in Eq. (3) is also called smooth parameter. Its selection impacts the performance of the kernel-based algorithm.

Weight Matrix. Based on the kernel function, we define the weight matrix. The weight matrix of a data set describes impacts of neighbors to each data point. Its rows and columns are all the data points. For each row, only columns corresponding to two groups of nearest neighbors are given weights, and other column are given 0. Then the matrix is not necessarily symmetric. Based on the weight matrix, a KNN graph is constructed.

Definition 2. *Let λ be a real number in $[0, 1]$. The weight matrix of data set $\mathcal{D} = \{\mathbf{x}_i\}_{1 \leq i \leq l+u}$ is defined as $\mathbf{W} = (w_{ij})_{(l+u) \times (l+u)}$, where*

$$w_{ij} = \begin{cases} K(\mathbf{x}_i, \mathbf{x}_j), & 1 \leq j \leq l \ \ and \ \ \mathbf{x}_j \in \{\mathbf{n}_{iq}\}_{1 \leq q \leq k_l}, \\ \lambda\, K(\mathbf{x}_i, \mathbf{x}_j), & l+1 \leq j \leq l+u \ \ and \ \ \mathbf{x}_j \in \{\mathbf{m}_{iq}\}_{1 \leq q \leq k_u}, \\ 0, & otherwise, \end{cases} \quad (4)$$

where $\{\mathbf{n}_{iq}\}_{1 \leq q \leq k_l}$ and $\{\mathbf{m}_{iq}\}_{1 \leq q \leq k_u}$ are $\mathbf{x}_i$'s neighbors from labeled data and unlabeled data.

The parameter λ, called influence factor of unlabeled data, is used to adjust unlabeled data's influence in the graph. If $\lambda = 0$, unlabeled data have no influence. If $\lambda = 1$, unlabeled data have the same influence as labeled data.

We row-normalize $\mathbf{W}$, i.e., adding row elements together and dividing each element by the corresponding row totals. Then we get row-normalized matrix $\mathbf{V}$. Obviously, the sum of each row's elements in $\mathbf{V}$ equals 1. We call $\mathbf{V}$ the *normalized weight matrix.* Considering $\mathcal{D} = \mathcal{D}_L \cup \mathcal{D}_U$, $\mathbf{V}$ consists of four blocks:

$$\mathbf{V} = (v_{ij})_{(l+u) \times (l+u)} = \begin{bmatrix} \mathbf{V}_{LL} & \mathbf{V}_{LU} \\ \mathbf{V}_{UL} & \mathbf{V}_{UU} \end{bmatrix}. \quad (5)$$

where $\mathbf{V}_{LL}, \mathbf{V}_{LU}, \mathbf{V}_{UL}$ and $\mathbf{V}_{UU}$ are l-by-l, l-by-u, u-by-l and u-by-u matrices, respectively. Superscripts L and U in $\mathbf{V}_{LU}$ mean rows from labeled data and columns from unlabeled data, respectively.

2.4   Recurrence Relation

Applying Eq. (1) to $\mathbf{x}_i$, we get estimation for $\hat{p}(c_r|\mathbf{x}_i)$ that is p_{ir}:

$$p_{ir} = \hat{p}(c_r|\mathbf{x}_i) = \frac{k_l(r) + k_u(r)}{k_l + k_u}. \quad (6)$$

Due to introduction of the kernel function, $k_l(r)$ and $k_u(r)$ are real numbers instead of integers. The sum of them can be estimated based on influence to $\mathbf{x}_i$ from other data points:

$$k_l(r) + k_u(r) = (k_l + k_u) \sum_{j=1}^{l+u} v_{ij}\, \hat{p}(c_r|\mathbf{x}_j) = (k_l + k_u) \sum_{j=1}^{l+u} v_{ij}\, p_{jr}. \quad (7)$$

Eq. (6) and (7) combine to give

$$p_{ir} = \sum_{j=1}^{l+u} v_{ij}\, p_{jr}. \quad (8)$$

Eq. (8) can be written in matrix: $\mathbf{P} = \mathbf{VP}$. Considering Eq. (2) and (5), Eq. (8) can be further written as:

$$\begin{bmatrix} \mathbf{P}_L \\ \mathbf{P}_U \end{bmatrix} = \begin{bmatrix} \mathbf{V}_{LL} & \mathbf{V}_{LU} \\ \mathbf{V}_{UL} & \mathbf{V}_{UU} \end{bmatrix} \begin{bmatrix} \mathbf{P}_L \\ \mathbf{P}_U \end{bmatrix}. \tag{9}$$

Since we do not consider the influence to a labeled data point from other points, we substitute the identity matrix $\mathbf{I}$ for $\mathbf{V}_{LL}$, and a zero matrix $\mathbf{0}$ for $\mathbf{V}_{LU}$. Eq. (9) becomes

$$\begin{bmatrix} \mathbf{P}_L \\ \mathbf{P}_U \end{bmatrix} = \begin{bmatrix} \mathbf{I} & \mathbf{0} \\ \mathbf{V}_{UL} & \mathbf{V}_{UU} \end{bmatrix} \begin{bmatrix} \mathbf{P}_L \\ \mathbf{P}_U \end{bmatrix}. \tag{10}$$

From Eq. (10), we obtained the recurrence relation in the data set:

$$\mathbf{P}_U = \mathbf{V}_{UL}\mathbf{P}_L + \mathbf{V}_{UU}\mathbf{P}_U. \tag{11}$$

3   Algorithm Derivation

3.1   Solution 1: Matrix Solution

From the recurrence relation in Eq. (11), we have the matrix solution

$$\mathbf{P}_U = (\mathbf{I} - \mathbf{V}_{UU})^{-1}\mathbf{V}_{UL}\mathbf{P}_L, \tag{12}$$

assuming that the inverse of matrix $(\mathbf{I} - \mathbf{V}_{UU})$ exists. This is the matrix solution.

After $\mathbf{P}_L$ and $\mathbf{V}$ are known, by Eq. (12), we obtain $\mathbf{P}_U$, the probability distributions of all unlabeled points over classes. The class label with the largest probability for each row of $\mathbf{P}_U$ is selected for classification decision. Assume $l \ll u$, the time complexity of this solution is $O(u^{2.376} + u^2l)$ which is the time for matrix inversion and multiplication, where u is the number of unlabeled data. This solution is appropriate for small and medium-size data sets.

Special Cases. If $\lambda = 0$, then $\mathbf{V}_{UU} = \mathbf{0}$ and Eq. (12) becomes $\mathbf{P}_U = \mathbf{V}_{UL}\mathbf{P}_L$, and the solution is downgraded to supervised learning — kernel-based weighted KNN. If $\lambda = 1$, the unlabeled data have the same influence as labeled data in the graph.

3.2   Solution 2: An Iterative Algorithm

In some cases the matrix $(\mathbf{I} - \mathbf{V}_{UU})$ in Eq. (12) is singular or approximately singular, causing that its inverse $(\mathbf{I} - \mathbf{V}_{UU})^{-1}$ does not exist (when $det(\mathbf{I} - \mathbf{V}_{UU}) = 0$) or approaches infinity. Then the matrix resolution fails or become imprecise. In addition, matrix solution is not efficient enough for large data sets.

We present an iterative algorithm that is more efficient and more flexible. Based on Eq. (8), we propose an iterative process to solve $\mathbf{P}_U$.

Iterative Process

- Initialization:
  - Normalized weight matrix $\mathbf{V}$ is initialized as in Section 2.3.
  - In class matrix $\mathbf{P}$, $\mathbf{P}_L$ is initialized as in Section 2.2. And $\mathbf{P}_U$ is initialized such that each row in it is an uniform distribution over all classes, i.e., each element is set to $\frac{1}{|C|}$.
- Repeat until convergence:
  - Update $\mathbf{P}$'s rows for unlabeled data according to Eq. (8). For all $l+1 \leq i \leq n$ $(n = l + u)$ and $1 \leq r \leq |C|$,

$$p_{ir}^{[t+1]} \leftarrow \sum_{j=1}^{n} v_{ij} p_{jr}^{[t]}. \tag{13}$$

where $p_{ir}^{[t+1]}$ is the new state and $p_{jr}^{[t]}$ is the current state.

Please note that when updating p_{ir}, we use previously computed p_{jr}'s as soon as they are available, instead of using old estimated p_{jr}'s.

Another reason that our algorithm converges quickly is that for each unlabeled point, there are always nearest neighbors from labeled data to influence its label. Then the rate of label propagation is fast. For most points, their labels can be stable quickly. That makes our algorithm faster than some previous graph-based iterative algorithms. In addition, considering $\mathbf{V}$ is a sparse matrix, each iteration's time complexity is actually $O(ku|C|)$ where $k = k_l + k_u$, instead of $O(nu|C|)$.

Convergence. We can prove that this iterative process is convergent, considering the nonnegative energy function $E = \frac{1}{2} \sum_{i=l+1}^{l+u} \sum_{r=1}^{|C|} \sum_{j=1}^{n} v_{ij} (p_{ir} - p_{jr})^2$ decreases monotonically in the process.

Flexibility. This algorithm is also flexible and can be combined with online learning. During online learning, the learner receives feedback and some unlabel data are labeled and added to labeled data. To address this problem, the algorithm can be slightly modified to allow $\mathbf{P}_L$ to change according to feedback.

3.3 Outlier Detection

In labeled data $\mathcal{D}_L$, there are possibly outliers, which could be labeled by mistake. Within the update process, we propose a method to detect outliers in $\mathcal{D}_L$. Different from Section 2.4, we consider the influence all over the data set, including both labeled data and unlabeled data. Assume the initial $\mathbf{P}_L$ is $\mathbf{P}_L^{[0]}$ and its estimation for outlier detection is $\hat{\mathbf{P}}_L$. From Eq. (9), we have

$$\hat{\mathbf{P}}_L = \mathbf{V}_{LL} \mathbf{P}_L^{[0]} + \mathbf{V}_{LU} \hat{\mathbf{P}}_U. \tag{14}$$

From Eq. (12), we have

$$\hat{\mathbf{P}}_U = (\mathbf{I} - \mathbf{V}_{UU})^{-1} \mathbf{V}_{UL} \mathbf{P}_L^{[0]}. \tag{15}$$

Algorithm 1. TKNN – transductive learning with nearest neighbors

---

Input:

λ: influence factor of unlabeled data.

C: the set of class labels $\{1, 2, \ldots, |C|\}$.

$\mathcal{D}_L$: set of labeled points, $\{\mathbf{x}_i\}_{1 \leq i \leq l}$.

Y_L: set of class labels for labeled points, $\{y_i\}_{1 \leq i \leq l}$ $(y_i \in C)$.

$\mathcal{D}_U$: set of unlabeled points, $\{\mathbf{x}_i\}_{l+1 \leq i \leq n}$ $(n = l + u)$.

k: number of nearest neighbors.

h: bandwidth for kernel method.

p_{thresh}: threshold of $\hat{p}(c_{y_i}|\mathbf{x}_i)$ for outlier detection.

Output:

Y_U: set of class labels for unlabeled points, $\{y_i\}_{l+1 \leq i \leq l+u}$.

Method:

1. Detects outliers $\mathcal{D}_O$.
2. $\mathcal{D}_L \leftarrow \mathcal{D}_L - \mathcal{D}_O$ and $l \leftarrow |\mathcal{D}_L|$.
3. $\mathcal{D}_U \leftarrow \mathcal{D}_L \cup \mathcal{D}_O$ and $u \leftarrow |\mathcal{D}_U|$.
4. Using λ, obtain row-normalized weight matrix $\mathbf{V} = (v_{i,j})_{n \times n}$ with four blocks denoted by $\mathbf{V}_{LL}, \mathbf{V}_{LU}, \mathbf{V}_{UL}$ and $\mathbf{V}_{UU}$.
5. Initialize class matrix $\mathbf{P} = (p_{i,r})_{n \times |C|}$ with two blocks denoted by $\mathbf{P}_L$ and $\mathbf{P}_U$.
6. **if** $\mathcal{D}_U$ is not too large and $det(\mathbf{I} - \mathbf{V}_{UU})$ is not too small **then**
7. $\mathbf{P}_U \leftarrow (\mathbf{I} - \mathbf{V}_{UU})^{-1}\mathbf{V}_{UL}\mathbf{P}_L$.
8. **else**
9.   {Iteratively update elements of $\mathbf{P}_U$ until convergence.}
10.   **repeat**
11.   **for all** $\mathbf{x}_i \in \mathcal{D}_U$ **do**
12.   **for** $r = 1$ to $|C|$ **do**
13. $p_{ir} \leftarrow \sum_{j=1}^{l+u} v_{ij}p_{jr}$
14.   **end for**
15.   **end for**
16.   **until** convergence
17. **end if**
18. Retrieve class labels $\{y_i\}_{l+1 \leq i \leq l+u}$ from class matrix $\mathbf{P}_U$.

Substitute $\hat{\mathbf{P}}_U$ in Eq. (14) with Eq. (15) and we have

$$\hat{\mathbf{P}}_L = \left[\mathbf{V}_{LL} + \mathbf{V}_{LU}(\mathbf{I} - \mathbf{V}_{UU})^{-1}\mathbf{V}_{UL}\right]\mathbf{P}_L^{[0]}. \qquad (16)$$

This is the estimation of $\mathbf{P}_L$ for outlier detection. In Eq. (16), $\mathbf{V}_{LL}$ reflects the influence between labeled data directly, and $\mathbf{V}_{LU}(\mathbf{I} - \mathbf{V}_{UU})^{-1}\mathbf{V}_{UL}$ reflects the influence between labeled data *through* unlabeled data, which is first introduced for outlier detection in this paper. Alternatively, in Eq. (14), $\hat{\mathbf{P}}_U$ can be obtained by the iterative process described in Section 3.2, which is more efficient for large data sets.

Looking for outliers, we compare each row of $\hat{\mathbf{P}}_L$ with the corresponding row of $\mathbf{P}_L^{[0]}$. Assume two corresponding rows are the row vector $\hat{\mathbf{p}}$ from $\hat{\mathbf{P}}_L$ and the row vector $\mathbf{p}_{[0]}$ from $\mathbf{P}_L^{[0]}$. Let a be the dot product of $\hat{\mathbf{p}}$ and $\mathbf{p}^{[0]}$. We know one component of $\mathbf{p}^{[0]}$ is 1 and others are 0. Then a is the component of $\hat{\mathbf{p}}$ for the class labeled in

$\mathbf{p}^{[0]}$. For example, if $\mathbf{p}^{[0]} = [0, 1, 0, 0]$ and $\hat{\mathbf{p}} = [0.1, 0.4, 0.3, 0.2]$, then $a = \hat{p}_2 = 0.4$. This is for one data point. For all labeled data, we have the column vector

$$\mathbf{a} = diag\left(\hat{\mathbf{P}}_L(\mathbf{P}_L^{[0]})^T\right), \tag{17}$$

where superscript T is matrix transpose, and $diag(\cdot)$ returns a vector that includes diagonal entries of a square matrix.

We decide if labeled data are outliers by vector $\mathbf{a}$. Assume a is a component of $\mathbf{a}$. If $a < p_{thresh}$, a's corresponding data point could have been mislabeled and is considered as an outlier. Generally we let the threshold p_{thresh} be $\frac{1}{|C|}$. The detected outliers are treated as unlabeled points; labeled and unlabeled data are re-partitioned before classification.

3.4   Algorithm

Based on our solution to the classification problem and outlier detection, we describe our algorithm called TKNN in Algorithm 1. We first detect outliers in labeled data and treat them as unlabeled. Then for a small or medium-size data set, we use matrix computation solution. And for a large data set, we use the iterative process solution.

4   Experiments

4.1   Data Sets

Three data sets are used in our experiments. (1) 2-rings. The 2-rings data set is generated to simulate the two interlock rings problem. 1000 data points are distributed around two rings in 3D space with Gaussian noise. Only two points are labeled, one for each class. (2) Digit1. This is from benchmark of [11], two classes, 241 dimensions, 1500 points. $Digit1$-10 means 10 points are labeled and $Digit$-100 means 100 points are labeled. (3) USPS. This is also from [11], also two classes, 241 dimensions, 1500 points. Two classes are imbalanced with relative sizes of 1:4. $USPS$-10 means 10 points are labeled and $USPS$-100 means 100 points are labeled.

Fig. 2. The convergence of TKNN iterative algorithm on the 2-rings data set. Two labeled points are marked with ● and ▲. Parameter configuration: $k_l = 1$, $k_u = 10$. From left to right: (1) after 1 iteration, accuracy 0.588; (2) after 10 iterations, accuracy 0.773; (3) after 30 iterations, accuracy 0.919; and (4) after 49 iterations, accuracy 1.0.

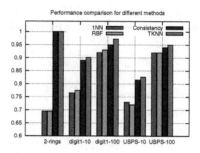

Fig. 3. Performance comparison of 1NN, RBF, Consistence method and TKNN

4.2   Experimental Results

First, we have to mention the parameter b. In Section 2.3 we introduced bandwidth h of the kernel function. The selection of h has impact to performance. But h ranges largely for different data sets. Then we define the bandwidth ratio b as the ratio of h and average distance of a data set. b is relatively stable and after its selection, h for a specific data is obtained.

Fig. 2 shows the convergence process of our TKNN iterative algorithm on the 2-rings data set. At the beginning of the process, points are randomly classified. It can be observed that during the process, the number of randomly classified points decreases and finally reaches zero, and all points are classified correctly, with accuracy 1.0. Fig. 4 (left) shows our algorithm converges faster than the consistency method [13]. Two reasons contribute to the efficiency. First, we considers labeled neighbors and unlabeled points can receive impact from labeled data at each iteration. Second, the newest data are immediately used to compute neighbors' impact to a point at each iteration.

We compare our algorithm to 1NN (KNN with $k = 1$), Radial Basis Function (RBF) and the consistency method as shown in Fig. 3. The Gaussian function is used for RBF. The best performance is recorded for each method. We can observe that when the number of labeled points is less (2-rings, Digit1-10 and USPS-10),

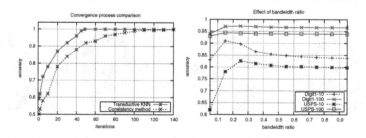

Fig. 4. **(left)** Convergence comparison of TKNN and consistence method on data set 2-rings. TKNN converges faster than the consistency method. **(right)** The effect of bandwidth to performance of TKNN. Here $k_l = 1$, k_u is 9 for Digit1 and 3 for USPS.

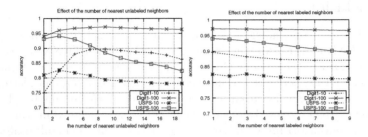

Fig. 5. (left) The effect of k_u of unlabeled data to performance of TKNN. Here $k_l = 1$ and b is 0.25 for Digit1 and 0.35 for USPS. **(right)** The effect of k_l of labeled data to performance of TKNN. Here $k_u = 9$, $b = 0.25$ for Digit1 and $k_u = 3$, $b = 0.35$ for USPS.

transductive learning improves more from supervised learning that is 1NN and RBF here. While both TKNN and the consistency method achieve 1.0 accuracy on the 2-rings data set, TKNN has the best performance on other data sets.

Fig. 4 (right) for b, Fig. 5(left) for k_u and Fig. 5(right) for k_l show effect of parameters. The performance is best when the bandwidth ratio $0.1 < b < 0.3$, the number of nearest unlabeled neighbors $3 \leq k_u \leq 10$ and the number of nearest labeled neighbors $k_l = 1$. The best value of k_u depends on data. The best value $k_l = 1$ agrees to human intuition that k_l/k_u should be proportional to ratio of number of l/u, the ratio of sizes of labeled data and unlabeled data.

5   Related Work

Our work is related to semi-supervised learning, especially graph-based learning. [4] introduced a method that take advantage of unlabeled data for classification tasks. After that, semi-supervised learning or trandsductive learning have been extensively researched. And graphs are introduced in semi-supervised learning [14]. Work in [10][13][15] is most related to our work. [10] proposed a semi-supervised learning based on a Gaussian random field model. Labeled and unlabeled data are represented in a weighted graph. The mean of the field is efficiently obtained using matrix methods or belief propagation. [13] proposed the consistency method that is a smooth solution which captures the intrinsic structure revealed by labeled and unlabeled points. Different from [10] and [13], we present an efficient iterative algorithm for large data sets and for circumstances where matrix computation fails because of the nonexistent inverse matrix. [15] introduced a two-stage approach. In the first stage, a model is built based on training data. In the second stage, the model is used to transform unlabeled data into weighted pre-labeled data set which is used in the classifier. Our work is not two-stage and influence between labeled and unlabeled data are considered at the same time. [9] proposed a graph-based approach, mixed label propagation, that explores similarity and dissimilarity simultaneously.

6 Conclusion

In KNN algorithm, only labeled data can be the nearest neighbors. In this paper, we extend KNN to a transductive learning algorithm called TKNN. We consider two groups of nearest neighbors. One from labeled data and the other from unlabeled data. We then derive the recurrence relation in a data set, considering that each point receives influence from neighboring points, labeled or unlabeled. A matrix solution is then proposed and a more efficient iterative algorithm is presented. We compare our algorithm with baseline algorithms. Experimental results show our algorithm are effective and efficient. Future work will be focused on more precise estimation of parameters and application on data with more than two classes.

Acknowledgments. This work is supported in part by the following NSF grants: IIS-0414981 and CNS-0454298.

References

1. Webb, A.R.: Statistical Pattern Recognition. John Wiley and Sons Ltd., Chichester (2002)
2. Bennett, K.P., Demiriz, A.: Semi-supervised support vector machines. In: NIPS (1998)
3. Kulis, B., Basu, S., Dhillon, I., Mooney, R.: Semi-supervised graph clustering: A kernel approach. In: ICML (2005)
4. Nigam, K., McCallum, A., Thrun, S., Mitchell, T.: Learning to classify text from labeled and unlabeled documents. In: AAAI (1998)
5. Nigam, K., McCallum, A.K., Thrun, S., Mitchell, T.: Text classification from labeled and unlabeled documents using em. Machine Learning 1(34) (1999)
6. Seeger, M.: Learning with labeled and unlabeled data. Inst. for Adaptive and Neural Computation, technical report (2001)
7. Sindhwani, V., Niyogi, P., Belkin, M.: Beyond the point cloud: from transductive to semi-supervised learning. In: ICML (2005)
8. Tang, W., Xiong, H., Zhong, S., Wu, J.: Enhancing semi-supervised clustering: A feature projection perspective. In: KDD (2007)
9. Tong, W., Jin, R.: Semi-supervised learning by mixed label propagation. In: AAAI (2007)
10. Zhu, X., Ghahramani, Z., Lafferty, J.: Semi-supervised learning using gaussian fields and harmonic functions. In: ICML (2003)
11. Chapelle, O., Schölkopf, B., Zien, A. (eds.): Semi-Supervised Learning. MIT Press, Cambridge (2006)
12. Tran, T.N., Wehrensa, R., Buydens, L.M.: Knn-kernel density-based clustering for high-dimensional multivariate data. Computational Statistics & Data Analysis 51(2), 513–525 (2006)
13. Zhou, D., Bousquet, O., Lal, T., Weston, J., Schölkopf, B.: Learning with local and global consistency. In: NIPS (2003)
14. Blum, A., Chawla, S.: Learning from labeled and unlabeled data using graph mincuts. In: ICML (2001)
15. Driessens, K., Reutemann, P., Pfahringer, B., Leschi, C.: Using weighted nearest neighbor to benefit from unlabeled data. In: Ng, W.-K., Kitsuregawa, M., Li, J., Chang, K. (eds.) PAKDD 2006. LNCS, vol. 3918, pp. 60–69. Springer, Heidelberg (2006)

An Improved Algorithm for Mining
Non-Redundant Interacting Feature Subsets

Chaofeng Sha[1], Jian Gong[1], and Aoying Zhou[2]

[1] School of Computer Science, Fudan University, China
{cfsha, jiangong}@fudan.edu.cn
[2] Shanghai Key Laboratory of Trustworthy Computing, ECNU, China
ayzhou@sei.ecnu.edu.cn

Abstract. The application of feature subsets with high order correlation in classification has demonstrates its power in a recent study, where non-redundant interacting feature subsets (NIFS) is defined based on multi-information. In this paper, we re-examine the problem of finding NIFSs. We further improve the upper bounds and lower bounds on the correlations, which can be used to significantly prune the search space. The experiments on real datasets demonstrate the efficiency and effectiveness of our approach.

1 Introduction

Pattern mining has been a focused theme in data mining research area with a large number of scalable methods proposed for mining various kinds of patterns, including frequent patterns [1], correlation patterns [3][13], and diverse patterns [11]. Those patterns have found broad applications in areas like association rule mining and classification [15].

In this paper, we re-examine the problem finding non-redundant interacting feature subsets from binary datasets, which has been proposed recently in [15]. The NIFSs are feature subset with higher order correlation, which is defined based on multi-information. Finding these high order correlations has important applications such as quantitative trait locus finding in genetics [15]. While a NIFS mining algorithm proposed in [15] used some bounds on multi-information to prune the search space, we found that there are some space to improve those bounds by careful re-examination of the multi-information.

The contribution of our paper can be summarized as follows: (1) By carefully examine the properties of multi-information, we prove serval new bounds on it. (2) We use these bounds to improve the pruning power of the mining algorithm. (3) We conduct experiments on two real datasets, which demonstrate the effectiveness of our new bounds.

The rest of this paper is structured as follows. Section 2 discusses related works. Section 3 presents formal definition, notation and problem description. Then we present some bounds of joint entropy that will be used to pruning the pattern searching space in Section 4. Section 6 describes our mining algorithm and Section 7 presents our experimental evaluation on real datasets. We conclude our work in Section 8.

Q. Li et al. (Eds.): APWeb/WAIM 2009, LNCS 5446, pp. 357–368, 2009.

2   Related Work

Mining frequent patterns is a matured problem in binary dataset processing. Since association rule mining introduced in [1], and then the Apriori algorithm proposed in [2], there are much work on frequent pattern mining, interest measure of association rule, constrained rule mining, and so on. [3] introduce the correlation pattern mining pattern, in which the authors used the χ^2 correlation measure, and this work is for the feature pairs. While [10] proposed all-confidence to measure the correlation of any itemset. The h-confidence proposed in [13] has the same form with all-confidence, but [13] studied more properties of their measure. Using information theory approach, [8] extended the correlation pattern mining problem to numeric datasets. The authors used normalized mutual information and all-confidence to measure the correlation of itemsets. But this work also focused on the correlation of pair of items.

[9] proposes joint entropy as a quality measure for itemsets, and serval efficient algorithms to mine those maximally informative k-itemsets. The main goal of their work is to select distinct items, as well as minimize redundancy within the resulted itemsets. [7] proposed to find those low-entropy sets, and introduced two low entropy trees. They discussed properties of their trees and proposed some mining algorithms. In addition, they also introduced high-scaled-entropy sets and high-normalized-entropy sets, but they have not proposed any algorithm to mining those patterns.

In [11], the authors defined subset coverage diversity, discussed the properties of this diversity measure, and proposed an algorithm to find minimum sample set with coverage diversity larger than a threshold ρ. The significance of their work is some pruning strategies to improve the efficiency of their enumeration algorithm.

In [12], the authors used mutual information to measure the similarity between two rows (elements). They proposed representative elements selection problem. There they treat binary dataset as joint probability distribution between rows and columns, and choose representative elements based on the proposed optimization criteria.

In [15], the authors explored the problem of finding non-redundant high order correlations in binary data. Both the algorithm in [15] and ours adopt depth-first search and use serval strategies to prune the pattern search space, which are based on some bounds on entropy.

3   Preliminaries and Problem Statement

In this section we introduce some notations we will use. A binary dataset D is a $N \times m$ binary matrix, where each column is a feature. We denote $F = \{X_1, \cdots, X_m\}$ as the whole feature subset. In our discussion, we treat each feature as a binary random variable, i.e., the domain of each feature $X_i (1 \leq i \leq m)$ is $D_{X_i} = \{0, 1\}$. For convenience, we use $X_1, X_2, \cdots$ to denote those random variables corresponding to features.

Definition 1. *The entropy of a random variable X, denoted as $H(X)$, is defined as*

$$H(X) = - \sum_{x \in D_X} p(x) \log p(x)$$

All logarithms are in base 2, and $0 \log 0 = 0$ by convention. It is known that $0 \le H(X) \le \log |X|$, with $H(X) = \log |X|$ only for the uniform distribution $P(X = x) = 1/|X|$ for all $x \in X$.

Definition 2. *The joint entropy of two random variables X and Y, denoted as $H(X, Y)$, is defined as*

$$H(X, Y) = - \sum_{x \in D_X} \sum_{y \in D_Y} p(x, y) \log p(x, y)$$

Definition 3. *The conditional entropy of a random variable Y given another variable X denoted as $H(Y|X)$, is defined as*

$$H(Y|X) = - \sum_{x \in D_X} \sum_{y \in D_Y} p(x, y) \log p(y|x)$$

Before defining the mining problem we are going to solve, we outline a property of entropy. In this paper, we treat a feature as a binary random variable, and an feature subset as a set of random variables. For more content of information theory, the reader is referred to [5,14].

Lemma 1. *For any two features X and Y, we have:*

$$H(X|Y) \le H(X)$$

Definition 4. *The multi-information of a set of features $\{X_1, \cdots, X_n\}$ is defined as*

$$C(X_1, \cdots, X_n) = \sum_{i=1}^{n} H(X_i) - H(X_1, \cdots, X_n)$$

We know that $H(X_1, \cdots, X_n) \le \sum_{i=1}^{n} H(X_i)$, with equality if and only if the random variables X_i are independent. Therefore the mutual information is non-negative. Larger multi-information corresponds to higher correlation, and small multi-information corresponds to higher diversity (independence). And we know that when $n = 2$, the multi-information of two features is equal to their mutual information.

Definition 5. *A set of features $\{X_1, X_2, \cdots, X_n\}$ is a strongly correlated if $C(X_1, X_2, \cdots, X_n) \ge \beta$, where $\beta > 0$ is a user defined threshold. In this case, $\{X_1, X_2, \cdots, X_n\}$ is called a Strong-correlated Feature Subset (SFS).*

Definition 6. *A set of features $\{X_1, X_2, \cdots, X_n\}$ is a weakly correlated if $C(X_1, X_2, \cdots, X_n) \le \alpha$, where $\alpha > 0$ is a user defined threshold. In this case, $\{X_1, X_2, \cdots, X_n\}$ is called a Weak-correlated Feature Subset (WFS).*

Definition 7. *A set of features* $\{X_1, X_2, \cdots, X_n\}$ *is a Non-redundant Interacting Feature Subset (NIFS) if the following two criteria are satisfied:*

1. $\{X_1, X_2, \cdots, X_n\}$ *is an SFS; and*
2. every proper subset $X' \subset \{X_1, X_2, \cdots, X_n\}$ *is a WFS.*

Problem Description. Given a binary dataset D, two thresholds $\alpha, \beta > 0$, the mining problem we are going to solve in this paper is to find all non-redundant interacting feature subsets from D.

4   New Bounds Based on Pairwise Correlations

Before presenting our new bounds, we introduce a conclusion from [6].

Lemma 2. *Let* $[n] = \{1, 2, \cdots, n\}$. *For any subset* A *of* $[n]$ *we denote* $H(X_A)$ *as the joint entropy* $H(X_i, i \in A)$. *For any* $1 \leq k \leq n$, *let*

$$H_k = \frac{1}{\binom{n-1}{k-1}} \sum_{A:|A|=k} H(X_A)$$

Then we have the following inequality:

$$H_1 \geq H_2 \geq \cdots \geq H_n.$$

And We can write the inequality $H_n \leq H_2$ explicitly as follows:

$$H(X_1, X_2, \cdots, X_n) \leq \frac{1}{n-1} \sum_{i<j} H(X_i, X_j) \tag{1}$$

Now we present a similar theorem as the property 3.1 from [11], there the conclusion is for subset coverage.

Theorem 1. *Given* $n \geq 3$ *random variables* $X_1, X_2, \cdots, X_n$, *a upper bound of* $H(X_1, \cdots, X_n)$ *is*

$$H(X_1, \cdots, X_n) \leq \frac{H(X_1, \cdots, X_{n-1}) + H(X_1, \cdots, X_{n-2}, X_n) + H(X_{n-1}, X_n)}{2} \tag{2}$$

Proof. By the chain rule for entropy, we can expand $H(X_1, X_2, \cdots, X_n)$ in two different ways.

$$H(X_1, \cdots, X_n) = H(X_1, \cdots, X_{n-1}) + H(X_n | X_1, \cdots, X_{n-1})$$

$$H(X_1, \cdots, X_n) = H(X_1, \cdots, X_{n-2}, X_n) + H(X_{n-1} | X_1, \cdots, X_{n-2}, X_n)$$

Summarization the left hand and right hand of the above two equations respectively, we have:

$$2H(X_1, \cdots, X_n) = H(X_1, \cdots, X_{n-1}) + H(X_n | X_1, \cdots, X_{n-1})$$
$$+ H(X_1, \cdots, X_{n-2}, X_n) + H(X_{n-1} | X_1, \cdots, X_{n-2}, X_n)$$

where,

$$H(X_n|X_1,\cdots,X_{n-1}) \leq H(X_n|X_{n-1})$$
$$H(X_{n-1}|X_1,\cdots,X_{n-2},X_n) \leq H(X_{n-1}|X_n)$$

Therefore,

$$H(X_n|X_1,\cdots,X_{n-1}) + H(X_{n-1}|X_1,\cdots,X_{n-2},X_n) \leq H(X_n|X_{n-1}) + H(X_{n-1}|X_n)$$
$$\leq H(X_n|X_{n-1}) + H(X_{n-1})$$
$$= H(X_{n-1},X_n)$$

According to the above inequalities, we have,

$$2H(X_1,\cdots,X_n) \leq H(X_1,\cdots,X_{n-1}) + H(X_1,\cdots,X_{n-2},X_n) + H(X_{n-1},X_n)$$

Which leads to the inequality (2).

Generally, using inequality (2) we can get the following lower bound on $C(X_1, \cdots, X_n)$,

$$C(X_1,\cdots,X_{n-1}) + C(X_1,\cdots,X_{n-2},X_n) + C(X_{n-1},X_n)$$
$$= \left(\sum_{i=1}^{n-1}H(X_i) - H(X_1,\cdots,X_{n-1})\right) + (H(X_{n-1}) + H(X_n) - H(X_{n-1},X_n))$$
$$+ \left(\sum_{i=1}^{n-2}H(X_i) + H(X_n) - H(X_1,\cdots,X_{n-2},X_n)\right)$$
$$= 2\sum_{i=1}^{n}H(X_i) - 2(H(X_1,\cdots,X_{n-1}) + H(X_1,\cdots,X_{n-2},X_n) + H(X_{n-1},X_n))$$
$$\leq 2\sum_{i=1}^{n}H(X_i) - 2H(X_1,\cdots,X_n) = 2C(X_1,\cdots,X_n)$$

Which leads to the following inequality:

$$C(X_1,\cdots,X_n) \geq \frac{C(X_1,\cdots,X_{n-1})+C(X_1,X_1,\cdots,X_{n-2},X_n)+C(X_{n-1},X_n)}{2}$$

where $\sum_{i<j}h_{i,j} = n/2$.

By recursively using the above inequality, we can establish a lower bound of the mutual information of an feature subset using only pairs of feature,

$$C(X_1,\cdots,X_n) \geq \sum_{i<j}h_{i,j} \cdot C(X_i,X_j) \qquad (3)$$

While [15] establishes the following lower bound, using the inequality (1):

$$C(X_1,X_2,\cdots,X_n) \geq \frac{1}{n-1}\sum_{i<j}C(X_i,X_j) \qquad (4)$$

As can be seen that the summarization of all coefficients of multi-information $C(X_i, X_j)$ in the right hands of both inequalities are $n/2$. Therefore we can show that the lower bound in inequality (4) is less than the the one in the inequality (3) by Lagrange multipliers method, which means that ours is more tighter.

On the other hand, due to the fact that for any pair of $i \neq j$ we have that $H(X_1, \cdots, X_n) \geq H(X_i, X_j)$, we can get an upper bounds on multi-information of feature subset $\{X_1, \cdots, X_n\}$ as follows [15]:

$$C(X_1, \cdots, X_n) \leq \sum_{i=1}^{n} H(X_i) - \max_{i \neq j} H(X_i, X_j) \qquad (5)$$

5   New Bounds Based on Mutual Information

5.1   Adding an Feature

In this section, we first present two propositions from [15], then propose an improvement on the bound given there.

Proposition 1. *Let X_1 and X_2 be two features in the dataset. If the Hamming distance between X_1 and X_2 is d, then*

$$0 \leq H(X_1, X_2) - H(X_1) \leq d \cdot H\left(\frac{1}{N}, \frac{N-1}{N}\right).$$

Proposition 2. *Let*

$$\Delta H = H(X_1, \cdots, X_n) - H(X_1, \cdots, X_{n-1})$$

If the minimum Hamming distance between X_n and $X_i(1 \leq i \leq n)$ is d, then

$$0 \leq \Delta H \leq d \cdot H\left(\frac{1}{N}, \frac{N-1}{N}\right).$$

Below we present some new bounds. Using the chain rule for entropy, we have:

$$H(X_1, \cdots, X_n) = H(X_1, \cdots, X_{n-1}) + H(X_n | X_1, \cdots, X_{n-1})$$

Using the fact that for $1 \leq i \leq n - 1$, $H(X_n | X_1, \cdots, X_{n-1}) \leq H(X_n | X_i)$, we get the following inequality:

$$H(X_1, \cdots, X_n) \leq H(X_1, \cdots, X_{n-1}) + \min_{1 \leq i \leq n-1} H(X_n | X_i) \qquad (6)$$

Therefore we know that the bound given in the inequality (6) is tighter than the one in the proposition 2, due to the fact that $H(X_2 | X_1) = H(X_1, X_2) - H(X_1)$. Based on this tighter upper bound, we have the following proposition.

Proposition 3. *(Adding Proposition) Let*

$$\Delta C = C(X_1, \cdots, X_n) - C(X_1, \cdots, X_{n-1}).$$

We have that,

$$\max_{1 \le i \le n-1} I(X_i; X_n) \le \Delta C \le H(X_n) \qquad (7)$$

Proof. We can rewrite ΔC as follows,

$$\begin{aligned}
\Delta C &= C(X_1, \cdots, X_n) - C(X_1, \cdots, X_{n-1}) \\
&= H(X_n) - (H(X_1, \cdots, X_n) - H(X_1, \cdots, X_{n-1}))
\end{aligned}$$

Due to the inequality (6), we have,

$$\begin{aligned}
\Delta C &\ge H(X_n) - \min_{1 \le i \le n-1} H(X_n | X_i) \\
&= \max_{1 \le i \le n-1} I(X_i; X_n)
\end{aligned}$$

The right hand of inequality (7) is trivial, due to the fact that $H(X_1, \cdots, X_n) - H(X_1, \cdots, X_{n-1}) \ge 0$.

5.2   Replacing a Feature

Using the chain rule for entropy again, we can expand $H(X_1, X_2, \cdots, X_n)$ in two different ways:

$$\begin{aligned}
H(X_1, X_2, \cdots, X_n) &= H(X_1, \cdots, X_{n-1}) + H(X_n | X_1, \cdots, X_{n-1}) \\
&= H(X_1, \cdots, X_{n-2}, X_n) + H(X_{n-1} | X_1, \cdots, X_{n-2}, X_n)
\end{aligned}$$

Using the fact that $H(X_{n-1} | X_1, \cdots, X_{n-2}, X_n) \ge 0$ and $H(X_n | X_1, \cdots, X_{n-1}) \le H(X_n | X_i)$ for any $1 \le i \le n-1$, we can get the following inequality:

$$H(X_1, \cdots, X_{n-2}, X_n) \le H(X_1, \cdots, X_{n-1}) + H(X_n | X_1, \cdots, X_{n-1}) \qquad (8)$$
$$\le H(X_1, \cdots, X_{n-1}) + \min_{1 \le i \le n-1} H(X_n | X_i) \qquad (9)$$

According to $H(X_n | X_1, \cdots, X_{n-1}) \ge 0$, we have,

$$H(X_1, \cdots, X_{n-2}, X_n) \ge H(X_1, \cdots, X_{n-1}) - H(X_{n-1} | X_1, \cdots, X_{n-2}, X_n)$$

Therefore using the fact that $H(X_{n-1} | X_1, \cdots, X_{n-2}, X_n) \le H(X_{n-1} | X_n)$ and $H(X_{n-1} | X_1, \cdots, X_{n-2}, X_n) \le H(X_{n-1} | X_i)$ for any $1 \le i \le n-2$, we have the following inequality:

$$H(X_1, \cdots, X_{n-2}, X_n) \ge H(X_1, \cdots, X_{n-1}) - \min \left(\min_{1 \le i \le n-1} H(X_{n-1} | X_i), H(X_{n-1} | X_n) \right)$$

$$(10)$$

Proposition 4. *(Replacing Proposition) Let*

$$\Delta C = C(X_1, \cdots, X_{n-2}, X_n) - C(X_1, \cdots, X_{n-2}, X_{n-1})$$

We have,

$$\max_{1 \le i \le n-1} I(X_i; X_n) - H(X_{n-1}) \le \Delta C \le H(X_n) - \max\left(\max_{1 \le i \le n-1} I(X_i; X_{n-1}), I(X_{n-1}; X_n)\right).$$

Proof. We can rewrite ΔC as follows:

$$\Delta C = H(X_n) - H(X_{n-1}) - (H(X_1, \cdots, X_{n-2}, X_n) - H(X_1, \cdots, X_{n-2}, X_{n-1}))$$

Using the inequality (9), we have that,

$$\Delta C \ge H(X_n) - H(X_{n-1}) - \min_{1 \le i \le n-1} H(X_n | X_i)$$
$$= \max_{1 \le i \le n-1} I(X_i; X_n) - H(X_{n-1})$$

On the other hand, using the inequality (10), we have that,

$$\Delta C \le H(X_n) - H(X_{n-1}) + \min\left(\min_{1 \le i \le n-1} H(X_{n-1} | X_i), H(X_{n-1} | X_n)\right)$$
$$= H(X_n) - \max\left(\max_{1 \le i \le n-1} I(X_i; X_{n-1}), I(X_{n-1}; X_n)\right)$$

6   The Mining Algorithm

In this section, we present our NIFS mining algorithm, which follows the framework in [15]. The novelty of mining algorithm are: (1) In the initialization step in NIFS miner (Figure 1), we do not need to calculate the Hamming distance between any pair of features. (2) In the DFS-Explore() sub-procedure we use the new bounds on joint entropy of feature subsets proposed in previous section to prune the pattern search space.

NIFS Mining Algorithm
Input: binary dataset D, thresholds α and β.
Output: all NIFSs in D.
  1.   calculate the entropy $H(X_i)$ of each feature $X_i \in D$, the joint entropy and mutual information of each pair of features in D.
  2.   **for** each node V at the first level of the search space **do**
  3.   DFS-Explore(X);

Fig. 1. NIFS mining algorithm

Procedure DFS-Explore(U)

```
1.      update ub(X) by inequality (1), lb(X) by inequality (5)
2.      update ub(X) and lb(X) by adding or replacing feature
3.      if lb(X) > α then
4.          if lb(X) ≥ β then
5.              if criterion of NIFS definition are satisfied then
6.                  output X;
7.          else
8.              if ub(X) ≥ β then
9.                  calculate H(X);
10.                 if C(X) ≥ β then
11.                     if criterion of NIFS definition are satisfied then
12.                         output X;
13.             DFS-Explore(the sibling of X);
14.     else
15.         if ub(X) ≤ α then
16.             DFS-Explore(the first child of X);
17.         else
18.             calculate C(X);
19.             if C(X) ≥ β then
20.                 if criterion of NIFS definition are satisfied then
21.                     output X;
22.                 DFS-Explore(the sibling of X);
23.             else
24.                 if C(X) ≤ α then
25.                     DFS-Explore(the first child of X);
26.                 else
27.                     DFS-Explore(the sibling of X)
```

Fig. 2. DFS-Explore

Before calculating the multi-information of a node in the search space, we first check its upper and lower bounds. The pairwise lower bound (inequality (3)) can be applied whenever the algorithm examines a new node. The adding proposition (Proposition 7) can be applied to the child nodes of a candidate feature subset and the two inequalities on replacing feature (Section 5.2) can be applied to the siblings.

The algorithm is performed in depth-first recursion [4]. Whenever NIFS miner finishes examining the current node X and its subtree, it proceeds to one of X's siblings, denoted by X'. The replacing inequalities in Section 5.2 can be applied to get upper and lower bounds on the multi-information of X'. Figure 1 and 2 show our NIFS miner. For more detail, the reader is referred to [15].

In the worse case, where the whole feature combinations should be checked, the running time of the above mining algorithm is exponential in the number of features. While the experimental results in Section 7 show that the running time of NIFS miner is quadratic in the number of features, which demonstrate the effectiveness of the pruning methods discussed in the previous section.

7   Performance Evaluation

In this section, we present experiments to evaluate the performance of our improved NIFS mining algorithm. While our work mainly focuses on improving the efficiency of the mining algorithm, we can also use the found feature subsets to improve the classification accuracy as in [15]. All experiments were run on a Windows XP machine with Intel Pentium4 2.4GHz CPU and 2GB RAM. Our algorithm is implemented using Java.

We report our experimental results on the performance of our NIFS miner in comparison with the one in [15]. And our experiments were performed on voting dataset from UCI machine learning repository. The voting data set contains 435 rows which correspond to 435 congressman, and 16 columns which correspond to 16 key votes. The votes can be 'yes' or 'no' and are denoted by 1 and 0.

In the first group of experiments, we test the scalability on the voting data set. We set $\alpha = 3$, $\beta = 3.1$, and number of column is 16. Figure 3(1) shows the shows the execution time of NIFS miner on the voting data set. As can be seen, the execution time of two algorithms increases with the increase of the number of rows. When we fix the number of row as 435, and still set $\alpha = 3$, $\beta = 3.1$. We get the time performance in varying number of column as Figure 3(2). As can be seen, the execution of two algorithm quadraticly increases with respect to number of column. But as anticipated, our approach has the better time performance than the one in [15].

In the following two experiments, we use the whole voting data set. Figure 4 shows the results by varying α while fix $\beta = 3.3$. As can be seen, the execution time increases significantly with the increase of α. While we fix $\alpha = 3.0$, we have the result as shown in Figure 4 where varies the threshold β. As can be can, in contrast, the execution time decreases significantly with the increase of β. Also as anticipated, our approach has the better time performance than the one in [15].

Figure 5 examines the effectiveness of our pruning strategies. The effects of pruning strategies based on bounds discussed in Section 4 are shown in

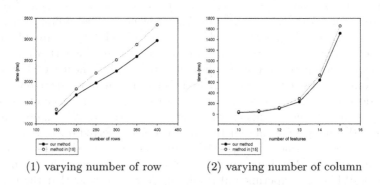

(1) varying number of row   (2) varying number of column

Fig. 3. Voting data: varying number of row and column, respectively

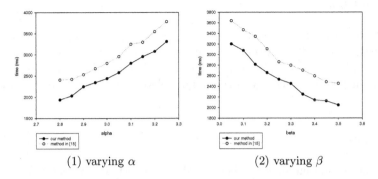

(1) varying α (2) varying β

Fig. 4. Voting data: varying α and β respectively

Figure 5(1). We see that the advantage is not significant just using the bounds based on pairwise correlation. But Figure 5(2) shows that the advantage of our approach increases with increases of the number of rows, using the bounds based on mutual information in Section 5. The bottom of Figure 5 shows the pruning effect of various bounds we proposed in previous sections.

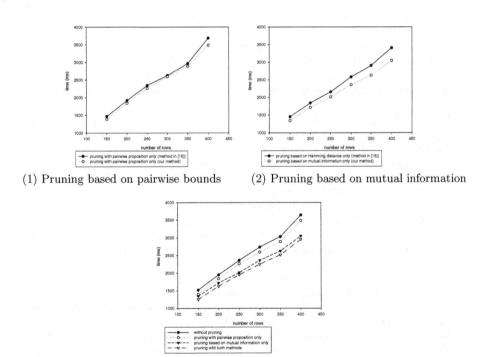

(1) Pruning based on pairwise bounds   (2) Pruning based on mutual information

Fig. 5. Voting data: Pruning based on various bounds

8 Conclusion

We re-examined the problem finding non-redundant interacting feature subsets from binary data. After presenting some improved bounds on the multi-information, we propose an improved algorithm to mine NIFSs. The experimental results justify the efficiency of the mining algorithm and effectiveness of new bounds.

References

1. Agrawal, R., Imielinski, T., Swami, A.: Mining Association Rules between sets of items in large databases. In: Buneman, P. (ed.) Proceedings of the 1993 ACM SIGMOD International Conference on Management of Data, Washington, DC, pp. 207–216. ACM Press, New York (1993)
2. Agrawal, R., Srikant, R.: Fast Algorithms for Mining Association Rules in Large Databases. In: VLDB 1994, pp. 487–499 (1994)
3. Brin, S., Motwani, R., Silverstein, C.: Beyond market baskets: generalizing association rules to correlations. In: Proceedings of 1997 ACM-SIGMOD International Conference on Management of Data (SIGMOD 1997) (1997)
4. Cormen, T.H., Leiserson, C.E., Rivest, R.L., Stein, C.: Introduction to Algorithms, 2nd edn. MIT Press, Cambridge (2001)
5. Cover, T., Thomas, J.: Elements of Information Theory. Wiley Series in Telecommunications. Wiley Interscience, Hoboken (1991)
6. Han, T.S.: Nonnegative entropy measures of multivariate symmetric correlations. Inform. Contr. 36, 133–156 (1978)
7. Heikinheimo, H., Hinkkanen, E., Mannila, H., Mielikainen, T., Seppanen, J.: Finding low-Entropy sets and trees from binary data. In: Proceedings of the 13th ACM SIGKDD International Conference on Knowledge Discovery and Data Mining (2007)
8. Ke, Y., Cheng, J., Ng, W.: Mining Quantitative Correlated Patterns Using an Information-Theoretic Approach. In: Eliassi-Rad, T. (ed.) Proceedings of the Twelfth ACM SIGKDD International Conference on Knowledge Discovery and Data Mining, pp. 227–236. ACM, Philadelphia (2006)
9. Knobbe, A., Ho, E.: Maximally informative k-itemsets and their efficient discovery. In: KDD 2006, pp. 237–244 (2006)
10. Omiecinski, E.R.: Alternative Interest Measures for Mining Associations in Databases. IEEE Transactions on Data Engineering 15(1), 57–69 (2003)
11. Pan, F., Roberts, A., McMillan, L., de Villena, F., Threadgill, D., Wang, W.: Sample selection for maximal diversity. In: Proceedings of the 5th IEEE International Conference on Data Mining (2007)
12. Pan, F., Wang, W., Tung, A.K.H., Yang, J.: Finding representative set from massive data. In: ICDM 2005, pp. 338–345 (2005)
13. Xiong, H., Tan, P., Kumar, V.: Hyperclique Pattern Discovery. Data Mining and Knowledge Discovery Journal 13(2), 219–242 (2006)
14. Yeung, R.W.: A first course in information theory. Springer, Heidelberg (2002)
15. Zhang, X., Pan, F., Wang, W., Nobel, A.: Mining non-redundant high order correlation in binary data. In: Proceedings of the 34th International Conference on Very Large Data Bases, Vienna, Austria, Auckland, New Zealand (2008)

AutoPCS: A Phrase-Based Text Categorization System for Similar Texts

Zhixu Li[1,2], Pei Li[1,2], Wei Wei[1,2], Hongyan Liu[3], Jun He[1,2],
Tao Liu[1,2], and Xiaoyong Du[1,2]

[1] Key Labs of Data Engineering and Knowledge Engineering, Ministry of Education, China
[2] School of Information, Renmin University of China, Beijing, China
`{lizx,lp,zauri,hejun,tliu,duyong}@ruc.edu.cn`
[3] Department of Management Science and Engineering, Tsinghua University, Beijing, China
`hyliu@tsinghua.edu.cn`

Abstract. Nearly all text classification methods classify texts into predefined categories according to the terms appeared in texts. State-of-the-art of text classification prefer to simply take a word as a term since it performs good on some famous datasets; some experts even pointed out that phrases don't improve or improve only marginally the classifiction accuracy. However, we found out that this is not always true when we try to categorize texts about similar topics in the same domain. With words only we can not categorize those texts effectively since they nearly share the same word set. Then we suppose the results might be improved if we also use phrases as terms. To testify our supposition, we propose our own phrase extraction way as well as select proper feature selection method and classifier by conducting experimental study on a data set which comes from paper abstracts in the field of *Databases*. Accordingly, we also develop a system called AutoPCS which can be used to help experts in choosing relevant topics for newly coming papers from a predefined topic list only by their abstracts.

Keywords: Text Classification, Phrase-based, BOP, Similar Texts.

1 Motivation

When there is a research paper submitted to a conference, some submission systems would ask its contributor to choose several relevant topics for the paper from a given predefined topic list so that the conference organization can arrange several corresponding reviewers to judge the paper. Those given predefined topics may cover most of the research areas in a certain domain. A snapshot of a manual relevant topic selection system can be seen in Fig. 1.

This topic selection process can be regarded as a manual text classification way. It might be simple and accurate if the researcher is familiar with most of the topics in the domain. But it becomes much unreliable and time-consuming if the researcher is a freshman of the research field who do not know most of those topics. In addition, there are also some conferences which would not ask contributors but some experts in

Q. Li et al. (Eds.): APWeb/WAIM 2009, LNCS 5446, pp. 369–380, 2009.

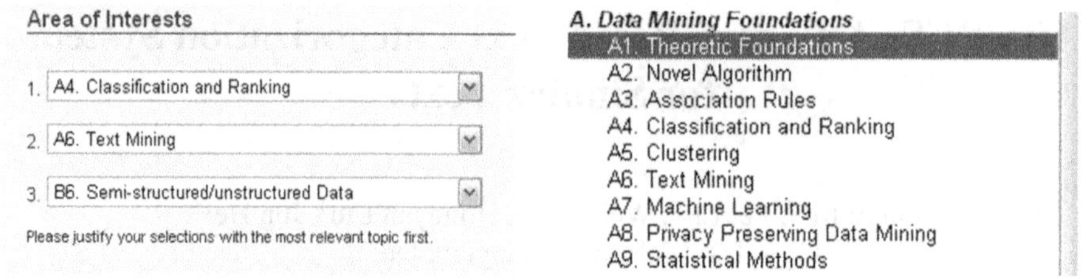

Area of Interests	A. *Data Mining Foundations*

(a) Relevant topics selection   (b) Predefined topic list

Fig. 1. Snapshot of the manual relevant topic selection in submission system of PAKDD2009

this domain to choose relevant topics for the paper. This might spend lots of time and they may disagree in whether or not to classify a document under a certain category. Besides, manual classification is incautious, which may also bring false classification. In order to solve these problems, we would like to provide some automatic relevant topics recommendation or checkout function to the manual categorization submission system by developing an **Auto**matic **P**ublication **C**ategorization **S**ystem, **AutoPCS**.

As a fundamental task in Information Retrieval and Data Mining, text categorization (or text classification) has been studied extensively in the past several decades. The classical approach to text categorization has so far been using a document representation in a word-based space, however, the main drawback of which is it destructs the semantic relations between words by using words in a phrase separately [7]. A classical example which has been proposed in [7] is *"White House"* or *"Bill Gates"*. Given a BOW (**B**ag-**O**f-**W**ords) of a document in which words *"bill"* and *"gates"* occur, one can suggests that the document is about accounting or gardening, but not about computer software. Whereas given a document representation that contains a phrase *"bill gates"*, the reader will hardly be mistaken about the topic of discussion.

These fairly obvious observations led researchers to an idea of enriching the BOW representation by phrases. **B**ag-**O**f-**B**igrams (pairs of consequent words) was proposed firstly in early 90s [8]. However, it always showed only marginal improvement or even a certain decrease. As far as we concerned, the word-based BOW is efficient enough for text classification because we are always classifying texts from different fields like *"Corporate/Industrial"*, *"Markets"* and *"Government Finance"* in *Reuters* dataset. Since each category has some solely-used domain words, adding phrases into the bag cannot improve much. But the situation becomes quite different when it comes to classifying **similar texts**. Similar texts means texts of similar topics which almost share the same word set, e.g. texts from two similar topics *"Data mining"* and *"Text Mining"*. The three words *"Data"*, *"Mining"* and *"Text"* are very common in text about either topic.

Since similar texts nearly share the same word set, it is difficult to classify them only by words (or BOW). But different topics have respective terms which are usually phrases; therefore, a phrase-based representation (We can call it as **B**ag-**O**f-**P**hrases, or simply BOP) is expected to be more efficient. In this paper we would like to overview some related works in recent years on the problem of using phrases for text classification. Then we try to use BOP in classifying similar texts in our dataset.

The rest of the paper is organized as follows: in Section 2 we describe the problem of text categorization; in Section 3 we briefly reviews related works on text classification in history, especially the most recent works on the problem of using phrases for text classification; in Section 4 we propose our BOP method of incorporating words and phrases in document representations; Section 5 presents our experimental study on classifying similar texts; Section 6 gives a introduction to the automatic new paper categorization system (AutoPCS) we developed for paper submission system. Finally we make a conclusion in Section 7.

2 Problem Statement

In order to better state the problem in text classification, we would like to give a simple formulation to text classification as follows. Assume we are given a training set:

$$T = \left\{ (d_i, c_j) \middle| d_i \in D, c_j \in C \right\} \tag{1}$$

In this formula (1), each document d_i belongs to a document set D and the label c_j is within a predefined set of categories $C = \{c_1, c_2, \dots, c_m\}$. Each (d_i, c_j) represents that d_i is labeled as c_j. The goal of text categorization is to devise a learning algorithm which can generate a classifier or a hypothesis $h: D \rightarrow C$ that can label unlabeled each document in D accurately, with the help of the training set T.

Designing a learning algorithm for text classification usually follows the classical approach in pattern recognition, where data instances (i.e. documents) first undergo a transformation of dimensionality reduction, and then a classifier learning algorithm is applied to the low-dimensionality representations. This transformation is also performed prior to applying the learned classifier to unseen instances.

The incentives in using dimensionality reduction techniques are to improve classification quality and to reduce the computational complexity of the learning algorithm and of the application of the classifier to unseen documents. Typically, dimensionality reduction techniques fall into two basic schemes: feature selection and feature generation. Feature selection can also be called as feature reduction, which tries to select the subset of features (words in text classification) that are most useful for text classification. In contrast, feature generation which can also be called as feature induction, tries to generate new features which are not necessarily words for representation.

After feature selection or feature generation, the next step is to choose a proper classifier. There have been lots of excellent algorithms proposed by researchers in this field such as Naïve Bayes, Bayes Networks, K Nearest Neighbors, Decision Trees, Decision Rules, Neural Networks, SVMs and so forth. Different classifier performs best in different situation.

3 Related Works

The classical approach to text categorization has so far been using a document representation in a word-based space which relies on classification algorithms that are trained in a supervised learning manner. In the early days of text categorization, classifier design has been significantly advanced [1] with lots of strong learning algorithms emerged such as [2], [3], [4]. Later, despite numerous attempts to introduce more sophisticated techniques for document representation, the simple minded independent word-based representation, known as bag-of-words (BOW), remains very effective. Indeed, to date the best multi-class, multi-labeled categorization results for the well-known Reuters-21578 dataset are based on the BOW representation [5] [6].

A sufficient effort has been expended on attempting to come up with a document representation which is richer than BOW. A widely explored approach is in using n-grams of words (or phrase) in addition to or in place of single words (or unigrams). However, after many years of unsuccessful attempts to improve the text categorization results by applying n-grams (usually n=2), many researchers agree that there might be a certain limitation in usability of phrases for text categorization. According to [7], this can probably be explained by two considerations: (a) the results achieved on these corpora are so high that they probably cannot be improved by any technique, because all the incorrectly classified documents are basically mislabeled; and (b) the corpora are "simple" enough so only a few extracted keywords can do the entire job of distinguishing between categories.

There are mainly two kinds of approaches to incorporate n-gram of words into the document representation: the first one applies n-grams together with unigrams, while the second one excludes unigrams from the representation and bases on n-grams only. However, in most cases the second approach leads to a certain decrease in the categorization results, while the first approach can potentially improve the results. This observation indicates that the simple BOW representation is powerful enough, so the classification results cannot be probably improved by replacing the BOW representation but only by extending it.

Now we give a particular presentation to the state-of-the-art of using n-gram of words into the document representation. [9] uses document representation based on Noun Phrases (obtained by a shallow parsing) and Key Phrases (the most meaningful phrases obtained by the Extractor system). The results achieved by either scheme are roughly the same as their baseline with BOW representation. [10] uses both unigrams and bigrams as document features and extract the top-scored features using various feature selection methods. Their results indicate that in general bigrams can better predict categories than unigrams. However, despite the fact that bigrams represent the majority of the top-scored features, the use of bigrams does not yield significant improvement of the categorization results while using the Rocchio classifier. [11] combines BOW and Bag-Of-Ngrams (BON) as document features. By n-grams the authors mean all continuous word sequences in texts. They use several common classifiers with the highest results obtained by the SVM classifier. [12] does feature induction with combination of single words and word pairs. The word pairs are of the Head/Modifier type, i.e. nouns are extracted with their modifiers. The authors show that using pairs without BOW, the results of both classifiers decrease, while using both pairs and BOW, the results are marginally above the BOW baseline. For extracting bigrams, [13] use the following method: first, they sort words according to their document frequency and consider only highly ranked words. Then they extract bigrams such that at least one of their components belongs to those highly ranked words. After that the authors filter the resulting bigrams according to their *tf·idf* and Mutual Information with respect to a category. One of few relatively successful attempts of using bigrams is demonstrated by [14], who propose a very sophisticated feature induction technique to improve the text categorization results on *Reuters* and *ComputerSelect* datasets. They apply a string distance measure which is similar to the String Kernel [15]. Basing on this measure the authors introduce a score according to which they rank bigrams. Then they extract highly ranked bigrams so that less than 1% of all bigrams are extracted. Using the SVM classifier, the authors achieve a significant improvement on the *ComputerSelect* dataset, while the improvement on the Reuters dataset is again statistically insignificant. Nevertheless, this result on Reuters is highly noticeable: 88.8% break-even point is clearly the state-of-the-art result. The success of this technique may be explained by the fact that documents of the Reuters dataset are very well structured (many of them are even not free text but tables) and the string similarity method used by the authors manages to capture this clear structure.

4 Phrase-Based Text Classification for Similar Texts

Now we propose our own method for classifying similar texts. Most of the time, we are classifying texts from totally different fields, each of which has some solely-used domain words, so the word-based BOW is effective enough for text classification. But the situation becomes quite different when it comes to classifying similar texts.

4.1 Similar Texts Analysis

Since similar texts are texts of similar topics which almost share the same word set, we cannot classify them only by words. In this situation, a phrase-based representation (BOP) is expected to be much more effective than BOW. Take the example we mentioned above: We have collected a small document set which contains 158 documents from two similar topics *"Data mining"* and *"Text Mining"*. There are 120 documents from *"Data Mining"* while other 38 from *"Text Mining"*. As it is performed in the table 1 below, the average number of the three words *"text"*, *"data"* and *"mining"* contained in documents of the two topics are not quite different, but the number of *"text mining"* in documents of *"Text Mining"* is apparently more than those of *"Data Mining"*. On contrast, the number of *"data mining"* in documents of *"Data Mining"* is much more than those of *"Text Mining"*.

Table 1. The average number of *"text"*, *"data"*, *"mining"*, *"text mining"* and *"data mining"* in the small document set which consists of 22 documents in *"Data Mining"* and 18 documents in *"Text Mining"*

	120 documents in *"Data Mining"*	38 documents in *"Text Mining"*
text	0.20	0.51
data	0.52	0.36
mining	0.50	0.41
text mining	0.08	0.36
data mining	0.42	0.05

The same problem also exists between some other similar topics such as *"Information Retrieval"* and *"Web Search"*, *"Information Security"* and *"Privacy and Trust"*, and so forth. There are still some other similar topics which may not have the same word set, but there is also a great intersection between their word sets. In this situation, BOW is not very applicable either. Therefore, in order to classify documents from similar topics more effectively, we propose to use phrases combined with words in classifying similar texts.

4.2 Using BOP in Text Categorization

It is necessary to explain that phrases we defined in bag-of-phrases (BOP) method are different from those in [10]. In [10] phrases are only Noun Phrases (obtained by a shallow parsing) and Key Phrases (the most meaningful phrases obtained by the Extractor system); while in our BOP method, phrases are frequently-used continuous word sequences (including a single word) in texts. In order to extract them from texts, we use an n-gram word sequence extractor which can get all word sequences no longer than n (including unigram), then only those word sequences which have appeared more than a *threshold* times in all texts are chosen as features in our BOP method.

After we get all potential phrases and words (word can also be seen as single-word phrase) as features for BOP, **feature selection** is necessary to reduce dimensionality and overcome the statistical sparseness of document representations. After feature selection we have to choose a suitable **classifier** which is also very influential to the results. Those previous attempts to incorporate phrases or n-grams in the past decade lead us to the following two strategies: (a) There are too much n-gram phrases (not words) in our phrase bags, it is necessary to make sure that the phrases we will use are all highly discriminative features. That means we should only choose those n-gram phrases which are "better" than all their components ("better" means more discriminative). Just like the phrase *"data mining"*, *which* can be chosen only if it is "better" than both *"data"* and *"mining"*; (b) In order to improve results that have been achieved, we should enrich the existing model rather than propose a new one. That implies that we prefer to choose feature selection method and classifier from existing ones [7].

Based on the two strategies above, we would like to find out which phrases are "better" than their components firstly. For each category we sort all the unigrams according to their *Mutual Information measure* with respect to the category [7]. Then we compare the rank of each n-gram phrase to its component unigram words. If the rank of n-gram phrase is better than all of its components, then it can be chosen, otherwise it should be removed from the phrase bag. After this process, the number of phrases left in the bag decrease sharply.

For the next step, we have to select a feature selection (FS) method for the phrases in the bag. There are lots of FS methods proposed. There are some class-independent measures such as *document frequency*. It is the simplest feature selection method which based on *Zipf's Law*. By this method, N most common phrases and those terms that appear in fewer than M documents (M usually 1 or 2) should be removed.

However, we prefer to use other methods which consider classes while selecting features such as *Information Gain* [16], *Chi-Square (CHI)* [17], *Gain Ratio* [18] and so forth.

1) *Information Gain,* which is also known as Expected Mutual Information, tries to keep only the terms distributed more differently in the training set of the categories.
2) *Chi-square* measures the lack of independence between a term t and a category c_i and can be compared to the chi-square distribution with one degree of freedom to judge extremeness.
3) *Gain Ratio* is commonly used as a splitting criterion in decision tree induction. It is defined whenever a data set is split into two or more subsets. The more a split helps create subsets with homogeneous classes, the better the gain ratio.

Since we don't know which FS method performs better for similar texts, we would like to use all of these methods and choose the most proper one. Later, we have to choose a proper classifier to learn the relationships between features and predefined categories. There have been lots of excellent algorithms proposed such as *Naïve Bayes* [19], *Bayes Networks* [20], *K Nearest Neighbors (KNN)* [21], *Decision Trees* [22], *SVMs* [23] and so forth.

1) *Naïve Bayes*: A very cheap but very successful classifier which assumes that all of the attribute values are independent given the class label.
2) *Bayes Networks*: This is a classifier which tries to improve the results of Naïve Bayes by relaxing the naïve Bayes independence assumption.
3) *KNN*: It formalizes the intuition that class of the unseen example is likely to be same as to the one of the closest known instances. Degree of similarity is to be defined according to a suitable criterion.
4) *Decision Trees*: The basic idea of decision tree is to classify texts through a decision tree. The most important advantage is their capability to break down a complex decision-making process into a collection of simpler decisions, thus providing a solution which is often easier to interpret.
5) *SVMs*: A SVM is an algorithm which computes the linear separation surface with maximum margin for a given training set. Only a subset of the input vectors will influence the choice of the margin; such vectors are called support vectors. When a linear separation surface does not exist, for example in presence of noisy data, SVMs algorithms with a slack variable are appropriate.

5   Experimental Study

Our experimental data set contains 917 abstracts from 15 different categories. These abstracts come from conference papers in *Databases* field from ACM Digital Library[1](ACMDL). To make this data set, we collected thousands of abstracts of conference papers from famous conferences of *Databases* in ACMDL, and then we label category for each abstract according to the name of session it belongs to, e.g. if

---

[1] ACM Digital Library: http://portal.acm.org/

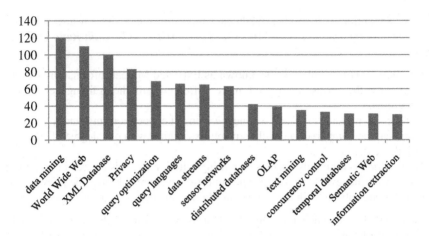

Fig. 2. Category name and the number of abstracts in each category of experimental data set

a paper belongs to a session named "Data Mining", then the category of its abstract can be labeled as "Data Mining". Since the limitation of our resources, although we find hundreds of different session names, most of them only contain no more than 10 abstracts. Therefore, we only have chosen some popular categories, each of which contains no less than 30 abstracts. All categories and the number of abstracts in each category of the data set are given in Fig. 2.

Firstly, we use an 3-gram word sequence extractor to extract all phrases no longer than 3 (including words), and only those word sequences which appear more than 3 times in all texts are chosen as features in our BOP method. Secondly, we remove those less discriminative phrases from the bag with the consideration we described in section 4.2. In order to make a comparison between BOP and BOW, we also use the classical BOW method to deal with the data sets with nearly the same first step and third step (The second step of BOP is not necessary for BOW)

In the third step, in order to choose a proper feature selection (FS) method for the phrases in the bag, we use some common FS methods including *Information Gain (IG), Chi-Square (CHI), Gain Ratio* respectively. After feature selection, we have to

Table 2. Accuracy of Naive Bayes Classifier

Feature Number		50	80	100	300	500	1000
Information Gain	BOP	0.6375	0.6308	**0.6414**	0.6282	0.5976	0.5631
	BOW	0.5878	0.5851	**0.5918**	0.5439	0.5173	0.4840
Chi-Square	BOP	0.6667	0.6467	0.6667	**0.6693**	0.6454	0.6069
	BOW	0.5904	0.6130	**0.6157**	0.5864	0.5439	0.5040
Gain Ratio	BOP	0.6247	0.6313	0.6260	**0.6366**	0.6340	0.5703
	BOW	0.5365	0.5551	**0.5418**	0.4861	0.4834	0.4436

Table 3. Accuracy of Bayes Networks Classifier

Feature Number		50	80	100	300	500	1000
Information Gain	BOP	0.6866	**0.6985**	0.6972	0.6960	0.6960	0.6959
	BOW	0.5851	**0.5851**	0.5851	0.5851	0.5851	0.5851
Chi-Square	BOP	0.6906	**0.6972**	0.6972	0.6959	0.6959	0.6959
	BOW	0.5851	**0.5851**	0.5851	0.5851	0.5851	0.5851
Gain Ratio	BOP	0.6830	**0.6910**	0.6910	0.6910	0.6910	0.6910
	BOW	0.5538	**0.5790**	0.5790	0.5790	0.5790	0.5790

Table 4. Accuracy of K Nearest Neighbors Classifier

Feature Number		50	80	100	300	500	1000
Information Gain	BOP	0.5046	0.4993	**0.5046**	0.4050	0.3293	0.2059
	BOW	0.4295	**0.4335**	0.4069	0.2939	0.2606	0.1676
Chi-Square	BOP	**0.5857**	0.5697	0.5339	0.4940	0.4117	0.2869
	BOW	**0.4934**	0.4601	0.4495	0.3404	0.3032	0.1941
Gain Ratio	BOP	0.5212	0.5358	**0.5371**	0.5027	0.4907	0.3912
	BOW	**0.5086**	0.4688	0.4462	0.3732	0.3559	0.3187

Table 5. Accuracy of Decision Trees Classifier

Feature Number		50	80	100	300	500	1000
Information Gain	BOP	0.2311	0.2311	0.2311	0.2311	0.2310	0.2310
	BOW	0.2181	0.2181	0.2181	0.2181	0.2181	0.2181
Chi-Square	BOP	0.2311	0.2311	0.2311	0.2311	0.2311	0.2311
	BOW	0.2181	0.2181	0.2181	0.2181	0.2181	0.2181
Gain Ratio	BOP	0.2308	0.2307	0.2307	0.2307	0.2307	0.2308
	BOW	0.2178	0.2179	0.2178	0.2178	0.2178	0.2178

Table 6. Accuracy of SVM Classifier

Feature Number		50	80	100	300	500	1000
Information Gain	BOP	0.5843	0.6096	**0.6162**	0.6082	0.5830	0.5206
	BOW	0.5559	0.5731	**0.5745**	0.5585	0.4854	0.3670
Chi-Square	BOP	0.6096	**0.6282**	0.6162	0.6003	0.5790	0.5139
	BOW	0.5559	**0.5665**	0.5625	0.5625	0.4907	0.3710
Gain Ratio	BOP	0.5889	**0.6167**	0.6141	0.5424	0.5133	0.3966
	BOW	**0.5193**	0.5113	0.5060	0.4117	0.3373	0.2364

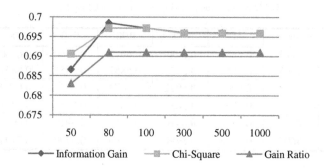

Fig. 3. The accuracy of the BOP + Bayes Networks classifier

choose a proper classifier from the following algorithms: *Naïve Bayes, Bayes Networks, K Nearest Neighbors, Decision Trees, SVMs* and so forth. The results with ten-fold cross-validation on the data set are given in the Table 2 to Table 6 and Fig. 3.

As we can see from Table 2 to Table 6, BOP performs better than BOW in most situations here, which proves that BOP is more effective for our data set. What's more, *Bayes Networks* performs much better than the other classifiers on our data set. it performs the best when there are only 80-100 words and phrases chosen in the *BOP* by using *Information Gain* feture selection method.

6   Introduction to AutoPCS

According to the experimental study in the last section, we decide to use *BOP + Information Gain + Bayes Networks* to do text classification for AutoPCS. Different from classifying unseen text into one of the predefined category, AutoPCS tries to find out three most relevant topics for each text. Therefore, we have to make a small adjustment to the classifier. We still use the BOP representation combining Information Gain feature selection method to get about 500 high-discriminative phrases and words in the bag. However, for an unseen text, the modified *Bayes Networks* classifier does not categorize it with only one category, but top three categories which ranked according to their relevant values.

AutoPCS is a domain-irrelevant tool. To have AutoPCS categorize new papers into three relevant topics according to its abstract effectively, it should be given a predefined list of topics in a certain domain, as well as a plenty of former accepted papers which have been categorized with their most relevant topics (only the most relevant **one** topic for each paper is enough) as training input. Since the topics of a certain domain are changing every year, the AutoPCS can be updated if only we add the newly accepted papers with their categorizations as additional input. A snapshot of AutoPCS is shown in Fig. 4.

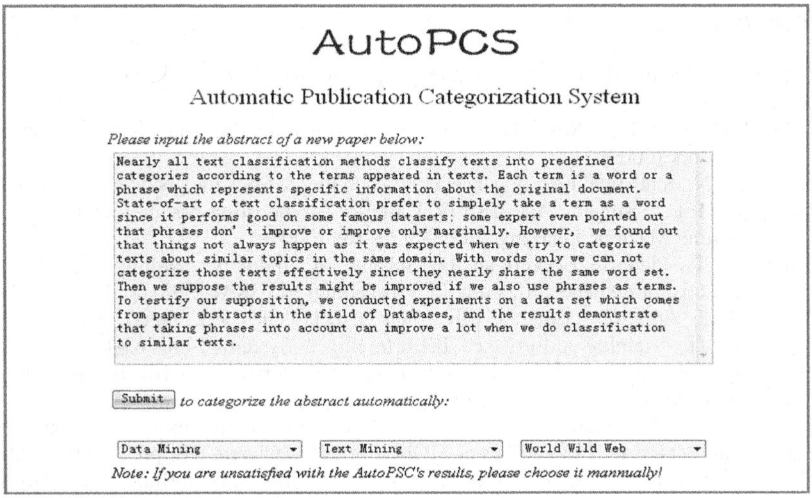

Fig. 4. A snapshot of AutoPCS

7 Conclusion

In this paper, we try to develop a tool named AutoPCS which can be used to help experts in choosing relevant topics for newly coming papers from a predefined topic list only by their abstracts. This topic selection process can be regarded as a text classification problem. Therefore, we tried to use classical BOW based text categorization method to help us but failed. The reason we concluded is that papers submitted to the same conference are all about a certain domain. Although concerning about different topics, they still share lots of common words. But different topics have their respective terms which are usually phrases, so a phrase-based representation (BOP) is expected to be more effective. In fact, a sufficient effort has been taken on attempting to use phrases or n-grams to represent a document which is richer than BOW. However, they are unsuccessful in improving the text categorization results on some famous data sets.

After an overview to related works in history, we design our own way to extract phrases from texts; through experimental study we find out proper feature selection method and classifier for our case. Finally, we decide to use *BOP + Information Gain + Bayes Networks* to do text classification for AutoPCS. Since AutoPCS has to label each new paper with three most relevant topics, we also have to make some adjustment to the output of our classifier. We believe that AutoPCS can be a good helper for conference organizations.

Acknowledgments

This work was supported in part by the National Natural Science Foundation of China under Grant No. 70871068, 70621061, 70890083, 60873017, 60573092 and 60496325.

References

1. Salton, G., McGill, M.: Introduction to Modern Information Retrieval. McGraw-Hill, New York (1983)
2. Duda, R.O., Hart, P.E., Stork, D.G.: Pattern Classification, 2nd edn. John Wiley & Sons, Inc., New York (2000)
3. Vapnik, V.N.: Statistical Learning Theory. John Wiley & Sons Inc., New York (1998)
4. Schapire, R.E., Singer, Y.: BOOSTEXTER: a boosting-based system for text categorization. Machine Learning 39(2/3), 135–168 (2000)
5. Dumais, S.T., Platt, J., Heckerman, D., Sahami, M.: Inductive learning algorithms and representations for text categorization. In: Proceedings of CIKM 1998, pp. 148–155 (1998)
6. Weiss, S.M., Apte, C., Damerau, F.J., Johnson, D.E., Oles, F.J., Goetz, T., Hampp, T.: Maximizing text-mining performance. IEEE Intelligent Systems 14(4), 63–69 (1999)
7. Bekkerman, R., Allan, J.: Using Bigrams in Text Categorization. CIIR Technical Report, University of Massachusetts at Amherst (2004)
8. Lewis, D.D.: An evaluation of phrasal and clustered representations on a text categorization task. In: Proceedings of SIGIR 1992, Kobenhavn, DK, pp. 37–50 (1992)
9. Scott, S., Matwin, S.: Feature engineering for text classification. In: Bratko, I., Dzeroski, S. (eds.) Proceedings of ICML 1999, Bled, SL, pp. 379–388 (1999)
10. Caropreso, M.F., Matwin, S., Sebastiani, F.: A learner-independent evaluation of the usefulness of statistical phrases for automated text categorization. In: Chin, A.G. (ed.) Text Databases and Document Management: Theory and Practice, pp. 78–102. Idea Group Publishing, Hershey (2001)
11. Zhang, D., Lee, W.S.: Question classification using support vector machines. In: Callan, J., Cormack, G., Clarke, C., Hawking, D., Smeaton, A. (eds.) Proceedings of SIGIR 2003, Toronto, CA, pp. 26–32 (2003)
12. Koster, C.H., Seutter, M.: Taming wild phrases. In: Sebastiani, F. (ed.) ECIR 2003. LNCS, vol. 2633, pp. 161–176. Springer, Heidelberg (2003)
13. Tan, C.M., Wang, Y.F., Lee, C.D.: The use of bigrams to enhance text categorization. Information Processing and Management 38(4), 529–546 (2002)
14. Raskutti, B., Ferra, H., Kowalczyk, A.: Second order features for maximising text classification performance. In: Flach, P.A., De Raedt, L. (eds.) ECML 2001. LNCS, vol. 2167, pp. 419–430. Springer, Heidelberg (2001)
15. Lodhi, H., Shawe-Taylor, J., Cristianini, N., Watkins, C.J.C.H.: Text classification using string kernels. In: Advances in Neural Information Processing Systems (NIPS), pp. 563–569 (2000)
16. Lewis, D.D., Ringuette, M.: A comparison of two learning algorithms for text categorization. In: Proceedings of SDAIR 1994, pp. 81–93 (1994)
17. Yang, Y., Pedersen, J.O.: A comparative study on feature selection in text categorization. In: Proceedings of ICML 1997, pp. 412–420 (1997)
18. Wiener, E.D., Pedersen, J.O., Weigend, A.S.: A neural network approach to topic spotting. In: Proceedings of SDAIR 1995, pp. 317–332 (1995)
19. Domingos, P., Pazzani, M.: On the optimality of the simple Bayesian classifier under zero-one loss. Machine Learning 29, 103–137 (1997)
20. Friedman, N., Geiger, D., Goldszmidt, M.: Bayesian network classifiers. Machine Learning 29, 131–163 (1997)
21. Yang, Y.: Expert network: Effective and efficient learning from human decisions in text categorization and retrieval. In: SIGIR 1994, pp. 13–22 (1994)
22. Yuan, Y., Shaw, M.J.: Induction of fuzzy decision trees. Fuzzy Sets and Systems 69, 125–139 (1995)
23. Joachims, T.: Text categorization with support vector machines: learning with many relevant features. In: Nédellec, C., Rouveirol, C. (eds.) ECML 1998. LNCS, vol. 1398, pp. 137–142. Springer, Heidelberg (1998)

A Probabilistic Approach for Mining Drifting User Interest

Pin Zhang, Juhua Pu, Yongli Liu, and Zhang Xiong

School of Computer Science and Technology,
Beihang University, Beijing 100083, China
{zhangpin,liuyl}@cse.buaa.edu.cn, {pujh,xiongz}@buaa.edu.cn

Abstract. Incremental approaches learn drifting user interests mainly from user feedbacks. Most of those existing approaches assume that data instances in user feedbacks are binary labeled. This paper presents a novel probabilistic approach that learns drifting user interests from numerically labeled feedbacks instead of binary labeled ones. The approach models user interests as a set of probabilistic concepts, considers numerical instance labels as probabilities that the user likes those instances, and uses feedbacks to update user interest models incrementally based on an exponential, recency-weighted average algorithm. Experimental results on different learning tasks show that the approach outperforms existing approaches in numerically labeled feedback environment.

Keywords: Incremental Learning, Concept Drift, Interest Model.

1 Introduction

Mining user interests is one of the essential processing phases of Web-based systems providing personalized services. It has become urgent for those systems of being able to track the users' interest as it changes over time [1]. Incremental approaches derived from concept drift learning area have been investigated to learn drifting user interests by many researchers. Those approaches use user feedbacks as labeled training data to track changes of user interests and to update user interest models incrementally.

In applications of personalized services, feedback values (i.e. data labels) representing degrees of user interest or user rankings may exist as continuous numbers (e.g. NewsDude [2]). The so-called numerically labeled feedbacks actually contain information more accurate than the binary labeled ones (i.e. labeled as Positive or Negative). However, to the best of our knowledge, none of the existing incremental approaches for concept/interest drift learning (Rocchio [3], MTDR [4], FLORA [5], etc.) take numerically labeled feedbacks into consideration, which results in no improvements in those approaches' performance when learning from numerically labeled feedbacks than from binary labeled ones.

Contributions: this paper presents PALDI, a novel Probabilistic Approach for Learning Drifting Interest from numerically labeled user feedbacks. The approach outperforms other existing incremental approaches when the feedback

Q. Li et al. (Eds.): APWeb/WAIM 2009, LNCS 5446, pp. 381–391, 2009.

values are in numerical forms and shows higher adaptability benefiting from being less influenced by system parameter selection.

2   Modeling User Interest

PALDI models user's interest as a set of interest concepts. An **interest concept** is a probabilistic concept [6]. Let X be a subset of the personalized content instances, an interest concept can be expressed as a real-valued function c : $X \rightarrow [0, 1]$. For any instance $x \in X$, $c(x)$ is interpreted as the probability that x is a positive example of interest concept c. Set X is called the instance domain of interest concept c. An interest concept is an abstraction to the features of a number of personalized content instances. The closer $c(x)$ is to 1, the more similar x is to the abstract feature of c. Thus an interest concept can be considered as a category that involves a set of instances in probability, while an instance belongs to one or more interest concepts in probability.

Given an instance x' and an interest concept $c : X \rightarrow [0, 1]$, we say x' is positive to c if $x' \in X$.

Let $C = \{c_1, c_2, \ldots, c_n\}$ be the set of interest concepts describing a certain user's interest and X_i be the instance domain of c_i in C, let $X_C = \bigcup_i X_i$ denote the union of X_i. Given $x \in X_c$, let $p_x = \{c_i | x \in X_i, c_i \in C\}$ denote the set of all the interest concepts in C whose instance domain contains x. Thus the user's interest to a certain personalized content instance can be expressed as conditional probability $\Pr_{p_x}(I|x)$ where I denotes random event "being interested". According to the law of total probability, we have:

$$\Pr_{p_x}(I|x) = \sum_{c_i \in p_x} \Pr(I|x, c_i) \times \Pr(c_i|x) \tag{1}$$

Since $x \in X_i$ is true for any $c_i \in p_x$, so we have $\Pr(I|x, c_i) = \Pr(I|c_i)$. Thus Eq. 1 can be written as:

$$\Pr_{p_x}(I|x) = \sum_{c_i \in p_x} \Pr(I|c_i) \times \Pr(c_i|x) \tag{2}$$

It can be seen from Eq.2 that the probability the user likes instance x depends on the probability the user likes interest concept c_i (i.e. $\Pr(I|c_i)$) and the probability instance x is positive to c_i (i.e. $\Pr(c_i|x)$ or $c_i(x)$). Obviously, conditional probability $\Pr(c_i|x)$ can be expressed as the feature similarity between c_i and x. Thus a user's interests can be modeled by maintaining the feature and interest probability of each interest concept. Consequently, we have the following definition: a **user interest model**, $UP = \{< D_1, w_1 >, < D_2, w_2 >, \ldots, < D_n, w_n >\}$ is a set of interest concepts where each interest concept c_i is a two-tuple consisting of a computable feature D_i and an interest weight w_i representing the probability that the user likes c_i.

According to the above definition of user interest model, the incremental learning of drifting user interest thereby refers to the incrementally updating process upon the interest weight w_i and interest concept feature D_i (i

$= 1, 2, \ldots, n$) of interest concepts in model UP using user feedbacks. A **user feedback** $UF = <x, d>$ is a two-tuple consisting of instance x and its feedback value (i.e. data label) d representing the probability that the user likes x. Thus the **incremental learning process** can be expressed as $w_i^{new} = \delta(w_i^{old}, UF)$ and $D_i^{new} = \phi(D_i^{old}, UF)$ ($i = 1, 2, \ldots, n$), where function δ and function ϕ refers to the expressions for calculating w_i and D_i respectively.

Given user interest model UP, according to Eq.2, the **predicted user's interest** to a new personalized content instance x'' can be given by:

$$
\begin{aligned}
\Pr(I|x'') &= \sum_{c_i \in p_{x''}} \Pr(I|c_i) \times \Pr(c_i|x'') \\
&= \sum_{i \in [1,n]} w_i \times Sim(D_i, x'')
\end{aligned}
\tag{3}
$$

where $Sim(D_i, x'')$ refers to the feature similarity between x'' and interest concept c_i.

In this paper, text-based representations are used as the computable description to the features of personalized content instances and interest concepts in consideration of the ubiquity of text information among different types of personalized contents (e.g. news [2], Web pages [7], TV programs [8], movies [9], etc.). Specifically, we use TF-IDF weighted term vectors to describe features of instances and interest concepts, and *cosine* similarity measure to calculate the feature similarity between instances and interest concepts [4].

3   ERWA-Based Learning Algorithm

Table 1 describes the general learning algorithm of PALDI. For each user feedback $< x_j, d_j >$ obtained by the system, the maximum similarity between instance x_j and each interest concept is first calculated. If the maximum similarity is less than the threshold θ, meaning that x_j is dissimilar to all of existing interest concepts, a new concept is created to describe the feature of x_j. The feature and interest weight of each interest concept in the user interest model are then updated using the above feedback. A threshold M is applied to limit the number of interest concepts in a single interest model to avoid the infinite increase of interest concepts when the variation in instances' features is very broad [4]. When the limit has been reached, no new concept will be created and existing concepts will be updated to describe the lately learned instance instead.

The incremental learning algorithm for updating interest weight w_i and concept feature D_i (i.e. Step 5 and 6 in Table 1) will be detailed in the rest of this section.

3.1   Learning of Interest Weights

According to the theorem of probability, the relation between the probability and the frequency of a random event e can be given by:

$$
\Pr(e) = \lim_{n \to \infty} \frac{n_e}{n_t}
\tag{4}
$$

Table 1. Learning algorithm of PALDI

Input: *model*: target interest model
c_i: the th interest concept in *model*,
D_i: feature descriptor of c_i,
$\{< x_1, d_1 >, < x_2, d_2 >, \ldots, < x_m, d_m >\}$: sequence of feedbacks,
θ: decision threshold constant $(0,1)$,
M: threshold number of interest concepts in *model*.
Output: modified *model*.
Function: $Size(model)$ returns number of interest concepts in *model*.
Steps:
1. **For** each x_j in $\{< x_1, d_1 >, < x_2, d_2 >, \ldots, < x_m, d_m >\}$ do
2.   let $s = Sim(D_i, x_j)$ such that $Sim(D_i, x_j) = \max_k\{Sim(D_k, x_j)\}$
3.   if $s < \theta$ and $Size(model) < M$ then
4.   create new concept $c' = < D', w' >$ with $D' = x_j$, $w' = d_j$
5.   **For** each c_i in *model* do
6. $w_i^{new} = \delta(w_i^{old}, < x_j, d_j >)$
7. $D_i^{new} = \phi(D_i^{old}, < x_j, d_j >)$
8.   **End for**
9. **End for**

Meaning that probability can be approximated by frequency when n is sufficiently large. Accordingly, the probability the user likes a certain interest concept can be approximated by the weighted average w_i of the probabilities the user likes the instances that are positive to the concept, in the sense that a interest concept involves a number of instances in probability. Besides, the impact of previously learned instances should be gradually weakened as user's interest changes over time [1]. Consequently, an exponential, recency-weighted average algorithm is introduced into PALDI to update w_i incrementally.

Exponential, recency-weighted average (ERWA) [10] is an iterative method for calculating weighted average. It adds the contribution of component r into the current average value Q_{old} to form the new average value Q_{new} with an expression like $Q_{new} = Q_{old} + \alpha(r - Q_{old})$. The method assures that the most recently added component holds the largest contribution to the overall average value (when updating rate α remains constant), which conforms to the goal of interest drift learning, i.e. reflecting the most recent changes of drifting user interests.

The selection of learning rate or updating rate (i.e. α in the above expression) is a key issue that impacts the performance of incremental learning approaches using ERWA. In literature [11], uniform updating rate was used for all feedbacks and the updating using a certain feedback $UF = < x, d >$ was only performed upon the very interest concept category most similar to x. The approach above resulted in a low computational complexity, but a limited flexibility as well. The updating based on the above approach is incomplete in condition that a certain instance may be positive to more than one interest concepts. In literature [4], a certain feedback was used to update multiple short term interest descriptors, which improved the completeness. Yet, uniform updating rate was used ignoring the various similarities between certain instance and different interest concepts. Based on ERWA method, we calculate interest weight w_i as:

$$w_i^{new} = \delta(w_i^{old}, < x_j, d_j >) = w_i^{old} + \beta_i^j(d_j - w_i^{old}) \tag{5}$$

where w_i^{old} and w_i^{new} are the value of w_i before/after the updating respectively, d_j refers to the feedback value of the jth feedback in feedback sequence, and $\beta_i^j \in [0, 1]$ represents the updating rate for w_i corresponding to the jth feedback.

We believed that the contribution of a certain instance's feedback value to the interest weight w_i should depend on the similarity between the instance and the interest concept as well as depend on the order of instances in the feedback sequence. Besides, the updating extent of interest model in a single learning cycle should differ among different users in consideration of the various changing speeds and ranges of users' interests. Consequently, the updating rate of PALDI for learning of w_i is determined by both the instance similarity factor r_i^x and the updating extent preference factor q (i.e. $\beta_i^j = qr_i^{x_j}$). The instance similarity factor r_i^x (i.e. conditional probability $\Pr(c_i|x)$ in Eq. 2) refers to the feature similarity between instance x and interest concept c_i, whereas the updating extent preference factor $q \in [0, 1]$ maintains the user preference of updating extent in a single learning cycle. The closer q is to 1, the faster the interest model is adapted to tracking interest drifts. By incorporating the above two factors, Eq.5 can be written as:

$$w_i^{new} = w_i^{old} + qr_i^{x_j}(d_j - w_i^{old}) \tag{6}$$

where x_j is the instance corresponding to the jth feedback.

In the actual learning situation, the updating extent preference factor q is to be set directly by the user, whereas the instance similarity factor r_i^x is calculated as follows: for each feedback $UF_j =< x_j, d_j >$, feature similarity between instance x_j and each interest concept in current user's interest model, $Sim(D_i, x_j)$ is measured successively. Then $r_i^{x_j}$ can be obtained by normalizing $Sim(D_i, x_j)$ as:

$$r_i^{x_j} = \frac{Sim(D_i, x_j)}{\sum_k Sim(D_k, x_j)} \tag{7}$$

As the similarity variety between a certain instance and different interest concepts has been considered during the incremental learning of each single interest weight, Eq.6 can be then applied uniformly to every interest concept in the interest model without thresholding the update scope of interest concepts, which reduces the number of system parameters and subsequently boosts the adaptability of the approach.

3.2   Learning of Concept Features

Corresponding to the uniform expression for learning the interest weights of interest concepts, a uniform algorithm for learning concept features should be designed and applied to all the interest concepts within a single model. In the sense that a interest concept could be expressed as a fuzzy set [12], the method for calculating category features of fuzzy sets may be adapted to the learning of interest concept features.

For a certain interest concept $c_i : X_i \rightarrow [0, 1]$, its feature D_i can be accordingly calculated as [12]:

$$
D_i = \frac{\sum\limits_{x_j \in X_i} [c_i(x_j)]^\gamma x_j}{\sum\limits_{x_j \in X_i} [c_i(x_j)]^\gamma} \tag{8}
$$

where exponent γ chosen from $(1, \infty)$ in advance determines the degree of fuzziness of the fuzzy set. It can be concluded from Eq.8 that interest concept feature D_i may also be calculated in the form of weighted average, of which the components are instance features and the weights are powers of instance-concept similarities. On the purpose of emphasizing the contribution of most recently learned instances, Eq.8 is transformed into an ERWA form as:

$$
\begin{aligned}
D_i^{new} &= \phi(D_i^{old}, < x_j, d_j >) \\
&= D_i^{old} + q(r_i^{x_j})^\gamma (x_j - D_i^{old})
\end{aligned} \tag{9}
$$

where the meaning of parameter q and $r_i^{x_j}$ remains the same as Eq.6.

4   Experimental Evaluation

The main objective of our experiments is to evaluate the prediction accuracy and recovery speed of PALDI in condition of drifting user interests. The performance of PALDI was compared with MTDR [4] in our experiment. The latter is an interest drift learning approach designed to track multiple target concepts. The MTDR learning approach uses Rocchio algorithm to update its long term interest descriptors, while the short term interest descriptors are updated using ERWA based algorithm with a uniform learning rate. According to the experimental results presented in literature [13], MTDR outperforms other existing interest drift learning approaches including Rocchio, FLORA, etc. The system parameters of MTDR were set to the same value as literature [4] in our experiment. The parameters of PALDI were set as follows: we set $q = 1$ to adapt the interest model as fast as possible and set $\gamma = 1.1$ to keep the updating rate of concept feature close to the rate of interest weight. Parameter θ and M was set to the same value as the corresponding parameter of MTDR due to the similar effect of the corresponding parameters between the two approaches.

In order to make MTDR approach able to learn user interest from numerically labeled feedbacks, each data label L_j in user feedbacks was first binarized to form a binary label B_j before being learned by the approach:

$$
B_j = \begin{cases} Positive, & L_j \geq 0.5 \\ Negative, & L_j < 0.5 \end{cases} \tag{10}
$$

4.1   Dataset

We derived our datasets from a subset of the Reuters-21578 data collection named ModApte split. The subset contains 9603 documents for the training set and 3299 documents for the test set. Each document in the subset has been assigned one or more topics. There are totally 1661 documents in the subset being assigned multiple topics, which conforms to our assumption that a single instance may be positive to multiple concepts. Documents were selected to form our set of personalized content instances from the topics used in the experiment of literature [4] (i.e. Trade, Coffee, Crude, Sugar and Acq). 100 documents (80 for training and 20 for test) that formed a tight cluster were selected from each of the five topics. These documents were pre-processed by removing stop words, stemming the remaining words, identifying bigrams and extracting them as individual terms, and counting term frequencies as other information retrieval processes did.

On the purpose of evaluating the proposed approach in the environment of numerically labeled feedbacks, data labels of the instances were generated according to the following method:

The user interests towards the content instances of a certain topic are assumed to be normally distributed [14]. The data labels of the instances of topic T are generated according to the distribution of $N(0.9, 0.0036)$ if T is assumed to be liked by the user; correspondingly, if T is assumed to be disliked, the labels are generated according to the distribution of $N(0.1, 0.0036)$. All the data labels are limited in the range of [0,1].

4.2   Learning Tasks

The effectiveness of PALDI to learn the evolution of user interests was measured by observing the performance of the approach on three learning tasks with different levels of difficulty. A learning task is a scenario that describes a user's interests that change over time. The scenario consists of a sequence of learning phases, each of which represents a stable period of the user's changing interests. A target topic group indicating the user's interests was assigned to each learning phase. Instance labels for both training data and test data were generated using the aforementioned method before each learning phase started. Each learning phase was divided into 10 evaluation cycles. Each evaluation cycle simulated an iteration of user's feedback, and consisted of (a) incrementally learning 10 documents (2 documents of each topic) selected randomly from training data, and (b) measuring the prediction accuracy of the interest model using the entire test dataset (100 documents). The final results of each learning task were averaged over 10 runs. The number of training documents in each evaluation cycle (i.e. 10) was determined by referring to the actual number of feedbacks that the system obtains in one user session of typical personalized services, so as to make the learning tasks close to realistic applications.

As the learning approaches were evaluated in the environment of numerical data labels, we used normalized mean absolute error (NMAE) to measure the

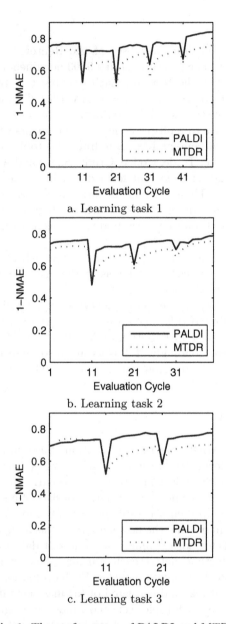

a. Learning task 1

b. Learning task 2

c. Learning task 3

Fig. 1. The performance of PALDI and MTDR

prediction accuracy by comparing predicted probability of user interest, $\Pr(I|x_j)$ with the actual generated feedback value d_j as:

$$NMAE = \frac{\sum\limits_{j=1}^{m} |\Pr(I|x_j) - d_j|}{m} \qquad (11)$$

Table 2. Learning tasks and evolution of target concepts over evaluation cycles

	Learning phase (Evaluation cycles)				
	1 (1-10)	2 (11-20)	3 (21-30)	4 (31-40)	5 (41-50)
Learning task 1	Trade	Coffee	Crude	Sugar	Acq
Learning task 2	Trade,	Coffee,	Crude,	Sugar,	-
	Coffee	Crude	Sugar	Acq	
Learning task 3	Trade,	Coffee,	Crude,	-	-
	Coffee,	Crude,	Sugar,		
	Crude	Sugar	Acq		

The learning tasks of the experiment and target topics of each learning phase of the tasks are summarized in Table 2.

4.3 Results

Figure 1 presents the experimental results, where part a, b, c are the result on learning task 1, 2, 3 respectively. The ordinate scale of the three figures is the prediction accuracy (1-NMAE). PALDI outperformed MTDR on all three learning tasks in both the recovery speed and the prediction accuracy within stable periods, which confirmed the design of PALDI. In comparison of the three figures, the prediction accuracy of PALDI within stable periods declined successively in learning task 1, 2, and 3 in response to the increasing learning difficulty of the tasks. Yet there are no obvious declines of its prediction accuracy in a single learning task as the times of interest drift accumulates. Thus it can be concluded that, by applying PALDI, relatively stable prediction accuracy can be obtained in the situation of drifting user interests.

Figure 2 shows the average prediction accuracy of the two approaches on different learning tasks. In all the three learning tasks, the average performance

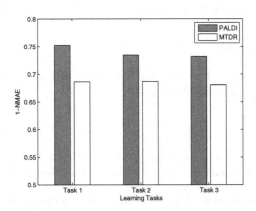

Fig. 2. Average performance of PALDI and MTDR

of PALDI is noticeably higher than MTDR, which further confirms that PALDI is more suitable for learning user interests from numerically labeled data.

5   Conclusions

This paper presented PALDI, a novel Probabilistic Approach for Learning Drifting Interest from numerically labeled user feedbacks. PALDI is superior to other existing incremental approaches for interest drift learning in the following aspects:

Processing of Feedbacks: By learning feedbacks as probabilities instead of as discretized labels, PALDI avoids the loss of information during discretization and thus improves the learning accuracy.

Presentation of Interest Models: PALDI models user interests as a set of probabilistic concepts and incorporates the impact of instance-concept similarity in learning and predicting users' interests, which improves the completeness of interest learning.

Learning Algorithm: PALDI incorporates ERWA-based algorithm to update the interest weights and concept features of interest concepts so as to emphasize the contributions of the most recently learned instances. Recovery speed of interest models is thus increased.

System Parameters: PALDI updates every interest concepts in user interest model with a uniform expression. By eliminating the threshold parameters for limiting the updating scope of concepts, the number of system parameters is reduced, and adaptability of the approach is thereby boosted.

Experimental results on different learning tasks show that the approach outperforms existing approaches in numerically labeled feedback environment.

In the future, we plan to experimentally evaluate the learning performance of PALDI with different parameter settings in detail and to quantify the impact of parameter settings to the performance.

Acknowledgement

This research is supported by National Science & Technology Pillar Program as part of the mega-project 2006BAB04A13.

References

1. Koychev, I., Schwab, I.: Adaptation to Drifting User's Interest. In: ECML Workshop: Machine Learning in New Information Age, Barcelona, Spain, pp. 39–45 (2000)
2. Billsus, D., Pazzani, M.J.: A Hybrid User Model for News Classification. In: The Seventh International Conference on User Modeling, pp. 99–108. Springer, Wien (1999)

3. Allan, J.: Incremental Relevance Feedback for Information Filtering. In: The Nineteenth International ACM-SIGIR Conference on Research and Development in Information Retrieval, pp. 270–278. ACM Press, Zurich (1996)
4. Widyantoro, D.H., Ioerger, T.R., Yen, J.: Learning User Interest Dynamics with a Three-Descriptor Representation. J. Am. Soc. Information Science 52(3), 212–225 (2001)
5. Widmer, G., Kubat, M.: Learning in the Presence of Concept Drift and Hidden Contexts. Machine Learning 23(1), 69–101 (1996)
6. Kearns, M.J., Schapire, R.E.: Efficient Distribution-free Learning of Probabilistic Concepts. In: 31st Annual Symposium on Foundations of Computer Science, St. Louis, MO, USA, vol. 1, pp. 382–391 (1990)
7. Ji, J., Liu, C., Yan, J., Zhong, N.: Bayesian Networks Structure Learning and Its Application to Personalized Recommendation in a B2C Portal. In: International Conference on Web Intelligence, pp. 179–184 (2004)
8. Kamahara, J., Asakawa, T., Shimojo, S., Miyahara, H.: A Community-Based Recommendation System to Reveal Unexpected Interests. In: The 11th International Multimedia Modeling Conference, pp. 433–438 (2005)
9. Good, N., Schafer, J., Konstan, J., Borchers, A., Sarwar, B., Herlocker, J., Riedl, J.: Combining Collaborative Filtering with Personal Agents for Better Recommendations. In: The Sixteenth National Conference on Artificial Intelligence (1999)
10. Sutton, R.S., Barto, A.G.: Reinforcement Learning: An Introduction. MIT Press, Cambridge (1998)
11. Singh, S., Shepherd, M., Duffy, J., Watters, C.: An Adaptive User Profile for Filtering News Based on a User Interest Hierarchy. In: 69th Annual Meeting of the American Society for Information Science and Technology, vol. 43 (2006)
12. Mika, S., Jain, L.C.: Introduction to Fuzzy Clustering. In: StudFuzz 205, pp. 1–8. Springer, Heidelberg (2006)
13. Hendratmo, W.D.: Concept drift learning and its application to adaptive information filtering, Texas A&M University (2003)
14. Hofmann, T.: Collaborative Filtering via Gaussian Probabilistic Latent Semantic Analysis. In: The 26th annual international ACM SIGIR, pp. 259–266. Association for Computing Machinery, New York (2003)

Temporal Company Relation Mining from the Web

Changjian Hu, Liqin Xu, Guoyang Shen, and Toshikazu Fukushima

14F, Bldg.A, Innovation Plaza, Building 1 Tsinghua Science Park
No 1, Zhongguancun East Road, Haidian District, Beijing 100084, China
{huchangjian, shenguoyang, fukushima}@research.nec.com.cn

Abstract. Relationships between companies and business events of company relation changing can act as the most important factors to conduct market analysis and business management. However, it is very difficult to track company relations and to detect the related events, because the relations change rapidly and continuously. Existing technologies are limited to analysis of non-temporal relations or temporal tracking web data in a single feature. Our work targets temporal and multiple-type relation mining. The proposed method automates relation instance extraction from the Web, temporal relation graph creation, and business event detection based on the created temporal graph. In the experiment, more than 70 thousand relation instances were extracted from 1.7 million English news articles, and a temporal relation graph of 255 companies from 1995 to 2007 was generated, and acquisition events and cases of competition relation evolution were detected. These results show our method is effective.

Keywords: Business Activity Analysis, Knowledge Acquisition, Graph Theory.

1 Introduction

Lots of business activity analysis applications have been developed until now. Some of them analyze only structured data of business activities within a company to find problems and to make decision in the company. Other applications analyze unstructured data as well within companies or from the Web to discover relations among companies, but those relation types are limited to single or a few. Therefore, the existing application can't give an overall view on business activities of a given company or multiple companies.

Our research aims to mine temporal company relations from the Web. It can guarantee the recall of relation discovery by collecting huge amount of data on the Web. Also it can create a global view of the business relation by tracking multiple kinds of relations (i.e., competition, cooperation, acquisition, incorporation, supply chain and stock share) between companies. After extracting temporal and multiple type relation instances to construct temporal relation graphs, trends of companies (even an industry) or useful and interesting business events can be detected as well. This is a new challenge in web information extraction.

The novel features of our approach are: 1) we deal with temporal relations among companies, which best reflect the trend of business, using data from the Web; 2) we

Q. Li et al. (Eds.): APWeb/WAIM 2009, LNCS 5446, pp. 392–403, 2009.

deal with multiple types of relations among companies, which give an overall view of business.

We give an introduction of related work in section 2. Our method is proposed in section 3. In section 4, the implementation and results are presented. Finally we conclude it in section 5.

2   Related Work

In this section, we discuss the related work from two perspectives: applications and algorithms.

2.1   Business Activity Analysis Application

In terms of the data source, we could classify business activity analysis applications into three types.

The first type uses the structured data within a company, but does not deal with relations between companies. For instance, Business Activity Monitoring (BAM) is software that aids in monitoring business activities [14]. Customer Relationship Management (CRM) collects customer information within a company and does data mining on this data to set up a market orientation production and management system [17]. Customer relation data is easier to get than company relation information, and structured data is simple to process.

The second type analyzes company relations based on the collected structured or semi-structured data of companies' activities. Hou et al. presented a company relationship analysis technology [7]. Grey et al. presented a supply-chain analysis method [6] which can be used in B2B marketplace. In an enlarged scale of industry which includes thousands of small companies, it becomes difficult to collect enough volume of detailed activity data.

The last type collects data from the Web. A typical application is user's opinion extraction system. This kind of system collects user's opinions from the Web and analyzes them [11][19]. Morinaga et al. proposed a temporal users' opinions analysis system, which only deals with a single simple relation between companies and products [15]. Liu et al. proposed a novel system for using the Web to discover comparable cases [13]. Liu et al. presented a web site comparison system [12]. These systems don't track and analyze temporal relationships or they only track a simple type of relation.

Our work belongs to the last type, a web data mining application. In the real world, multiple types of relations do exist at the same time [3], and they always have impact to each other. Besides, they always change rapidly with time. Therefore, temporal comprehensive mining of multiple types of company relation mining is required.

2.2   Information Analysis and Graph-Based Analysis

In company relation mining, a company's relations can be modeled as a graph. How to create a graph according to data sources and how to do analysis on this graph are two main topics for this kind of algorithm.

How to create a graph depends on what kind of data this system is using. Some algorithms extract relations from un-structured data like text and HTML pages. This is a traditional information extraction problem. The problem definition and a basic survey were given in [5][2]. According to the survey, relation extraction is still a challenging problem in terms of the accuracy. Some other algorithms mine useful information from the structured data such as event log, transaction data, etc. [1][8][10]. These approaches don't deal with the temporal graph.

How to do company relation analysis is a conventional graph analysis problem. Social network analysis is a typical method. It includes a set of techniques for identifying and representing patterns of interaction among social entities. It can be applied in a variety of applications [20][18] which have a sociometric characteristics. This kind of conventional social network analysis technology tends to compute the centrality, density, and distance of the nodes in a graph [4]. But it lacks the advanced analytic feature of temporal event detection.

In terms of how to create a graph, we developed not only technologies to extract relation instances but also new technologies to create temporal relation graphs from the relation instances with filtering and interpolation. In terms of how to analyze the graph, we developed new technologies to detect important events on the temporal graph, while the existing technologies mainly handle a non-temporal graph.

3 Proposed Method

3.1 Overview

We realize the multiple-type relation extraction and the temporal relation analysis in the following steps. (1) The first step is to extract temporal relation instances from the Web. Here multiple types of relation are handled. (2) Then the temporal graphs are created based on these relation instances. (3) And the significance of each company at a specific time is calculated based on its temporal graph, and comprehensive relations are constructed. Now, each graph consists of weighted links corresponding to company relationships, and weighted nodes corresponding to companies. (4) Business events are detected by applying a rule-based method to the temporal graphs. The architecture of our method is shown in Figure 1.

3.2 Relation Instance Extraction

In general, a news article contains three sections: title, content and publishing time, Figure 2 shows an example.

The part of content displayed with red color font contains the information of competition relations between companies. This part means that "Oracle" has "competition" relation with "Red Hat", "Novell" and "Microsoft" on "2007-01-24". We call them three relation instances. Each relation instance has the following parts: Company A, Company B, Relation Type X, and Date t. We use $RI(A, B, X, t)$ to denote it. The date is very important. It is the key point to do temporal analysis. We define a template for HTML page to extract such information.

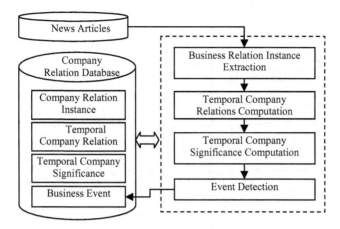

Fig. 1. Architecture of our method

We set some candidates of relation types, and ask people to look in news corpus to see which type appears more frequently. Finally we decide six types of relation, competition, cooperation, acquisition, incorporation, supply chain and stock share, are best ones. They represent an overall view on business.

The technology we used to extract the relation instance is basically a NLP-rule-based method. But a rule-based method has its disadvantages. It requires a lot of human labor to define the rules and cannot extract those instances which don't match any pre-defined rules. To overcome it, we integrate a machine learning method into our system. Figure 3 shows our method to extract relation instance. At first, the input news sentence is segmented into words, and part of speech (POS) of the words is tagged. The SVM-based classifier called "SVM sifter" is designed to select the candidate sentences of relation description. All selected sentences are called "instance candidates".

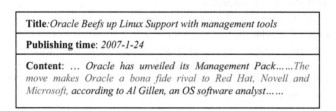

Fig. 2. An example of news article

The NLP-rule-based extractor (RE) performs syntactic analysis on the instance candidates and then uses pre-defined rules to match the text. In the very first step, input news sentence has already been segmented and tagged with POS. Here all words are first matched with a predefined keyword list. The keyword list includes company names, addresses, industry names and key verbs. All matched words

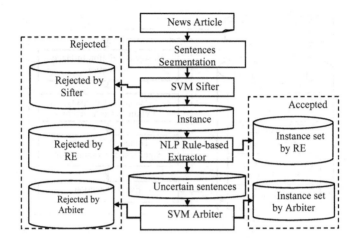

Fig. 3. Relation instance extraction

or word combinations are tagged as entities. Then the labeled entities are matched with every rule. The instance candidates could be classified into three parts: The first part contains no subjects (company names) and matches none of the pre-define rules, and it is discarded directly. The second part matches one or more rules, and it is collected as the accepted relation instance. The third part contains subjects but does not match any rule. They will be output to the next step.

The SVM arbiter reads the third part of output of RE. And it determines whether the input sentence contains a relation instance. The conventional rule-based method gives high precision and low recall, while SVM arbiter amends recall in this case.

3.3 Temporal Relation Graph Creation

Before creating the temporal relation graph based on the extracted instances, unreliable instance should be filtered out. There are several reasons: Firstly, some relation instances are false due to the error in our extractor. Secondly, we use publishing time as the time of a relation instance, but sometimes this is not the actual relation occurring time. Finally, there are conflictions among relation instances.

Multiple types of relation were considered in our work. And some type of relation has impact on others. So this feature can be used to filter out some un-reliable instances. This is an advantage of multiple type of relation.

3.3.1 Relation Instance Filtering
We use $S(A, B, X, t)$ to denote the strength of the relation between company A and company B with relation type X at time t, and calculate it using the following formula:

$$S(A,B,X,t) = s_{A,B,X}(t) = \sum_{i=1}^{C_t} ms(n_i) \tag{3.1}$$

Where C_t is the number of relation instances on A, B, X within time t, n_i is the ith relation instance, $ms(n_i)$ is the matching score of this relation instance in the news article.

There are two kinds of relation: 1) event-like relation, which can only occur once, and 2) continuous relation, which can last for a long time. Assuming X is event-like relation, if more than one relation instances are detected, e.g., both $R(A, B, X, t_1)$ and $R(A, B, X, t_2)$ $(t_1 < t_2)$ exist, we take the value $S'_{A,B,X}(t)$ calculated using Formula 3.2 as the new relation strength of this kind of relation.

$$s'_{A,B,X}(t_1) = s_{A,B,X}(t_1) + s_{A,B,X}(t_2)$$
$$s'_{A,B,X}(t_2) = 0 \tag{3.2}$$

Next we solve the direction conflict for event-like relation. We select the relation direction with larger strength as the real relation direction and delete the other relation direction.

Then we filter all relation, including both event-like relation and continuous relation, with a threshold. The relation strength which is smaller than a threshold is set to 0. Here the thresholds for event-like relation and continuous relation are different.

Finally we solve the conflict between special relation types. If an event-like relation X_1 and a continuous relation X_2 can not occur at the same time t_0, we compare $s_{A,B,X_1}(t_0)$ with $\sum_{t > t_0} s_{A,B,X_2}(t)$. If $s_{A,B,X_1}(t_0) > \sum_{t > t_0} s_{A,B,X_2}(t)$, we filter out all relation X_2 after time t_0; otherwise we filter out relation X_1.

3.3.2 Temporal Relation Interpolation

For the continuous relation, the relation always continues for a time period. But from the relation instances, we can only see some relation occur at some specific time but not occur at other time. The interpolation could be done to guarantee the continuous characteristic. If a relation doesn't exist at a time t_n, we will try to use the relation before t_n, and the relation after t_n, to do interpolation. A simple linear or exponential interpolation just works.

Now we have each type of temporal relation. These temporal relations could be used to create temporal graphs. Each company corresponds to a node of a graph, and each relation corresponds to a link of a graph. And the weight of a link is equal to the strength of the related relation.

3.3.3 Comprehensive Relation Calculation

It is not a good way to display all the relations in a single graph because it will be very complex. We could build a comprehensive relation between all companies. The comprehensive relation between two companies shows how deeply they relate to each other in this community in the sense of business. We simply use weighted company relation strength of each relation type and sum them up. The weight for every type of relation could be calculated with the probability of each type of relation. And in case not enough data exists, an experience value could be used. Here only continuous relation types are calculated.

3.4 Business Event Detection

With temporal relation graph on each type of relation and temporal comprehensive relation graph, the trend in companies and industry are more likely to be discovered.

3.4.1 Company Significance Calculation

In this paper, we define the significance of a company as the importance of a company to this business community. In a business relation network, a company has larger significance if it has more relations with other companies. We use the significance of a node in this graph to represent the significance of a company.

We use PageRank [16] to compute the significance. Assuming the adjacent matrix of a graph is A, we can compute the largest eigenvector of matrix $A + \lambda E$ where E is the matrix consists of all ones. The ith element in the eigenvector represents the significance of ith company. And we normalize it to let the biggest one always be equal to 1.

The graph we used is the comprehensive relation graph. The significance changes over time, so we compute the significances of all companies at every time unit. Then all the significances of a company form a temporal significance chart.

3.4.2 Rule-Based Business Event Detection

Business events depend highly on the prior knowledge. Each domain has its own definition of event mining. We use a basic rule-based method to do event detection. Rules are compiled by people with prior business knowledge. The example rules are shown as follows:

1) For Company A, if $\dfrac{S_A(t_1) - S_A(t_0)}{t_1 - t_0} > Th_1$, then A developed rapidly from t_0 to t_1

2) For Company A, if $\dfrac{S_A(t_0) - S_A(t_1)}{t_1 - t_0} > Th_2$, then A had a problem from t_0 to t_1.

The simplest event is acquisition and merging event, which could be detected directly through analyzing the acquisition and incorporation relation strength.

4 Implementation and Experiment

4.1 System Implementation

255 companies, which produce/sell computer and electronics products, were collected. The aliases of each company were prepared manually. For instance, "3M", "MMM" are aliases of 3M Corporation. By search engines, 1.8 million web pages, which were published from January 1st, 1995 to July 1st, 2007 and contain company names, were analyzed. More than 100 templates were defined to extract text of content, the title of every news article and the time stamp. And finally about 1.7 million of articles were successfully extracted.

More than 4,000 sentences were tagged for training SVM sifter and SVM arbiter, and total 146 rules were created by human to build relation extraction module. Gate [9] was applied in rule based extraction, and SVM-Light as SVM training tool. Regarding its extraction accuracy, closed test was done on the tagged samples. And the overall accuracy for all kinds of relation is from 60% to 80%. Finally 4864, 40057, 8069, 1952, 14108, 1764 instances for cooperation, competition, acquisition, incorporation, supply chain and stock share were detected respectively.

Month and year were used as time units to build the temporal graph. In order to aid visual comparison in the graph, relation strength was normalized from absolute value to a period of 0 to 1.

4.2 Extraction and Mining Results

4.2.1 Acquisition Event Detection Result

We detected the following acquisition events in Table 1. Each row in Table 1 shows an acquisition event. The first company acquired the second company.

Table 1. Acquisition event

No.	Company	Company	Time
1	3Com	USR	1997-02-17
2	AMD	ATI	2003-06-23
3	EMC	BMC	2002-05-21
4	Corel	Borland	2002-2-8
5	Cisco	Linksys	2003-3-20
6	EMC	Dantz	2004-10-11
7	EMC	RSA	2005-10-6
8	Seagate	Maxtor	2004-7-26
9	Veritas	Seagate	1999-6-25
10	Symantec	Veritas	2002-1-30
11	Toshiba	Westinghouse	2001-2-5
12	Lenovo	IBM	2002-2-17
13	LexMark	IBM	1998-7-27
14	Ricoh	IBM	2002-2-6
15	Corel	Novell	1996-1-31
16	Novell	CNet	2000-1-25

Only one error, Novell/CNet acquisition, is in this set. This is caused by the falsely extracted acquisition relation instance.

There are two types of acquisition relationships. One is that one company acquires the other company entirely. The other is that one company acquires parts of the other company. We compare the significances of these two companies to tell the type.

The significance of 3Com on Feb. 1997 is 0.1698. The significance of US Robotics on Feb. 1997 is 0.0003. From the significance chart of 3Com and US Robotics shown in Figure 4, we could easily know that 3Com acquired US Robotics entirely. The vertical blue line shown in Figure 4 indicates the acquisition date. All acquisition case 1~11 in Table 1 follow this rule.

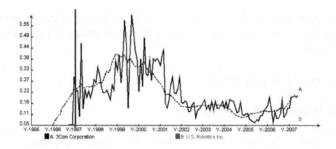

Fig. 4. Company significance comparison between 3Com and US Robotics

In Figure 5, IBM always has much larger significance than Lenovo. After the acquisition event date 2002-2-17, the significance of IBM did not decrease. So we can definitely say Lenovo acquired a department of IBM. From Figure 5, we also know that when Lenovo acquired IBM's department in 2002, IBM kept its leader status in this industry but Lenovo spent years to fit in this overseas acquisition. All cases 12~15 in Table 1 are this kind of acquisition relationship.

4.2.2   Complicated Event Detection Example

With all the data we get, more complex events could be detected. But prior knowledge is required to obtain it. Here we give some as an example.

We show a part of the competition relation graph in 1995, 1996 and 1997 respectively in Figure 6~8. Each node in the figure represents a company, and each link corresponds to a competition relation between two companies. The number labeled on a link is the normalized competition relation strength. We select some companies from the total 255. In each graph, the nodes in the left frame don't have any other links to other companies which do not appear in this graph. The companies shown in the right part of the figure are some kernel companies we selected from the total graph. They may have other links to other companies which do not appear in this graph. We could display the whole graph and detect the same event. But the graph is too large to be displayed in a page. So what we show here is the part of the graph.

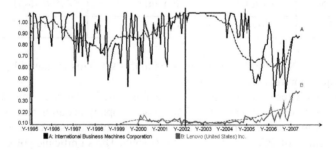

Fig. 5. Company significance comparison between Lenovo and IBM

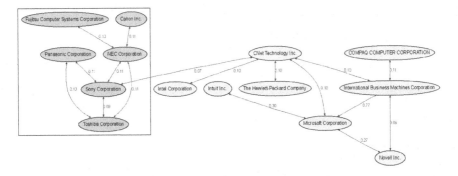

Fig. 6. Part of the company competition graph in 1995

The nodes in the left frame in Figure 6 form a relatively independent group. These companies are all Japanese companies. Here all news we collected is English news which mainly reports about company activities in countries where English is the native language, especially in the 1990's. We could infer that in 1995 these Japanese companies don't compete with companies in the western world. They tend to be a close group. In 1996, they remain as a group but some of them begin to compete with companies in the western world. The precedents are Sony, Toshiba and NEC. In 1997, the original members are divided into two parts. The first one remained in this group. This group evolved into a group producing camera related products. Other Japanese companies escaped this small group successfully and became involved in the whole big community. But Panasonic is an exception. Though it belongs to the big community, it only has competition relation with the original group member.

In this procedure, Sony, Toshiba are more open than other companies. This predicted their future development opportunity in some sense.

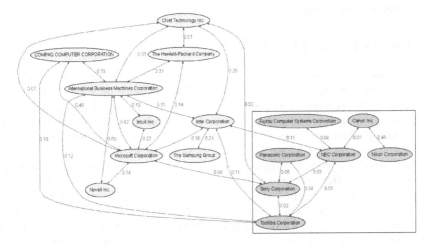

Fig. 7. Part of the company competition graph in 1996

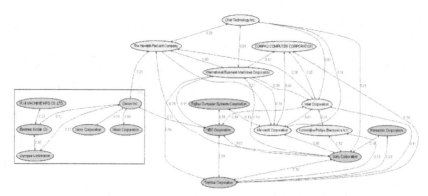

Fig. 8. Part of the company competition graph in 1997

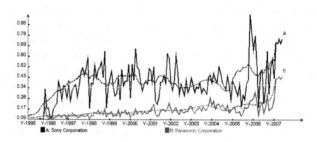

Fig. 9. Company significance comparison between Sony and Panasonic

The significance comparison between Sony and Panasonic is shown in Figure 9. We could see that though their starting points are almost same in 1995 in the western world market, Sony developed much better than Panasonic subsequently in western world market, especially from 1995 to 1997.

5   Conclusions and Future Work

In this paper, we studied the problem of temporal company relation mining. Our work focuses on multiple types of relation and temporal relation analysis. Those are new features in this area. We proposed a new approach of creating temporal relation graphs from the Web. Based on the temporal graphs we created, we developed a new approach on business event mining. The experiments showed that this temporal relation mining approach can be applied on large amounts of data. The results showed some interesting cases.

We started the work in temporal company relation mining area. It is worth to do more work on it. Here, we give some future directions.

Our method is validated, and is effective in extracting useful information (i.e., company relations, business events) from news articles. But we believe it can be applied to other sources (i.e., BLOG data, reports, etc.). And it will result in higher information extraction accuracy by using multi-source information fusion.

We plan to do automated industry and sub-industry clustering based on the relation graphs. Because of the data we collected, the companies are not easy to be clustered based on their industry. All companies are those who produce/sell electronics or computer related product. They tend to form a tight community. Next we will work on more companies from diverse industries.

The proposed method can be easily expanded to support multiple languages. Actually, it has been applied to several other languages (i.e., Chinese, Japanese) and generated similar result as English. With multi-lingual support, it enables to build a more global view on the business environment.

Acknowledgements

We thank Hong Wu, Genqing Wu and Jianming Jin for their initial work on this paper.

References

1. van der Aalst, W.M.P., Song, M.S.: Mining social networks: Uncovering interaction patterns in business processes. In: Desel, J., Pernici, B., Weske, M. (eds.) BPM 2004. LNCS, vol. 3080, pp. 244–260. Springer, Heidelberg (2004)
2. Appelt, D.: An introduction to information extraction. Artificial Intelligence Communications 12(3), 161–172 (1999)
3. Bengtsson, M., Kock, S.: Cooperation and competition in relationships between competitors in business networks. Journal of Business and Industrial Marketing 14(3), 17–19 (1999)
4. Chung, F.R.: Spectral Graph Theory. American Mathematical Society, Providence, Rhode Island (1997)
5. Cunningham, H.: Information Extraction, Automatic. Encyclopedia of Language and Linguistics (2005)
6. Grey, W., Olavson, T., et al.: The role of e-marketplaces in relationship-based supply chains: A survey. IBM Systems Journal 44(1) (2005)
7. Hou, X., Liu, T., et al.: Connection Network Intelligence (2006), http://www.research.ibm.com/-jam/CNI_Jam_Demo_PPT_v1.5.pdf.2006
8. http://www.buddygraph.com/
9. http://gate.ac.uk/
10. http://www.metasight.co.uk/
11. Hu, M., Liu, B.: Mining and summarizing customer reviews. ACM SIGKDD, 168–177 (2004)
12. Liu, B., Zhao, K., et al.: Visualizing Web Site Comparisons. In: WWW 2002 (2002)
13. Liu, J., Wagner, E., et al.: Compare&Contrast: Using the Web to Discover Comparable Cases for News Stories. In: WWW 2007, pp. 541–550 (2007)
14. McCoy, D.W.: Business Activity Monitoring: Calm Before the Storm. Gartner Research, ID: LE-15-9727 (2002)
15. Morinaga, S., Yamanishi, K., et al.: Mining Product Reputations on the Web. In: KDD 2002, pp. 341–349 (2002)
16. Page, L., Brin, S.: The PageRank Citation Ranking: Bringing Order to the Web. Stanford Technical Report (1998)
17. Rygielski, C., Wang, J.C., et al.: Data mining techniques for customer relationship management. Technology in Society 24, 483–502 (2002)
18. Scott, J.: Social Network Analysis. Sage, Thousand Oaks (1992)
19. Tateishi, K., Ishiguro, Y., et al.: A Reputation Search Engine that Collects People's Opinions by Information Extraction Technology. IPSJ Transactions on Databases 22, 115–123 (2004)
20. Wasserman, S., Faust, K.: Social Network Analysis: Methods and Applications. Cambridge University Press, Cambridge (1994)

StreetTiVo: Using a P2P XML Database System to Manage Multimedia Data in Your Living Room

Ying Zhang[1], Arjen de Vries[1,2], Peter Boncz[1], Djoerd Hiemstra[3], and Roeland Ordelman[3]

[1] Centrum voor Wiskunde en Informatica, Amsterdam, the Netherlands
[2] Delft University of Technology, Delft, The Netherlands
[3] University of Twente, Enschede, The Netherlands
{zhang,arjen,boncz}@cwi.nl, {hiemstra,ordelman}@cs.utwente.nl

Abstract. *StreetTiVo* is a project that aims at bringing research results into the living room; in particular, a mix of current results in the areas of Peer-to-Peer XML Database Management System (P2P XDBMS), advanced multimedia analysis techniques, and advanced information retrieval techniques. The project develops a plug-in application for the so-called Home Theatre PCs, such as set-top boxes with MythTV or Windows Media Center Edition installed, that can be considered as programmable digital video recorders. StreetTiVo distributes compute-intensive multimedia analysis tasks over multiple peers (i.e., StreetTiVo users) that have recorded the same TV program, such that a user can search in the content of a recorded TV program *shortly* after its broadcasting; i.e., it enables *near real-time* availability of the meta-data (e.g., speech recognition) required for searching the recorded content. Street-TiVo relies on our P2P XDBMS technology, which in turn is based on a DHT overlay network, for distributed collaborator discovery, work coordination and meta-data exchange in a volatile WAN environment. The technologies of video analysis and information retrieval are seamlessly integrated into the system as XQuery functions.

1 Introduction

Things are changing in the living room, under the TV set: TV is going digital and consumer electronics gets networked. The so-called "set-top" boxes that are needed for digital television have appeared in the houses of many; and, many of these set-top boxes are rather powerful computers connected to the Internet, running Windows Media Center or its open-source MythTV equivalent. A related trend is the increasing demand for multimedia information access. Presently, "ordinary" people own hundreds of gigabytes of multimedia data, resulting from their digital photo cameras, hard-disk video recorders, etc. However, searching multimedia files is usually restricted to simple look up in the meta-data of files, such as file names and (human-edited) descriptions. More advanced multimedia retrieval requires highly compute-intensive pre-processing of the data

Q. Li et al. (Eds.): APWeb/WAIM 2009, LNCS 5446, pp. 404–415, 2009.

(e.g., speech recognition and image processing). As a rough estimation, it takes a moderate computer more than one order of magnitude more time to derive the auxiliary data that would enable better search facilities.

The idea of *StreetTiVo* is to unite the computing power of those Media Center devices (peers) in the living rooms. By distributed and parallel execution of compute-intensive multimedia analysis tasks on multiple peers, near real-time indexing of the content can be provided using just the ordinary hardware available in the network. StreetTiVo divides the Media Center devices into groups and assigns each group a short time slice of a recording (e.g., ten seconds), to run multimedia analysis tools on those time slices only. Thus, the peers form virtual digital streets and are virtual neighbours of each other. Note that StreetTiVo uses the P2P concept in a strictly legal way, as it is not used to distribute the video files themselves. Users can only watch the content that they have recorded themselves. What is exchanged by StreetTiVo are only the results of multimedia analysis of those videos, i.e., generated meta-data.

Summarizing, the goal of the StreetTiVo project is: *unite Media Center devices using P2P technologies to cooperatively run compute intensive multimedia analysis applications just in everybody's living room so that they could produce results in near real-time.*

Imagine for example, if only a tiny fraction of the millions of people recording the Champions League soccer competition would participate in StreetTiVo! Useful media analysis tools would include the transcription of the text spoken by the presenters during the match, but also the cross-media analysis to recognize goals and other exciting moments (e.g., from audio volume and/or camera motion patterns). After each group of media centers has exchanged its partial analysis results with the other groups (that recorded the same match), all StreetTiVo participants obtain the complete set of automatically derived annotations, so that meta-data could be used for direct entry to the most exciting moment(s) of the game, or, the automatic generation of a summary including all highlights.

In this paper we describe the P2P data management approach of StreetTiVo in Section 2. The current system architecture is presented in Section 3. We briefly describe the three major components: XRPC, ASR and PF/Tijah. This first version of StreetTiVo was chosen to be simple, so that we could quickly demonstrate the cooperation between XRPC, ASR, PF/Tijah in a non-trivial application setting. A distributed architecture is adopted, in which all StreetTiVo users (i.e., clients) are managed by a central server. In Section 4, we explain the envisioned P2P model and discuss the challenges on our way to make StreetTiVo a truly P2P application. We conclude in Section 5.

2   P2P Data Management

From a network communication perspective, DHT-based overlays have gained much popularity in both research projects and real-world P2P applications [1,2,3,4,5,6,7,8,9,10,11,12]. DHT networks have proven to be efficient and scalable (guarantees $O(logN)$ scalability) in volatile WAN environments [6,13].

From a data management perspective, XML has become the de facto standard for data exchange over the Internet, and XQuery the W3C standard for querying XML data. The infrastructure of StreetTiVo is therefore provided by *MonetDB/XQuery** [14], a P2P XML DBMS that supports distributed evaluation of XQuery[15] queries over DHT networks [1,3,4,6]. Each peer runs an XDBMS, *MonetDB/XQuery* [16], to manage its local data. Communication among peers is done by remote execution of XQuery functions using *XRPC* [17,18], a simple XQuery extension for Remote Procedure Calls (RPC) that enables *efficient* distributed querying of *heterogeneous* XQuery data sources.

The current implementation of StreetTiVo as XQuery expressions applies a Dutch automatic speech recognition system (*ASR*) [19] to the recorded programmes, and provides a full-text retrieval service of the ASR output by employing the MonetDB/XQuery extension *PF/Tijah* [20]. When a TV program is broadcast, all participants (peers that are recording this TV program) jointly extract the (Dutch) spoken text using the ASR component, and exchange local partial ASR results with each other. ASR produces XML documents containing speech texts and some meta-data, for example, start/end timestamp of a sentence in the video file. Each participant stores *all* resulting XML documents of the recording in its local MonetDB/XQuery that can then be queried using PF/Tijah. The meta-data of the retrieved sentences provide sufficient information for the StreetTiVo GUI to display only the desired video fragment.

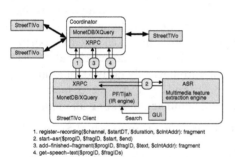

1. register-recording($channel, $startDT, $duration, $cIntAddr): fragment
2. start-asr($progID, $fragID, $start, $end)
3. add-finished-fragment($progID, $fragID, $text, $cIntAddr): fragment
4. get-speech-text($progID, $fragIDs)

```
⟨recording progID="bbc1_20080425_200000"⟩
  ⟨participants⟩
    ⟨participant host="x.example.org" /⟩
    ⟨participant host="y.example.org" /⟩
  ⟨/participants⟩
  ⟨fragments⟩
    ⟨fragment fragID="1" start="0" end="10"⟩
      ⟨owner host="x.example.org" /⟩⟨/fragment⟩
    ⟨fragment fragID="2" start="10" end="25"⟩
      ⟨assignee host="y.example.org" /⟩⟨/fragment⟩
    ⟨fragment fragID="3" start="25" end="30"⟩
      ⟨assignee host="x.example.org" /⟩⟨/fragment⟩
    ⟨fragment fragID="4" start="30" end="38" /⟩
  ⟨/fragments⟩
⟨/recording⟩
```

Fig. 1. StreetTiVo architecture: client-server model

Fig. 2. Information maintained for each recording

3 StreetTiVo Architecture

The current version of StreetTiVo uses a simple client-server network model, as shown in Figure 1. In this setup, we assume that the coordinator is a reliable host, while the clients join and leave unpredictably (similar to the early work of [21]). While our next step will be to replace this model with a more sophisticated DHT-based P2P model, we first detail the current implementation.

Each peer runs a MonetDB/XQuery server and communicates with other peers via XRPC, by sending SOAP XRPC request/response messages. The central StreetTiVo coordinator is responsible for registration of recordings, and the generation and distribution of ASR tasks. For each recording, the coordinator maintains a list of participating peers and a list of tasks (called fragments). A recording is divided into short fragments (usually several seconds) to be analyzed parallelly by the participants. For each fragment, the coordinator maintains if it is being processed by a peer (i.e., it has an assignee), or if its speech text is already available (i.e. it has an owner). All meta-data are in XML format and stored in MonetDB/XQuery.

Example 1. The XML snippet in Figure 2 shows the recording element for the TV program bbc1_20080425_200000. The *progID* is determined by channel, date and start time. The TV program is recorded by two peers and is divided into 4 fragments. The attributes start and end of a fragment indicate the relative start/end timestamps of the fragment in the video file. Fragments are assigned to the participants in the order they register, so initially fragments 1 and 2 are assigned to the hosts x.example.org and y.example.org, respectively. So far, x.example.org has finished analyse fragment 1 (indicated as the owner of the fragment) and has been assigned a new job fragment 3. The host y.example.org is still processing fragment 2. Since there are no more participants, fragment 4 is waiting to be assigned.

All interfaces between coordinator, clients and the local ASR engine have been defined as XQuery functions. A small Java program implements the XQuery function (start-asr() in Figure 1) that triggers the ASR engine. The interaction between one StreetTiVo client and the coordinator is shown in details. Collaborative speech recognition works as follows.

Step ①. When a TV program is scheduled for recording, the StreetTiVo client sends an XRPC request to the coordinator to execute the function register-recording(). The coordinator responds with a fragment, which has not been processed by any participants, and inserts the client as an assignee into the fragment element. For reliability reason, each fragment is assigned to multiple clients.

If the request is the first registration for a TV program, the coordinator first needs to generate ASR tasks. An easy way to do this is to divide the whole recording into equal sized fragments. To get high quality ASR result, the ASR segmenter should be used, which is able to filter out the audio that do not contain speech and generates fragments accordingly (see Section 3.2). Information provided by the ASR segmenter ensures that the ASR speech recognizer produces more accurate results. However, the better quality comes at a high cost of speed, because the coordinator must record the TV program itself and run ASR segmenter *afterwards*. So, there is a trade-off between speed and quality.

Step ②. Upon receipt of the response from the coordinator (in *Step* ①), the StreetTiVo client starts its local ASR engine to analyze the fragment specified in the response message, by calling the interface function start-asr().

Step ③. After the ASR engine has finished analyzing a fragment, the StreetTiVo client reports this by calling add-finished-fragment() on the coordinator and passing among others the retrieved text as parameter.

Step ④. After having finished one task, the StreetTiVo client is expected to request a new task (**get-job()**), until the coordinator responds with an empty task, which might mean that there are sufficient number of assignees for each fragment, or that the coordinator has received the ASR results of all fragments.

Step ⑤. If there are no new tasks, the StreetTiVo client waits for some predefined time to give the other participants the opportunity to finish their ASR tasks, and then attempts to retrieve the speech text of the missing fragments. The StreetTiVo client can directly ask the coordinator for the missing fragments (**get-speech-text()**), since all ASR results are also stored at the coordinator, but the preferred procedure is to just call the coordinator's **get-fragments()** function (not shown) to find out what participant owns which ASR result, and retrieve the fragments' texts from those nodes.

Once the ASR results are locally available, StreetTiVo users can search for video fragments by entering keywords in the GUI. The keywords are subsequently translated into Tijah queries. PF/Tijah returns matching sentences ranked by their estimated probability of relevance.Each sentence is tagged with its relative start/end timestamps in the video file, this way, the desired video fragments can be retrieved. In summary, StreetTiVo has three major components: XRPC takes care of communication among peers, ASR provides video analysis functions, and PF/Tijah enables retrieval of video fragments using keywords. In the remainder of this section, we give a brief overview of these techniques.

3.1   XRPC: Distributed XQuery Processing

XQuery 1.0 [15] only provides a *data shipping* model for querying XML documents over the Internet. The built-in function fn:doc() fetches an XML document from a remote peer to the local server, where it subsequently can be queried. The recent W3C Candidate Recommendation *XQuery Update Facility* (XQUF) [22] introduces a built-in function fn:put() for remote storage of XML documents, which again implies data shipping. There have been various proposals to equip XQuery with *function shipping* style distributed querying abilities [23,24,25]. On the syntax level, we consider our XRPC proposal an incremental development of these. XRPC adds RPC to XQuery in the most simple way: adding a destination URI to the XQuery equivalent of a procedure call (i.e. function application).

Remote function applications in XRPC take the XQuery syntax: execute at {Expr}{FunApp (ParamList)}, where *Expr* is an XQuery xs:string expression that specifies the URI of the peer on which the function *FunApp* is to be executed. As a running example, we assume a set of XQuery DBMS (peers) that each store a movie database in an XML document filmDB.xml with contents similar to:

```
⟨films⟩
    ⟨film⟩⟨name⟩The Rock⟨/name⟩⟨actor⟩Sean Connery⟨/actor⟩⟨/film⟩
    ⟨film⟩⟨name⟩Green Card⟨/name⟩⟨actor⟩Gerard Depardieu⟨/actor⟩⟨/film⟩
⟨/films⟩
```

We assume an XQuery module film.xq stored at the host example.org that defines a function filmsByActor():

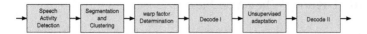

Fig. 3. Overview of the ASR decoding system

```
module namespace file= "films" ;
declare function film:filmsByActor($actor as xs:string) as node()*
  { doc( "filmDB.xml" )//name[../actor=$actor] };
```

With XRPC, we can execute this function on a remote peer, e.g. x.example.org, to get a sequence of films in which Sean Connery plays in the film database stored on the remote peer.

```
import module namespace f= "films" at "http://example.org/film.xq" ;
⟨films⟩ { execute at { "xrpc://x.example.org" } {f:filmsByActor( "Sean Connery")} } ⟨/films⟩
```

which yields: ⟨films⟩⟨name⟩The Rock⟨/name⟩⟨/films⟩. We introduce here a new xrpc:// network protocol, accepted in the destination URI of execute at, to indicate a peer's ability of handling XRPC queries. The generic form of such URIs is xrpc://⟨host⟩[:port] [/[path]], where xrpc:// is the network protocol, ⟨host⟩[:port] identifies the remote peer, and [/[path]] is an optional local path at the remote peer.

The SOAP XRPC Protocol. The design goal of XRPC is to create a distributed XQuery mechanism with which *different* XQuery engines at different sites can jointly execute queries. This implies that our proposal also encompasses a *network protocol*, SOAP XRPC, which uses the Simple Object Access Protocol (SOAP) [26] (i.e. XML messages) over HTTP. XML is ideal for distributed environments (think of character encoding hassles, byte ordering), XQuery engines are perfectly equipped to process XML messages, and an XML-based message protocol makes it trivial to support passing values of any type from the XQuery Data Model [27]. The choice for SOAP brings as additional advantages seamless integration of XQuery data sources with web services and Service Oriented Architectures (SOA) as well as AJAX-style GUIs. The complete specification of the SOAP XRPC protocol can be found in [17]. Here we show, as an example, the XRPC request message that should be generated for the query above:

```
⟨?xml version="1.0" encoding="utf-8"?⟩
⟨env:Envelope xmlns:xrpc="http://monetdb.cwi.nl/XQuery"
              xmlns:env="http://www.w3.org/2003/05/soap-envelope"
              xmlns:xs="http://www.w3.org/2001/XMLSchema"
              xmlns:xsi="http://www.w3.org/2001/XMLSchema-instance"
              xsi:schemaLocation="http://monetdb.cwi.nl/XQuery
                                  http://monetdb.cwi.nl/XQuery/XRPC.xsd" ⟩
  ⟨env:Body⟩
    ⟨xrpc:request xrpc:module="films" xrpc:method="filmsByActor" xrpc:arity="1" xrpc:iter-count="1"
                  xrpc:updCall="no" xrpc:location="http://x.example.org/film.xq" ⟩
      ⟨xrpc:call⟩
        ⟨xrpc:sequence⟩
          ⟨xrpc:atomic-value xsi:type="xs:string"⟩Sean Connery⟨/xrpc:atomic-value⟩
        ⟨/xrpc:sequence⟩
      ⟨/xrpc:call⟩
    ⟨/xrpc:request⟩
  ⟨/env:Body⟩
⟨/env:Envelope⟩
```

3.2   ASR: I Know What You Said

Automatic Speech Recognition (ASR) supports the conceptual querying of video content and the synchronization to any kind of textual resource that is accessible, including other annotations for audiovisual material such as subtitles. The potential of ASR-based indexing has been demonstrated most successfully in the broadcast news domain. Typically large vocabulary speaker independent continuous speech recognition (LVCSR) is deployed to this end.

The ASR system deployed in StreetTiVo was developed at the University of Twente and is part of the open-source SHoUT speech recognition toolkit[1]. Figure 3 gives an overview of the ASR decoding system. Each step provides the input for the following step. The whole process can be roughly divided into two stages. During the first stage, the *Speech Activity Detection* (SAD) is used to filter out the audio parts that do not contain speech. This step is crucial for the performance of the ASR system to avoid that it tries to recognize non-speech audio that is typically found in recorded TV programs such as music, sound effects or background noise with high volume (traffic, cheering audience, etc). After SAD, the system tries to figure out 'who spoke when', a procedure that is typically referred to as speaker diarization. In this step, the speech fragments are split into segments that only contain speech from one single speaker. Each segment is labelled with its corresponding speaker ID. Next, for each segment the vocal tract length (VTLN) warping factor is determined for vocal tract length normalisation. Variation of vocal tract length between speakers makes it harder to train robust acoustic models. In the *SHoUT* system, normalisation of the feature vectors is obtained by shifting the Mel-scale windows by a certain warping factor during feature extraction for the first decoding step.

After having cleaned up the input sound and gained sufficient meta information, speech recognition can be started in the second stage. Decoding is done using the HMM-based Viterbi decoder. In the first decoding iteration, triphone VTLN acoustic models and trigram language models are used. For each speaker, a first best hypothesis aligned on a phone basis is created for unsupervised acoustic model adaptation. Optionally, for each file a topic specific language model can be generated based on the input of first recognition pass. The second decoding iteration uses the speaker adapted acoustic models and the topic specific language models to create the final first best hypothesis aligned on word basis. Also, for each segment, a word lattices is created. A more detailed description of each step can be found in [19].

3.3   PF/Tijah: XML Text Search

PF/Tijah [20] is another research project run by the University of Twente with the goal to create a flexible environment for setting up search systems by integrating MonetDB/XQuery that uses the Pathfinder compiler [28] with the Tijah

---

[1] For information on the use of the SHoUT speech recognition toolkit see http://wwwhome.cs.utwente.nl/~huijbreg/shout/index.html

XML Information Retrieval (IR) system [29]. The main features supported by PF/Tijah include the following:

- Retrieving arbitrary parts of textual data, unlike traditional IR systems for which the notion of a document needs to be defined up front by the application developers. For example, if the data consist of scientific journals one can query for complete journals, journal issues, single articles, sections from articles or paragraphs *without* adapting the index or any other part of the system configuration;
- Complex scoring and ranking of the retrieved results by means of so-called *Narrowed Extended XPath* (NEXI) [30] queries. NEXI is a query language similar to XPath that only supports the descendant and the self axis step, but that is extended with a special about() function that takes a sequence of nodes and ranks those by their estimated probability of relevance to the query;
- PF/Tijah supports incremental indexing: when new ASR fragments are added to the database, their text will be automatically indexed by PF/Tijah, *without* the need to re-index the entire database from scratch;
- search combined with traditional database querying, including for instance joins on values. As an example, one could search for programmes mentioning "football" that are broadcast on the same channel as programmes mentioning "crime".

StreetTiVo inserts fragments containing the transcripts of ASR whenever they are available. Therefore, they will not be nicely grouped per programme in the database, nor will they be in chronological order. The combination XQuery and NEXI text search enables StreetTiVo to search matching fragments, combine the fragments with the same programme identifier, combine their scores (or take the score of the best matching fragment), rerank programmes by the scores of their fragments, and display the matching programmes along with their best matching fragments: all of this is done in one query.

4   Next Steps

Our next step in the development of StreetTiVo is to replace the client-server model with MonetDB/XQuery* [14], which integrate the P2P data structure Distributed Hash Tables (DHTs) into XQuery (see Figure 4). A DHT [1,3,4,6] provides *(i)* robust connectivity (i.e., it tries to prevent network partitioning), *(ii)* high data availability (i.e., prevent data loss if a peer goes down by automatic replication), and *(iii)* a

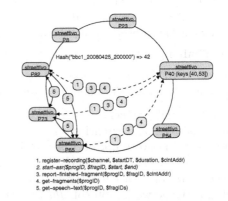

Fig. 4. StreetTiVo architecture: P2P model

scalable $(key, value)$ storage mechanism with $O(log(N))$ cost complexity, where N is the number of peers in the network. A number of P2P database prototypes have already used DHTs [5,7,8,10,11].

In a StreetTiVo system using a DHT model, peers are managed by a DHT ring. There is no single StreetTiVo coordinator. All peers are unreliable and each can be both a coordinator and a client, thus, each peer must additionally support the functions provided by the coordinator, as discussed in Section 3. The process to collectively speech extraction using ASR is similar as in the client-server model, except several small changes:

Example 2. Figure 4 shows an example scenario, in which three peers N65, N73 and N82 will record the TV program with *progID*="bbc1_20080425_200000". All participants use the same hash function to calculate the hash of the *progID*, which is 42 here. Since the peer P40 is responsible for keys [40, 53], it is chosen as the coordinator for this TV program. The participating peers register the recording at P40 (Step ①). Generating ASR tasks is still done by the (temporary) coordinator, P40. When a peer finished an ASR task, it reports this at the coordinator (Step ③), but *without* sending the extracted text. All participants will repeat steps ④, ② and ③ to process more fragments, until there are no more ASR tasks. Finally, the participants exchange the ASR results by first finding the owner of each fragments from the coordinator (Step ⑤), and then retrieving the missing ASR results from each other (Step ⑤).

Note that, in the DHT model, the availability of the recording meta-data (i.e., the recording elements maintained by a coordinator as discussed in Section 3) is guaranteed thanks to the automatic replication facility provided by the underlying DHT network, that is, all data on a peer managed by the DHT network are replicated on the peer's predecessors and successors. Thus, if peer P40 would fail in our example, all request with key 42 would be routed by the DHT network to P23 or P54. Also note that, the ASR results are not managed by the DHT network. Basically, all StreetTiVo peers can retrieve these data, but only the peers that have recorded the particular TV program can display the video. The availability of the ASR results is affected by the number of participants (more participants $\Rightarrow$ higher availability). However, this is not a crucial issue, since every StreetTiVo peer is able to run ASR on a missing fragments.

The challenges in integration of XQuery and DHT are:

(i) how a DHT should be exploited by an XQuery processor,
(ii) if and how the DHT functionality should surface in the query language.

In MonetDB/XQuery*, we propose to avoid any additional language extensions, but rather introduce a new dht// network protocol, accepted in the destination URI of fn:doc(), fn:put() and execute at. The generic form of such URIs is dht://dht_id/key, where dht:// is the network protocol, dht_id is the ID of the DHT network to be used. Such an ID is useful to allow a P2P XDBMS to participate in multiple (logical)

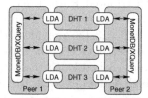

Fig. 5. Tight coupling

DHTs simultaneously (see Figure 4). The *key* is used to store and retrieve values in the DHT.

In the architecture shown in Figure 4, we run the DHT as a separate process called the Local DHT Agent (LDA). Each LDA is is connected to one DHT *dht_id*. We propose a *tight coupling* between the DHT network and the XDBMS [14], in which each DHT peer uses its local XDBMS to store the data (i.e. XML documents) and the local XDBMS uses its underlying DHT network to route XQuery queries to remote peers for execution (i.e. pass XRPC requests to the LDA). A positive side-effect of this tight coupling is that the DBMS gets access to the information internal to the P2P network. This information (e.g. peer resources, connectivity) can be exploited in query optimization. To realize this coupling, we need to extend the DHT API (**put()** and **get()**) with one new method: xrpc (*key*, q, m, $f(ParamList)$) : item()*, where $f(ParamList)$ is the XQuery function that is to be executed on a remote DHT peer determined by *key*. The parameters q and m specify XQuery module, in which the function f_r is defined and the location of the module file. With this method, an XRPC call on a peer p_0 to a dht://dht_x/key_y URI is handled as follows:

1. The XRPC request($q, m, f, ParamList$) is passed to the Local DHT Agent lda_0^x of p_0, which in turn passes the request to the DHT network dht_x.
2. The DHT dht_x routes the request using the normal DHT routing mechanism to the peer p_i responsible for key_y.
3. When the LDA lda_i^x on p_i receives such a request, it performs an XRPC call containing the same request to the MonetDB/XQuery instance on p_i.
4. When lda_i^x receives the response message, it transports the response back via dht_x to the query originator p_0.

Use Cases. Below we show how two main StreetTiVo functions can be implemented as XQuery module functions, which then can be executed using XRPC and the tightly coupled DHT semantics.

(i) Collaborator Discovery. In StreetTiVo, every TV program has a unique identifier *progID*, and for each recorded TV program a recording element is maintained by the peer responsible for the key *hash(progID)* with lists of participants and fragments. If a peer is going to record the TV program "bbc1_20080425_200000", it should register the recording at the coordinator of this TV program. This can be done by the following XRPC call:

```
import module namespace stv = "streettivo" at "http://example.org/stv.xq";

let $key := hash("bbc1_20080425_200000"),
    $dst := fn:concat("dht://dht_x/", $key)
return execute at {$dst} {stv:register-recording(bbc1, "2008-04-25T20:00:00", "1H", "x.example.org")}
```

(ii) Distributed Keyword Retrieval. Assume a StreetTiVo user wants to search in today's newscast "bbc1_20080425_20-0000", he/she has recorded, for video fragments that were about the situation in Tibet, but the ASR results are not completely available (yet) on his/her local machine. Then the search request might be sent to other StreetTiVo peers that have recorded the same newscast.

The following pseudo-code first retrieves the list of fragments from the coordinator, and then sends a search request to each peer that owns ($frags//owner/@host) the ASR results of a fragment:

```
import module namespace stv = "streettivo" at "http://example.org/stv.xq";

let $key := hash("bbc1_20080425_200000"),
    $dst := fn:concat("dht://dht_x/", $key)
    $frags := execute at {$dst} {stv:get-fragments(bbc1, "2008-04-25T20:00:00"}
return for $p in $frags//owner/@host return
        execute at {$p} {stv:search("situation in Tibet")}
```

5   Conclusion

In this paper, we have described StreetTiVo, a P2P Information Retrieval system that enables near real-time search in video contents by just using existing hardware in the living rooms to collectively run compute-intensive video analysis video content analysis tools.

Thanks to its implementation in a high-level declarative database language, it is straightforward to extend StreetTiVo with other types of functionality. We plan to complement the current media analysis with image processing techniques to automatically detect celebrities in news broadcasts or goals in soccer matches. Maybe even more interesting is that StreetTiVo users can also easily share their own human made annotations. For example, people usually schedule a recording several minutes before/after the start/end of the to be recorded TV program, to prevent missing part of the program. If just one StreetTiVo user has annotated the exact start/end timestamp of the TV program, the information can be shared in the platform with other StreetTiVo users who have recorded the same program, and the unnecessary parts of their recordings can be removed transparently.

Acknowledgement

The authors would like to thank Marijn Huijbregts for creating the SHoUT ASR toolkits, Erwin de Moel for developing the GUI for Windows Vista Media Center, Henning Rode and Jan Flokstra for their contribution to PF/Tijah, and Niels Nes, Matthijs Mourits and Roberto Cornacchia for their help in organizing the digital video recording. MultimediaN is funded by the Dutch government under contract BSIK 03031.

References

1. Aberer, K.: P-Grid: A Self-Organizing Access Structure for P2P Information Systems. In: CoopIS (2001)
2. Rowstron, A.I.T., Druschel, P.: Pastry: Scalable, decentralized object location, and routing for large-scale peer-to-peer systems. In: Guerraoui, R. (ed.) Middleware 2001. LNCS, vol. 2218, p. 329. Springer, Heidelberg (2001)
3. Ratnasamy, S., Francis, P., Handley, M., Karp, R., Schenker, S.: A Scalable Content-Addressable Network. In: SIGCOMM (2001)
4. Stoica, I., et al.: Chord: A Scalable Peer-to-peer Lookup Service for Internet Applications. In: SIGCOMM (2001)

5. Huebsch, R., et al.: Querying the Internet with PIER. In: VLDB (2003)
6. Rhea, S.C., Geels, D., Roscoe, T., Kubiatowicz, J.: Handling Churn in a DHT. In: USENIX Annual Technical Conference, General Track (2004)
7. Rao, W., Song, H., Ma, F.: Querying XML data over DHT system using xPeer. In: Jin, H., Pan, Y., Xiao, N., Sun, J. (eds.) GCC 2004. LNCS, vol. 3251, pp. 559–566. Springer, Heidelberg (2004)
8. Bonifati, A., Chang, E.Q., Lakshmanan, A.V.S., Ho, T., Pottinger, R.: HePToX: marrying XML and heterogeneity in your P2P databases. In: VLDB (2005)
9. Rhea, S., Godfrey, B., Karp, B., et al.: OpenDHT: a public DHT service and its uses. In: SIGCOMM (2005)
10. Huebsch, R., Chun, B.N., et al.: The Architecture of PIER: an Internet-Scale Query Processor. In: CIDR (2005)
11. Karnstedt, M., Sattler, K.U., et al.: UniStore: Querying a DHT-based Universal Storage. Technical report, EPFL (2006)
12. Zhao, B.Y., et al.: Tapestry: A Resilient Global-scale Overlay for Service Deployment. IEEE J-SAC 22(1) (January 2004)
13. Rhea, S., et al.: Fixing the Embarrassing Slowness of OpenDHT on PlanetLab. In: USENIX WORLDS 2005 (2005)
14. Zhang, Y., Boncz, P.: Integrating XQuery and P2P in MonetDB/XQuery*. In: EROW (January 2007)
15. Boag, S., et al.: XQuery 1.0: An XML Query Language W3C Candidate Recommendation, June 8 (2006)
16. Boncz, P., et al.: MonetDB/XQuery: A Fast XQuery Processor Powered by a Relational Engine. In: SIGMOD (June 2006)
17. Zhang, Y., Boncz, P.: XRPC: Interoperable and Efficient Distributed XQuery. In: VLDB (September 2007)
18. Zhang, Y., Boncz, P.: Distributed XQuery and updates processing with heterogeneous XQuery engines. In: SIGMOD (2008)
19. Huijbregts, M., Ordelman, R., de Jong, F.: Annotation of heterogeneous multimedia content using automatic speech recognition. In: Falcidieno, B., Spagnuolo, M., Avrithis, Y., Kompatsiaris, I., Buitelaar, P. (eds.) SAMT 2007. LNCS, vol. 4816, pp. 78–90. Springer, Heidelberg (2007)
20. Hiemstra, D., Rode, H., van Os, R., Flokstra, J.: PFTijah: text search in an XML database system. In: OSIR (August 2006)
21. de Vries, A., Eberman, B., Kovalcin, D.: The design and implementation of an infrastructure for multimedia digital libraries. In: IDEASapos (July 1998)
22. Chamberlin, D., et al.: XQuery Update Facility (W3C Working Draft 11 (July 2006)
23. Onose, N., Siméon, J.: XQuery at your web service. In: WWW (2004)
24. Re, C., et al.: Distributed XQuery. In: IIWeb (September 2004)
25. Thiemann, C., Schlenker, M., Severiens, T.: Proposed Specification of a Distributed XML-Query Network. CoRR cs.DC/0309022 (2003)
26. Mitra, N., Lafon, Y.: SOAP Version 1.2 Part 0: Primer W3C Recommendation, June 24 (2003), http://www.w3.org/TR/2003/REC-soap12-part0-20030624
27. Fernández, M., et al.: XQuery 1.0 and XPath 2.0 Data Model (XDM) W3C Recommendation, January 23 (2007), http://www.w3.org/TR/xpath-datamodel
28. Grust, T., Sakr, S., Teubner, J.: XQuery on SQL Hosts. In: VLDB (2004)
29. List, J., et al.: Tijah: Embracing information retrieval methods in XML databases. Information Retrieval Journal 8(4), 547–570 (2005)
30. O'Keefe, R.A., Trotman, A.: The simplest query language that could possibly work. In: INEX (2004)

Qualitative Spatial Representation and Reasoning for Data Integration of Ocean Observational Systems[*]

Longzhuang Li[1], Yonghuai Liu[2], Anil Kumar Nalluri[3], and Chunhui Jin[3]

[1] Dept. of Computing Sciences, Texas A&M Uni.-Corpus Christi, Corpus Christi,
TX 78412, USA
Longzhuang.Li@tamucc.edu
[2] Dept. of Computer Science, Uni. Of Wales, Aberystwyth, Ceredigion, SY 23 DB, UK
yyl@aber.ac.uk
[3] Dept. of Computing Sci. Texas A&M Uni.-CC, Corpus Christi, TX, USA

Abstract. Spatial features are important properties with respect to data integration in many areas such as ocean observational information and environmental decision making. In order to address the needs of these applications, we have to represent and reason about the spatial relevance of various data sources. In the paper, we develop a qualitative spatial representation and reasoning framework to facilitate data retrieval and integration of spatial-related data from ocean observational systems, such as in situ observational stations in the Gulf of Mexico. In addition to adopt the state-of-the-art techniques to represent partonomic, distance, and topological relations, we develop a probability-based heuristic method to uniquely infer directional relations between indirectly connected points. The experimental results show that the proposed method can achieve the overall adjusted correct ratio of 87.7% by combining qualitative distance and directional relations.

Keywords: Data integration, qualitative spatial representation and reasoning.

1 Introduction

Historically, environmental, hydrographic, meteorological, and oceanographic data have been collected by numerous local, state, and federal agencies as well as by universities around the Gulf of Mexico (GOM). The GOM observing systems are organized into two groups pertaining to (1) in situ observations (see Figure 1), and (2) satellite observations. Each in situ observation contains a number of stations and each station (represented as a point in Figure 1) provides public Web data source access [7]. Nowadays, users have to manually interact with these large collections of Internet data sources, determine which ones to access and how to access, and manually merge results from different observational stations. Without an adequate system and personnel for managing data, the magnitude of the effort needed to deal with such large and complex data sets can be a substantial barrier to the GOM research.

---

[*] Research is supported in part by the National Science Foundation under grant CNS-0708596 and CNS-0708573.

Q. Li et al. (Eds.): APWeb/WAIM 2009, LNCS 5446, pp. 416–428, 2009.

We intend to answer complex queries by developing a mediator-based data integration system to access the underlying distributed and heterogeneous Web sources of various locations. The integration system should be able to not only answer the traditional keyword-based queries, but also take the consideration of the spatial terms/concepts in the search. The examples of the spatial-aware queries are:

- The water level in *Houston* coastal area (see Figure 1).
- The wind speed in the area close to *TABS F* (see Figure 1).
- The water temperature to the south of *TCOON* 068 (see Figure 1).

Every GOM in situ observational station contains geospatial information including addresses and geographic references ((x, y)-coordinates). This information can be exploited and used to provide spatial awareness to integration systems. Yet most people use place names to refer to geographical locations, and will usually be entirely ignorant of the corresponding coordinates. In the paper, our focuses are to investigate the methods that represent the spatial features of the underlying data sources for qualitative spatial reasoning. We employ a qualitative notion instead of a quantitative notion of spatial information because the chosen representation does not need to worry about the real data and measurements. Verbal descriptions are typically not metrically precise, but sufficient for the task intended. Imprecise descriptions are necessary in query languages where one specifies some spatial properties such as the wind temperature not far away from *Corpus Christi* coast.

In order to support expressive spatial queries as well as to carry out qualitative spatial reasoning in an intuitive way, four relations have been proposed in [13]: *Partonomic relations*, *Distance*, *Topological relations*, and *Directional relations*. To the best of our knowledge, the BUSTER system [12,13,15] is the only implemented data integration system that supports two spatial features, partonomy and neighborhood graphs, in query answering and data retrieval. But the BUSTER system does not incorporate the distance and directional relations. Their research efforts have concentrated on specific aspects of spatial reasoning in isolation from each other, partially due to the lack of a comprehensive spatial reasoning framework.

Often partonomic, distance, topological, and directional information is available about objects in a scene and it is necessary to use a combination of the two or more types of spatial information in order to infer new facts that could not be inferred by considering individual type of relations in isolation. In the paper, we consider all of the above four spatial features. In particular, the framework utilizes inferences on combined spatial information about distance and directional relations. Unlike the BUSTER system where the build of neighborhood graph is straightforward because it deals with regions with the natural border, we have to utilize the Delaunay triangulation to determine the direct neighbor of point objects with no obvious border (see Figure 1). In addition, we propose a probability-based heuristic method to uniquely infer the directional information of indirectly connected points.

2   Related Work

Techniques for representing and reasoning about qualitative spatial information have been extensively studied in *Artificial Intelligences* (AI). The efficient spatial problem solving in this line of research depends on the abstraction of spatial details. Three levels of spatial abstraction have been identified in terms of topology, ordinal and metrics [13], which correspond to the topological, directional and distance relations defined in Section 1, respectively.

The spatial abstraction and reasoning approaches developed in AI research are inadequate for the field of information retrieval and data integration because we have to explicitly represent and reason spatial relevance as well as retrieve conceptually relevant data. Although it is possible to use OWL (Web Ontology Language) to describe as well as reason about spatial properties, OWL has very limited build-in inference capabilities, and an extension of the formal semantics to integrate spatial relations into OWL turns out to be undecidable [8]. A better solution is to evaluate the queries with conceptual and spatial criteria separately. As a result, *Stuckenschmidt* et al. [13] proposed to represent the spatial information in four relations: partonomic, topological, directional, and distance relations. But the four relations have never been applied to a real system and the scenarios handled are relative simple in [13] with the four relations considered separately, while the proposed probability-based heuristic method takes into the consideration the combination of distance and directional relations. BUSTER system [12,13,15] is the only implemented system that supports the partonomic and neighborhood relations, which are derived from the geographic tessellation of homogeneous decomposition. Spatial relevance is determined by the weighted sum of the partonomic and neighborhood relations. In addition, a system of eight calculus RCC-8 is widely used in GIS applications to describe the region connection relations [2,4,11].

In the rest of the paper, we use Texas area as an example to present the framework and the same idea can be applied to other GOM US states, such as Florida.

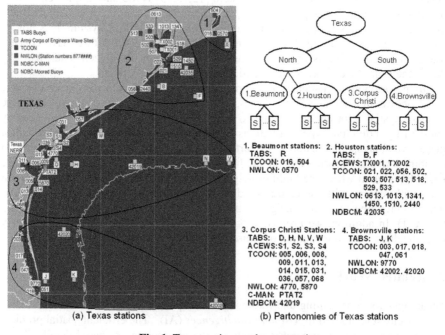

(a) Texas stations   (b) Partonomies of Texas stations

Fig. 1. Texas stations and partonomies

3   Qualitative Spatial Representation and Reasoning

In this section, we propose a framework to qualitatively represent the four spatial relations: partonomic, distance, topological, and directional relations.

3.1   Partonomic Relation

We consider areas $A_1,...,A_n$ as a partition of a larger area A if (1) $A_1 \cup ... \cup A_n = A$ and (2) $A_i \cap A_j = 0$ for all $i \neq j$ from $(1,....,n)$ [13]. Partonomies are created by recursively applying the above partition method to divide an area, such as A_1, to multiple smaller areas $A_{11},...,A_{1m}$. Figure 1(a) represents station partonomies in Texas area from north to south: 1. *Beaumont*, 2. *Houston*, 3. *Corpus Christi*, and 4. *Brownsville*, and Figure 1(b) shows the corresponding decomposition tree, where S represents a station or a buoy. Each partonomy contains several different types of stations, and one type of station can appear in more than one partonomy. For example, *TABS*, *TCOON*, and *NWLON* stations can be found in all four partonomies, and the partonomy *Corpus Christi* includes *TABS*, *Army Corps of Engineers Wave Sites* (*ACEWS*), *TCOON*, *NWLON*, and *NDBC Moored* (*NDBCM*) buoys. For more details about stations in each area, please refer to Figure 1(b).

This partition is sort of coarse and other finer partitions may be created based on the application needs. The simple queries can be answered by partonomic information including the water level in the *Houston* area or in the south Texas, etc.

3.2   Distance Relation

The Texas observational systems can be classified to three categories: coast, offshore, and ocean based upon their distance from the land. The coastal stations are the ones that are installed on the shore and are just a little bit stick out to the water from the land. Figure 2 shows that there are four kinds of costal stations: *ACEWS*, *TCOON*, *NWLON*, and *NDBC C-MAN*. The offshore stations usually are the buoy systems that are not close to the land but within the continental shelf. All of the *TABS* and part of *NDBCM* buoys fall in this category (see Figure 1). The ocean stations are those far away from the land and are outside of the continental shelf. The only station in Figure 2 that can be put in this category is *NDBCM Buoy 42002* in the partonomy *Brownsville*.

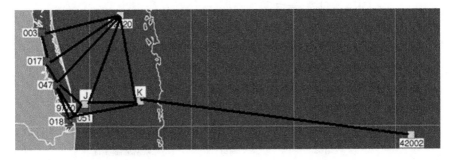

Fig. 2. Station neighborhood graph in *Brownville* area

Correspondingly, we adopt from [5] the five-step distance with five symbols: *very close* (*VC*), *close* (*C*), *medium* (*M*), *far* (*F*), and *very far* (*VF*). The concept of distance in [5] is defined as the straight line length between two point objects. The distance between two nearest coastal stations, between one costal station and one offshore station, and between one offshore station and one ocean station in one area are usually defined as *very close* or *close*, *medium*, and *far*, respectively. For example, in Figure 2 the distance is *very close* from *TCOON* 051 to *TCOON* 018, *close* from *TCOON* 017 to *TCOON* 047, *medium* from *TCOON* 051 to *TABS K*, and *far* from *TABS K* to *NDBC* 42002. The last symbol *very far* is used to describe the distance that is farther than *far*, for example, between a Texas coastal station and a Florida coastal station (not shown in Figure 2).

Based upon the above distance concept, the distance addition operation is defined as $X + Y = succ(X, Y)$ if $X=Y$, otherwise $X + Y = max(X, Y)$. For instance, $VC + VC = C$, $C + C = M$, $C + M = M$, and $C + F = F$, *etc*. The operations obey the associative law. A real example for $C + C = M$ is that in Figure 2 *TCOON* 003 indirectly links to 047 through 017, so the distance between 003 and 047 is *M*. The neighborhood graph is introduced in Section 3.3 and Figure 2 is a subgraph of Figure 3(e).

The simple queries can be answered by combining partonomic and distance relations including the water level *close* to *TABS J*, and the salinity not far away (*medium*) from *TCOON* 047, etc (see Figure 2).

3.3 Topological Relation

Topology is useful in information integration and retrieval because many spatial modeling operations require only topological information instead of coordinates (latitude and longitude). In GIS, the mostly considered topological relationships are neighborhood (what is next to what), containment (what is enclosed by what), and proximity (how close something is to something else). Because there is no containment relationship between any observational sites, we only consider the neighborhood and

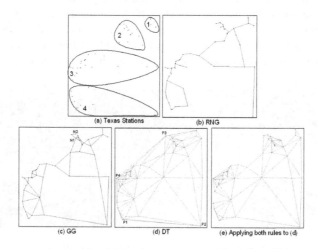

Fig. 3. Neighborhood graphs of Texas stations

proximity relationships in the paper. To be more specific, proximity graphs are exploited to represent the neighborhood of a set of points in the Euclidean plane. The neighborhood graph of the ocean observational stations is a graph $G = (N, E)$, where N represent stations and E the imaginary edges between two neighboring stations.

We now look at four types of neighborhood relations, minimum spanning tree (MST), relative neighborhood graph (RNG) [14], Gabriel graph (GG) [6], and Delaunay triangulation (DT) [1]. A point in MST has the least number of neighbors while a node in DT has the most neighbors. In other words, for a given set of points N, each neighborhood graph (the graph giving for each point its set of neighbors) in the following list is a subgraph of the subsequent one [9]: $MST \subseteq RNG \subseteq GG \subseteq DT$. Stronger requirement of neighborhood makes it relatively difficult for two points to be neighbors. In the paper, we do not consider MST neighborhood relation because each point has the least neighbors in the above four neighborhood relations.

The Texas stations and the corresponding neighborhood graphs of RNG, GG, and DT are shown in Figure 3(a), 3(b), 3(c), and 3(d), respectively. It is easy to tell from Figure 3(b) that the RNG graph is not a good candidate for our application because most nodes only have two direct neighbors. Too few direct neighbors complicate the queries in which the user is requesting the information around a station. Although more edges have been added in Figure 3(c), the GG graph still has the similar not-enough-direct-neighbor problem. In the same area we expect that the closely neighboring nodes are connected. For example, both coast stations N_1 and N_2 are located in *Houston* area but N_1 is not directly linked to N_2. On the other hand, the DT graph Figure 3(d) generates some extra unnecessary neighbors. The problems in Figure 3(d) are two folds: (1) a coastal station should not connect to an ocean station because the environmental and weather situation in one area is quite different from that of another remote area. For example the coastal station P_1 should not be the direct neighbor of the ocean station P_2. (2) A coastal station in one region should not link to the coastal station in another region. The reason is that the coastal stations are usually classified as C (*close*) in terms of distance and the correlation between two coastal stations in different regions is low. For example, P_3 in *Houston* should not link to P_4 in *Corpus Christi*.

As a result, an intermediate neighborhood graph between GG and DT will suit our application needs. There are two ways to achieve this goal: one way is to add more edges to produce an enhanced GG graph and another way is to delete unnecessary connections from DT to create a reduced DT graph. We have chosen the latter one because it is relatively easy and straightforward to implement. We have designed two rules to eliminate extra edges: (1) coastal stations in one area do not connect to coastal stations in another area; (2) a coastal station does not link to an ocean station. The first rule is proposed to add the partonomic factor to the neighborhood graph, and the second rule is exploited to handle the connection of far away stations. The results of applying both rules are shown in Figure 3(e), respectively.

3.4 Directional Relation

Qualitative direction is a function between two points in the plane that maps onto a symbolic direction. The n different symbols available for describing the directions are given as a set C_n. The value of n depends on the specific system of direction used, e.g., C_4={*North, East, South, West*} or more extensively C_8={*North, Northeast, East, Southeast, South, Southwest, West, Northwest*} (see Figure 4(a)). In the paper, we employ the cone-shaped direction system C_8. For instance, the direction *Northeast* refers to a cone-shaped section with the angular degree [22.5°, 67.5°]. This type of direction system results in the feature that the allowance area for any given direction increase with distance [5].

In the qualitative directional reasoning, we should be able to reason directional information of indirectly connected points, or the composition of two directional relations to derive a new directional relation. A typical composition is like: given B is east of A and C north of B, what is the directional relationship between A and C (see Figures 4(b), 4(c), and 4(d))? The composition table of the eight cone-shaped cardinal direction relation is shown in Table 1 [5,10]. For 64 compositions in Table1, only 8 can be inferred exactly with one answer, 8 (represented by *) may be any of the eight directions, and the remaining 48 generate two possible answers, three possible answers, or four possible answers. The reason is that the mere directional relationships do not provide enough information to infer the result of a composition. For example, C may be east (see Figure 4(b)), northeast (see Figure 4(c)), or north (see Figure 4(d)) of A depending on directions and distances between A and B and between B and C. In our application domain, we need to provide a unique and precise answer for each user's spatial query instead of two, three, or four possible answers. Next, we first present a probabilistic solution [3] to reason a unique composition answer for the scenarios, such as Figures 4(c), where the two distances, A to B and B to C, are the same, e.g., *medium*. Then we propose the heuristics to improve the above probabilistic method for the cases where two composition distances vary. For example, AB is *medium*, and BC is *close* and *far* in Figures 4(b) and 4(d), respectively.

Table 1. The composition results for eight cone-shaped cardinal direction relations

	N	S	E	W	NE	NW	SE	SW
N	N	*	N,NE,E	N,NW,W	N,NE	N,NW	N,NE,E,SE	N,NW,W,SW
S	*	S	S,SE,E	S,SW,W	S,SE,E,NE	S,SW,W,NW	S,SE	S,SW
E	E,NE,N	E,SE,S	E	*	E,NE	E,NE,N,NW	E,SE	E,SE,S,SW
W	W,NW,N	W,SW,S	*	W	W,NW,N,NE	W,NW	W,SW,S,SE	W,SW
NE	NE,N	NE,E,SE,S	NE,E	NE,N,NW,W	NE	NE,N,NW	NE,E,SE	*
NW	NW,N	NW,W,SW,S	NW,N,NE,E	NW,W	NW,N,NE	NW	*	NW,W,SW
SE	SE,E,NE,N	SE,S	SE,E	SE,S,SW,W	SE,E,NE	*	SE	SE,S,SW
SW	SW,W,NW,N	SW,S	SW,S,SE,E	SW,W	*	SW,W,NW	SW,S,SE	SW

Dehak et. al. [3] proposed a probabilistic method to uniquely determine the angular relationship γ between two points, A and C, giving the angular value α from A to B and the angular value β from B to C. Under the assumption of uniform distribution for all points in a circular region and no knowledge of the point coordination information (longitude and latitude), γ is calculated as:

$$r = \begin{cases} \dfrac{\alpha+\beta}{2} & \text{if } |\beta-\alpha| \in \left[2k\pi, 2k\pi+\dfrac{\pi}{2}\right] \\[2mm] \pi+\dfrac{\alpha+\beta}{2} & \text{if } |\beta-\alpha| \in \left[2k\pi+\dfrac{3\pi}{2},(2k\pi+2)\pi\right] \\[2mm] \dfrac{\varphi(\alpha,\beta)}{\sigma(\alpha,\beta)} & \text{if } |\beta-\alpha| \in \left(2k\pi+\dfrac{\pi}{2},2k\pi+\dfrac{3\pi}{2}\right), \quad \alpha<\beta, \ |\beta-\alpha|\neq\pi \\[2mm] \pi+\dfrac{\varphi(\alpha,\beta)}{\sigma(\alpha,\beta)} & \text{if } |\beta-\alpha| \in \left(2k\pi+\dfrac{\pi}{2},2k\pi+\dfrac{3\pi}{2}\right), \quad \alpha>\beta, \ |\beta-\alpha|\neq\pi \\[2mm] \alpha+\dfrac{6\pi}{7} & \text{if } \beta=\alpha+\pi, \ \text{or} \qquad\qquad \alpha-\dfrac{6\pi}{7} \ \text{if } \alpha=\beta+\pi \end{cases} \tag{1}$$

with

$$\varphi(\alpha,\beta) = -4((2\beta-\alpha-\pi)^2 - \beta^2 + 1)\cos(\alpha-\beta) + 5\cos(3\alpha-3\beta) - \cos(5\alpha-5\beta)$$
$$- (12\alpha-20\beta+8\pi)\sin(\alpha-\beta) + (8\alpha-10\beta+5\pi)\sin(3\alpha-3\beta)$$
$$- (2\beta-\pi)\sin(5\alpha-5\beta),$$

and

$$\sigma(\alpha,\beta) = 8(\alpha-\beta+\pi)\cos(\alpha-\beta) + 8\sin(\alpha-\beta) - 2\sin(3\alpha-3\beta) - 2\sin(5\alpha-\beta)$$

Equation (1) works well if two distances (AB and BC) are similar.

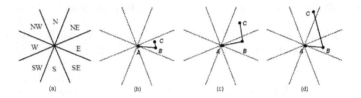

Fig. 4. The eight direction system and a composition example when $|\beta-\alpha| = \dfrac{\pi}{2}$

In our application domain, we have the knowledge of cardinal directions, such as north or southwest, between two points, but the real angular information between two points is unknown. So we have to employ the average angular degree of a cone-shaped section to represent the angle of a direction. For example, the direction north-east is between $\dfrac{\pi}{8}$ and $\dfrac{3\pi}{8}$ and is represented by the average angle $\dfrac{\pi}{4}$. In this case, the values of α and β can only be 0, $\dfrac{\pi}{4}$, $\dfrac{\pi}{2}$, $\dfrac{3\pi}{4}$, π, $\dfrac{5\pi}{4}$, $\dfrac{3\pi}{2}$, or $\dfrac{7\pi}{4}$ for eight directions, respectively. As a result, Equation (1) only returns two types of γ values when $|\beta-\alpha|$ is in the first or fourth quadrant: γ is right on the delimiter of two cone-shaped sectors when $|\beta-\alpha| = \dfrac{\pi}{4}$ or $\dfrac{7\pi}{4}$, or γ is the average angle of a cone-shaped sector when $|\beta-\alpha| = 0, \dfrac{\pi}{2},$ or $\dfrac{3\pi}{2}$. For instance, $\alpha = 0$, $\beta = \dfrac{\pi}{4}$, then $\gamma = \dfrac{\alpha + \beta}{2} = \dfrac{\pi}{8}$, which is the degree that separates east and northeast. In the above example, to determine a direction for AC, we may use some heuristics or randomly pick one if the two distances, AB and BC, are the same.

On the other hand, the two distances can vary widely, for example, AB is *medium* and BC is *far* in Figure 4(d). If we do not take the distance into consideration, the result of Equation (1) may drift away from the real angle between A and C. To handle the above disparate distance problem, we propose heuristics to improve the performance of Equation (1) according to the $|\beta - \alpha|$ value and length of AB and BC. Next, we discuss the heuristics proposed for various scenarios.

Heuristics 1: When $|\beta - \alpha| = \dfrac{\pi}{2}$ or $\dfrac{3\pi}{2}$, AC is in the same direction with the longer one of AB and BC, or AC is in the same direction with the middle section between AB direction and BC direction if AB and BC are of same length. For example, in Figure 4, $\alpha = 0$, $\beta = \dfrac{\pi}{2}$, and $\gamma = \dfrac{\pi}{4}$ according to Equation (1). Suppose the distance of AB is *medium*, C is east of A if BC is *short* (see Figure 4(b)), C is northeast of A if BC is *medium* (see Figure 4(c)), or C is north of A if BC is *far* (see Figure 4(d)).

Heuristics 2: When $|\beta - \alpha| = \dfrac{\pi}{4}$ or $\dfrac{7\pi}{4}$, AC is in the same direction with the longer one of AB and BC, or AC is in the same direction with BC if AB and BC are of same distance. For example, B is east of A and C is northeast of B, then $\alpha = 0$, $\beta = \dfrac{\pi}{4}$, and $\gamma = \dfrac{\pi}{8}$ according to Equation (1). Suppose the distance of AB is *medium*, C is east of A if BC is *short* (see Figure 5(a)), or C is northeast of A if BC is *medium* or *far* (see Figure 5(b) and 5(c)).

Heuristics 3: When $|\beta - \alpha| = \dfrac{3\pi}{4}$ or $\dfrac{5\pi}{4}$, the direction of AC is determined by the result of Equation (1) if AB and BC are of same length. Otherwise, γ is increased/decreased by $\dfrac{\pi}{8}$ if AB is shorter than BC, or γ is increased/decreased by $\dfrac{\pi}{2}$ if AB is longer than BC. The reason we do this is because usually the value γ falls in the third or fourth direction from AB direction. For example, in Figure 10, B is east of A and C is northwest of B, then $\alpha = 0$, $\beta = \dfrac{3\pi}{4}$, and $\gamma \approx 1.95$, which is about $112°$ and north of A according to Equation (1). Suppose the distance of AB is *medium*, in

Figure 6(a) C is east of A because BC is *short* and γ is decreased by $\dfrac{\pi}{2}$ to 0.38 (about 22°), while in Figure 6(d) C is northwest of A since BC is *far* and γ is increased by $\dfrac{\pi}{8}$ to 2.34 (about 134°). Figures 6(b) and 6(c) show two possible locations of C when AB and BC are of same length.

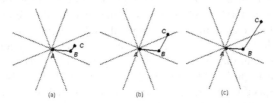

Fig. 5. Composition results when $\left|\beta - \alpha\right| = \dfrac{\pi}{4}$

Another heuristics we utilize is when one of AB and BC is *very close* and the length of AB and BC varies, the direction of AC is inferred the same as the longer one of AB and BC. For example, in Figure 2, the distance between $TCOON$ 051 and $TABS$ J is *close* and the distance from $TCOON$ 051 to $TCOON$ 018 is *very close*, then the direction from $TABS$ J to $TCOON$ 018 is the same as that from $TABS$ J to $TCOON$ 051, which is southwest.

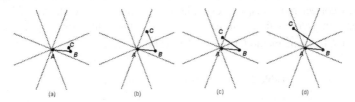

Fig. 6. Composition results when $\left|\beta - \alpha\right| = \dfrac{3\pi}{4}$

4   Experiments

In the experiments, the effectiveness of the proposed heuristics method is tested based on the Texas in situ stations (see Figure 1). The compared three methods are all based on the probabilistic method in [3]: the first one is called *exact probabilistic* method uses the exact angular degree, the second one is called *approximate probabilistic* method which does not use the exact α and β values and employs the average angular degree to represent each direction in Equation (1), and the last one is called *heuristic* method which is developed from the second method and is enhanced with heuristics as proposed in Section 3. In the *approximate probabilistic* method a direction is randomly selected if γ is the delimiter of two directions, while in the heuristic method the second direction anti-clockwise is chosen.

Twenty stations are randomly selected from the four areas of Texas partonomies: 2 from *Beaumont*, 6 from *Houston*, 8 from *Corpus Christi*, and 4 from *Brownville*. Among the above selected stations, 9 are from coast, 10 from offshore, and 1 from ocean (see Table 2). The total numbers of tested compositions for the chosen stations are 97, 153, and 11, respectively.

The overall results are shown in Table 3. Correct ratio (CR) is calculated as (*correct infer (CI)*)/(*total compositions*). For example, CR of exact probabilistic method in coastal area is 69/97=71.1%. Gap score (GS) is the directional difference between the inferred result and the real result. For example, the GS is 1 if the inferred direction is east and the real direction is northeast. In Table 3, the heuristic method has slightly better performance than that of exact probabilistic method and is significantly better than the approximate probabilistic method in terms of CR in coastal and offshore area. To be more specific, the heuristic method achieves the CR of 74.2% and 75.2%, exact probabilistic method 71.1% and 69.9%, and approximate probabilistic method 60.8% and 59.5%, respectively in coastal and offshore area. From the perspective of GS, the heuristic method is the best among three methods with most GS as 1. The result generated for the ocean area is not conclusive because we only have one ocean station with 11 compositions in Texas, which is *NDBC* 42002 (see Figure 2).

On the other hand, sometimes there is more than one path to reach from one station to another station in the composition operation, so the final CR can be adjusted better to some extent. For example, in Figure 2 *TABS J* can reach *TCOON* 018 through *TCOON* 047 or through *TCOON* 051. Both composition results will be counted corrected even if only one composition can successfully reason that *TCOON* 018 is southwest of *TABS J*. In Table 3, ACI, AGS, and ACR represent adjusted correct infer, adjusted gap score, and adjusted correct ratio, respectively. The performance of overall ACRs is at least 13% higher than overall CRs for all three methods. Without the exact angular degree of α and β, the heuristic method achieves the overall ACR as 87.7%.

Table 2. Selected stations and compositions

	Coast			Offshore		Ocean
	TCOON/NWLON	C-MAN	ACEWS	TABS	NDBCM	NDBCM
Stations	5	1	3	7	3	1
Compositions	48	18	31	91	62	11
Total Compositions	97			153		11

Table 3. Overall composition results

	Exact Probabilistic			Approx. probabilistic			Heuristic		
	Coast	Offshore	Ocean	Coast	Offshore	Ocean	Coast	Offshore	Ocean
CI	69	107	10	59	91	10	71	114	9
GS	41	68	1	45	71	1	33	46	2
CR	71.1%	69.9%	90.9%	60.8%	59.5%	90.9%	74.2%	75.2%	81.8%
Overall CR	71.3%			61.3%			74.3%		
ACI	89	133	10	78	111	10	86	133	10
AGS	15	28	1	20	45	1	12	22	1
ACR	91.8%	86.9%	90.9%	80.4%	72.5%	90.9%	88.7%	86.9%	90.9%
Overall ACR	88.9%			76.2%			87.7%		

Table 4. Composition results when $\alpha \neq \beta$

	Exact Probabilistic		Approximate probabilistic		Heuristic	
	VSL	VDL	VSL	VDL	VSL	VDL
compositions	114	82	114	82	114	82
CI	73	48	64	31	71	58
GS	61	49	61	55	54	27
CR	64.0%	58.5%	56.1%	37.8%	62.3%	70.7%

But results in Table 3 do not really reflect the usefulness and effectiveness of the proposed heuristics for the cases where the length of AB and BC vary because in the tested 261 (97+153+11) compositions there are 65 compositions where AB and BC have the same direction ($\alpha = \beta$), 114 compositions where $\alpha \neq \beta$ and AB and BC are of the same length (represented as VSL), and only 82 compositions where $\alpha \neq \beta$ and the length of AB and BC diff (represented as VDL). The heuristics proposed in Section 3 are for VDL-type compositions. As we know from Table 1, AC simply has the same direction as AB and BC when $\alpha = \beta$ and no equation or heuristics is needed at all. So 65 compositions with $\alpha = \beta$ should be excluded from consideration. In Table 4, three methods obtain close performance for VSL-type compositions and their CRs are 64.0%, 56.1%, and 62.3 %, respectively. As for VDL-type compositions, the heuristic method significantly outperforms other two methods. Especially, the CR of approximate probabilistic method is only 37.8% which is way too low. The proposed heuristics improves the performance of the exact probabilistic method and approximate probabilistic method by 12.2% and 32.9%, respectively.

5   Conclusion

In the paper, we developed a framework to represent and reason the spatial relationship qualitatively for ocean observational systems. The framework considers four spatial relations, partonomic, distance, topological, and directional relation as well as the effects of their combinations. With only qualitative distance and directional information, the proposed probability-based heuristic method obtains very good results. The proposed framework can be utilized to represent spatial features of observational systems in other areas or other domains. In our future work, we are going to further improve the accuracy of the probabilistic method and investigate the methods to represent and reason the spatial relations between two areas and between a point and an area.

References

1. de Berg, M., van Kreveld, M., Overmars, M., Schwarzkopf, O.: Computational Geometry: Algorithms and Applications. Springer, Heidelberg (2000)
2. Cohn, A.G., Renz, J.: Qualitative spatial representation and reasoning. In: Handbook of Knowledge Representation. Elsevier, Amsterdam (2008)
3. Dehak, S., Bloch, I., Maitre, H.: Spatial reasoning with incomplete information on relative positioning. IEEE Trans. on Pattern Analysis and Machine Intelligence 27(9), 1473–1484 (2005)

4. Duckham, M., Lingham, J., Mason, K., Worboys, M.: Qualitative reasoning about consistency in geographic information. Information Sciences 176(6), 601–627 (2006)
5. Frank, A.: Qualitative Spatial Reasoning: Cardinal Directions as an Example. Twenty Years of the Intel. Journal of Geographical Information Science and Systems (2006)
6. Gabriel, K.R., Sokal, R.R.: A new statistical approach to geographic variation analysis. Systematic Zoology 18, 259–270 (1969)
7. GCOOS, http://www-ocean.tamu.edu/GCOOS/gcoos.htm
8. Haarslev, V., Lutz, C., Moeller, R.: Foundations of spatioterminological reasoning with description logics. In: Principles of Knowledge Representation and Reasoning (1998)
9. Jaromczyk, J.W., Toussaint, G.T.: Relative neighborhood graphs and their relatives. Proc. IEEE 80(9), 1502–1517 (1992)
10. Ligozat, G.: Reasoning about cardinal directions. Journal of Visual Languages and Computing 9, 23–44 (1998)
11. Liu, Y., Zhang, Y., Gao, Y.: GNet: a generalized network model and its applications in qualitative spatial reasoning. Information Sciences 178(9), 2163–2175 (2008)
12. Schlieder, C., Vogele, T., Visser, U.: Qualitative spatial representation for information retrieval by gazetteers. In: Montello, D.R. (ed.) COSIT 2001. LNCS, vol. 2205, p. 336. Springer, Heidelberg (2001)
13. Stuckenschmidt, H., Harmelen, F.V.: Information Sharing on the Semantic Web. Springer, Heidelberg (2004)
14. Toussaint, G.T.: The relative neighborhood graph of a finite planar set. Pattern Recognition 12, 261–268 (1980)
15. Visser, U.: Intelligent Information Integration for the Semantic Web. LNCS, vol. 3159. Springer, Heidelberg (2004)

Managing Hierarchical Information on Small Screens

Jie Hao[1], Kang Zhang[1,2], Chad Allen Gabrysch[1], and Qiaoming Zhu[2]

[1] Dept. of Computer Science, University of Texas at Dallas, Richardson,
TX 75080-3021, USA
{jie.hao,kzhang,cag016500}@utdallas.edu
[2] School of Computer Science and Technology, Soochow University, Suzhou, China 215006
qmzhu@suda.edu.cn

Abstract. This paper presents a visualization methodology called Radial Edgeless Tree (RELT) for visualizing and navigating hierarchical information on mobile interfaces. RELT is characterized by recursive division of a polygonal display area, space-filling, maximum screen space usage, and clarity of the hierarchical structure. It is also general and flexible enough to allow users to customize the root location and stylize the layout. The paper presents the general RELT drawing algorithm that is adaptable and customizable for different applications. We demonstrate the algorithm's application for stock market visualization, and also present an empirical study on an emulated implementation with a currently used cell phone interface in terms of their performances in finding desired information.

Keywords: Hierarchy visualization, mobile interface, tree drawing, aesthetic layout, stock market visualization.

1 Introduction

With the increasingly powerful processing and rendering capabilities of modern computers and display technology, sophisticated visualization techniques have been widely applied to various application domains. There are numerous examples of application data, such as financial statistics, remote-sensing results, real-time data, gene sequences, etc, that can be intuitively represented and analyzed using information visualization tools.

Numerous types of modern-day information are structured hierarchically, such as product catalogs, Web documents, computer file systems, organizational charts, etc. Visualization of hierarchical data improves not only the understanding of the data and the relationship among the data items, but also the way of manipulating and searching for desirable data items.

As a simple form of graph, a tree represents a hierarchical structure that provides efficient content search. Research has been conducted on transforming a graph to a tree representation [1]. The ability of visualizing and navigating hierarchical information, usually represented as a tree, is urgently needed and becomes an emerging challenging research topic.

With the increasing number of users of mobile devices, such as PDAs and cellular phones, the need for searching and navigating time-critical data on small screens

Q. Li et al. (Eds.): APWeb/WAIM 2009, LNCS 5446, pp. 429–441, 2009.
© Springer-Verlag Berlin Heidelberg 2009

becomes apparent. Various approaches have been proposed for browsing the Web information on small screen devices [3,4] and some with innovative interaction techniques [5,17]. Although many visualization techniques have been proposed for viewing tree structures, little progress has been made on maximizing the space usage for displaying hierarchical structures on small screens.

To achieve the objective of effective visualization of trees on small screens with space optimization, we proposed an approach, called *Redial EdgeLess Tree* (RELT) [7], for visualizing and navigating hierarchical information on mobile devices such as a PDA and cellular phones. RELT implements a simple area partitioning algorithm which minimizes the computational cost. By arranging a set of tree nodes as non-overlapped polygons adjacent in a radial manner, RELT maximally utilizes the display area while maintaining the structural clarity.

This paper generalizes RELT with the following additional characteristics, and reports new experimental results:

- The root can locate anywhere on the viewing space; and
- A given hierarchical structure can be represented and navigated in either concentric or multi-centered mode depending on the viewer's preference or the application requirement.

The remainder of this paper is organized as follows. Section 2 reviews the current hierarchy visualization techniques. Section 3 presents the generalization concepts and a corresponding general layout algorithm. Section 4 applies the RELT to stock visualization. Section 5 reports the experimental results on the performance of an emulated implementation of RELT compared with that of an existing cell phone interface. Finally, Section 6 summarizes the paper and discusses the remaining challenges.

2 Related Work

Various approaches have been investigated for visualizing hierarchical information and can be generally classified into **Connection** and **Space-filling** categories. The **Connection** approaches draw the hierarchical information in a node-edge diagram. Each information object is visualized as a node. An edge depicts the relationship between two connected nodes. All the connection approaches essentially focus on two major layout issues: how to locate the nodes, and how to connect the related nodes?

The method by Reigold and Tilford [11] is possibly one of the best known techniques which simply locate child nodes below their common ancestors. H-trees [6,8,15] positions the child nodes in the vertical or horizontal direction of their common ancestors. Concentric circles are created on a radial tree [6,8] with nodes placed on them according to the nodes' depths in the hierarchy. The elegance of this technique is in keeping adjacent branches from overlapping. Researchers have also developed hierarchy information visualization in 3D, such as Cone Trees [8,12,13], and Balloon Views [8,9] by projecting the Cone Tree onto a plane.

The **Space-filling** approaches are generally divided into two techniques, rectangular space-filling such as Tree Maps [8,10,16], and radial space-filling such as Information Slices [2] and InterRing [18]. Without edges between nodes, space-filling techniques use the position and adjacency to express the parent-child relationship. For example, by mapping a tree structure onto nested rectangles, enclosure is applied to express the parent-children relationship in Treemap. Space-filling techniques are gaining attention in the recent years.

Information Slices [2] is a radial space-filling technique which uses a series of semi-circular discs to visualize hierarchical information. Each disc represents multiple levels of a hierarchy and the number of levels can be user defined. The child nodes on the same level are fanned out according to their size. InterRing is another radial space-filling method which can be constructed by the several rules [18].

Both the connection and the space-filling approaches are effective methods for hierarchy information visualization. Although the effectiveness varies according to the data's properties, connection approaches are essentially advantageous in revealing the information structure while space-filling approaches are good at maximizing the space usage by getting rid of connecting edges.

The above approaches, when simply adapted to mobile interfaces, usually cannot show the structure as effective as on desktop screens. Connection approaches are usually chosen to be adapted for mobile interfaces due to their desirable property of showing a clear structure. Magic Eye View and Rectangular View [19] are typical examples. However, the drawback, i.e. uneconomic space-usage, is as distinct as their advantage. The biggest limitation of these approaches for mobile interfaces is their loss of legibility or the overall sense of a hierarchy when reduced in size.

Despite their economical screen usage, space filling methods have their own problems in the process of adaptation. For InterRing [18] and Information Slices [2], efficiency of space usage depends significantly on how balance the tree is. The branches whose depths are much bigger than the others greatly decrease in the performance of space usage. Moreover, these two methods visually express a node by a part of cirque on which it is difficult to label. Tree Maps fully utilize the display space. The hierarchical structure on a Tree Map is, however, hidden behind the enclosure relationships.

In summary, none of the above hierarchy visualization methods can achieve both structural clarity and economic space usage on mobile devices. Next section describes the RELT methodology that aims at achieving both goals.

3   General Radial Edgeless Trees

3.1 Methodology and Terminology

Assume a university's web site to be visualized and navigated on a mobile screen, as depicted in Figure 1, where schools of "Management", "Engineering", and "Science" each have a few departments. The root ("University") locates at the top-left corner of the screen as shown in Figure 2(left). It is sometimes desirable to view the overall structure with the root at the center of the screen. More generally, we should allow the root to be displayed anywhere on the screen, depending on the user's preference and the requirement of the application domain. With the generalized RELT, it is straightforward to display a layout with the root at the bottom-left corner (Figure 2(right)) or one of the other corners.

With a rooted tree like the university structure, a center-rooted layout can be visualized as in Figure 3(left), where each level-2 node, i.e. a school, shares a border with the root, and each level-3 node shares a border with its level-2 parent. We will call this type of center-rooted layout as *concentric*, or simply *CC*. As shown in Figure 3(right) with the university example, departments locate under the school they belong to, which in turn locates adjacent to the university.

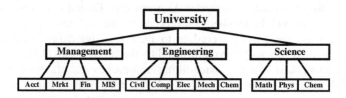

Fig. 1. A university Web structure

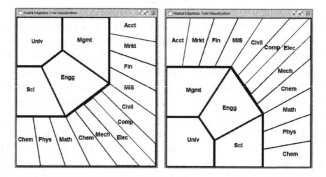

Fig. 2. University structure visualized on RELT: RELT layout with root at top-left (left); An alternative layout with root at bottom-left (right)

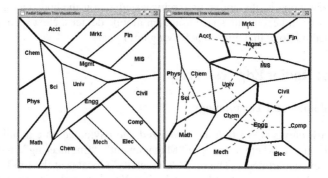

Fig. 3. Center-rooted: Concentric layout (left); Multi-centric layout (right)

Another type of center-rooted layout is to consider the sub-area of each child node as a center-rooted sub-tree, as illustrated in Figure 3(b). We will call this later type of layout as *multi-centric*, or simply *MC*. In this case, the schools form the sub-centers with their departments around them. In fact, different layout types may be used for different levels of details whenever desirable.

Given any of these layouts, the user is able to navigate down and up the hierarchy, namely zoom in and out, to view and upper lower level structures. For example, to view the detailed information on the Department of Computer Engineering, the user can select "Comp" in any of the top-level views in Figures 2 and 3. By selecting "Comp" in Figure 2(a) that has the root at the top-left, we obtain the zoom-in view, where the entire display space is filled by the sub-tree rooted at "Comp". Therefore with a single user interaction, several levels of hierarchy (five in the university example) can be reached.

How a node is selected depends on which interaction technique is supported for navigation. A touch-screen interface (by either a finger or a stylus) would apparently suit the RELT methodology very well. A single touch on a node would serve the selection of the node. With a key or button based interaction support, direction buttons may be assigned to the moving directions within a predefined radial range, and/or up and down the hierarchical levels, plus a confirmation button, as we experimented on a cell phone emulator to be discussed in Section 5.

In general, the number of levels on a single display can be determined based on the application, or maximized to the extent that the polygonal nodes have enough space for text labels. If color coding can easily discriminate and identify multiple groupings and levels of information, each tree node may be drawn by just a few colored pixels and thus maximize the size of the hierarchy being visualized. We therefore consider the RELT methodology to be generally scalable.

In discussing the concept and terminology on how to **draw** the general RELT layout, we will refer to a tree node as a *vertex* that occupies a polygonal area in the drawing space. A *rooted tree $T = (V, E, r)$* is a tree with a vertex as the *root, $r \in V$*, and, only one path can exist from r to any other vertex. V is the vertex set and E is the edge set. All *leaf* vertices form a subset of V, denoted as $L \subset V$. For two directly connected vertices, the one closer to the root is the parent and the other is the child. In a typical node-edge type of drawing, a directed edge $(v, u) \in E$ is used to express the parent-child relationship, where v is the parent and u is the child. In a RELT drawing, the relationship between a parent and its children is shown by their adjacency on a radial sector and/or positions relative to each other.

In general, there are three types of vertices in V for a RELT layout:

1. $v = r$: $T(v)$ represents the tree itself covering the entire display space.
2. $v \neq r$ AND $v \notin L$: $T(v)$ represents the sub-tree rooted at v and contains the polygonal area that covering all the descendents of v's vertices.

3. $v \in L$: $T(v)$ covers the area for only v itself.
   If $v \notin L$, i.e. type 1 or 2, then the set of v's children can be expressed by:
   $$CV = \{u \mid (v, u) \in E\} \qquad (1)$$
   CV_i is the i_{th} child of v.

The display space is recursively divided into several non-overlapping polygonal areas each visually representing a vertex. Three types of "area" are defined for the general RELT:

1. $WA(v)$ represents the entire area under vertex v.
2. $OA(v)$ is the area occupied by v itself.
3. $DA(v)$ is the area occupied by all of v's descendants.

3.2 Rules and Algorithm

A radial edgeless tree can be constructed by applying the following four general rules for space allocation:

1. Every $T(v)$ has $WA(v)$ where $T(v)$ should be drawn. $WA(v)$ is assigned to v by its parent. $WA(r)$ is the entire display area.
2. How $OA(v)$ is determined inside $WA(v)$ depends on the layout type *(LT)* specified by the user.
3. If v is a non-leaf vertex, it divides its area $DA(v)$ to be distributed to its children.
4. The area size of a leaf vertex is proportional to one of the vertex's properties, such as the weight.

A radial edgeless tree can be represented as a function RELT(V, r, LT) in which V is the vertex set, r specifies the root vertex, and LT indicates the layout type. The user-provided LT parameter determines whether the hierarchy is viewed concentric, i.e. the CC layout, or children surrounding their parents that form multiple sub-centers, i.e. the MC layout, as introduced in Section 3.1. The RELT algorithm is presented in pseudocode below.

```
Algorithm RELT
Input: vertex set V, root r, layout type LT
Begin
Set root location          // User decides root location
DFS                        // Depth first search to traverse the tree
if vertex v is not yet processed then
  if vertex is a non-leaf
        OA(v) = createOA(v, LT); //calculate the area occupied by v: see Appendix
        if v has more than one child
            DA(v) = WA(v) − OA(v)
            DF (DA(v)) -> WA(CV)   // v distributes DA(v) to its children
        else                       // v has only one child
            DA(v) = WA(v)          // DA(v) belongs to the only child
  else                             // leaf vertex
            WA(v) = OA(v) = DA(v)
end
```

Each vertex v may be assigned a corresponding variable $w(v)$ which we call the *weight* of v. One way to calculate the value of $w(v)$ is to use a weight function WF:

$$w(v) = WF\ (w\ (CV_1),\ w\ (CV_2)...\ w\ (CV_n)) \qquad (2)$$

Equation (2) basically says that, the value of $w(v)$ depends on the overall weight of v's children. Different weight functions may be defined to meet different application requirements (see Section 4.2 for the weight function used to visualize the stock market). Figures 3 through 5 in this paper are all generated using the following simple weight function:

1. If $v \in L$, $w(v) = WF(1) = 1$;
2. Otherwise, assuming the given v has n children, then

$$w(v) = WF(1, w(CV_1), w(CV_2)...w(CV_n)) = 1 + \sum_{i=1}^{n} w(CV_i)$$

A leaf vertex v has no children, and thus requires no distribution. Only its own area $OA(v)$ needs to be calculated. For a non-leaf vertex v, after $OA(v)$ is constructed, v's remaining area, i.e. $DA(v)$, is distributed to its children. The distribution is determined by a distribution function $DF()$. Here, we propose $DF()$ that partitions the distribution area based on the size of $DA(v)$ and the children's weights:

$$DF(DA(v), w(CV_1), w(CV_2)...w(CV_n)) \rightarrow WA\ (CV) \quad (3)$$

$DF()$ also allocates the position of $WA\ ()$. For a vertex v, in our implementation, $DF()$ positions $WA()$ of v's children in clockwise order.

We now discuss the complexity of the RELT algorithm. As discussed in Section 3.1, a given tree $T = (V, E, r)$ has n vertices which are divided into three types. RELT applies depth first search to traverse the tree. Table 1 illustrates the vertex types and their corresponding operations used in the RELT algorithm. All the operations in the right column can be computed in linear time. Therefore the overall time complexity is $O(n)$, where n is the number of vertices.

In summary, RELT recursively partitions the display area into a set of non-overlapping polygons. Because every part of screen is utilized, economic screen estate is achieved. Parent-child relationships are explicitly represented by adjacent relationships, and thus the structural clarity is maintained.

Table 1. Vertex types and corresponding operations

Vertex Type	Operation
$v=r$	*createOA*(**v, LT**) **DA(v) = WA(v) – OA(v)** **If |CV| > 1, *DF*(DA(v)) → WA(CV)**
$v{\neq}r$ AND $v{\notin}L$	*createOA*(**v, LT**) **DA(v) = WA(v)** **If |CV| > 1, *DF*(DA(v)) → WA(CV)**
$v \in L$	**WA(v) = OA(v) = DA(v)**

4 Application in Stock Visualization

The continuing improvement in the computational and graphic capabilities for mobile devices is making timely retrieval of time-critical data a possibility. Stock marketing on mobile devices is one of the appealing and useful activities. Applying hierarchy visualization and navigation techniques to monitoring the market trend and instant changes could significantly reduce the browsing time. The current practice is to classify securities into different sectors which are further divided into specific industries, as illustrated in Figure 4.

Fig. 4. A typical stock market classification

4.1 The Current Treemap Approach

The most popular stock market visualization tools currently in use is the treemap approach, called portfolio map [14]. Treemap is a space-constrained and rectangular space-filling technique for visualizing hierarchical information. None of the other approaches emphasizes the economic usage of the screen estate. Figure 5 visualizes the "10 stocks now 799". Traditional stock coloring indicates stock price performance. Green stands for price up and red for down, black indicates

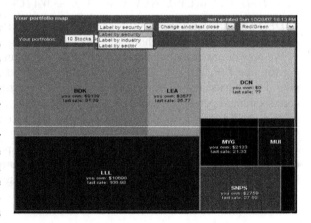

Fig. 5. Treemap for stock market visualization

an unchanged price, and gray means that users do not own any shares. The size of rectangle indicates the market capitalization of the corresponding company.

"How does the market look like today" is the essential question that a stock visualization tool needs to answer. What makes this question really tough is that a good answer must contain both macro and micro views. Neither simply answering the market as a whole is up, nor answering a certain stock is down will satisfy the user. A good answer may be like: the whole market rallied but technology stock is down, especially the software industry stock; "SNPS", however, performs well among the software stocks. None of the current approaches including Treemaps can offer simultaneous display of both a big picture and individual stock details.

4.2 The RELT Solutions

RELT offers alternative solutions that overcome the major drawbacks of the approaches discussed above. Assume we would like to emphasize the price change of each individual stock as well as the capitalization of the corresponding company. We can use the color scheme similar to the one used for the Treemap in Figure 5, and make the size of the polygonal vertex for each stock a function of the company's capitalization. The corresponding weight function can then be customized as follows:

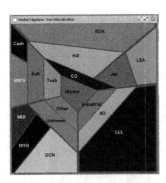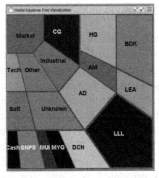

Fig. 6. A RELT visualization of stock market with (left) the center-rooted concentric layout and (right) the root at the top-left

1. If $v \in L$, $w(v) = WF(Cpt(v)) = Cpt(v)$, where $Cpt(v)$ is the capitalization of the company represented by v;
2. Otherwise, assuming the given v has n children, then $w(v) = WF(1000, w(CV_1)$,

$$w(CV_2)...w(CV_n)) = 1000 + \sum_{i=1}^{n} w(CV_i)$$

It is generally more difficult to tell the difference of the sizes between two arbitrarily shaped polygons than between two regular shapes such as rectangles. We therefore use a large constant (1000) in $w(v)$ above, instead of 1 as in Section 3.2, to emphasize the size difference, i.e. the difference between the companies' capitalizations. Giving more weight to a non-leaf vertex v would contrast the relative proportion of $DA(v)/WA(v)$. The results are visualized in Figure 6, that includes a layout view with the root at the center and at the top-left corner.

We believe that RELT outperforms the Treemap in visualizing the tree structure, since RELT shows multiple levels on a single display while the Treemap shows only one level, as illustrated in Figure 5. This also implies that only one level of hierarchy may be reached (i.e. zoom it or out) by a user interaction with the Treemap. Multiple user interactions are needed to navigate from the high-level market overview to individual stock (company) performances.

5   Emulated Comparison with a Current Cell Phone Interface

In order to evaluate whether the RELT methodology provides a more efficient interface for navigation, this section presents an experimental comparison of our implementation on an emulator with one of the currently used cell phone interfaces, Sprint PCS Vision Phone®. The comparison is performed for the RELT center-rooted layout since the Sprint top-level interface looks similar and provides the navigation structure from the center as pictured in Figure 7. The hierarchical information to be navigated by both interfaces is the Sprint cell phone's 5-level functions and services as described below.

5.1 Sprint Interface

The Sprint interface has 9 top level service
categories, as shown in Figure 7(a). There
are totally 43 second level sub-categories
under the 9 services, and each second level
sub-category may include 0 to 6 third
level items or sub-sub-categories. A third
level sub-sub-category may include a few
items at the fourth level. The last level is
the 5th level. The entire hierarchical structure can be represented as a tree that has
about 150-250 nodes. Some of the nodes at
levels 2 through 5 are leaves, which should
be the cell phone functions that the user
wishes to arrive at in the shortest time, or
with the fewest user interactions.

5.2 Emulated RELT

RELT has been implemented on a SUN
cell phone emulator as in Figure 7(right)

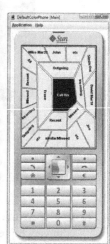

Fig. 7. Sprint PCS Vision Phone®(left)
and RELT emulator (right)

that reads in the Sprint hierarchy structure as an XML file. The default selection is at
the root, i.e. "Menu", at the beginning, and the selected node is highlighted in the
inverse color as shown in Figure 7(right). Each screen display shows several levels of
hierarchy (3 in this experiment), and a selection on a lower level node will make that
node a root, whose child and grand-child nodes are then displayed as illustrated in the
two examples in Figure 8. The left image shows when the level 2 node "Messaging"
becomes the root, and the right one shows the level 3 node "Sound" to be the root.

The text labels are arranged to maximize the space usage while avoiding cluttering,
particularly the overlap with the polygon edges. We use a simple approach in labeling
the almost arbitrarily oriented polygons. The approach finds the center point as well
as the longest side of the polygon to be labeled. It then rotates the text label around
the text's midpoint to the same angle as the longest side of the polygon. The label is
then placed in the polygon such that its midpoint coincides with the polygon's midpoint. Though this approach may not generate perfectly aesthetic labeling, it is simple
and fast with quite satisfactory display results as shown.

5.3 Results Comparison

To make a fair comparison, we have also implemented the Sprint interface on a computer screen with the same size as that of the RELT interface. Interactions with the
interfaces are through mouse clicks. We conducted an empirical study on the performance of each interface on the given example of hierarchical information (i.e.
Sprint cell phone functions).

Eighteen Computer Science graduate students were involved in this study as the
subjects, of whom half performed first on Sprint and then the RELT interface and the

other half performed on RELT followed by Sprint. The subjects were asked to perform the following 9 different actions on both interfaces,

1. Find if you called Mike on March 23;
2. Find whether David has called you on March 10;
3. Find the Settings of the Receiver Volume of the Speaker;
4. Go to Stopwatch Lap2;
5. Find the Settings of the Power-off Tone;
6. Launch Tetris game;
7. View the Picture Album;
8. Find the Settings of the Messaging Signature; and
9. Find the Directory Services.

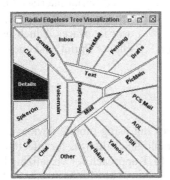 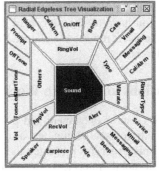

Fig. 8. Views of levels 2, 3 and 4 (left) and levels 3, 4 and 5 (right) on RELT

A touch-screen interface is assumed for both Sprint and RELT, the user can touch directly on the screen to select desired functions without pushing any buttons. Each "touch" is simulated by a mouse click on the selected node. We compared the number of clicks (touches) on the screen performed and total time in seconds taken for each action on both Sprint and RELT interfaces, as shown in Figure 9. An action is considered complete after the node for the corresponding action is touched. The numbers in the figures for both RELT and Sprint are averaged over the 18 subjects.

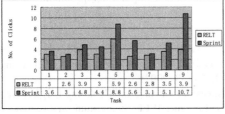 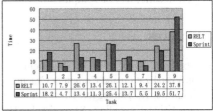

No. of Clicks	1	2	3	4	5	6	7	8	9
RELT	3	2.6	3.9	3	5.9	2.6	2.8	3.5	3.9
Sprint	3.6	3	4.8	4.4	8.8	5.6	3.1	5.1	10.7

Time	1	2	3	4	5	6	7	8	9
RELT	10.7	7.9	26.6	13.4	26.1	12.1	9.4	24.2	37.8
Sprint	18.2	4.7	13.4	11.3	25.4	13.7	5.5	19.5	61.7

Fig 9. Number of clicks performed (left) and total time (in sec) taken (right)

In Figure 9(left), RELT shows superior performance over Sprint with 0.3 to 6.8 fewer touches on average for the 9 actions. Figure 9(right) shows the mixed results on the speed of performing the 9 actions. Although on average six actions (verses three) were completed in shorter times on Sprint than on RELT, the total averaged time difference is very close, i.e. 18.16 (Sprint) vs 18.89 (RELT). We observed that those who performed faster on Sprint had been using the similar cell phone interface and thus were fairly familiar with the Sprint interface (as evidenced below), while the RELT interface is totally unfamiliar to all the subjects.

We also asked the subjects several general questions in a questionnaire. When asked "In your experience in using cell phones, finding a piece of information on the current cell phone is ...", 6 subjects chose "easy", 7 chose "OK but improvements are needed", 2 with no response, and 3 felt unsatisfactory. Seventeen subjects responded positively to the question "Have you used the provided Sprint interface before, or are you familiar with this particular interface?", with only one negative. In response to the question "If you have to choose one of the two user interfaces, which one you will choose?" 10 subjects reported to choose RELT, 7 choose Sprint, and one did not indicate any preference.

6 Conclusions and Future Work

The ubiquitous data management and processing using mobile devices create a new era for information visualization, and also multiple challenges for effective user interface design. This paper has presented the RELT approach for visualizing hierarchical information on mobile interfaces. Two issues, economical space usage and clear hierarchical structure have been effectively addressed in RELT. The algorithm has been implemented to generate all the RELT figures in this paper. It has a linear time complexity and can be easily adapted to suit various applications. We have demonstrated this adaptability by visualizing the stock market performance, where the size of each leaf vertex is made a function of the corresponding company's capitalization.

The RELT display and interactivity are also ported to a cell phone emulator platform in order for us to have a meaningful comparison with some existing cell phone interfaces. Having experimented a real-world hierarchical information on the RELT emulated interface and the Sprint PCS interface, we found that RELT outperforms Sprint for the given example. This is of course only a preliminary experiment. More empirical evidences need to be established before general conclusions can be made. Our immediate future work is to conduct comprehensive usability study with varied navigation schemes, in comparison with Sprint and other popular cell phone interfaces.

References

1. Abello, J., Korn, J.: MGV: A System for Visualizing Massive Multi-digraphs. IEEE Trans. on Visualization and Computer Graphics 8(1), 21–38 (2002)
2. Andrews, K., Heidegger, H.: Information Slices: Visualising and Exploring Large Hierarchy Using Cascading, Semi-circular Discs. In: Proc. IEEE Symposium on Information Visualization, pp. 9–12 (1998)
3. Arase, Y., Hara, T., Uemukai, T., Nishio, S.: OPA Browser: A Web Browser for Cellular Phone Users. In: Proc. ACM UIST 2007, Newport, USA, pp. 71–80 (2007)

4. Baudisch, P., Xie, X., Wang, C., Ma, W.-Y.: Collapse-to-Zoom: Viewing Web Pages on Small Screen Devices by Interactively Removing Irrelevant Content. In: Proc. ACM UIST 2004, Santa Fe, USA, pp. 91–94 (2004)

5. Dontcheva, M., Drucker, S.M., Wade, G., Salesin, D., Cohen, M.F.: Summarizing Personal Web Browsing Sessions. In: Proc. ACM UIST 2006, Montreux, Switzerland, pp. 115–124 (2006)

6. Eades, P.: Drawing Free Trees. Bulletin of the Institute for Combinatorics and Its Applications, 10–36 (1992)

7. Hao, J., Zhang, K.: RELT – visualizing trees on mobile devices. In: Qiu, G., Leung, C., Xue, X.-Y., Laurini, R. (eds.) VISUAL 2007. LNCS, vol. 4781, pp. 344–357. Springer, Heidelberg (2007)

8. Herman, I., Melançon, G., Marshall, M.S.: Graph Visualization in Information Visualization: a Survey. IEEE Trans. on Visualization and Computer Graphics, 24–44 (2000)

9. Jeong, C.S., Pang, A.: Reconfigurable Disc Trees for Visualizing Large Hierarchical Information Space. In: Proc. of IEEE Symposium on Information Visualization, pp. 19–25 (1998)

10. Johnson, B., Shneiderman, B.: Tree-maps: A Space-filling Approach to the Visualization of Hierarchical Information Structures. In: Proc. 1991 IEEE Visualization, pp. 284–291 (1991)

11. Reingold, E.M., Tilford, J.S.: Tidier Drawing of Trees. IEEE Trans. on Software Engineering 7(2), 223–228 (1981)

12. Robertson, G.G., Mackinlay, J.D., Card, S.K.: Cone Trees: Animated 3D Visualizations of Hierarchical Information. In: Proc. ACM CHI 1991 Human Factors in Computing Systems Conference, April 28 - June 5, pp. 189–194 (1991)

13. Robertson, G.G., Card, S.K., Mackinlay, J.D.: Information Visualization Using 3D Interactive Animation. C ACM 36(4), 57–71 (1993)

14. Portfolio Map, http://www.smartmoney.com/

15. Shiloach, Y.: Arrangements of Planar Graphs on the Planar Lattices, PhD Thesis, Weizmann Institute of Science, Rehovot, Israel (1976)

16. Shneiderman, B.: Treemaps for Space-Constrained Visualization of Hierarchies, http://www.cs.umd.edu/hcil/treemap-history/index.shtml

17. Wobbrock, J.O., Forlizzi, J., Hudson, S.E., Myers, B.A.: WebThumb Interaction Techniques for Small-Screen Browsers. In: Proc. ACM UIST 2002, Paris, France, pp. 205–208 (2002)

18. Yang, J., Ward, M.O., Rundensteiner, E.A.: InterRing: An Interactive Tool for Visually Navigating and Manipulating Hierarchical Structures. In: Proc. IEEE Symposium on Information Visualization, pp. 77–84 (2002)

19. Yoo, H.Y., Cheon, S.H.: Visualization by information type on mobile device. In: Proc. Asia-Pacific Symposium on Information Visualization, vol. 60, pp. 143–146 (2006)

An Approach to Detect Collaborative Conflicts for Ontology Development

Yewang Chen, Xin Peng, and Wenyun Zhao

School of Computer Science, Fudan University, Shanghai 200433, China
{061021061,pengxin,wyzhao}@fudan.edu.cn

Abstract. Ontology has been widely adopted as the basis of knowledge sharing and knowledge-based public services. However, ontology construction is a big challenge, especially in collaborative ontology development, in which conflicts are often a problem. Traditional collaborative methods are suitable for centralized teamwork only, and are ineffective if the ontology is developed and maintained by mass broadly distributed participators lacking communications. In this kind of highly collaborative ontology development, automated conflicts detection is essential. In this paper, we propose an approach to classify and detect collaborative conflicts according to some mechanisms: 1) impact range of a revision, 2) semantic rules, and 3) heuristic similarity measures. Also we present a high effective detecting algorithm with evaluation.

1 Introduction

Ontology has been widely adopted as the basis of knowledge sharing and knowledge-based public services. In this kind of knowledge-based systems, a rich and high-quality ontology model is the premise of satisfactory knowledge services. Therefore, besides ontology-based knowledge service, it is also important to get support for ontology creation and maintenance. In fact, ontology construction itself is a big challenge, especially when the ontology is expected to have relevance and value to a broad audience[1].

In most cases, the ontology is constructed manually by centralized teams, which is a complex, expensive and time-consuming process (e.g. [10]). In recent years, some methods and tools for collaborative ontology construction are also proposed which meet the requirements of public ontologies having relevance and value to a broad audience better. For these systems, it is essential to keep knowledge in consistency.

However, collaborative developments are often accompanied by conflicts. In traditional way, collaborative conflicts are usually handled by mechanisms of locking or branching/merging provided by version management systems (e.g. CVS). The conflicts occur when two developers revise the same file in a same phase. However, in ontology, the basic units are concepts and relationships between them, without explicit file-based artifact. Therefore, conflicts in this environment much more likely exist. The rich semantic relationships in ontology model exacerbate the situation, since different parts of the ontology model have much more complicated impacts on each other. On the other hand, in large-scale collaboration-based ontology development, there are many participators, most of them are not aware of the existences of

Q. Li et al. (Eds.): APWeb/WAIM 2009, LNCS 5446, pp. 442–454, 2009.

their co-workers at all, let alone communicate enough, this exacerbates the problem. Therefore, it is obvious that effective conflicts detection and resolution are essential for large-scale collaborative ontology development. For these, in this paper, we propose an approach to classify and detect three kinds of collaborative conflicts according to some mechanisms. Also we present a high effective detecting algorithm with evaluation.

The remainder of this paper is organized as follows. Section 2 discusses related work. Section 3 addresses the overview of our framework. Section 4 proposes an approach to detect collaborative conflicts. Section 5 shows experiments with evaluation. The last section is our conclusion.

2 Related Works

KAON [5] focuses on that changes in ontology can cause inconsistencies, and proposes deriving evolution strategies in order to maintain consistencies. Protégé [8] is an established line of tools for knowledge acquisition, which is constructed in an open, modular fashion. OntoEdit [7] supports the development and maintenance of ontologies, with an inferencing plugin for consistency checking, classification and execution of rules. OntoWiki [3] fosters social collaboration aspects by keeping track of changes, allowing comment and discuss every single part of a knowledge base, enabling to rate and measure the popularity of content and honoring the activity of users. OILEd [12] is a graphical ontology editor that allows the user to build ontologies with FACT reasoner [2] to classify ontologies and check consistency.

As far as conflicts detection and resolve are concerned, some works, such as [5] provides concurrent access control with transaction oriented locking, Noy[8] provides discussion thread for users to communicate. In some cases, even rollback. SCROL[11] provides a systematic method for automatically detecting and resolving various semantic conflicts in heterogeneous databases with a dynamic mechanism of comparing and manipulating contextual knowledge of each information source. Cupid [6] is an approach uses a thesaurus to identify linguistic matching and structure matching. Peter Haase[4] discusses the consistent evolution of OWL ontologies, and presents a model considering structural, logical, and user-defined consistency.

Almost all current ontology construction tools provide functionality for consistency checking. But based on our humble knowledge, none of them distinguishes collaborative conflicts from logical inconsistency. It is helpful to differentiate them from each other, because some collaborative conflicts may not result in logical inconsistency, but violate user defined rules or other exception (we will explain in section 3.3). Furthermore, concurrent access control with transaction oriented locking as database is not suitable for ontology, for the rich semantic relationships among ontology entities. Therefore, we need new approach to overcome these deficiencies.

3 Overview of Our Approach

3.1 System Framework

Fig.1 shows a framework of our collaborative ontology building tools. It adopts B/S architecture, and uses OWL[9] as ontology language. There are two parts, one is private workspace, each user has his own private workspace, and the other is public workspace. When a user logins to his own private workspace successfully, he could select an ontology segment from server. All revisions the user performs will be transferred into a Command-Package and stored until submission. In the public workspace, there are 5 main processes: 1) a command package pool stores all packages received from each user. 2) For every period, the collaborative conflicts checking process, which includes three sub-processes, i.e. hard soft and latent conflicts checking. All conflicts will be handled by Conflicts Handler, in this paper we omit the detail of the process. 3) The 2nd checking is consistency checking, which deals with logical inconsistency, and all inconsistency will be handled by another handler which is also omitted in this paper. 4) After 2 proceeding checking, some mistakes still remain for they can not be found automatically. Therefore, experts' review is necessary. 5) Lastly, all correct revisions will be executed to update ontology base. In this paper, we focus on collaborative conflicts detection.

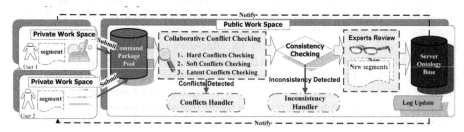

Fig. 1. Overview of Collaborative Ontology Building Process

3.2 Ontology Commands

In our method, we provide a set of commands named Onto-Command for users to modify the ontology content, each of which is composed of an operation and some operands (ontology entity). Table 1 lists a part of atomic commands.

Definition 1(OC:Onto-Command). OC=<*Name,E,V*> is a function s.t. E$\rightarrow$E', where, *Name* is the command name, *E* is an ontology entity set , *V* is the value of parameter.

Our system differentiates these commands into three kinds:

Table 1. A Part of Ontology Commands

Name	Parameter (E)
AddClass/DelClass	(aClass)
AddSubClass/DelSubClassRelation/DelEquClass/DelSameClass	(aClass,supClass)
AddDataProperty/AddObjProperty/DelObjProperty	(aProp)
AddDifferentClassRelation/AddEquivalentClass/AddSameClass	(aClass1,aClass2)
AddDifferentIndividual/AddEquivalentIndividual	(aInd1,aInd2)

$$OCKind(oc) = \begin{cases} DEL, oc \in \{DelClass, DelIndividual, DelRestriction, DelProperty\} \\ ADD, oc \in \{ AddClass, AddIndividual, AddRest, AddProperty \} \\ MOD, otherwise \end{cases}$$

where, *DEL* means commands of this kind is for deleting entity, *ADD* is for adding, and *MOD* is for modifying.

3.3 Collaborative Conflicts and Logical Inconsistency

As mentioned in the last paragraph of section 2, some conflicts may not always result in ontology logical inconsistency, but violate user-defined rules or get result unexpected. For example, fig.2(a) shows an original ontology segment, which depicts the relation between *Animal* and *Plant* as well as their subclasses. If one participator adds *Eat* object property between *Rat* and *Apple,* meanwhile the other participator moves *Apple* to be a subclass of *Computer,* as showed in fig.2(b). There is no inconsistency at all in fig.2(b), but what the 1st participator wants is '*Rat eats a kind plant*', instead of '*Rat eats a kind Computer*'. Therefore, in our opinion, collaborative conflicts should be handled different from logical inconsistency.

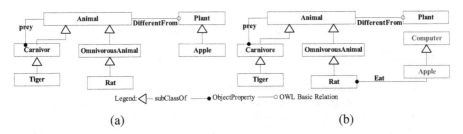

Fig. 2. (a) Original Ontology Segment. **(b)** Result of Accepting All Revisions.

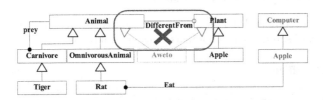

Fig. 3. An Example of Logical Inconsistency

Definition 2(Collaborative Conflicts). We call two revisions performed by different participators on the same ontology entity conflict with each other, if one changes the semantic of the entity, while the other still uses its original semantic.

Definition 3 (Logical Inconsistency). We call a revision on ontology O, is inconsistent with a set of consistency conditions K, iff $\exists k \in K$ such that O' does not satisfies the consistency condition k(O'), where O' is the result from O by the revision.

For example, suppose one participator adds a concept *Aweto*, and then makes it as a subclass of both *Animal* and *Plant* as fig.3 shows. It is obvious that these changes breaks the logical condition '*Animal owl:differentFrom Plant*'.

4 Collaborative Conflicts Detection

In this section we address how to detect collaborative conflicts. Firstly we introduce Command Impact Range, and base on it we classify 3 kinds of collaborative conflicts, and finally propose two algorithms with comparison between them.

4.1 Command Impact Range

Because subtle semantic relationship exists among entities in ontology, change one entity may produce chain reaction. This means that for each revision, there is a set of entities may be impacted. Therefore, for each atomic Onto-Command we define a unique **IR(Impact Range)** value, table 2 lists some of them.

Look back on fig.2(b), for example, if a user uses *DelClass* command *oc* to delete class '*Carnivore*'. According to table 2, **IR**(*oc*)={ *Carnivore, prey, Tiger* }.

Table 2. A Part of IRs Value

OC Name	E	IR(oc)
DelClass	(aClass)	Return $E \cup$ {o| o.Domain=aClass } $\cup$ { o| o.Range=aClass } $\cup$ ...
DelInstance	(aIndividual)	Return $E \cup$ { ins| ins is the same as aIndividual }
DelProperty	(aProperty)	Return $E \cup$ {r| r.actOn(aProperty)= true, r is Restriction }
AddObjProperty	(class1,class2)	Return $E \cup$ {class| (aClass.directSubclassOf(aClass2)}...

4.2 Three Kinds of Collaborative Conflicts

In our method, according different criterions, we classify three kinds of collaborative conflicts, i.e. hard soft and latent conflicts. They are following.

4.2.1 Hard Conflicts

Hard conflicts are easy to detect according to the impact range and command type. Before go on we will introduce two functions, which are used to judge whether it is possible for two commands conflict with each other.

OTC (OC Type Conflicts Table) is a function table returns whether it is possible for two commands (oc and oc') conflict with each other. As table 3 lists, F/T is false/true, which means that it is impossible/possible for oc directly conflicts with oc'.

In our opinion, if two participators perform the same operation on the same entity with same value, we will unite two operations as one. Therefore, there is no conflict if both participators perform commands to delete or add the same class. i.e., *OTC(DEL,DEL)=F* and *OTC(ADD,ADD)=F*.

Table 3. Onto-Command Type Conflicts Rules

OTC(oc,oc')		OCKind(oc')		
		MOD	ADD	DEL
OCKind(oc)	MOD	**OTC_MOD**(oc,oc')	F	T
	ADD	F	F	T
	DEL	T	T	F

Table 4. A Part of OTC_MOD Conflicts Rules

oc	oc'	OTC_MOD(oc,oc')
AddEquivalentIndividual	AddDifferentIndividual	If (oc.E=oc'.E) T Else F
AddEquivalentClass	AddDifferentClassRelation	If (oc.E=oc'.E) T Else F
AddEquivalentClass	AddDisjointWithRelation	If (oc.E=oc'.E) T Else F
AddEquivalentProperty	AddDifferentPropertyRelation	If (oc.E=oc'.E) T Else F

OTC_MOD (MOD OC Conflicts Table) is a function table returns whether it is possible for two *MOD* commands (i.e. OCKind(*oc*) =*MOD*) directly conflict with each other. Table 4 lists a part of them.

For example, if one participator use command *AddEquivalentClass* oc1 to make *class1* and *class2* equivalent, while the other use command *AddDifferentClass* oc2 to make *class1* and *class2* different, then *OTC_MOD(oc1,oc2)=T*.

Definition 4 (Hard Conflicts '@'). Given 2 commands *oc* and *oc'* performed by different users, *oc@oc'* if **OTC**(*oc,oc'*) $\wedge$ (IR(oc) $\cap$ IR(oc')) $\neq$ NULL .

For example, look back on fig.2(b), if user U1 performs a *DelClass* command oc for delete class *Carnivore*, and user U2 performs *AddSubClass* command oc' to add a subclass *Lion* to *Carnivore*. In this case, DR(oc, oc') = IR(oc) $\cap$ IR(oc') ={*Carnivore*} and OTC(oc,oc') holds, then oc@oc', that means oc conflicts with oc' directly.

4.2.2 Soft Conflicts

In some cases, there is nothing wrong if both revisions happen alone, but would result in inconsistency if both were accepted. Therefore, we define **Semantic-Rule-Set** which includes a set of semantic rules, in order to deal with these cases. Table 5 lists a part of these rules.

Definition 5 (Semantic Rules). Given an entity *e* and two commands *oc* and *oc'*, where, e $\in$ IR(oc) $\cap$ IR(oc'). **SEM** is a semantic rule judges whether *e1* is inconsistent with *e2*, where *e1* and *e2* originate from *e* by executing *oc* and *oc'* respectively.

Table 5. A Part of Semantic Rules in Semantic-Rule-Set

SEM	Description
Same VS Different	Check whether there is an entity is the same as e1 but different from e2
Same VS DisjointWith	Check whether there is an entity is the same as e1 but disjoint with from e2
Functional Same VS Different	Check whether there is an entity is the same as e1 but different from e2 according the principle of functional object property

For example, ***Functional Same VS Different*** is one semantic rule, whose principle is that for a functional object property **P** in OWL, **Y=Z** if **P(X, Y)** and **P(X, Z)**. Therefore,

$$SEM_{functional\&diff}(e,oc,oc') = \begin{cases} 0, \exists c, s.t.\ FunctionalSameAs(c,e1) \wedge c.differentFrom(e2) \\ 0, \exists c, s.t.\ FunctionalSameAs(c,e2) \wedge c.differentFrom(e1) \\ 1,\ otherwise \end{cases},$$

where,

    (a) differentFrom means '*owl:differentFrom*' or '*owl:disjointWith*'.

    (b) *e1* and *e2* originate from *e* by executing *oc* and *oc'* respectively.

    (c) $FunctionalSameAs(c1,c2) = \begin{cases} 1, \exists p, \exists c, s.t.,\ p(c, c1) \wedge p(c, c2),\ where,\ p\ is\ a\ functional\ property \\ 0,\ otherwise \end{cases}$

Definition 6 (Soft Conflicts '#'). Given 2 commands oc and oc' performed by different users, oc#oc' if $\neg$ OTC(oc,oc') $\wedge$ ($\exists$SEM, $\exists$e, s.t. SEM(e,oc,oc')=0), where, $e \in IR(oc) \cap IR(oc')$, SEM $\in$ **Semantic-Rule-Set**.

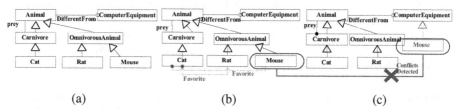

Fig. 4. (a) Another Segment. **(b)**Result Revised By U1. **(c)** Result Revised By U2.

For example, fig. 4 (a) shows another ontology segment. If participator U1 adds a functional object property *Favorite* from *Cat* to *Rat* and *Mouse* by command *oc*, according to the principle of functional object property, we can infer that *Rat* is the same as *Mouse*. Meanwhile participator U2 moves *Mouse* to be a subclass of *ComputerEquipment* that is different from *Animal* by command *oc'*. We can infer that it is also different from *Rat*. i.e. $SEM_{functional\&diff}(e, oc, oc') = 0$, and $\neg$ OTC(oc,oc') therefore, *oc#oc'*. Fig.4 (b) and (c) show the result.

4.2.3 Latent Conflicts

Different from hard and soft conflicts, which can be detected explicitly, some conflicts are not so easy to judge. Look back on the first example in section 3.3, suppose that the second participator only inserts new concept *Fruit* between *Apple* and *Plant*, instead of moving *Apple* to be a subclass of *Computer*, then the result is acceptable. Fig. 5 shows the comparison, where (a) shows the acceptable result, and (b) shows unacceptable case. In this case, it is not easy to judge whether the changes on concept *Apple* is acceptable or not, for lacking absolute standard. Therefore, heuristic measures must be adopted to deal. In our system, we define a set of similarity measures, and differentiate them into two types. Table 6 lists a part of them.

Two Types of Heuristic Similarity Measures:

(1) **HSemSIM**: calculates similarity based on semantic information.
(2) **HStruSIM**: calculates similarity based on the structure information.

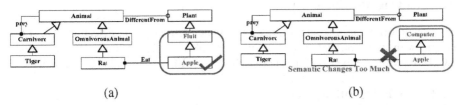

(a)   (b)

Fig. 5. (a) Acceptable Result. (b) Unacceptable Result.

Table 6. A Part of Characteristic and Heuristic Similarity Measures

Measure type	Measure	Heuristic matcher Description
Semantical	Entity Name	Compare the name or label of e1 and e2
	Class Property	Compare matched property of e1 and e2
	Property Domain & Range	Compare matched domain and range of e1 and e2
	Class Restriction	Compare matched restriction of e1 and e2
	Restriction On Property	Compare matched property of e1 and e2
Structural	Direct Sup/Subproperty	Compare matched direct sup/subproperty of e1 and e2
	Direct Sup/Subclass	Compare matched direct sup/subclass of e1 and e2
	Depth Distance	Compare hierarchy distance of e1 and e2

For example, **Depth Distance** is one structural measure that calculates similarity between two ontology classes or two object properties, based on the fact that the deeper the two entities locate in one hierarchy, the higher the similarity is. Therefore,

$$HStruSim_{DepthDistance}(e1,e2) = \frac{2 \times depth(LCA(e1,e2))}{depth(e1) + depth(e2)}$$

where, LCA(e1,e2) gets the nearest ancestor of e1 and e2.

Similarity Aggregation: for all measures, **OntoSIM** is given:

$$OntoSIM(e1,e2) = \begin{cases} 0, OntoSemSIM(e1,e2) < t_1 \\ 1, OntoSemSIM(e1,e2) > t_2 \\ \lambda_1 OntoSemSIM(e1,e2) + \lambda_2 OntoStruSIM(e1,e2), otherwise \end{cases}$$

where, $OntoSemSIM(e1,e2) = Min(HSemSim_0(e1,e2), HSemSim_1(e1,e2), ... HSemSim_N(e1,e2))$

$$OntoStruSIM(e1,e2) = \frac{\Sigma_{i=0}^{i=M} W_i \times HStruSim_i(e1,e2)}{\Sigma_{i=0}^{i=M} W_i}$$

$$t_1 = 0.3, \quad t_2 = 0.95, \quad \lambda_1 = 0.6, \quad \lambda_2 = 0.4$$

where, N is the number of semantical matchers. M is the number of structural matchers. W_i is the weight of each individual structural measure.

Definition 7 (Latent Conflicts'!!'): Given 2 commands oc and oc' performed by different users. $oc!!oc'$ if $\neg OTC(oc,oc') \wedge \exists e \in (IR(oc) \cap IR(oc'))$, s.t. $OntoSIM(e1,e2)<t$, $t \in [0,1]$, where $e1$ and $e2$ originate from e by executing oc and oc' respectively.

4.3 Checking Algorithm

This section presents two algorithms for collaborative conflicts detection. The 1st is simple with high complexity that is unacceptable. The 2nd is high effective named CCD. We will give their performance comparison with experiments in section 5.3.

4.3.1 Simple Algorithm

Algorithm 1 lists the simple algorithm. It scans command-pool by 2 loops and gets all possible command pairs. For each pair, we do three kinds of conflicts checking, and each detected conflict will be added to appropriate conflicts set.

Algorithm 1. Simple Conflicts Detector Pseudo Code

Input Data: All Commands in **Command-Pool**

Result: three conflicts sets with initialization *hardConflicSet={}*, *softConflictSet={}*, *latentConflictSet={}*

0   scan **Command-Pool** by 2 loops and get all possible pair $<c,c'>$ where c *is different from* c'

1   **for** each command pair $<c,c'>$ **do**

2   calculate intersection entity set $S := (\mathbf{IR}(c) \cap \mathbf{IR}(c'))$

3   **if** ($OTC(c, c')$ and S is not empty) *hardConflicSet* $:= hardConflicSet \cup \{<c,c'>\}$

4   **for** each entity e in S **do**

5   **for** each semantic rule *SEM* in **Semantic-Rules-Set do**

6   **if** (SEM (e,c,c') is zero) *softConflictSet* $:= softConflictSet \cup \{<c,c'>\}$

7   **end**

8   **if** (***OntoSIM***(c.execute(e) , c'.execute(e))$<T$) *latentConflictSet* $:= latentConflictSet \cup \{<c,c'>\}$

9   **end;**

10 **end ;**

Complexity Analysis. Because 1) in line 0, it scans pool by two loops which makes complexity $O(n^2)$. 2) The number of entities that one command may impact is finite (by our statistic, none exceeds 100), i.e. $O(IR)=O(C)$. 3) The number of *SEMs* in Semantic-Rules-Set and the number of heuristic similarity measures in OntoSIM are both finite. Therefore, the total complexity is $O(n^2)$, this is unacceptable.

4.3.2 High Effective Collaborative Conflicts Detection Algorithm (CCD)

In order to detect conflicts effectively, we define a structure STRU_CON_SET (see fig.6). Each instance of this type holds an entity e, and a command set *IRSet*, where, $\forall oc \in IRSet$ s.t. $e \in IR(oc)$, i.e. *IRSet* stores all commands whose impact range includes e. All instances are stored and sorted by e in Sorted-List v.

Theorem 1: $\forall oc, oc' \in scs.IRSet$, $oc@oc'$ if $OTC(oc,oc')$ holds.

Theorem 2: $\forall oc, oc' \in scs.IRSet$, $oc\#oc'$ if $\neg OTC(oc,oc) \wedge \exists SEM, s.t. SEM(scs.e, oc, oc') = 0$.

Theorem 3: $\forall oc, oc' \in scs.IRSet$, $oc!!oc'$ if $\neg OTC(oc,oc') \wedge OntoSIM(e1,e2)<t$, where $e1$ and $e2$ are produced from e by executing oc and oc' respectively.

Algorithm 2 presents a high effective checking algorithm named **CCD**. It scans Command-Pool only once, for each command c in the pool, we compare it with candidate commands that may conflict with c in Sorted-List v by binary searching. The remainder is similar with Algorithm 1.

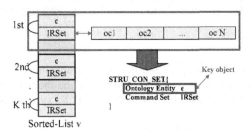

Fig. 6. STRU_CON_SET Data Structure

Algorithm 2. CCD Pseudo Code
Input Data: All Commands in **Command-Pool**
Result: three conflicts sets with initialization *hardConflicSet*={}, *softConflictSet*={}, *latentConflictSet*={}
0   create and init a **Sorted-List<STRU_CON_SET > v.**
1   **for** each command c in **Command-Pool do**
2   **for** each entity e in **IR**(c) **do**
3   **STRU_CON_SET** *scs* :=do binary search in v by e, if not found create a new and insert into v
4   **for** each command c' in. ***IRSet*** of *scs* **do**
5   **if** (*OTC(c, c')*) **do** *hardConflictSet* := *hardConflicSet* ∪ {<c,c'>} and get next c'
6   **for** each semantic rule *SEM* in **Semantic-Rules-Set do**
7   **if** (*SEM* ($e,c,\ c'$) is zero) *softConflictSet* := *softConflictSet* ∪ {<c,c'>} and get next c'
8   **if** (***OntoSIM***(c.execute(e) , c'.execute(e))<*T*) *latentConflictSet*:= *latentConflictSet* ∪ {<c,c'>}
9   **end**
10   *scs.**IRSet*** := *scs.**IRSet*** ∪ {oc}
11   **end**
12 **end**

Complexity Analysis. As stated in Algorithm 2, in line 1 it scans the pool by a loop, whose complexity is $O(n)$. In line 3 the binary retrieval's complexity is $O(Log_2 (n))$. Other loops' complexities are all $O(C)$, just the same as Algorithm 1. Therefore, the total complexity of CCD is $O(n)*O(Log_2 (n))$. In the worst case is $O(n^2)$. The cost is the extra spending of v whose space complexity is $O(n)$.

5   Case Study and Experimental Design

We developed our system by Java1.5 and MySql. Client runs on web, and the server was conducted on Intel Centrino Duo T2400 1.83GHz PC with 2GB RAM, running WindowsXP sp2. Table 7 lists the detail of the experiment data, which comes from FAO (Food and Agriculture Organization).

Table 7. Experiment Data for Collaborative Conflicts Detection

Entity type	Number	Entity type	Number	Entity type	Number
Class	82	TransitivePropertiy	14	FunctionalProperty	13
Individual	233	FunctionalProperty	13	InverseFunctionalProperty	9
Restriction	113	Data-Property	105	SymmetricProperty	12
ObjectProperty	212	Data-Range	176	InverseFunctionalProperty	9

5.1 Experiment 1

In order to evaluate, we compare the manually determined real conflicts(R) against the detected result P returned by **CCD**, and determine the true positives, I, as well as false positives, F=P-I. Then we have: (1) **Recall**=$|I|/|R|$, (2) **Precision** =$|I|/|P|$.

5.1.1 Hard and Soft Conflicts Detection Result

This experiment is to evaluate the hard and soft conflicts checking algorithm, fig. 7 (a) and (b) show the result.

Analysis. From fig.7 we can infer, 1) the algorithm is good enough for detecting soft conflicts. 2) As far as the hard conflicts are concerned, at the start of X axes, both precision and recall change radically, because the data for analysis is not enough. 3) With number increasing, the precision/recall of hard conflicts converge 95%/94%, and we consider the constant is the real data we need.

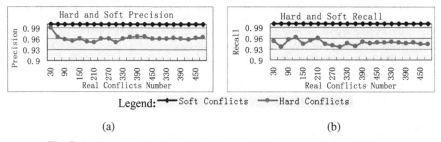

(a)   (b)

Fig. 7. (a) Hard and Soft Precision Graph. **(b)** Hard and Soft Recall Graph.

5.1.2 Latent Conflicts Detection Result

This experiment is to analysis the impact of heuristic matchers on latent conflicts detection result. The threshold is 0.65. Firstly there are few matchers in **OntoSIM**, and then add a number, finally we add enough. Fig. 8 (a) and (b) show the result.

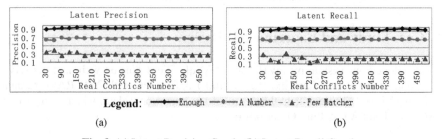

(a)   (b)

Fig. 8. (a) Latent Precision Graph. **(b)** Latent Recall Graph.

Analysis. From the two figures, we can infer, 1) at the start of X axes, the result of *'Few Matcher'* changes radically. We also believe that is because of the data for analysis is not enough, 2) with more and more conflicts coming, the results converge to a constant, and we consider the constant is the real data we need, too. 3) *'Enough Matchers'* produces higher precision and recall than *'Few Matcher'* and *'A Number of Matchers'*. Therefore, we can say that heuristic matchers play an important role in the collaborative conflicts detecting process.

5.2 Experiment 2

In this experiment, we adopt 'Enough Matchers' in OntoSIM with threshold 0.65, and make a running time comparison between **Simple** and **CCD** algorithm. Table 8 and fig.9 present the performance comparison.

Analysis. 1) It clearly shows that with commands increasing, the running time of CCD increases flatly, while the performance of optimized simple algorithm is far worse than CCD. 2) According to the analysis in section 4.3.1 and 4.3.2, the running time of Simple algorithm should be (N/ Log $_2$ (N)) times of the time of CCD. From the last two lines of table 8, we can see that the experiment result complies with it perfectly. 3) Notice that the running time of Simple algorithm does not increase by the way of N^2 parabola, this is because we optimize the checking algorithm.

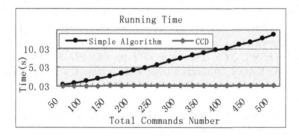

Fig. 9. Running Time Comparison between Simple Algorithm and CCD

Table 8. Running Time and Comparison between Two Algorithms (time is in seconds)

Comparison	N (Commands Number)						
	200	250	300	350	400	450	500
Running time of Simple(s)	4.6079	5.9016	7.4763	8.9776	10.1837	11.492	13.869
Running time of CCD (s)	0.1605	0.1746	0.1976	0.2134	0.2192	0.2237	0.2374
Time(Simple)/Time(CCD)	28.70	33.79	37.81	42.05	46.44	51.36	58.41
N/ Log $_2$ (N)	26.16	31.51	36.46	41.41	46.27	51.06	55.76

6   Conclusion and Future Work

Ontology construction itself is a big challenge, especially in large-scale collaboration-based ontology development, because, 1) in ontology, the basic units are concepts and relationships between them, without explicit file-based artifact, which makes conflicts much more likely exist; 2) Most of participators are not aware of the existences of their co-workers at all, let alone have enough sufficient communications. Therefore, it is obvious that effective conflicts detection and resolution are essential for large-scale collaborative ontology development.

In this paper, we propose a novel approach to deal with collaborative conflicts in our ontology construction system. The main contributions are: 1) differentiate collaborative conflicts from logical inconsistency. 2) Classify three kinds of collaborative conflicts, i.e. hard conflicts soft conflicts and latent conflicts. 3) Propose a formal method to detect collaborative conflicts with an effective algorithm and evaluation.

In the future, we will adapt the method to be a real time agent for more efficiency.

Acknowledgements. This work is supported by the National High Technology Development 863 Program of China under Grant No. 2007AA01Z179.

References

1. Noy, N.F., Chugh, A., Alani, H.: The CKC Challenge: Exploring Tools for Collaborative Knowledge Construction. IEEE Intelligent Systems 23(1), 64–68 (2008)
2. Horrocks, Sattler, U., Tobies, S.: Practical reasoning for expressive description logics. In: Ganzinger, H., McAllester, D., Voronkov, A. (eds.) LPAR 1999. LNCS, vol. 1705. Springer, Heidelberg (1999)
3. Auer, S., Dietzold, S., Riechert, T.: OntoWiki – A tool for social, semantic collaboration. In: Cruz, I., Decker, S., Allemang, D., Preist, C., Schwabe, D., Mika, P., Uschold, M., Aroyo, L.M. (eds.) ISWC 2006. LNCS, vol. 4273, pp. 736–749. Springer, Heidelberg (2006)
4. Haase, P., Stojanovic, L.: Consistent evolution of OWL ontologies. In: Gómez-Pérez, A., Euzenat, J. (eds.) ESWC 2005. LNCS, vol. 3532, pp. 182–197. Springer, Heidelberg (2005)
5. Bozsak, E., et al.: KAON - towards a large scale semantic web. In: Bauknecht, K., Tjoa, A.M., Quirchmayr, G. (eds.) EC-Web 2002. LNCS, vol. 2455, p. 304. Springer, Heidelberg (2002)
6. Madhavan, J., Bernstein, P.A., Rahm, E.: Generic Schema Matching with Cupid. In: Proc. VLDB Conf., pp. 49–58 (September 2001)
7. Sure, Y., Erdmann, M., Angele, J., Staab, S., Studer, R., Wenke, D.: OntoEdit: Collaborative ontology development for the semantic web. In: Horrocks, I., Hendler, J. (eds.) ISWC 2002. LNCS, vol. 2342, p. 221. Springer, Heidelberg (2002)
8. Noy, N.F., Sintek, M., Decker, S., Crubezy, M., Fergerson, R.W., Musen, M.A.: Creating Semantic Web Contents with Protege 2000. IEEE Intelligent Systems 16(2), 60–71 (2001)
9. Web—ontology working group. OWL Web ontology Language Overview, http://www.w3.org/TR/2003/PR-owl-features-20031215
10. Hwang, S.-H., Kim, H.-G., Yang, H.-S.: A FCA-based ontology construction for the design of class hierarchy. In: Gervasi, O., Gavrilova, M.L., Kumar, V., Laganá, A., Lee, H.P., Mun, Y., Taniar, D., Tan, C.J.K. (eds.) ICCSA 2005. LNCS, vol. 3482, pp. 827–835. Springer, Heidelberg (2005)
11. Ram, S., Park, J.: Semantic conflict resolution ontology (SCROL): an ontology for detecting and resolving data and schema-level semantic conflicts. IEEE Transactions on Knowledge and Data Engineering 16(2) (February 2004)
12. Bechhofer, S., Horrocks, I., Goble, C., Stevens, R.: OilEd: A reason-able ontology editor for the semantic web. In: Baader, F., Brewka, G., Eiter, T. (eds.) KI 2001. LNCS, vol. 2174, p. 396. Springer, Heidelberg (2001)

Using Link-Based Content Analysis to Measure Document Similarity Effectively

Pei Li[1,2], Zhixu Li[1,2], Hongyan Liu[3], Jun He[1,2], and Xiaoyong Du[1,2]

[1] Key Labs of Data Engineering and Knowledge Engineering, Ministry of Education, China
[2] School of Information, Renmin University of China, Beijing, China
{lp,lizx,hejun,duyong}@ruc.edu.cn
[3] Department of Management Science and Engineering, Tsinghua University, Beijing, China
hyliu@tsinghua.edu.cn

Abstract. Along with a massive amount of information being placed online, it is a challenge to exploit the internal and external information of documents when assessing similarity between them. A variety of approaches have been proposed to model the document similarity based on different foundations, but usually they are not applicable for combining internal and external information. In this paper, we introduce a link-based method into content analysis, which is based on random walk on graphs. By defining similarity as the meeting probability of two random surfers, we propose a computational model for content analysis, which can also be integrated with external information of documents. Empirical study shows that our method achieves good accuracy, acceptable performance and fast convergent rate in multi-relational document similarity measuring.

Keywords: link graph, content analysis, document similarity.

1 Introduction

Document similarity needs to be measured in a variety of applications for clustering, filtering, sorting and retrieving, etc. For example, in a personalized digital library, the computing of document similarity is the foundation of collaborative filtering [1]. But along with astonishing amount of information being placed online, the computation of document similarity encounters great challenges in two aspects: (1) the complexity of internal and external information of a document; (2) the large scale of document amount. For this reason, the ability to measure similarity between documents in an accurate and efficient way is a key determinant for many applications.

A variety of approaches have been proposed to model document similarity based on different foundations. Some traditional approaches calculate similarity according to document contents (especially document-term relationship), such as Vector Space Model [2], n-gram measures [3] and Latent Semantic Analysis [4], etc. Recently, by exploiting link structure of objects, some methods focusing on link-based object ranking are proposed by researchers [5, 6, 7]. If viewing documents as nodes and relationships among documents as edges, document similarity can be measured by these link-based object ranking methods with the contents of documents ignored.

Q. Li et al. (Eds.): APWeb/WAIM 2009, LNCS 5446, pp. 455–467, 2009.

In this paper, we propose a new approach by using link-based content analysis to measure document similarity effectively. This approach takes advantage of document-term relationship and builds a link graph among documents. Then link analysis is imported to assess the similarity between documents. There are plenty of link analysis methods introduced in [12]. Our approach propagates the similarity between documents with a certain transition probability, and has a theoretical foundation based on random walk theory [9]. Moreover, internal and external information of documents can be combined effectively using our method, which is not applicable for most similarity measuring methods mentioned above.

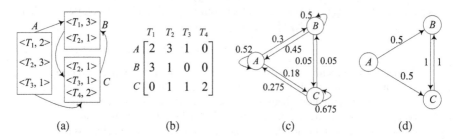

Fig. 1. (a) Documents and citation relationship. (b) Document-term relationships. (c) Transition probability based on contents. (d) Transition probability based on references.

As Figure 1(a) shows, the content of a document can be modeled as a collection of *<Term, TF>* pairs, where *TF* is term frequency in the corresponding document. Using some simple statistics, we obtain a matrix describing relationship between documents and terms as seen in Figure 1(b). Transition probability based on contents between documents is derived from relationship matrix, via normalization step and matrices product step (explained in Section 3.2 and shown in Figure 1(c)). External information is also considered (discussed in Section 4). From the citation relationship shown in Figure 1(a), we can get transition probability shown in Figure 1(d). Given a ratio of importance between contents and citation relationship in measuring the similarity, we can combine these two kinds of transition probabilities as a whole, which is the input of similarity computation. Based on random walk theory, we define similarity as the meeting probability of two surfers. Given a document-pair (a, b), its similarity is determined by similarities of all its direct-connected document-pairs and transition probabilities from (a, b) to these pairs (details can be found in Section 3.3).

The internal information (that is content) and external information (namely, outside links) of documents can be combined effectively using the same computational model. Experiments on real datasets are conducted to test the accuracy and performance of our link-based content analysis. Observation on convergence indicates that similarity result of the first iteration is acceptable for most cases.

The rest of the paper is organized as follows. Section 2 surveys related work. In Section 3, we introduce link-based content analysis. Afterwards, Section 4 describes how to combine internal and external information. The results of experiments are shown in Section 5 and this study is concluded in Section 6.

2 Related Works

Measuring pair-wise document similarity has been extensively studied for decades, with lots of methods proposed. These methods can be roughly divided into several types according to the document information they focus on. Generally, related works of document similarity measuring can be considered from three views: content analysis, link analysis and their combination.

(1) Content analysis

Analysis of content (or internal information) is a traditional information retrieval task. Early methods treat a document as a bag of words and calculate cosine similarity according to the *tf-idf* weight [13], such as Vector Space Model and its variations [2, 14]. Considering the sequence of terms, *n*-grams [3] are introduced to gauge the similarity. A more complicated approach is Latent Semantic Analysis [4], which maps each document and term vector into a lower dimensional space associated with concepts. Aslam et al. [15] propose an information-theoretic measure for document similarity in an axiomatic manner, which is a different research route from others.

(2) Mining links of documents

Some documents (especially web pages) have rich outside links (or called external information). Viewing documents as nodes and external relationships as edges, the corpus of documents can be modeled as a graph, and exploiting this link structure may be one of the best ways to measure document similarity. There are multiple link-based object ranking methods. *SimRank* [5] provides a wonderful definition for similarity on a link graph and takes $O(N^2)$ time for each iteration. The intuitive underlying model of *SimRank* is "random surfer-pairs", a concept derived from random walk theory. Xi et al. [6] use a Unified Relationship Matrix (URM) to represent a collection of heterogeneous objects and their relationships, and compute the similarity iteratively over the URM. Besides, Yin et al. [7] proposed a hierarchical structure called *SimTree* to represent similarities between objects in a compact way. These link-based object ranking methods usually involve iterative computation and ignore the inside attributes of an object.

(3) Combining content and outside links

Researchers have introduced techniques for combining link-based and content-based methods to improve the accuracy of web document classification [16]. Multiple models are developed. Jin et al. [17] introduce a probabilistic model that integrates content matching and link information in a single unified framework to improve retrieval. Zhu et al. [18] design an algorithm that carries out a joint factorization on both the linkage adjacency matrix and the document-term matrix, and represents web pages in a low-dimensional space.

The intuitive model of our method is based on random walk on graphs, which is a special Markov Chain [10]. We apply link-based idea to content analysis, and integrate it with outside links naturally, thus making it different from other methods.

3 Link-Based Content Analysis

In this section we introduce link-based content analysis method.

3.1 Modeling the Content of a Document

Before modeling the content of a document, a cleaning step is performed to remove stop-words, and stem leftover words using the Porter Stemmer algorithm [19]. The stop-words list is taken from SMART Retrieval System [20], and a complete manifest can be found in [21].

We model the content of a document as a set of *<Term, TF>* pairs, where *TF* is the frequency of this term occurring in this document. Formally, let *D* denote the content of a document. Supposing that *D* contains *n* different terms, we describe *D* by

$$D = \{< Term_1, TF_1 >, \cdots, < Term_n, TF_n >\} \tag{1}$$

For a corpus of documents, we give simple definitions for some terminologies which will be used in the rest of this paper.

Definition 1. (**Document Vector**) The document vector of a corpus is an ordered array of all document objects in this corpus. Let *DV* denote it.

Definition 2. (**Term Vector**) The term vector of a corpus is an ordered array of all *different* terms in this corpus. We signify it by *TV*.

Obviously, the relationship between document vector and term vector can be modeled as a weighted bipartite graph with two disjoint sets corresponding to documents and terms. A relationship matrix can be deduced from this bipartite graph, with |*DV*| rows and |*TV*| columns, where |*DV*| denotes the number of documents in *DV* and analogously for |*TV*|. Let *R* denote relationship matrix.

We take a corpus example of documents shown in Figure 2(a) for the convenience of following analysis. Note that term T_4 only occurs in document *C*. We present the bipartite graph corresponding to corpus example in Figure 2(b). The relationship matrix between documents and terms is easy to obtain in Figure 2(c).

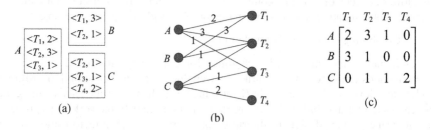

Fig. 2. (a) A corpus example of documents. (b) Weighted bipartite graph. (c) Corresponding relationship matrix *R*.

3.2 Transition Probability between Documents

From the viewpoint of Random Walk Theory, given a directed unweighted graph *G(V, E)*, supposing there is a random surfer standing on node *a*, he has identical possibility to visit each node at which he can arrive on next step, and *zero* possibility to other nodes. If it is a weighted graph, the possibility will be adjusted according to edge weight. This possibility is entitled "transition probability" by researchers.

Treating every document in *DV* as a node and transition probability of each pair of documents as an edge, the corpus of documents can be viewed as a graph. However, the transition probability between two documents can't be ascertained directly. The reason is, from the view of a random surfer in document *A*, he do not know any internal information (the content) of another document *B*. He only knows the terms in *A*. So our solution is using terms as the bridge between two documents.

A normalization step is performed before computing of transition probability. From the probability theory, transition probabilities of a node to all other nodes should sum to 1. For a matrix *M*, we use M_N to denote the normalized matrix. The terminology "*transition matrix*" is defined by researchers to represent the normalized probability matrix from object sets O_1 to O_2, and we use $T(O_1, O_2)$ to denote it. Moreover, given the *i*-th element *a* in O_1 and the *j*-th element *b* in O_2, let $P(a, b)$ denote the element in the *i*-th row and the *j*-th column of $T(O_1, O_2)$.

Base on the above discussions, the transition matrix from documents to terms can be described as $T(DV, TV) = R_N$, and the transition matrix from terms to documents can be described as $T(TV, DV) = (R^T)_N$, where *R* is given in Section 3.1 and R^T is the transpose of *R*.

$$
\begin{array}{c}
\begin{array}{ccc}
A & B & C
\end{array} \\
\begin{array}{c}
A \\ B \\ C
\end{array}
\left[
\begin{array}{ccc}
0.52 & 0.3 & 0.18 \\
0.45 & 0.5 & 0.05 \\
0.275 & 0.05 & 0.675
\end{array}
\right]
\end{array}
$$

(a)   (b)

Fig. 3. (a) The computation model of $P(A, B)$. (b) Transition matrix of the corpus example.

We model the computation of transition probability from document *A* to *B* as two steps (shown in Figure 3(a)). The first step is a random walk from document *A* to some term T_n with transition probability $P(A, T_n)$, and the second step is another random walk from term T_n to document *B* with transition probability $P(T_n, B)$. Serial steps mean a multiplication. Considering all terms including T_n, we take

$$P(A,B) = \sum_{n=1}^{|TV|} P(A,T_n) \cdot P(T_n, B) \tag{2}$$

to represent the transition probability from document *A* to *B*, where |*TV*| is the number of different terms in term vector *TV*.

Considering the transition matrix of documents, supposing *A* and *B* are the *i*-th and the *j*-th documents in *DV* respectively, we get $P(A, B) = T_{ij}(DV, DV)$, $P(A, T_n) = T_{in}(DV, TV)$ and $P(T_n, B) = T_{nj}(TV, DV)$. Thus we have the following theorem.

Theorem 1. The transition matrix of documents, $T(DV, DV)$, can be computed by

$$T(DV, DV) = T(DV, TV) \cdot T(TV, DV) \tag{3}$$

Proof. From Equation (3) we can get $T_{ij}(DV,DV) = \sum_{n=1}^{|TV|} T_{in}(DV,TV) \cdot T_{nj}(TV,DV)$, which can be also obtained from Equation (2).

Taking the corpus shown in Figure 2(a) as an example, we can obtain $T(DV, TV)$ and $T(TV, DV)$ using relationship matrix R shown in Figure 2(c). The transition matrix of documents is computed by Equation (3) and shown in Figure 3(b) with another view in Figure 1(c). Note that the transition probability from a document to itself is not equal to *zero*, which means the random surfer has possibility to stay in his current position on next step.

3.3 Assessing Similarity

Let $Sim(A, B)$ denote the similarity between object A and B in a link graph. If $A = B$, we define $Sim(A, B) = 1$, otherwise we define $Sim(A, B)$ as the meeting probability of two random surfers starting from A and B respectively (similar definition is found in [5]). It is easy to get $Sim(A, B) = Sim(B, A)$ according to this definition.

$$\begin{array}{c c c c}
 & A & B & C \\
A & \begin{bmatrix} 1.0 & 0.585 & 0.524 \\ B & 0.585 & 1.0 & 0.486 \\ C & 0.524 & 0.486 & 1.0 \end{bmatrix}
\end{array}$$

(a)   (b)

Fig. 4. (a) The computation model of $Sim(A, B)$. (b) $S(DV)$ of the corpus example.

In our method, the meeting probability (or similarity) is reinforced step by step. Each step of random surfers means a re-distribution of meeting probabilities, and results of the $(k+1)$th step are based on results of the k-th step. Assessing similarity $Sim(A, B)$ is an iterative process. We model the $(k+1)$th iteration of $Sim(A, B)$ (we denote it by $Sim_{k+1}(A, B)$) in Figure 4(a). Supposing document A has n outgoing edges and B has m outgoing edges, there are n×m document-pairs needing to be considered. Let d be a decay factor (usually $d = 0.8$) and we represent $Sim_{k+1}(A, B)$ by

$$Sim_{k+1}(A,B) = d \cdot \sum_{i=1}^{n} \sum_{j=1}^{m} P(A,A_i) \cdot Sim_k(A_i, B_j) \cdot P(B, B_j), \text{ where } A \neq B \tag{4}$$

For the convenience of derivation, we take the objects not in $\{A_1, A_2, ..., A_n\}$ or $\{B_1, B_2, ..., B_m\}$ into consideration. Since transition probabilities from A or B to these objects are zero, they make no contribution to $Sim_{k+1}(A, B)$. Denoting similarity matrix of the k-th iteration by $S_k(DV)$ and supposing A and B are the p-th and the q-th documents in DV respectively ($p \neq q$), we get

$$(S_{k+1}(DV))_{pq} = d \cdot \sum_{i=1}^{|DV|} \sum_{j=1}^{|DV|} T_{pi}(DV,DV) \cdot (S_k(DV))_{ij} \cdot T_{jq}(DV,DV) \tag{5}$$

Based on Equation 5, we give a theorem as follows.

Theorem 2. The $(k+1)$th iteration of similarity matrix $S_{k+1}(DV)$ can be computed by

$$S_{k+1}(DV) = d \cdot T(DV, DV) \cdot S_k(DV) \cdot (T(DV, DV))^T + M \tag{6}$$

where $(T(DV, DV))^T$ is the transpose of transition matrix $T(DV, DV)$, and M is a correction matrix making every element on diagonal of $S_{k+1}(DV)$ to be 1.

Proof. In fact, Equation (6) is the matrix form of Equation (5).

A notice here is the initialization of $S_k(DV)$ when $k = 0$. Since we can't foreknow the similarity between two objects before iteration, it is reasonable to simply define $(S_0(DV))_{ij} = 1$ for $i = j$, and $(S_0(DV))_{ij} = 0$ for $i \neq j$. Hence, $S_0(DV)$ is an identity matrix and is symmetrical, which ensures $Sim(A, B) = Sim(B, A)$ on all iterations.

Our similarity computation method can be viewed as an extension of *SimRank* [5] on directed weighted graph. The naive method of *SimRank* is only applicable for undirected unweighted graph. Besides, the starting points of our method and *SimRank* are different. *SimRank* measures the Expected-*f* Meeting Distance on a graph, while our method evaluates the meeting probability of two random surfers. *SimRank* gets an iterative formula similar to Equation (4), and the mathematic proof of convergence given in the Appendix of [5] is also adaptable for our method with a few changes.

Let's consider the example shown in Figure 2(a). Using the transition matrix shown in Figure 3(b), we can compute its similarity matrices easily by Equation (6), and the convergent result $S(DV)$ (via 10 iterations) is presented in Figure 4(b).

We summarize the major steps of link-based content analysis method as follows.

1. Preprocessing. Remove stop-words from document contents and stem the left words (Section 3.1).
2. Obtain relationship matrix R between documents and terms using statistical technologies (Section 3.1).
3. Compute transition matrix of documents $T(DV, DV)$ by Equation (3) (Section 3.2).
4. Initialization. Set $k = 0$ and $S_0(DV)$ to be a $|DV|$-by-$|DV|$ identity matrix.
5. For the $(k+1)$th iteration, compute $S_{k+1}(DV)$ using Equation (6) (Section 3.3).
6. If $S_{k+1}(DV)$ is not convergent when compared with $S_k(DV)$, let $k = k+1$ and then jump to step 5; else return $S_{k+1}(DV)$.

3.4 Complexity Analysis

For simplicity, we suppose $|DV| = n$ and $average(|D|) = m$ in a corpus of documents, and the average outgoing edge number of a node is d. The time cost on our link-based content analysis can be roughly divided into two parts: (1) Computation of transition matrix. According to Equation (2), the time consumed by transition matrix computation is $O(mn^2)$. (2) Iterative computation of similarity matrix. According to Equation (4), supposing the number of iterations is k, time complexity for iterative computation is $O(kd^2n^2)$. In real datasets, m and d are usually constants, and $k < 10$ in most cases.

There are some methods for improving performance of *SimRank*, such as pruning *SimRank* [5], fingerprint *SimRank* [11], etc. These methods are also suitable for improving the performance of our link-based content analysis, but it is not the key point of this paper. The aim of our work is to introduce a link-based method into content analysis, and combine internal and external information (introduced in Section 4) for more accurate similarity measuring. In addition, the convergence feature studied in Section 5.3 indicates that the result is acceptable when $k = 1$.

4   Combining Internal and External Information

Relationships between documents can be explored from different aspects, for instance references, publication date, authors, and so on. The best way to describe external information of documents is using a link graph. Extensive studies have been performed on exploiting the link structure such as web graph. In this section, we focus on how to integrate external information into link-based content analysis.

4.1   Integration of Transition Matrices

Usually, there is more than one kind of relationships between documents, and we only consider the ones independent with each other. For example, references and authors are independent, while references and "cited-by" are not. A relationship matrix R can be obtained according to each relationship. Similar with link-based content analysis in Section 3.2, the transition matrix between documents can be described as $T(DV, DV)$ $= R_N$. Taking documents shown in Figure 1(a) as an example, transition probabilities of citation relationship are shown in Figure 1(d).

For a corpus of documents, supposing there are K different relationships between documents, we can get K different transition matrices $T_1(DV, DV)$, $T_2(DV, DV)$, ..., $T_K(DV, DV)$. Considering the transition matrix computed by link-based content analysis (let $T_0(DV, DV)$ denote it), we give different weights to these $K+1$ transition matrices and compute a weighted mean matrix. That means the integrated transition matrix $T(DV, DV)$ can be calculated by the following formula.

$$T(DV, DV) = \sum_{i=0}^{K} w_i \cdot T_i(DV, DV)$$

(7)

where w_i is the weight of $T_i(DV, DV)$ and $\sum_{i=0}^{K} w_i = 1$. Then, similarity matrix $S(DV)$ is computed iteratively using Equation (6).

4.2   Estimating Weights

In Equation 7, the weights are usually determined by experts. That is not a perfect solution, because different people have different evaluations of importance. Inspired by machine learning methods such as decision tree and neural networks, we design an approach to learn weights from training dataset.

Usually we take a portion of classified documents (e.g. 10%) as training dataset. For a transition matrix $T_i(DV, DV)$, we get corresponding similarity matrix $S(DV)$ by Equation 6 and cluster documents based on $S(DV)$. Correct classified documents are counted (we denote the sum by C_i) and let $accuracy(T_i)$ be the ratio between C_i and $|DV|$. The weight of $T_i(DV, DV)$ can be estimated by

$$w_i = accuracy\ (T_i)\bigg/ \sum_{x=0}^{K} accuracy\ (T_x) \qquad (8)$$

5 Empirical Study

We have described a method for link-based content analysis and a solution to combine internal and external information. In this section, the accuracy and iterative process of similarity propagation will be tested, compared with (1) *VSM* [2], a traditional content-based method. We implement it strictly following this equation

$Sim(D_1, D_2) = \cos\theta = \sum_{k=1}^{|TV|} W_{1k} \times W_{2k} \bigg/ \sqrt{(\sum_{k=1}^{|TV|} W_{1k}^2)(\sum_{k=1}^{|TV|} W_{2k}^2)}$, where W is *tf-idf* weight of a

term in a document. (2) *SimRank* [5], a link-based approach on graph.

The similarity score between two objects is hard to ascertain without performing extensively user studies. ACM Computing Classification System [22] (CCS) is a credible subject classification system for Computer Science, which provides an identification of similar papers by organizing these papers in the same category. Note that CCS is only a rough evaluation of similarity. Our dataset is crawled from three categories in ACM CCS, and contains 5469 documents with the ineffectual papers removed. There are three kinds of information in this dataset: (1) Document-term information. Terms are extracted from ABSTRACT of each paper. This relationship is used for content analysis. (2) Reference information. If document A refers to document B, there is an edge from A to B, so we can construct a citation relationship which is a directed weighted graph. (3) Author information. If A and B share the same author, an edge between A and B exists. Thus author relationship can be described as an undirected weighted graph. More details are listed in Table 1.

All experiments are performed on a PC with a 1.86G Intel Core 2 processor, 2 GB memory, and Windows XP Professional. All algorithms are implemented using Java.

Table 1. Details of each relationship

Information	Details (5469 documents in total)
document-term	11,312 different terms; each document contains 43.9 terms on average, and a term exists in 21.2 documents on average
reference	29,304 references in total; 5.4 references per paper
author	7238 different authors; each author appears in 2.2 documents and there are 1.7 authors per document on average

5.1 Similarity Evaluating

In this section we report experiments to examine the accuracy of our method, compared with others. For the given ACM dataset, we compute the similarity matrix of documents in five different ways. To be specific, based on document-term relationship, we obtain two similarity matrices using *VSM* and our link-based content analysis respectively. Another two are obtained by using *SimRank* on reference relationship and author relationship. At last, by combining internal and external information, we obtain a holistic similarity measuring of this dataset.

We use PAM [8] to cluster papers into groups based on similarity matrix. Comparing these groups with CCS categories, we define accuracy as the maximal ratio between the number of correct classified papers and the total number of papers.

Table 2. Accuracy of different approaches

Approach	Max Accuracy
VSM	0.4452
Link-Based Content	0.5072
SimRank (reference)	0.5009
SimRank (author)	0.4876
Combination	0.5681

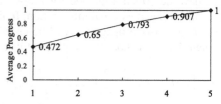

Fig. 5. Average progress of each iteration

Table 2 shows the accuracy of different approaches. The accuracy of *VSM* is not very good compared with other methods, mainly because *VSM* behaves weak in large amount of short documents (e.g. web pages). Anyway, our link-based content analysis gets comparable accuracy with *SimRank*. Then, we compute weights via accuracy according to Equation 8, and obtain a combination of internal and external information by Equation 7. The accuracy of this combination is 56.81% and higher than others, which means a more accurate similarity measuring.

To observe the reinforcement of similarity in iterative process of our link-based content analysis, we take $average(Sim_k(A,B)/Sim(A,B))$ to denote the average progress of each document-pair on the k-th iteration. In Figure 5, the similarity score increases iteration by iteration, and the rising amplitude of each iteration indicates the effect of this iteration to final convergent similarity.

5.2 Performances

We have discussed issues about time complexity in Section 3.4. Supposing there are n documents and a document has m terms on average, *VSM* takes $O(mn^2)$ time and for the worst case, time complexity is $O(|TV|n^2)$. In comparison, our link-based content analysis takes $O(kd^2n^2)$ time, where d is the average number of outgoing edges.

Table 3. Performances of different approaches

Approach	Time/Iteration (sec)	Iteration Num	Total (sec)
VSM	1665	1	1665
Link-Based Content	5858	5	29289
SimRank (reference)	1101	9	9908
SimRank (author)	1030	9	9270
Combination	5519	5	27593

Table 3 lists performances of different approaches. We can see link-based content analysis takes longer time than *VSM* and other link-based methods. That is because viewing terms as bridges between documents, outgoing edge number of a document is usually big. As we have said before, there are some technologies to improve the performance of our method. In next subsection, convergence feature indicates that we only need to perform the first iteration when using link-based content analysis. Moreover, our link-based content analysis can be combined with external information such as references and authors. Transition matrix of this combination is dense too, which results in a performance similar to link-based content analysis.

5.3 Convergence Feature

Iterative process is common for link-based methods based on random walk on graphs. In this set of experiments, the convergence rate will be measured from two aspects: maximum difference and accuracy. If let $Sim_k(A, B)$ denote the similarity score on the k-th iteration, maximum difference M_k can be described as $max(|Sim_{k+1}(A,B) - Sim_k(A,B)|)$, where (A, B) is an arbitrary pair of documents. When M_k is less than the tolerance factor of convergence (e.g. 0.001), iterative process will stop.

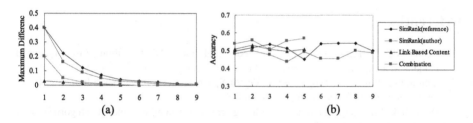

Fig. 6. (a) Maximum differences of different approaches. (b) Accuracy in iterative process.

We get maximum difference and the accuracy of each link-based method on each iteration shown in Figure 6(a) and Figure 6(b) respectively. Comparing with other methods, link-based content analysis has remarkable advantage on the first iteration. In Figure 6(a) the maximum difference M_1 of link-based content analysis is 0.03, which means the increase of similarity score on latter iterations is very limited. The reason is that dense links between documents can accelerate similarity propagation and result in faster convergence rate (but also consume more time on each iteration).

Based on this feature, we can draw a conclusion that in most cases, similarity computed by the first iteration of link-based content analysis is acceptable for most applications. That usually means $Sim_1(A,B) \approx Sim(A,B)$ and we can replace $Sim(A,B)$ by $Sim_1(A,B)$ to avoid expensive computing cost of iterative process.

6 Conclusion

In this paper, we introduce a link-based method to content analysis, and by exploiting document-term relationship, we propose a link-based content analysis to measure document similarity iteratively. Moreover, our link-based content analysis can be combined with external information to obtain a more accurate similarity measuring.

The contributions of our work are summarized as follows.

- We introduce a link-based method into content analysis research. Traditionally, link analysis and content analysis are mutual noninterference, and they differ in research routes and theory models. Our method is a link-based content analysis based on random walk on graphs.
- The internal and external information of documents can be combined effectively using the same computational model. That means our method not only suits for content analysis (utilizing internal information), but also is applicable for utilizing external information such as references, authors and publication date, etc.

Acknowledgments

This work was supported in part by the National Natural Science Foundation of China under Grant No. 70871068, 70621061, 70890083, 60873017, 60573092 and 60496325.

References

1. Renda, M.E., Straccia, U.: A Personalized Collaborative Digital Library Environment: a model and an application. Information Processing and Management 41(1), 5–21 (2005)
2. Salton, G., Wong, A., Yang, C.S.: A Vector Space Model for Automatic Indexing. Communications of the ACM 18, 613–620 (1975)
3. Damashek, M.: Gauging similarity with n-grams: Language-independent categorization of text. Science 267, 843–848
4. Deerwester, S.C., Dumais, S.T., Landauer, T.K., Furnas, G.W., Harshman, R.A.: Indexing by latent semantic analysis. Journal of the American Society of Information Science 41(6), 391–407 (1990)
5. Jeh, G., Widom, J.: SimRank: A measure of structural-context similarity. In: SIGKDD 2002, pp. 538–543 (2002)
6. Xi, W., Fox, E.A., Fan, W., Zhang, B., Chen, Z., Yan, J., Zhuang, D.: SimFusion: measuring similarity using unified relationship matrix. In: SIGIR 2005, pp. 130–137 (2005)
7. Yin, X., Han, J., Yu, P.S.: Linkclus: Efficient clustering via heterogeneous semantic links. In: VLDB 2006, pp. 427–438 (2006)
8. Kaufman, L., Rousseeuw, P.J.: Finding Groups in Data: an Introduction to Cluster Analysis. John Wiley & Sons, Chichester (1990)
9. Lovasz, L.: Random walks on graphs: a survey. In: Combinatorics, Paul Erdos is Eighty, vol. 2, pp. 1–46, Keszthely, Hungary (1993)
10. Kallenberg, O.: Foundations of Modern Probability. Springer, New York (1997)
11. Fogaras, D., Racz, B.: Scaling Link-Based Similarity Search. In: WWW 2005, pp. 641–650 (2005)
12. Getoor, L., Diehl, C.P.: Link mining: A survey. In: SIGKDD 2005 Explorations, vol. 7(2), pp. 3–12.
13. Salton, G., Buckley, C.: Term-weighting approaches in automatic text retrieval. Information Processing & Management 24(5), 513–523 (1988)

14. Hammouda, K.M., Kamel, M.S.: Phrase-based Document Similarity Based on an Index Graph Model. In: ICDM 2002, pp. 203–210 (2002)
15. Aslam, J.A., Frost, M.: An Information-theoretic Measure for Document Similarity. In: SIGIR 2003, pp. 449–450 (2003)
16. Calado, P., Cristo, M., Moura, E.S., Ziviani, N., Ribeiro-Neto, B.A., Goncalves, M.A.: Combining link-based and content-based methods for web document classification. In: CIKM 2003, pp. 394–401 (2003)
17. Jin, R., Dumais, S.: Probabilistic Combination of Content and Links. In: SIGIR 2001, pp. 402–403 (2001)
18. Zhu, S., Yu, K., Chi, Y., Gong, Y.: Combining content and link for classification using matrix factorization. In: SIGIR 2007, pp. 487–494 (2007)
19. Porter, M.: An algorithm for suffix stripping. Program, vol. 14(3), pp. 130–137 (1980), http://www.tartarus.org/~martin/PorterStemmer
20. Salton, G.: The SMART Retrieval System—Experiments in Automatic Document Processing. Prentice-Hall, Inc., Upper Saddle River (1971)
21. The stop-words list, http://members.unine.ch/jacques.savoy/clef/englishST.txt
22. ACM Computing Classification System, http://portal.acm.org/ccs.cfm

Parallel Load Balancing Strategies
for Tree-Structured Peer-to-Peer Networks

Yaw-Huei Chen and Yu-Ren Ju

Department of Computer Science and Information Engineering
National Chiayi University, Chiayi, Taiwan, ROC
ychen@mail.ncyu.edu.tw, yurenju@gmail.com

Abstract. A logical balanced tree structure can be overlaid on a peer-to-peer (P2P) network to support both exact match and range queries. Load balancing mechanisms are needed for handling skew problems in a tree-structured network. Traditional load balancing operations, such as sequentially probing for helper nodes and readjusting the tree structure after moving a node, may incur high costs. This research develops two new parallel load balancing strategies: vicinity load balancing and virtual group load balancing. The former adopts parallel transmission techniques to distribute the load of an overloaded node over a set of neighboring nodes, and the latter redirects the load of an overloaded node to a lightly loaded region. Simulation results indicate that these strategies are efficient not only in time and message, but also in the work load of participant nodes.

Keywords: Peer-to-peer network, load balancing, balanced tree, parallel transmission.

1 Introduction

A peer-to-peer (P2P) network is an internet communication model which allows a group of users to share the computing power, bandwidth, and other resources of participant computers. These autonomous computers form a self-organizing overlay network on the internet so that they can share resources without centralized control. There are two classes of P2P overlay networks: unstructured and structured [11]. In an unstructured P2P system, the overlay network does not form any particular topology and the peers can freely join the system by following some simple rules [5], [7]. Because of the unstructured nature, flooding is the most feasible mechanism for sending a query in the network. However, if there are a large number of peers in the system, the costs of locating an unpopular item using flooding are very high.

On the other hand, various topologies, including mesh [14], [16], ring [4], [15], multi-dimensional space [13], XOR metric [12], list [1], and tree [2], [3], [8], [9], [10], have been proposed to organize the peers in a structured P2P overlay network. Because every data item stored in the network is assigned a unique key and the key is mapped to a peer according to the overlay topology, a data item can be easily found using the key. Many of the structured P2P network systems use a distributed hash table (DHT) as a substrate to support a hash-table based insertion and retrieval of data

Q. Li et al. (Eds.): APWeb/WAIM 2009, LNCS 5446, pp. 468–479, 2009.

items. These DHT-based systems are scalable with respect to network size and efficient in terms of overlay routing hops for searching a data item. However, since they use hash tables to map data items, DHTs support only exact match queries. One effective means for supporting range based data partitioning and retrieval is to build a balanced tree overlay network.

BATON is a balanced tree overlay network proposed for a P2P system [8]. The overlay network is based on a binary balanced tree structure, where each tree node is maintained by a peer in the network. Upper level nodes in a tree structure are more frequently accessed because a message needs to be routed through a common ancestor before it can reach a node at the same level. In order to reduce the access load of nodes near the root, BATON maintains both vertical and horizontal routing information. Therefore, each node in the tree structure stores not only links to its parent and children nodes, but also links to its adjacent nodes and some selected nodes at the same level. The tree overlay structure also acts as a distributed index structure. BATON assigns to each node in the tree a range of index values according to the inorder traversal sequence. Initially we would like to distribute the data items evenly across all nodes. However, the data distribution may become uneven after data items are inserted into and/or deleted from the system. In addition, certain data ranges may be frequently accessed because they contain some popular data items. These situations cause undesired data and execution skew problems, where most queries are executed on only a few nodes.

In order to remedy these skew problems, we need efficient strategies to perform load balancing. NbrAdjust and Reorder are two basic load balancing operations proposed in the literature [6]. The NbrAdjust operation moves a portion of data from the overloaded node to its neighbor. The Reorder operation first tries to find a lightly loaded helper node and then does the load balancing in three steps: (1) Moving all data from the helper node to its neighbor; (2) Logically inserting the empty helper node as a neighbor to the overloaded node; (3) Splitting the load of the overloaded node between the two neighboring nodes. BATON implements its load balancing strategies based on these two operations [8]. On the other hand, P-Ring applies the same principle but improves the load balancing performance by keeping a set of helper nodes handy [4]. However, these load balancing strategies are basically sequential algorithms, which try to balance the load of the overloaded node with the helper nodes one at a time. This is an expensive approach. Furthermore, if we have a highly skewed situation, the problem is getting worse because the overloaded node and its neighbors are already busy with the overwhelming demands and cannot help much in executing the load balancing algorithms. Thus, we need parallel load balancing strategies that do not rely on active participation of the overloaded nodes.

This paper proposes two parallel load balancing strategies for tree-structured P2P networks: vicinity load balancing and virtual group load balancing. The former strategy first probes neighboring nodes of the overloaded node to find possible helper nodes and then distributes the load over them. The latter strategy tries to locate a lightly loaded region and then redirects the load to the region. Because we send messages in parallel, it takes less time and requires smaller number of messages to find the potential helper nodes. The time complexity of this step is in the order of $O((\log_2 n)^2)$ and the message complexity is in the order of $O(n)$, where n is the number of nodes participated in this step. Similarly, parallel propagation of range and routing information reduces the costs of load distribution. In this step, the time complexity is in the order of $O(\log_2 N)$ and the message complexity is in the order of $O(N)$, where N is the number of tree nodes. In addition, we intentionally simplify the role played by the initiator of the algorithm so that the overloaded node does not need to pay too much attention to the load balancing process. Simulation results indicate that the proposed strategies perform well and are better than the BATON load balancing methods in most cases.

The rest of the paper is organized as follows. Section 2 provides necessary background information about BATON. Section 3 presents the two new parallel load balancing strategies. Section 4 describes the simulation results. Finally, Section 5 concludes this paper.

2 BATON Network

BATON (BAlanced Tree Overlay Network) is a balanced tree structure overlay network designed for P2P networks [8]. Each node in the tree handles a range of index values, which are assigned to the nodes according to an inorder traversal of the tree. In addition to the basic parent-child link information, each node in the tree structure also maintains links to its adjacent nodes in the inorder traversal sequence and some selected nodes at the same level. Links to the selected nodes at the same level are stored in a left routing table and a right routing table, which contain the links to nodes that are 2^i nodes away from the node ($i = 0, 1, 2, \ldots$). An example of the BATON tree structure is shown in Fig. 1.

Because BATON is a balanced tree structure, we can search for an exact index value in $O(\log N)$ steps, where N is the number of tree nodes. A node is free to join or leave a P2P system, which may cause imbalance in a BATON tree. Therefore, we need to balance the tree after node addition or deletion by adjusting the tree structure and updating the routing tables. It takes $O(\log N)$ messages for performing both of these operations on a BATON tree.

Each node in a BATON tree manages a range of index values. The computational load on a node is the number of queries issued against the data items managed by the node. Therefore, we can share the load of an overloaded node with its adjacent nodes by adjusting the data range between them. However, if the adjacent nodes are also heavily loaded, this range adjustment operation may propagate to other nodes. This sequential load balancing process will incur high costs. An alternative way of doing load balancing is to use a helper node. We first try to find a lightly loaded helper node using neighbor routing tables. Then we pass the load of the helper node to its adjacent node, remove the helper node from its original position, and insert it as a new adjacent node of the overloaded node. Finally, we split the load of the overloaded node between the node and its newly inserted empty adjacent node. Because moving the helper node may cause imbalance in the tree structure, we need to adjust the tree structure and update the routing tables. If the load is too heavy to be balanced using only one helper node, we need to repeat this process. Although the costs of each operation are in the order of $O(\log N)$, repeatedly readjusting the tree structure may incur very high costs. Neither sequentially propagating load through neighboring

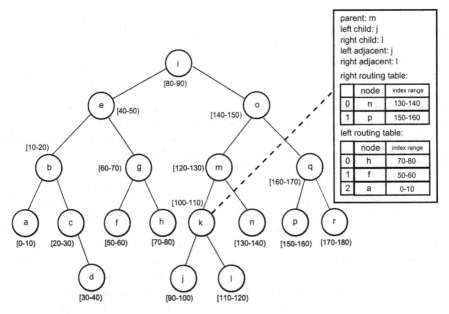

Fig. 1. An example of index architecture and range values in a BATON tree structure

nodes nor repeatedly readjusting the tree structure is an effective way of doing load balancing in a tree-structured network. Thus, we need load balancing strategies that can efficiently distribute the extra load over multiple helper nodes in parallel.

3 Parallel Load Balancing Strategies

In order to effectively handle a highly skewed dataset, we propose two parallel load balancing strategies: vicinity load balancing and virtual group load balancing. The vicinity load balancing strategy distributes the heavy load over a set of neighboring nodes which manages a continuous range of the index values. On the other hand, the virtual group load balancing strategy moves the heavy load to another lightly loaded region. The proposed strategies perform well because of using three basic techniques: (1) Combining multiple messages for the same target node into a single message; (2) Sending messages in parallel; (3) Reducing the work load of the overloaded node when doing load balancing. We present these strategies in the following subsections.

3.1 Vicinity Load Balancing

The basic concept of the vicinity load balancing strategy is to use multiple neighboring nodes to share the heavy load. This strategy consists of two phases: probing for participant nodes and sending update information to affected nodes. Because the index values are partitioned into continuous ranges, we need to probe neighboring nodes to find enough logically adjacent nodes for sharing the heavy load. Without loss of generality, we only probe the nodes on the right side of the overloaded node, but the algorithm can be extended to search both sides with similar costs and results. Sending messages to all nodes in the right routing table of the overloaded node can probe a large number of nodes. However, this is wasteful of resources if only a few neighboring nodes are needed. We propose a round-based probing approach, which will double the total number of nodes searched at the same level after each round.

Probing Phase. Probing nodes in the routing table cannot cover every node because the routing table only contains links to nodes that are 2^i nodes away from the node. In order to probe 2^i nodes, we send messages to the first $(i+1)$ nodes in the right routing table in parallel. Among these nodes, we dedicate the last one as a leader node, which will collect all response messages and decide whether the next round of probing is necessary. When the jth node in the routing table receives the probing message, if $j > 2$ and the node is not the last node, it will send probing messages to the first $(j-1)$ nodes in its own right routing table. In addition, the node will send a probing message to the first node in its left routing table if $j > 2$. Fig. 2 illustrates the three steps of probing 16 nodes. In the first step, node 0 sends probing messages to the first five nodes in its right routing table: node 1, node 2, node 4, node 8, and node 16, and notifies them that node 16 is the leader node of this round. Because node 4 is the 3rd node in node 0's routing table and it is not the last node in this round, node 4 needs to send probing messages to the first 2 nodes in its own right routing table. Similarly node 8 needs to send probing messages to nodes 9, 10, and 12 as shown in step 2. In addition, nodes 4, 8, and 16 also send probing messages to nodes 3, 7, and 15, respectively. This process continues until every node receives the probing message. In the mean time, node 16 will receive response messages from all nodes participated in this round and decide whether it should initiate the next round of probing.

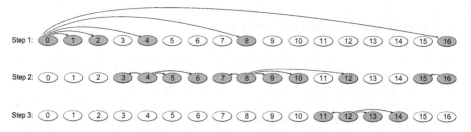

Fig. 2. Probing 16 nodes at the same level

This probing process only propagates messages among nodes at the same level. Since index values are assigned to nodes according to the inorder traversal sequence, we need to consider nodes at other levels in the probing phase. Note that at least one of the two adjacent nodes of any internal node in a BATON tree is a leaf node [8]. Therefore, if the overloaded node is an internal node, we can use its right adjacent leaf node to initiate the probing process. On the other hand, if the internal overloaded node does not have a right adjacent node, we can initiate the probing process from the overloaded node itself. Thus, the probing messages are propagated at a level that is mainly composed of leaf nodes. However, some nodes at this level may still have child nodes. In this case, we need to probe these child nodes, too. For probing internal nodes, it is necessary to probe the right adjacent node of each probed leaf node. If the intended recipient node does not exist, we will have a missing node situation. For example, as shown in Fig. 1, node d can relay the message through its parent node c to node f, because if a node x contains a link to another node y then the parent node of node x must also contain a link to the parent node of node y [8]. The execution time of probing m nodes at the same level is $T(m) = T(m/2) + ((\log_2 (m/2) - 1)$. The general solution to this equation is $T(m) = 3 + ((k - 1)(k - 2)/2)$, where $m = 2^k$, $k > 2$, and $T(4) = 2$. Other operations such as probing adjacent nodes and child nodes only take constant time. Thus, the time complexity of the probing process is in the order of $O((\log_2 n)^2)$ and the message complexity is in the order of $O(n)$, where n is the number of nodes probed.

Updating Phase. When the probing phase finishes, the last leader node has received response messages from all the nodes probed. It can distribute the heavy load over the participant nodes by reassigning their index values. Therefore, the leader node needs to notify the participant nodes about their new range values. For those nodes that contain links to the participant nodes in their routing tables must change their routing tables accordingly. The leader node also needs to notify these affected nodes about the new range values. In order to reduce the number of messages, we propose a method that combines messages sent to the same node into a single message and transmits messages to multiple nodes in parallel.

The leader node first divides participant nodes into buckets, each of which contains a set of consecutive nodes at the same level. For example, as shown in Fig. 3, the left subtree of the root node contains all participant nodes, which are divided into five different buckets because of the missing right child node of node 4. Secondly, the leader node passes all range values and routing tables of the nodes in the same bucket to the first node in the bucket.

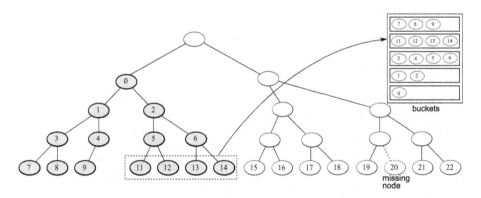

Fig. 3. An example of participant nodes and the buckets they form

Knowing all range values and routing tables of the nodes in the bucket, the first node can combine the update information for a node into a single message. However, if the first node sends out all update messages at the same time, the node will be overwhelmed by this task because of the large number of messages. Thus, the node will divide the affected nodes into groups and notify them in parallel. For example, as shown in Fig. 4, the bucket of nodes 11, 12, 13, and 14 is divided into five groups of consecutive nodes. Nodes 12, 13, 15, 19, and 21 become seed nodes of these five groups. This arrangement makes sure that each node will receive only one update message. Thus, the first node in the bucket sends messages to all seed nodes, and then the seed nodes initiate the updating process in its own group. This process of the example bucket is illustrated in Fig. 5. Because the group length is in the order of $O(N)$ and the message can be transmitted to all nodes in a group in $O(\log N)$ steps, the time complexity of this process is in the order of $O(\log N)$. On the other hand, because each node receives only one message, the message complexity of this process is in the order of $O(N)$, where N is the number of tree nodes.

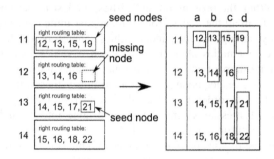

Fig. 4. Seed nodes and corresponding consecutive groups in an example bucket

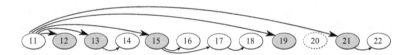

Fig. 5. Updating affected nodes through seed nodes

3.2   Virtual Group Load Balancing

Virtual group load balancing strategy moves the heavy load to another lightly loaded region. The main advantage of this approach is that the neighboring nodes of the overloaded node do not share the heavy load, so they can continue to work on their tasks. Redirecting the heavy load to a virtual group is very efficient if we have an overloaded area instead of a single overloaded node. However, the drawback of this approach is that we need to modify the original query processing and index updating algorithms to incorporate the existence of the virtual group.

In the first step of the virtual group load balancing process, the overloaded node sends messages to all nodes in its routing table and finds out the node that has the lightest load. Then, the lightly loaded node will establish a virtual group to share the heavy load by going through the above mentioned probing phase and updating phase of the vicinity load balancing process. For example, as shown in Fig. 6, node 9 is the overloaded node and node 73 is the lightest loaded node in node 9's routing table. Node 9 redirects its load to node 73, and node 73 forms a virtual group marked by the dotted rectangle to share the load.

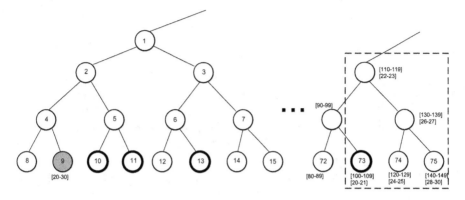

Fig. 6. An example of virtual group load balancing.

4 Performance Evaluation

This section presents the simulation results of vicinity load balancing, virtual group load balancing, and sequential adjusting strategies. We did compare the performance between the two methods used in BATON: adjusting with an adjacent node and inserting an empty helper node. One set of experiment results indicated that the number of messages sent in the latter approach was 30 times larger than the former approach. Because the costs of readjusting tree structure were much higher than the costs of other operations, we will not present the simulation results of this approach. The simulation programs were written in Python and run on a Linux system. We used a randomly generated BATON tree with 1000 nodes as the experiment data set. The experiment results were average values of 500 different experiments.

We first compared the three load balancing strategies used in a system with a single overloaded node. In the experiments we assigned each node a default 0.8 unit load, and then varied the load of the single overloaded node from 2 to 17 units. The load balancing strategies shared the heavy load with helper nodes and ensured that the load of every node was within a 1.2 units load threshold. Fig. 7 and Fig. 8 indicate that the total messages sent and time units used in vicinity and virtual group strategies are much smaller than the sequential adjusting method when the load of the overloaded node is heavy. As shown in Fig. 9, the number of participant nodes in the two parallel methods is larger than the sequential method because the round-based probing phase may probe more nodes than necessary.

Then, we compared the three load balancing strategies used in a system with multiple overloaded neighboring nodes. We assigned 3 units of load for all overloaded nodes and varied the number of overloaded nodes from 1 to 19 in these experiments. Fig. 10 and Fig. 11 indicate that the costs of the virtual group load balancing strategy are the lowest among the three methods tested. In addition, as shown in Fig. 12, the number of participant nodes in the virtual group load balancing strategy is the smallest because it does not probe heavily loaded nodes.

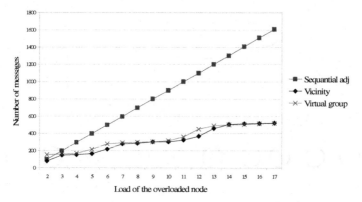

Fig. 7. The total number of messages sent in a single overloaded node system

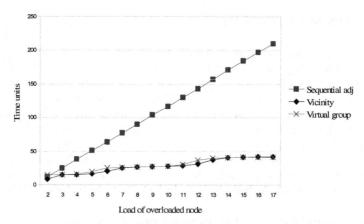

Fig. 8. Time units used in a single overloaded node system

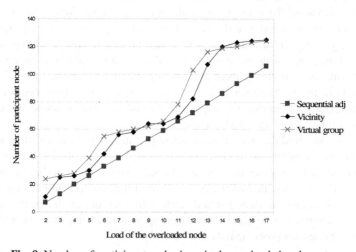

Fig. 9. Number of participant nodes in a single overloaded node system

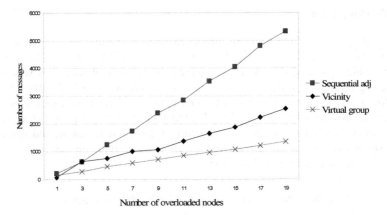

Fig. 10. The total number of messages sent in a multiple overloaded nodes system

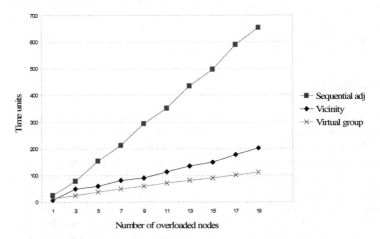

Fig. 11. Time units used in a multiple overloaded nodes system

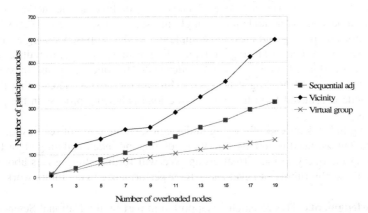

Fig. 12. Number of participant nodes in a multiple overloaded nodes system

4.1 Working Load

We use the number of messages sent by each node to represent the working load of the node that participates in a load balancing process. This experiment assigned a single 5 units overloaded node and measured the number of messages sent in each node using the three strategies. It is shown in the following table that the sequential adjusting method causes more nodes to send larger number of messages. Another set of experiments indicated that the overloaded node and its neighboring nodes sent more messages using the sequential adjusting strategy. Note that we did not include the costs of data migration in the experiments because they are similar in the three strategies after readjusting the index ranges.

Table 1. Number of nodes at different levels of working load

	Number of messages sent per node		
	Low (1 to 10)	Medium (11 to 20)	High (21 to 30)
Sequential adj	99	5	8
Vicinity	46	4	0
Virtual group	53	6	0

5 Conclusions

In tree-structured P2P networks, traditional load balancing operations either use an existing adjacent node or insert a helper node as an adjacent node to share the heavy load of the overloaded node. This approach is adequate for handling the data skew problem caused by data insertion and deletion, but this sequential approach may trigger a sequence of load balancing operations in case of execution skew. In order to handle both skew problems, we propose two parallel load balancing strategies for tree-structured P2P networks. The vicinity load balancing strategy, which consists of a probing phase and an updating phase, tries to find enough neighboring nodes to share the load in a round-based fashion. On the other hand, the virtual group load balancing strategy moves the heavy load to a lightly loaded region instead of moving helper nodes to the overloaded node. Because we combine all the messages that are intended to be sent to one node into a single message and send the messages in parallel, the time complexity and message complexity of the proposed strategies are lower than the sequential load balancing strategies. Furthermore, in order to reduce the work load of the overloaded node, we assign leader nodes to collect response messages and initiate next round of probing if necessary. Simulation results indicate that the overloaded node is relieved from the work of the load balancing process in the proposed strategies.

Moving indexes instead of moving nodes can save the costs of readjusting the tree structure, but we need to modify query processing and index updating algorithms for redirecting the query to the virtual group. We will further study issues about implementing these algorithms and evaluating their performance in the future work.

Acknowledgments. This research is supported in part by the National Science Council of ROC under grant NSC97-2221-E-415-006.

References

1. Aspnes, J., Shah, G.: Skip Graphs. In: Proceedings of the 14th Annual ACM-SIAM Symposium on Discrete Algorithms (2003)
2. Chou, J.C.-Y., Huang, T.-Y., Huang, K.-L., Chen, T.-Y.: SCALLOP: A Scalable and Load-Balanced Peer-to-Peer Lookup Protocol. IEEE Transactions on Parallel and Distributed Systems 17(5), 19–26 (2006)
3. Crainiceanu, A., Linga, P., Gehrke, J., Shanmugasundaram, J.: Querying Peer-to-Peer Networks Using P-Trees. In: Proceedings of the 7th International Workshop on the Web and Databases (2004)
4. Crainiceanu, A., Linga, P., Machanavajjhala, A., Gehrke, J., Shanmugasundaram, J.: P-Ring: An Efficient and Robust P2P Range Index Structure. In: Proceedings of the ACM SIGMOD International Conference on Management of Data (2007)
5. Clarke, I., Sandberg, O., Wiley, B., Hong, T.W.: Freenet: A distributed anonymous information storage and retrieval system. In: Federrath, H. (ed.) Designing Privacy Enhancing Technologies. LNCS, vol. 2009, p. 46. Springer, Heidelberg (2001)
6. Ganesan, P., Bawa, M., Garcia-Molina, H.: Online Balancing of Range-Partitioned Data with Applications to Peer-to-Peer Systems. In: Proceedings of the 30th International Conference on Very Large Data Bases (2004)
7. Gnutella, The Gnutella Protocol Specification v0.4
8. Jagadish, H.V., Ooi, B.C., Vu, Q.H.: BATON: A Balanced Tree Structure for Peer-to-Peer Networks. In: Proceedings of the 31st International Conference on Very Large Data Bases (2005)
9. Jagadish, H.V., Ooi, B.C., Tan, K.-L., Vu, Q.H., Zhang, R.: Speeding up Search in Peer-to-Peer Networks with a Multi-way Tree Structure. In: Proceedings of the ACM SIGMOD International Conference on Management of Data (2006)
10. Jagadish, H.V., Ooi, B.C., Vu, Q.H., Zhang, R., Zhou, A.: VBI-Tree: A Peer-to-Peer Framework for Supporting Multi-Dimensional Indexing Schemes. In: Proceedings of the 22nd International Conference on Data Engineering (2006)
11. Lua, E.K., Crowcroft, J., Pias, M., Sharma, R., Lim, S.: A Survey and Comparison of Peer-to-Peer Overlay Network Schemes. IEEE Communications Surveys & Tutorials 7(2), 72–93 (2005)
12. Maymounkov, P., Mazières, D.: Kademlia: A Peer-to-Peer Information System Based on the XOR Metric. In: Proceedings of the 1st International Workshop on Peer-to-Peer Systems (2002)
13. Ratnasamy, S., Francis, P., Handley, M., Karp, R., Shenker, S.: A Scalable Content-Addressable Network. In: Proceedings of the 2001 Conference on Applications, Technologies, Architectures, and Protocols for Computer Communications (2001)
14. Rowstron, A., Druschel, P.: Pastry: Scalable, decentralized object location, and routing for large-scale peer-to-peer systems. In: Guerraoui, R. (ed.) Middleware 2001. LNCS, vol. 2218, pp. 329–350. Springer, Heidelberg (2001)
15. Stoica, I., Morris, R., Karger, D., Kaashoek, M.F., Balakrishnan, H.: Chord: A Scalable Peer-to-Peer Lookup Service for Internet Applications. In: Proceedings of the 2001 Conference on Applications, Technologies, Architectures, and Protocols for Computer Communications (2001)
16. Zhao, B.Y., Huang, L., Stribling, J., Rhea, S.C., Joseph, A.D., Kubiatowicz, J.D.: Tapestry: A Resilient Global-Scale Overlay for Service Deployment. IEEE Journal on Selected Areas in Communications 22(1), 41–53 (2004)

A Verification Mechanism for Secured Message Processing in Business Collaboration

Haiyang Sun[1], Jian Yang[1], Xin Wang[2], and Yanchun Zhang[2]

[1] Department of Computing, Macquarie University,
Sydney, NSW2109, Australia
{hsun,jian}@ics.mq.edu.au
[2] School of Computer Science and Mathematics, Victoria University,
Melbourne, Victoria, Australia
xin@csm.vu.edu.au, Yanchun.zhang@vu.edu.au

Abstract. Message processing can become unsecured resulting in unreliable business collaboration in terms of authorization policy conflicts, for example, when (1) incorrect role assignment or modification occurs in a partner's services or (2) messages transferred from one organization are processed by unqualified roles in other collaborating business participants. Therefore, verification mechanism based on access policies is critical for managing secured message processing in business collaboration. In this paper, we exploit a role authorization model, Role-Net, which is developed based on Hierarchical Colored Petri Nets (HCPNs) to specify and manage role authorization in business collaboration. A property named *Role Authorization Based Dead Marking Freeness* is defined based on Role-Net to verify business collaboration reliability according to partners' authorization policies. An algebraic verification method for secured message processing is introduced as well.

Keywords: Role Authorization, Reliability Verification, Secured Message Processing, Hierarchical Colored Petri Net.

1 Introduction

Business collaboration is about coordinating the flow of information among organizations and linking their business processes into a cohesive whole. Emerging web service and business process technologies have provided a technology foundation for cross-organization business collaboration [1]. However, security concerns are crucial for the success of collaborative business and become one of the main barriers that prevent widespread adoption of this new technology.

Role Based Access Control (RBAC) as an access control mechanism has been widely accepted in the business world [2]. In RBAC, users are assigned with roles to process messages or perform tasks [3, 4]. However, in business collaboration environment, role assignments or modifications are more complicated and prone to error because different parties and services are involved in. For example, incorrect role assignment or modification may occur in any parties' services,

Q. Li et al. (Eds.): APWeb/WAIM 2009, LNCS 5446, pp. 480–491, 2009.

or messages transferred from one organization may be processed by unqualified roles in other collaborating business partners. Therefore, verification mechanism for such variation of role authorization is critical to manage secured message processing in business collaboration.

1.1   A Motivating Example

Let us take an example of Bank Home Loan Application (see Fig. 1) which involves four collaborating parties: **Customer**, **Bank**, **Credit Rating Agency** and **Property Agent**. The loan application process begins with receiving an application from a **Customer**. The **Bank** will then cooperate with **Credit Rating Agency** to check the **Customer**'s credit level. The **Property Agent** will also be involved to assess the value of the property. The risk assessment will be executed at the **Bank** side when the results for the credit check and value assessment are returned. The eligible applications are approved while the unacceptable high risk applications are rejected.

Due to space limit, we only present in this paper the role based authorization policies of **Bank** which are separated into two categories according to the types of message transfer:

- **Intra-organization message transfer**: When a message is transferred between services within the same organization, individual service must be assigned with roles in order to process the message according to authorization

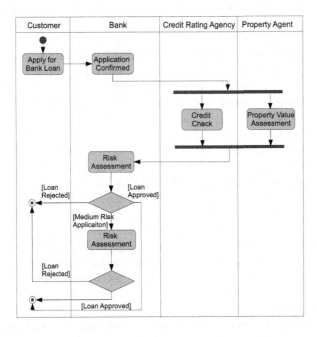

Fig. 1. Bank Home Loan Application Activity Diagram

policies. For example, the role of `Risk Administrator` in **Bank** is assigned to *Risk Assessment* service to deal with huge-amount loan application.

- **Cross-organization message transfer**: Since all involving organizations in business collaboration are peers with equal rights, their internal authorization policies can not be totally revealed to each other. However, individual organizations need to send their collaborators the message being processed as well as the role requirement to enforce that the message can be dealt with by a qualified role of the collaborators. For instance, the **Bank** may require the role of `Credit Check Manager` to process the credit check requirement message at *Credit Check* service of **Credit Rating Agency** to protect client's privacy.

1.2 Role Authorization Policy Conflicts

We can observe from the above motivating scenario that collaboration authorization control and enforcement are governed by the authorization policies of collaborating organizations. Any authorization policy conflicts within or across organization can lead to unsecured message processing in business collaboration as follows:

- **Incorrect role assignment or modification in a service within one organization**: Any message should be associated with a set of required roles before it can be processed in a service, and any service should be assigned with a set of available roles for processing different types of messages. After a message has been imputed in a service, the required role need to be checked against the set of available roles. If the two types of roles are not matched, e.g. the available role assignment or modification is incorrect, then we can conclude that the authorization policy has conflict. For instance, let us assume that there is a *Value Assessment* service after *Credit Check* service in **Bank** to evaluate **Customer**'s collateral, and the service is assigned as the exclusive role of `Value auditor` to process message. However, if `value audit manager` is the required role for processing this message, then no matched role exists in *Value Assessment* service, and the business process will be suspended due to authorization policy conflict. Let us look at another example in relation to role modification. The `risk administrator` and `risk officer` are two available roles assigned to *Risk Assessment* service to process message. The `risk officer` can process small-amount loan applications while the `risk administrator` can deal with both huge- and small-amount loans. However, when one huge-amount loan application is being processed in **Bank**, the role of `risk administrator` may be removed from *Risk Assessment* service at runtime. Therefore, there does not exist any matched role that the *Risk Assessment* service can be assigned to handle the huge-amount loan application.
- **Unexpected role access in collaborating business partner**: Business partners are peers with their own authorization policies that are agnostic to each other. Therefore, without central control, it is difficult to guarantee that

the message is processed by the expected roles in business partners' service. For instance, due to confidential reason for corporate clients, message sent from *Credit Check Requirement* service at **Bank** side may require to be processed by the role of `Credit Check Manager` at *Credit Check* service in **Credit Rating Agency**. However, if the message is handled and modified by an unexpected role in **Credit Rating Agency**, e.g. a `general credit check officer`, then **Bank** may not accept the credit check result.

In summary, in collaborative business environment, organization does not only require the correct role assignment to process messages for its own services, but also the right role to handle messages it passes to its collaborating partners. In order to guarantee that the messages transferred within or across organizations can be processed by the qualified roles and the role assignment or modification to specific service at design time or runtime is proper, each message transferred between services within one organization or in various partners need to carry the information of the required roles. Moreover, each service should be assigned with a set of available roles that can deal with different messages. If the required roles are matched with the available roles, then the service can process the message as an operational role that is selected from the intersection of the two types of role set. Based on the operational role and organizational authorization policy, the set of required roles for next service are generated. If the two types of roles are not matched at specific service, then authorization policy conflicts can be detected. According to this mechanism, the authorization policies can then be coordinated to enforce secured message processing in business collaboration.

Even with the above assumed setting, message processing can still become un-secured resulting in unreliable business collaboration in terms of authorization policy conflicts occurred within or across organizations. Therefore, model and techniques are required to verify business collaboration reliability in terms of authorization policies for secured message processing. In our project, a Role-Net is developed based on Hierarchical Colored Petri Net [5]. It has following features: (1) A Role-Net is separated into two layers, which correspond to local organization and its collaborators respectively. (2) A *Role* is modeled as a RO-Token in Role-Net. Its movement among consecutive transitions thereby models the role assignment at specific services, which consequently generates a role flow. However, before a *Place*, the RO-Token is called **Operational Role** which represents the role that processes and modifies the message at previous service; while after a *place* the RO-Token represents **Required Role** which is used to describe the roles required in the next service. (The detailed petri net terminology such as *Token, Place* and *Transition* will be explained in section 2.1). (3) There are two types of tokens moved within Role-Net: AO-Token and RO-Token, each of which correspond to application message and role. These two types of tokens are dynamically combined and moved together in Role-Net. The correlation of the two types of tokens can guarantee that the desired message can only be processed by the specific service as the designated role.

In our previous work [6], the structure and the execution policy of Role-Net were introduced. In this paper, we will focus on providing a verification mechanism for secured message processing in business collaboration. A property named *Role Authorization Based Dead Marking Freeness* is defined based on Role-Net to verify business collaboration reliability. An algebraic verification method for secured message processing is also introduced.

The rest of paper is organized as follows. In section 2, the specification of Role-Net is briefly described. The reliability property based on Role-Net is presented in section 3 associated with the algebraic verification method. Discussion on related work is introduced in section 4 while conclusion and future work are presented in section 5.

2   Specifications of the Role Authorization Model - Role-Net

2.1   Structure of Role-Net

In Role-Net (see Fig. 2), *Places* denoted as circles are used to model the state of business collaboration and provide function to transfer Role-Tokens from representing *Operational role* to *Required role* according to organizational authorization policies. *Transitions* depicted as black bars are used to model services. The *flow relations* in Role-Net which link the transitions and places in Fig. 2 represent the execution order of services in business collaboration.

Role-Net, as mentioned above, is separated into two layers to model the inter-organizational role authorization in business collaboration. The upper layer is used to model the business process in local organization and the lower layer is used to simulate the projection of local organization's view on its collaborators' processes. In Fig. 2, we give an example of Role-Net of **Bank** from our motivating scenario. The upper layer models loan application process at **Bank** side while lower layer simulates the projection of **Bank** on **Credit Rating Agency**'s Role-Net and **Property Agent**'s Role-Net. They are linked by Refinement Function at the transitions which are represented as *Credit Check Requirement* Service and *Property Value Assessment Requirement* Service respectively. a...j in the figure are AO-Tokens which indicate the messages transferred in the business collaboration. $r_1 \ldots r_7$ and $r_0^\varepsilon \ldots r_5^\varepsilon$ including R_1^ε and R_2^ε are RO-Tokens while $r_1 \ldots r_7$, R_1^ε, and R_2^ε represent operational roles in the services of local organization and the services of collaborators, and the rest are required roles. We present the formal definition of Role-Net's two layers based on Hierarchical Colored Petri Nets as follows:

Definition 1. *The upper layer of organization G_i's Role-Net is a tuple ρ_{Gi}^{upper} = $(P_{Gi}^{upper}, T_{Gi}^{upper}, F_{Gi}^{upper}, \Gamma_{Gi}^{upper}, \Delta_{Gi}^{upper}, \Theta_{Gi}^{upper}, \Omega_{Gi}^{upper})$, Where:*

- P_{Gi}^{upper} *is a set of places in upper layer of G_i's Role-Net which graphically are represented as circle in Fig. 2.*
- T_{Gi}^{upper} *is a set of transitions graphically represented as dark bars in Fig. 2. $t_{Gi}^\tau \in T_{Gi}^{upper}$ is a set of empty transitions for distributing and collecting tokens, e.g., t_2 and t_7 in Fig. 2. $P_{Gi}^{upper} \cap T_{Gi}^{upper} = NULL$.*

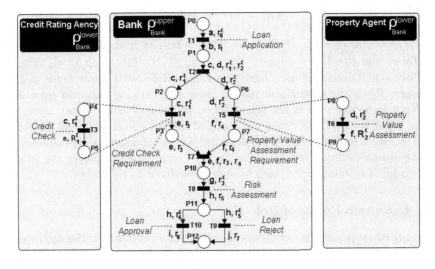

Fig. 2. Role-Net of Bank

- $F_{Gi}^{upper} = (p^u \times V \times t^u) \cup (t^u \times V \times p^u)$ is the flow relation between places and transitions, where $p^u \in P_{Gi}^{upper}$, $t^u \in T_{Gi}^{upper}$, and V is the sets of variables $V = \{x, y, ...\}$ to represent the tokens.
- Γ_{Gi}^{upper} $(p^u,\ a,\ r) \to Boolean$ is a correlation function to evaluate the relationship of RO-Token and AO-Token at specific place, where $a \in AO\text{-}Token$, $r \in RO\text{-}Token$, and $p^u \in P_{G_i}^{upper}$. Γ_{Gi}^{upper} guarantees that the AO-Token can only be moved with assigned RO-Tokens at specific places.
- Δ_{Gi}^{upper} $(p^u,\ a,\ r) \to r^\varepsilon$ is a function to change RO-Token from representing the operational role r that accessed the AO-Token at previous transition to the required roles r^ε which are needed by the next transition, according to role authorization policies, where $p^u \in P_{G_i}^{upper}$, $a \in AO\text{-}Token$, $r, r^\varepsilon \in RO\text{-}Token$.
- Θ_{Gi}^{upper} $(t^u,\ \varphi,\ r^\varepsilon) \to Boolean$ is comparison function, where $t^u \in T_{Gi}^{upper}$, $r^\varepsilon \in RO\text{-}Token$, and φ is a threshold variable representing the role element selected from the set γ. γ is the set of available roles that are permitted to process and modify the AO-Token in the transition at upper layer of G_i's Role-Net. The TRUE result of Θ_{Gi}^{upper} function reflects the existence of qualified roles for specific transition t.
- Ω_{Gi}^{upper} $(t_\beta^u, a) \to L$ is refinement function on transition t_β^u to connect Role-Net's lower layer, where $t_\beta^u \in T_{G_i}^{upper}$ represents transition including link between two layers of Role-Net. $L = \{g(x),\ \{e(x), \rho_{G_i}^{lower}, r(x)\}\}$ $x \in V$. $g(x)$ is a function to decide which $\rho_{G_i}^{lower}$ shall be initiated in other collaborators. $e(x)$ and $r(x)$ are the guard functions of corresponding lower layer to evaluate whether or not the $\rho_{G_i}^{lower}$ is available to initiate and exit.

Definition 2. The lower layer of organization G_i's Role-Net is a tuple ρ_{Gi}^{lower} $= (P_{Gi}^{lower},\ T_{Gi}^{lower},\ F_{Gi}^{lower},\ \Gamma_{Gi}^{lower},\ \Psi_{Gi}^{lower}\ )$, Where:

- P_{Gi}^{lower}, T_{Gi}^{lower}, F_{Gi}^{lower}, Γ_{Gi}^{lower} are as same as relevant elements in upper layer of Role-Net.
- $\Psi_{Gi}^{lower}(t^l,\ a,\ r^\varepsilon) \rightarrow (b,\ R^\varepsilon)$ is a switch function to transfer the value of AO-Token and RO-Token, where $t^l \in T_{Gi}^{lower}$, a, $b \in AO$-Token $(a \neq b)$, r^ε, $R^\varepsilon \in RO$-Token. r^ε is required roles transferred from Role-Net's upper layer to lower layer. R^ε is the operational role in lower layer and is returned from Role-Net's lower layer to upper layer. When R^ε arrives at the upper layer, it is an input to Θ_{Gi}^{upper} to detect the qualified role set. Since any modification on AO-Token and RO-Token in collaborators is unknown to organization G_i, the switch function in G_i's lower layer is only used to transfer the value of two types of tokens after they have been modified in collaborator.

2.2   Execution Policy of Role-Net

There are two types of tokens that are operated within a Role-Net: the Application-Oriented Token (AO-Token) and the Role-Oriented Token (RO-Token). The AO-Token will move together with the relevant RO-Token to correlate the message flow and role flow. The execution policies of Role-Net are described as follows:

- **Token at Place**
    1. Each RO-token is correlated to a specific AO-token. If a RO-Token and an AO-Token are received separately, the *Place* will abandon the token as an unexpected role or message respectively.
    2. **(i)** Before *Places* in upper layer, the RO-Token r represents the *Operational role* that has processed the correlated message at previous *Transition*. After *Places* in upper layer, the RO-Token r^ε will represent the *Required role* which will be required by the next *transition*. **(ii)** The place in lower layer is used to receive the AO-Token and RO-Token from upper layer, and return the two correlated tokens together to upper layer after they are processed by the services of the collaborating partners.

- **Token at Transition**
    - *Transition in upper layer of organization G_i's Role-Net*
        1. If the transition has link between upper layer and lower layer of a Role-Net, the AO-Token and RO-Token r^ε (as *Required roles*) will move together to the lower layer of Role-Net as cross-organizational message transfer. The transition in upper layer will identify the two types of tokens when they are returned.
        2. Detecting qualified roles is implemented when *Required role* r^ε arrives at transition in upper layer with AO-Token where no link between lower layer and upper layer in this transition or returned RO-Token R^ε arrives at the transition with AO-Token from lower layer under the condition that link between lower layer and upper layer exists in this transition. **(i)** Each transition in upper layer of organization G_i's Role-Net has a set of available roles γ which are

qualified to process message in this transition. However, depending on the properties of message and role authorization policies, all or part of them may not be authorized to process message at runtime. Therefore, a threshold φ is dynamically decided at runtime by choosing roles from γ at each transition. If the transition in upper layer of Role-Net has link to lower layer, then φ is selected from R^ε and is used to verify whether the AO-Token is modified by the *Required role* r^ε at collaborator's Role-Net. **(ii)** If r^ε's element equals to the threshold φ, the role in threshold will be moved to the set ϱ as qualified role to process the message in this *transition* and the threshold will be degraded for the next role in γ. The comparison will continue until all role elements in required roles r^ε and available roles γ (or R^ε) have been dealt with. **(iii)** If ϱ is not empty, then the role elements in this set will be authorized the permission to process the messages in this service. The RO-Token thus represents the role that actually processes the message and is moved with AO-Token together to the next places. If ϱ is empty, then there is no matched role to deal with this message at this service. The process will be suspended due to the role authorization runtime error.

- *Transition in lower layer of organization G_i's Role-Net*
  1. Transition at lower layer of organization G_i's Role-Net is used to transfer the value of AO-Token and RO-Token. It means that the local organization G_i is agnostic to its collaborator's internal process, including how to deal with the message AO-Token and which role is assigned to the collaborator's service to process the message. These modifications on AO-Token and RO-Token are implemented according to collaborator's own authorization policies, and transition in the lower layer of local organization G_i's Role-Net can only identify and exchange the result of modification on tokens.

3   Detecting Authorization Policy Conflicts

In this section, we define a reliability property named as *Role Authorization Based Dead Marking Freeness* based on Role-Net to verify authorization policy based business collaboration reliability for secured message processing. An algebraic approach to detect authorization policy conflicts is also presented in this section.

The *labeled transitive matrix* L^*_{BP} [7] used in Petri-Net expresses the relationship between •t and t• based on transition t (•t is the set of pre-places of a transition t while t• represents the set of post-places of a transition t). However, it does not elaborate the role authorization relationship between •t and t•. We extend the transitive matrix by associating role authorization impact called *role-embedded transitive matrix* and use it to verify the *Role Authorization Based Dead Marking Freeness* property.

We firstly describe the syntax of two types of matrix algebraic operators $\diamond$ and $\triangle$ which are used to generate *role-embedded transitive matrix* and verify the *Role Authorization Based Dead Marking Freeness* property. The grammar of definition follows BNF-like notation:

$$M ::= M_1 \diamond M_2 | M_1 \triangle M_2$$

Where:

- $M_1 \diamond M_2$: Given an n×m matrix M_1, an m×n matrix M_2, and an n×n matrix M_3 where $M_1 = [c_{ij}]$, $M_2 = [d_{ji}]$, and $M_3 = [e_{ii}]$ (i=1..n, j=1..m). Then

$$M_3 = M_1 \diamond M_2 \Rightarrow [e_{ii}] = \bigcup_{j=1}^{m} ([(c_{ij})] \cap [d_{ji}])$$

- $M_1 \triangle M_2$: Given an n×m matrix M_1, an m×n matrix M_2, and an n×n matrix M_3 where $M_1 = [c_{ij}]$, $M_2 = [d_{ji}]$, and $M_3 = [e_{ii}]$ (i=1..n, j=1..m). Then

$$M_3 = M_1 \triangle M_2 \Rightarrow [e_{ii}] = \bigcap_{j=1}^{m} ([(c_{ij})] \cap [d_{ji}])$$

The proposed algebraic operators guarantee that each result of an operation on matrix is still a matrix to which we can again apply algebraic operators. Here below we formally define how these two operators can be used.

Definition 3. *A role-embedded transitive matrix:*

$$L_{BP}^{Ro} = (A^-)^T \diamond Diag(\gamma_1, \gamma_2, ...\gamma_n) \diamond A^+$$

$$\Downarrow$$

$$[c_{jg}] = \bigcup_{k=1}^{n} (\bigcup_{i=1}^{n} ([(a_{ij}^-)^T] \cap [b_{ik}]) \cap [a_{kg}^+])$$

where

- γ_h *(h=1,2,...,n) is the available role set of each transition, or Ξ if the transition is empty transition t_{Gi}^τ. $Diag(\gamma_1, \gamma_2, ...\gamma_n) = [b_{ik}]$ is n×n matrix (i,k=1..n). Ξ represents the set indicating **all** role elements in business collaboration.*
- $A^- = [a_{ij}^-]$ *and $A^+ = [a_{kg}^+]$ are n×m matrix (i,k=1..n, j,g=1..m). T means transpose matrix. x∈P, y∈T. (n transitions, m places)*

$$a_{ij}^- = \begin{cases} \Xi \ (x,y) \ In \ \rho_{G_i}^{upper/lower} \\ \emptyset \ (x,y) \ Not \ In \ \rho_{G_i}^{upper/lower} \end{cases}$$

$$a_{ij}^+ = \begin{cases} \Xi \ (y,x) \ In \ \rho_{G_i}^{upper/lower} \\ \emptyset \ (y,x) \ Not \ In \ \rho_{G_i}^{upper/lower} \end{cases}$$

- $[c_{jg}]$ *is role-embedded transitive matrix. $L_{BP}^{Ro} = [c_{jg}]$ is then m×m matrix.*
- $\Downarrow$ *means "refer to" for detailed matrix computing method.*

Marking of Petri-Net based model is an allocation of tokens to the places of the net formally defined as a function M: $P \rightarrow R^{|P|}$, where $R^{|P|}$ is $|P| \times 1$ vector. The marking reflects the state of the petri net after each transition firing. In a marking k, if a token in place p, then $M_k(p)=1$, otherwise $M_k(p)=0$. A Petri-Net based model is called *Dead Marking Free* if there do not exist places having no enabled transition. It means that the change of marking from state k to state k+1 is not impeded by the absence of a transition firing during the execution of the model. Hence, we extend the property *Dead Marking Freeness* by embedding role authorization impact called *Role Authorization Based Dead Marking Freeness*. This reliability property is used in Role-Net to verify authorization policy based business collaboration reliability by detecting whether there exist places having no enabled transition caused by the errors of role authorization at design time and runtime. Here we present the formal definition of this property.

Definition 4. *A Role-Net is authorization based dead marking free* **if**

$$\forall M_k^{*Ro}, \exists M_k^{*Ro}(w) = S_w^{Ro} \neq \emptyset \; w = 1..m$$

where:

- $M_k^{*Ro}=[S_w^{Ro}]$ *(w=1..m) is named* RO-Token transitive marking *which is used to detect whether the transition is qualified to facilitate the change of* RO-Token marking, e.g., $M_8^{*Ro} = \begin{bmatrix} \emptyset & \emptyset & \emptyset & \emptyset & \emptyset & \emptyset & \emptyset & r_3^\varepsilon \cap \gamma_5 & \emptyset \end{bmatrix}$ *in upper layer of* **Bank**'s Role-Net in Fig. 2. RO-Token transitive marking *is calculated as:*

$$M_k^{*Ro} = M_{k-1}^{Ro} \triangle L_{BP}^{Ro} \;\; \Rightarrow \;\; [S_w^{Ro}] = \bigcap_{i=1}^{m}([Q_i] \cap [c_{ig}])$$

Where:
- $M_{k-1}^{Ro} = [Q_i]$ *(i=1..m) is called* RO-Token marking *which indicates the state of Role-Net from the movement of RO-Token point of view, e.g.,* M_7^{Ro} *in upper layer of* **Bank**'s Role-Net in Fig. 2 is $\begin{bmatrix} \emptyset & \emptyset & \emptyset & \emptyset & \emptyset & \emptyset & r_3^\varepsilon & \emptyset & \emptyset \end{bmatrix}$.
- L_{BP}^{Ro} *is role-embedded transitive matrix.* $L_{BP}^{Ro}=[c_{ig}]$ *(i,g=1..m).*
- $\Rightarrow$ *means "refer to" for detailed matrix computing method.*

Here we present an example on how *Role-Token marking* and *RO-Token transitive marking* work together to detect authorization policy conflicts. (See upper layer of **Bank**'s Role-Net in Fig. 2) Let us assume:

$$K = 8, \; M_{k-1}^{Ro} = M_7^{Ro} = \begin{bmatrix} \emptyset & \emptyset & \emptyset & \emptyset & \emptyset & \emptyset & r_3^\varepsilon & \emptyset & \emptyset \end{bmatrix}$$

$$L_{BP}^{Ro} = \begin{bmatrix} \emptyset & \gamma_1 & \emptyset & \emptyset & \emptyset & \emptyset & \emptyset & \emptyset & \emptyset \\ \emptyset & \emptyset & \Xi & \emptyset & \Xi & \emptyset & \emptyset & \emptyset & \emptyset \\ \emptyset & \emptyset & \emptyset & \gamma_3 & \emptyset & \emptyset & \emptyset & \emptyset & \emptyset \\ \emptyset & \emptyset & \emptyset & \emptyset & \emptyset & \emptyset & \Xi & \emptyset & \emptyset \\ \emptyset & \emptyset & \emptyset & \emptyset & \emptyset & \gamma_4 & \emptyset & \emptyset & \emptyset \\ \emptyset & \emptyset & \emptyset & \emptyset & \emptyset & \emptyset & \Xi & \emptyset & \emptyset \\ \emptyset & \emptyset & \emptyset & \emptyset & \emptyset & \emptyset & \emptyset & \gamma_5 & \emptyset \\ \emptyset & \emptyset & \emptyset & \emptyset & \emptyset & \emptyset & \emptyset & \emptyset & \gamma_6 \cup \gamma_7 \\ \emptyset & \emptyset & \emptyset & \emptyset & \emptyset & \emptyset & \emptyset & \emptyset & \emptyset \end{bmatrix}$$

$M_8^{*Ro} = M_7^{Ro} \bigtriangleup L_{BP}^{Ro} = \left[\emptyset\ \emptyset\ \emptyset\ \emptyset\ \emptyset\ \emptyset\ \emptyset\ r_3^\varepsilon \cap \gamma_5\ \emptyset \right]$, then $\exists$ w=8,

$$S_8^{Ro} = \begin{cases} r_3^\varepsilon \cap \gamma_5 & \text{Role exists} \\ \emptyset & \text{NO role exists} \end{cases}$$

Let us assume that the **Bank** expects the *Risk Assessment* service as the role of `risk administrator` to process the loan case due to huge-amount loan application. $r_3^\varepsilon=\{$"`Risk Administrator`"$\}$ and $\gamma_5= \{$"`Risk Officer`", "`Risk Administrator`", "`Bank Manager`"$\}$. There exists S_8^{Ro} which is not empty. The Role-Net of **Bank** is *role authorization based dead marking free*. However, if the `risk officer` is the only available role for this service, then $\gamma_5=\{$"`Risk Officer`"$\}$ and $S_8^{Ro}=\emptyset$. The runtime role authorization error will be detected resulting in the suspension of business collaboration execution in this service.

Consequently, we can conclude that unsecured message processing in business collaboration in terms of authorization policy conflicts can be modeled and detected in Role-Net as follows: (1) **Incorrect role assignment or modification in a service within one organization** can then be seen as the inconsistent design on available role set γ in specific service where the qualified role is not initially set up in γ, or as incorrect runtime modification on available role set γ which leads to $r_i^\varepsilon \cap \gamma = \emptyset$; (2) **Unexpected role access in collaborating business partner** can be detected as the returned RO-Token R_i^ε is not equivalent to the required role in r_i^ε.

4   Related Work

Research has been done in the area of role authorization in business collaboration. We shall look into some of the representative work.

Paci et al [8] proposed an approach to decide whether the choreography can be implemented by a set of web services according to the match of access control policies and credential disclosure policies in various organizations. However, this work only addresses the access control in design time. The security breaches in run time are not discussed, which is a common scenario in enterprize world.

The authors in [9] focused on extending the specification WS-BPEL with role authorization constraints in business collaboration. A language named BPCL to express these constraints based on XACML and an algorithm to check the consistency of the access control model were developed. However, the authors did not elaborate how these role authorization constraints can facilitate the authorization policy based verification for secured message processing at runtime.

Liu and Chen [10] developed another extended RBAC model, WS-RBAC. Three new elements were introduced into the original RBAC model, namely enterprize, business process and web services. However, this model was designed for organizational behavior level web services security. Reliability verification on message level was not investigated by the authors.

In a summary, existing approaches are insufficient in: (1) describing role authorization in business collaboration with regard to the organization's peer nature; (2) detecting role authorization errors and verifying collaboration reliability for secured message processing.

5   Conclusion and Future Work

Message processing can become unsecured resulting in unreliable business collaboration in terms of authorization policy conflicts. In this paper, we exploit a role authorization model (Role-Net) to provide a verification mechanism by introducing a reliability property named as *Role Authorization Based Dead Marking Freeness* and an algebraic verification method.

Currently we are working on how to dynamically determine the required roles for each service according to authorization policies at runtime. A Role-Net simulator will be developed as well to implement the verification approach.

References

[1] Papazoglou, M., Georgakopoulos, D.: Service-Oriented Computing. Communications of the ACM 46(10), 25–28 (2003)

[2] Wang, X., Zhang, Y., Shi, H., Yang, J.: BPEL4RBAC: An Authorisation Specification for WS-BPEL. In: Bailey, J., Maier, D., Schewe, K.-D., Thalheim, B., Wang, X.S. (eds.) WISE 2008. LNCS, vol. 5175, pp. 381–395. Springer, Heidelberg (2008)

[3] Sandhu, R.S., Coyne, E., Feinstein, H., Youman, C.: Role-based Access Control Models. IEEE Computer 29(2), 38–47 (1996)

[4] Ferraiolo, D., Cugini, J., Kuhn, R.: Role Based Access Control: Features and Motivations. In: Proceedings of ACSAC (1995)

[5] Girault, C., Valk, R.: Petri Nets for Systems Engineering: A Guide to Modeling, Verification, and Applications. Springer, Heidelberg (2003)

[6] Sun, H., Wang, X., Yang, J., Zhang, Y.: Authorization Policy Based Business Collaboration Reliability Verification. In: Proceedings of ICSOC, pp. 579–584 (2008)

[7] Song, Y., Lee, J.: Deadlock Analysis of Petri Nets Using the Transitive Matrix. In: Proceedings of the SICE Annual Conference, pp. 689–694 (2002)

[8] Paci, F., Ouzzani, M., Mecella, M.: Verification of Access Control Requirements in Web Services Choreography. In: Proceedings of the IEEE International Conference on Service Computing, pp. 5–12 (2008)

[9] Bertino, E., Crampton, J., Paci, F.: Access Control and Authorization Constraints for WS-BPEL. In: Proceedings of the IEEE International Conference on Web Services, pp. 275–284 (2006)

[10] Liu, P., Chen, Z.: An Access Control Model for Web Services in Business Process. In: Proceedings of the 2004 IEEE/WIC/ACM International Conference on Web Intelligence, pp. 292–298 (2004)

Semantic Service Discovery by Consistency-Based Matchmaking*

Rajesh Thiagarajan, Wolfgang Mayer, and Markus Stumptner

Advanced Computing Research Centre
University of South Australia
{cisrkt,mayer,mst}@cs.unisa.edu.au

Abstract. Automated discovery of web services with desired functionality is an active research area because of its role in realising the envisioned advantages of the semantic web, such as functional reuse and automated composition. Existing approaches generally determine matches by inferring subsumption relationships between a request and a service specification, but may return poor results if service profiles are overspecified or provide only partial information. We present a two-staged consistency-based matchmaking approach where services that potentially match the request are identified in the first stage, these services are queried for concrete information, and finally this information is used to determine the matches. We evaluate our matchmaking scheme in the context of the *Discovery II and Simple Composition* scenario proposed by the SWS Challenge group. Preliminary evaluation shows that our approach is robust while handling overspecified profiles, does not return false positives, and is able to handle partial information in service and requirements specification.

1 Introduction

Automated discovery of web services with desired functionality is an active research area because it is crucial for realising the envisioned advantages of the semantic web, such as functional reuse and automated composition. This scenario is expected to be frequently encountered in E-Commerce and application integration scenarios, but also in e-Science with Semantic Grid technology, with enterprises or end user applications attempting to compose complex interactions from offered services in a variety of application contexts: brokering, purchasing, supply chain integration, or experiment assembly.

A web service discovery request typically consists of a functional requirement such as *book a hotel room*, a set of additional requirements such as *price < 100*, input and output (IO) requirements such as *one of the service inputs should be of type CreditCard*, Quality of Service (QoS) and other non-functional requirements. For a given web service discovery request, matchmaking is the process of discovering a set of web services from one or more repositories such that all the matched services provide the required functionality and satisfy the constraints imposed by the request. Earlier approaches to

---

* This work was partially supported by the Australian Research Council (ARC) under grant DP0988961.

Q. Li et al. (Eds.): APWeb/WAIM 2009, LNCS 5446, pp. 492–505, 2009.

address the matchmaking problem adopted a keyword-based search mechanism [1]. In the web services context, keyword search schemes are ineffective in practice because of either the absence of sufficient textual descriptions or the ambiguity in the semantics of the textual descriptions [2], and formalisms to precisely represent service functionality are desired. OWL-S [3] attempts to address this problem with an approach based on the OWL family of Description Logics.

Several OWL-S based matchmaking approaches have been proposed for service discovery [4,5,6,7,8,9,10]. The approaches fall under two major categories: Input-Output (IO)-based and description-based. IO-based approaches [4,5,6] select matches by determining the relationship between the IO profile of a request and the IO profile of a service. While IO-based matching allows to eliminate some services from consideration, in general, a large number of inappropriate services may remain because services with identical IO signatures yet different functionality cannot be differentiated. Description-based approaches [7,8,10,9] adopt a more rigorous logical inference process to eliminate false positives by inferring subsumption relationships between a request and a service specification. However, services with functionality contradicting the request may be returned, and a matching service may be ranked poorly if its description does not completely satisfy a request [9,11,12,13].

In this paper, we present a consistency-based matchmaking scheme to address these issues. Our matchmaking scheme is a two-stage process, where in the first stage, a set of conceptual service profiles that seem to match the request are shortlisted, and subsequently, in the second stage, the concrete information about these services is obtained by querying the services directly to determine whether the services do meet the requirements or not.

Our matchmaking process (shown in Figure 1) utilises both the general service descriptions and detailed information obtained from services. The conceptual description of a service holds information that represents terminological or general knowledge about it. For example, an airline service might be described as *"sells airline tickets between Australian cities for less than $100"*. The concrete description of a service holds instance-level information. For example, the same airline service might provide information such as *"sells tickets between Adelaide and Melbourne for $69"*. The concrete information of a service are specialisations of its conceptual description.

Given a set of service descriptions and a discovery request, our matchmaking approach identifies the set of matching services profiles in two stages: first, we determine the set of conceptual descriptions that are consistent with the request (*potential matches*), and second, we restrict the selected candidates to those that indeed offer concrete descriptions that are consistent with the request (*concrete matches*). Unlike existing proposals, where only the conceptual knowledge is used to determine matching services, both conceptual and instance-level information are utilised in matchmaking. As a result, only the services that indeed satisfy the request are returned.

We evaluate our consistency-based matchmaking scheme in the context of the *Discovery II and Simple Composition* scenario proposed by the SWS Challenge group. We show that our matchmaking scheme (1) is robust with respect to overspecified service profiles, (2) always returns functional semantic matches (no false positives), and (3) is able to handle unspecified values in the service specifications.

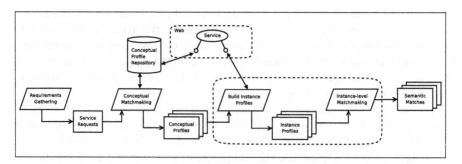

Fig. 1. Consistency-based Matchmaking Process

The rest of the paper is organised as follows. The sales portal scenario used for evaluation is presented in Section 2. Related description-based matchmaking proposals and their shortcomings are discussed in Section 3. Our consistency-based matchmaking scheme is detailed in Section 4. In Section 5 we evaluate our approach in the context of the sales portal scenario. In Section 6, an overview of other related work is presented, followed by a summary of our contributions and an outline of future directions in Section 7.

2   A Sales Portal Scenario

We adopt the *Discovery II and Simple Composition*[1] scenario proposed by the SWS Challenge group as a platform to evaluate our consistency-based matchmaking scheme. The sales portal proposed in the scenario has two vendors (*Rummage* and *Hawker*) who sell computers and accessories. The sales processes of these vendors are exposed as web services, where sales services offer a comprehensive description of the available products. In the adopted scenario, Rummage offers 3 services and Hawker offers 1 service. The services, 4 representative requests, and expected matches for the requests in the sales portal scenario are presented in Table 1.

3   Description-Based Matchmaking: A Discussion

Description-based semantic matchmaking is the process of identifying a set of services whose descriptions satisfy the requirements in a discovery request. A semantic service description (also referred to as a *service profile*) is a representation of a service's properties such as functionality, attributes, restrictions on attribute values etc., usually authored in a declarative formalism such as OWL DL and $\mathcal{SHOIQ}(\mathbf{D})$. In general, a service description contains conceptual and instance-level information. The conceptual layer holds information that represents terminological or general knowledge about a service, while the concrete layer holds instance-level or specific information about the service offerings. Both conceptual knowledge and concrete knowledge are essential

---
[1] http://sws-challenge.org/wiki/index.php/Scenario:_Discovery_II_and_Simple_Composition

Table 1. Services and Queries in the Sales Portal Scenario

Services	Description
S01	Notebook (GTIN: 1), MacBook 13', $1099.0, 13' flat screen, 1.83 Intel Duo Core, 512 MB, 60 GB, color: white, Vendor: Rummage
S02	Notebook (GTIN: 2), MacBook 13', $1699.0, 13' flat screen, 2.0 Intel Duo Core, 1 GB, 100 GB, color: black, No OS supplied, Vendor: Rummage
S03	Notebook (GTIN: 3), MacBook 13', $1099.0, 13' flat screen, 1.83 Intel Duo Core, 512 MB, 60 GB, color: white, Windows XP, 56K Modem, Vendor: Hawker
S04	Web cam (GTIN: 4), iSight, VGA 640x480, $149.0, Vendor: Rummage

Requests	Description
Q01	Services that sell MacBook 13', for $1099.0 with 13' flat screen, with Intel Duo Core 1.83, 60 GB HDD, 512 MB RAM,and white in colour (*Expected matches: S01 and S03*).
Q02	Services that sell Notebooks for less than $1600.0 (*Expected matches: S01 and S03*)
Q03	Services that sell Notebooks with pre-installed Windows XP (*Expected match: S03*).
Q04	Services that sell Notebooks that don't have a pre-installed OS (*Expected match: S02*).

to model service descriptions. Matchmaking processes in description-based schemes depend heavily on the completeness and quality of the service description. Description logics provide the two abstractions in the form of *TBox* (terminological box) and *ABox* (assertive box). Since most DL based approaches operate only on the conceptual level, instance information must be lifted to concepts and predicates. However, this approach does not scale well to large, detailed profiles [9,12].

3.1 Formal Matchmaking

Subsumption is the predominant inference mechanism for semantic matchmaking in most of the description-based approaches [7,5,8]. In general multiple degrees of match for a given request Q and service S are considered (in order of preference): *exact* ($S \equiv Q$), *plug-in* ($Q \sqsubseteq S$), *subsumed* ($S \sqsubseteq Q$), *intersection* ($\neg(S \sqcap Q \sqsubseteq \bot)$) and *mismatch* ($S \sqcap Q \sqsubseteq \bot$). Even though the approach improves upon pure IO matchmaking, additional information in service profiles and the Open World Assumption (OWA) can

Table 2. Sales Portal Scenario: OWL DL Observations

	S01	S02	S03	S04
Q01	SUBSUME (+)	DISJOINT (+)	INTERSECTION ($\oplus$)	DISJOINT (+)
Q02	SUBSUME (+)	DISJOINT (+)	SUBSUME (+)	INTERSECTION (−)
Q03	INTERSECTION (−)	INTERSECTION (−)	SUBSUME (+)	INTERSECTION (−)
Q04	PLUG-IN (−)	PLUG-IN (+)	PLUG-IN (−)	PLUG-IN (−)

(+) Consistent, (−) Inconsistent, and ($\oplus$) Consistent but poorly rated

lead to counter-intuitive results. Hence, false positives may be returned and profiles that provide the required functionality may be rated poorly. Table 2 shows the results of DL matchmaking applied to the scenario in Section 2. We use $(+)$ to indicate that the observations are consistent with the expectations listed in Table 1. For instance, the query Q01 is rightly matched with S01, S02, and S04. The inconsistent results are denoted as $(-)$. Instances where the matches are poorly rated are denoted with $(\oplus)$. The results support the following observations:

Intersection Matches. *Subsumed* matches are too strong in situations where a service description contains more information than what is requested[2]. Hence, the *intersection match* $(\neg(S \sqcap Q \sqsubseteq \bot))$ is required to select profiles that are similar to the request, but do not satisfy the requirements precisely. The intuition behind this type of match is that if a profile and request overlap, the service may be similar to the one requested. However, undesirable service profiles may be returned: Consider the combination Q02 and S04. Even though S04 offers only Web cams it is matched with the request Q02 from a Notebook sales service because the disjointness between Web cams and Notebooks cannot be derived under OWA. Such false positives also arise with the following pairs (Q03,S01), (Q03,S02), (Q03,S04), (Q04,S01), (Q04,S03), and (Q04,S04). Conversely, intersection matches cannot be ignored, since it is unlikely that a single profile will suit all requests (such as (Q01,S03) where S03 provides a modem that is not requested in Q01). Rather, intersection matches must be considered selectively to avoid false positives.

Overspecified Profiles. The existence of additional information in a service profile that is not in a request will result in poor ranking [7]. We refer to such profiles as being *overspecified* profiles for the given request. For example, S01 is preferred over S03 for Q01 in Table 2. While both S01 and S03 match the query Q01, S03 matches only by intersection because S03 offers more information about its product than requested. This limitation is counter-intuitive since vendors prefer to furnish more information about their services to match a wide variety of requests.

Unspecified Values. The OWA in DL matchmaking approaches implies that properties that are not constrained in a profile can take on any value and is therefore a potential intersection match to a query that requires that property. As a result, queries such as Q04 are not handled properly and result in false positives such as (Q04,S01), and (Q04,S03). The limitation of this approach is that there is no difference between a situation where an attribute has no value to that where an attribute does not exist at all.

4   Consistency-Based Matchmaking

Our consistency-based matchmaking is a two staged approach where in the first stage a set of conceptual profiles that could match the request are identified based on conceptual information, and subsequently in the second stage the concrete services are queried to determine whether a service indeed meets the requirements.

---

[2] Similarly, *plug-in* matches are too strong in cases where the request is overspecified.

Formally, every web service is augmented with two profiles: a *conceptual* profile and an *instance* profile. The conceptual profile represents the terminological knowledge about a service which is used to determine potential matches. The *instance* profile represents knowledge about the concrete capabilities of a service. We assume that the instance profile can be obtained by querying the service directly. The purpose of dividing a service's profile into two distinct parts is partly to ease the effort to describe a service and partly to support efficient discovery while avoiding undesired matches.

Our matchmaking process (shown in Figure 1) begins with a user formulating a service request for desired functionality and other requirements, such as QoS attributes. The consistency-based matchmaking scheme consists of two stages:

Step 1. Identify potential matches
  - Conceptual profiles that are consistent with the request are selected from a repository. The conceptual profiles are used to limit the search process to a few likely candidates that potentially match the request.

Step 2. Filter those services that do not satisfy the request
  - The instance profile of each candidate service identified in Step 1 is built by querying the respective services for concrete information.
  - Each of the gathered instance profiles is compared with the request to assess if the service can satisfy the request. Concrete profiles that are not consistent with the requirements are considered mismatches.

We refer to the instance profile building process described above as the *on-line* approach. Alternatively, we are also able to operate with a database of instance profiles in an off-line setting. Due to amount and volatility of information, most often it is not practical to map all the instance-level details onto the conceptual domain. Conversely, it is not possible to completely ignore the conceptual knowledge in favour of instance-level knowledge because the conceptual knowledge captures inherent relationships that may go unnoticed if only the instances are considered. Since all the concrete information available about a service can be queried, it can be assumed that the instance profile is complete. This allows us to apply the Closed World Assumption (CWA) where a particular service that cannot be shown to definitely match a request is assumed to be a mismatch.

The consistency-based approach offers the following benefits.

  - Only the services that match both the conceptual and the instance profile are returned, thereby eliminating potential false positives. The additional instance level check ensures that the returned matches only include the services that can be shown to offer the requested service.
  - The clear distinction of the conceptual and instance profiles potentially reduces the likelihood of a profile being overspecified. Moreover, values at the instance level need not be transformed into conceptual entities, and the conceptual profiles can be kept general.
  - The approach is potentially more efficient than a pure instance level checking scheme because only the services that have been identified as potential matches on the conceptual level are checked for consistency.

- The distinction between conceptual and instance profiles also provides the opportunity to use different formalisms that are best suited for each level. For example, variants of logic (DL, FOL) could be used as the formalism to model at the conceptual level, while on-line querying or instance databases could be used to gather instance level information.

4.1   Matchmaking as a Constraint System

Our consistency-based matchmaking scheme is modelled as a constraint system in which request and service profiles are represented as constraints. Note that the matchmaking scheme as such is generic and could be implemented using other methods that support logical entailment. We define a common template called *service constraint* to model both requests and service profiles.

Definition 1 (Service Constraint). *A service constraint p is defined as a triple $p = \langle A, D, C \rangle$ where $A = \{a_1, \dots, a_n\}$ is a set of service attributes, $D = \{D_{a_1}, \dots, D_{a_n}\}$ is the set of domains such that D_a is the set of allowed values of a , and C is a set of constraints specifying the conceptual properties of a service. A service constraint $p' = \langle A', D', C' \rangle$ specialises p iff $A \subseteq A'$ and $C \subseteq C'$ and $\forall d \in D, d' \in D' : d' \subseteq d$.*

Let $\perp$ denote the infeasible constraint representing either an empty profile that does not offer any service or a request that cannot be fulfilled.

Example 1 (BudgetNotebooks). A conceptual profile for a service that sells notebook computers for at-most \$1500 could be

$A = \{category, provider, price, name, gtin, screen, processor, memorySize, hddSize, modem, color, os\}, D = \{D_{category} = \{Notebook, Webcam\}, \dots, D_{price} = \mathbb{R}\}$
$C = \{category = Notebook, price \leq 1500.0\}$

Example 2 (Q02). A service request for services that sell notebooks for \$1600.0 or less could be

$A = \{category, price\}, D = \{D_{category} = \{Notebook, Webcam\}, D_{price} = \mathbb{R}\}$
$C = \{category = Notebook, price < 1600.0\}$

To access advertised services, we assume the existence of a registry (e.g., UDDI [1] or semantic registry [14]) that publicises service profiles. We define both conceptual and instance profiles in our abstract representation of a service registry called *Environment*.

Definition 2 (Environment). *An environment E is defined as a tuple $\langle S, E_C, E_I \rangle$ where, S is a set of concrete services, E_C is the set of conceptual service profiles for services in S, and E_I denotes the set of instance level profiles over S. Let $suppliers(p) \subseteq S$ denote the services that conform to instance-level profile $p \in E_I$.*

The environment that hosts the sales portal discussed in Section 2 is shown in Figure 2. *EService* is the generic profile that all the other profiles specialise. A profile is specialised by adding constraints. For example in Figure 2, the *BudgetNotebooks* profile specialises the profile *NotebookSales* by restricting the value of the attribute price to at-most \$1500.

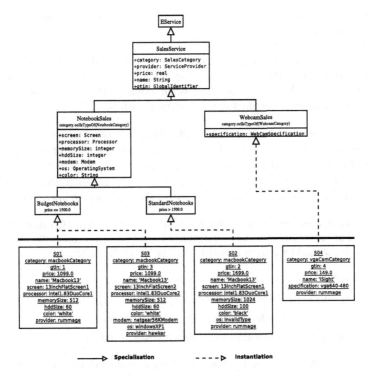

Fig. 2. Environment: Sales Portal Scenario

In our approach, conceptual profiles allow complex constraints over its attributes, whereas specialisation into instance profiles is limited to equality predicates to assign values to attributes. Therefore, a conceptual profile can be specialised into another conceptual or instance profile whereas all the instance profiles are specialised only by ⊥. We refer to the specialisation of a conceptual profile into an instance profile as *instantiation*. For example in Figure 2, the instance profile *S01* instantiates the conceptual profile *BudgetNotebooks*.

Example 3 (S01). An instance profile that specialises the conceptual profile from Example 1 is $C = \{gtin = 1, price = 1099.0, name = Macbook13, memorySize = 512, screen = 13InchFlatScreen1, processor = Intel1.83DuoCore1, hddSize = 60, color = white, provider = rummage\}$

Consistency at the Conceptual Level. The conceptual profiles that are consistent with the request are shortlisted in this step. We use logical entailment as a means to determine whether a service profile matches a request or not. Our consistency check is formalised as a Constraint Satisfaction Problem (CSP). A CSP is defined as follows:

Definition 3 (CSP). *A CSP is defined as a triple* $\langle X, V, R \rangle$ *where*

- *X is a set of variables* $\{x_1, x_2, \ldots, x_n\}$
- *for each* $x \in X$, *V_x denotes the set of allowable values for variable v*

– R is a set of constraints that need to be satisfied for any given assignment to the variables in X.

A CSP is *satisfiable* if there exists at least one valid assignment for all the variables in X. A service constraint can be treated as a CSP where service attributes are the constraint variables $(X = A)$ with appropriate domains $(V = D)$, and constraints $(R = C)$. A consistent conceptual profile is defined as follows:

Definition 4 (Consistent Conceptual Profile). *Given a service request $Q = \langle A, D, C \rangle$ and an environment $E = \langle S, E_C, E_I \rangle$, $ccp_E(Q)$ denotes the set of conceptual profiles from E_C that are consistent with Q: $ccp_E(Q) = \{\langle A', D', C' \rangle \mid \langle A', D', C' \rangle \in E_C \wedge A' \cap A \neq \emptyset \wedge$ the CSP $\langle A \cup A', D \dot{\cup} D', C \cup C' \rangle$ is satisfiable} where $D \dot{\cup} D'$ denotes the union combination[3] of corresponding domains in both operands.*

If there is at least one solution for the CSP then $S \in ccp_E(Q)$ otherwise $S \notin ccp_E(Q)$. For example, the CSP formulated to check for consistency between *BudgetNotebooks* and *Q02* from Examples 1, 2 is $\langle V, D, R \rangle$, where

$$X = \left\{ \begin{array}{l} category, provider, name, gtin, screen, processor, \\ memorySize, hddSize, modem, color, os, price \end{array} \right\},$$

$$V = \{V_{category} = \{Notebook, Webcam\}, \ldots, V_{price} = \mathbb{R}\},$$

$$R = \{category = Notebook, price \leq 1500.0, price < 1600.0\}.$$

In this step only the conceptual profiles that have common attributes with the request and that have non-conflicting constraints are selected. This is similar to the *intersection matches* for selecting overlapping profiles. The consistency check can be performed using a constraint solver with flexible instance reasoning such as ILOG's JConfigurator [15], or Seimens COCOS [16], or can be carried out using a (DL) reasoner such as Racer.

A profile S being in $ccp_E(Q)$ implies that the instance profiles that specialise S are potential matches for Q. All the concrete profiles that specialise any profile in $ccp_E(Q)$ are candidates for the second stage in our matchmaking process.

Consistency at the Instance Level. The candidate services identified in the previous step are checked further to determine whether their instance profiles satisfy the requirements in the request. Similar to the previous step, instance-level consistency is formulated as a CSP. This requires an additional restriction on some domains in the instance profile to accommodate the CWA for overspecified requests. Specifically, for a request $\langle A, D, C \rangle$ and a matching instance profile $\langle A', D', C' \rangle$, we write $\hat{D}$ to refer to the set of domains for attributes in A such that $\hat{D}_a = \emptyset$ if $a \in A \setminus A'$ else $\hat{D}_a = D_a \cup D'_a$. We now formally define the consistent instance profile.

Definition 5 (Consistent Instance Profile). *Given an environment E, a request $Q = \langle A, D, C \rangle$, and a set of candidate instance profiles $P = \{S_1, \ldots, S_n\}$, where $P \subseteq E_I$, $cip_E(Q, P)$ denotes the set of instance profiles that are consistent with Q: $cip_E(Q, P) = \{\langle A', D', C' \rangle \mid \langle A', D', C' \rangle \in E_I \wedge$ the CSP $\langle A \cup A', \hat{D}, C \cup C' \rangle$ is satisfiable}.*

---

[3] Union combination denotes the union of corresponding domain values of every attribute.

If there is at least one solution for the CSP then $S \in cip_E(Q, P)$; otherwise $S \notin cip_E(Q, P)$. In this step only the instance profiles with attribute values that definitely meet the restrictions in the request are selected. Matches for a given request Q are: $\cup_{S \in cip_E(Q,P)}$, $supplier(S)$. For example, consistency check between the instance profile $S01$ and $Q02$ from Examples 3, 2 is modelled as a CSP $\langle V, D, R \rangle$ where

$$X = \left\{ \begin{array}{l} category, provider, name, gtin, screen, processor, \\ memorySize, hddSize, modem, color, os, price \end{array} \right\},$$

$$V = \{V_{category} = \{Notebook, Webcam\}, \ldots, V_{price} = \mathbb{R}\},$$

$$R = \{category = Notebook, price = 1099.0, \ldots, price < 1600.0\}.$$

The consistency test at the instance level may alternatively be implemented in a number of ways, including querying over a database of instance profiles, logic inference in FOL, and on-line querying of services. Since all the information about the service is rigorously tested in this step to ascertain whether the service definitely meets the request, the matches returned in this step supersede the ones derived through pure conceptual reasoning (see Section 5).

4.2   On the Interpretations of Instance Profiles

All the attributes A of the concrete profiles $S = \langle A, C \rangle \in E_I$ are bound to concrete values from the defined range using equality constraints. If no equality relationship is defined over an attribute then this implies the absence of any information about that attribute. Under the CWA, the meaning of unspecified attribute is that such an attribute does not exist. For example in Figure 2, the *os* property of the service $S01$ does not have a value associated with it hence $S01$ does not offer the *os* property ($V_{os} = \emptyset$). That is, if a query restricts an *unknown* attribute of a profile then the service is not a match for the query. For example, $S01$ is not a match for the query $Q03$ that restricts the attribute *os*. It has been shown that under the CWA more accurate web service matches can be obtained than with the general open semantics [9,12].

Table 3. Characteristics of Consistency-based Matchmaking Scheme

(a) Interpretation of *instance* Profiles

Attributes in Profile		Existent	Non-existent
Attribute Value	**Query Restriction**		
Known	Value	(+)	(−)
	Open	(+)	(+)
	Existence	(−)	(+)
Unknown	Value	(−)	(−)
	Open	(+)	(+)
	Existence	(−)	(+)

(+) Potential Match   (−) Mismatch

(b) Sales Portal Scenario: Observations

Requests	S01	S02	S03	S04
Q01	(+)	(−)	(+)	(−)
Q02	(+)	(−)	(+)	(−)
Q03	(−)	(−)	(+)	(−)
Q04	(−)	(+)	(−)	(−)

(+) Match   (−) Mismatch

Inspired by prior research in the database domain to provide additional semantics for *null* values, we explore the possibility of explicitly specifying that an attribute *does not exist*. In our approach, an attribute whose value is *InvalidType* implies that the attribute is not offered. *InvalidType* is an universal type whose instances can be assigned to attributes irrespective of their prescribed data type. For example in Figure 2, the *os* attribute of the service *S02* is assigned to the special type *InvalidType* hence it is understood that *S02* does not offer the property *os*. Service requests may be modelled to take advantage of this modelling style in order to search for services that do not offer certain properties. For example in Table 1, the query *Q04* searches for services that do not offer the *os* property.

Table 3a presents the possible interpretations of instance profiles. For example in *S02*, the value of the attribute *os* is *InvalidType* which implies that *S02* does not offer *os*. Therefore, *S02* is a potential match for a query like *Q04* that requires the absence of the *os* attribute.

5   Evaluation: Matchmaking in the Sales Portal Scenario

The results of the evaluation of our consistency-based matchmaking scheme are presented in this section. The services offered by Rummage and Hawker in the sales portal scenario were modelled within the abstract environment that holds conceptual and concrete service details. The environment used in the evaluation is presented in Figure 2. The service discovery requests from Table 1 were modelled within our framework. For each of these requests, the consistency-based matchmaking scheme was applied. The candidate profiles that are selected in the conceptual matchmaking stage are the following:

Q01: {*BudgetNotebooks*}; Concrete profiles → {*S01,S03*}
Q02: {*BudgetNotebooks*}; Concrete profiles → {*S01,S03*}
Q03: {*BudgetNotebooks,StandardNotebooks*}; Concrete profiles → {*S01,S03,S02*}
Q04: {*BudgetNotebooks,StandardNotebooks*}; Concrete profiles → {*S01,S03,S02*}

Subsequently, the respective concrete profiles were queried to isolate the services that do match the requests. The results are presented in Table 3b.

Our results support that following observations:

Absence of False Positives. All results are consistent with the expected results as specified in Table 1; there are no false positives because of the rigorous testing at concrete layers of the service descriptions. Therefore, the false positives (Q02,S04), (Q03,S01), (Q03,S02), (Q03,S04), (Q04,S01), (Q04,S03), (Q04,S04) from Table 2 are not present in our approach.

Overspecified Profiles. For the query Q01, subsumption-based matchmakers penalise the profile S03 for providing a more detailed description. In our approach, S01 and S03 both match equally well since the attributes that are not restricted by the service requests are ignored. Therefore, excess information that is not restricted in a request does not play any role in the matching process.

Unspecified Values. The flexible modelling paradigm in our approach allows OWA, CWA and reified representation (*InvalidType*) to restrict the existence of certain attributes in queries. For example, query Q04 restricts the existence of the OS attribute. The matchmaking between Q04 and the service S01 is interpreted as: *"attribute 'os' with an unknown value exists in the profile but the query restricts its existence"*. As shown in Table 3a, a mismatch is returned in this case. Profiles S02 and S03 are also checked in a similar fashion[4].

Our preliminary evaluation suggests that our consistency-based matchmaking approach is robust while handling additional information in service specifications, always returns functional semantic matches (no false positives), and supports unspecified values.

Our consistency-based matchmaking process has been implemented within the Model-Driven Architecture (MDA) framework to support the (semi-)automatic composition of web services [17,18]. The combination of MDA and multi-level consistency-based matchmaking facilitates a more interactive composition process (if so desired).

6   Related Work

Other than the DL based approaches mentioned in the introduction, the following works are relevant in the context of this paper:

Our work is similar to [19] in that both the approaches advocate querying instance level information about the services to improve accuracy. However, different matchmaking schemes are employed. In [19], a set-theoretic matchmaking approach is used to assess suitability of a service with respect to a request. In this scheme, a service profile (or a goal) is modelled as "intentions", that is, a universally (or existentially) quantified expression over a set of *relevant objects* (service capabilities), where the quantification determines whether a profile matches a goal. The main drawback of this approach is that explicit intention modelling is required; the absence of such information reduces the problem to subsumption reasoning and may result in counter-intuitive matches (see Section 3). There is no explicit consideration of gathering intention and its representation in [19]. Conversely, our approach does not require any additional information besides the service profiles.

In our previous work [18], an interactive CSP-based algorithm that composes services by checking consistency between requirements and service specifications was presented. This work assumes the existence of a database of instance profiles prior to composition. Here, we present the characteristics of the consistency-based matchmaking scheme especially in situations where concrete details about the service capabilities cannot be predicted prior to matchmaking.

Carman and Serafini [20] propose interleaving planning and execution of service composition. The execution of services produces information that can be used to

---

[4] Note that S04 is not considered at this stage because the profile *WebcamSales* is not selected in the first stage.

extend and revise a plan to ensure a request is met. This approach does not scale to handle all services because executing a service may have permanent effects and compensation actions may be required. Moreover, the relationship between any two service interfaces have to be specified prior to planning. In our approach, no such pre-specification is required because consistency between services and requests is assessed during matchmaking by gathering instance level details when required.

7 Conclusion

We presented a consistency-based matchmaking approach where services that potentially provide the requested functionality are identified, candidate services are queried for more information to isolate those services that indeed meet the request to prune undesired results. Our evaluation shows that our matchmaking scheme is robust while handling over-specified profiles, always returns matches where every individual match provides the requested functionality, avoids false positives, and is able to handle omitted attributes effectively. Further work in this direction includes developing a ranking mechanism based on an optimisation scheme where attributes are qualified with weights and extending our work to situations where a service offers only a part of the required functionality. We also plan to incorporate our matchmaking process within a framework for semi-automated web service composition.

References

1. Bellwood, T., et al.: UDDI Version 3.0. Technical report, UDDI.org. (July 2002)
2. Klusch, M., Fries, B., Sycara, K.P.: Automated semantic web service discovery with owls-mx. In: Proc. AAMAS (2006)
3. OWL-S: Semantic markup for web services (2004),
   http://www.w3.org/Submission/OWL-S/
4. Paolucci, M., Kawamura, T., Payne, T.R., Sycara, K.P.: Semantic matching of web services capabilities. In: Horrocks, I., Hendler, J. (eds.) ISWC 2002. LNCS, vol. 2342, pp. 333–347. Springer, Heidelberg (2002)
5. Kawamura, T., Blasio, J.A.D., Hasegawa, T., Paolucci, M., Sycara, K.P.: Public Deployment of Semantic Service Matchmaker with UDDI Business Registry. In: McIlraith, S.A., Plexousakis, D., van Harmelen, F. (eds.) ISWC 2004. LNCS, vol. 3298, pp. 752–766. Springer, Heidelberg (2004)
6. Grønmo, R., Jaeger, M.C.: Model-Driven Semantic Web Service Composition. In: APSEC 2005, pp. 79–86 (December 2005)
7. Li, L., Horrocks, I.: A software framework for matchmaking based on Semantic Web technology. In: Proc. of WWW, Budapest, pp. 331–339 (May 2003)
8. Wang, H., Li, Z.: A Semantic Matchmaking Method of Web Services Based On $\mathcal{SHOIN}^{+}(D)*$. In: Proc. of IEEE APSCC, Guangzhou, China, pp. 26–33 (December 2006)
9. de Bruijn, J., Lara, R., Polleres, A., Fensel, D.: OWL DL vs. OWL flight: conceptual modeling and reasoning for the semantic Web. In: Proc. of WWW 2005 Conf., Chiba, Japan (May 2005)
10. Patel-Schneider, P.F., Horrocks, I.: A comparison of two modelling paradigms in the semantic web. J. Web Sem. 5(4), 240–250 (2007)

11. Balzer, S., Liebig, T., Wagner, M.: Pitfalls of OWL-S: a practical semantic web use case.. In: Proc. ICSOC 2004, New York, NY, USA, pp. 289–298 (November 2004)
12. Grimm, S., Motik, B., Preist, C.: Matching Semantic Service Descriptions with Local Closed-World Reasoning. In: Sure, Y., Domingue, J. (eds.) ESWC 2006. LNCS, vol. 4011, pp. 575–589. Springer, Heidelberg (2006)
13. Thiagarajan, R., Stumptner, M.: A Native Ontology Approach for Semantic Service Descriptions. In: Proc. of AOW 2006, Hobart, Australia, pp. 85–90 (2006)
14. Nguyen, K., Cao, J., Liu, C.: Semantic-enabled organization of web services. In: Zhang, Y., Yu, G., Bertino, E., Xu, G. (eds.) APWeb 2008. LNCS, vol. 4976, pp. 511–521. Springer, Heidelberg (2008)
15. Albert, P., Henocque, L., Kleiner, M.: Configuration Based Workflow Composition. In: Proc. of ICWS 2005 (July 2005)
16. Stumptner, M., Friedrich, G., Haselböck, A.: Generative constraint-based configuration of large technical systems. AIEDAM 12(4) (1998)
17. Thiagarajan, R., Stumptner, M., Mayer, W.: Semantic Web Service Composition by Consistency-based Model Refinement. In: Proc. of IEEE APSCC, Tsukuba Science City, Japan (December 2007)
18. Thiagarajan, R., Stumptner, M.: Service Composition With Consistency-based Matchmaking: A CSP-based Approach. In: Proc. ECOWS (2007)
19. Zaremba, M., Vitvar, T., Moran, M.: Towards optimized data fetching for service discovery. In: Proc. ECOWS, Halle, Germany (November 2007)
20. Carman, M., Serafini, L.: Planning For Web Services the Hard Way. In: SAINT Workshops, Orlando, FL, USA (January 2003)

A Fast Heuristic Algorithm for the Composite Web Service Selection

Rong Wang[1], Chi-Hung Chi[1], and Jianming Deng[2]

[1] School of Software, Tsinghua University, Beijing, China
[2] College of Software Engineering, SouthEast University, Nanjing, China

Abstract. Composite Web Service selection is one of the most important issues in Web Service Composition. During the selection process, while the decision making during the selection process is much easy in the term of the functional properties of Web Service, it is very difficult in terms of the non-functional properties. In this paper, we investigate the problem of composite Web Service selection. We propose the utility function to be the evaluation standard as a whole by considering all QoS parameters of each component service based on the definition in [16]. We map the multi-dimensional QoS composite Web Service to the multi-dimensional multi-choice knapsack (MMKP). And we propose a fast heuristic algorithm with $O(nlm+nl\lg n)$ complexity for solving the problem.

1 Introduction

The rapid development of internet results in more and more Web Services published on the internet. Together with the e-business development, a single service often cannot satisfy the functional needs in most practical situation. To complete a service of complex function often requires a batch of atomic service to work together. So Web Service Composition (WSC) has become the core Service-Oriented Computing technology. And the composite Web Service selection is the one of the most important issues in Web Service Composition. What Web Service selection cares about is mainly on how to select atomic service so that the resulting composite service satisfies both the functional and QoS requirements. As more and more web services are published on the internet, the number of Web Services with same or similar function properties but different QoS properties grows sharply. This fact makes a challenge for the composite Web Service selection: to give an optimal or approximate choice solution in real time. While the decision making during the selection process is much easy in the term of functional properties, it is very difficult in term of the non-functional properties. This triggers substantial amount of research in composite Web Service selection, such as in [18, 10, 7, 17, 19, 2, 9], in which they mainly focus on the non-functional properties, i.e. QoS of Web Services.

In this paper, we study the problem of the composite Web Service selection. We propose the utility function to be the evaluation standard as a whole, considering all QoS parameters of each component service based on the definition in [16]. We map the composite Web Service selection model with multi-dimensional QoS constraints and the *general flow structure* to the multi-dimensional multi-choice knapsack (MMKP) and propose a fast heuristic solution with $O(nlm+nl\lg n)$ complexity.

Q. Li et al. (Eds.): APWeb/WAIM 2009, LNCS 5446, pp. 506–518, 2009.

The rest of this paper is organized as follows. Section 2 reviews related research to our work. Section 3 proposes the utility function definition with QoS attributes. We also describe a QoS-based service selection model and transform the model to MMKP and MCKP in Section 3. Section 4 discusses a heuristic algorithm F-HEU with $O(nlm+nllgn)$ complexity to solve the MMKP. Finally, we conclude the paper in Section 5.

2 Related Work

Majority of existing methods on composite Web Service selection can be divided into two groups, local optimal and global optimal.

For local optimal methods, EFlow[1] and CMI[6] focus on optimizing service se-lection at a task level, i.e. a single task. However, under the most practical case, the global optimal case can satisfy the service request, while such local strategies cannot tackle the global constraints and optimal problem for the composite services. Thus, the global optimal solution is the main objective of composite Web Service selection. Based on the constraints' dimensions, the global selection problems can be divided into four categories:

- Selection problem with single QoS constrain and a sequential flow structure;
- Selection problem with multi-dimension QoS constraints and a sequential flow structure;
- Selection problem with single QoS constraint and a general flow structure; and
- Selection problem with multi-dimensional QoS constraint and a general flow structure.

[16] maps the second category of the selection problem to a multi-dimension multi-choice knapsack (MMKP) at first, then proposes a heuristic algorithm WS-HEU based on the algorithm M-HEU [11]. The worst case running time of M-HEU, a heuristic based on the aggregate resource consumption, is $O(mn^2(l-1)^2)$. Here, n is the number of tasks, l the number of candidate services in each service class (assumed constant for convenience of analysis) and m the constraint dimensions. The algorithm C-HEU in [12] is a heuristic algorithm by constructing convex hull [5] to solve the MMKP, with complexity being $O(nlm+nllgl+nllgn)$. The optimality achieved by this algorithm is between 88% and 98%. There are other algorithms [1, 13, 14] for MMKP.

A few algorithms are also proposed to solve the fourth category of the selection prob-lem. GODSS in [15][15] is an algorithm for global optimal solution of dynamic Web service selection based on genetic algorithm. The running time of GODSS is $O(((m+n+L)N2+(m+n)NN^*)T)$. Here m is the number of the objective functions, n the number of constraints, N the number of services in a service class, N^* the size of assis-tant population, L the number of tasks, T the number of repetitions. [8] presents a fast method for quality driven composite Web services selection. The workflow partition strategy is to divide an abstract workflow into two sub-workflows so that the number of candidate services that should be considered is decreased. Then it utilizes a mixed inte-ger linear programming for solving the service selection problem. In [17], the selection problem with multi-dimension QoS constraints and a general flow structure is mapped to the 0-1 Integer Programming (0-1 IP) problem. And the paper proposes WFlow

Algorithm which finds a feasible solution at first and then trying to optimize the solution through upgrades and downgrades. WFlow is with $O(N^2(l-1)^2 m)$ complexity. In the next sections, we will map the fourth service selection problem with multi-dimension QoS constraints and a general flow structure to MMKP and present a heuristic algorithm to solve it.

3   QoS-Based Service Selection Model

This paper focuses on four workflow structures (shown in Figure 1). They are sequential, parallel, conditional and loop structure. The QoS model based on these four workflow patters will be presented, and this QoS-based service selection model will be mapped into MMKP.

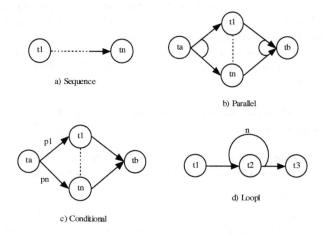

Fig. 1. Four Workflow Structures

Table 1. QoS Aggregation Rules

	Sequence	Conditional	Parallel	Loop
Time	$\sum_{i=1}^{n} Time(t_i)$	$\sum_{i=1}^{n} p_i \times Time(t_i)$	$\underset{i=1}{\overset{n}{Max}}(Time(t_i))$	$n \times Time(t_i)$
Cost	$\sum_{i=1}^{n} Cost(t_i)$	$\sum_{i=1}^{n} p_i \times Cost(t_i)$	$\sum_{i=1}^{n} Cost(t_i)$	$n \times Cost(t_i)$
Probability	$\prod_{i=1}^{n} Probability(t_i)$	$\sum_{i=1}^{n} p_i \times Probability(t_i)$	$\prod_{i=1}^{n} Probability(t_i)$	$(Probability(t_i))^n$

3.1   QoS Model for Composite Web Service

The QoS attributes of Web Service include response time, service time, availability, reliability, service cost and loss probability etc. The QoS information of atomic services is already given by the QoS broker or service registry. The QoS of a composite

Web service is calculated by aggregating the QoS of constituent Web services. Here depending on the computing nature, the QoS attributes are divided into four groups [8]: 1) time, such as response time; 2) cost, such as service cost; 3) probability, such as reliability; and 4) capacity, such as throughput. In this paper, only main three groups are taken into account. The rules for aggregating the QoS of workflow patterns are described in the Table 1 [8].

3.2 Definition of Utility Function

In general, the QoS attributes can be divided into two groups, the positive attribute (the higher, the better) and the negative attribute (the lower, the better). Based on the definition in [16], the utility function is defined as follows:

Definition 1 (Utility Function). Suppose there are x positive QoS attributes and y negative QoS attributes. The utility function for a service s_{ij} is defined as:

$$F_{ij} = \sum_{\alpha=1}^{x} \omega_\alpha * \left(\frac{q_{ij}^\alpha - \mu_i^\alpha}{\sigma_i^\alpha} \right) + \sum_{\beta=1}^{y} \omega_\beta * \left(\frac{\mu_i^\beta - q_{ij}^\beta}{\sigma_i^\beta} \right) + C, \tag{1}$$

where ω_α and ω_β are the weights for each QoS attribute $(0 < \omega_\alpha, \omega_\beta < 1)$, and $\sum_{\alpha=1}^{x} \omega_\alpha + \sum_{\beta=1}^{y} \omega_\beta = 1 \cdot \mu_i$ and σ_i are the average and standard deviation of the values of QoS attributes for all candidates in service class Si. C is a constant to ensure the value of F_{ij} be non-negative. For example, set $C = \sum_{\beta=1}^{x} \omega_\beta * \left(\dfrac{\max\limits_{j=1}^{l}\left(q_{ij}^\beta\right) - \mu_i^\beta}{\sigma_i^\beta} \right)$.

3.3 Web Service Selection Model

Here we describe a Web Service selection model, which is to choose services s_{ij} from the service candidate class S_i for Task t_i in the workflow W, maximize the sum of utility of the chosen services and at the same time make the QoS of the workflow W with chosen services satisfy the QoS requirement.

- W is a workflow, which consists of four structures mentioned above. And T is the set of tasks in W, i.e. T={t_1,...,t_n}, $t_i \in W$.
- S is the set of service classes, S={S_1,...,S_n}. And Si includes all services with similar function that t_i requires and different QoS, Si = {s_{i1},...s_{iNi}}. That is to say Si is the service candidate class for Task t_i.
- R is the QoS requirement.
- The QoS vector of service s_{ij} is [q_1,...q_m], and these attributes are divided into three groups: time, cost and probability.
- The utility u_{ij} of s_{ij} can be calculated by Equation 1.
- The function $Time(t_1,..., t_n,w)$ calculates the time value of workflow W based on the QoS aggregation rules in Table 1. $Cost(t_1,..., t_n,w)$ and $Probability(t_1,..., t_n,w)$ are to calculate the cost and probability of the workflow respectively.

So the Web Service model can be formulated as follows:

$$Max(\sum_{i=1}^{n}\sum_{j=1}^{N_i} x_{ij} * u_{ij})$$ (2a)

$$s.t. \ Time(\sum_{j=1}^{N_1} t_{1j} \cdot x_{1j}, .., \sum_{j=1}^{N_n} t_{nj} \cdot x_{nj}, W) \leq R_{time}$$ (2b)

$$Cost(\sum_{j=1}^{N_1} c_{1j} \cdot x_{1j}, ..., \sum_{j=1}^{N_n} c_{nj} \cdot x_{nj}, W) \leq R_{cost}$$ (2c)

$$Probability(\sum_{j=1}^{N_1} p_{1j} \cdot x_{1j}, ..., \sum_{j=1}^{N_n} p_{nj} \cdot x_{nj}, W) \geq R_{probability}$$ (2d)

$$\sum_{j=1}^{N_i} x_{ij} = 1, \quad and \ x_{ij} = \begin{cases} 0, & s_{ij} \ not \ chose \\ 1, & s_{ij} \ is \ chose \end{cases}$$

3.4 Mapping to MMKP

To map the selection model to MMKP, the aggregate of QoS of constituent Web services must be linearized. Depending on the three QoS attributes (time, cost and probability), three steps are needed to linearize the computing of the two categories of QoS attributes.

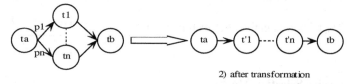

Fig. 2. Transformation of Conditional Structure

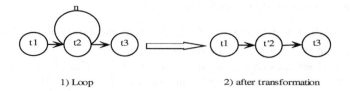

Fig. 3. Transformation of Loop Structure

Step 1: Linearize the Time of Computing.

a)   Transform the conditional and loop structure to the sequence, as shown in Figure 2 and Figure 3.

According to the transformation in Figure 2, there is some computing transformation:

> t'_i.time=p_i*t_i.time
> Each service s_{ij} in the service class S_i for t_i will be transformed to s'_{ij} with s'_{ij}.time=p_i*s_{ij}.time (j=1,…,l)

According to the transformation in Figure 3, there is some computing transformation:

> t'_2.time=$n*t_2$.time
> Each service s_{2j} in the service class S_2 for t_2 will be transformed to s'_{2j} with s'_{2j}.time=$n*s_{2j}$.time (j=1,…,l)

b) Add additional attributes based on the path in the simplified graph.

Now there are only parallel and/or sequence structures in the graph. Calculate the number of the paths in the graph and number them $Path_i$ (i=1,…,j). Add attribute a^k (k=1,…,j) for $Path_i$. Then take following transformation:

> $ti.a^k$ = ti.time, where ti belongs to $Path_k$
> $ti.a^k$ = 0, where ti doesn't belong to $Path_k$
> Make s_{ij} in S_i to have the corresponding change according to t_i.

So the constraints on time will be transformed into the constraints on a^k, k=1,…,x. It can be formulated as follows:

$$\sum_{i=1}^{n}\sum_{j=1}^{Ni} a_{ij}^k * x_{ij} \leq R_{time} , k=1,…,j \tag{3}$$

Step 2: Linearize the computing of cost.

Transform the conditional structure to sequence just as we do in linearizing the computing of time and set

> t'_i.cost=p_i*t_i.cost
> Each service s_{ij} in service class S_i for t_i will have the corresponding change with s'_{ij}.time=p_i*s_{ij}.time (j=1,…,l)

So the constraints on cost will be formulated as follows:

$$\sum_{i=1}^{n}\sum_{j=1}^{Ni} cost_{ij} * x_{ij} \leq R_{cost} \tag{4}$$

Step 3: Linearize the computing of probability.

a) Transform the parallel and loop structure to the sequence, as is shown in Figure 4 and 5.

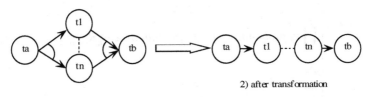

1) Parallel   2) after transformation

Fig. 4. Transformation of Parallel Structure

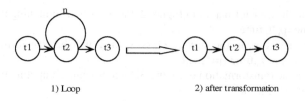

1) Loop   2) after transformation

Fig . 3-5. Transformation of Loop Structure

According to the transformation in Figure 4, there is no transformation in computing. According to the transformation in Figure 5, there is some computing transformation:

➤ t'$_2$.probability=(t$_2$.probability)^n
➤ Each service s$_{2j}$ in the service class S$_2$ for t$_2$ will be transformed to s'$_{2j}$ with s'$_{2j}$.probabilty=(s$_{2j}$.probability)^n , where j=1,...,N$_i$.

b)   Add additional attributes based on the path in the simplified graph
     Now there are only conditional or sequence structure in the graph, such as the graph shown in Figure 6.

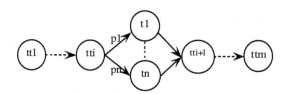

Fig. 6. A Simplified Graph After the Operation Described Above

So the computing of probability of Figure 6 can be formulated as follows:

$$\prod_{i=1}^{m} tt_i.probability * \sum_{i=1}^{n} (p_i * t_i.probability) \geq R_{probability} \tag{5a}$$

$$\Leftrightarrow \begin{cases} \prod_{i=1}^{m} tt_i.probability * p_i * t_1.probability \geq \omega_1 R_{probability} \\ \qquad \cdots\cdots\cdots\cdots \\ \prod_{i=1}^{m} tt_i.probability * p_i * t_n.probability \geq \omega_n R_{probability} \\ \sum_{i=1}^{n} \omega_i = 1 \end{cases} \tag{5b}$$

$$\Leftrightarrow \begin{cases} \sum_{i=1}^{m} \log(tt_i.probability) + \log(p_1) + \log(t_1.probability) \geq \log \omega_1 + \log(R_{probability}) \\ \qquad\qquad \cdots\cdots\cdots\cdots \\ \sum_{i=1}^{m} \log(tt_i.probability) + \log(p_n) + \log(t_n.probability) \geq \log \omega_n + \log(R_{probability}) \\ \sum_{i=1}^{n} \omega_i = 1 \end{cases} \qquad (5c)$$

Relax the inequality in Formula (5c) and make further adjustment, we have the followings:

$$\begin{cases} -n\sum_{i=1}^{m} \log(tt_i.probability) - \sum_{i=1}^{n} \log(t_i.probability) \leq -\sum_{i=1}^{n} \log \omega_i + \\ \qquad\qquad\qquad\qquad \sum_{i=1}^{n} \log(p_i) - n\log(R_{probability}) \qquad (5d) \\ \sum_{i=1}^{n} \omega_i = 1 \end{cases}$$

Then we can set ω_i, i=1,...,n based on the historical information to make the solutions of the probability constraint in Formula (5a) to be included in the solution of the inequality in Formula (5d) and at the same time the non-solutions of the probability constraint are excluded from the solutions of the inequality in Formula (5d). For example set $\sum_{i=1}^{n} \log \omega_i = \sum_{i=1}^{n} \log(p_i)$. Based on the Formula (5d), the probability attributes of services and $R_{probability}$ will be changed correspondingly as follows:

1. Set
   $s'_{ij}.probability = -n\log(s_{ij}.probability)$, where $s_{ij} \in$ the service candidate class of tt_i

2. Set
   $s'_{ij}.probability = -\log(s_{ij}.probability)$, where $s_{ij} \in$ the service candidate class of t_i

3. Set $R'_{probability} = -\sum_{i=1}^{n} \log \omega_i + \sum_{i=1}^{n} \log(p_i) - n\log(R_{probability})$

After a series of operations, the Web Service selection model is transformed into the MMKP shown below:

$$Max(\sum_{i=1}^{n} \sum_{j=1}^{N_i} x_{ij} * u_{ij}) \qquad (6a)$$

$$\sum_{i=1}^{n} \sum_{j=1}^{Ni} a_{ij}^{k} * x_{ij} \leq R_{c}^{k}, \quad k = 1, \ldots, m$$

such that   (6b)

$$\sum_{j=1}^{Ni} x_{ij} = 1, \quad and \quad x_{ij} \in \{0,1\}$$

3.5 Mapping to MCKP

Here we will map the multi-dimensional multi-choice knapsack (MMKP) to multi-choice knapsack (MCKP). We use the vector ω as a slack variable to agglomerate the constraints into one single constraint. So MCKP can be expressed as

$$Max(\sum_{i=1}^{n} \sum_{j=1}^{Ni} x_{ij} * u_{ij})$$ (7a)

$$\sum_{i=1}^{n} \sum_{j=1}^{Ni} r_{ij} * x_{ij} \leq \sum_{k=1}^{m} \omega^{k} R_{c}^{k},$$

such that $$r_{ij} = \sum_{k=1}^{m} \omega^{k} a_{ij}^{k}$$ (7b)

$$\sum_{j=1}^{Ni} x_{ij} = 1, \quad and \quad x_{ij} \in \{0,1\}$$

It is obviously that if the choice X={a_{11},...,a_{nNi}} does not satisfy the constraints of MCKP, X will definitely not satisfy the constraints of the MMKP.

4   Heuristic Selection Algorithm for MMKP

In this section, we will propose a new heuristic selection algorithm to solve MMKP. Then we will analyze its computational complexity.

4.1   A Heuristic Selection Algorithm

Below is the heuristic algorithm that we propose to solve the MMKP. The pseudo-code of the algorithm presented below is preceded by the definitions of some variables used in the algorithm.

snf:	a Boolean variable. If *snf* is ture, current_choice will not be feasible.
penalty:	a slack variable to transform multidimensional QoS to single dimension.
isfeasible( ):	a function returning true if the solution satisfies the constraints and false otherwise.
Utility( ):	a function returning the total utility value earned from the selection choice.

initial_penalty( ): return the initial *penalty* vector.
adjust_penalty( ): return the *penalty* vector based on the current resource usage.

Begin Procedure Adjust_selected_item (p, current_choice, snf)

1.   temp←current_choice
2.   update temp with p
3.   fp←isfeasibility(temp)
4.   if(fp==ture)
5.   update current_choice with p
6.   update u_c←utility(current_choice)
7.   update snf←false   // set signal of finding feasible solution in this iteration
8.   return
9.   else
10.   if snf
11.   update current_choice with p;
12.   return
End Procedure

Begin Procedure initial_penalty( )

/* Calculate the initial penalty vector. */
1.   sumR←vector summation of QoS vectors of each item in each group.
2.   avg ← sumR/(n*l)
3.   q← avg /|avg|*m;
4.   return q.
End Procedure

Begin Procedure adjust_penalty(q)

/* Update the penalty vector using information about the QoS of the current choice */
1.   q ←apply appropriate formula on vector q, total resource vector and available resource vector.
2.   return q;
End Procedure

Begin Procedure F-HEU()

/* Find solution for MMKP. Services in each service class are in increasing order of the value of the utility */
1   current_choice←the item with lowest utility value from each group
2   if isfeasible(current_choice)
3   u_c←utility(current_choice)   // set the current utility value if the current choice is feasible
4   else
5   u_c←0   // set the current utility value zero
6   endif
7   penalty = initial_penalty( )

```
8     for repeat←1 to 3 do              // only three iterations for finding solution
9         saved_choice←current_choice    // save the current solution
10        u ←u_c                         //save the current utility
11        initialize current_choice, u_c, and snf
12        for each service class Sᵢ in MMKP do
13            Transform each QoS vector of each service to single dimension using
                 vector penalty .
14            Chᵢ ← pick_up_efficient_points (Sᵢ') // pick up the frontiers.
15            Seᵢ ← point_to_segment(Chᵢ) // link points to produce segments and
                                            record their slope.
16        endfor
17        Se ← Combine(Seᵢ , n)      // combine Seᵢ (i=1,…,n) in a special way
18        If ( no solution for MCKP )         // verify the combination of the service
                                                  with lowest r
19            return Solution not found    // from each service class
20        endif
21        for each segment in Se do
22           p1, p2 ←The items associated with the segment.
23           adjust_selected_item (p1, current_choice, snf)
24           adjust_selected_item (p2, current_choice, snf)
25        endfor
26        if Utility(current_sol) < u then        // new solution is inferior than the saved
                                                      one
27           current_sol←saved_sol
28           u_c ← u
29        endif
30        penalty←adjust_penalty(penalty)
31     endfor
32     if u_c == 0 then
33        return Solution Not found
34     else
35        return current_choice
36     end
End Procedure.
```

4.2 Computational Complexity

Now we analyze the worst-case complexity of the heuristic algorithm F-HEU. First we assume that there are n service classes, l services in each service class and m-dimension QoS vector. It is obvious that the running time of initial_penalty( ) and adjust_penalty() is of $O(nlm)$ and $O(m)$ respectively. The time complexity of the procedure Adjust_selected_item () is of $O(m)$. Lines #12–#16 iterates n times. So the running time of Line #15 is of $O(nlm)$. The running time of the procedure pick_up_efficient_points() is of $O(l)$, so that of Line #17 is of $O(nl)$. The worst running time of the procedure point_to_segment() is of $O(l)$, so that of Line #18 is of $O(nl)$. The worst running time of Line #19 is of $O(nllgn)$. Lines #21–#25 iterates at most $O(nl)$ times. So the worst running time of Line #25,26 is $O(nlm)$. So the overall worst running time of the procedure F-HEU can be calculated as follows:

$$O(nlm) + O(nlm) + O(nl) + O(nl) + O(n(l-1)\lg n) + O(n(l-1)m) + O(m) = O(nlm + nl\lg n)$$

The running time of F-HEU is $O(nlm + nl\lg n)$.

5 Conclusion

In this paper, we study the problem of the composite Web Service selection. We propose the utility function to be the evaluation standard as a whole, by considering all QoS parameters of each component service based on the definition in [16]. We map the composite Web Service selection model with multi-dimension QoS constraint and *a general flow structure* to MMKP and MCKP. Then a heuristic algorithm F-HEU with $O(nlm + nl1\lg n)$ complexity is proposed to solve the MMKP.

Acknowledgement

This work is supported by the 863 project #2007AA01- Z122 & project #2008AA01Z12, the National Natural Science Foundation of China Project #90604028, and 973 project #2004CB719406.

References

[1] Drexel, A.: A simulated annealing approach to the multi-constraint zero-one knapsack problem. Annuals of Computing 40, 1–8 (1988)

[2] Ardagna, D., Pernici, B.: Global and local qoS guarantee in web service selection. In: Bussler, C.J., Haller, A. (eds.) BPM 2005. LNCS, vol. 3812, pp. 32–46. Springer, Heidelberg (2006)

[3] Lee, C.: On QoS management, PhD dissertation. School of Computer Science, Carnegie Mellon University (August 1999)

[4] Casati, F., Shan, M.C.: Dynamic and Adaptive Composition of EServices. Information Systems 26(3), 143–162 (2001)

[5] Cormen, T.H., Leiserson, C.E., Rivest, R.L., Stein, C.: Introduction to Algorithms, 2nd edn. MIT Press, Cambridge (2002)

[6] Georgakopoulos, D., Schuster, H., Cichocki, A., Baker, D.: Managing Process and Service Fusion in Virtual Enterprises. Information System, Special Issue on Information System Support for Electronic Commerce 24(6), 429–456 (1999)

[7] Issa, H., Assi, C., Debbabi, M.: QoS-aware middleware for Web services composition - A qualitative approach. In: Proceedings - International Symposium on Computers and Communications, Proceedings - 11th IEEE Symposium on Computers and Communications, ISCC 2006, pp. 359–364 (2006)

[8] Jang, J.-H., Shin, D.-H., Lee, K.-H.: Fast Quality Driven Selection of composite Web services. In: Proceedings of ECOWS 2006: Fourth European Conference on Web Services, pp. 87–96 (2006)

[9] Clabby, J.: Web Services Explained: Solutions and Applications for the Real World. Prentice Hall PTR, Englewood Cliffs (2002)

[10] Zeng, L., Benatallah, B., Ngu, A.H.H., Dumas, M., Kalagnanam, J., Chang, H.: QoS-Aware Middleware for Web Services Composition. IEEE Transactions on Software Engineering 30(5), 311–327 (2004)

[11] Akbar, M.M., Manning, E.G., Shoja, G.C., Khan, S.: Heuristic solutions for the multiple-choice multi-dimension knapsack problem. In: Alexandrov, V.N., Dongarra, J., Juliano, B.A., Renner, R.S., Tan, C.J.K. (eds.) ICCS-ComputSci 2001. LNCS, vol. 2074, p. 659. Springer, Heidelberg (2001)

[12] Mostofa, A.M., Sohel, R.M., Kaykobad, M., Manning, E.G., Shoja, G.C.: Solving the Multidimensional Multiple-Choice Knapsack Problem by Constructing Convex Hulls. Computers and Operations Research 33(5), 1259–1273 (2006)

[13] Magazine, M., Oguz, O.: A Heuristic Algorithm for Multi-Dimensional Zero-One Knapsack Problem. European Journal of Operational Research 16(3), 319–326 (1984)

[14] Khan, S.: Quality Adaptation in a Multi-Session Adaptive Multimedia System: Model and Architecture. PhD Dissertation. Department of Electrical and Computer Engineering, University of Victoria (1998)

[15] Liu, S.-L., Liu, Y.-X., Zhang, F.: Dynamic Web Services Selection Algorithm with QoS Global Optimal in Web Services Composition. Journal of Software 18(3), 646–656 (2007)

[16] Yu, T., Zhang, Y., Lin, K.-J.: Efficient algorithms for Web services selection with end-to-end QoS constraints. ACM Transactions on the Web 1(1), 6 (2007)

[17] Yu, T.: Quality of Service (QoS) in Web Services Model, Architecture and Algorithms, http://lsdis.cs.uga.edu/lib/download/ CSM+QoS-WebSemantics.pdf

[18] Tsesmetzis, D., Roussaki, I., Sykas, E.: QoS-Aware Service Evaluation and Selection. European Journal of Operational Research 191(3), 1101–1112 (2008)

[19] Gao, Y., Na, J., Zhang, B., Yang, L., Gong, Q.: Optimal Web Services Selection Using Dynamic Programming. In: Proceedings of 11th IEEE Symposium on Computers and Communications, pp. 365–370 (2006)

Intervention Events Detection and Prediction in Data Streams[*]

Yue Wang[1], Changjie Tang[1], Chuan Li[1,**], Yu Chen[1],
Ning Yang[1], Rong Tang[1], and Jun Zhu[2]

[1] The Database and knowledge Engineering Lab, Computer School of Sichuan University
[2] China Birth Defect Monitoring Centre, Sichuan University
{wangyue,tangchangjie,lichuan}@cs.scu.edu.cn

Abstract. Mining interesting patterns in data streams has attracted special attention recently. This study revealed the principles behind observations, through variation of intervention events to analyze the trends in the data streams. The main contributions includes: (a) Proposed a novel concept *intervention event*, and method to analyze streams under intervention. (b) Proposed the methods to evaluate the impact of intervention events. (c) Gave extensive experiments on real data to show that the newly proposed methods do prediction efficiently, and the rate of success is almost reach 92.6% recall in adaptive detection for intervention events in practical environment.

Keywords: Intervention Events, Data stream, Detection, Prediction.

1 Introduction

With the rapid development of modern sensor device technology, huge data are generated every day in forms of stream, Data streams have raised challenges to data mining for their characters. Among these challenge tasks, mining pattern form streams with intervention is specially difficult but meaningful.

Motivating example. Consider the traffic prediction on the freeway: There is a loop-sensor located on the Glendale on-ramp to the 101-North freeway in Los Angeles. The sensors are located around stadium in order to inspect the traffic stream near stadium. By investigating the variation in stream, people detect the game event occurred in the stadium as illustrated in Fig.1.

In Fig.1, with the time evolving, the traffic data are varying temporally. The data distribution in time window is following some interesting rules. The shapes of traffic distributions in $date_1$ and $date_2$ are similar. Check $date_2$ and $date_3$, some obvious difference between them, the difference will raises an alarm indicating some event happened. The event monitors will check associated media news and conclude that there is a game play during $date_3$.

---

[*] Supported by the National Science Foundation of China under Grant No. 60773169, the 11th Five Years Key Programs for Sci. &Tech. Development of China under grant No. 2006BAI05A01, and the Youth Foundation of Computer School, Sichuan University.
[**] Corresponding author.

Q. Li et al. (Eds.): APWeb/WAIM 2009, LNCS 5446, pp. 519–525, 2009.
© Springer-Verlag Berlin Heidelberg 2009

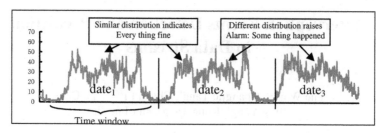

Fig. 1. The traffic observation data collected by the loop-sensor

The above principle about intervention can be applied in many fields, such as: birth defect analysis, nuclear leak detection, malfunction detection, and so on. Mining this new knowledge brings special challenges: How to detect intervention events adaptively in stream? When intervention event's status changing, how to analyze its impacts? How to apply the mining results of intervention events?

2 Related Work

Researchers proposed many (traditional) mining methods to deal with the problems in data streams [1, 2, 3, 4, 5, 6, 7, 8, 9]. The literatures [10, 11] gave frameworks based on regression analysis to deal with incidents detective and trends analysis. Due to the limitation of regression, it only analyzes changes in a local scope. Thus regression is not widely applied in mining data stream. Literature [12] applied statistical methods to mine the data stream, but it needs prior distribution judging. This limits the application objects and causes high computation cost. Literature [13] proposed a high-order model "Concept Drift" to reveal the rule behind the varying of data. However, current researches on "concept drift" usually needs prior methods to model over patterns as its concepts, and then to monitor the varying of data stream. For those researches on "concept drift" ignore the backward direction. This would bring new problems, such as: How to use these detected patterns to analysis the trends of the data? How to evaluate the change rate of these patterns, and how to predict their impact to the data streams? To tackle these problems, a series of novel strategy was proposed in this paper.

3 Problem Statements

According to the motivating example, we have the following assumption:

Stability Assumption. During the process of a data stream, if no outside intervention exits, the stream will retain a stable status; in another words, unstable status means intervention events happened.

3.1 Basic Definitions and Terminologies

To describe the concept and principle formally, the basic notations and descriptions are listed as Tab. 1.

Table 1. Symbols

Symbols	Descriptions and Examples
d	Describes the size of time window. Example: process start time is t_1, end time is t_2, then $d=t_2-t_1$.
W_t	The t-th time window, t refers to the time in this study
N_w	The number of data item in window W, or abbr. as N when w is understood.
SqureSum	SqureSum is Sum of Square between groups, it defined by: $$SqureSum(Seq1, Seq2) = \frac{1}{n}\sum_{i=1}^{k} x_i^2 . - (\sum_{i=1}^{k} x_i)^2 / kn$$

Complexion Entropy (C_E) is borrowed from thermodynamics. It describes molecules' distribution status among chambers. It is applied to analysis the distribution status of stream in t-th Window (W_t). We call the newly distribution-describe variance as Complexity Entropy (C_E). It can be calculated as:

$$C_E(t) = \log(N! / \prod_{i=1}^{k}(N_i!))$$ (E1)

Observation 1. Consider an intervention observation system. Denote observation distribution at time t as Distr (t): Then there exists threshold h such that, SquareSum $(S_{CE}(t1), S_{CE}(t_2)) <= h$ implies $Distr(t1) \approx Distr(t2)$.

Definition 1 (Sequence significant difference $P_s(t)$). Let $S = (X_1, X_2,...X_t..)$ be a temporal sequence, $S_{CE}(t)$ be sequence of C_E value in date t , and N(t) be the value of SquareSum computation between 2 neighbor data items, then $\mathbf{Inf}(t_i) = N(t+1)/N(t)$ is called inflation ratio at time t; the **Sequence significant difference** of S, denoted as $\mathbf{P_s}(t)$ (or abbr. **P (t)** when S is understood) is defined as follows:

$N(t) = SquareSum(S_{CE}(t), S_{CE}(t+1))$ (E2)

$$Inf(t) = \begin{cases} N(t)/N(t-1) & \text{if } N(t)>N(t-1); \\ N(t-1)/N(t) & \text{if } N(t)<=N(t-1); \end{cases}$$ (E3)
(inflation ratio)

Definition 2 (Intervention Event and intensity). Let P (t) be the sequence significant difference value of observation sequence O(t) (t=0,1,2,…).Suppose that intervention observing system consists of two sequences: (a) observation sequence (refer to the input data stream); (b) intervention events sequence. Assume the variation of stream is caused by the varying of intervention events sequence.

(1) (**Intervention event) Let** k>0 be a predefined threshold by experiments. **We say** intervention event e happen at t if P (t) >k ;

(2) The **Intervention Intensity** at time t of intervention event E, **denoted as I (t),** is defined as:

$I(t) = P(t) - P(t-1)$ (E5)

Definition 3. The Intervention Events Scheduler (IES) is defined as a 4-tuple, (O (t), Q, P, C, S(t)), where:

(1) O (t) = {item$_1$, item$_2$, item$_3$,…item$_t$} is its observation sequence, and t is the tick of its observed time and its value in a finite set.
(2) Q={q$_1$,q$_2$,…q$_n$} is a finite set of intervention events;
(3) M$_t$ is the transition matrix between intervention events;
(4) M$_C$ is the confusion transition matrix between intervention events and observation, confusion matrix is concept from HMM which uses to describe the relations between intervention events and observation;
(5) S(t)={event$_1$, event$_2$, event$_3$…event$_t$} is the state sequence of intervention(happened or not, binary sequence), and t is the time tick.

Observation 2. In IES model, state sequences S(t) is a Markov chain.

4 Algorithms for Intervention Events

This section proposes three algorithms to implement the **Intervention Events Scheduler** (IES) model, and all the algorithms are designed in one data scan style. The main structure of IES is listed in Fig.2 (a).

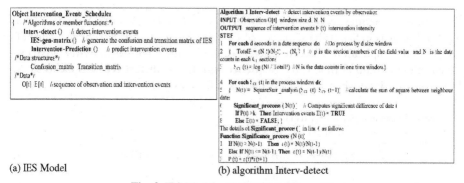

(a) IES Model   (b) algorithm Interv-detect

Fig. 2. IES Model and algorithm Interv-detect

Algorithm **Interv-detect** describes the detection step in IES. As it is listed in the Fig. 2 (b), the key idea of **Interv-detect** is as follows: Computes the sum of square of the sequence of C$_E$; Calculates the sequence of significant degree P (t); Detects the happening of intervention events by contrasting the P$_s$ (t) with threshold.

Algorithm **IES-gen-matrix** is listed in Fig. 3 (a), and it generates the parameters needed by IES model, including: states transition matrix, confusion matrix between states and observations. Algorithm **Intervention-Prediction** gives the methods to predict intervention events and the details are listed in the Fig. 3(b).

Algorithm 2 IES-gen-matrix	Algorithm 3 Intervention-Prediction
INPUT Observation sequence C[t] intervention events sequenceE[t]	**INPUT** Parameters of IES
OUTPUT IES Model with its parameters. (Q I F C s)	**OUTPUT** Intervention intensity I Intervention impact w future states
STEP	**STEP**
1 int confusion_matrix[m][n] //m intervention events status n observation	1 int states_matrix[m][t]
2 int w_counter = 0 // window counter	2 int pattern_path[t]
3 For each t do //data streams arrive with the time ticking	3 Initial_states(states_matrix[m][t])
4 If events_happened()	4 StatesMatrix_Gene(states_matrix[m][t] patterr_path[t])
5 Then Confusion_Matrix_generation(confusion_matrix[m][n])	
6 w_counter++	**Function Initial_states(states_matrix[m][t])**
7 If w_counter>d Then {w_counter =0 Break; }	1 For each i in [0 m] do
	2 states_matrix[i][0] = Initial_Probability[i]* confusion_matrix[i][0]
Function Confusion_Matrix_generation(confusion_matrix[m][n])	
1 For each i in [0 n] do	**Function StatesMatrix_Gene(states_matrix[m][t] patterr_path[t])**
2 { For each i in [0 n] do	1 int MaxProb_state = MaxProb(states_matrix[m][0])
3 If C[t] belong to [Current_Min (C[t]) + i*(Current_Span (C[t]))/n	2 For each i in [0 t] do
4 Current_Min (C[t]) + (i+1)*(Current_Span (C[t]))/n]	3 { patterr_path[i] = ArgMax(MaxProb_state confusion_matrix[m][i])
5 Then confusion_matrix[ ][i]++; }	4 MaxProb_state = Max(MaxProb_state * transition_matrix[m][i]
	* confusion_matrix[m][t])
	5 For each _ in [0 m] do
	states_matrix[ ][i]=states_matrix[ ][i-1]* transition_matrix[m][i]
	* confusion_matrix[m][t]; }

(a) algorithm interv-detect   (b) algorithm IES-gen-matrix

Fig. 3. Algorithms to detect intervention events and to generate confusion matrices

5   Experiments and Discussions

The experiments are conducted on a Linux machine with a 1.7GHZ CPU and 512 Mb main memories. The experiments are based on the data sets "Dodgers" [14] downloaded from UCI Machine Repository. It is generated by a loop sensor around the Glendale on ramp for the 101 North freeways in Los Angeles. There are totally 50,400 traffic records, and 67 events.

5.1 Experiment

Three experiments were carried out in this section: In the experiment 1, the correctness and completeness of our algorithms are analyzed. The final result of IES analysis is listed in the Fig. 4. The entire observed traffic data set (from April to October) was processed: We test different detective threshold k in order to give guideline about setting k's value: when k is 30, the detective accuracy is 65.4%, when k is 12, the detective accuracy is 92.6%, and when k is set to 11, the entire 81 events were detected correctly. The total processing time is within 800 milliseconds. The more, process efficiency is inverse correlated to the window size d.

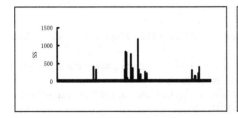

(a) Result of SquareSum analysis   (b) Result of P (t) computation

Fig. 4. Results of intervention events detection from 4/12/2005 ~ 10/1/2005

In experiment 2, half of the detected events in last experiment were used as training, and the rest were used to verify the results of prediction. The transition and confusion matrix obtained in our experiment is listed in the Table.3:

Table 2. The basic configuration of experiments

Experiment purpose	To verify the correctness of the give algorithms
C_E section number p	5
Window	10
Detective Threshold k	11, 30, 50
Time range	4/12/2005 ~ 10/1/2005

Table 3. Results of matrices generation

Transition_Matrix0:	Confusion_Matrix0:
0.33, 0.11	0.0160, 0.0299, 0.0145, 0.00243, 0
0.11, 0.45	0.392, 0.431, 0.114, 0, 0

Experiment 2 uses half of the events with their events to train algorithm, and the whole processing time costs 550 milliseconds. During the process, it updates parameters of IES model incrementally for every input observation data, and the prediction accuracy was 61.4%.

6 Conclusions and Future Work

This paper provides methods to mine intervention events in data stream. It is very useful to support intervention decision making in many fields, such as birth defect analysis, nuclear leak detection, and malfunction detection, etc. The research for mining intervention rule is just in its beginning. A lot of researches are to be done, such as, investigate the multi-status conditions about intervention events; analyze complex problems caused by combining of difference intervention events. These traits will be added to IES in the next research report; they will be especially useful in analyzing the complex data stream problems with interventions. Interested readers may refer to our full paper [15].

References

1. Gaber, M.M., Zaslavsky, A., Krishnaswamy, S.: Mining Data Stream: A Review. SIG-MOD record 34(2) (2005)
2. Teng, W.-G., Chen, M.-S., Yu, P.S.: A regression-based temporal pattern mining scheme for data streams. In: Proc. 29th VLDB Conference (2003)
3. Chen, Y., Dong, G., Han, J., Wah, B.W., Wang, J.: Multi-Dimensional Regression Analysis of Time-Series Data Streams. In: Proc.VLDB (2002)
4. Guha, S., Mishra, N., Motwani, R., O'Callaghan, L.: Clustering data streams. In: Proc. FOCS, pp. 359–366 (2000)
5. Aggarwal, C.C., Han, J., Wang, J., Yu, P.S.: A framework for clustering evolving data streams, pp. 81–92 (2003)
6. Ganti, V., Gehrke, J., Ramakrishnan, R.: Mining Data Streams under Block Evolution. SIGKDD Explorations (2002)

7. Ding, Q., Ding, Q., Perrizo, W.: Decision Tree Classification of Spatial Data Streams Using Peano Count Trees. In: Proc. ACM Symposium on Applied Computing (2002)
8. Chen, S., Wang, H., Zhou, S., Yu, P.S.: Stop Chasing Trends: Discovering High Order Models in Evolving Data. In: ICDE 2008 (2008)
9. Wang, H., Yin, J., Pei, J., Yu, P.S., Yu, J.X.: Suppressing Model Overfitting in Mining Concept-Drifting Data Streams. In: SIGKDD (2006)
10. Cai, Y.D., Clustter, D., Pape, G., Han, J., Welge, M., Auvil, L.: MAIDS: Mining Alarming Incidents from Data Streams. In: Proc. ACM SIGMOD (2004)
11. Curry, C., Grossman, R., Lockie, D., Vjcil, S., Bugajski, J.: Detecting Changes in Large Data Sets of Payment Card Data: A Case Study. In: Proc. SIGKDD (2007)
12. Ihler, A., Hutchins, J., Smyth, P.: Adaptive Event Detection with Time-Varying Poisson Process. In: Proc. ACM SIGKDD international conference (2006)
13. Wang, H., Fan, W., Yu, P.S., Han, J.: Mining concept-drifting data streams using ensemble classifiers. In: SIGKDD (2001)
14. http://archive.ics.uci.edu/ml/datasets/Dodgers+Loop+Sensor
15. Wang, Y., Tang, C., Yang, N., Chen, Y., Li, C., Tang, R., Zhu, J.: Intervention Events Detection and Prediction in Data Streams (full version of the paper),
    http://cs.scu.edu.cn/~wangyue/InterventionFullpaper.pdf

New Balanced Data Allocating and Online Migrating Algorithms in Database Cluster[*]

Weihua Gong[1], Lianghuai Yang[1], Decai Huang[1], and Lijun Chen[2]

[1] College of Information Engineering, Zhejiang University of Technology,
Hangzhou China, 310014
[2] School of Electronics Engineering and Computer Science, Peking University,
Beijing China, 100871

Abstract. An improved range data distribution method is proposed, which is suitable for both the homogeneous and heterogeneous database cluster with consideration of full use of different computing resources of nodes. In order to avoid the problem of load imbalance caused by the hot accessing, an online migrating algorithm is presented during the parallel processing. The experimental results show that the improved range partition method and the rebalancing strategy of online migrating algorithm not only significantly improve the throughput of database cluster but also keep the balanced state well. At the same time, the cluster system has achieved better scalability.

Keywords: Data distribution, Heterogeneous database cluster, Over-loaded nodes, Load balancing.

1 Introduction

The system balance and performance improvement in database cluster mainly includes two aspects: the balanced data distribution and the workload balance. Although several kinds of dynamic load balancing methods are proposed in [2,5,7], they are closely related to the data distribution approaches. Some other works have discussed the issues of data distribution algorithms in distributed environments[1,3, 9,11]. In another point of view, the heterogeneity is unavoidable in database cluster. Different nodes have different computing capabilities due to a variety of CPU calculation power, memory capacity and disk IO bandwidth etc. Some works have focused on the parallel processing in shared memory DBMS[5], and some mainly considered the I/O parallelism in parallel disk systems[4,10] or the I/O load balancing in heterogeneous file system by Jose[8]. Others have only taken into account the calculation power of each CPU in the heterogeneous cluster[6,7]. Only when all the computing resources of each node are fully used, the cluster system can obtain highly parallel processing performance.

---

[*] Supported by the National High-Tech Research and Development Plan of China under Grant(863) No.2007AA01Z153, also Supported by the Natural Science Foundation of Zhejiang Province No.Y1080102 and the School Scientific Research Founds of ZJUT No.X1038109.

Q. Li et al. (Eds.): APWeb/WAIM 2009, LNCS 5446, pp. 526–531, 2009.

In this paper, we mainly discuss an improved data distribution method based on the widely-used Range partition algorithm, which means distributing the data non-uniformly or asymmetrically among the nodes in cluster according to the different computing capabilities of homogeneous or heterogeneous nodes. For the problem of the hot data accessing in future time, we further design a specific on-line algorithm through migrating the hot data so as to share the workloads with the under-loaded nodes, which can dynamically re-balance both of the distribution of the data and the workloads in the cluster system.

The rest of this paper is organized as follows. Section 2 proposes an improved data distribution algorithm. Section 3 introduces how the online migrating algorithm rebalances the database cluster system because of the hot data accessing. Section 4 provides system architecture and analyzes the experimental results. Section 5 offers concluding remarks.

2   An Improved Data Distribution Algorithm

When initializing the data allocation in the database cluster, we assume that the whole data in the cluster system are called global data and all the global data have the same accessing probability. The global data are allocated in hybrid mode and classified into two types: replicated data and partitioned data. We will mainly focus on developing an improved data distribution method about the global partition data in this paper.

For the partitioned data, we propose an Improved-Range (IR) partition algorithm which is not only fit for the symmetric distribution in homogeneous cluster but also fit for the asymmetric distribution in heterogeneous cluster. The IR partition scheme is a weighted Range partition method according to the different comprehensive compute capability of heterogeneous nodes. That means $D^P = \sum_{i=1}^{n} D_i$, where $D_1 \cap D_2 \cdots \cap D_n = \Phi$. The range of every interval D_i is closely related to the compound weighted compute value of different nodes, and the cardinality of D_i can be evaluated with $|D_i| = \alpha_i \times |D^P|$, and $\alpha_i = \dfrac{W_i}{W_T}$, where the numerator W_i is the compound weighted compute value of single node i, while the denominator W_T is the sum of the weighted compute value of all nodes in database cluster, that is

$$W_T = \sum_{i=1}^{n} W_i \tag{1}$$

Since the compound compute capability in a single node is mainly associated with three types of computing resources mentioned above, the weighted compute value of some node is given by

$$W_i = e_1 * W_i^C + e_2 * W_i^R + e_3 * W_i^I \tag{2}$$

Where W_i^C, W_i^R and W_i^I denote the CPU speed, available memory and IO access rate weights of the ith node taken up in the cluster respectively, and the formula must satisfy the constraint that is $\sum_{j=1}^{3} e_j = 1$, which combines three different types of re-

sources into a uniform evaluation value.

3 The Process of Dynamic Workload Balancing

Although the IR data distribution scheme maybe works well in both homogeneous and heterogeneous cluster under the assumption of the same accessing probability, the data accessing rate of different nodes often changes indefinitely as increasing of the workloads in the real applications. After longer running time, the cluster system may incur some over-loaded spots because of the hot data accessing. In order to achieve load balancing in database cluster, we can use a bivariate function to distinguish the states of over-loaded and under-loaded nodes with formula (3), the two variables ti and xi denote the response time from some node and the number of transactions dispatched to the node respectively.

$$f(t_i, x_i) = \sqrt{\left(1 - \frac{t_i}{\bar{t}}\right)^2 + \left(1 - \frac{x_i}{\bar{x}}\right)^2} \tag{3}$$

To solve the problem of the data skew, we propose a dynamic on-line data migrating algorithm. We consider the over-loaded node as the source node (S) and the node set that will receive the migrated hot data are looked as the destination nodes (D). Before the migrating procedure, the cluster system should timely get both of the source node S come from the $max\{ f(t_i, x_i) | f(t_i, x_i) > \sqrt{2}, 1 \le i \le n\}$, and the destination set D got from the set $\{ \min_{j=1}^{m} \{ f(t_i, x_i) \} | 0 \le t_i \le 2\bar{t}, 0 \le x_i \le 2\bar{x}, 1 \le i \le n\}$. Obviously, all the status functions in the destination set D should satisfy the constraint $f(t_i, x_i) \le \sqrt{2}$ if there exist m minimal nodes. The online hot data migrating procedure is just as illustrated below:

Procedure On-line Migrate(S, D)

Step 1.   statistic the hot data D_{hot} and partition $\theta_i = D_{hot} /(m+1)$, $(1 < i \le m)$;

Step 2.   lock the hot data of the source node, queue the transactions which need to access the migrating data;

Step 3.   replicate every part of partitioned hot data to corresponding Di which receive the amount of θ_i hot data;

Step 4. log the changes of distribution into meta data of scheduler;

Step 5. delete the hot data of S which have been migrated;

Step 6. unlock the hot data and take up the transactions from the queue to execute;

Step 7. check if the system is balanced, then exit the procedure;

Step 8. else loop to step 1.

4 Experiments and Performance Analysis

In the database cluster, we develop a middleware to apply the balance methods of data and workloads distribution described in section 2 and section 3. The architecture of database cluster system shows in figure 1. To simplify the measurement of the compound weighted value of the nodes, we only simulate the type of OLTP requests. As a result, we can assume the vector $e(0.3,0.2,0.5)$, each W_i will have the same weight for different resources in some application. We can focus on the data distribution and the performance of the cluster system.

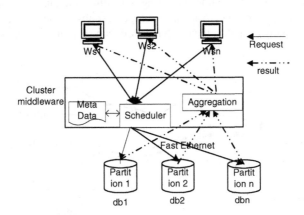

Fig. 1. Cluster middleware architecture

The ideal geometric increasing of throughput as increasing of nodes has been shown in Fig 2, which is an ideal case marked with dotted line without considering any extra cost such as communication cost. Whereas, in the actual test the throughput goes up almost linearly along the trends of dotted line when applied our improved-range partition method in the database cluster, which has pictured with solid square line. But for the traditional range partition method, the max transaction throughput of it grows slowly with increasing of nodes and the cluster will quickly achieve the saturation point because of hot access. Even though more nodes are added to the database cluster, the scale up of the database cluster is still finite.

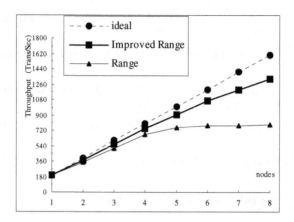

Fig. 2. Transaction throughput with Improved Range(IR) partition method

In the graph 3, the balanced curve denotes the middleware system applied with our online migrating algorithm and the unbalanced curve is on the contrary. In the way of range distribution, the cluster system gets to unbalanced workloads in shorter time and the average response time of transaction processing increases slowly since that more and more transactions are blocked in waiting queue. When applying the IR data distribution method and the online migrating algorithm, the cluster system can avoid hot accessing for longer time. And the average response time of transaction is decreasing as increasing of nodes, the cluster system can scale up to more nodes.

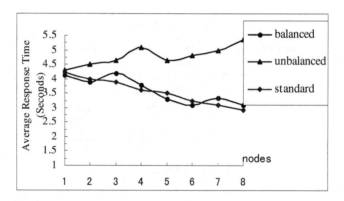

Fig. 3. Average response time

5   Conclusion

The performance improvement of database cluster is mainly related to the data distribution strategy and the workload allocation scheme. In this paper, we have proposed an Improved-Range (IR) data distribution method, which is not only suitable for the uniform data distribution in the homogeneous cluster, but also suitable for the non-uniform data distribution in the heterogeneous cluster. For the problem of load imbalance caused by the data skew during the parallel processing, we present an online migrating algorithm to move the hot data to the under-loaded nodes so as to make full use of its computing resources.

Our experimental results show that the improved range partition method does keep the balanced state well. At the same time, the throughput of the database cluster is significantly improved compared to the traditional range partition method. And the rebalancing strategy of online migrating algorithm can improve the performance of parallel processing and decrease the average response time of transaction as increasing the nodes. As a result, the cluster system achieves better scalability.

References

[1] Nguyen, K.Q., Thompson, T., Bryan, G.: An enhanced hybrid range partitioning strategy for parallel database systems. In: Proceedings of the Eighth International Workshop on Database and Expert System Applications, pp. 289–294. IEEE Computer Society, Los Alamitos (1997)

[2] Hababeh, I.O., Ramachandran, M., Bowring, N.: A high-performance computing method for data allocation in distributed database systems. The Journal of Supercomputing 39, 3–18 (2007)

[3] Wang, J., Tsutaya, Y., Segawa, N., et al.: Approaches to balancing data load of shared-nothing clusters and their performance comparison. In: Proceedings of the 9th International Conference on Parallel and Distributed Systems, pp. 293–299. IEEE Computer Society, Los Alamitos (2002)

[4] Perez, J.M., Garcia, F., Carretero, J., Calderon, A., Sanchez, L.M.: Data allocation and load balancing for heterogeneous cluster storage systems. In: Proceedings of the 3rd IEEE/ACM International Sympo-sium on Cluster Computing and the Grid (CCGRID 2003), pp. 718–723 (2003)

[5] Hirano, Y., Satoh, T., Inoue, U., Teranaka, K.: Load balancing algorithms for parallel database processing on shared memory multiprocessors. In: Proceedings of 1st Parallel and Distributed Information Systems, pp. 210–217 (1991)

[6] De Giusti, A.E., Naiouf, M.R., De Giusti, L.C., Chichizola, F.: Dynamic load balancing in parallel processing on non-homogeneous clusters. Journal of Computer Science and Technology 5(4), 272–278 (2005)

[7] Rahm, E., Marek, R.: Analysis of dynamic load balancing strategies for parallel shared nothing database systems. In: Proceedings of 19th Conference on VLDB, pp. 182–193 (1993)

[8] Scheuermann, P., Weikum, G., Zabback, P.: Data partitioning and load balancing in parallel disk systems. The VLDB Journal (7), 48–66 (1998)

[9] Dewitt, D., Gray, J.: Parallel database system: the future of high performance database systems. Communication of ACM 33(6) (1992)

[10] Beynon, M.D., Kurc, T., Catalyurek, U., Chang, C., Sussman, A., Saltz, J.: Distributed processing of very large datasets with DataCutter. Parallel Computing 27(11), 1457–1478 (2001)

[11] Zhu, F., Sun, X., Salzberg, B., Hvasshovd, S.-O.: Supporting load balancing and efficient reorganization during system scaling. In: Proceedings of the 19th IEEE International Parallel and Distributed Processing Symposium(IPDPS 2005) (April 2005)

An Evaluation of Real-Time Transaction Services in Web Services E-Business Systems

Hong-Ren Chen

Department of Digital Content and Technology, National Taichung University,
Taichung 403, Taiwan

Abstract. The use of services-oriented computing by enterprises has increased noticeably, with the aim of speeding up application processes and in response to changing e-business needs. This paper addresses the efficient processing of e-business real-time transaction services using the emerging web services technology and considers a services-oriented computing environment in which multiple service providers provide identical real-time transaction services. An efficient mechanism called the sensorial value-planned mechanism for web services (SVM-WS) is proposed for differentially treating e-business real-time transaction services according to their importance, such as inquiry and purchase requests. The results of a performance evaluation of an experimental setup for a web services e-business system demonstrate the feasibility of the scheme.

1 Introduction

Web services technology provides a new computing model for greatly accelerating application processing and responding to changing business needs within and across enterprises. Web services transactions are increasingly becoming key additions to business processes that access multiple web services, which need to combine various web services transactions. As fundamental components of e-commerce and enterprise application integration as well as on-demand computing, web services are expected to greatly enhance the promise of distributed computing [1, 2]. The demand for real-time transaction services has been increasing conspicuously in many web services e-business applications. The current trend is for multiple service providers with common interfaces to provide the same web services over the Internet [3, 4]. This implies that web services can be dynamically determined and executed, which would make their execution more flexible and robust. The Internet purchasing example illustrated in Fig. 1 involves the following three web services: (1) credit validation, (2) inventory management, and (3) customer accounting [5]. Each of these web services must be executed in real-time. A web service transaction is executed by the web services transaction model (WSTM) selecting an available web services and interpreting the script that defines the process. The services aggregator is responsible for managing various types of web services and communicating with the universal web services registry (UDDI) to update the service directory, which is a central registry of service providers whose methods and interfaces are listed in this registry.

In this paper, we consider a services-oriented computing environment in which identical real-time transaction services are provided by multiple service providers, one

Q. Li et al. (Eds.): APWeb/WAIM 2009, LNCS 5446, pp. 532–537, 2009.
© Springer-Verlag Berlin Heidelberg 2009

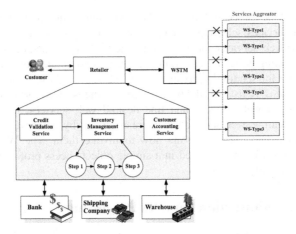

Fig. 1. The business process of the web services Internet purchasing [1, 6, 7]

of which can be chosen during the run-time phase to execute transaction services on the basis of minimizing the execution cost. A novel value-based scheduling strategy called sensorial value-planned mechanism for web services (SVM-WS) is proposed here for differentially treating e-business real-time transaction services according to their importance, such as inquiry and purchase requests. The essential objective of our approach is to maximize the profit by preferentially serving high-priority services based on the principles of (1) cost-effectiveness, by exploring adaptive web services from different service providers to ensure communication quality and reduce the dependency over remote transaction services, (2) differential profit treatment, by preferentially executing the purchase requests to be completed while trying to execute as many requests for inquiry services as possible, and (3) variable value setting involving the dynamic determination of mean values based on the system load for each inquiry and purchase transaction services in real-time to satisfy diverse data needs. Simulation results demonstrate that the SVM-WS can deliver good performance in a web services e-business system.

The remainder of this paper is organized as follows: Section 2 details the web services e-business system, Section 3 presents real-time transaction services processing and the SVM-WS method. Section 4 describes the simulation model and performance results, and conclusions are drawn in Section 5.

2 Web Services E-Business System

In a web services computing environment, a communication network interconnects multiple sites. Each site contains a transaction generator, a web services selection manager, a communication interface, a scheduling policy, a cache manager, and a resource manager. The transaction generator is responsible for generating the workload for each site. The arrival of data and/or messages at a site is assumed to be independent of the arrivals at other sites. Real-time web services transactions are classified into inquiries and purchases, where the former involves only read operations on requested data items. To effectively execute a web services transaction, the actions of web services needed by transactions must be processed through WSTM. While interpreting the transaction script, the web services selection manager selects the appropriate web services from among multiple web services for execution. The real-time scheduler is based on the transaction scheduling policy described in Section 3. A communication interface at each site is responsible for sending/receiving data or messages to/from other sites. The universal web services registry (UDDI) is used by the service provider and service requester to publish web services and find desired web services, respectively. Once applications are built as web services, they are described by a web services description language (WSDL) that tells requesters how to invoke them. The WSDL and XML metadata of the services are wrapped in a simple object access protocol (SOAP) message that can be sent to the UDDI.

3   Real-Time Transaction Services Processing

The design of the scheduling strategy for real-time transaction services processing is driven by its real-time constraints. Generally, such e-business systems will execute the transaction assigned with the highest execution priority based on a particular formula. I briefly describe the following transaction services scheduling strategy using the notations as listed below.

T_i	a real-time transaction service in the system
$C_{es}(T_i)$	estimated communication delay of transaction service T_i
$C_{ep}(T_i)$	elapsed communication delay of transaction service T_i
$C_{rr}(T_i)$	remaining communication delay of transaction service T_i, obtained from $C_{es}(T_i)$-$C_{ep}(T_i)$
$E_{es}(T_i)$	estimated execution time of transaction service T_i
$E_{ep}(T_i)$	elapsed execution time of transaction service T_i
$E_{rr}(T_i)$	remaining execution time of transaction service T_i, obtained from $E_{es}(T_i)$ - $E_{ep}(T_i)$
$P(T_i)$	priority of transaction service T_i
$S(T_i)$	slack time of transaction service T_i
$V(T_i)$	mean value of transaction service T_i
W	weight of adjusted value and urgency of transaction service.

3.1   Earliest Deadline

The earliest deadline strategy (ED) assigns high priorities to transaction services with early deadlines [7]. All (sub)transaction services have the same deadlines. The priority assignment formula is given by:

$$P(T_i) \leftarrow \frac{1}{D(T_i)}$$

3.2   Highest Value

The disadvantage of ED is that it does not consider the values of transactions. Assigning high priorities to transactions with high values is called Highest Value (HV) [8].

As ED, all (sub)transaction services have the same value. The priority assignment formula is given by:

P(Ti) ← V(Ti)

3.3 Highest Reward and Urgency

In contrast to ED, HV focuses on completing transaction services with high values. However, the urgency of transaction service is not considered. To eliminate the disadvantage, Highest Reward and Urgency (HRU) considers both the deadline and value as design factors. It gives a high priority to a transaction service with high value and shortest remaining execution time, and the priority assignment formula is given by:

$$P(T_i) \leftarrow \frac{V(T_i)}{E_{rt}(T_i)} - S(T_i) * W$$

HRU considers the reward ratio of scheduling a transaction service and provides an adjustable policy for various system load conditions [9, 10].

3.4 Sensorial Value-Planned Mechanism for Web Services

Scheduling strategies for web services computing environments that consider a differentiated real-time data services for e-business applications have not been discussed previously. The enormous popularity of using the Internet has increased the demand for real-time transaction services in many web services e-business applications, such as searching for airfares and buying flight tickets. The proposed SVM-WS method as shown in Fig. 2 can differentially treat e-business real-time transaction services based on their class, such as browsing, searching, and payment requests.

Algorithm SVM-WS Mechanism

Input: a real-time transaction service T_i, $\forall T_i \in \{T_i,..., T_n\}$
Output: the priority of T_i

Begin

$Exec\_Cost_{ws}(T_i) \leftarrow$ calculated by the selection strategy for multiple web services providers

$$P(T_i) = \frac{Factor_{dynamic} \times V(T_i)}{E_{rt}(T_i)} - \frac{S(T_i)}{Exec\_Cost_{ws}(T_i)} \text{ , for } \begin{cases} Factor_{dynamic} < 1 \text{ if } T_i \in \text{ inquiry requests} \\ Factor_{dynamic} \geq 1 \text{ if } T_i \in \text{ purchase requests} \end{cases}$$

End-Begin

Fig. 2. Pseudo code of the SVM-WS method

4   Simulation Model and Performance Evaluation

This section describes the simulation model expanded from [8-11] that was developed using the popular open-source discrete-event SimJava package to evaluate the performance of the proposed method [12]. Most of workload parameters are similar to these values used in previous studies [7-9, 11]. We compared the transaction scheduling policies of the ED, HV, HRU and SVM-WS under various conditions. In this

experiment of increasing load, we varied the arrival rate from 20 real-time web services transaction services/second (abbreviated as real-time services/sec) to 120 ser-
vices/sec in increasing steps of 20 in order to model different system loads. As shown
in Figs. 3a and 3b, the performance order based on the *MissRatio* and *LossRatio* metrics is SVM-WS > HRU > HV > ED regardless of a normal load or a high load. The
excellent performance of the SVM-WS results from the following reasons. The cost-
effective scheme is employed in the SVM-WS to satisfy the execution quality of service and concern considering the minimal execution cost for web services since such
a remote web service requires expensive cost for accessing data objects. However, the
value-based priority assignment strategy also depends on the value of transaction
service. The highest utilization of different profit treatment will preferentially give the
high priority to purchase services. Hence, an inquiry request will have a better chance
to be executed completely under the adjustment of the reflecting value of transaction
service determined by the various value-setting scheme based on the different system
load. In addition, I also observed narrowly the impact of the increasing arrival rate of
real-time services for inquiry and purchase requests.

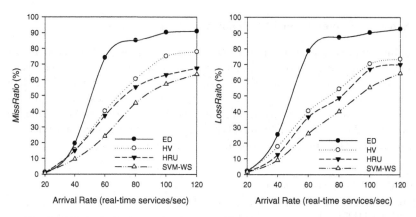

Fig. 3a. *MissRatio* for increasing load   **Fig. 3b.** *LossRatio* for increasing load

5   Conclusions

Web services technology is becoming an important topic for building distributed
applications based on open data communication and standard data formats. Web services transaction services are increasingly becoming key additions to business processes that access multiple web services, which need to combine various web services
transaction services. This paper has addressed the efficient processing of e-business
real-time transaction services using the emerging web services technology. Our
approach exploits the current capability of the services-oriented computing environment where the same real-time transaction services are provided by multiple service
providers, with the proposed SWM-WS mechanism differentially treating e-business
real-time transaction services according to their importance, such as inquiry and purchase requests.

Acknowledgement

This research was partially supported by National Taichung University and the National Science Council in Taiwan through Grant NSC96-2221-E-007-143.

References

[1] Kerger, H.: Web services conceptual architecture. IBM Software Group (2001)

[2] Kapsalis, V., Charassis, K., Georgoudakis, M., Nikoloutsos, E., Papadopoulos, G.: A SOAP-based system for the provision of e-services. Computer standards & Interfaces 26, 527–541 (2004)

[3] Ibach, P., Tamm, G., Matthias, H.: Dynamic Value Webs in Mobile Environments Using Adaptive Location-Based Services. In: IEEE Proceedings of International Conference on System Sciences, pp. 1–9 (2005)

[4] Chafle, G., Dasgupta, K., Kumar, A., Mittal, S., Srivastava, B.: Adaptation in Web Service Composition and Execution. In: IEEE international Conference on Web Services, pp. 549–557 (2006)

[5] Cram, C.M.: E-commerce concepts: illustrated introductory. Course Technology, Boston (2001)

[6] Chen, H.-R.: An evaluation of concurrency control protocols for web services oriented E-commerce. In: Li Lee, M., Tan, K.-L., Wuwongse, V. (eds.) DASFAA 2006. LNCS, vol. 3882, pp. 530–540. Springer, Heidelberg (2006)

[7] Abbott, R., Garcia-Molina, H.: Scheduling real-time transactions: a performance evaluation. ACM Trans. Data. Sys. 17, 513–560 (1992)

[8] Haritsa, J.R., Carey, M.J., Livny, M.: Value-based scheduling in real-time database systems. VLDB J. 2(2), 117–152 (1993)

[9] Tseng, S.M.: Design and analysis of value-base scheduling policies for real-time database systems. PhD. Thesis, National Chiao Tung University, Taiwan (1997)

[10] Chen, H.R., Chin, Y.H.: An adaptive scheduler for distributed real-time database systems. Info. Scie.: an Intl. J. 153, 55–83 (2003)

[11] Agrawal, R., Carey, M.J., Livny, M.: Concurrency control performance modeling: alternative and implications. ACM Trans. Data. Sys. 12(4), 609–654 (1987)

[12] Fishwick, P.A.: SimJava: Java-Based Simulation Tool Package. University of Florida (1992)

Formal Definition and Detection Algorithm for Passive Event in RFID Middleware

Wei Ye[1,2,3], Wen Zhao[2,3], Yu Huang[2,3], Wenhui Hu[2,3],
Shikun Zhang[2,3], and Lifu Wang[2,3]

[1] School of Electronics Engineering and Computer Science,
Peking University, Beijing 100871, China
[2] National Engineering Research Center for Software Engineering,
Peking University, Beijing 100871, China
[3] Key Laboratory of High Confidence Software Technologies, Ministry of Education,
Beijing 100871, China
cactus.ye@pku.edu.cn, owen@sei.pku.edu.cn

Abstract. In RFID middleware, passive event refers to one kind of composite event, some of whose constituent sub-events do not occur under certain condition. To process complex RFID business logic, we should be able to define a wide variety of passive event types and perform efficient event detection. In this paper, we design an event definition language which well supports specifying complex hierarchical passive event, and propose a detection algorithm with some optimization techniques for recognizing passive event. We finally compare our work with Esper in detection performance.

1 Introduction

RFID system consists of four components: tag, interrogator, middleware and application. RFID middleware, which is also named RFID edge server, is a fundamental infrastructure which facilitates data and information communication between automatic identification physical layer and enterprise applications. In RFID middleware, we can devise RFID business logic into hierarchical composite event and automate business processing as event detection based on CEP (Complex Event Processing) [1] technology.

In complex event processing, *passive event* refers to one kind of composite event, some of whose constituent sub-events do not occur under certain condition. In contrast to passive event, we define *active event* as another kind of event which can be detected just by waiting for notice of upcoming event. Primitive event is the simplest active event. To process complex RFID business logic, in addition to active event types, we must also be able to define a wide variety of passive event types with rigorous semantics. Meanwhile, in passive event detection process, RFID middleware has to perform efficient queries on event's absence in runtime context besides waiting for notice of primitive event occurrence. Though most work on event processing proposes *negation operator* to address passive event issues [3,4,5,6,7,8,9], we argue that it needs more in-depth exploration in RFID event processing research.

Q. Li et al. (Eds.): APWeb/WAIM 2009, LNCS 5446, pp. 538–543, 2009.

Our contribution is an expressive and rigorous event definition language which especially well supports specifying passive event and an efficient event detection algorithm based on Petri net structure with a suit of optimization technology for recognizing passive event. Our work is based on a prototype of RFID middleware PKU RFID Edge Server (PRES), a pure Java middleware which is compatible with ALE [2]. The remainder of this paper is organized as follows. We first present an overview of RFID event in section 2. Section 3 and 4 is a formal presentation of RFID event definition language and event detection algorithm respectively. Experiment and discussion is demonstrated in section 5, followed by a conclusion in section 6.

2   RFID Event Preliminaries

RFID Event is the specification of significant occurrence in or around RFID middleware. There are two kinds of RFID event: *primitive event* and *composite event*. When a tag passes through the range of transmission, or through the gateway of a RFID reader, it generates a series of "tag is seen" events. These events are instantaneous and atomic, which form main primitive event stream flowing into RFID middleware. Another kind of primitive event is messages sent by extern business application while executing a particular step of business process. Composite event types are aggregated event types that are created by combining other primitive or composite event types using a specific set of event operators such as disjunction, conjunction, sequence, etc [10]. In this paper, we will use capital letters like E and A to represent (composite or primitive) *event type*, while use corresponding lowercase letters e and a to represent *event instance*. The event time of event instance e is a time interval represented by $[e \uparrow, e \downarrow]$. In a clear context, we will say *event* to represent *event type*. We use subscript to distinguish different instances of the same event type like e_1 and e_2 and use expression like $E(t, t')$ to represent an instance of E occurs over the interval $[t, t']$. For simplicity, we assume event time is discrete.

3   Event Definition Language

Most event definition language will propose a set of event operators like *conjunction, disjunction* and *sequence*. However, there is still no standard language exists for describing all composite events possible to detect. Starting from common composite events abstract from business logic of RFID applications like monitoring and supply-chain management, we propose an event definition language for PRES, which aims to provide a well support for passive event. Since we mainly focus on the definition and detection of passive event, we divide event operators into *active event operator* and *passive event operator*, instead of dividing them into *temporal event operator* and *logic event operator*. Note that complex passive event type definition may be embedded with both active event operator and active event operator.

Though our concern is passive event, we present four active event operators for the sake of completeness of event definition language. Meanwhile, passive event can be defined by combining primitive event using both passive event operator and active event operator. Formal semantics of active event operators is defined as following:

- $E = A\&B : E(t,t') \Leftrightarrow \exists t_1, t_2, t_1', t_2'(A(t_1, t_2) \wedge B(t_1', t_2') \wedge t = min(t_1, t_1') \wedge t' = max(t_2, t_2'))$
- $E = A|B : E(t,t') \Leftrightarrow A(t,t') \vee B(t,t')$
- $E = A \rightarrow B : E(t,t') \Leftrightarrow \exists t_1, t_1'(A(t,t_1) \wedge B(t_1', t') \wedge t_1 < t')$
- $E = w(A,T) : E(t,t') \Leftrightarrow t' - t \leq T \wedge A(t,t')$

Passive event operator is used to express absence of some event under some certain conditions. No matter defining negation operator as binary or ternary operator, the essence is imposing stronger constraints on event absence. From the perspective of event detection, these constraints actually tell event detection engine when to check the absence of the corresponding event. Thus, we need a more flexible mechanism to specify constraint on event absence. We propose six passive event operators for RFID application.

We assume there is a start timestamp t_0, and define $Phi(T)$ as a sequence of timestamp, where $Phi(T, i) = t_0 + T \bullet i$. Using $\partial(A, t, t')$ to denote that $\forall t_1, t_1' : t \leq t_1' \leq t' \Rightarrow \neg A(t_1, t_1')$, we present formal semantics of passive event operators as following:

- $E = A\ B : E(t,t') \Leftrightarrow \partial(A, t, t') \wedge B(t, t')$
- $E = before(A, B, T) : E(t,t') \Leftrightarrow \exists t_1(t - t_1 = T \wedge \partial(A, t, t_1) \wedge B(t_1, t'))$
- $E = after(A, B, T) : E(t,t') \Leftrightarrow \exists t_1(t' - t_1 = T \wedge \partial(A, t1, t') \wedge B(t, t1))$
- $E = between(A, B, C) : E(t,t') \Leftrightarrow \exists t_1, t_2(t_2 - t_1 > 0 \wedge \partial(A, t_1, t_2) \wedge B(t, t_1) \wedge C(t_2, t'))$
- $E = between(A, T_1, T_2) : E(t,t') \Leftrightarrow \partial(A, t, t') \wedge t = T_1 \wedge t' = T_2$
- $E = between(A, T) : E(t,t') \Leftrightarrow \exists i > 0(t = \Phi(T, i-1) \wedge t' = \Phi(T, i) \wedge \partial(A, t, t'))$

Take $E = between(A, B, C)$ as an example, given B occurs in the time interval $[t, t_1]$ and A occurs in the time interval $[t_2, t']$, if $t_1 < t_2$ and there is no A instances whose end timestamp fall into the time interval $[t_1, t_2]$, we say E occurs and the time interval of this event instance is $[t, t']$.

4   Event Detection

Since it is rather natural to represent complex event using Petri net because of its hierarchy and concurrency properties, we propose a Petri net structure-based event detection algorithm. It is straightforward to use *place* to represent event type, use *transition* to represent event operator and use *token* to represent event instance. However, using P/T system or colored Petri net [11] to model complex passive event will make the net structure quite complicated. Thus, instead of designing a new net system extended from P/T system or colored Petri net, we

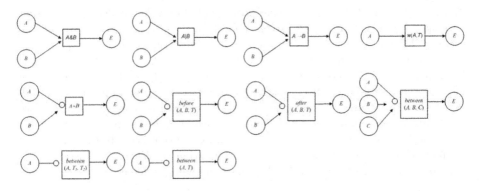

Fig. 1. Petri net-like structures for each event operator

propose a set of Petri net-like structures based on ten event operators, which are illustrated as in figure 1.

Structures shown in figure 1 are basic components of more complicated structure for hierarchical composite event and we call them *basic net*. Note that in basic net, we use *inhibit arc* to connect the event (place) which needs to check absence. We call this kind of event *negated event*. For example, in composite event $E = between(A, B, C)$, A is a negated event. For each composite event, we transform event definition to an abstract grammar tree, which is finally transformed to a Petri net-like structure. If a place has no pre transitions, we call it *input place*; and if a place has no post transitions, we call it *output place*.

In a net structure, input place represents primitive event, output place represents composite event. Each kind of transition is associated with a special firing rule, according which composite (or sub-composite) event instances is generated. So based on Petri net-like structure, to describe event detection algorithm, we just need to describe how each kind of transition play the token game.

The detection process is like this: input places wait for notification of primitive event instances, and transitions play token games according to corresponding firing rule; once a token is generated in output place, a composite event instance is detected. So, the whole detection algorithm is actually divided into ten kinds of firing rules associated with event operators. Firing rules includes not only condition checking on tokens but also query plans. Each inhibit arc associated with passive event operator requires a query plan to check absence of the corresponding event.

To optimize query plan, we introduce the following two query functions, where N represents non-negative integer.

♦ $\alpha : PE \rightarrow \{N, -1\}$. $\alpha(A)$ retrieves the end timestamp of latest A instance. If there is no A instance, $\alpha(A)$ returns -1;
♦ $\beta : PE \times N \rightarrow N$. $\beta(A, t)$ retrieves the number of A instances before a timestamp t.

While playing the token game, each place will hold a token collection represent corresponding event instances. Token collection usually can not be shared because of different effect of consumption policy. Since tokens in negated place will not be consumed while transition firing, we can optimize detection efficiency by sharing token collection of the same negated place type. Token collection of negated place is used to check absence of event, so there is no need to keep all negated event instances. Among the six basic net of passive event operators, we need only keep the latest negated event instance in detection process except operator $before(A, B, T)$ given that B is a composite event type. In that situation, we need to keep all A instances for query function β. When a B instance b is detected and $\beta(A, b \uparrow) = (A, b \uparrow +T)$, one E instance is detected.

5   Experiment

To evaluate our work, we make a comparison with Esper [8] which is an open source component for CEP and ESP (Event Stream Processing) applications. Note the comparison is only about CEP part of Esper. Equivalence of semantics of event type is a foundation of performance comparison, since the number of detected event instance may vary enormously for two similar but not equivalent event types. However, defining equivalent active event types is much easier, so in this paper we make a performance comparison with Esper based on active event types. For passive event types, we checked event processing time versus the number of primitive events. The hardware environment for performance evaluation is Intel Pentium IV CPU at 2.4GHz and 1GB RAM .The operation system is Windows XP Professional with JDK 1.6.0_05. We generate primitive event at a rate of 1000 per seconds, and process 20 event types in parallel each time. Experiment result is illustrated in figure 2 and figure 3. Experiment result shows that our algorithm is efficient for both active event type and passive event type. A more matured comparison strategy for passive event is one of our future works.

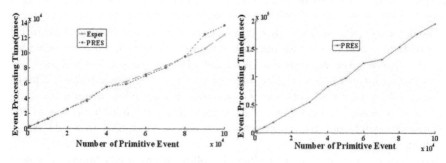

Fig. 2. Active event processing time   **Fig. 3.** Passive event processing time

6   Conclusion

Passive event is one common kind of event in RFID middleware, while its definition faces some logic difficulties and expressiveness problems. Aiming at expressing a wide variety of RFID business logic with rigourous semantics, we design an event definition language embedded with a set of passive event operators and active event operators, of which semantics is specified formally. Our language especially well supports specifying complex hierarchical passive event. We propose a Petri net structure-based event detection algorithm with efficient query plan organization for passive event. Experiment shows that detection algorithm is efficient and scalable. Technology presented in this research has integrated into our Java-based RFID middleware.

References

1. Luckham, D.C., Frasca, B.: Complex Event Processing in distributed system. Stanford University Technical Report CSL-TR-98-754 (March 1998)
2. EPCglobal Application Level Events (ALE) Specification Version 1.0, Technical Report, EPC Global Inc. (September 2005)
3. Wu, E., Diao, Y., Rizvi, S.: High-Performance Complex Event Processing over Streams. In: Proceeding of SIGMOD 2006, pp. 407–418 (2006)
4. Wang, F., Liu, S., Liu, P., Bai, Y.: Bridge physical and virtual worlds: complex event processing for RFID data streams. In: Ioannidis, Y., Scholl, M.H., Schmidt, J.W., Matthes, F., Hatzopoulos, M., Bhm, K., Kemper, A., Grust, T., Böhm, C. (eds.) EDBT 2006. LNCS, vol. 3896, pp. 588-607. Springer, Heidelberg (2006)
5. Gatziu, S., Dittrich, K.R.: Detecting Composite Events in Active Databases Using Petri Nets. In: Workshop on Research Issues in Data Engineering: Active Database Systems (1994)
6. Motakis, I., Zaniolo, C.: Formal Semantics for Composite Temporal Events in Active Database Rules. Journal of Systems Integration 7, 291–325 (1997)
7. Chakravarthy, Mishra, D.: Snoop: an Expressive Event Specification Language for Active Databases. Data Knowl. Eng. 14(1), 1–26 (1994)
8. Esper Reference Documentation, Technical Report (2007),
   http://esper.codehaus.org/esper-1.0.0/doc/reference/en/pdf/
   esper_reference.pdf
9. Carlson, J., Lisper, B.: An Event Detection Algebra for Reactive Systems. In: EMSOFT 2004, September 27–29 (2004)
10. David Luckham: Event Processing Glossary (2007),
    http://complexevents.com/?p=195
11. Chongyi, Y.: Principals and Application of Petri Nets. Publishing House of Electronics Industry, Beijing (2005)

Distributed Benchmarking of Relational Database Systems

Murilo R. de Lima,[1] Marcos S. Sunyé[1], Eduardo C. de Almeida[2],
and Alexandre I. Direne[1]

[1] Federal University of Paraná, C3SL,
Curitiba, Brazil
{mrlima,sunye,alexd}@c3sl.ufpr.br
http://www.inf.ufpr.br
[2] Université de Nantes, LINA,
Nantes, France
eduardo.almeida@univ-nantes.fr
http://www.lina.univ-nantes.fr

Abstract. Today's DBMS are constantly upon large-scale workloads
(e.g., internet) and require a reliable tool to benchmark them upon a
similar workload. Usually, benchmarking tools simulate a multi-user
workload within a single machine. However, this avoids large-scale
benchmarking and also introduces deviations in the result. In this
paper, we present a solution for benchmarking DBMS in a fully
distributed manner. The solution is based on the TPC-C specification and aims to simulate large-scale workloads. We validate our
solution through implementation and experimentation on three open-
source DBMS. Through experimentation, we analyze the behavior of
all systems and show how our solution is able to scale up the benchmark.

Keywords: Database, Benchmarking, Distributed System.

1 Introduction

Performance analysis and workload measurement appear as important issues in
Computer Science, especially in database technology. Along the years researchers
have been proposing different ways to benchmark Database Manager Systems
(DBMS). The TPC-C appears as the most popular specification for benchmarking transactional DBMS.

Today's transactional DBMS are constantly upon large-scale workloads (e.g.,
internet) and require a reliable tool to benchmark them upon a similar workload.
Usually, benchmarking tools based on TPC-C simulate a multi-user workload
within a single machine. However, this avoids large-scale benchmarking and also
introduces deviations in the results. Moreover, most of these tools are developed
for a particular DBMS. This means that the benchmark logic needs to be adapted
to the DBMS features and sometimes needs to be implemented directly in it.
Optimizations can be implemented to achieve requirements of the DBMS. For

Q. Li et al. (Eds.): APWeb/WAIM 2009, LNCS 5446, pp. 544–549, 2009.

example, by using stored procedures, the source code will be compiled and stored in the DBMS memory and this could affect the evaluation's result.

Indeed, developing a tool driven to a particular DBMS yields to two problems: (i) it is impossible to evaluate a wide list of DBMS. (ii) the workload tends to be unreal.

In this paper, we present a solution for benchmarking DBMS in a fully distributed manner that also aims to simulate large-scale workloads. Our solution called TPCC-C3SL is a distributed system developed in Java that complies with the TPC-C specification.

TPCC-C3SL was developed to be executed in a fully distributed way yielding scalability to the benchmark. Furthermore, its architecture allows the injection of failures (e.g., network failures) and overload situations during benchmarking.

To be platform independent, this tool was developed as open source software and makes extensive use of patterns and several *frameworks* for object persistence and record log tracks.

The rest of this article is organized as follows. Section 2 shows some related work. Section 3 describes the architecture of TPCC-C3SL. Section 4 presents the validation of our tool through benchmarking some open-source DBMS and Section 5 concludes the text.

2   Benchmarks Tools

In general, the DBMS vendors implement their own benchmark tool [17] and make the code available to the community in the TPC website. Also available on the internet are some open-source implementations such as OSDL-DBT2 [16], TPCC-UVa [8] and BenchmarkSQL [15]. However, these tools should be more carefully viewed under two issues.

Firstly, they simulate a multi-user environment in a single node where each emulated user executes as a new Thread. However, this is not scalable since this node can run out of resources quickly whenever benchmarking is carried out with a large number of emulated users. A better approach should comprehend a distributed manner: one machine executes the DBMS and other machines emulate the users.

Secondly, they are tied up to a specific OS and/or DBMS (e.g., implemented as stored procedures) and cannot be used elsewhere. We claim that this may affect the benchmarking results since the code is implemented directly in the DBMS and, therefore, can take advantage of the internal DBMS functions that are already optimized. A fair approach should comprise an open-source, DBMS-independent benchmarking tool.

3   TPCC-C3SL

TPCC-C3SL benchmarks in a fully distributed manner. Assuming that one wants to simulate 128 concurrent users but has only 64 machines, it suffices to replicate TPCC-C3SL over all 64 machines and simulate two users within

```
package br.ufpr.tpccufpr.vo;

@Entity
@Table(name = "WAREHOUSE")
public class Warehouse implements Serializable {
    private Long id;

    @Id
    @Column(name = "W_ID")
    public Long getId() {
        return id;
    }
    public void setId(Long id) {
        this.id = id;
    }
}
```

Fig. 1. Java class with JPA Annotations

each machine. Likewise, if 128 or more machines are available it is possible to have only one simulated user per machine.

Furthermore, TPCC-C3SL allows final users to evaluate any DBMS without changing the source code. To accomplish this task we used an object-relational mapping framework [10]. This framework enables the usage of meta-data included in the source code, known as annotations. The annotations permit the execution of *Data Definition Language* (DDL) and *Data Manipulation Language* commands against the DBMS without using any implicit *Structured Query Language* SQL instruction. An example of annotation can be seen in the Figure 1.

The annotation @Entity instructs the DBMS to create (or update) a new table if it does not exist and @Table defines that the name of this table will be WAREHOUSE. The annotations over the getId() method regards the id attribute. @Column indicates that a column must be created in this table and the name of this column will be W_ID. @Id defines that this column will be a primary key.

As a TPC-C implementation, the main functionality of our system is to continuously execute the 5 transaction types defined in the specification during the benchmark execution. The TPC-C specifies a fixed frequency distribution (FD) for the number of transactions that must be executed. An additional feature that we have added is the configurable weighted transactions. This enables users to adjust the transaction's FD in order to adapt the benchmark to their particular needs. But it should be noted that modifying a transaction's frequency distribution may violate the TPC-C specification.

4   Distributed Benchmarking

To validate our tool, we benchmarked three open source DBMSs: Post-greSQL [14], MySQL [18] and a pure Java DBMS, the Apache Derby [19]. We considered two different scenarios: (1) all simulated users submit transactions from the same machine and (2) all simulated users submit transactions from different machines. The first scenario is called *Server*, a centralized approach, and the second one *Multi-Server*, a distributed approach.

In the *Server* scenario only one machine was used. It was an Intel Xeon Quad Core processor at 2 GHz, 2 GB RAM and two 250GB SCSI HD configured for RAID-0 running Ubuntu 8.04 LTS Server Edition for 64bit processors. In *Multi-Server* we used 5 other machines to run the simulated users. All of them shared the file system (with 5TB) and ran the same OS (Debian 4.1.1-21 -Linux version 2.6.24.7). Table 1 shows the characteristics of these machines.

Table 1. Client machines configuration

Name	Processor	Memory (in GB)
macalan	Dual-Core AMD Opteron(tm) Processor 2220 at 2800 MHz	20
caporal	AMD Opteron(tm) Processor 242 at 1600 MHz	4
talisker	AMD Opteron(tm) Processor 242 at 1600 MHz	4
cohiba	Intel(R) Xeon(TM) at 3.60GHz	4
bolwmore	Dual-Core AMD Opteron(tm) Processor 2220 at 2800 MHz	32

We remark that in the *Server* scenario, the base machine had much more computing power than all five machines used in the *Multi-Server* scenario. Figure 2 presents the results of the PostgreSQL's benchmark[1].

Each scenario was executed 4 times. The first execution simulated 10 concurrent users and the following executions simulated 30, 50 and 100 concurrent users. During the execution, the simulated users submitted several transactions in parallel against the DBMS.

The set of all completed transactions per minute is called *tpmC* by the TPC-C. In the more restricted execution, i.e., 10 to 30 concurrent users, it has been observed that the tpmC is higher in the centralized approach. However, when 50 concurrent users are simulated, the distributed approach exhibits better performance than the centralized approach. This trend gets more evident when 100 concurrent users are simulated. The tpmC obtained in the *Server* scenario is only half of the one in the *Multi-Server* scenario.

---

[1] For space reasons, we present only the PostgreSQL's result. The results of MySQL and Apache Derby showed a similar behavior.

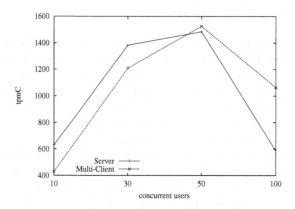

Fig. 2. PostgreSQL's benchmark

As expected, this was due to the *Server* scenario configuration, since a single machine is responsible both for handling the entire set of simulated users and executing the DBMS. During the experiment, the machine ran out of computing resources, affecting the overall performance. By contrast, the *Multi-Server* scenario scaled up correctly and avoided the overload of the DBMS machine with simulated users.

5   Conclusion

In this paper we presented a new tool for benchmarking DBMS, called TPCC-C3SL, which is based on the TPC-C specification.

TPCC-C3SL allows the distribution of users across multiple processing nodes in order to simulate a multi-user environment. This allows a considerable number of transactions to be submitted in parallel to simulate a large-scale environment (e.g. internet). The results showed that the way the transactions are submitted affects the DBMS's performance. As a result, we suggested that for benchmarking large-scale user access, the performance evaluation of DBMS should be primarily supported by a *Multi-Server* method that deals with large-scale workloads.

We also showed that users can adjust the TPCC-C3SL for their needs. Our tool allows users to modify the transaction's frequency distribution to better reflect their needs.

TPCC-C3SL was also developed using the three-layer concept: presentation, business logic and persistence. Based on these ideas, we believe to have taken a step further towards the generalization of TPC-C benchmarking.

References

1. Bitton, D., DeWitt, D.J., Turbyfill, C.: Benchmarking Database Systems A Systematic Approach. In: VLDB, pp. 8–19 (1983)
2. DeWitt, D.J., Levine, C.: Not just correct, but correct and fast: a look at one of Jim Gray's contributions to database system performance. SIGMOD Record 37(2), 45–49 (2008)

3. DeWitt, D.J.: The Wisconsin Benchmark: Past, Present, and Future. The Benchmark Handbook (1993)
4. Datamation: A Measure of Transaction Processing Power, pp. 112–118 (1985)
5. Transactional Processing Process Council, http://www.tpc.org
6. Serlin, O.: The History of DebitCredit and the TPC. The Benchmark Handbook (1993)
7. TPC Benchmark C, http://www.tpc.org/tpcc
8. Llanos, D.R.: TPCC-UVA: An Open-Source TPC-C Implementation for Global Performance Measurement of Computer Systems. ACM SIGMOD Record 35(4), 6–15 (2006)
9. de Almeida, E.C.: Estudo de viabilidade de uma plataforma de baixo custo para data warehouse. UFPR, Master Thesis (2004).
10. Hibernate - Relational Persistence for Java and .NET, http://www.hibernate.org
11. The Java Persistence API - A Simpler Programming Model for Entity Persistence, http://java.sun.com/developer/technicalArticles/J2EE/jpa
12. JSR-000220 Enterprise JavaBeans 3.0, http://jcp.org/aboutJava/communityprocess/final/jsr220
13. IBM Smalltalk Tutorial, http://www.inf.ufsc.br/poo/smalltalk/ibm/tutorial/oop.html
14. PostgreSQL, http://www.postgresql.org
15. BenchmarkSQL, http://sourceforge.net/projects/benchmarksql
16. OSDL-DBT2, http://sourceforge.net/projects/osdldbt
17. TPC-C Result Highlights, http://www.tpc.org/tpcc/results/tpcc_result_detail.asp?id=107022701
18. MySQL, http://www.mysql.com
19. The Apache DB Project, http://db.apache.org/derby
20. JDBC Drivers, http://developers.sun.com/product/jdbc/drivers

Optimizing Many-to-Many Data Aggregation in Wireless Sensor Networks

Shuguang Xiong and Jianzhong Li

Department of Computer Science and Technology,
Harbin Institute of Technology, Harbin, China
n2xiong@gmail.com, lijzh@hit.edu.cn

Abstract. Wireless sensor networks have enormous potential to aid data collection in a number of areas, such as environmental and wildlife research. In this paper, we study the problem of how to generate energy efficient aggregation plans to optimize many-to-many aggregation in sensor networks, which involve a many-to-many relationship between the nodes providing data and the nodes requiring data. We show that the problem is NP-Complete considering both routing and aggregation choices, and we present three approximation algorithms for two special cases and the general case of the problem. By utilizing the approximation algorithms for the minimal Steiner tree, we provide approximation ratios on the energy cost of the generated plan against optimal plans.

Keywords: Wireless Sensor Network, Data Aggregation, Energy Efficiency, In-Network Execution.

1 Introduction

As an important way for people to acquire data from physical world, wireless sensor network (WSN) has received much concern from research society [1,3,6]. One of the most common usage of a WSN is data collection/aggregation. Because of limited power of a sensor, energy efficiency is the primary concern.

Recently, a type of data aggregation in sensor networks, namely *many-to-many data aggregation*, has been presented [2]. In many-to-many data aggregation, there are multiple nodes that provide data (i.e., their readings), which are called *producers*, while another set of nodes require the aggregated data from part of the producers, which are called *consumers*. The relationship between consumers and producers is many-to-many, i.e., one producer may be needed in computing aggregates acquired by multiple consumers, and one consumer may require data aggregated from multiple producers.

In this paper, we study the problem of how to generate energy-efficient aggregation plans for many-to-many aggregation in a sensor network. Different from the problem in [2] restricting that the multiple routing trees are fixed, this paper focuses on a new problem in which the input only includes the network topology, the producer and consumer sets and their supply and demand relations. The output, an aggregation plan, consists of two parts: the first one is the set of routing tree rooted at a producer and spans all consumers of this producer, and the second part is the forwarding strategies adopted by each node in the routing

Q. Li et al. (Eds.): APWeb/WAIM 2009, LNCS 5446, pp. 550–555, 2009.

trees, i.e., the decision to forward the aggregated data or to forward the data separately when two data r_1 and r_2 arrive at it.

The contributions of this paper are listed as follows. First, we focus on the many-to-many aggregation problem in WSN with both routing choices and aggregation schemes that is considered by few works, and show the hardness of the problem, i.e., the problem is NP-Complete. Second, we propose three approximation algorithms to get close to the optimal solution for three different cases of the problem, in which the two cases are commonly-used special cases. Finally, we proved the ratio bounds of the three algorithms.

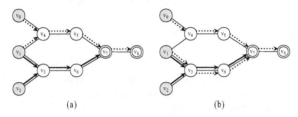

(a)   (b)

Fig. 1. The energy costs in different routing strategies. v_7 requires $v_0 + v_1$, and v_8 requires $v_1 + v_2$. (a) The cost is 9 since there is one piece of data transmitted on each edge. (b) The cost is 10. Note that v_1 sends its data through $v_1 v_3$ only once, and the values of v_1 and v_2 are transmitted through $v_3 v_6$ and $v_6 v_7$.

2   Problem Statement

A sensor network is abstracted as a connected, undirected graph $G(V, E)$, where V indicates the set of sensors and E stands for the links of the sensors. Let $|V|$ and $|E|$ denote the number of sensors and links, respectively. Each sensor $v_i \in V$ has one reading r_i, which is a real number, i.e., $r_i \in \mathbb{R}$.

The class of aggregate functions we concern meets the following requirement. For any function f in the class, $f : 2^{\mathbb{R}} \to \mathbb{R}$, and $f(S) = f(\{f(A), f(B)\})$ for arbitrary real number sets S, A and B that $A \cup B = S$ and $A \cap B = \varnothing$.

Given a set of consumers C ($C \subseteq V$) that consists of n sensors, $\forall c_i \in C$, c_i requires an aggregation result $f(P_{c_i})$, where P_{c_i} is a given set of sensors called the producer set of c_i. Denote the set of producers as $P = \cup_{i=1}^{n} P_{c_i}$ and let the number of producers is m, we can analogously define the consumer set C_{p_i} for $p_i \in P$, to which p_i provide its reading. Clearly, $C = \cup_{i=1}^{m} C_{p_i}$.

Definition 1 (Aggregation plan). *Given a connected graph $G(V, E)$, the set of consumers $C = \{c_1, c_2, \ldots, c_n\}$ and the set of producers $P = \{p_1, p_2, \ldots, p_m\}$, an aggregation plan $\mathcal{A}$ consists of a set of routing trees and the forwarding strategies of all the nodes on these trees. $\mathcal{A}$ is valid if $\forall c_i \in C$, c_i is able to calculate $f(P_{c_i})$ according to the data it receives.*

Figure 1 shows two aggregation plans with different routing strategies. In this example, v_7 and v_8 require $v_0 + v_1$ and $v_1 + v_2$, resp.

Definition 2 (The Optimization Problem for Many-to-Many Aggregation). *The Optimization problem for many-to-many Aggregation (in short, the*

OA problem) problem is to find out an optimal aggregation plan $\mathcal{A}_{opt}$, so that $E(\mathcal{A}_{opt})$ is minimized among all the valid aggregation plans.

The OA problem is NP-Complete, however, because its special case with one consumer is NP-Complete. In this case, since every producer has a directed path to the consumer, the consumer and the producers are connected. Denote the number of edges of a minimal Steiner tree $MST(P \cup C)$ that spans $P \cup C$ as $|MST(P \cup C)|$, and $E(\mathcal{A}_{opt}) \geq |MST(P \cup C)|$. On the other hand, one can construct an aggregation plan $\mathcal{A}$ based on $|MST(P \cup C)|$ so that $E(\mathcal{A}) = |MST(P \cup C)|$. Hence the special case is equivalent to a MST problem on the graph, which is known as NP-Complete [8], so the OA problem is NP-Complete.

3   Solutions

In this paper, we focus on the approximation algorithms to the OA problem. Before our study for the general case, we discuss two special cases of the problem and propose two constant-approximated algorithms, resp. We do not give the proofs of the theorems due to page limitation.

3.1   Special Case I

This special case restricts that all the consumers require the aggregated result $f(V)$ of the entire network. To achieve the optimality, algorithm 1 employs the mechanism that first aggregates the data of the whole network to a randomly selected consumer c_0 via a spanning tree of the network, and then disseminate the data to all the consumers along the paths of a tree that spans the consumers.

Algorithm 1. Optimal aggregation I

INPUT: Network $G(V, E)$, consumers C
OUTPUT: An optimal aggregation plan $\mathcal{A}_{opt}$ for special case I
1: construct a spanning tree T_{c_0} of G rooted at c_0
2: v forwards $f(Subtree_v)$ to its parent on T_{c_0}, $\forall v \in V$
3: construct $MST(C)$ rooted at c_0
4: v forwards $f(V)$ to its children on $MST(C)$ for all v involved in $MST(C)$
5: return $\mathcal{A}$ that consists of T_{c_0}, $MST(C)$ and the forwarding strategies

---

Theorem 1. *Algorithm 1 outputs an optimal aggregation plan $\mathcal{A}_{opt}$ for special case I, i.e., $E(\mathcal{A}_{opt}) = |V| - 1 + |MST(C)|$.*

To our best knowledge, the lowest ratio bound of the polynomial-time algorithms on graph MST problem is $\beta = 1 + \frac{\ln 3}{2} \approx 1.55$ [7], so Algorithm 1 can employ this approximation algorithm to reduce computational complexity. Denote the energy cost of the tree generated by the approximation algorithm be $|MST'(C)|$, it means that $|MST'(C)| \leq \beta \cdot |MST(C)|$. Because $|V| - 1 \geq MST(C)$, the energy cost of a plan $\mathcal{A}_{app}$ generated by the approximation algorithm is bounded by

$$\frac{E(\mathcal{A}_{app})}{E(\mathcal{A}_{opt})} = \frac{|V| - 1 + |MST'(C)|}{|V| - 1 + |MST(C)|} \leq \frac{|V| - 1 + \beta \cdot |MST(C)|}{|V| - 1 + |MST(C)|} \approx 1.275$$

3.2 Special Case II

We continue to study a more general case, in which either $P_{c_i} \cap P_{c_j} = P_{c_i} = P_{c_j}$ or $P_{c_i} \cap P_{c_j} = \varnothing$ for arbitrary consumer c_i and c_j. Hence, $\forall P_{c_i}$, there is a set of consumers C_i that require $f(P_{c_i})$. Let a *cluster* of nodes be $\mathcal{D}_i = P_{c_i} \cup C_i$ and the number of distinct clusters $n_c \leq n$, and denote the n_c clusters with their producer sets and consumer sets as $\mathcal{D}_1(P_1, C_1), \mathcal{D}_1(P_1, C_1), \ldots, \mathcal{D}_{n_c}(P_{n_c}, C_{n_c})$.

The key idea of the approximation algorithm for special case II is that constructing two routing trees for each cluster independently. The first routing tree of $\mathcal{D}_i$ spans P_i, and the value $f(P_i)$ is obtained by a producer v_i in the tree, while the second one spans $C_i \cup \{v_i\}$ so that $f(P_i)$ can be disseminated to all the nodes in C_i from v_i. The pseudo code is shown in Algorithm 2.

Algorithm 2. Near-optimal aggregation II

INPUT: Network $G(V, E)$, consumers C and producer sets $\{P_{c_i}\}$, $1 \leq i \leq n$
OUTPUT: An aggregation plan $\mathcal{A}$ for special case II
1: aggregate $P \cup C$ into n_c distinct clusters $\mathcal{D}_1(P_1, C_1), \mathcal{D}_1(P_1, C_1), \ldots, \mathcal{D}_{n_c}(P_{n_c}, C_{n_c})$

2: **for** $i = 1$ to n_c **do**
3:     construct $MST(P_i)$ rooted at v_i that spans P_i
4:     v forwards $f(Child_v)$ to its parent on $MST(P_i)$, $\forall v \in MST(P_i)$
5:     construct $MST(C_i \cup \{v_i\})$ rooted at v_i
6:     v forwards $f(P_i)$ to its children on $MST(C_i \cup \{v_i\})$, , $\forall v \in MST(C_i \cup \{v_i\})$
7: **end for**
8: **return** $\mathcal{A}$ with $MST(P_i)$, $MST(C_i \cup \{v_i\})$ and the forwarding strategies

---

Theorem 2. *Algorithm 2 outputs an aggregation plan $\mathcal{A}$ for special case II with approximation ratio $\frac{E(\mathcal{A})}{E(\mathcal{A}_{opt})} \leq 2$, where $\mathcal{A}_{opt}$ is an optimal aggregation plan.*

Note that if Algorithm 2 employs Robinsy's approximation algorithm [7] in finding out a minimal Steiner tree to generate an aggregation plan $\mathcal{A}'$, it is easy to see that Algorithm 2 runs in polynomial time and $\frac{E(\mathcal{A}')}{E(\mathcal{A}_{opt})} \leq 2 + \ln 3$.

3.3 General Case

In the general case, the consumers can have arbitrary producers and vice versa. We propose an efficient algorithm to get close to the optimality for this case. The basic idea is to divide the consumers and producers into *groups*, and then perform data aggregation on these groups independently. Nodes in the same producer group send their readings or partially aggregated data to one producer in the group, while nodes in the same consumer group share the data that has been aggregated on a specific consumer.

Theorem 3. *Algorithm 3 generates an aggregation plan $\mathcal{A}$ that its energy cost $\mathcal{A}$ is bounded by $\frac{E(\mathcal{A})}{E(\mathcal{A}_{opt})} \leq 2 \cdot \min\{n_p, n_c\}$.*

Similarly, Algorithm 3 can utilize the approximation algorithm in [7] to reduce the computational complexity. In this case $\frac{E(\mathcal{A})}{E(\mathcal{A}_{opt})} \leq (2 + \ln 3) \cdot \min\{n_p, n_c\}$.

Algorithm 3. Near-optimal aggregation III

INPUT: Network $G(V, E)$, consumers C with $\{P_{c_i}\}$, $1 \le i \le n$, producers $P = \cup_{i=1}^{n} P_{c_i}$
OUTPUT: An aggregation plan $\mathcal{A}$ for special case III

1: aggregate P into n_p groups $gp_1, gp_2, \ldots, gp_{n_p}$
2: aggregate C into n_c groups $gc_1, gc_2, \ldots, gc_{n_c}$
3: **if** $n_c > n_p$ **then**
4:     **for** $i = 1$ to n_p **do**
5:         construct $MST(gp_i)$ rooted at v_{gp_i}
6:         v forwards $f(Child_v)$ to its parent on $MST(gp_i)$, $\forall v \in MST(gp_i)$
7:         construct $MST(S_{gp_i})$ rooted at v_{gp_i}
8:         v forwards $f(gp_i)$ to its children on $MST(S_{gp_i})$, $\forall v \in MST(S_{gp_i})$
9:     **end for**
10: **else**
11:     **for** $i = 1$ to n_c **do**
12:         construct $MST(S_{gc_i})$ rooted at v_{gc_i}
13:         v forwards $f(Child_v)$ to its parent on $MST(S_{gc_i})$, $\forall v \in MST(S_{gc_i})$
14:         construct $MST(gc_i)$ rooted at v_{gc_i}
15:         v forwards $f(S_{gc_i} \setminus \{v_{gc_i}\})$ to its children on $MST(gc_i)$, $\forall v \in MST(gc_i)$
16:     **end for**
17: **end if**
18: return $\mathcal{A}$ that consists of the routing trees and the forwarding strategies

---

4   Related Work

Efficient data gathering and aggregation in wireless sensor networks has been long studied in many articles [2,4,3,10,6,14], which is defined as the process of aggregating the data from multiple sensors to eliminate redundant transmission and provide fused information to the base station [10].

Existing data aggregation techniques can be categorized into several types. One of them is cluster/grid-based approaches [11,12,13], while another type of works focus on the efficient algorithms for constructing routing trees in aggregation [5,15,6]. The concept of many-to-many data aggregation is proposed by Silberstein et al. [2]. They consider the optimization problem for many-to-many aggregation relying on given multiple routing trees with optimal solutions.

Numerous approximation algorithms has been proposed for the classical Steiner tree problem on graph, in which [7] has the best approximation ratio $1 + \frac{\ln 3}{2}$ to our best knowledge. The hardness of the problem is proved in [8].

5   Conclusion and Future Works

We study the OA problem which requires an efficient aggregation plan for many-to-many data aggregation in sensor networks, and show that the problem is NP-Complete considering both routing and aggregation strategies. We then propose three approximation algorithms to handle three different cases of the problem. For the two special cases, we prove that the proposed algorithms have constant approximation ratios, and for the general case, the approximation ratio is no larger than $(2 + \ln 3) \cdot \min\{n_p, n_c\}$.

Next, we plan to evaluate the efficiency of the algorithms by experiments, and since n_c and p_c may be large in the worst cases, one possible future work will seek for tighter approximation ratios.

References

1. Akyildiz, I.F., Su, W., Sankarasubramaniam, Y., Cayirci, E.: Wireless Sensor Networks: a Survey. Computer Networks 38(4), 393–422 (2002)
2. Silberstein, A., Yang, J.: Many-to-Many Aggregation for Sensor Networks. In: IEEE ICDE (2007)
3. Madden, S., Szewczyk, R., Franklin, M., Culler, D.: Supporting Aggregate Queries Over Ad-Hoc Wireless Sensor Networks. In: IEEE Workshop on Mobile Computing Systems and Applications (2002)
4. Silberstein, A., Munagala, K., Yang, J.: Energy-Efficient Monitoring of Extreme Values in Sensor Networks. In: ACM SIGMOD (2006)
5. Goel, A., Estrin, D.: Simultaneous Optimization for Concave Costs: Single Sink aggregation or Single Source Buy-at-bulk. In: ACM Symposium on Discrete Algorithms (SODA) (2003)
6. Wu, Y., Fahmy, S., Shroff, N.: On the Construction of a Maximum-Lifetime Data Gathering Tree in Sensor Networks: NP-Completeness and Approximation Algorithm. In: IEEE INFOCOM (2008)
7. Robinsy, G., Zelikovskyz, A.: Improved Steiner Tree Approximation in Graphs. In: ACM SODA (2000)
8. Karp, R.M.: Reducibility Among Combinatorial Problems. In: Miller, R.E., Thatcher, J.W. (eds.) Complexity of Computer Computations, pp. 85–103 (1972)
9. Koch, T., Martin, A.: Solving Steiner Tree Problems in Graphs to Optimality. Networks 32(3), 207–232 (1998)
10. Rajagopalan, R., Varshney, P.: Data Aggregation Techniques in Sensor Networks: A Survey. Communication Surveys and Tutorials 8(4), 48–63 (2006)
11. Heinzelman, W., Chandrakasan, A., Balakrishnan, H.: An Application-Specific Protocol Architecture for Wireless Microsensor Networks. IEEE Trans. on Wireless Communications 1(4), 660–670 (2002)
12. Younis, O., Fahmy, S.: HEED: A Hybrid, Energy-Efficient, Distributed Clustering Approach for Ad Hoc Sensor Networks. IEEE Trans. on Mobile Computing 3(4), 660–669 (2004)
13. Vaidhyanathan, K., Sur, S., Narravula, S., Sinha, P.: Data Aggregation Techniques in Sensor Networks. Technical Report (2004)
14. Intanagonwiwat, C., Govindan, R., Estrin, D.: Directed Diffusion: A Scalable and Robust Communication Paradigm for Sensor Networks. In: ACM MOBICOM (2000)
15. Khan, M., Pandurangan, G.: A fast distributed approximation algorithm for minimum spanning trees. In: Dolev, S. (ed.) DISC 2006. LNCS, vol. 4167, pp. 355–369. Springer, Heidelberg (2006)

Bag of Timestamps: A Simple and Efficient Bayesian Chronological Mining

Tomonari Masada[1], Atsuhiro Takasu[2], Tsuyoshi Hamada[1], Yuichiro Shibata[1], and Kiyoshi Oguri[1]

[1] Nagasaki University, 1-14 Bunkyo-machi, Nagasaki, Japan
{masada,hamada,shibata,oguri}@cis.nagasaki-u.ac.jp
[2] National Institute of Informatics, 2-1-2 Hitotsubashi, Chiyoda-ku, Tokyo, Japan
takasu@nii.ac.jp

Abstract. In this paper, we propose a new probabilistic model, *Bag of Timestamps (BoT)*, for chronological text mining. BoT is an extension of latent Dirichlet allocation (LDA), and has two remarkable features when compared with a previously proposed *Topics over Time (ToT)*, which is also an extension of LDA. First, we can avoid overfitting to temporal data, because temporal data are modeled in a Bayesian manner similar to word frequencies. Second, BoT has a conditional probability where no functions requiring time-consuming computations appear. The experiments using newswire documents show that BoT achieves more moderate fitting to temporal data in shorter execution time than ToT.

1 Introduction

Topic extraction is an outstanding agenda item in practical information management. Many researches provide efficient probabilistic methods where topics are modeled as values of latent variables. Further, researchers try to avoid *overfitting* via Bayesian approaches. Latent Dirichlet allocation (LDA) [5] is epoch-making in this research direction. In this paper, we focus on documents with *timestamps*, e.g. weblog entries, newswire articles, patents etc. *Topics over Time (ToT)* [9], one of the efficient chronological document models, extends LDA with per-topic beta distributions defined over the time interval normalized to $[0, 1]$. However, since beta distribution parameters are directly estimated with no corresponding priors, ToT may suffer from overfitting to temporal data. In general, we can observe overfitting to temporal data in two ways. First, same topics are assigned to word tokens from documents having similar (i.e., close) timestamps even when those word tokens relate to dissimilar semantical contents. Second, different topics are assigned to word tokens from documents having dissimilar timestamps even when those word tokens relate to semantically similar contents. We will later show that ToT suffers from the latter case in our experiments.

In this paper, we propose *Bags of Timestamps (BoT)* as a new probabilistic model for Bayesian topic analysis using temporal data. Our approach is similar to mixed-membership models [6], where reference data of scientific publications are treated in a Bayesian manner along with word frequency data. In BoT, we

Q. Li et al. (Eds.): APWeb/WAIM 2009, LNCS 5446, pp. 556–561, 2009.

attach an array, called *timestamp array*, to each document and fill this array with tokens of document timestamps. Further, a Dirichlet prior is also introduced for timestamp multinomials, not only for word multinomials. We will show that BoT can realize more moderate fitting to temporal data than ToT.

2   Previous Works

Recent researches provide efficient and elaborated document models for utilizing temporal data. In Dynamic Topic Models [4], a vector is drawn, at each position on time axis, from a normal distribution conditioned on the previously drawn vector. This vector is used to obtain topic and word multinomials. However, normal distribution is not conjugate to multinomial, and thus inference is too complicated. On the other hand, Multiscale Topic Tomography Models [8] segment the given time interval into two pieces recursively and construct a binary tree whose root represents the entire time span and each internal node represents a shorter time interval. Leaf nodes are associated with a Poisson distribution for word counts. These two researches do not explicitly discuss overfitting to temporal data. In this paper, we focus on this problem and propose a new probabilistic model, *Bag of Timestamps (BoT)*. We can compare BoT with Topics over Time (ToT) [9], because both are an extension of LDA. ToT has a special feature that it can model continuous time with beta distributions. However, beta distribution parameters are estimated in a non-Bayesian manner. This may lead to overfitting to temporal data. BoT discards sophisticated continuous time modeling and takes a simpler approach. We attach an array, called *timestamp array*, to each document and fill timestamp arrays with tokens of document timestamps. Each timestamp token is drawn from a per-topic multinomial in the same manner as word tokens. Therefore, our approach is similar to mixed-membership models [6] where references of scientific articles are generated along with word tokens within the same LDA framework. We can introduce a Dirichlet prior not only for per-topic word multinomials but also for per-topic timestamp multinomials. Therefore, we can expect that overfitting to temporal data will be avoided. Further, BoT requires no time-consuming computations of gamma functions and of power functions with arbitrary exponents, which are required by ToT.

However, ToT may avoid overfitting with likelihood rescaling [10]. Here we introduce notations for model description. z_{ji} is the latent variable for the topic assigned to ith word token in document j. n_{jk} is the number of word tokens to which topic k is assigned in document j, and n_{kw} is the number of tokens of word w to which topic k is assigned. A symmetric Dirichlet prior for per-document topic multinomials is parameterized by α, and a symmetric Dirichlet prior for per-topic word multinomials is parameterized by β. t_j is the observed variable for the timestamp of document j. In case of ToT, t_j takes a real value from $[0, 1]$. ψ_{k1} and ψ_{k2} are two parameters of a beta distribution for topic k. By applying likelihood rescaling for beta distribution, the conditional probability that z_{ji} is updated from k' to k in Gibbs sampling for ToT

is $\frac{n_{jk}+\alpha-\Delta_{k=k'}}{\sum_k n_{jk}+K\alpha-1} \cdot \frac{n_{kx_{ji}}+\beta-\Delta_{k=k'}}{\sum_w n_{kw}+W\beta-\Delta_{k=k'}} \cdot \left\{ \frac{(1-t_j)^{\psi_{k1}-1}t_j^{\psi_{k2}-1}}{B(\psi_{k1},\psi_{k2})} \right\}^\tau$ where $\Delta_{k=k'}$ is 1 if

$k = k'$, and 0 otherwise. τ is a parameter for likelihood rescaling and satisfies $0 \leq \tau \leq 1$. When $\tau = 1$, we obtain the original ToT. When $\tau = 0$, ToT is reduced to LDA. Therefore, we can control the degree of fitting to temporal data by adjusting τ. BoT will be also compared with ToT after likelihood rescaling in our experiments.

3   Bag of Timestamps

We modify LDA to realize an efficient chronological document modeling and obtain BoT as described in the following. First, a topic multinomial is drawn for each document from a corpus-wide symmetric Dirichlet prior parameterized by α. Second, for each document, topics are drawn from its topic multinomial and assigned to the elements of both word array and timestamp array. y_{js} is the latent variable for the topic assigned to sth timestamp token in document j. Let n_{jk} be the number of both word and timestamp tokens to which topic k is assigned in document j. Third, we draw words from word multinomials in the same manner with LDA, where per-topic word multinomials are drawn from a corpus-wide symmetric Dirichlet prior parameterized by β. Fourth, we draw timestamps from timestamp multinomials, where per-topic timestamp multinomials are also drawn from another corpus-wide symmetric Dirichlet prior paremeterized by γ. Then the conditional probability that z_{ji} is updated from k' to k is $\frac{n_{jk}+\alpha-\Delta_{k=k'}}{\sum_k n_{jk}+K\alpha-1}$ $\cdot \frac{n_{kx_{ji}}+\beta-\Delta_{k=k'}}{\sum_w n_{kw}+W\beta-\Delta_{k=k'}}$ as in case of LDA [7]. Further, let n_{ko} be the number of tokens of timestamp o to which topic k is assigned. Then the conditional that y_{js} is updated from k' to k is $\frac{n_{jk}+\alpha-\Delta_{k=k'}}{\sum_k n_{jk}+K\alpha-1} \cdot \frac{n_{ky_{js}}+\gamma-\Delta_{k=k'}}{\sum_o n_{ko}+O\gamma-\Delta_{k=k'}}$. BoT is more than just introducing timestamps as new vocabularies because two distinct Dirichlet priors are prepared for word multinomials and timestamp multinomials.

By using document timestamps, we determine observed configuration of timestamp arrays as follows. We assume that all documents have a timestamp array of the same length L. The timestamp arrays of documents having timestamp o are filled with $L/2$ tokens of timestamp o, $L/4$ tokens of $o-1$, and $L/4$ tokens of $o+1$, where $o-1$ and $o+1$ are two timestamps adjacent to o along time axis. When a document is placed at either end of the given time interval, we leave $L/4$ elements of its timestamp array empty. Obviously, this is not the only way to determine configuration of timestamp arrays. However, we use this configuration in this paper for simplicity.

4   Experiments

We conduct experiments to reveal differences between BoT and ToT. The number of topics is fixed to 64. α is set to $50/K$, and β is 0.1 for both ToT and BoT [7]. γ, playing a similar role with β, is set to 0.1. We test three timestamp array lengths: 32, 64, and 128. In this order, dependency on temoral data gets stronger. We abbreviate these three settings as BoT32, BoT64, and BoT128. In ToT, we test 0.5 and 1.0 for τ. These two settings are referred to by ToTsqrt and ToTorig. Preliminary experiments have revealed that 300 iterations in Gibbs sampling are enough for all settings. We use the following three data sets.

"MA" includes 56,755 Japanese newswire documents from Mainichi and Asahi newspaper web sites (`www.mainichi.co.jp`, `www.asahi.com`). Document dates range from Nov. 16, 2007 to May 15, 2008. While we collapse the dates into 32 timestamps for BoT, the dates are mapped as is to $[0, 1]$ for ToT. We use MeCab [1] for morphological analysis. The number of word tokens and that of unique words are 7,811,336 and 40,355, respectively. It takes nearly 90 minutes (resp. 105 minutes) for 300 iterations in case of BoT64 (resp. BoT128) on a single core of Intel Q9450. The same number of iterations require 135 minutes for both ToTorig and ToTsqrt. "S" consists of 30,818 Korean newswire documents from Seoul newspaper web site (`www.seoul.co.kr`). Document dates range from Jan. 1 to Dec. 31 in 2006. We map the dates to real values from $[0, 1]$ for ToT, and to 32 timestamps for BoT. KLT version 2.10b [2] is used for Korean morphological analysis, and we obtain 5,916,236 word tokens, where 40,050 unique words are observed. The execution time of BoT64 (resp. BoT128) is 60 minutes (resp. 70 minutes) for 300 iterations. Both ToTorig and ToTsqrt require 100 minutes. "P" includes 66,050 documents from People's Daily (`people.com.cn`). The dates range from May 1 to 31 in 2008. We use a simple segmenter prepared for Chinese word segmentation bakeoff [3] with its prepared dictionary. So far as comparison between BoT and ToT is concerned, we think that this segmenter is enough. "P" includes 41,552,115 word tokens and 40,523 unique words. We map the dates to $[0, 1]$ for ToT and discretize them into 24 timestamps for BoT. The execution time of 300 iterations is about 360 minutes (resp. 400 minutes) for BoT64 (resp. BoT128). Both ToTorig and ToTsqrt require 690 minutes.

Fig. 1 includes six graphs showing the results for "MA" under various settings. The horizontal axis represents time, and the vertical axis shows the percentage of the assignments of each topic to word tokens from the documents at each position on time axis. Every graph has a plot area divided into 64 regions. Each region corresponds to a distinct topic. When a region occupies a larger area, the corresponding topic is assigned to more word tokens. Top left graph is the result for LDA. While the graph shows poor dynamism in time axis direction, this is not a weakness. LDA is efficient in distinguishing topics enduring over time (e.g. stock news, weather forecasts, news of a person's death). However, LDA cannot control the degree of fitting to temporal data. Top right, middle left, and middle right graphs are obtained by BoT32, BoT64, and BoT128, respectively. More intensive temporal dynamism appears in middle right than in top right. We can say that the degree of fitting to temporal data is controlled by adjusting timestamp array length. Bottom left and bottom right graphs show the results for ToTorig and ToTsqrt, respectively. Each region has a smooth contour, because ToT can model continuous time. However, many topics are localized on narrow segments of time axis. Namely, same topics are rarely assigned to word tokens from distant positions on time axis. With respect to ToTorig (bottom right), only 7~20 topics among 64 are observed at every position of time axis. In contrast, by using ToTsqrt (bottom left), we can find 44~51 topics at each position of time

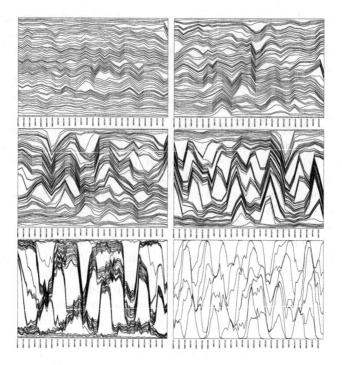

Fig. 1. Graphs of topic distribution changes for MA dataset

axis. However, many of these 44~51 topics are assigned to very few word tokens. Likelihood rescaling seems to result in partial success. For both "S" and "P", we have obtained similar results.

We can also estimate the degree of overfitting to temporal data as follows. For each topic k, we sort words by n_{kw}. Many of top-ranked words may represent semantical content of the corresponding topic. We select 100 top-ranked words for each k. As the number of topics is 64, we have 64 lists of 100 top-ranked words. Further, for each word, we count the number of topics whose top-ranked word lists include the word. For example, when we apply LDA to "MA", the word "sakunen (last year)" appears in 25 among 64 lists. This kind of words may not help in focusing on a specific semantical content. In contrast, "bouei (military defence)" appears only in two among 64 lists. When many different topics are assigned to a word focusing on a specific content, poor topic extraction will result. For "P", "aoyun (Olympic games)" appears only in four among 64 lists for both LDA and BoT64. However, when we use ToTorig, "aoyun" appears in 17 among 64. This suggests that ToT assigns different topics to word tokens only because they appear in documents far apart in time. This corresponds to the second overfitting case described in Section 1. Table 1 includes other examples.

We can give another evidence. For "MA", 64 lists of 100 top-ranked words consists of 3,189 and 2,934 unique words when we use LDA and BoT64, respectively. However, when we use ToTorig, only 1,177 unique words are observed.

Table 1. Examples of words presenting the difference between BoT and ToT

dataset	word	LDA	BoT64	ToTsqrt	ToTorig
MA	Minshu-tou (a political party)	5	5	8	16
(Japanese)	Ilaku (Iraq)	1	1	1	6
S	Lee Seung-Yeob (a baseball player)	1	1	1	7
(Korean)	Hannara (a political party)	2	3	3	14
P	Sichuan (a province)	16	20	28	32
(Chinese)	Shenghuo (Olympic Flame)	10	10	17	21

Namely, the same word appears in many different lists of top-ranked words. For both "S" and "P", we have obtained similar results. ToT tends to assign different topics to word tokens from documents having dissimilar timestamps even when they relate to a similar semantical content. ToT may be efficient when the given document set rarely includes outstanding contents ranging over a long period of time. However, when we are interested in both the semantical contents ranging over a long period of time and those localized on a small portion of time axis, BoT is a better choice, because BoT can respect both semantical similarity and temporal similarity, at least for Chinese, Japanese, and Korean documents.

5   Conclusion

In this paper, we propose a new probabilistic model to realize an efficient and intuitive control of the degree of fitting to temporal data of documents. As future work, we plan to model timestamp array lengths also in a Bayesian manner to realize more flexible control of the degree of fitting to temporal data.

References

1. http://mecab.sourceforge.net/
2. http://nlp.kookmin.ac.kr/HAM/kor/
3. http://www.sighan.org/bakeoff2005/
4. Blei, D., Lafferty, J.: Dynamic Topic Models. In: Proc. of ICML 2006, pp. 113–120 (2006)
5. Blei, D., Ng, A., Jordan, M.: Latent Dirichlet Allocation. Journal of Machine Learning Research 3, 993–1022 (2003)
6. Erosheva, E., Fienberg, S., Lafferty, J.: Mixed-membership models of scientific publications. Proc. Natl. Acad. Sci. 101(suppl. 1), 5220–5527 (2004)
7. Griffiths, T., Steyvers, M.: Finding Scientific Topics. Proc. Natl. Acad. Sci. 101(suppl. 1), 5228–5235 (2004)
8. Nallapati, R., Cohen, W., Ditmore, S., Lafferty, J., Ung, K.: Multiscale Topic Tomography. In: Proc. of KDD 2007, pp. 520–529 (2007)
9. Wang, X., McCallum, A.: Topics over Time: A Non-Markov Continuous-Time Model of Topical Trends. In: Proc. of KDD 2006, pp. 424–433 (2006)
10. Wang, X., Mohanty, N., McCallum, A.: Group and Topic Discovery from Relations and Text. In: Proc. of LinkKDD 2005, pp. 28–35 (2005)

Efficient Hybrid Password-Based Authenticated Group Key Exchange*

Shuhua Wu and Yuefei Zhu

Department of Networks Engineering,
Zhengzhou Information Science Technology Institute,
Zhengzhou 450002, China
wushuhua726@sina.com.cn

Abstract. This paper presents an efficient solution to the group key exchange problem in the password-based scenario. Our scheme can combine existing group protocols to get a hybrid group key exchange protocol which is efficient in terms of both computation and communication when the number of group users is very large. Our solution allows users to securely join and leave the group at any time—the so-called dynamic case. Moreover, we prove its security in the random oracle model.

Keywords: password authenticated, group key exchange, provably secure.

1 Introduction

A group key exchange protocol allows a group of users to exchange information over public network to agree upon a common secret key from which a session key can be derived. This common session key can later be used to achieve desirable security goals, such as authentication, confidentiality and data integrity. Secure virtual conferences involving up to one hundred participants is an example.

Due to the usefulness of such protocols, several papers have attempted to design secure group key exchange protocols. In order to protect against an active adversary who may inject messages, impersonate one or more of the parties, or otherwise control the communication in the network, these protocols need incorporate some authentication mechanism to be authenticated ones. The most classical way to add authentication to key exchange protocols is to sign critical message flows. Unfortunately, such techniques require the use of complex infrastructures to handle public keys and certificates. One way to avoid such infrastructures is to use passwords for authentication. Humans directly benefit from this approach since they only need to remember a low-quality string chosen from a relatively small dictionary (e.g. 4 decimal digits). However, since passwords are easily-guessed strings, many password-based systems are vulnerable to replay attack or dictionary attacks [1]. To design a secure password-based system is a precise task that has attracted many cryptographers.

---

* This work was partially supported by a grant from the National High Technology Research and Development Program of China (863 Program) (No. 2007AA01Z471).

Q. Li et al. (Eds.): APWeb/WAIM 2009, LNCS 5446, pp. 562–567, 2009.

During the last decades, the design of 2-party password-based authenticated key establishments has been explored intensively [2,3,4,5]. Nonetheless, very few group key exchange protocols have been proposed with password authentication. In [6,7], Bresson et al. firstly showed how to adapt their group Diffie-Hellman protocols to the password-based scenario. However, as the original protocols on which they are based, the total number of rounds is linear in the number of players, making their schemes impractical for large groups. More recently, several constant-round password-based group key exchange protocols have been proposed in the literature by Abdalla et al. [8,9], by Dutta and Barua [10], and by Kim, Lee, and Lee [11]. All of these constructions are based on the Burmester and Desmedt protocol [12,13]. However, as the original protocols on which they are based, if the group is large, then such solutions will involve a large number of computations for each participant, also making their schemes impractical for large groups. Other protocols, such as the Dutta and Barua's protocol for authenticated key exchange [14], do consider the group setting, but not in the password-based scenario.

In this paper, we propose an efficient solution to the group key exchange problem in the password-based scenario. It is very attractive while the number of group users is very large. We deal with the problem using ideas similar to those used in the Dutta and Barua's protocol [14]. Our solution consists of two stages. In the first stage, all group users are divided into several subsets and each user executes a regular group key exchange among its user subset. In the second stage, the users that represent the subsets authenticate each other and help to establish a new session key so that this value is known and only known to the members of the whole group. Our protocol has several advantages. First, our scheme can combine efficient regular group protocols to get a hybrid group key agreement protocol which is efficient in terms of both computation and communication. Second, our solution allows users to securely join and leave the wireless group at any time—the so-called dynamic case—[7]. Third, our solution is password-only authenticated: users need remember only short passwords, and no cryptographic key(s) of any kind. Finally, we can prove its security in the random oracle model[21].

2   Our Hybrid Protocol

In this section, we introduce our hybrid password-based group key exchange protocol. It combines a 2-party password-based key exchange, a regular password-based group key exchange and a $\mathcal{MAC}$-based key exchange. Our scheme can combine existing regular group protocols to get a hybrid group key agreement protocol which is efficient in terms of both computation and communication.

2.1   Description of the Hybrid Protocol

We assume all group users are divided into several subsets and subset has a representative. And passwords are pairwise shared among the users in each subset and among the representatives. Our solution consists of two stages. In the first stage, each user executes a regular group key exchange among its user subset. In the second stage, each user executes a $\mathcal{MAC}$-based key exchange in an authenticated way under the aid of the representative of its subset to establish a new session key so that this value is known and only known to the members of the whole group. For simplicity, and without loss of generality, we assume all group users are divided into two subsets: user subset A and B. Also we assume U_A is the representative of A and U_B the representative of B. We denote by $\mathcal{H} : G \rightarrow \{0,1\}^l$ is a random oracle with l the security parameter.

The protocol runs as follows:

1. Users execute a regular group key exchange $\mathcal{P}_1$ among its subset A or B to establish a secure common agreed key $sk_A = (x, k_1)$ and $sk_B = (y, k_2)$ respectively, where x, y represents the higher parts and k_1, k_2 represents the rest parts of sk_A and sk_B respectively. Note sk_A and sk_B are only known to the members in subset A and B respectively.

2. The two representatives help to execute a $\mathcal{MAC}$-based key exchange between A and B in an authenticated way. At first, each player U chooses a random nonce N_U and broadcasts $T_U = (U, N_U)$. Then U_A and U_B execute a 2-party password-based key exchange $\mathcal{P}_2$ to establish a secure common agreed key k_3. Afterward, U_A computes $X = g^x$ and the $\mathcal{MAC}$ digest of the string X under the key k_3: $\alpha_1 = \mathrm{MAC}_{k_3}(N_{AB}\|X)$, and sends (X, α_1) to U_B, where N_{AB} is the concatenation of such N_U that U is either U_A or U_B. Similarly, U_B also computes $Y = g^y$ and $\alpha_2 = \mathrm{MAC}_{k_3}(N_{AB}\|Y)$ and sends (Y, α_2) to U_A. Upon receiving the message from the other side, each representative verifies the the $\mathcal{MAC}$ digest is valid in the straight way. If α_2 is valid, U_A proceeds to compute the $\mathcal{MAC}$ digest of the string Y under the key k_1: $\beta_1 = \mathrm{MAC}_{k_1}(N_A\|Y)$ and broadcasts (Y, β_1) among A, where N_A is the concatenation of all such N_U that U is a member in A. If α_1 is valid, U_B proceeds to compute $\beta_2 = \mathrm{MAC}_{k_2}(N_B\|X)$ and broadcasts (X, β_2) among B, where N_B is the concatenation of all such N_U that U is a member in B. Upon receiving the message from the representative, each user checks that the $\mathcal{MAC}$ digest is valid. If it is valid, he proceeds to complete the $\mathcal{MAC}$-based key exchange to establish a secret key g^{xy}. Please note, to guarantee the security, the session key is set to be the hash (random oracle) of the secret key g^{xy}.

The solution can be easily extended to the more general case where all group users are divided into more than two subsets. Actually, we can combine two subsets as one and set a user to be their representative when the members in the two subsets have established a common agreed key. In this case, we can combine all the subsets into one by repeating the operations of the second stage after each user has executed a regular group key exchange $\mathcal{P}_1$ among its subset. Our proposal is essentially a solution to tree-based group key agreement in password-based setting. The procedure in combination can be designed in an optimized way so that a user is enabled to join or leave the group efficiently. The algorithm of the combination designed in non-password-based setting [14] still works in our case.

Furthermore, we can prove our hybrid password-based key authenticated protocol described above is secure in the random oracle model. The security mode is that used in [22]. Due to limitation of the paper length, the complete proof is omitted here.

2.2 Practical Considerations

Our protocol has been considered a lot from practical prospective.

First, our solution assumes that all group users are divided into several subsets in the first stage. This requirement is reasonable because, in real life, the membership of the group is not built once and for all but is built incrementally as the network topology evolves [15,16,17].

Second, our scheme can combine efficient regular group protocols to get a hybrid group key agreement protocol which is efficient in terms of both computation and communication. Since executing the existing regular group protocols among all the users may be expensive from computation or communication point of view when the number of users in the whole group is large, our solution applies the regular group key exchange operation on the subsets of small size. Besides, these operations can be done in parallel. In real life, the participants may be part of several pools at the same time and, therefore, may run multiple key exchanges in parallel. Our solution will be especially efficient in this case. Moreover, in our case, both the computation cost and communication cost are linear in the logarithm of the group size if the size of each subset is upper bounded by a constant. Since our proposal is essentially a solution to tree-based group key agreement in password-based setting, the costs largely depend on the depth of the tree. None of the existing password-based solutions has the attractive feature. For some of them, such as [6,7], the total number of rounds is linear in the group size. For some of them, such as the one proposed by M. Abdalla et al. [9] , the amount of the most expensive part of computation by each user is linear in the group size.

Third, our solution is also straightward to allows users to securely join and leave the group at any time —the so-called dynamic case—[7]. When a user joins or leaves the group, the related subset, say A, re-establish a fresh session key $sk'_A = (x', k'_1)$ among the subset A and sends $g^{x'}$ to the subset B in an authenticated way. In the end, the group will accept a new session key $g^{x'y}$ among the whole group. We note that re-running the protocol from scratch is always possible, and hence the goal of such operations is to provide an efficient means to update the existing session key into a new one. Other constant round protocols in the literature, such as the one proposed by M. Abdalla et al. [18] and the two in [19,20] do not have the attractive feature. In addition, some optimization could also be achieved on how to update the session key sk_A in the first stage but it depends on the underlying protocol.

Finally, our solution is password-only authenticated protocol: users need remember only a short password, and no cryptographic key(s) of any kind. However, some previous protocols, such as the protocol proposed recently in [9], assume that each participant must store a pair of secret and public keys; in some sense, this obviates the reason for considering password-based protocols in the first place: namely, that human users cannot remember or securely store long, high-entropy keys.

3   Conclusion

We have presented an efficient hybrid password-based authenticated group key exchange protocol. Our protocol has been considered much more from practical prospective. It is very attractive while the number of group users is very large. Furthermore, we proved its security in the random oracle model.

References

1. Bellovin, S.M., Merritt, M.: Encrypted key exchange: password-based protocols secure against dictionary attacks. In: Proc. of 1992 IEEE Computer Society Symp. on Research in security and Privacy, pp. 72–84 (May 1992)
2. Bresson, E., Chevassut, O., Pointcheval, D.: Security proofs for an efficient password-based key exchange. In: ACM CCS 2003, pp. 241–250. ACM Press, New York (2003)
3. Bresson, E., Chevassut, O., Pointcheval, D.: New security results on encrypted key exchange. In: Bao, F., Deng, R., Zhou, J. (eds.) PKC 2004. LNCS, vol. 2947, pp. 145–158. Springer, Heidelberg (2004)
4. Abdalla, M., Pointcheval, D.: Simple password-based encrypted key exchange protocols. In: Menezes, A. (ed.) CT-RSA 2005. LNCS, vol. 3376, pp. 191–208. Springer, Heidelberg (2005)
5. Abdalla, M., Chevassut, O., Pointcheval, D.: One-time verifier-based encrypted key exchange. In: Vaudenay, S. (ed.) PKC 2005. LNCS, vol. 3386, pp. 47–64. Springer, Heidelberg (2005)
6. Bresson, E., Chevassut, O., Pointcheval, D.: Group diffie-hellman key exchange secure against dictionary attacks. In: Zheng, Y. (ed.) ASIACRYPT 2002. LNCS, vol. 2501, pp. 497–514. Springer, Heidelberg (2002)
7. Bresson, E., Chevassut, O., Pointcheval, D.: A Security Solution for IEEE 802.11's Ad-hoc Mode: Password-Authentication and Group-Diffie-Hellman Key Exchange. International Journal of Wireless and Mobile Computing 2(1), 4–13 (2007)
8. Abdalla, M., Bresson, E., Chevassut, O., Pointcheval, D.: Password-based group key exchange in a constant number of rounds. In: Yung, M., Dodis, Y., Kiayias, A., Malkin, T.G. (eds.) PKC 2006. LNCS, vol. 3958, pp. 427–442. Springer, Heidelberg (2006)
9. Abdalla, M., Pointcheval, D.: A scalable password-based group key exchange protocol in the standard model. In: Lai, X., Chen, K. (eds.) ASIACRYPT 2006. LNCS, vol. 4284, pp. 332–347. Springer, Heidelberg (2006)
10. Dutta, R., Barua, R.: Password-based encrypted group key agreement. International Journal of Network Security 3(1), 30–41 (2006), http://isrc.nchu.edu.tw/ijns
11. Kim, H.-J., Lee, S.-M., Lee, D.H.: Constant-round authenticated group key exchange for dynamic groups. In: Lee, P.J. (ed.) ASIACRYPT 2004. LNCS, vol. 3329, pp. 245–259. Springer, Heidelberg (2004)
12. Burmester, M., Desmedt, Y.: A secure and efficient conference key distribution system. In: De Santis, A. (ed.) EUROCRYPT 1994. LNCS, vol. 950, pp. 275–286. Springer, Heidelberg (1995)

13. Burmester, M., Desmedt, Y.: A secure and scalable group key exchange system. Information Processing Letters 94(3), 137–143 (2005)
14. Dutta, R., Barua, R.: Dynamic Group Key Agreement in Tree-Based Setting. In: Boyd, C., González Nieto, J.M. (eds.) ACISP 2005. LNCS, vol. 3574, pp. 101–112. Springer, Heidelberg (2005)
15. Agarwal, D., Chevassut, O., Thompson, M.R., Tsudik, G.: An integrated solution for secure group communication in wide-area networks. In: Proc. of 6th IEEE Symposium on Computers and Communications, pp. 22–28. IEEE Computer Society Press, Los Alamitos; also Technical Report LBNL-47158, Lawrence Berkeley National Laboratory
16. Amir, Y., Kim, Y., Nita-Rotaru, C., Schultz, J., Stanton, J., Tsudik, G.: Secure group communication using robust contributory key agreement. IEEE Transactions on Parallel and Distributed Systems 15(5), 468–480
17. Rodeh, O., Birman, K.P., Dolev, D.: The architecture and performance of the security protocols in the ensemble group communication system. ACM Trans. on Information and System Security 4(3), 289–319 (August)
18. Abdalla, M., Bresson, E., Chevassut, O., Pointcheval, D.: Password-based group key exchange in a constant number of rounds. In: Yung, M., Dodis, Y., Kiayias, A., Malkin, T.G. (eds.) PKC 2006. LNCS, vol. 3958, pp. 427–442. Springer, Heidelberg (2006)
19. Dutta, R., Barua, R.: Password-based encrypted group key agreement. International Journal of Network Security 3(1), 30–41 (2006), http://isrc.nchu.edu.tw/ijns
20. Lee, S.-M., Hwang, J.Y., Lee, D.H.: Efficient password-based group key exchange. In: Katsikas, S.K., López, J., Pernul, G. (eds.) TrustBus 2004. LNCS, vol. 3184, pp. 191–199. Springer, Heidelberg (2004)
21. Bellare, M., Rogaway, P.: Random oracles are practical: A paradigm for designing efficient protocols. In: ACM CCS 1993: 1st Conference on Computer and Communications Security, Fairfax, Virginia, USA, pp. 62–73. ACM Press, New York (1993)
22. Abdalla, M., Fouque, P.-A., Pointcheval, D.: Password-based authenticated key exchange in the three-party setting. In: Vaudenay, S. (ed.) PKC 2005. LNCS, vol. 3386, pp. 65–84. Springer, Heidelberg (2005)

Optimal K-Nearest-Neighbor Query in Data Grid[*]

Yi Zhuang[1,2], Hua Hu[1,3], Xiaojun Li[1], Bin Xu[1], and Haiyang Hu[1]

[1] College of Computer & Information Engineering, Zhejiang Gongshang University, P.R. China
[2] Zhejiang Provincial Key Laboratory of Information Network Technology, P.R. China
[3] Hangzhou Dianzi University, P.R. China
{zhuang,huhua,lxj,xubin,hhy}@zjgsu.edu.cn

Abstract. The paper proposes an optimal distributed k Nearest Neighbor query processing algorithm based on Data Grid, called the opGkNN. Three steps are incorporated in the opGkNN. First when a user submits a query with a vector V_q and a number k, an iDistance[3]-based vector set reduction is first conducted at data node level in parallel. Then the candidate vectors are transferred to the executing nodes for the refinement process in which the answer set is obtained. Finally, the answer set is transferred to the query node. The experimental results show that the performance of the algorithm is efficient and effective in minimizing the response time by decreasing network transfer cost and increasing the parallelism of I/O and CPU.

1 Introduction

In high-dimensional databases, a k-Nearest-Neighbor(k-NN) query is a popular and expensive query operation which is to find the k most similar objects in the database with respect to a given object. Although a considerable amount of related work has been carried out on multi- or high-dimensional indexing and k-NN query in high-dimensional spaces [4], they all focus on a centralized (i.e., single-PC-based) k-NN query which do not scale up well to large volume of data. Therefore the design of high performance k-NN query methods becomes a critically important research topic.

Grid technology, especially data grid, provides a new and high performance computing infrastructure [1] which motivates us to speed up a query performance by using grid techniques. Compared with a CGkNN query algorithm proposed in [5], the paper considers an optimal grid- based distributed k-NN query(opGkNN) that mainly focuses on the optimal enabling techniques to further improve the query efficiency.

As discussed in [5], for the grid-based similarity query, the challenges include two main aspects: 1) *The instability of data grid environment*, and 2) *The heterogeneity of*

---

[*] This paper is partially supported by the Program of National Natural Science Foundation of China under Grant No. 60873022; The Program of Natural Science Foundation of Zhejiang Province under Grant No. Y1080148; The Key Program of Science and Technology of Zhejiang Province under Grant No. 2008C13082; The Open Project of Zhejiang Provincial Key Laboratory of Information Network Technology; The Key Project of Special Foundation for Young Scientists in Zhejiang Gongshang University.

Q. Li et al. (Eds.): APWeb/WAIM 2009, LNCS 5446, pp. 568–575, 2009.

data grid environment. So in this paper, to address the above challenges, based on the CG*k*NN [5], we propose an optimal distributed *k*-NN(opG*k*NN) query processing technique in data grid. It includes two optimal enabling techniques such as *an adaptive load balancing scheme* and *a histogram-based minimal query radius estimation*. Extensive experiments indicate that this method is better than the CG*k*NN [5]. The contributions of this paper are as follows:

- We introduce an optimal Grid-based *k*-Nearest-Neighbor(opG*k*NN) query algorithm including three enabling techniques to further improve search efficiency of a *k*-NN query.
  - Extensive experimental studies are conducted to validate its query efficiency.

The rest of paper is organized as follows. Preliminary work is given in Section 2. In Section 3, two enabling techniques, viz., the *adaptive load balancing scheme* and the *histogram-based estimation of minimal query radius* are introduced to facilitate a fast *k*-NN search over data grid. In Section 4, we propose an opG*k*NN query processing algorithm. In Section 5, we perform comprehensive experiments to evaluate the efficiency of our proposed method. We conclude the paper in Section 6.

2 Preliminaries

Definition 1. *A data grid is a graph which is composed of Node and Edge, formally denoted as G=(N,E), where N refers to the set of nodes, E refers to a set of edges representing the network bandwidths for data transfer.*

Definition 2. *The nodes in a data grid, formally denoted as $N=Nq+Nd+Ne$, can be logically divided into three categories: the query node(Nq), data nodes(Nd) and executing nodes(Ne), where Nd is composed of α data nodes(N_d^i) and Ne is composed of β executing nodes(N_e^j), where N_d^i is the i-th data node, N_e^j denotes the j-th executing node, for $i \in [1,\alpha]$ and $j \in [1,\beta]$ (cf. Figure 1).*

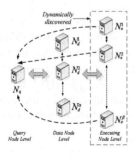

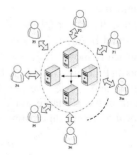

Fig. 1. The architecture of a data grid environment   **Fig. 2.** An example of query polling

Symbols	Meaning	Symbols	Meaning
Ω	a set of vectors	$d(V_i, V_j)$	the distance between two vectors
V_i	the i-th vector and $V_i{\in}\Omega$	Ω'	the candidate vector set
D	the number of dimensions	Ω''	the answer vector set
n	the number of vectors in Ω	$|\bullet|$	the number of vectors in $\bullet$
V_q	a query vector user submits	α	the number of the data nodes
$\Theta(V_q, r)$	the query sphere with centre V_q and radius r	β	the number of the executing nodes

As defined in Definition 2, in a data grid, the functionalities of the query node, the α data nodes and the β executing nodes are same to that of [5]. The list of symbols used throughout the rest of this paper is first summarized in Table 1.

3 Optimal Enabling Techniques

To better facilitate efficient k-Nearest-Neighbor query processing in data grid environment, in this section, based on the enabling techniques proposed in [5], we introduce two optimal techniques including *a dynamic load balancing scheme*(including a query-polling(QP)-based method to estimate processing capacity of each node and a optimal start-distance(SD)-based data partition scheme) and *a histogram-based estimation of minimal query radius*, respectively. The purpose of them is to minimize the transfer cost and maximize the query parallelism.

3.1 Dynamic Load Balancing Scheme in Data Nodes

• A QP-Based Approach to Estimate the Processing Capacity of Each Node

For many existing parallel database systems, the study of data partition is critically important to the efficiency of parallel query processing. Different from some traditional parallel systems, the data grid is a heterogeneous environment in which the processing capacity of each node is different. It is relatively hard to accurately model a processing capacity of a node. So we propose a query-polling(QP)-based approach to deal with this challenge.

Specifically, in our QP-based approach, the processing capacity of each data node can be measured by the average corresponding time of a same query from different users' ends, which is shown in Figure 2. Formally, given m users and x nodes, let the total corresponding time of the i-th node from the j-th user be T_{ij}, where $i{\in}[1,x]$ and $j{\in}[1,m]$. Thus for m users and x nodes, we can obtain a table called the U_ser-T_ime(UT) table.

Based on the UT table, for the i-th node, its processing capacity (denoted as ρ_i) is reciprocal to the average corresponding query processing time, which is shown below:

$$\rho_i \propto \frac{1}{\frac{1}{m}\sum_{j=1}^{m} T_{ij}} = \frac{m}{\sum_{j=1}^{m} T_{ij}} \qquad (1)$$

That is, as shown in Eq.(1), the average corresponding time can be modeled by a function of ρ_i:

Based on the above analysis, for the i-th node, the percentage of its processing capacity can be derived as follows:

$$Per(i) = \frac{1\big/\sum_{j=1}^{m} T_{ij}}{1\big/\sum_{j=1}^{m} T_{1j} + 1\big/\sum_{j=1}^{m} T_{2j} + \ldots + 1\big/\sum_{j=1}^{m} T_{xj}} \tag{2}$$

where $i \in [1,x]$ and $j \in [1,m]$.

Algorithm 1. Query polling Algorithm

Input: Ω: the vector set, the x nodes, m users;
Output: ρ_i: the processing capability of x nodes;
1. **for** i:=1 to x **do**
2.   **for** j:=1 to m **do**
3.   send a query to the i-th node to perform the query respectively from the j-th user's end;
4.   record the corresponding time T_{ij};
5.   the processing capacity of the i-th node (ρ_i) can be obtained by Eq. (1);
6.   the percentages of the processing capacity of each node can be obtained by Eq. (2);

- **An Optimal SD-Based Data Partition Scheme**

As mentioned in [5], to maximize the parallelism of vector set reduction at data node level, based on the query-polling technique mentioned above, an optimal *start-distance(SD)*-based data partition scheme is proposed, which is suitable for the optimal data partition in the heterogeneous nodes since the processing capacity of each node is different.

Algorithm 2. Vector allocation in data nodes

Input: Ω: the vector set, the α data nodes;
Output: $\Omega(1$ to $\alpha)$: the placed vectors in data nodes;
1. The high-dimensional space is equally divided into α slices in terms of start-distance;
2. **for** j:=1 to α **do**
3.   the $|n \times Per(j)|$ vectors($\Omega(j)$) are randomly selected from the each sub-range of start distance respectively;
      /* $Per(j)$ is defined in Eq. (2) */
4. $\Omega(j)$ is deployed in the j-th data node;
5. **end for**

3.2 Histogram-Based Estimation of MQR

As we know, k-NN query is a time-consuming operation which starts with a small query sphere, then incrementally enlarges its radius, till k nearest neighbor vectors are found. It is possible that a k-NN query will be progressively improved if we can estimate a minimal query radius(MQR) for some k in k-NN queries as accurate as possible. Therefore we propose a novel *histogram-based approach* to estimate the value of MQR for opGkNN queries.

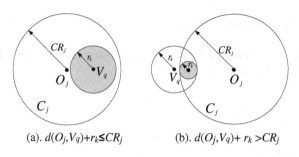

(a). $d(O_j,V_q)+r_k \leq CR_j$ (b). $d(O_j,V_q)+ r_k > CR_j$

Fig. 3. Radius estimation for query $\Theta(Vq,r_k)$

Specifically, in high-dimensional spaces, for any cluster C_j that intersects with the query $\Theta(Vq,r_k)$, we estimate the approximate number of vectors that can be retrieved by the query based on Eqs.(3-4). Assuming a radius r_k, for any intersecting cluster C_j, there are two cases to be considered: 1) the inequality $d(O_j,Vq)+r_k \leq CR_j$ holds, or 2) cluster C_j partially intersects with the range query $\Theta(Vq,r_k)$.

In the first case, we estimate the number of vectors retrieved by the query $\Theta(Vq,r_k)$ from cluster C_j:

$$num(\Theta(V_q,r_k)) = |C_j| * \frac{Vol(\Theta(V_q,r_k))}{Vol(\Theta(O_j,CR_j))} = |C_j| * \left(r_k / CR_j\right)^d \tag{3}$$

In the second case, the query sphere $\Theta(Vq,r_k)$ partially intersects with a cluster C_j, the number of vectors retrieved by the query from cluster C_j is estimated based on a smaller radius that corresponds to the intersection and is approximated by $r_k' = \frac{CR_j + r_k - d(O_j,V_q)}{2}$. The estimated number of vectors retrieved from C_j is:

$$num(\Theta(V_q,r_k) \cap \Theta(O_j,CR_j)) = |C_j| * \frac{Vol(\Theta(V_x,r_k'))}{Vol(\Theta(O_j,CR_j))} = |C_j| * \left(\frac{CR_j + r_k - d(O_j,V_q)}{2CR_j}\right)^d \tag{4}$$

The query area in Figure 3(b) is represented by a shaded circle that is contained in the cluster, which is smaller than the real intersection, and leads to an underestimation of the number of vectors contained, therefore an overestimation of radius r_k.

Algorithm 3. k-NN radius estimation
Input: Vq, k, S: Maximum histogram bin of any cluster C_j
Output: Estimated radius r_k
1. $sum \leftarrow 0$; $r_k \leftarrow rB$
2. **for** $i=1$ to S **do**
3.   **for** each cluster C_j **do**
4. $r_k \leftarrow i*rB$
5.   **if** $(d(O_j,Vq)+r_k \leq CR_j)$ **then** $sum \leftarrow sum+|C_j| * \left(r_k / CR_j\right)^d$;
6.   **if** $(d(O_j,Vq)+ r_k > CR_j)$ **then** $sum \leftarrow sum+|C_j| * \left(\frac{CR_j + r_k - d(O_j,V_q)}{2CR_j}\right)^d$;
7.   **if** $(sum>k)$ **and** $flag$(contained)=TRUE **then return** r_k , break;
8.   **if** $flag$(intersected)=TRUE **and** $(sum>k)$ **then return** r_k , break;

4 The opGkNN Query Algorithm

As mentioned above, in essence, an opGkNN query is completed by iteratively invoking the range queries based on data grid. Similar to the CGkNN [5], the algorithm is composed of three stages: 1). *Global vector set reduction(i.e., GVReduce)*, 2). *Hash mapping*, and 3). *Refinement (i.e., Refine)*, which are detailed in [5]. Algorithm 4 shows the detailed steps of our proposed opGkNN algorithm. It is worth mentioning that steps 5-9 are executed in parallel.

Algorithm 4. opGkNNSearch(Vq, k)
Input: a query vector Vq, k
Output: the query result S
1. $r \leftarrow 0$, $\Omega' \leftarrow \Phi$, $\Omega'' \leftarrow \Phi$; /* initialization */
2. a query request is submitted to the query node Nd;
3. the minimal query radius (R_{min}) is obtained by the histogram;
4. $r \leftarrow R_{min}$;
5. $\Omega' \leftarrow$ **GVReduce**(Vq, r);
6. the β nodes are discovered by the *GRMS* as the executing nodes;
7. the candidate vector set Ω' is sent to the β executing nodes by hash mapping;
8. $\Omega'' \leftarrow$ **Refine**(Ω', Vq, r);
9. the answer vector set Ω'' is sent to the query node Nq;
10. **if** ($\| \Omega'' \| > k$) **then** /* if the number of answer vectors is larger than k */
11.   remove the ($\| \Omega'' \| - k - 1$) vectors from Ω'' which are farthest to Vq
12. **else if** ($\| \Omega'' \| < k$) **then**
13.   **while** ($\| \Omega'' \| < k$)
14. $r \leftarrow r + \Delta r$;
15. $\Omega' \leftarrow$ **GVReduce**(Vq, r);
16.   the candidate vectors in Ω' are disseminated to the β executing nodes by hash mapping;
17. $\Omega'' \leftarrow$ **Refine**(Ω', Vq ,r);
18.   the answer vector set Ω'' is sent to query node Nq;
19.   **if** ($\| \Omega'' \| > k$) **then** remove the ($\| \Omega'' \| - k - 1$) vectors from Ω'' which are farthest to Vq
20. **return** Ω''

5 Experimental Results

In this section, to verify the efficiency of the proposed method, we conduct two simulation experiments to demonstrate the performance of the opGkNN method. We have implemented the iDistance[3]-based vector set reduction algorithm in C language and the index is deployed at the data node level. B$^+$-tree is used as a single-dimensional index structure. All experiments are run in a local network. We use two data sets as experimental data: (i). The color histogram data from *UCI KDD Archive* [6], which includes 68040 32-D color histogram features, the value range of each dimension is between 0 and 1; (ii). The computer randomly generated 100-D 5,000,000 vectors which are uniformly distributed, with each dimension's value ranging between 0 and 1.

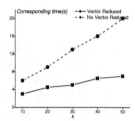
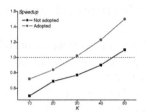

Fig. 4. Effect of VSR   **Fig. 5.** Effect of Histogram

5.1 Effect of VSR

In the experiment, we study the effect of vector set reduction(VSR) on the performance of the k-NN query processing. Two methods are performed and compared for this purpose. The first does not adopt the VSR algorithm and the second uses this algorithm. Figure 7 shows that when k is fixed, the total response time using the vector set reduction is superior to that of no VSR. Meanwhile, with the increase of k, the performance gap becomes larger since the candidate vectors which are filtered by the iDistance[3] dramatically reduces the network transfer cost, therefore the time T_r for VSR is far less than the response time of network transfer, and T_r can be neglected.

5.2 Effect of Histogram

In the last experiment, we study the effect of histogram on the performance of distributed k-NN query processing. The speedup refers to the ratio of the response time for a single-PC- based k-NN query to that of Grid-based one. We also use the second data set as the experimental data and comparatively perform the two experiments by using two schemes: *adopted* and *not adopted*. In Figure 5, the speedup gap between two schemes is increasing rapidly with the increase of k, it is expected that the *histogram*-based estimation of minimal query radius can effectively boost up the query performance by significantly reducing the loop time of range searches.

6 Conclusions

In this paper, based on the CGkNN [5], we have presented a optimal Grid-based k-Nearest- Neighbor(opGkNN) query processing, which specifically caters for the different bandwidth of nodes in the data grid. Two optimal enabling techniques, viz., adaptive load balancing scheme which includes the *query-polling(QP)-based processing capacity estimation of each node* and the *start-distance(SD)-based data partition policy*, and the histogram-based estimation of query radius are proposed to reduce the query response time. The experimental studies demonstrate the validity of the proposed opGkNN method.

References

[1] Foster, I., Kesselman, C.: The Grid: Blueprint for a New Computing Infrastructure. Morgan Kaufmann, San Francisco (1998)

[2] Böhm, C., Berchtold, S., Keim, D.: Searching in High-dimensional Spaces: Index Structures for Improving the Performance of Multimedia Databases. ACM Computing Surveys 33(3) (2001)

[3] Jagadish, H.V., Ooi, B.C., Tan, K.L., Yu, C., Zhang, R.: iDistance: An Adaptive B+-tree Based Indexing Method for Nearest Neighbor Search. TODS 30(2), 364–397 (2005)

[4] Berchtold, S., Bohm, C., Braunmuller, B., Keim, D.A., et al.: Fast Parallel Similarity Search in Multimedia Databases. In: SIGMOD, pp. 1–12

[5] Zhuang, Y., Zhuang, Y.T., Li, Q., Wu, F.: Speeding Up Similarity Queries over Large Chinese Calligraphic Character Databases Using Data Grid. In: GCC, pp. 225–236 (2007)

[6] UCI KDD Archive (2002), http://www.kdd.ics.uci.edu

A Routing Scheme with Localized Movement in Event-Driven Wireless Sensor Networks

Fang-Yie Leu, Wen-Chin Wu, and Hung-Wei Huang

Database and Network Secuity Lab, Department of Computer Science,
Tunghai University, Taiwan
leufy@thu.edu.tw, g96350038@thu.edu.tw

Abstract. In wireless sensor networks (WSNs), energy is one of the most important resources that should be economically used. But most routing approaches employed by WSNs are hop-by-hop relay schemes, which cause sensors along a routing path not only collecting its environmental data, and then sending the data to base station, but also relaying data received from its neighbors toward base station. This will result in an unbalanced energy consumption problem. In the paper, we propose a routing approach, named Routing Assistant Scheme with Localized Movement (RASLM) which deploys mobile sensors to help relay of packets for nodes along an active routing path so as to prevent them from dying much earlier than others. This scheme can also stabilize the routing path to ensure that the sensed data can be sent to the base station safely and smoothly. Experimental results show that the RASLM can effectively improve the system lifetime in event-driven wireless sensor networks.

Keywords: RASLM, wireless sensor networks, event-driven, mobile nodes, energy hole problem.

1 Introduction

A WSN often consists of a huge number of sensor nodes. But, sensor power is limited. To overcome power constraint of sensor nodes so as to prolong a WSN's system lifetime and balance workload of its nodes, the ability of movement of sensor nodes or base station (BS) has become one of the hottest research topics nowadays.

Many studies on WSNs have tried to prolong a network's lifetime by balancing power consumption among static sinks or nodes. In a uniformly distributed and homogeneous node environment, if static nodes or static sinks are used, energy hole near base station can not be avoided [1]. Wu and Chen [4] proposed an approach to solve energy hole problem.

To follow the trend of sensor network development, in this paper, we propose a routing approach, named Routing Assistant Scheme with Localized Movement (RASLM) for a uniformly distributed event-driven WSN. This scheme deploys mobile nodes to help relay of packets for nodes along an active routing path in such an environment. The propose is to prolong lifetime of a node N on the active routing path that has suffered from relaying too many packets. The mobile nodes located near N

Q. Li et al. (Eds.): APWeb/WAIM 2009, LNCS 5446, pp. 576–583, 2009.

when necessary can move toward N and share relaying burden with N so as to mitigate $N's$ energy consumption speed. Experimental results show that RASLM can effectively prolong lifetime of a WSN for a uniformly distributed event-driven WSN. In the following, we use energy consumption rate and workload interchangeably, even though someone defines them differently.

The rest of the paper is organized as follows. Section 2 introduces background and related work of the paper. In section 3, we explain the preliminaries used in this study. Section 4 presents the detail of RASLM scheme. The performance of the proposed scheme is evaluated in section 5. Section 6 concludes this paper and addresses our future work.

2 Background and Related Work

In this section, we classify studies relevant to mobility of wireless sensor networks into two categories, those based on mobile sinks (base stations), and those based on mobile nodes.

2.1 Mobile Sinks

Many researchers have tried to improve a WSN's lifetime by using single or multiple mobile sinks. Wu et al. [2] proposed a method called Dual Sink which combines mobile and static sinks to change routing destination. A node forwards packets to one of the sinks based on distance between it and the sinks. This approach truly extends the lifetime of a wireless sensor network, but too much overhead is required to maintain the routing path. Wang et al. [5] proposed an optimal predefined mobility trajectory of the base station to balance the energy consumption of sensors in event-driven WSNs. However, the two approaches are all developed on static sensor environments.

2.2 Mobile Sensor Nodes

Unlike the sink mobility, there are a lot of studies trying to solve unbalanced energy consumption problem for WSNs. Yang et al. [3] proposed a local load balance solution by modifying the Hungarian-method-based optimal solution, and integrating with movement assisted sensors to redeploy their positions into a balanced grid. Leu et al. [6] adjusted node energy reserved for sending packets generated by itself, its direct out-neighbors and other outer coronas to prolong system life.

3 Preliminaries

Energy consumption in an event-driven environment and in a message-evenly-generated environment is different in that in an event-driven environment nodes collect environmental data only when they are triggered by external events or by user commands. When not triggered or requested to relay packets, they do nothing but listen to its upstream nodes. In a message-evenly-generated environment, nodes generate packets periodically with the same sensing rate. In both environments, nodes close to base station should relay packets for those far away, consequentially resulting in an energy hole.

In the following, we will use nodes and sensors interchangeably and analyze energy a sensor in an event-driven but uniformly distributed environment may consume.

3.1 Event-Driven Environments

In such a network, sensors, e.g., nodes A, B, D and W shown in Fig. 1, might be triggered at the same time. They forward data packets/messages to the base station through a hop-by-hop routing approach. Node A forwards its sensing data to the base station. The packets it generates need to be relayed six times before they can arrive at the base station. Node C relays packets for, e.g., nodes A and B. Node D, in addition to transmitting its packets, also relays packets generated by nodes A and B. Node B does the same, but it only relays packets for node A. The workload and power consumption of nodes A, B, C, D and E are different. Obviously, nodes close to base station will exhaust their energy far faster than those far away from the base station. An active routing path (an active path for short) P in a WSN is a routing path established when P's source node, e.g., node A or node W, is triggered by its surrounding events. The source node then broadcasts a route request packet. After the events are solved or removed, and the source node stops sending packets, P will be in an inactive state.

3.2 Energy Hole Problem and Non-surveillant Zones

Basically, in an event-driven WSN, if nodes triggered by events, e.g., nodes A and B, continue to transmit their sensing data to the base station through one or several routing paths, nodes on the path but near the base station, like nodes E and F in Fig.1, will die soon, producing a non-surveillant zone (NSZ). Situation becomes worse when nodes E and F are articulation nodes. An articulation node is a node that joins at least two upstream paths and at least one downstream path. This might be very dangerous for critical monitoring missions, such as fire monitoring and detection, and chemical/nuclear reaction detection. So, to prolong the lifetime of an event-driven WSN, we have to first reduce energy consumed by articulation nodes. The nodes are very often not far away from the base station.

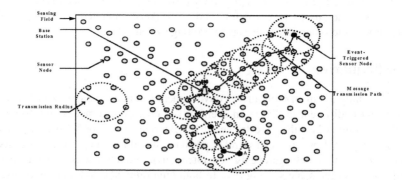

Fig. 1. An event-driven based wireless sensor network in which nodes are randomly distributed

In this study, a node computes its energy consumption rate periodically. A node N with a very high energy consumption rate will soon exhaust its energy. However, if N's nearby "duty-free" nodes can share burden of relaying packets with it, N's lifetime and of course the WSN's lifetime can be effectively prolonged.

3.3 Mobility Models

To reduce the probability of forming NSZ or energy holes in an event-driven WSN, we assume that when events occur, all its sensors are able to move to accurate positions by themselves. We use the mobility model of sensors similar to that defined by Chellappan et al. in [7] to migrate sensors. In this model, sensor mobility is something like a flip, and the authors assumed that a sensor could move from its current position to a new location by using the power of propellers, fuels, coiled springs unwinding during flips, external agents launching sensors, etc, implying this sensor's movement consumes no battery energy of its own.

4 Proposed Protocol

In the following, we would like to introduce how RASLM prolongs lifetime of an articulation node on an active path.

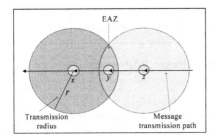
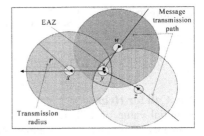

Fig. 2. Node y's EAZ is the intersection region of node x's and z's communication ranges

Fig. 3. Node y's EAZ is the intersection region of nodes x's, w's and z's communication ranges

4.1 Effective Assistant Zone

In RASLM, when a node N on an active path realizes that its energy consumption rate is higher than a predefined threshold δ, it broadcasts a Request-Assist Packet (RAP) to request its neighbors to come to relay packets for it. The effective assistant zone (EAZ) of a node, e.g., node y, as shown in Fig. 2, is defined as the intersection region of communication ranges of node y's immediate upstream and immediate downstream nodes, i.e., nodes z and x, respectively. Node y's EAZ is the only region that its neighbors coming to help it can communicate with both its immediate upstream and downstream nodes. Fig.3 show another example.

4.2 Broadcasting RAPs

A RAP packet as shown in Fig.4 consists of five main fields. The first is RAP-ID which comprises two sub-fields, sender ID and serial-number. The former shows who sends the packet, whereas the latter is a number used to discriminate different RAPs. Each time when a node broadcasts a new RAP, it increases its serial number by one to ensure the uniqueness of the RAP-ID. The second field is the sender's coordinates with which an assistant node can realize where it should move to. The third field is TTL (time to live) value which is used to control the broadcast range of a RAP packet. When the energy consumption rate of a node is higher, it broadcasts a RAP packet with a higher TTL so that many more nearby nodes can receive the packets and come to help the packet sender. The fourth field is a list of direct neighbors which records the sender's immediate upstream and downstream nodes. An assistant node on deciding to help the sender of a received RAP packet must keep its antenna working while moving toward the sender. Once they can receive all the radio signals sent by those nodes listed in the RAP received, it can realize that it is now in the EAZ, thus stopping its movement. This is helpful in shortening an assistant node's moving distance, consequentially saving its moving energy. The last field is timestamp T_s which is used to tell receiving nodes that this "call for help" will time out at timestamp T_s+T_m, where T_m, a predefined time period, is time interval of a call-for-help round.

RAP-ID		Sender's Coordinates	TTL	List of direct neighbors	Timestamp
Sender ID	Serial-No.				

Fig. 4. Format of a Request-Assist Packet (RAP) packet

4.2.1 Call for Help

The initial value of TTL field is 1, which can be adjusted according to system requirements. The value can be either statically or dynamically set. When the network scale is small, broadcasting RAPs with static TTL, which is fixed to a constant, is not harmful to system lifetime. But, when the network scale is large, TTL value should be adjusted according to actually need of the relay nodes. On each call for help, the sender node S evaluates its own workload $W(S)$ by using the formula $W(S)=Data\_Length \cdot (u \cdot e_1 + j \cdot e_2) / T_m \cdot (N_A + 1)$, where Data_Length is the length of a packet that a sensor sends out, u is number of sensor node along a routing segment between S and the source node of the routing path, $j=u-1$, T_m is time interval of a call-for-help round, and N_A is number of assistant nodes. If the burden is still higher than its threshold, it broadcasts a RAP with a $TTL = TTL + n$ to call for many more assistant nodes, where n, a system-size parameter and $n \geq 0$, is an integer whose value is related to node density of the sensing field and the energy consumption rate of S.

Values of $TTL + n$ range between 1 and $\left\lceil l/r \right\rceil$, where l is the side length of underlying sensing field, and r is the transmission radius of a sensor node. When a RAP with TTL=1 is broadcasted, the maximum number of assistant nodes that can come to help S is $m \cdot \dfrac{\pi \cdot r^2}{l^2}$ nodes, where m is number of nodes distributed to the sensing field. When the TTL increases to i, $0 > i \geq \left\lceil \dfrac{l}{r} \right\rceil$, maximum number of nodes that may come increases to $m \cdot \dfrac{\pi \cdot i^2 r^2}{l^2}$. If $h\%$ of nodes on receiving RAPs for TTL=i can come to help S, the new workload of S, $W'(S) = \dfrac{W(S)}{(m \cdot \frac{\pi \cdot (ir)^2}{l^2} \cdot \frac{h}{100})} \Bigg/ (m \cdot \frac{\pi \cdot (qr)^2}{l^2} \cdot \frac{h}{100}) = W(S) \cdot \dfrac{q^2}{i^2}$ in average, where $W(S)$ denotes the previous workload of S at $TTL = q$, and $0 \leq h \leq 100$. If $W(S)' \leq \delta$, we can derive that $i \geq q \sqrt{\dfrac{W(S)}{W'(S)}}$, and then $n = i - TTL_{old}$. The new RAP would be broadcasted with $TTL_{new} = TTL_{old} + n = q + n$.

5 Experiments and Discussion

Two experiments were performed in this study. The first evaluates how nodes are distributed after RASLM is enabled in WSN. The second measured lifetime of a WSN given different node-failure rates.

5.1 Sensor Nodes Distributed in RASLM System

In Fig. 5a, four nodes are triggered by events. They establish four routing paths connected to the base station. Node 0 is denoted as the base station. The topology generated after active nodes sent out RAPs with TTL=1, and non-active nodes moved to the chosen EAZs is shown in Fig. 5b. We can see that the RASLM scheme called for sufficient assistant nodes. However, some senders, e.g., nodes 27 and 16, have called for too many assistant nodes, causing unnecessary energy consumption for node movement.

5.2 System Lifetime Measured at Different Node-Failure Rates

In the second experiment, we measured the system life given different node-failure rates to further verify the effectiveness of RASLM. In Fig.6, we can see system lifetime grows when node failure rate increases. When the event-triggered ratio is 10%, the system lifetimes that had been shown were only at node-failure rates 10%, 20% and 30%, because in the whole field only small portions of nodes participated in packet generating and packet relaying. Other nodes did not consume their energy. That is why there were no available values which are higher than 30% at event-triggered ratio 10%.

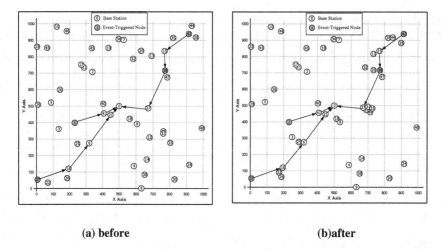

(a) before   **(b)after**

Fig. 5. The network topology and position change of assistant nodes before and after the experiments. The experiment was performed in the 1000×1000 m^2 sensing field in which 49 nodes (excluding base station) were distributed and event-triggered ratio is 5%.

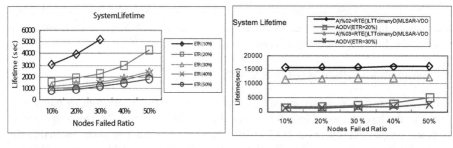

Fig. 6. System lifetime measured at different node-failure rates with different event-triggered ratio (ETR)

Fig. 7. System lifetime measured at different node-failure rates with event-triggered ratios at 20% and 30% compared to AODV-RASLM (Dynamic TTL)

Fig. 7 shows system lifetime measured at different node-failure rates from 10% to 50% when the event-triggered ratios are at 20% and 30%. We can see that AODV with RASLM has actually improved system lifetime of a WSN compared that with pure AODV routing, because the RASLM scheme not only called for assistant nodes, but also evenly shared the burden of relaying job. That means RASLM can balance the energy consumption along the routing path even some heavy burden nodes exist.

6   Conclusions and Future Research

In the paper, we have addressed the energy hole problem which may occur both in message-evenly-generated sensing environment, and event-driven wireless sensor networks. This problem may cause some other problems, such as fire alarms can not be sent to administrators, and collected data can not arrive at its destination database. Therefore, we propose a routing support scheme named RASLM which can prolong lifetimes of uniformly distributed event-driven WSNs by using mobile nodes that share relay burden with nodes along an active path so as to prevent these nodes from exhausting energy quickly. The latter can ensure that the sensed data to be sent to base station safely and smoothly.

In the future, we would like to analyze the reliability and performance models for RASLM so users can predict how reliable the system is and what the behavior the system may perform is.

References

1. Wu, X., Chen, G., Das, S.K.: On the Energy Hole Problem of Non-uniform Node Distribution in Wireless Sensor Networks. In: Proc. of IEEE International Conference on Mobile Ad-hoc and Sensor Systems, pp. 180–187 (October 2006)
2. Wu, X., Chen, G.: Dual-Sink: Using Mobile and Static Sinks for Lifetime Improvement in Wireless Sensor Networks. In: Proc. of International Conference on Computer Communications and Networks, pp. 1297–1302 (August 2007)
3. Yang, S., Wu, J., Dai, F.: Localized Movement-Assisted Sensor Deployment in Wireless Sensor Networks. In: Proc. of IEEE International Conference on Mobile Ad-hoc and Sensor Systems, pp. 753–758 (October 2006)
4. Wu, X., Chen, G.: Avoiding Energy Holes in Wireless Sensor Networks with Non-uniform Node Distribution. IEEE Transaction on Parallel and Distributed Systems 19(5), 710–720 (2008)
5. Wang, B., Xie, D., Chen, C., Chen, C., Ma, J., Cheng, S.: Employing Mobile Sink in Event-Driven Wireless Sensor Networks. In: Proc. of IEEE International Conference on Vehicular Technology, pp. 188–192 (May 2008)
6. Leu, F.Y., Li, G.C., Wu, W.C.: An Autonomous Energy-Aware Routing Scheme: a Supplementary Routing Approach for Path-Preserving Wireless Sensor Networks. In: Proc. of the IFIP Annual Mediterranean Ad Hoc Networking Workshop, pp. 49–60 (June 2008)
7. Chellappan, S., Bai, X., Ma, B., Xuan, D., Xu, C.: Mobility limited flip-based sensor networks deployment. IEEE Transactions on Parallel and Distributed Systems 18(2), 199–211 (2007)

Short Text Clustering for Search Results

Xingliang Ni[1,2,3], Zhi Lu[2], Xiaojun Quan[2], Wenyin Liu[2,3], and Bei Hua[1,3]

[1] Dept. of Computer Sci. and Tech., University of Sci. and Tech. of China, Hefei, China
[2] Department of Computer Science, City University of Hong Kong, HKSAR, China
[3] Joint Research Lab of Excellence, CityU-USTC Advanced Research Institute, Suzhou, China
xlni@mail.ustc.edu.cn, {xiaoquan,csliuwy}@cityu.edu.hk,
bhua@ustc.edu.cn

Abstract. An approach to clustering short text snippets is proposed, which can be used to cluster search results into a few relevant groups to help users quickly locate their interesting groups of results. Specifically, the collection of search result snippets is regarded as a similarity graph implicitly, in which each snippet is a vertex and each edge between the vertices is weighted by the similarity between the corresponding snippets. *TermCut*, the proposed clustering algorithm, is then applied to recursively bisect the similarity graph by selecting the current *core* term such that one cluster contains the term and the other does not. Experimental results show that the proposed algorithm improves the *KMeans* algorithm by about 0.3 on *FScore* criterion.

Keywords: Clustering, Short Text Clustering, Graph Partitioning.

1 Introduction

Currently, the number of web pages in the World Wide Web has increased exponentially and the search engine has been proved to be a successful technique to retrieve useful information from the huge number of web pages. However, as more and more web pages are being accumulated day by day, a search engine usually returns a huge number of results corresponding to a single query. Consequently, users may spend a lot of time on browsing the results list to find their expected pages. For instance, given the query of *milk powder*, the number of web pages Google returns can reach up to 2,390,000. In addition, different web sites often publish the web pages with similar or even the same content. This phenomenon is especially common in the news domain where an event is usually reported on different web sites. Hence, it is necessary to reorganize the search results to facilitate users' browsing.

Many researchers have employed the technique of text clustering to reorganize the search results. The key step of the text clustering method [1, 2, 3] is to compute the similarity between two documents, and many similarity metrics heavily rely on the co-occurrence of terms between the two documents. However, the search results are usually composed of short text snippets, such as titles, which make the traditional similarity metrics not suitable for the clustering task of search results owing to the limited number of common terms among them.

Q. Li et al. (Eds.): APWeb/WAIM 2009, LNCS 5446, pp. 584–589, 2009.

Intuitively, short texts are usually clustered into groups according to some core terms. For instance, given three sentences:

S1: *How many products does Google have?*
S2: *How many products does Yahoo! have?*
S3: *Whether Yahoo! is a good web company?*

S2 and S3 are usually clustered into a group because they are describing the different profiles of the same term *Yahoo!*, even if S2 shares more common terms with S1 than S3. This shows that the *core* terms play an important role in short text clustering.

In this paper, we propose a short text clustering method which can be used to cluster search results. The proposed algorithm, *TermCut*, is to recursively bisect clusters according to a *core* term in each iteration. The *core* terms are found based on minimizing the *Mcut* criterion [2], in which the collection of search results is treated as a similarity graph. A comparative study with the *KMeans* algorithm as baseline is conducted to evaluate our method's clustering accuracy. Experimental results indicate that *TermCut* obtains a higher accuracy than the baseline.

The rest of the paper is organized as follows. *TermCut*, the proposed algorithm, is presented in detail in Section 2. In Section 3 we explain the evaluation criteria and present the experimental results. Section 4 concludes our work and discusses future work.

2 The Proposed Algorithm

2.1 Text Model

Throughout this paper, we use n, M and K to denote the number of documents, the number of terms and the number of clusters, respectively, in the entire corpus. The vector-space model is employed to represent short texts, and each short text d is considered to be a vector in this model. In particular, we employ the TF-IDF term weighting model. Since TF of each term in a short text is usually very low, it can be smoothed to be 1 or 0, as Zhang and Lee [4] do in their method. Specifically, if a term appears in the short text, the TF of this term is smoothed to be 1; and 0 otherwise. In addition, the dot product is used to compute the similarity between texts.

2.2 *Mcut* Criterion

The *Mcut* criterion [2] is a graph based criterion function which models the collection of texts as a graph. It measures the quality of clusters according to the min-max clustering principle—minimizing the inter-cluster similarity and maximizing the intra-cluster similarity. If a collection of short texts C is clustered into K clusters $\{C_1, C_2, ..., C_k, ..., C_K\}$, their *Mcut* is defined to be Eq. (1).

$$\text{Mcut}(C_1, C_2, \cdots, C_k, \cdots, C_K) = \sum_{k=1}^{K} \frac{\sum_{d_i \in C_k} \sum_{d_j \notin C_k} sim(d_i, d_j)}{\sum_{d_s, d_t \in C_k} sim(d_s, d_t)} \tag{1}$$

The similarity between a pair of short texts is usually not accurate owing to the limited co-occurring terms between two short texts. The *Mcut* criterion can weaken the impact of the sparseness of short texts by summing up all the pair-wise similarities within a cluster. However, the *Mcut* criterion has a computational complexity of $O(n^2)$ considering that the number of clusters and the average number of terms in each short text are small and independent of n. Next, we further propose an efficient method to compute the *Mcut* with a complexity of $O(n+M)$. In the case of short texts, the numerator part of the *MCut* can be calculated as Eq. (2).

$$\sum_{d_i \in C_k} \sum_{d_j \notin C_k} sim(d_i, d_j) = \sum_{d_i \in C_k} \sum_{d_j \notin C_k} \sum_{t \in d_i, t \in d_j} idf(t) \cdot idf(t)$$

$$= \sum_{d_i \in C} \sum_{d_j \in C} \sum_{m=1}^{M} idf(t_m)^2 \cdot [t_m \in d_i] \cdot [t_m \in d_j] \cdot [d_i \in C_k] \cdot [d_j \in C - C_k]$$

$$= \sum_{m=1}^{M} (idf(t_m)^2 \cdot \sum_{d_i \in C} [t_m \in d_i, d_i \in C_k] \cdot \sum_{d_j \in C} [t_m \in d_j, d_j \in C - C_k]) \tag{2}$$

We use $F(t_m, C_k)$ and $F(t_m, C-C_k)$ to denote the frequency of t_m in C_k and the frequency of t_m in $C-C_k$ respectively. Hence, Eq. (2) can be further simplified as:

$$\sum_{d_i \in C_k} \sum_{d_j \notin C_k} sim(d_i, d_j) = \sum_{m=1}^{M} idf(t_m)^2 \cdot F(t_m, C_k) \cdot F(t_m, C - C_k) \tag{3}$$

On the other hand, the denominator part of the *Mcut* takes into account the intra-similarity. It sums up all the pair-wise similarities between the short texts in C_k, which can be calculated as:

$$\sum_{d_s, d_t \in C_k} sim(d_s, d_t) = \sum_{m=1}^{M} idf(t_m)^2 \cdot F(t_m, C_k)^2 \tag{4}$$

Using Eq. (3) and Eq. (4), the *Mcut* criterion function can be deduced into Eq. (5).

$$Mcut(C_1, C_2, \cdots, C_k) = \sum_{k=1}^{K} \frac{\sum_{m=1}^{M} idf(t_m)^2 \cdot F(t_m, C_k) \cdot F(t_m, C - C_k)}{\sum_{m=1}^{M} idf(t_m)^2 \cdot F(t_m, C_k)^2} \tag{5}$$

The value of $idf(t_m)$, $F(t_m, C_k)$ and $F(t_m, C-C_k)$ can be computed by scanning all the short texts in the corpus one time in $O(n)$ time and the computational complexity of *Mcut* is $O(M)$. Hence, the complexity of *Mcut* can be reduced to $O(n+M)$.

2.3 The *TermCut* Algorithm

The basic idea of the proposed algorithm, named *TermCut*, is to find the *core* term which can minimize the *Mcut* in each iteration. One cluster is then divided into two clusters—one cluster containing the current *core* term and the other not. The pseudo-code of *TermCut* algorithm is given in Fig. 1.

Algorithm TermCut

1. Initialize the clusters list to contain a single cluster with all the short texts
2. **Repeat**
3.   Select a cluster C_k from the clusters list randomly
4.   Init *CoreTerm:=null*
5.   *CoreTerm*:=FindCoreTerm(C_k)
6.   **If** *CoreTerm<>null*
7.   Divided the cluster C_k into two clusters according to *CoreTerm*
8.   Register the two generated clusters into the clusters list
9.   **End If**
10. Remove the cluster from the clusters list
11. **Until** No cluster left in the clusters list

Fig. 1. The pseudo-code of *TermCut* algorithm

In the *TermCut* algorithm, the clustering process begins with a single cluster containing all the short texts. We then find the *core* term in each iteration based on minimizing the *Mcut* criterion. If we find a *core* term for a cluster, the original cluster is bisected and replaced by the two new generated clusters from the clusters list. Otherwise, if any *core* term for a cluster cannot be found, the cluster is simply regarded as an atomic cluster. This process continues until there is no cluster in the clusters list. The function, *Find-CoreTerm*, called in Step 5 in *TermCut* in Fig. 1 is used to find the *core* term to bisect the original cluster. The pseudo-code of *FindCoreTerm* function is shown in Fig. 2.

In the *FindCoreTerm* function, we generate two clusters C_{k1} and C_{k2} according to each term t in the given cluster C_k, where C_{k1} covers all the texts containing term t and C_{k2} covers the rest texts in C_k not containing t. The *Mcut* criterion for the new set of clusters is calculated, and the term which minimizes the *Mcut* is selected as the *core* term candidate. If the decrease between the *Mcut* criterion value of the original set of clusters and the minimized *Mcut* is above a *threshold*, the candidate is returned as the

---

Function FindCoreTerm(Cluster C_k)

1. Init CoreTerm:=null, *MinMcut*:=maxValue, *OriginalMcut:= Mcut(C_1,...,C_k...,C_K)*
2. **For** each term t in the cluster C_k
3.   Generate two clusters C_{k1} and C_{k2} from C_k where C_{k1} with all the texts containing t and
        C_{k2} with the rest of texts in C_k
4.   **If** *Mcut($C_1,C_2,...,C_{k1},C_{k2},...,C_K$)< MinMcut*
5.   *CoreTerm:=t*
6.   *MinMcut:= Mcut($C_1,C_2,...,C_{k1},C_{k2},...,C_K$)*
7.   **End If**
8. **End For**
9. **If** *OriginalMcut-MinMcut >threshold*
10. **Return** *CoreTerm*
11. **else**
12. **Return** *null*
13. **End If**

Fig. 2. The pseudo-code of *FindCoreTerm* Function

core term; otherwise, the *threshold* is not satisfied and we simply return *null* to denote no *core* term can be found. The *threshold* is set to a negative number for starting the clustering process, because according to Eq. (1) the initial value of *Mcut* is 0.

3 Experiments

In this section, we conduct two experiments to verify the effectiveness of the proposed method. The first one is clustering the search results returned by Google. The second one is clustering questions, which are typical short text documents.

3.1 Search Result Clustering

We build up a web application which can extract the search results from Google including links and titles corresponding to a given query. These search results are then clustered using *TermCut* according to their corresponding titles. The resulting clusters are labeled as their corresponding *core* terms, and the only cluster without any *core* term is labeled as the term *others*. Here we define the cluster labeled with the term *others* as *impure* cluster and the rest clusters as *pure* clusters. Any *pure* clusters containing only one search result are merged into the *impure* cluster.

We hire a three-expert group and randomly choose 50 Chinese queries for testing. A *pure* cluster is determined as *false* if it contains any incorrect search result judged by experts; and *true* otherwise. Our method for this test generates 6.84 *pure* clusters for each query in average; whereas 1.38 clusters and 5.46 clusters are judged to be *false* and *true* respectively. The manual evaluation shows that our method reports a relatively good performance for a practical search engine.

3.2 Question Clustering

We conduct a comparison between the *TermCut* and the *KMeans* algorithm using the ground truth test dataset of questions selected from our system *BuyAns* [6]. In our system, each question is a short text, which is used to simulate the snippets a search engine could return. In total, 453 Chinese questions are selected and manually classified into 4 coarse-grain and 13 fine-grain classes. To test the effectiveness of the proposed *TermCut* method, the *FScore* measure [1, 3] is employed. We utilize the Macro Averaged *FScore* and Micro Averaged *FScore* [5] of the entire clustering result to evaluate a clustering method's performance.

Performance comparison between our algorithm and *KMeans* algorithm is shown in Table 1 where the *threshold* of the *TermCut* algorithm is set to -1.5. The input cluster number of *KMeans* is 13 which is the same with the number of the fined-grain classes of the ground truth dataset. Since the *KMeans* algorithm we implemented selects the initial centroid randomly, the result shown in Table 1 is the average result of running 100 times.

Table 1. Performance comparison of *TermCut* and *Kmeans* as the baseline

	MicroFScore	*MacroFScore*
TermCut	0.77	0.68
KMeans (K=13)	0.49	0.31

The above experimental results exhibit that *KMeans* performs badly on clustering short text, like the search result snippets, whereas the proposed method *TermCut* performs much better on these data collections.

4   Conclusion

We propose a short text clustering method to cluster search results. An algorithm, *TermCut*, is presented to find the *core* terms based on minimizing the Mcut and generate clusters according to the *core* terms. Experiments show that the proposed algorithm can help efficiently and effectively organize the search results.

In the future, we will use even big corpora to fully test the proposed method and investigate its performance with external semantic knowledge (such as Wikipedia).

Acknowledgement

The work described in this paper was fully supported by a grant from City University of Hong Kong (Project No. 7002336).

References

1. Zhao, Y., Karypis, G.: Evaluation of hierarchical clustering algorithms for document data-sets. In: Proc. 11th international conference on information and knowledge management, pp. 515–524 (2002)
2. Ding, C., He, X., Zha, H.: A min-max cut algorithm for graph partitioning and data clustering. In: Proc. international conference on data mining, pp. 107–114 (2001)
3. Larsen, B., Aone, C.: Fast and effective text mining using linear-time document clustering. In: Proc. 5th ACM SIGKDD International Conference on Knowledge Discovery and Data Mining, pp. 16–22 (1999)
4. Zhang, D., Lee, W.S.: Question classification using support vector machines. In: Proc. 26th annual international ACM SIGIR conference on Research and development in information retrieval, pp. 26–32 (2003)
5. Hachey, B., Grover, C.: Sequence modelling for sentence classification in a legal summarisation system. In: Proc. ACM symposium on applied computing, pp. 292–296 (2005)
6. Buyans (2008), http://www.buyans.com

Load Shedding for Shared Window Join over Real-Time Data Streams

Li Ma[1,2], Dangwei Liang[3], Qiongsheng Zhang[1], Xin Li[4], and Hongan Wang[2]

[1] School of Computer Science and Communication Engineering,
China University of Petroleum, Dongying 257061, P.R. China
[2] Institute of Software, Chinese Academy of Sciences, Beijing 100190, P.R. China
[3] Geophysical Research Institute of Shengli Oil Field,
China Petroleum & Chemical Corporation, Dongying 257000, P.R. China
[4] Shandong University, Ji'nan 250101, P.R. China
{mali, zqsheng}@hdpu.edu.cn

Abstract. Join is a fundamental operator in a Data Stream Management System (DSMS). It is more efficient to share execution of multiple windowed joins than separate execution of everyone because the former saves a part of cost in common windows. Therefore, shared window join is adopted widely in multi-queries DSMS. When all tasks of queries exceed maximum system capacity, the overloaded DSMS fails to process all of its input data and keep up with the rates of data arrival. Especially in a time-critical environment, queries should be completed not just timely but within certain deadlines. In this paper, we address load shedding approach for shared window join over real-time data streams. A load shedding algorithm LS-SJRT-CW is proposed to handle queries shared window join in overloaded real-time system effectively. It would reduce load shedding overhead by adjusting sliding window size. Experiment results show that our algorithm would decrease average deadline miss ratio over some ranges of workloads.

Keywords: Data stream, load shedding, quality of service, query deadline.

1 Introduction

Data stream is defined as a real-time, continuous, ordered sequence of data items [1]. One primary characteristic of Data Stream Management System (DSMS) is that its data is active and its query is passive. It is impossible to control the arrival rates and contents of the stream data. When the arrival rates are bursty, resources in DSMS, including CPU and memory, are not enough to meet all requirements of the current system, and then the system becomes overloaded. Therefore, load shedding which is to drop redundant tuples is a new challengeable issue. Join is a fundamental operation for relating information from different data streams in streaming systems. The execution of multiple queries over shared windows separately wastes the system resources, because the common join's execution between these queries will be repeated. Implementing multiple queries shared join in a single execution plan avoids such redundant

Q. Li et al. (Eds.): APWeb/WAIM 2009, LNCS 5446, pp. 590–596, 2009.

processing [8]. The shared join operation processes every window once in a join execution and produces multiple output data to the next operators in the operator network. The shared join is at the very core of the query plan in a continuous query system. Therefore, it should actively take effective measures to avoid its system degradation sharply when such a system is overloaded.

In this paper, we propose a load shedding algorithm for shared window join to efficiently handle huge fluctuant overload in real-time DSMS. The algorithm realizes queries' execution by adjusting processing window sizes, and only drops tuples in the largest window during overload. Experiment results show that this algorithm has better performance.

2 Related Works

Load shedding over data streams is discussed broadly since 2002. According to the ways of dropping tuples, it is classified into random load shedding and semantic load shedding [3]. From the perspective of system implementation architecture, it is classified into load shedding based on closed loop control, such as [9], and load shedding based on opened loop control [2-7]. From the perspective of the query plan, it can be classified into load shedding aimed at non-join characteristics [2, 3] and load shedding aimed at join characteristics [4-7]. From the perspective of real-time property, all above approaches don't consider real-time property. There are some load management approaches in real-time DSMS, such as [10] which presented a two-level load management method aimed at periodic queries. But these approaches don't specially consider the situation of shared join operation over real-time data streams.

The shared execution of multiple queries over data streams could optimize the system resources utilization, so it is a good query optimization way [12]. Hammad et al [8] discussed the scheduling strategy of shared join queries under normal workload. At present, the study of load shedding for shared window join is rarely. The literature [7] firstly discussed this problem, and proposed two load shedding strategies (uniform strategy and small window accuracy strategy) to improve query throughput. But these strategies are not suitable for time-critical situation where queries should be completed within their deadlines.

3 Preliminaries

Sliding window is a recent segment of data stream. It is classified into time-based sliding window being discussed in this paper and count-based sliding window [1]. Our target is to provide firm real-time query services, in which overdue results have no value for the system. There are two input streams S_1 and S_2, and N real-time binary join queries shared one window join. CPU restriction is one of the reasons for memory restriction [6], so we only consider that the system has limited CPU and unlimited memory.

Compared with the unary operators, such as select and project, join is a high overhead operator in a data stream processing system. Shared window join could reduce the number of join probe, and improve system execution efficiency. For example, in intelligent system building, plenty of sensors monitor the temperature and smoke viscosity of each room. A monitoring system receives sensor data continuously, and produces alarming and statistic information for users. There are two queries.

Q_1: SELECT S_1.Timestamp, S_1.RoomID, S_1.Temp, S_2.Smoke FROM S_1 [Range 6 seconds], S_2[Range 6 seconds] WHERE S_1.RoomID=S_2.RoomID AND S_1.Temp>40 AND S_2.Smoke>0.03 DEADLINE 5 seconds;

Q_2: SELECT S_1.RoomID, AVG (S_1.Temp), AVG (S_2.Smoke) FROM S_1[Range 10 minutes], S_2[Range 10 minutes] WHERE S_1.RoomID=S_2.RoomID GROUP BY S_1.RoomID DEADLINE 2 minutes;

The above two queries would waste system resources if they are separately executed. Because these queries are all join operations on stream S_1 and S_2, and their join conditions are same, implementing them in a single shared execution plan would avoid redundant processing. As shown in Fig.1, the processing of the shared query plan performs the shared window join operator firstly and then routes the output to the select operator (σ) of Q_1 and the aggregate operator (avg) of Q_2.

4   LS-SJRT-CW Algorithm

A new LS-SJRT-CW algorithm (Load Shedding for Shared window Join over Real-Time data streams based on Changeable Window size) is proposed to handle overload of shared window join. This algorithm adjusts practical processing window size according to sample ratios for all queries, and finally drops tuples only in the largest sliding window. The basic idea of adjusting window size has been mentioned in [13]. Different from them, this paper mainly discusses load shedding algorithm based on adjusting the processing window size. The LS-SJRT-CW does not need to buffer the dropped tuples, and then reduces the overhead of load shedding.

■   **Sample Ratio**

The data sample ratio is used to measure the percentage of input tuples processed to produce query results. We allow the users to specify the relationship between output quality and sample ratio by a two-dimensional QoS graph [11] which is shown in Fig.2. For simplicity, the QoS graph is shown as a piece-wise linear function of sample ratio with one critical point, at which a query reaches its tolerable quality (TQ) with the least sample ratio. When a DSMS is overloaded, the sample ratio of each current query could be decided by the two-dimensional QoS graph according to users' quality requirements. Assume that SR_i is the sample ratio of query Q_i, and n_i ($n_i>0$) denotes the number of operators in the query plan of Q_i. O_{ij} (Assume that $j=1$ in this paper) denotes the j-th operator of query Q_i, and is the current scheduled join operator. There are $m = r \cdot UnitTime$ (*UnitTime* denotes one unit time) tuples in its one side input queue. When there is no other join operator in this query plan, RT (Remain Time) of one side join direction is estimated as follows.

$$RT_{ij} = \sum_{p=j}^{n_i}(m \cdot c_{ip} \cdot \prod_{k=j}^{p-1}s_{ik})$$

(1)

Where c_{ip} is the execution cost of operator O_{ip}, and s_{ik} is the selectivity of operator O_{ip}. The sample ratio of one join direction of the binary join operator O_{ij} is computed as $SRO_{ij} = \sqrt{(D_i - T_{ij})/(2 \cdot RT_{ij})}$, where T_{ij} is the wasted time.

The sample ratio P_i of the window segment Seg_i could be computed by SRO_{ij}

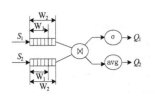

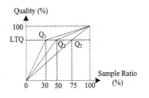

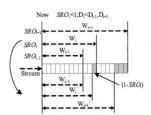

Fig. 1. Shared execution of multi-windows join

Fig. 2. QoS graph based on sample ratio

Fig. 3. Architecture of the LS-SJRT-CW

■ **Adjusting Window Size**

Assume that the deadline D_i of query Q_i is the shortest one in all deadlines. The system set up the sample ratio of query Q_i, i.e., $SR_i (SR_i \leq 1)$ to ensure that the deadline D_i of query Q_i is in the range of quality. $(SR_i < 1)$ means that we should sample tuples in window $W_i (W_i = W_{i-1} + Seg_i)$. If the window size of query Q_{i+1} is larger than that of query Q_i, the results of window W_{i+1} share these of window W_i. Because of dropping some tuples on query Q_i, the precision of query Q_{i+1} would be also reduced. However, the query Q_{i+1} may have ability to process some tuples dropped by Q_i within its deadline. The architecture of the LS-SJRT-CW is shown in Fig.3.

W_i' denotes the adjusted sliding window. After getting the sample ratio of the join operator, we could compute the new window size T_i' based on the initiation window size T_i.

$$T_i' = (SRO_i) \cdot T_i \quad (1 \leq i \leq N) \tag{2}$$

From the point of intuition, this approach has little overhead to release system workload by adjusting the window size according to the sample ratio of join operator.

■ **Algorithm Description**

When the system is overloaded, it should be shed after making sure $SR_i (1 \leq i \leq N)$ according to the QoS graph. The LS-SJRT-CW algorithm is described as follows.

Step1. Compute the new processed window size $T_i' (1 \leq i \leq N)$ of each query and each Seg_i after adjusting it.

Step2. Change all window sizes in accordance with each T_i', and finally drop redundant tuples in the largest window.

Step3. Recover the initiation window size immediately once detecting that the system is normal.

The main drawback of the LS-SJRT-CW algorithm is that its output tuples aren't strictly processed on the basis of the window size specified by users.

5 Experiment

We carry out many simulated experiment results to evaluate our proposed algorithm. Join is fully matched in the experiment, and its overhead is proportional to the window size. We choose the small window first algorithm (LS-SWF for short) proposed in [7] to compare with our algorithm. The system is under unstable workload, and the maximum workload is about 1.5 times maximum system capacity.

■   **Evaluation Metrics**

In our experiments, we use Average Miss Ratio (*AMR*) as the metric of evaluating the algorithm's performance.

$$AMR = \sum_{i=1}^{N} \Delta m_i \bigg/ \sum_{i=1}^{N} \hat{m}_i \tag{3}$$

Where $\hat{m}_i$ and Δm_i are the count of output tuples of the *i*-th query and the count of those missing their deadlines respectively, and N is the count of queries. Those dropped tuples are part of tuples that missed their deadlines when computing the *AMR*. Therefore, the *AMR* reflects the system throughput. The smaller the *AMR*, the larger the throughput, and vice versa.

■   **Experiment Results**

There are three kinds of query deadline distributions in this experiment. (1) the shortest deadline distributes in a small window, denotes as **set₁**; (2) the shortest deadline distributes in a medium window, denotes as **set₂**; (3) the shortest deadline distributes in a large window, denotes as **set₃**.

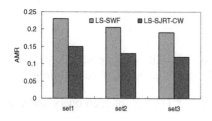 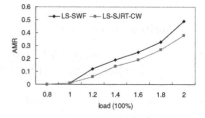

Fig. 4. *AMR* under different deadline distribution   **Fig. 5.** *AMR* under different workload

The LS-SWF and the LS-SJRT-CW are compared on *AMR* shown in Fig.4 and Fig.5. Although LS-SWF is a better load shedding algorithm for shared window join, it is not suitable for real-time situation. Therefore, its *AMR* is larger than one of the LS-SJRT-CW under the same deadline distribution. The LS-SJRT-CW would decrease *AMR* under greater changeable workloads, that is, the number of output tuples that are not missed deadlines is more. The number of effective window segments is little in the query distribution set_3 where the query with minimum deadline distributes in maximum window, so scheduling overhead is decreased. Thus, the *AMR* of the LS-SJRT-CW in set_3 are smaller than these in set_1 and set_2. Gradient of each curve is larger and larger with increasing workload in Fig.5. When the system is seriously overloaded, the LS-SWF shows poor performance. This is why the slope of the LS-SWF is greater than that of the other algorithm.

6 Conclusion

The problem of load shedding for shared window join over real-time data streams is described in this paper. An algorithm LS-SJRT-CW based on adjusting sliding window size is proposed to handle overload of shared window join effectively. The LS-SJRT-CW algorithm would decrease *AMR*. For future work, we will experiment further with different workloads and configurations. In addition, we plan to study adaptive join and asymmetric shared window join.

References

1. Golab, L., Ozsu, M.T.: Issues in data stream management. SIGMOD Record 32(2), 5–14 (2003)
2. Babcock, B., Datar, M., Motwani, R.: Load shedding for aggregation queries over data streams. In: Proceedings of ICDE, Boston, USA, pp. 350–361 (2004)
3. Tatbul, N., Cetintemel, U., Zdonik, S., Cherniack, M., Stonebraker, M.: Load shedding in a data stream manager. In: Proceedings of VLDB, Berlin, Germany, pp. 309–320 (2003)
4. Srivastava, U., Widom, J.: Memory-limited execution of windowed stream joins. In: Proceedings of VLDB, Toronto, Canada, pp. 324–335 (2004)
5. Gedik, B., Wu, K.L., Yu, P., Liu, L.: Adaptive load shedding for windowed stream joins. In: Proceedings of the 14th ACM international conference on Information and knowledge management (CIKM), Bremen, Germany, pp. 171–178 (2005)
6. Ayad, A., Naughton, J., Wright, S., Srivastava, U.: Approximating streaming window joins under CPU limitations. In: Proceedings of ICDE, Atlanta, Georgia, p. 142 (2006)
7. Yan, Y., Jin, C.Q., Cao, F., Wang, H.J., Zhou, A.Y.: Load shedding for shared window joins over data streams. Journal of Computer Research and Development 41(10), 1836–1841 (2004) (in Chinese)
8. Hammad, M.A., Franklin, M.J., Aref, W.G., Elmagarmid, A.K.: Scheduling for shared window joins over data streams. In: Proceedings of VLDB, Berlin, Germany, pp. 297–308 (2003)
9. Tu, Y.C., Liu, S., Prabhakar, S., Yao, B.: Load shedding in stream databases: a control-based approach. In: Proceedings of VLDB, Seoul, Korea, pp. 787–798 (2006)

10. Wei, Y., Prasad, V., Son, S.H., Stankovic, J.A.: Prediction-based QoS management for real-time data streams. In: Proceedings of RTSS, Rio de Janeiro, Brazil, pp. 344-358 (2006)
11. Li, X., Ma, L., Li, K., Wang, K., Wang, H.A.: Adaptive load management over real-time data streams. In: Proceedings of the 4th International Conference on Fuzzy Systems and Knowledge Discovery (FSKD), Hainan, China, pp. 719–725 (2007)
12. Madden, S., Shah, M.A., Hellerstein, J.M., Raman, V.: Continuously adaptive continuous queries over streams. In: Proceedings of SIGMOD, Madison, USA, pp. 49–60 (2002)
13. Motwani, R., Widom, J., Arasu, A., et al.: Query processing, resource management, and approximation in a data stream management system. In: Proceedings of CIDR, Asilomar, USA, pp. 245–256 (2003)

A Sliding-Window Approach for Finding Top-k Frequent Itemsets from Uncertain Streams

Xiaojian Zhang[1] and Huili Peng[2]

[1] Department of Computer Science, Henan University Finance & Economics,
Zhengzhou 450002, China
[2] Department of Education, Henan Radio & Television University,
Zhengzhou 450008, China
{xjzhang82,phl81}@gmail.com

Abstract. The analysis and management of uncertain data has attracted a lot of attention recently in many important applications such as pattern recognition and sensor network. Frequent itemset mining is often useful in analyzing uncertain data in those applications. However, previous works just focus on the static uncertain data instead of uncertain streams. In this paper, we study the problem of mining top-k FIs in uncertain streams. We propose an efficient algorithm, called UTK-FI, based on sliding-window and Chernoff bound techniques for finding k most frequent itemsets of different sizes. Experimental results show that our algorithm performs much better than many established methods in uncertain streams environment.

1 Introduction

Top-k Frequent itemset (top-k FI) mining in certain data streams has been thoroughly studied. However, previous approaches almost coped with the certain data streams that contain precise data. Actually, a number of applications today need to analyze streaming data that is uncertain, such as medical data managing and mobile object tracking. These applications has attracted a lot of attention recently [1, 2]. Since large amounts of uncertain streaming data could arrive at high speed, there are two main challenges to handle it, the first one stems from the limited memory and CPU recourse, the other is how to deal with the exponential blowup in the number of FIs. While streaming data is highly time sensitive, people are generally more interested in the most recent transactions than those in the past, and dated uncertain transactions are no longer interesting. In this situation, sliding window model can be used to deal with these recent transactions directly.

Most previous algorithms like Lossy Counting algorithm [3] for processing data streams are controlled by an error parameter ε set by the users. However, the setting of ε is quite subtle, which often leads to a dilemma. ε too big yields inaccurate results and makes the number of itemsets large while setting it too small leads to excessive memory consumption. To solve the above questions, we propose an algorithm, called UTK-FI (Top-K Frequent Itemset mining in Uncertain streams), for finding top-k FIs

Q. Li et al. (Eds.): APWeb/WAIM 2009, LNCS 5446, pp. 597–603, 2009.

in a sliding window, and we apply Chernoff bound technique [4] to reset ε that is not given by user and neither fixed with the window sliding, and design an increasing expected support function to compute the potentially expected support for each itemset. Our experiments show that UTK-FI algorithm is significantly efficient.

2   Preliminaries

An uncertain stream S is a continuous sequence of transactions, $\{T_1, T_2, ..., T_n, ...\}$. A transaction T_i ($T_i \subset S$) contains a number of items, and each item x in T_i is associated with a probability $P_{Ti}(x)$, which expresses the possibility that x exists in T_i. A possible world model in [5] can be adopted to interpret the uncertain streams. Each probability $P_{Ti}(x)$ with an item x deduces two possible worlds, pw_1 and pw_2. In pw_1, item x exists in T_i, and in pw_2, item x does not exist in T_i. Each possible world pw_i is associated with an existence probability, denoted as $P(pw_i)$ that the possible world happens, then we can get $P(pw_1)=P_{Ti}(x)$ and $P(pw_2)=1-P_{Ti}(x)$.

A sliding window is a window that slides forward for every time unit with a fixed size, w. We use t_τ and $Sw_\tau = <t_{\tau-w+1},..., t_\tau>$ to denote the current time unit and the current sliding window, respectively. Let $W_{tl} = \{pw_1, pw_2, ...\}$ be the set space of all possible worlds in t_l. Let $\Phi_{i,j}$ be the set of items in the ith transaction T_i in pw_j over t_l, $|T_{tl}|$ be the number of transactions that arrive in t_l, and $P(pw_j)$ be the probability of pw_j. Assuming that all items in each sliding window are independent, we have

$$P(pw_j) = \prod_{i=1}^{|T_{tl}|} \left(\prod_{x \in \Phi_{i,j}} P_{T_i}(x) \times \prod_{y \notin \Phi_{i,j}} (1 - P_{T_i}(y)) \right). \tag{1}$$

We define $tran(Sw)$ as the set of transactions over a sliding window Sw, $|tran(Sw)|$ as all the number of transactions in $tran(Sw)$, W_{Sw} as the worlds space that consists of all items in $tran(Sw)$ over Sw, and $|W_{Sw}|$ as the number of possible worlds in W_{Sw}. For an uncertain stream, we do not determine whether a transaction contains X, since X is associated with a probability. We can depend on the possible worlds model to propose the notion of expected support count of an itemset over Sw, denoted as $E\_S(X, Sw)$.

$$E\_S(X, Sw) = \sum_{j=1}^{|W_{Sw}|} P(pw_j) \times S(X, pw_j) = \sum_{i=1}^{|tran(Sw)|} \prod_{x \in X} P_{T_i}(x), \tag{2}$$

where $S(X, pw_j)$ denotes the absolute support count of X with respect to pw_j, $pw_j \in W_{Sw}$.

Given a support threshold, σ ($0 < \sigma \le 1$), we define X as a frequent itemset over Sw if $E\_S(X, Sw) \ge \sigma |tran(Sw)|$. Given σ and X, X is a top-k FI if there exists no more than (k-1) FIs whose expected support count is higher than that of X.

3   Increasing Expected Support Function

In uncertain streams, we cannot obtain the true support count of X, but has to make an approximation. Many methods often use an error parameter ε to estimate the support of X. However, the setting of the parameter is quite subtle, which often leads to a dilemma. Since all itemsets whose expected support count over time unit t_l is less than $r\sigma|tran(t_l)|$ are discarded, we define the potentially expected support count (denoted as $PE\_S(X, t_l)$) of the itemsets as follows.

$$PE\_S(X,t_l) = \begin{cases} 0 & \text{if } E\_S(X, t_l) < r\sigma|tran(t_l)| \\ E\_S(X,t_l) & \text{otherwise.} \end{cases} \tag{3}$$

We can define the expected support count of X over time unit t_l, as $E\_S(X, t_l)$.

$$E\_S(X,t_l) = \sum_{j=1}^{|W_{t_l}|} P(pw_j) \times S(X, pw_j) = \sum_{i=1}^{|T_{t_l}|} \prod_{x \in X} P_{T_i}(x), \tag{4}$$

where $S(X, pw_j)$ denotes the support count of X with respect to pw_j, for $pw_j \in W_{t_l}$, and $|W_{t_l}|$ denotes the number of possible worlds in the set W_{t_l} over t_l.

Clearly, we can get the potentially expected support count over the current window Sw_τ (denoted as $PE\_S(X, Sw_\tau)$), which is defined as the following formula.

$$PE\_S(X,Sw_\tau) = \sum_{i=\tau-w+1}^{\tau} PE\_S(X,t_i). \tag{5}$$

Based on Chernoff bound method, we do not use the fixed error $\varepsilon = r\sigma$ to control the mining process, where r is the relaxation ratio, but $\varepsilon_n = [(2\sigma ln(2/\delta)/n)]^{1/2}$, where δ is the reliability parameter, n is the number of transactions observed in each window.

Support σ, δ are the two given parameters, let $Sw_R = <t_{R-w+1},\ldots, t_\tau>$ be the most recent R time units in Sw_τ, where $1 \le R \le w$. The increasing expected support function of an itemset, denoted as $IE\_S(R)$, is defined as the following formula.

$$IE\_S(R) = \left\lceil \left(\sigma - \sqrt{\frac{2\sigma \ln(2/\delta)}{|tran(Sw_R)|}} \right) \times r \mid tran(Sw_R) \mid \right\rceil. \tag{6}$$

An X is a potentially frequent itemset (denoted as PFI) if $PE\_S(X, Sw_R) \ge IE\_S(R)$. Given σ, δ, and a PFI X, if there exists no more than $(k-1)$ PFIs whose potentially expected support count is higher than that of X, then X is defined as a potential top-k FI, denoted as ptop-k FI. Support F and C are the set of ptop-k FIs over t_τ and Sw_τ, respectively. Hence, the task of finding ptop-k FIs over an SW is to update C with F.

4   UTK-FI Algorithm

Our algorithm utilizes a Top-K Prefix tree, called TK_Ptree, to store all the ptop-k FIs. Each entry in TK_Ptree presents an itemset X, which contain three fields: *item*, t_{id} and $PE\_S(X)$. *Item* denotes the last item of X. t_{id} registers the ID of a time unit, in which X is inserted the tree. $PE\_S(X)$ presents the potentially expected support count of X. For the first sliding window $Sw_{first} = <t_1,\ldots, t_{w+1}>$, we find all the top-$k$ FIs from

Table 1. The UTK-FI Algorithm

ALGORITHM UTK-FI $(S, w, \sigma, r, \delta, k)$
$\qquad$ S: uncertain stream; $\quad \sigma$: minimum support; $\quad k$: user-specified value
$\qquad$ r: relaxation ratio; $\quad \delta$: reliability parameter; $\quad w$: window size

---

1. The TK_Ptree T has been initialized according to the Sw_{first}
2. $F \leftarrow \varnothing$, $C \leftarrow \varnothing$
3. **foreach** arriving t_τ such that $t_\tau \subset Sw_\tau$ **do**
4. $\quad$ find all ptop-k FIs X over time unit t_i
5. $\quad$ **if** X belongs to T **then**
6. $\quad\quad$ update the $PE\_S(X)$ of each entry in F
7. $\quad\quad$ $PE\_S(X) \leftarrow PE\_S(X) + PE\_S(X, t_\tau)$
8. $\quad\quad$ $C \leftarrow F \cup C$
9. $\quad\quad$ **if** $((\tau - t_{id}(X) + 1 < w)$ **and** $(PE\_S(X) < IE\_S(\tau - t_{id}(X) + 1)))$ **or** $((\tau - t_{id}(X) + 1 \geq w)$ **and** $(PE\_S(X) < IE\_S(w)))$ **then**
10. $\quad\quad\quad$ remove the entry containing X and its descendants from T
11. $\quad$ **if** X does not belong to T **then**
12. $\quad\quad$ generated a new entry of form $(X, \tau, PE\_S(X))$
13. $\quad\quad$ $t_{id}(X) \leftarrow \tau$
14. $\quad\quad$ $PE\_S(X) \leftarrow PE\_S(X, t_\tau)$
15. **foreach** expiring $t_{\tau-w+1}$ such that $t_{\tau-w+1} \subset Sw_\tau$ **do**
16. $\quad$ find all ptop-k FIs X over time unit $t_{\tau-w+1}$
17. $\quad$ **if** $(X$ such that $X \subset T)$ **and** $((\tau - t_{id}(X) + 1 \geq w)$ **then**
18. $\quad\quad$ $PE\_S(X) \leftarrow PE\_S(X) - PE\_S(X, t_{\tau-w+1})$
19. $\quad$ remove the entries with $PE\_S(X) = 0$ in F
20. output C with k FIs if $PE\_S(X) > \sigma |tran(w)|$

$tran(t_i)$, where $t_i \subset Sw_{first}$. According to these top-k FIs, we initially construct the TK_Ptree in the following way: For each mined X, if X does not belong to TK_Ptree, recursively we can create a new entry of form $(X, i, PE\_S(X))$. On the other hand, if X belongs to the tree, we should insert $PE\_S(X, t_i)$ into $PE\_S(X)$.

Table 1 presents the UTK-FI algorithm. After initializing the tree, we should slide the window, and process the added transactions in the arriving time units and deleted transactions in the expiring time units. For an expiring time unit $t_{\tau-w+1}$, if X satisfies the condition in line 17, its support must be subtracted from $PE\_S(X)$. For an arriving time unit t_τ, if X meets the condition in line 9, we should remove it from the TK_Ptree, otherwise, we insert the $PE\_S(X, t_\tau)$ into the $PE\_S(X)$.

5   Experiments

We perform experiments on a PC computer with 2.0 GHz Pentium 4 CPU, 1.0 GB main memory. We compare UTK-FI with UTK-LC algorithm that is an improved Lossy Counting algorithm for finding Top-K FIs on Uncertain streams. We generate a synthetic dataset, *T15I4D100K*, applying IBM synthetic dataset generator [6]. During generating *T*15*I*4*D*100K. We randomly assign probability values 0.9, 0.8, 0.7, 0.6, 0.5, 0.4, 0.3, and 0.2 to the items of each transaction, respectively. In these experiments, the default parameters setting is: w =100,000, r=0.1 and δ=0.01.

Table 2. Pecision and Recall forT15I4D100K with varying ε

ε (%)	UTK-FI		UTK-LC	
	Precision (%)	Recall (%)	Precision (%)	Recall (%)
0.005	100.00	100.00	98.00	99.00
0.01	100.00	100.00	90.00	91.00
0.03	100.00	100.00	87.00	84.00
0.04	100.00	100.00	81.00	80.00
0.05	99.00	98.00	78.00	75.00
0.07	98.00	98.00	70.00	71.00

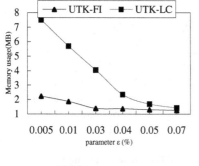

(a) Memory usage

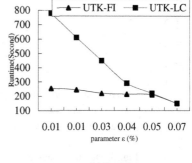

(b) Runtime

Fig. 1. Memory usage and Runtime for T15I4D100K with varying ε

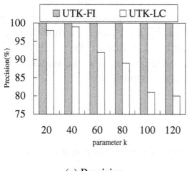

(a) Precision

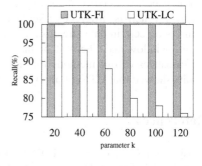

(b) Recall

Fig. 2. Precision and Recall for T15I4D100K with varying k

We first study how ε affects the two algorithms. Table 2 shows that UTK-FI algorithm can achieve over 98% precision and recall. However, the precision and the recall of UTK-LC descends clearly. Especially, when ε=0.07%, UTK-LC only obtain 70% precision and 71% recall. Fig.1 (a) and (b) show that the memory usage and the runtime of UTK-FI remain nearly invariable when changes ε from 0.005 to 0,07, but that of UTK-LC are changed largely. From the runtime viewpoint, UTK-FI is about 2.5 times faster than UTK-LC.

We evaluate how the parameter k impacts on the performances of UTK-FI and UTK-LC. Fig. 2 (a) and Fig. 2 (b) show that UTK-FI algorithm can obtain 100% precision and 100% recall, and UTK-LC can only achieve average 85% recall and 84% precision by varying k from 20 to 120. Fig. 3 (a) and Fig. 3 (b) show that the number of ptop-k FIs kept by UTK-LC is approximately 3 times larger than that of UTK-FI, and UTK-FI is nearly 3 times faster than UTK-LC.

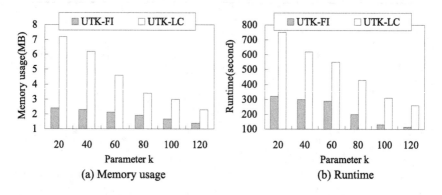

(a) Memory usage   (b) Runtime

Fig. 3. Memory usage and Runtime for T15I4D100K with varying k

6   Conclusion

In this paper, we studied the problem of finding top-k frequent itemsets on uncertain streams, which is different in semantics from the past proposals of top-k frequent itemsets on certain data streams. A novel algorithm, UTK-FI based on time-sensitive sliding window and Chernoff bound was developed and evaluated. We evaluated the performances of our methods through the above experiments, and the experimental results showed that UTK-FI algorithm is efficient and feasible.

References

1. Aggarwal, C.C., Yu, P.S.: A Framework for Clustering Uncertain Data Streams. In: Proc. of 34th ICDE, Cancún, México, pp. 150–159 (2008)
2. Cormode, G., Garofalakis, M.: Sketching Probabilistic Data Streams. In: Proc. of ACM SIGMOD, Beijing, China, pp. 281–292 (2007)
3. Manku, G., Motwani, R.: Approximate Frequency Counts over Data Streams. In: Proc. of the 28th VLDB, Hong Kong, China, pp. 346–357 (2002)

4. Chernoff, H.: A Measure of Asymptotic Efficiency for Tests of a hypothesis Based on the Sum of Observations. The Annals of Mathematical Statistics 23(4), 493–507 (1952)
5. Zimányi, E., Pirotte, A.: Imperfect information in relational databases. In: Uncertainty Management in Information Systems, pp. 35–88 (1996)
6. Agrawal, R., Srikant, R.: Fast Algorithms for Mining Association Rules in Large Databases. In: Proc. of the 20th ICDE, pp. 487–499. Morgan Kaufmann, San Francisco (1994)

A Novel ID-Based Anonymous Signcryption Scheme

Jianhong Zhang, Shengnan Gao, Hua Chen, and Qin Geng

College of sciences, North China University of Technology,
Beijing 100144, P.R. China
jhzhang@ncut.edu.cn

Abstract. Signcryption is a novel cryptographic primitive. In this work, by combining ring signature with ID-based signcryption , we build a security model of anonymous signcryption and have proposed an ID-based anonymous signcryption scheme based on the bilinear pairings to adapt to the anonymous setting. And we prove that our proposed scheme is secure and satisfies the confidentiality and unforgeability. By comparison our scheme with the other two schemes, we show that our scheme is more efficient in terms of computational cost and the length of ciphertext.

Keywords: anonymous signcryption, security proof, ID-based ring signature.

1 Introduction

Message security and the sender's identity authentication for communication in the open channel is a basic and important technology of internet. In 1997, Zheng [1] first proposed a new cryptographic primitive: signcryption, which can perform digital signature and public key encryption in a logic step, at lower computational costs and communication overheads than the above sign-then-encrypt way. Since then, there are many signcryption schemes proposed.

It is only recently that a formal security proof model[5] is formalized providing security proof for Zheng[1] in the random oracle model. By combining ID-based cryptology and signcryption, Malone-Lee proposed a first ID-based signcryption scheme. But Libert and Quisquater[9] pointed out that Malone-Lee's scheme [6] is not semantically secure. Chow *et al*[8] proposed an ID-based signcryption scheme that can provide both public verifiability an forward security. In 2003, Boyen [6] proposed a secure identity-based signcryption scheme with ciphertext anonymity and provable secure in the random oracle model. Their security proof model is slightly different from that of [5] which includes the ciphertext anonymity. In 2004, Libert and Quisquater [7] modified Boyen's security proof model to non-dentity based signcryption scheme and proposed a signcryption scheme. Unfortunately, Tan showed that the scheme did not satisfy the above properties in [10]. Recently, Duan *et.al* presented ID-based signcryption scheme [11] in the multi-user setting, and Ma *et.al* gave a short signcryption scheme [12]. Up to now, the most efficient ID-based signcryption

Q. Li et al. (Eds.): APWeb/WAIM 2009, LNCS 5446, pp. 604–610, 2009.

scheme[13] was proposed by Barreto *et al* and the security of the scheme was based on recently studied computational assumptions: **the** $q-$**BDH Inversion problem**.

In some cases, we hope to realize anonymous authentication and encryption. For example, the famous Watergate scandal in the history of the United States. An anonym sends some information to Woodward and Bernstein and proves the information authenticity. Only Woodward and Bernstein can obtain these information. We call the technology which is able to anonymous authentication and encryption, as anonymous signcryption.

Ring signcryption is an important technology to realize anonymous signcryption. And some ring signcryption schemes [3,4] were proposed. But these scheme are not efficient. To raise efficiency of anonymous signcryption, In this work, by combining ring signature scheme idea and ID-based signcryption idea, we propose an ID-based anonymous signcryption scheme to make a user anonymously produce a ciphertext for the recipient on half of the group, and show that the scheme is secure in the random oracle model. The security of the scheme is based on DDH problem and the CDH problem.

2   Preliminaries

Let $\mathbb{G}_1$ be a cyclic additive group generated by the generator P, whose order is a prime q, and $\mathbb{G}_2$ be a cyclic multiplicative group of the same prime order q. We assume that the discrete logarithm problem (DLP) in both $\mathbb{G}_1$ and $\mathbb{G}_2$ are hard. An admissible pairing $e : \mathbb{G}_1 \times \mathbb{G}_1 \longrightarrow \mathbb{G}_2$, which satisfies the following three properties:

- Bilinear: If $P, Q \in \mathbb{G}_1$ and $a, b \in Z_q^*$, then $e(aP, bQ) = e(P, Q)^{ab}$;
- Non-degenerate: There exists a $P \in \mathbb{G}_1$ such that $e(P, P) \neq 1$;
- Computable: If $P, Q \in \mathbb{G}_1$, one can compute $e(P, Q) \in \mathbb{G}_2$ in polynomial time.

We note the modified Weil and Tate pairings associated with supersingular elliptic curves are examples of such admissible pairings. The security of the ID-based signcryption scheme discussed in this paper is based on the following security assumption.

Definition 1. *Given two group $\mathbb{G}_1$ and $\mathbb{G}_2$ of the same prime order q, a bilinear map $e : \mathbb{G}_1 \times \mathbb{G}_1 \longrightarrow \mathbb{G}_2$ and a generator P of $\mathbb{G}_1$, the Decisional Bilinear Diffie-Hellman problem (DBDHP) in $(\mathbb{G}_1, \mathbb{G}_2, e)$ is to decide whether $h = e(P, P)^{abc}$ given (P, aP, bP, cP) and an element $h \in \mathbb{G}_2$.*

Definition 2 (Computational Diffie-Hellman (CDH) Assumption). *Let $\mathcal{G}$ be a CDH parameter generator. Given (P, aP, bP) in $\mathbb{G}_1$, It is hard to compute abP*

Lemma 1. *[**Ring Forking Lemma**] Consider a generic ring signature scheme with security parameter k. Let $\mathcal{A}$ be a probabilistic polynomial time Turning machine which receives as input the digital identities of users in a set L and other public data; the machine $\mathcal{A}$ can ask q queries to random oracle.*

Assume that $\mathcal{A}$ produces, within time bound T and with non-negligible probability of success $\varepsilon \geq \frac{7V_{q,n}}{2^k}$, a valid ring signature $(L', m, R_1, \cdots, R_n, h_1, \cdots, h_n, \delta)$ for some ring $L' \subset L$ of n user.

Then, in time $T' \leq 2T$ and with probability $\varepsilon' \geq \frac{\varepsilon^2}{66V_{q,n}}$, we obtain two valid ring signatures $(L', m, R_1, \cdots, R_n, h_1, \cdots, h_n, \delta)$ and $(L', m, R_1, \cdots, R_n, h'_1, \cdots, h'_n, \delta')$ such that $h_j \neq h'_j$, for some $j \in \{1, \cdots, n\}$ and $h_i = h'_i$ for all $i = 1, 2, \cdots, n$ such that $i \neq j$.

3   Formal Model of ID-Based Anonymous Signcryption

An ID-based anonymous signcryption scheme consists of the algorithms <**Setup, Key Extract, Anonymous-Signcrypt, Unsigncrypt**>. Malone-Lee[6] define the security requirements for ID-based signcryption schemes. These security requirements include indistinguishable against adaptive chosen ciphertrext attacks and unforgeability against adaptive chosen message attacks. In the following, we modify his definitions to adapt for our ID-based anonymous signcryption scheme.

Definition 3. (Confidentiality.) An ID-based anonymous signcryption scheme is indistinguishable against adaptive chosen ciphertext attacks property(IND-IDAS-CCA2), if no polynomially bounded adversary has a non-negligible advantage in the following game.

Setup. The challenger $\mathcal{C}$ runs **Setup** algorithm with a security parameter k and sends the system parameters to the adversary $\mathcal{A}$.

Phase 1. $\mathcal{A}$ performs a series of queries in an adaptive fashion. The following queries are allowed:

Key extraction queries: $\mathcal{A}$ chooses an identity ID. $\mathcal{C}$ computes private key $s_{ID} = Extract(ID)$ to response to $\mathcal{A}$.

Anonymous-signcryption queries: $\mathcal{A}$ produces a signer list $L = (ID_1, \cdots, ID_n)$, the recipient identity ID_B and a message m. $\mathcal{C}$ randomly selects a user ID_i as the signer and computes $\delta = Anonymous-signcrypt(m, L, S_{ID_i}, ID_B)$ and sends δ to $\mathcal{A}$, where S_{ID_i} is the secret key of the user with identity ID_i.

Unsigncryption queries: given a ciphertext δ. The challenger $\mathcal{C}$ runs **Unsigncrypt** algorithm on input $(\delta, s_{ID_B}, L = (ID_1, \cdots, ID_n))$ and returns its output to $\mathcal{A}$, where S_{ID_B} is the secret key of the receiver ID_B.

Challenge. At the end of phase 1, $\mathcal{A}$ generate two equal length plaintext m_0, m_1 and a signer list $L = (ID_1, \cdots, ID_n)$ and the recipient ID_B on which he wants to be challenged. He cannot have query the private key of ID_B as the first phase. The challenge flips $b \in \{0, 1\}$, then computes $delta^* = Anonymous - signcryption(s_{ID_1}, \cdots, s_{ID_n}, ID_B, m_b)$ to return it to the adversary $\mathcal{A}$.

Phase 2. $\mathcal{A}$ can ask a ploynomially bounded number of queries adaptively again as in the first phase. But he cannot make a key extraction query on ID_B and cannot make an unsigncryption query on δ^*.

Output. $\mathcal{A}$ outputs a bit b' and wins the game if $b' = b$.

The advantage of $\mathcal{A}$ is defined as $Adv\mathcal{A} = |\ 2P[b' = b] - 1\ |$, where $P[b' = b]$ denotes the probability that $b' = b$.

Definition 4. (Unforgeability). An ID-based anonymous signcryption scheme is existential unforgeable against adaptive chosen message attack (EUF-IDAS-CMA) if no polynomially bounded adversary has a non-negligible advantage in the following game.

1. The challenger $\mathcal{C}$ runs the **Setup** algorithm to produce system parameters with a security parameter k, and sends the system parameters to the adversary $\mathcal{F}$.
2. $\mathcal{F}$ makes a number of queries to the challenger. The attack may be conducted adaptively, and allows the same **Key Extract**, **Signcrypt** and **Unsigncrypt** queries as the Definition 3.
3. Finally, $\mathcal{F}$ produces a new triple $(\delta^*, L = (ID_{R_1}^*, \cdots, ID_{R_n}^*), ID_B^*)$ where L is the signer list and ID_B^* is the identity of the recipient.
4. $\mathcal{F}$ wins the game if (I) the signcryptext δ^* is a valid ciphertext under the ring L^* and the recipient with the identity ID_B and (II) the query $(\delta^*, L = (ID_{R_1}^*, \cdots, ID_{R_n}^*))$ has not been submitted by the adversary as public input to **Signcrypt** oracle.

The advantage of $\mathcal{F}$ is defined as the probability $Adv(\mathcal{F}) = Pr[\mathcal{F}\ win\ the\ above\ game]$.

4   Our Proposed Scheme

In this section, we give a novel ID-based anonymous signcryption scheme, which is a combination of ring signature and signcryption scheme. The details of our scheme are as follows:

Setup: given a security parameter k, the PKG chooses bilinear map groups $(\mathbb{G}_1, \mathbb{G}_2)$ of prime order $q > 2^k$, $e : \mathbb{G}_1 \times \mathbb{G}_1 \to \mathbb{G}_2$ and a generator $P \in_R \mathbb{G}_1$. It then chooses a master $s \in_R Z_q$, a system-wide public key $P_{pub} = sP$. Let H_1, H_2, H_3 and H_4 be four cryptographic hash functions where $H_1 : \{0,1\}^* \to \mathbb{G}_1$, $H_2 : \mathbb{G}_2 \to \{0,1\}^n$,$H_3 : \{0,1\}^* \times \mathbb{G}_2 \to Z_q$ and $H_4 : \{0,1\} \times \mathbb{G}_1 \times \{0,1\} \to \mathbb{G}_1$. The public parameters are

$$(\mathbb{G}_1, \mathbb{G}_2, P, e, H_1, H_2, H_3, H_4, P_{pub}, k)$$

Extract: for a given user's identity string $ID \in \{0,1\}^*$, the algorithm responses as follows: Compute $Q_{ID} = H_1(ID) \in \mathbb{G}_1$ and Set the user's private key $S_{ID} = s \cdot Q_{ID}$, where s is a master secret of PKG.

Anonymous Signcrypt: given a message m, a receiver's identity ID_B, and Let $L = (ID_{A_1}, ID_{A_2}, \cdots, ID_{A_n})$ denote n users' identity. For a user with identity ID_s, it executes the following steps:

1. For $i = 1, \cdots, n$ and $i \neq s$, randomly select $x_i \in Z_q$ to compute $R_i = x_i P$.
2. randomly choose $x_s \in Z_q$ to compute $\omega = e(P_{pub}, \sum_{j=1}^{n} x_j Q_{ID_B})$ and set $R_s = x_s P - \sum_{j=1, j \neq s}^{n} (H_3(R_i, \omega) Q_{A_j} + R_i)$, where $Q_{A_j} = H_1(ID_{A_j})$.
3. compute $c = H_2(\omega) \oplus m$ and $U = \sum_{i=1}^{n} x_i P$
4. compute $S = \sum_{i=1}^{n} x_i H_4(L, U, m) + x_s P_{pub} + H_3(R_s, \omega) \cdot S_{ID_{A_s}}$

Finally, the resultant ciphertext on the message m is $\delta = (c, S, U, R_1, \cdots, R_n)$.

Unsigncrypt: given $\delta = (c, U, S, R_1, \cdots, R_n)$, and the identity list L in the ring, where $L = (ID_{A_1}, \cdots, ID_{A_n})$. The receiver with identity ID_B can compute as follows to recover and verify the message:

1. compute $\omega = e(U, S_{ID_B})$ and $m = H_2(\omega) \oplus c$
2. accept the message if and only if the following equation holds $e(S, P) = e(U, H_4(L, U, m)) \cdot e(P_{pub}, \sum_{j=1}^{n} (R_i + H_3(R_i, \omega) \cdot Q_{A_j}))$

Security Analysis: In the following, we prove that the proposed anonymous signcryption scheme is secure in the random oracle model.

Theorem 1. *(Confidentiality) Assume that an IND-IDAS-CCA adversary $\mathcal{A}$ has an advantage ϵ against our anonymous signcryption scheme when running in time τ, asking q_{h_i} queries to random oracles $H_i (i = 1, 2, 3, 4)$, q_{se} anonymous-signcryption queries and q_{us} queries to the unsigncryption oracle. Then there is an algorithm $\mathcal{B}$ to solve the DBDH problem with probability .*

Theorem 2. *Let us assume that there exists an adaptively chosen message and identity attack $\mathcal{F}$ making q_{h_i} queries to random oracles H_i (i=1,2,3,4), q_e queries to key extract oracle and q_s queries to the signcryption oracle. If $\mathcal{F}$ can produce a forged ring signcryption with non-negligible probability. Then, there exists an algorithm $\mathcal{B}$ is able to solve the CDH problem.*

Efficiency Analysis: In the following,n, we analyze efficiency of the proposed scheme by comparing our scheme with recent schemes in literatures [3,4]. To reduce the signature length on supersingular curves, we instantiate pairing-based schemes using Barreto-Naehrig curve and adopt the embedding degree 6 and a large prime p with the size 160bits. Then, to achieve 1024 bit DL difficulty, the sizes of $\mathbb{G}_1$ and $\mathbb{G}_2$ elements need 171 bits and 1026 bits, respectively. In the following Table 1, the efficiency of our scheme is shown by comparison with that of [4,3]. In Table1, Let P_s be scalar multiplication on the group $\mathbb{G}_1$, P_e be pairings computation, P_m be multiplication operator in group $\mathbb{G}_2$.

Table 1. Comparison of our proposed scheme with [3]'s scheme and [4]'s scheme

Scheme Size		Unsigncryption	Signcryption
[3]	171*2+n*1186*n bits	$n(2P_s + P_m) + 3P_e$	$(3n + 2)P_s + (n + 2)P_e + 1P_m$
[4]	(n+2)*171+160*n bits	$2nP_s + 3P_e$	$(3n + 2)P_s + 1P_e$
ours	(n+2)*171+160 bits	$nP_s + 4P_e$	$3nP_s + 1P_e$

5   Conclusions

Anonymous signcryption is a useful tool to realize anonymity technolgy. In this paper, by combining ring signature with ID-based signcryption scheme, we build security model of ring signcryption and give a detail instance based on the bilinear pairings to adapt to anonymous setting. And the proposed scheme is provably secure and satisfies the two primitive properties of signcryption: confidentiality and unforgeability. Finally, Comparing our scheme with the other two schemes [3,4], we show that our scheme is more efficient in terms of computation cost and the size of ciphertext.

Acknowledgement

This work is supported by Natural Science Foundation of China (NO:60703044, 90718001), The Beijing Nova Programma (NO:2007B001), the PHR, Program for New Century Excellent Talents in University(NCET-06-188)and The Beijing Natural Science Foundation Programm and Scientific Research Key Program of Beijing Municipal Commission of Education (NO:KZ2008 10009005)

References

1. Zheng, Y.: Digital signcryption or how to achieve cost (signature & encryption) << cost(signature)+cost(encryption). In: Kaliski Jr., B.S. (ed.) CRYPTO 1997. LNCS, vol. 1294, pp. 165–179. Springer, Heidelberg (1997)
2. Zhang, G., Wang, S.: A Certificateless Signature and Group Signature Schemes against Malicious PKG. In: AINA 2008, pp. 334–341 (2008)
3. Huang, X.Y., Su, W., Mu, Y.: Identity-based ring signcryption scheme: cryptographic primitives for preserving privacy and authenticity in the ubiquitous world. In: Safavi-Naini, R., Seberry, J. (eds.) ACISP 2003. LNCS, vol. 2727, pp. 649–654. Springer, Heidelberg (2003)
4. Zhang, M., Yang, B., Chen, Y., Zhang, W.: Efficient secret authenticatable anonymous signcryption scheme with identity privacy. In: Yang, C.C., Chen, H., Chau, M., Chang, K., Lang, S.-D., Chen, P.S., Hsieh, R., Zeng, D., Wang, F.-Y., Carley, K.M., Mao, W., Zhan, J. (eds.) ISI Workshops 2008. LNCS, vol. 5075, pp. 126–137. Springer, Heidelberg (2008)
5. Baek, J., Steinfeld, R., Zheng, Y.: Formal proofs for the security of signcryption. In: Naccache, D., Paillier, P. (eds.) PKC 2002. LNCS, vol. 2274, pp. 80–98. Springer, Heidelberg (2002)
6. Boyen, X.: Multiprupose identity-based signcryption: A Swiss ary knife for identity-based cryptology. In: Boneh, D. (ed.) CRYPTO 2003. LNCS, vol. 2729, pp. 383–399. Springer, Heidelberg (2003)
7. Libert, B., Quisquater, J.: Efficient signcryption with key privacy from gap diffie-hellman groups. In: Bao, F., Deng, R., Zhou, J. (eds.) PKC 2004. LNCS, vol. 2947, pp. 187–200. Springer, Heidelberg (2004)
8. Tan, C.-H.: On the signcryption Scheme with key privacy. IEICE Trans. Fundamentals E88-A, 1093–1095

9. Libert, B., Quisquater, J.: A new Identity based signcryption schemes from pairings. In: IEEE infomation theory workshop 2003, pp. 155–158 (2003)
10. Chow, S.M., Yiu, S.M., Hui, L., Chow, K.: Efficient forward and provably secure ID-based signcryption Scheme with public verifiability and public ciphertext authenticity. In: Lim, J.-I., Lee, D.-H. (eds.) ICISC 2003. LNCS, vol. 2971, pp. 352–369. Springer, Heidelberg (2004)
11. Duan, S., Cao, Z.: Efficient and provably secure multi-receiver identity-based signcryption. In: Batten, L.M., Safavi-Naini, R. (eds.) ACISP 2006. LNCS, vol. 4058, pp. 195–206. Springer, Heidelberg (2006)
12. Ma, C.: Efficient short signcryption scheme with public verifiability. In: Lipmaa, H., Yung, M., Lin, D. (eds.) Inscrypt 2006. LNCS, vol. 4318, pp. 118–129. Springer, Heidelberg (2006)
13. Barreto, P.S.L.M., Libert, B., McCullagh, N., Quisquater, J.-J.: Efficient and provably-secure identity-based signatures and signcryption from bilinear maps. In: Roy, B. (ed.) ASIACRYPT 2005. LNCS, vol. 3788, pp. 515–532. Springer, Heidelberg (2005)

Pseudo Period Detection on Time Series Stream with Scale Smoothing

Xiaoguang Li[1,*], Long Xie[1], Baoyan Song[1], Ge Yu[2], and Daling Wang[2]

[1] School of Information,
Liaoning University, Shenyang, China
{xgli,bysong}@lnu.edu.cn,
long.xie.lnu@gmail.com
[2] College of Information Science and Engineering,
Northeastern University, Shenyang, China
{yuge,wangdaling}@ise.neu.edu.cn

Abstract. In this paper a novel technique is proposed to detect the pseudo period on time series streams. To address the memory limitation, a new summary method called *scale smoothing* is proposed for its scale smoothing property on data compression. Based on such summary data, a prune-based detection algorithm is designed to probe the pseudo period more efficiently. With the extensive experiments on the real and synthetic data sets, our method outperforms the wavelet-like methods.

Keywords: time series stream, pseudo period, scale smoothing, period detection.

1 Introduction

The periodicity is an instinct of many time series streams in real domains, such as SAT data, so period detection is a quite important and basic issue to the analysis of streams. For example, once people find the period of SAT data, they can use the historical data to forecast weather more accurately. But the real data tend to be noisy. As a result the data seem to repeat in a certain period, whereas a tiny difference exists between pair of data in a periodic interval. This kind of time series stream is called *pseudo periodical time series stream*.

For period detection of such streams, there are two problems: one is that the memory is limited; the other is that the data typically arrive at high rates. For the first, we adopt a DWT-like summary, called *scale smoothing*. While to solve the second a basic approach is proposed, and a prune-based detection approach is designed to compute the period approximately but efficiently enough.

Contribution:

- To address the infinite of time series stream, a new compressed method, *scale smoothing*, is introduced. It provides a lower compress ratio change rate than DWT-like methods to improve the utilization rate of memory;
- Two novel period detection algorithms, BPDA and RPDA, based on the scale smoothing are designed. BPDA can detect the pseudo period accurately

---

* Supported by National Science Foundation 60703068 and 60873068.

Q. Li et al. (Eds.): APWeb/WAIM 2009, LNCS 5446, pp. 611–616, 2009.

but less efficiently than RPDA. The detection by RPDA benefits the efficiency but suffers a bit error.

- Extensive experiments on the real and synthetic data sets show our algorithms scale smoothing, BPDA and RPDA are efficient and effective.

The remainder of this paper is organized as follows: Sect.2 briefly discusses related work; Sect.3 provides background material; Sect.4 discusses scale smoothing method to construct the summary; Sect.5 introduces the pseudo period detection algorithm BPDA and RPDA; Sect.6 reports the experimental results; finally, we summarize the paper in Sect.7.

2   Related Work

There are two critical problems for our research:(1) compression of time series streams; (2) efficient period detection algorithm.

To compress the time series streams, there are many methods ([1] is a suvery by DeVore), including DFT, DCT, PAA [2], DWT, histogram[3] and sampling [4,5]. Among these synopsises, wavelet has been found extremely useful and popular in stream processing field[6,7]. In [8] Anna give a novel method, called *IncDWT*, to get the B-term[9] wavelet coefficients by on-pass of the stream. It is infeasible for the period detection on stream data because of the reconstruction of raw data. [10] proposed a DWT-like approach of just using the consecutive average coefficients to approximate the stream, and detect the period. But for such DWT-like summary, at each time the data are compressed to the half size, that is half of the memory is unused and the compress ratio sharply becomes high.

To solve the issue of detection period, there are also some methods, such as, in the literature [5], the author studied the issue of *relax trend*, which is similar to ours, but the *sketch* cannot handle the stream data. While in the literature [11] the authors used DFT and Circular Autocorrelation to detect *accurate periodicity* in time series, and their methods can process stream data, but they do not consider the finite of memory. Recently, Lv-an in [12], introduced the concept of *pseudo period*, but that paper focused on how to manage patterns, not pseudo period detection.

3   Background

Definition 1 (time series stream). *A time series stream* $X = (x_0, \ldots, x_i, \ldots, x_n)$ *is a series of infinite items, where* $n \to \infty$, x_i *is an item at timestamp* i *and* $x_i \in \mathbb{R}$.

$X[i]$ denotes the i-th item of X, i.e., $X[i] = x_i$, and $X[i \ldots j]$ denotes a subsequence of X started at x_i and ended at x_j.

A similarity measure function $\mathscr{F}$ is used to evaluate the similarity degree of two fragments, where the higher the degree is, the more similar the fragments are.

Definition 2 (pseudo period). *Let* X *be a time series stream, we say* T *is pseudo period of* $X \iff \mu_{\mathscr{F}}(T) = \frac{1}{\lceil n/T \rceil} \sum_{i=1}^{\lfloor \frac{n}{T} \rfloor} (\mathscr{F}(X[0 \ldots n - iT], X[iT \ldots n])$

is maximum and $\mu_{\mathscr{F}}(\mathcal{T}) > \theta$ for $\mathcal{T} \in [1 \ldots \lfloor \frac{n}{2} \rfloor]$, where n is the amount of X's arrival items and θ is a threshold.

Definition 3. *[13,14,15][autocorrelation] Let X be a series, then the autocorrelation $auto\_corr(d)$ is defined as:*

$$auto\_corr(d) = corr(X[0 \ldots n - d], X[d \ldots n]);$$ (1)

where

$$corr(X, Y) = \frac{\sum_i [(X[i] - \mu_X) * (Y[i] - \mu_Y)]}{\sqrt{\sum_i (X[i] - \mu_X)^2} * \sqrt{\sum_i (Y[i] - \mu_Y)^2}}$$

and μ_X (μ_Y) denotes the average of sequences X (Y).

4   Scale Smoothing

Definition 4 (scale). *The scale of a summary of a stream X is the ratio of the size of X and the size of the summary:*

$$scale = \frac{|X|}{|Summary(X)|}$$ (2)

where $Summary(X)$ denotes the summary of X, and $| \bullet |$ denotes the size of $\bullet$.

Usually, with a fix data size, the larger the scale is, the most the information lost. So a novel compressed method, *scale smoothing*, is introduced to construct a stream's summary when the data volume is far beyond the limitation of memory.

Scale smoothing uses the average of $X[i * k, \cdots, (i + 1) * k - 1]$ to be the i-th item of scale k, denoted by x_i^k. Eq.3 can be used to estimated x_i^{k+1}'s evaluation $\hat{x}_i^{k+1}$. And $\hat{x}_W^k$, the last one of k-scale, may be only paretically used, so the remainder of its should be added with $k + 1 - (k * W) \mod (k + 1)$ new arrivals to generate an item at $(k + 1)$-scale.

$$\hat{x}_i^{k+1} = \frac{k - (i \mod k)}{k + 1} \hat{x}_{i+[i/k]}^k + \frac{(i \mod k) + 1}{k + 1} \hat{x}_{i+1+[i/k]}^k$$ (3)

where $i = 0, 1, \ldots, \lfloor \frac{k}{k+1}(W - 1) \rfloor$, and $[\bullet], \lfloor \bullet \rfloor$ are the integral part and floor of $\bullet$ respectively. For the new arrivals, their average is appended to the window.

5   Period Detection Algorithm

5.1   Mean Autocorrelation Measure

Due to the fact that if $\mathcal{T}$ is a period of a stream, if slipping the stream to $\mathcal{T}$-length clips, then those clips should also be correlate, Eq.1 is modified from Eq.4.

$$auto\_corr(d) = \frac{1}{\lfloor \frac{n}{d} \rfloor} \sum_{j=1}^{\lfloor \frac{n}{d} \rfloor} corr(X[0 \ldots d - 1], X[j * d \ldots (j + 1) * d - 1])$$ (4)

Eq.4 is called *mean autocorrelation*, which is adopted in the following, while the old one is *basic autocorrelation*.

5.2   Basic Period Detection Algorithm

The idea of *Basic Period Detection Algorithm* (BPDA) is directly from the definition of Def.2: whenever a new value added to the window, compute the $\mu_{auto\_corr(d)}$ for each possible vale, and select the value of d at the maximum of $\mu_{auto\_corr(d)} \geq \theta$, to be the period of this moment.

Theorem 1. *The time complexity of scale smoothing is $O(W)$, while the complexity of period detection is $O(W^2)$, so the time complexity of BPDA is $O(n + W^2 \ln(1 + n/W))$.*

5.3   Reduced Period Detection Algorithm

As shown by Theorem 1, the BPDA is inefficient because the computation of $auto\_corr(d)$ is so costly. So a *Reduced Period Detection Algorithm* (RPDA) is put forward with two improvements against BPDA.

Lemma 1. *If $\mu_{auto\_corr}(d) > \theta$, then it must hold that:*

$$\sum_{i=1}^{p} auto\_corr(i * d) > p - (1 - \theta) * \left[\frac{n}{2d}\right] \tag{5}$$

where $p \in [1, n/(2 * k)]$.

Lemma 2. *If $auto\_corr(d) > \theta_1$, then the below must be hold:*

$$\sum_{i=1}^{j} corr(X[0 \ldots d-1], X[i * d \ldots (i+1) * d - 1) > j - (1 - \theta_1) * \left[\frac{n}{d}\right] \tag{6}$$

where $j \in \left[1, \frac{n}{d}\right]$, and $\theta_1 = p - (1 - \theta) * \lfloor \frac{n}{2d} \rfloor - \sum_{i=1}^{p-1} auto\_corr(i * d)$, p and θ come from lemma 1.

The proof of lemma 1 and 2 is omitted for the limited space. So we can use lemma 1 and 2 to terminate the computing of $\mu_{auto\_corr}(d)$, when $\exists p \in [1 \ldots [n/(2d)]]$ that Eq.5 dose not be hold, and $\exists j \in \left[1, \frac{n}{d}\right]$ that Eq.6 dose not be hold.

Observation 1. *If a time series stream X has a pseudo period T, then the autocorrelation of X will be summit near the $T, 2 * T, \cdots$, and low at others.*

According to Observation 1, we reexamine our basic period detection algorithm by skipping some low probabilistic periods. The main idea is that we maintain a candidate set storing the periods with high probability. Whenever a new item add to the window, only the candidates will be tested. And the candidates set need refresh periodically, such as every time the scale changed.

   To address the high complexity of computing the autocorrelation, we proposes a computation pruning approach to terminate the process of computation earlier.

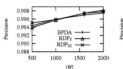

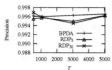

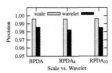

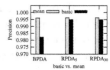

(a) $\mathcal{D}_1$ (b) $\mathcal{D}_2$ (a) n (b) $|W|$

Fig. 1. Stability of autocorrelation   **Fig. 2.** Time Complexity

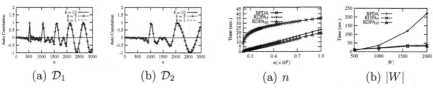

Fig. 3. Effectiveness   **Fig. 4.** Effectiveness   **Fig. 5.** Scale vs.   **Fig. 6.** Basic vs.
under $|W|$ under T Wavelet   Mean

6   Performance Evaluation

Our experiments are written by C++, and run on a PC with 256M RAM,2.00GHz CPU, Windows® XP. We use both synthetic and real data sets to test our algorithms' performance. The four synthetic data sets we used are all sinusoidal data with amplitude $A = 10$, $N(0, 0.2A)$ Gauss noise, and the periods are $531, 1031, 3031$ and 5031 respectively. The real data sets are monthly SAT[16] data of Shenyang, China from 1951 to 2004, and a sunspot daily data from 1945 to 2007.

Fig.1 shows the autocorrelation function at different scale with $T = 531$ (Fig.1(a)) and $T = 1031$ (Fig.1(b)). As illustrated by Fig.1, the autocorrelation at different scale are almost the same as the original without the smoothing.

As Fig.2(a) illustrated, with the increasing of n, the runtime of all algorithms increase. The time reduced by RDPA is remarkable; RDPA's is nearly 39% of BDPA. And the lazy refreshing reduces the process time. With the increase of window's size W, the running time of our period detection algorithms are increase, Fig.2(b) is the result.

The precision is defined as $precision = 1 - \frac{1}{n}\sum \frac{|T-\hat{T}|}{|T|}$, where T is the data's real period and $\hat{T}$ is determined by our algorithms.

As shown by Fig.3, with the increasing of W, the precision of our methods are all increased. And the period T has only a slightly influence to the precision. Fig.4 shows that T may cause precision be higher. The curves in Fig.4 are first down and than raise up is that when T is small we can finding the period at low scale, which means we have a more exact result, with the increasing of T the finding scale is also increasing, so the precision is down, but when the period is larger, a bigger denominator increases the precision.

We use the scale smoothing and wavelet method to compress the time series stream respectively, Fig.5 is the result. As Fig.5 shown, our scale smoothing time series stream compress method is better than wavlet. As shown in Fig.6, mean autocorrelation is better than basic's. The basic at BPDA is too low, that because it report some 4s, which are false positives.

7 Conclusion

In this paper, to detect *pseudo period* of time series streams with limited memory, a novel data stream approximate algorithm, *scale smoothing* is introduced. And then a novel pseudo period detection algorithm BPDA and its improvement RPDA are proposed. The massive experiments on synthetic and real data sets show that the algorithms are efficiency and effectiveness. In the future, we will study the properties of mean autocorrelation, and make the bound of lemma 1 and lemma 2 tighter, to make our pseudo period detection more efficiency.

References

1. DeVore, R.: Nonlinear approximation. Acta Numerica 7, 1–99 (1998)
2. Palpanas, T., Vlachos, M., et al.: Online Amnesic Approximation of Streaming Time Series. In: ICDE (2004)
3. Guha, S., Koudas, N., et al.: Approximation and streaming algorithms for histogram construction problems. ACM TODS 31, 396–438 (2006)
4. Manku, G.S., Rajagopalan, S., et al.: Random Sampling Techniques for Space Efficient Online Computation of Order Statistics of Large Datasets. In: SIGMOD (1999)
5. Indyk, P., Koudas, N., et al.: Identifying Representative Trends in Massive Time Series Data Sets Using Sketches. In: VLDB (2000)
6. Percival, D.B., Walden, A.T.: Wavelet Methods for Time Series Analysis. Cambridge University Press, Cambridge (2000)
7. Sarkar, T.K., Su, C., et al.: A tutorial on wavelets from an electrical engineering perspective. I. Discrete wavelet techniques. Antennas and Propagation Magazine 20, 49–68 (1998)
8. Gilbert, A.C., Kotidis, Y., et al.: One-Pass Wavelet Decompositions of Data Streams. IEEE TOKDE 15, 541–554 (2003)
9. Guha, S., Harb, B.: Approximation Algorithms for Wavelet Transform Coding of Data Streams. IEEE TOIT 54, 811–830 (2008)
10. Papadimitriou, S., Brockwell, A., et al.: Adaptive, unsupervised stream mining. The VLDB J. 13, 222–239 (2004)
11. Vlachos, M., Yu, P., et al.: On Periodicity Detection and Structural Periodic Similarity. In: SDM (2005)
12. Tang, L.-a., Cui, B., et al.: Effective Variation Management for Pseudo Periodical Streams. In: SIGMOD (2007)
13. Priestley, M.B.: Spectral analysis and time series. Academic Press, New York (1982)
14. Mark Last, A., Kandel, A., et al.: Data Mining in Time Series Databases. World Scientific, Singapore (2004)
15. Shumway, R.H., Stoffer, D.S.: Time Series Analysis and Its Applications With R Examples, 2nd edn. Springer, Berlin (2006)
16. Met. Data Services of LiaoNing Provice, P.R.China, http://data.lnmb.gov.cn

An Affinity Path Based Wireless Broadcast Scheduling Approach for Multi-item Queries[*]

Wei Li, Zhuoyao Zhang, Weiwei Sun[**], Dingding Mao, and Jingjing Wu

School of Computer Science, Fudan University
Shanghai, China
{li_wei,zhangzhuoyao,wwsun,maodingding,wjj}@fudan.edu.cn

Abstract. Data broadcasting is accepted as an effective solution to disseminate information to a large number of clients in wireless environments. Data broadcast scheduling attracts extensive attention, especially for multi-item queries. However, most previous works focusing on this issue do not take data replication in broadcast schedule into consideration, which leads to poor performance when the access probability of data items is skewed. In this paper, we proposed an efficient schedule algorithm called the affinity path approach. It firstly creates paths based on affinity of different data items, and then broadcast the paths based on the square root rule with data replication. Experiments prove that our approach reduces the average access time significantly.

Keywords: broadcast scheduling, affinity path, multi-item queries.

1 Introduction

The rapid development of wireless communication technology leads to the increasing concern to the important issue of effective information delivery in wireless environment. Due to the asymmetry in the wireless communication environment and the scalability of broadcast, which can make the information available simultaneously to a large number of clients, the data broadcasting has become an attractive and efficient solution for data dissemination [1, 2].

Lots of researchers have proposed their methods on designing broadcast schedules for multi-item queries. [3] introduces a scheduling method for multi-item queries called QEM. The measure Query Distance (QD) is defined, which shows the coherence degree of a query's data set in a schedule. [4] presents the Modified-QEM, which releases the restriction that the QD of previously expanded queries cannot be changed. Actually, a "move" action could be executed so that the QD of the new coming query is optimized and the better performance may be able to achieve. A broadcast data clustering approach for multi-item queries is proposed in [5]. Firstly, the data affinity between two data records is computed representing the degree of their referencing together in a query. Secondly, combine the pair with the highest data

---

[*] This research is supported in part by the National Natural Science Foundation of China (NSFC) under grant 60503035 and SRF for ROCS, SEM.
[**] Corresponding author.

Q. Li et al. (Eds.): APWeb/WAIM 2009, LNCS 5446, pp. 617–622, 2009.

affinity to form a segment. It forms a broadcast schedule finally by merge segments with the highest segment affinity in sequence.

In this paper, we investigate an efficient broadcast scheduling approach called *the affinity path approach*, which can get much lower access time than the previous works especially when the access probability of data items is skewed. In the affinity path approach, we first find out the affinity of every pair of data items, and then we create affinity paths according to the data affinity by greedy algorithms. These affinity paths are considered as the atomic data item and are scheduled based on the square root rule [6]. The experiment results show that the affinity path approach could attain a considerable improvement.

The organization of the rest part is as follows. We first present a formal definition of the broadcast scheduling for multi-item queries problem in section 2, and then describe the affinity path approach in section 3. The performance evaluation is illustrated in Section 4, and finally all of the above comes to a conclusion in section 5.

2   Preliminaries

We first introduce some notations which will be used in the rest of the paper.

d_i : a data item in the database.
N : the number of data in the database.
D : the set of database, where $D=\{ d_1, d_2, \ldots\ldots d_N\}$.
q_i : a query issued by a mobile client which access at least two data items.
M : the number of queries in Q.
Q : the set of all queries, where $Q=\{ q_1, q_2, \ldots\ldots q_M\}$.
$QDS(qi)$: the set of data items that q_i accesses (Query Data Set).
$freq(q_i)$: the reference frequency of the query q_i.
σ : the broadcast schedule.

The problem of data scheduling in this paper is to find a broadcast schedule σ which minimizes total access time (*TAT*) [7], denoted by

$$TAT(\sigma) = \sum_{k=1}^{M} freq(q_k)AT^{avg}(q_k,\sigma)/\sum_{k=1}^{M} freq(q_k) \tag{1}$$

Where $AT^{avg}(q_k,\sigma)$ is the average access time of the query q_i based on σ.

The data affinity of two data items denotes the degree that they are required together in a query. We introduce the definition of the data affinity between two data items d_i and d_j, which is defined in [5]:

$$aff(d_i,d_j) = \begin{cases} \displaystyle\sum_{\forall q_k \in Q} \frac{has(q_k,d_i)\times has(q_k,d_j)\times freq(q_k)}{|q_k|} & if\ i \neq j \\ 0 & otherwise \end{cases} \tag{2}$$

Where $has(q_k,d_i)=1$ if $d_i \in q_k$ and $has(q_k,d_i)=0$ otherwise.

Actually, the affinity of two data items is the sum of frequencies of queries that access both of them. The affinity of a pair of data items will be high if they are required together by many queries.

3 The Affinity Path Approach

3.1 Motivation

Compared with single-item queries, scheduling for multi-item queries is more complex because we must consider the access frequency of both data item and request. If we design the scheduling algorithm without considering the relation of data items accessed by a query, there may exist many half-satisfied queries, which means some of the data items they require are satisfied while some are not. As a result, the average access time is prolonged greatly. On the other hand, if we focus mainly on the access frequency of queries and satisfy the query with all the data items it requires at a time, the other request must wait for a long time once we decide to satisfy a query with a large *QDS*. We can see the consideration of both data items and requests in the definition of data affinity. So it is a reasonable parameter used in our approach.

In most of previous studies of broadcast scheduling for multi-item queries, researchers do not consider data replication which means each data item appears exactly once in a periodic cycle of broadcast schedule. "Replication or non-replication" is meaningless to users, because it does not affect users' access protocol. We try to make a tradeoff between the data items and requests by take data replication into consideration.

3.2 An Example

Let us take an example to show the process of the affinity path approach. Suppose that there are 12 data items and 11 queries, and the frequency values and data sets of queries are as Table 1.

We first construct a weighted graph that contains 12 vertexes, each vertex denotes a data item. There is an edge between two data items if the data affinity between the two data items is greater than 0. The weight of the edge between d_i and d_j is $aff(d_i, d_j)$.

Then, we create the paths. First, we select the edge with the largest weight as the initial of a path, here is the $edge(d_1, d_3)$, and delete it from E, Then we extend the path with edges, which is selected in E. It must satisfy the following tow conditions: 1) it must contain the vertex d_1, and 2) it has the largest weigh among all the candidates satisfying condition 1). It is $edge(d_1, d_2)$ in this example. We delete that edge in E and extend the path, now the path is (d_2, d_1, d_3). After that, we continue to extend the path in the same way from the endpoint of the edge with the larger weight until the other endpoint of the selected edge is already exist in the path. In this example, when we expand the path to $(d_8, d_5, d_2, d_1, d_3, d_6)$, we will find the largest edge from d_8 is (d_8, d_1), it is noticed that the other vertex d_1 already exists in the path, then we stop extending the current path, delete all the edge between any two vertexes in the path from E, and calculate the average weight of the path. After that we continue to create another path in the same way. For the above example, we can create seven paths which are listed in Table 2.

At last, we consider each path as a whole data item with the same size, and then we broadcast the path base on the square root rule, where the probability of a path refers to its average weight. We do not take the length of path into consideration, because we have considered the affect of length in calculating average weight of path.

Table 1. The set of queries   **Table 2.** The set of paths

query	QDS	frequency
q_1	d_1, d_2	3
q_2	d_1, d_3	3
q_3	d_1, d_3, d_4	3
q_4	d_1, d_2, d_4, d_5	2
q_5	d_1, d_3, d_6, d_8	1
q_6	d_1, d_4, d_7, d_9	1
q_7	d_2, d_5, d_{10}	2
q_8	d_2, d_3, d_6, d_{11}	2
q_9	d_4, d_7, d_{12}	1
q_{10}	d_5, d_8	2
q_{11}	d_2, d_6, d_9	1

Number	Average weight	Data set of path
P_1,	1.00971	$d_6, d_3, d_1, d_2,$ d_5, d_8
P_2,	0.715452	d_4, d_1, d_7
P_3,	0.445453	$d_5, d_{10}, d_2, d_4,$ d_3, d_{11}
P_4,	0.343443	d_{11}, d_6, d_9
P_5,	0.333333	d_9, d_2
P_6,	0.305556	d_7, d_{12}, d_4, d_9
P_7,	0.25	d_9, d_1

3.3 Algorithm

We describe the affinity path approach by three steps as follows:

Step 1: construct graph
We construct a graph $G = (V, E)$, where V is the set of all data items and each vertex denote a data item, every pair of vertexes have a edge with a weight which equals to the affinity of the corresponding pair of data items. The algorithm is described as follows.

Input: The set of data items $D = \{ d_1, d_2, \ldots\ldots d_N \}$.
Output : a graph $G = (V, E)$.

Algorithm:
1. for each data item d_i, create a corresponding vertex v_i in the graph;
2. if($aff(d_i, d_j) > 0$){
3.   create a edge $e(d_i, d_j)$ between v_i and v_j;
4. $weight(e(v_i, v_j)) = aff(d_i, d_j)$;
5. }

Step 2: create paths
First, we select the edge with the largest weight in E as the initial of the path and delete the edge from E. Then we extend that path from the endpoint v_i with the larger value of frequency by adding a edge in E which contains the same vertex v_i. This edge must be the one with largest weight among all the candidates and will be deleted from E once selected. After that, we continue to extend the path from the endpoint of the path in the similar way until the two vertexes of the selected edge are both included in the path. The difference is that each time we choose the endpoint v_i with larger weight of the corresponding edge as the vertex to extend Then we delete all the edges in E between the vertexes that selected to the path and calculate the average weight for the next step. We continue to create other paths until no edge exists in G. The algorithm is described as follows.

Input: the weighted graph G.
Output: the *PathSet* containing sets of affinity paths with average weight.

Algorithm:
1. initiate *PathSet* = empty;
2. while(*E* is not empty){
3.   find edge $e(v_a, v_b)$ in *E* with largest weight;
4.   initiate *Plist* = { v_a, v_b};
5.   delete $e(v_a, v_b)$ from *E*;
6.   set v_t = the vertex with larger value of frequency between v_a and v_b;
7.   while(*E* is not empty){
8.   find the edge $e(v_t, v_s)$ in *E* with largest weight;
9.   if(v_s already exist in *Plist*) break;
10.   else extend *Plist* with $e(v_t, v_s)$;
11.   delete $e(v_t, v_s)$ in *E*;
12.   set v_t = the endpoint of the edge with the larger weight between the head and tail edge in the *Plist*;
13.   }
14.   add *Plist* into *PathSet*;
15.   calculate the average weight of *Plist*;
16. }

Step 3: broadcast path
We consider each path as a whole data item with the same size, and then we broadcast the path based on the square root rule [6], where the probability of a path refers to its average weight.

4   Experiments and Evaluation

In this section, we compare our affinity path approach with the *data clustering* approach proposed in [5]. In our experiment, the parameter of Zipf distribution's default value is 1. The default value of both number of query types and data items is 500. The access time of the affinity path approach is the average access time evaluated after broadcasting 100,000 data items which is long enough to generate a stable result. Fig.1 depicts the effect of Zipf parameter. The figure indicates that when the Zipf parameter increases the average access time of the affinity path approach falls sharply. The reason for this result is that when the access probability of data items skewed, data items in queries with a higher access probability will broadcast frequently in the affinity path approach, while in the data clustering approach all the data have a broadcast interval equal to the length of periodic cycle, that is each data items appearance exact once in a broadcast cycle. So, the affinity path approach is especially suitable for scheduling multi-item queries with skewed access probability.

The effect of number of data items is showed in our second experiment. The result of experiment with number of data items varies from 200 to 1200 is illustrated in Fig.2. With the increase of number of data items for the data clustering approach, the length of periodic cycle is increasing linearly and the average access time is increasing sharply because the time to access a data item is proportional to the length of periodic cycle. While in the affinity path approach, the performance of average access time is less affect by increase of number of data items because of data replication.

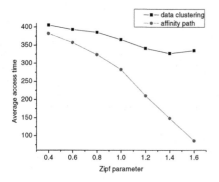
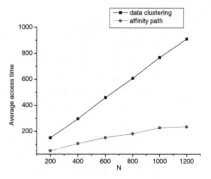

Fig. 1. Effect of Zipf parameter   **Fig. 2.** Effect of number of data items

5   Conclusion

In this paper, we have studied the issue of broadcast scheduling for multi-item queries. Most previous works are based on the consumption that each data item appears exactly only once in a periodic cycle. However, such limitation is meaningless to users' access and brings poor performance.

To find an efficient solution, we loosen the restriction that each data item appears exactly once in a periodic cycle. In the affinity path approach, we create affinity paths based on the affinity between different data items and broadcast these paths according to the square root rule. Through experiment results, we show that the affinity path approach performs much better than the previous approaches, especially when the access possibility of data items is skewed.

References

1. Imielinski, T., Viswanathan, S., Badrinath, B.R.: Data on air: organization and access. IEEE Transactions on Knowledge and Data Engineering 9(3) (1997)
2. Xu, J., Lee, D., Hu, Q., Lee, W.C.: Data broadcast. In: Handbook of wireless networks and mobile computing. John Wiley & Sons, Chichester (2002)
3. Chung, Y., Kim, M.: QEM: A scheduling method for wireless broadcast. In: DASFAA 1999 (1999)
4. Lee, G., Yeh, M.-S., Lo, S.-C., Chen, A.L.P.: A strategy for efficient access of multiple data items in mobile environments. In: MDM 2001 (2001)
5. Chung, Y., Kim, M.: A wireless data clustering method for multipoint queries. Decision Support Systems 30(4) (2001)
6. Vaidya, N., Hameed, S.: Scheduling data broadcast in asymmetric communication environment. ACM Journal of Wireless Networks 5, 171–182 (1999)
7. Tsaih, D., Wu, G., Wang, C., Ho, Y.: An efficient broadcast scheme for wireless data schedule under a new data affinity model. In: Kim, C. (ed.) ICOIN 2005. LNCS, vol. 3391, pp. 390–400. Springer, Heidelberg (2005)

Lightweight Data Mining Based Database System Self-optimization: A Case Study

Xiongpai Qin, Wei Cao, and Shan Wang

Key Laboratory of Data Engineering and Knowledge Engineering, School of Information,
Renmin University of China, MOE, Beijing 100872, P.R. China
qxp1990@sina.com, {caowei,swang}@ruc.edu.cn

Abstract. With the system becoming more complex and workloads becoming more fluctuating, it is very hard for DBA to quickly analyze performance data and optimize the system, self optimization is a promising technique. A data mining based optimization scheme for the lock table in database systems is presented. After trained with performance data, a neural network become intelligent enough to predict system performance with newly provided configuration parameters and performance data. During system running, performance data is collected continuously for a rule engine, which chooses the proper parameter of the lock table for adjusting, the rule engine relies on the trained neural network to precisely provide the amount of adjustment. The selected parameter is adjusted accordingly. The scheme is implemented and tested with TPC-C workload, system throughput increases by about 16 percent.

Keywords: database self-optimization, rule engine, neural network predictor.

1 Introduction

Optimizing the DBMS with manual work gets more and more difficult due to the following facts. Firstly, it is time consuming to analyze a lot of performance data before tuning the DBMS. Secondly, it is difficult for DBA to continuously monitors and analyzes fluctuating workload and to react to it properly. In addition, embedded DBMS usually has on DBA at all. Lastly, there is not enough number of experienced database management experts, and the cost for the manpower is prohibitive. Database research community is trying to build database systems [1] with the ability of self-managing and self-optimization. DBMS could perform the work of parameters setting, optimization, system healing, and protecting the system against malicious attacks by itself, without much human intervention.

We present a lightweight data mining based database system self-optimization scheme in this paper. The lock table is used as the case for study. The training module feeds historical lock table performance data to a neural network to get it trained. The neural network learns from the data through a lightweight data mining process. After trained, the neural network is intelligent enough to predict lock table and system performance according to lock table parameters. A rule engine makes decisions on which parameter to be adjusted during system running, and gets quantitative hints from the neural network predictor to get the job done.

Q. Li et al. (Eds.): APWeb/WAIM 2009, LNCS 5446, pp. 623–628, 2009.
© Springer-Verlag Berlin Heidelberg 2009

2 Related Works

Database self-optimization techniques can be categorized into two classes: Global Tuning & Optimization, and Local Tuning & Optimization [2].

Local Tuning & Optimization tries to solve specific database tuning problems such as data distribution, index selection, buffer replacement policies choice, multi programming level setting and so on [1]. On the other hand, Global Tuning & Optimization focuses on the construction of a brand-new database system with the ability of self-optimization. The DBMS can automatically maintain a subtle equilibrium among the resources used by individual system components to achieve the best overall performance. Paper [3] and paper [4] propose database systems with self-tuning ability. Traditionally successful DBMS systems enjoy a large number of customers and possess of a large code base, radical modifications to the code base is not a good idea. As mentioned in [4], these systems need to cope with the complexity of integrating self-tuning functions into the already existed system code base. No practical systems using this approach exist until now except a prototype is presented in [5]. Other than rewriting the code base of the DBMS, another approach is to view the database system as a black box. The tuning component is decoupled from target database system cleanly, there is no need to integrate the self-tuning capability into database system code base. During DBMS running, the tuning module continuously monitors the performance of the system [6], and tunes some critical parameters that are externally accessible for better performance [7]. In [8] an agent-based local self-tuning scheme for index selection is proposed, the experiment result shows its feasibility. In [9] Benoit used a decision tree to make tuning decisions, several iterations of parameter adjusting are needed. Our work differs from previous work in that a neural network based predictor is used to avoid several iterations of tuning.

3 Data Mining Based DB Self-optimization

3.1 Lock Table Performance Behavior

Modeling lock table running behavior is the precondition of tuning.1) Lock Granularity: DBMS commonly provides various lock granularities for various concurrencies. Lock granularity directly affects system concurrency. When lots of users access the system concurrently, locks of small granularity are usually used to avoid locking conflicts and increase system concurrency. When memory is in short, the system will elevate some tuple-level locks to a table-level lock, to make room for new lock requests. Table level locks, which are large in granularity, often lead to decreased concurrency and poor system throughput. 2) Lock Table Size and Memory Consumption: In real life when memory budget is limited, Arbitrarily extending lock table size leads to negative effects because reduced buffer size decreases cache hit during data accesses. 3) Solving the Deadlock Problem: DBMS periodically detects deadlocks, if the time interval is too long, maybe some deadlocks cannot be handled timely, transaction response time will get postponed. If the time interval is too short, DBMS maybe do some futile work in deadlock detection, system process power is wasted and performance get decreased. 4) Locking Time Out Parameter: We can set a parameter, namely locking time out for SQL execution. When the transaction cannot get the necessary locks before time out, the transaction is aborted. With some transactions aborted, other blocked transactions can move on, system throughput is ensured. Above-mentioned parameters should be adjusted according to workload fluctuations, data access conflicts, and resource consumption status.

3.2 ABLE Toolkit and Its Application

The self-optimization scheme for lock table is implemented using the Java language, the ABLE toolkit (Agent Building & Learning Environment)[10] is used as the underlying framework and running platform.

We develop a rule engine to solidify expert knowledge on lock table tuning, the rules come from DBA's experience and the database-tuning guide, and are expressed using ABLE rule language [10]. According to current performance data, some parameter may be chose to be adjusted according to the rules. A neural network based predictor is also implemented to help determine the amount of parameter adjustment.

3.3 Self-Optimization System Implementation

【1】 **System Architecture.** Figure 1 shows the architecture the self-optimization system. 1) Performance Data Collecting Module is responsible for gathering lock table settings and performance metrics. During training, the data is used to train the neural network; during running, the data is handed over to the rule engine for further processing. 2) Training Module uses the performance data to train the neural network to make it capable of predicting. 3) Rule Engine uses rules in the rule set to make decisions on which parameter to be adjusted. 4) The function of the Neural Network based Predictor is to predict lock table performance with provided lock table parameters. To make the neural network capable of the job, it is trained beforehand as mentioned before.

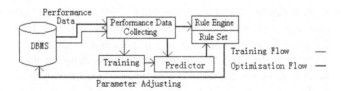

Fig. 1. Self-Optimization System Architecture

【2】 **Performance Metrics for the Lock Table.** Two performance metrics can be used to indicate the performance of the lock table, namely *Deal Lock Rate*, and *Average Lock Wait Time*. We define the *Deal Lock Rate* to be the number of deadlocks that occur for every 10,000 transactions. The *Average Lock Wait Time* is calculated from the amount of time spent waiting for locks and the number of locks granted. The two performance metrics can be obtained directly from DBMS management API or calculated from the data obtained from the API. The two metrics maybe conflict with each other, one of them should be discarded.

【3】 **Prepare Training Data Set.** A serial of experiments are conducted to find out the relationship between *Deadlock Check Time* and lock table performance. The experiment result is shown in figure 2. Figure 2 shows the *Deadlock Rates* and *Average Lock Wait Times* corresponding to various *Deadlock Check Time* intervals. The X-arises of the figure represents *Deadlock Check Time* intervals, ranges from 10000ms to 1000ms. The Y-axis on the left side represents *Deadlock Rate*, and the Y-axis on the right side represents system throughput (*Transactions per Minutes*) and *Average Lock Wait Times*.

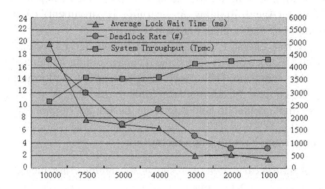

Fig. 2. Lock Table & System Performance vs. Deadlock Check Time

From figure 2, we can see that the relationship between *Deadlock Check Time* and *Deadlock Rate* is not a monotone relationship. It is hard to find out a commonly acceptable deadlock rate for various OLTP workloads. We choose *Average Lock Wait Time* to be the major lock table performance indicator. When deadlock check time interval is below 3000ms, the curve of *Average Lock Wait Time* and the curve of system throughput flatten out. For OLTP workloads, we choose 1000ms as a threshold for *Average Lock Wait Time*. When *Average Lock Wait Time* is greater than 1000ms, it is necessary to adjust lock table parameters to achieve better performance.

The relationships between other lock table parameters and lock table performance are also determined through experiments. The parameters include *Lock Table Size, SQL Lock Time Out, and Max Locks*. When experimenting on specific parameter concerned, other parameters are set to default values, and the concerned parameter is varied by small steps from the low bound of its domain to the high bound, lock table performance and system throughput are measured and recorded. We do not try to experiment with every possible parameter combinations, trained predictor should tell what performance the system will get when using specific parameter combination.

【4】 **Training the Neural Network -- Lightweight Data Mining.** After enough performance data is collected, it is used to train the neural network. The data scheme for training is < *Lock Table Size, Deadlock Check Time, Lock Time Out, Max Locks, Average Lock Wait Time*>, the preceding 4 parameters are input values, and the fifth parameters is an output value. 7000 training samples and 3000 testing samples are prepared for training and testing respectively.

The predictor is a back propagation neural network. We use a 3 layer neural network to ensure the precision of predicting as well as low cost of running, the numbers of neurons in input layer, hidden layer and output layer of the network are 4, 6, 1 respectively. After trained, the neural network changes its internal structure and becomes more intelligent to predict lock table performance according to provided parameters. The result of the testing shows that, within the tolerable error bound of $\pm 5\%$, the network predicts correctly.

【5】 **Tuning the Lock Table.** During DBMS running, performance data collected is handed over to the rule engine to decide whether any parameter should be adjusted. The rule engine provides the predictor with various parameter values increased by small steps, the parameter setting for the best predicted-performance is chose for actual parameter adjustment. The rule engine evaluates three rules in sequence. 1) When tuple level locks are elevated to table level lock, it indicates that the lock table size is too small, the rule engine tries to increase lock table size. 2) When *Average Lock Wait Time* is greater than the preset threshold (i.e. 1000 ms), the rule engine try to decrease deadlock detection interval or SQL lock wait time to reduce lock wait time. The two parameters can both be used to cut down the *Average Lock Wait Time*, the policy is that adjusting *Deadlock Check Time* twice for every adjustment of *Lock Time Out*. 3) When the lock table usage percentage is below the preset threshold, the rule engine tries to decrease the size of lock table to release memory for other using, e.g. database page buffer.

4 Experiments

The experiment is conducted on a HP Proliant DL 380 G4 with two Intel Xeon 3600MHz CPUs, and 5GB RAM. The operating system is Windows 2003. We use IBM DB2 version 7.2 as the experimenting DBMS. 10 warehouses' of data is populated into the database and the TPC-C workload is run on the DBMS.

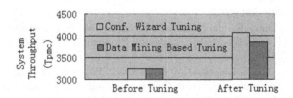

Fig. 3. Comparison of Optimization Effects

The experiment result is shown in figure 3, the system throughput (*Transaction per Minute*) increase by 15.85 percent, from 3250 to 3765, a considerable increase. From the figure, we can also see that the tuning effect of our self-optimization scheme is inferior to IBM Configuration Wizard. Besides the lock table, IBM Configuration Wizard tunes other critical performance parameters, thus it gets more performance improvement. The result encourages us to extend the scheme to support other DBMS sub systems optimization to achieve more overall performance improvement.

5 Conclusions

A data mining based database system self-optimization scheme for lock table is presented. A neural network is trained with historical performance data of the lock table is used as a predictor. The self-optimization process is driven by a rule engine, which chooses proper parameters to be adjusted during runtime, and gets quantitative hints from the neural network predictor to accomplish parameter adjusting. With the help of predictor, the self-optimization system avoids several rounds of tuning and evaluating. Experiment result demonstrates the feasibility of the scheme.

References

1. Weikum, G., et al.: Self-tuning database technology and information services: from wishful thinking to viable engineering. In: Proceedings of VLDB, Hong Kong, China, pp. 20–31 (2002)
2. Milanés, A.Y., Lifschitz, S.: Design and Implementation of a Global Self-tuning Architecture. In: Simpósio Brasileiro de Bancos de Dados, Uberlândia, MG, Brazil, pp. 70–84 (2005)
3. Martin, P., Powley, W., Benoit, D.: Using reflection to introduce self-tuning technology into dbmss. In: Proceedings of IDEAS, Coimbra, Portugal, pp. 429–438 (2004)
4. Chaudhuri, S., Weikum, G.: Rethinking database system architecture: Towards a self-tuning risc-style database system. In: Proceedings of VLDB, Cairo, Egypt, pp. 1–10 (2000)
5. Harizopoulos, S., Ailamaki, A.: A case for staged database systems. In: Proceedings of CIDR, Asilomar, CA, USA, pp. 121–130 (2003)
6. Bigus, J.P., Hellerstein, J.L., Jayram, T.S., Squillante, M.S.: Auto tune: A generic agent for automated performance tuning. In: Proceedings of International Conference and Exhibition on the Practical Application of Intelligent Agents and Multi-Agent Systems (PAAM), Manchester, UK, pp. 33–52 (2000)
7. Martin, P., Powley, W., Li, H., Romanufa, K.: Managing database server performance to meet QoS requirements in electronic commerce systems. International Journal on Digital Libraries 3(4), 316–324 (2002)
8. Costa, R., Lifschitz, S., de Noronha, M., Salles, M.: Implementation of an Agent Architecture for Automated Index Tuning. In: Proceedings of SMDB, Tokyo, Japan, pp. 1215–1223 (2005)
9. Benoit, D.: Automatic Diagnosis of Performance Problems in Database Management Systems. PhD. Dissertation, School of Computing, Queen's University, Canada (2003)
10. Bigus, J.P., et al.: ABLE: A Toolkit for Building Multi-agent Autonomic Systems. IBM System Journal 41(3), 250–371 (2002)

Efficient Detection of Discords for Time Series Stream

Yubao Liu, Xiuwei Chen, Fei Wang, and Jian Yin

Department of Computer Science of Sun Yat-Sen University, Guangzhou, 510275, China
{yubaoliu,xiaowei,issjianyin}@mail.sysu.edu.cn,
wilderwang@gmail.com

Abstract. Time discord detection is an important problem in a great variety of applications. In this paper, we consider the problem of discord detection for time series stream, where time discords are detected from local segments of flowing time series stream. The existing detections, which aim to detect the global discords from time series database, fail to detect such local discords. Two online detection algorithms are presented for our problem. The first algorithm extends the existing algorithm HOT SAX to detect such time discords. However, this algorithm is not efficient enough since it needs to search the entire time subsequences of local segment. Then, in the second algorithm, we limit the search space to further enhance the detection efficiency. The proposed algorithms are experimentally evaluated using real and synthesized datasets.

Keywords: Time discord, Online detection, Time series stream.

1 Introduction

Time discord is such subsequence that is the least similar to other sequence of time series. The detection of discords in time series data is an important problem in a great variety of applications, such as space shuttle telemetry, mechanical industry, biomedicine, and financial data analysis [1-3]. In real applications, time series data often has the form of data streams in which time series data points continuously arrives. For example, cardiogram time series of certain a patient often continuously flow during his/her each treatment.

In this paper, we consider the problem of discord detection for time series stream, where time discords are detected from local segments of flowing time series stream. The existing methods, which aim to detect the global discords from the entire time series database but a segment of time series, fail to detect such local discords. In addition, the existing methods need to capture the entire time series before detecting time discords and that is impossible in the case of flowing time series stream environment. In this paper, two online detection algorithms are presented. The first algorithm BFHS (stands for Brute Force HOT SAX) is simple and intuitive. It extends the existing algorithm HOT SAX to local segments of time series stream to detect the local discords. However, this solution is not efficient enough since it needs to search all the possible time subsequence of local time series segment during each detecting. Then, in the second algorithm DCD (stands for Detection of Continuous Discords), we limit the search space to further enhance the detection efficiency. The proposed algorithms are

Q. Li et al. (Eds.): APWeb/WAIM 2009, LNCS 5446, pp. 629–634, 2009.

experimentally evaluated using real and synthesized datasets. There are some other related works [4-6]. Different from our work, these works mainly focus on finding of time discord or anomaly from static time series database. The problem of text data streams clustering is studied in [7]. The concept-drifting mining problem is studied in [8]. The aggregate query over data streams is studied in [9]. Different from these work, our work focus on time series streams but general data stream or text stream.

2 Basic Notations and Definitions

Some basic time series notations are similar to the previous works [2-4]. We denote a time series as $T = t_1,\ldots,t_m$ that is an ordered set of m real-valued variables. A subsequence C of T of length m is a sampling of length $n \leq m$ of contiguous position from p, that is, $C = t_p,\ldots,t_{p+n-1}$ for $1 \leq p \leq m-n+1$. We can extract all possible subsequences of length n from a given time series T by sliding a widow of size n across T. We use function $Dist$ to measure the distance between two given time series

M and C, the Euclidean distance is used, that is, $Dist(M,C) = \sqrt{\sum_{i=1}^{n}(m_i - c_i)^2}$. It

is easy to know the function $Dist$ is symmetric, that is, $Dist(C, M) = Dist(M, C)$.

Definition 1. *Non-Self Match*: Given a time series T, containing a subsequence C of length n beginning at position p and a matching subsequence M beginning at q, if $| p - q| \geq n$, we say M is a non-self match of C and their distance is $Dist(M, C)$.

Definition 2. *Time Series Discord*: Given a time series T, the subsequence D of length n beginning at position l is said to be the discord of T if D has the largest distance to its nearest non-self match. That is, $\forall$ subsequences C of T, non-self matches M_D of D, and non-self matches M_C of C, $\min(Dist(D, M_D)) > \min(Dist(C, M_C))$.

We use time series buffer B to contain the local segment of flowing time series. In general, at certain a time point t, suppose B contains a time series of length n, $B \equiv T = t_1,\ldots,t_n$, and t_{new} is the next arriving time series data point. At next time $t+1$, the time series in B is changed as $B \equiv T' = t_2,\ldots, t_n, t_{new}$. In general, the length of time series buffer is specified in advance. We use $(m, n)(t)$ to denote the time subsequence whose position is m of time series buffer and length is n at time t. Then $(p-1, n)(t+1)$ and $(p, n)(t)$ are the same time subsequence for a given times series stream. The time discord found from B at time t is the local discord at time t and denoted as $LD(l, n)(t)$.

Definition 3. *Non-Similar Distance*: Given a time series T, for any subsequence P of T, Q is the nearest non-self match of P, the distance from P to Q is the non-similar distance of P.

The non-similar distance of local discord is the distance between local discord and its nearest non-self match. We use the vector V of length n to record the non-similar distances of n historical local discords. Assume the vector at time t is $V=d_1, d_2, \ldots, d_n$, then d_1 is the non-similar distance of the local discord at time $t-n$, d_n is the non-similar distance of the local discord at time $t-1$, and if the non-similar distance of the local discord at time t is d_{new}, the vector V at time $t+1$ is changed as $V= d_2, \ldots, d_n, d_{new}$.

Definition 4. *Available Discords*: Assume the local discord at time t is $LD(l, n)(t)$, and the local discord at time $t+1$ is $LD(k, n)(t+1)$. We say $LD(k, n)(t+1)$ is available if the following conditions hold (a) $k \neq l$, and (b) *valid* $(dist, V)$ = true, where *dist* denotes the non-similar distance of $LD(k, n)(t+1)$.

The condition (a) assures the local discord at time $t+1$ is not same to the local discord at time t. The condition (b) avoids generating too many trivial local discords. The purpose of function *valid* is to determine the interesting discords through the historical non-similar distances of local discords. Our problem is to find the available discords from the given time series stream.

3   The Presented Detection Algorithms

3.1   Online Detection Algorithm Framework

The online detection framework in Fig.1 firstly initializes some basic data structures and related variables, such as, the time series buffer and the vector V etc. Then the time series data points are added into time series buffer B. Next, the function *FindDiscord* is firstly called to find the local discord from B. The non-similar distance of the local discord, *dist*, is added into V. Next, we read the new time series data points from the flowing stream. After that, we continue to find the local discord from the updated time series in B by calling the *FindDiscord* function. If the local discord is available and different with the previous local discord (step 9), then we output the found local discord. Next, the position of the local discord and its non-similar distance are saved (steps 12 and 13). The loop continues until the time series stream is stopped. The *valid* function is defined as *valid*$(dist, V) \equiv dist >$ mean$(V)*$ *threshold*, where mean(V) function is the mean value of vector member, *threshold* and the number of vector member n are specified in advance.

A simple solution for *FindDiscord* function is to use the existing algorithm such as HOT SAX to find the discord from time series buffer. Such *FindDiscord* function is adopted in BFHS algorithm. But, BFHS algorithm is low since HOT SAX needs to search the entire local segment while detecting each time.

3.2   DCD Algorithm

In DCD algorithm, we adopt a new FindDiscord function, which is shown in Fig. 2.

Definition 5. *Small match*: For any subsequence P of a given time series T, Q is the non-self match of P, if $Dist(P, Q) < dist$, where *dist* is the non-similar distance of current local discord of time series, then Q (P) is a small match of P (Q).

Next, let observe the possible changes of local discords at two continuous time point. Assume current time is t. The position of current local discord is l. The position of the nearest non-self match of current local discord is *lnn*. Then at next time $t+1$, we have two cases: (a) The non-similar distance of the local discord at time $t+1$ is smaller than that of the local discord at time t, (b) The non-similar distance of the local discord at time $t+1$ is larger than that of the local discord at time t. For case (a), we have two sub-cases. (a_1) The time series subsequence containing the arriving data points is the nearest non-self match of the local discord at time t (i.e. $LD(l, n)(t)$), and the non-similar distance of $LD(l, n)(t)$ may be smaller than that of other time subsequences. Then the local discord at time t may be not local discord at time $t+1$. (a_2) The position of the current local discord at time t is equal to 1 (i.e. $LD(1, n)(t)$). Then the local discord will be changed as a non-discord subsequence at time $t+1$ because the first data point of $LD(1, n)(t)$ will be deleted with the flowing of time series stream. For case (b), due to deleting some time series points of time series buffer, there may be some subsequences whose non-similar distances are larger than that of the local discord at time t. These subsequences may become the new local discord at time $t+1$. For case (a), the non-similar distance of local discord at time $t+1$ will be smaller than that of the local discord at time t, we need to search all possible subsequences of time series buffer to find the new local discord at time $t+1$. For case (b), we can reduce the search space using the following lemma 1. Due to the limit of space, the proof of lemma1 is omitted here.

Algorithm 1: Online Detection Framework
Input: A given time series stream and the length of discord is n
Output: The local discords
1.   Initialize ();
2.   Read the time series data points into B;
3. $[prev\_loc, dist] = FindDiscord(B, n)$;
4.   Output the local discord $LD(prev\_loc, n)(t)$; // t is the current time;
5.   Add $dist$ into V;
6.   While (the time series stream is not stopped) do
7.   Read the next data point into B;
8. $[loc, dist] = FindDiscord(B, n)$;
9.   If $loc \mathrel{!=} prev\_loc\text{-}1$ && $valid(dist, V)$ then
10.   Output the local discord $LD(loc, n)(t)$; // t is the current time;
11.   End
12. $prev\_loc = loc$;
13.   Add $dist$ into V;
14.   End // end of while

Fig. 1. The description of online detection framework

Lemma 1. Assume current time is t, $B \equiv T = t_1, t_2, ..., t_m$, the current local discord is $LD(l, n)(t)$ and its non-similar distance is $dist$. Then at next time $t+1$, assume the arriving data point is t_{new} and local discord is $LD(k, n)(t+1)$. If $l \neq 1$ and $Dist(t_l, ..., t_{l+n-1}, t_{m-n+2}, ..., t_m, t_{new}) > dist$ (i.e. the above case (b)), we have: if $k \neq l-1$ (i.e. the local discord at time $t+1$ is not same to the local discord at time t) and $k \neq m-n+1$ (i.e., the local discord at time $t+1$ is not the time series subsequence containing the arriving data point), then $(k+1, n)(t)$ (i.e. $t_{k+1}, ..., t_{k+n}$) and $(1, n)(t)$ (i.e. $t_1, ..., t_n$) are small match.

In Fig.2, *Candidate* denotes the possible time subsequences could be the local discords. The *Search* function is similar to the discord discovery function of existing HOT SAX algorithm and is to quickly find the small matches of time subsequences. Due to the limit of space, the *Search* function is omitted here.

Algorithm 2: *FindDiscord*
Input: *B*: the time series buffer at time *t+1*; *n*: the length of discord;
Output: The position and non-similar distance of local discord at time *t+1*.
1.   read the next data point t_{new};
2. $currDist=Dist(t_{loc},...,t_{loc+n-1},\ t_{m-n+2},...,\ t_m, t_{new})$; /* *loc*: the position of local discord at time *t*; *dist*: the non-similar distance of local discord at time *t*. */
3.   if *currDist < dist*   // The case (a₁)
4.   *Candidate* = 1:|B|-n+1;
5.   else if *loc*=1   // The case (a₂)
6.   *Candidates* = 1:|B|-n+1;
7.   else   // The case (b)
8.   *Candidates* = {The small match of subsequence $(1, n)(t)$}∪ {The local discord at time *t*}∪{The subsequence $(m-n+1, n)(t+1)$ };
9.   [*loc, dist*] = *Search (Candidates, n, B)*;

Fig. 2. The description of FindDiscord function

4   The Experimental Evaluation

The experimental computer system hardware is AMD Atholon 4400 CPU and 1G memory and the OS is Microsoft Windows XP. The algorithms are implemented using Mathlab 7.4.0.

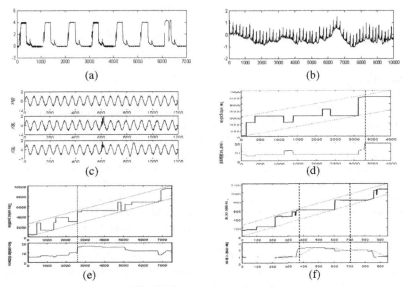

Fig. 3. The experimental results

The first real dataset is the sensor dataset shown in Fig. 3(a). This dataset contains about 7000 data point. The period of this dataset is not fixed and is about 1000. From this figure, we can see that the seventh period has an anomaly. In this test, the length of discord is 256, the length of time series buffer is 3000, and the arguments of valid function are: *n*=600 and *threshold*=2. The running results of our algorithms on this dataset are shown in Fig. 3(d). In this figure, the above part shows the found discords, included in two dashed biases, at each time point. The biases indicate the changing trend of time series of sensor dataset between 1000 and 7000. The changes of non-similar distance are given in the below part of this part. The largest changing point is indicated by a dashed vertical line. From this line, we can see that the largest non-similar distance appears at position 6053 of time series at time point 3300. Compared with Fig. 3(a), we can see that this indicated point is matched with the anomaly. The second real cardiogram dataset is shown in Fig. 3(b) and contains about 10000 data point. The running results of our algorithms on this dataset are in Fig. 3(e). In this test, the length of discord is 256 and the length of time series buffer is 2500. The arguments of valid function are: $n=600$ and *threshold*=1.5. The third dataset is synthesized and shown in Fig. 3(c). The results for this dataset are in Fig. 3(f). The arguments of valid function are: $n=100$ and *threshold*=1.05.

5 Conclusions

In this paper, we consider the problem of local time discord detection for time series stream. Two online detection algorithms BFHS and DCD are presented for our problem. The proposed algorithms are experimentally evaluated using real and synthesized datasets. How to determine the algorithm parameters such as valid function arguments n and *threshold* etc. according to the detailed application and enhance the detection algorithm efficiency is our future work.

Acknowledgments. This paper is supported by the NSFC (60703111, 60773198).

References

1. Fu, A., Keogh, E., Lau, L.Y.H., Ratanamahatana, C.A.: Scaling and time warping in time series querying. In: Proc. VLDB, pp. 649–660 (2005)
2. Keogh, E., Lin, J., Fu, A.: HOT SAX: efficiently finding the most unusual time series subsequence. In: Proc. ICDM, pp. 226–233 (2005)
3. Bu, Y., Leung, T.-W., Fu, A., Keogh, E., Pei, J., Meshkin, S.: WAT: Finding Top-K Discords in Time Series Database. In: Proc. SDM, pp. 449–454 (2007)
4. Yankov, D., Keogh, E., Rebbapragad, U.: Disk Aware Discord Discovery: Finding Unusual Time Series in Terabyte Sized Datasets. In: Proc. ICDM 2007 (2007)
5. Chan, K.-P., Fu, A.: Efficient time series matching by wavelets. In: Proc. ICDE, pp. 126–133 (1999)
6. Fu, A., Leung, O., Keogh, E., Lin, J.: Finding time series discords based on haar transform. In: Li, X., Zaïane, O.R., Li, Z. (eds.) ADMA 2006. LNCS, vol. 4093, pp. 31–41. Springer, Heidelberg (2006)
7. Liu, Y., Cai, J., Yin, J., Fu, A.: Clustering text data streams. Journal of Computer Science and Technology 23, 112–128 (2008)
8. Wang, H., Yin, J., Pei, J., Yu, P.S., Yu, J.X.: Suppressing model over-fitting in mining concept-drifting data streams. In: Proc. KDD, pp. 736–741 (2006)
9. Qin, S., Qian, W., Zhou, A.: Approximately processing multi-granularity aggregate queries over data streams. In: Proc. ICDE (2006)

SLICE: A Novel Method to Find Local Linear Correlations by Constructing Hyperplanes[*]

Liang Tang[1], Changjie Tang[1], Lei Duan[1], Yexi Jiang[1], Jie Zuo[1,**], and Jun Zhu[2]

[1] School of Computer Science, Sichuan University
610065 Chengdu, China
{tangliang,tangchangjie}@cs.scu.edu.cn
[2] National Center for Birth Defects Monitoring
610041 Chengdu, China

Abstract. Finding linear correlations in dataset is an important data mining task, which can be widely applied in the real world. Existing correlation clustering methods may miss some correlations when instances are sparsely distributed. Other recent studies are limited to find the primary linear correlation of the dataset. This paper develops a novel approach to seek multiple local linear correlations in dataset. Extensive experiments show that this approach is effective and efficient to find the linear correlations in data subsets.

Keywords: Data Mining, Linear Correlation, Principal Component Analysis.

1 Introduction

Linear correlations reveal the linear dependencies among several features in a dataset. Finding theses correlations is an interesting research topic. For example, in sensor network, a latent pattern, which is the linear correlation of multiple time series data received, can be used to detect the data evolution [8]. Principal component analysis (PCA) is able to capture linear correlations in a dataset [4]. PCA assumes that all instances in a dataset are in the same correlation. However, instances collected from the real world may have different characteristics, so the linear dependencies among features may be different in different data subsets.

Some clustering methods are developed to find data clusters in subspaces [2, 3, 5, 9]. Correlation clustering methods, such as 4C, try to seek clusters in a linear correlation [6, 7]. The first step of these methods is to generate ε–neighborhoods first. However, it is hard to generate high quality ε–neighborhoods to find linear correlations, when the distribution of instances in the correlations is sparse. In [1], a method called CARE is proposed to find local linear correlations. This method adopts PCA to analyze the linear correlations on both feature subsets and instance subsets. CARE focuses on finding the primary linear correlations from the whole dataset. It is not suitable to find linear correlations that exist in data subsets.

---

[*] This work was supported by the National Natural Science Foundation of China under grant No. 60773169 and the 11th Five Years Key Programs for Sci. &Tech. Development of China under grant No. 2006BAI05A01.
[**] Correspondence author.

Q. Li et al. (Eds.): APWeb/WAIM 2009, LNCS 5446, pp. 635–640, 2009.

The study of this paper focuses on finding multiple linear correlations in data subsets. We refer such correlations only existed in the subset of dataset as *local linear correlations*. The main challenge for us is that the number of data subsets is so large that it is hard to enumerate all subsets in reasonable runtime. Thus, we design a method to search linear correlations by constructing hyperplanes. The time complexity of our proposed method is polynomial.

Our Contributions. (1) analyzing the limitations of applying current methods on finding linear correlations in data subsets; (2) developing a heuristic algorithm SLICE (significant local linear correlation searching) to find multiple local linear correlations in data subsets. The basic idea of our method is using a heuristic to construct hyperplanes that represent linear correlations; (3) conducting extensive experiments to show that SLICE is effective to find correct correlations in both synthetic and real-world datasets.

2 Significant Local Linear Correlation Searching

Firstly, we use the definition of strongly correlated feature subset proposed in [1] to measures both accuracy and significance of the local correlation.

Definition 1 (Significant Local Linear Correlation). Given a data matrix D containing M tuples, each tuple has d features. Let C_D be the covariance matrix of D, and $\{\lambda_i\}$ ($1 \leq i \leq d$) be the eigenvalues of C_D, where $\lambda_1 \leq \lambda_2 \leq \ldots \leq \lambda_d$, The data subset S $= \{x_{j1}, \ldots, x_{jm}\}$ ($0 \leq j_t \leq M$, $t = 1, 2, \ldots, m$) is in a *significant local linear correlation* if following conditions are true:

$$f(S,k) = \sum_{t=1}^{k} \lambda_t / \sum_{t=1}^{N} \lambda_t \leq \varepsilon \tag{1}$$

$$m / M \geq \delta \tag{2}$$

where ε, δ and k ($k \leq N$) are user defined parameters. The meanings of these user defined parameters are as same as in [1]. The computation cost would be very large if each subset is enumerated and tested to make sure it is in a significant local linear correlation or not. Given a data matrix with M tuples, there are total 2^M data subsets to be tested, which is impractical for real-world applications.

Our SLICE (significant local linear correlation searching) adopts a heuristic to find the data subset within a significant local linear correlation. Suppose the dataset has M tuples and d features, the heuristic searching begins with an initial seed (d tuples), and let S be the set of the d tuples, then S absorbs the next "best" tuple in rest tuples iteratively until the value of $f(S, k)$ becomes larger than ε, which is introduced in Definition 1. If $|S|$ satisfies the requirement in Definition 1, the linear correlation established on S can be regarded as a *significant local linear correlation*. The concept of the "best" of a tuple will be discussed later in this section. We introduce the definition of distance from a tuple to a hyperplane firstly.

Definition 2 (Distance from a tuple to a hyperplane). Given a d dimensional data subset S and a tuple x. The distance from x to the hyperplane established on S, denoted as $d(x, S)$, is defined as follows.

$$d(x,S) = f(S \cup \{x\}, k) - f(S, k) \tag{3}$$

where $f(S, k)$ is defined in Definition 1.

According to Definition 1 and Definition 2, we can see that if the distance from a tuple to the hyperplane is small, the increase of $f(S, k)$ would be small after S absorbs this tuple. So, in the process of searching, the tuple with the minimal distance to the hyperplane is the "best" tuple to be absorbed. Our approach to find the "best" tuple is efficient and incremental. The computational complexity of updating the covariance matrix of S in each iteration is $O(d^2)$ (d is the dimensionality). Considering the computational complexity of PCA, the time complexity of finding the minimum distance between a tuple and the hyperplane is $O(d^2+d^3)$. Thus, finding the "best" tuple in each step costs $O(n(d^2+d^3))$ at most.

The second key point of SLICE is how to arrange the initial seeds to start the searching to find all significant local linear correlations. If the number of features is d and the size of dataset is M, there are C_M^d different initial seeds to be tested. It is easy to see that the computational cost is large when the dimensionality is large. Fortunately, we can use a pruning strategy to set the initial seeds for searching. Then the whole process can converge very fast. SLICE does not choose the tuples as initial seeds which have been absorbed by a hyperplane again. Algorithm 1 describes the implementation details of SLICE.

Algorithm 1. SLICE $(D, \varepsilon, \delta, k)$

Input: D: a d-dimensional data set D, ε, δ, k: user defined parameters.
Output: SS: data subsets that are in a *significant local linear correlation*.
Begin
1    $M \leftarrow |D|, C \leftarrow D, SS \leftarrow \varnothing$
2   **while** $|C| > d$
3       $seed \leftarrow \varnothing$
4   **while** $|seed| < d$
5          $seed \leftarrow seed \cup \text{RANDOM}(C, d)$
6   **end**
7       $S \leftarrow \text{SEARCH}(D, seed, \varepsilon, \delta, k)$
8   **if** $|S| \geq M \cdot \delta$ **then**
9          $SS \leftarrow SS \cup \{S\}$
10   **end**
11     $C \leftarrow (C - S)$
12 **end**
End.

The time complexity of Algorithm 1 is $O(t \cdot n(d^2+d^3))$, where t is the times of invoking searching method. In the worst case, $t = n/d$. If ε is small, the number of tuples got in Subroutine SEARCH(D, $seed$, ε, δ, k) is small. In this case, t would be large. So we can see that t is associated with parameter ε. Our experimental results show that t is less than n in most cases.

3 Experimental Study

To evaluate the performance of SLICE, we test it on several synthetic datasets and a real world dataset. SLICE is implemented using Matlab 7.6. The experiments are performed on a 1.8GHz PC with 2G memory running Windows Server 2003 operating system. The characteristics of experimental datasets are presented as follows.

- **Synthetic datasets:** we generate 500 distinct datasets. These synthetic datasets are categorized to 5 different groups. For example, "D300F4C3" means each dataset in this group has 300 tuples with 4 features, and contains 3 predefined local linear correlations.
- **Real world dataset:** we use NBA statistics dataset[1] to test the performance of SLICE. We use all 458 players' statistical scores in season 2006..

In order to evaluate the effectiveness of algorithms, we conduct SLICE on these 500 synthetic datasets. We compare the discovered correlations with predefined correlations in datasets. If SLICE finds all predefined correlations, we mark this run as a *success*. We set $k=1$. The values of ε, δ and *success* rate[2] are list in Table 1. From Table 1, we can conclude that our SLICE has high probability to reach all linear correlations in these synthetic datasets.

Table 1. Results for testing on synthetic datasets

D300F3C3	D400F3C4	D600F3C4	D600F4C4	D500F4C5
$\varepsilon =0.0006$	$\varepsilon =0.0006$	$\varepsilon =0.0006$	$\varepsilon =0.0001$	$\varepsilon =0.0001$
$\delta =0.3$	$\delta =0.2$	$\delta =0.2$	$\delta =0.2$	$\delta =0.18$
Rate=100%	Rate=95%	Rate=99%	Rate=97%	Rate=100%

Next, we compare SLICE with 4C and CARE, since 4C and CARE are mostly recent works close to ours to the best of our knowledge.

a) Comparison with algorithm 4C: 4C is a kind of correlation clustering algorithm [6]. This algorithm has to generate $\varepsilon -$ neighborhoods at first. In the synthetic datasets, each correlation intersects with another correlation. Therefore, for this kind of datasets, the generated $\varepsilon -$ neighborhoods contain tuples in different correlations that would mislead searching the correct correlation clusters. Due to the space limit, we only give the results in dataset D300F3C3 here (Figure 1). Similar results are found in other datasets.

b) Comparison with algorithm CARE: In this experiment, we use CARE to find linear correlations in tuple subset. Figure 2 illustrates the hyperplane found by CARE in dataset D300F3C3. Based on this result, we can see the main limitation of CARE is that it is only capable to find the primary linear correlation of the dataset. Due to the space limit, we only give the results in dataset D300F3C3 here (Figure 2). Similar results are found in other datasets.

---

[1] http://sports.espn.go.com/nba/statistics
[2] We define the *success* rate is the percent of SLICE finds all predefined correlations over 500 datasets.

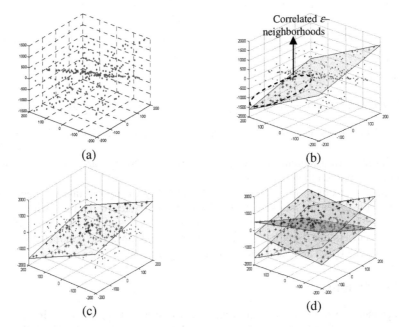

Fig. 1. (a) D300F3C3 dataset. **(b)** Correlation found by 4C in D300F3C3 dataset. **(c)** Correlation found by CARE in D300F3C3 dataset. **(d)** Correlations found by SLICE in D300F3C3 dataset.

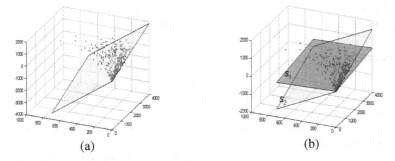

Fig. 2. (a) Correlations found by CARE in NBA dataset. **(b)** Correlations found by SLICE in NBA dataset.

As the authors of [1] had demonstrated that CARE can better results than 4C, we just compare our SLICE with CARE in NBA dataset. We use the same parameters in CARE and SLICE ($\varepsilon = 0.0006$, $\delta = 0.5$, $k = 1$). Table 2 lists the linear correlations discovered by CARE and SLICE. In common sense, different players have different

Table 2. *Success* rates of SLICE for each group of datasets

CARE	$0.090645 * minutes - 0.976344 * assists - 0.196303 * rebounds = 0.796832$
SLICE	$0.157510 * minutes - 0.898574 * assists - 0.409580 * rebounds = 8.273140$ $0.112557 * minutes - 0.121851 * assists - 0.986146 * rebounds = -8.695694$

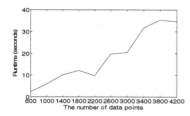

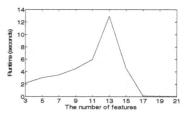

Fig. 3. Varying data size   **Fig. 4.** Varying feature size

responsibilities in a match. We believe the results of SLICE are much closer to the real world compared the result of CARE.

In order to evaluate the efficiency and scalability of algorithms, we generate 10 datasets with different data sizes. Moreover, we generate 10 datasets with different dimensionalities to evaluate the scalability of SLICE on feature size. All these synthetic datasets has 4 predefined local linear correlations. The parameter $k = 1$, and $\varepsilon = 0.0001$, $\delta = 0.2$. The results are showed in Figure 3 and Figure 4.

4 Conclusions

Finding linear correlations in dataset has many real world applications. In this paper, we propose a method to find local linear correlations in data subsets. The full work of this paper could be seen in the place[3]. In future, we plan to integrate user interests in our method to find interesting local linear correlations. Furthermore, developing a method that can guarantee to find all correct linear correlations in polynomial time complexity in each execution is a challenging work.

References

1. Zhang, X., Pan, F., Wang, W.: CARE: Finding Local Linear Correlations in High Dimensional Data. In: The 24th IEEE International Conference on Data Engineering (ICDE), pp. 130–139 (2008)
2. Aggarwal, C., Yu, P.: Finding Generalized Projected Clusters in High Dimensional Spaces. In: ACM SIGMOD 2000, pp. 70–81 (2000)
3. Aggarwal, C., Wolf, J., Yu, P.: Fast Algorithms for Projected Clustering. In: ACM SIGMOD 1999, pp. 61–72 (1999)
4. Jolliffe, I.: Principal Component Analysis. Sprinter, New York (1986)
5. Agrawal, R., Gehrke, J., Gunopulos, D., Raghavan, P.: Automatic Subspace Clustering of High Dimensional Data for Data Mining Applications. In: ACM SIGMOD 1998, pp. 94–105 (1998)
6. Bohm, C., Kailing, K., Kroger, P., Zimek, A.: Computing Clusters of Correlation Connected Objects. In: ACM SIGMOD 2004, pp. 455–466 (2004)
7. Achtert, E., Bohm, C., Kriegel, H.-P., Kroger, P., Zimek, A.: Deriving Quantitative Models for Correlation Clusters. In: ACM KDD 2006, pp. 4–13 (2006)
8. Papadimitriou, S., Sun, J., Faloutsos, C.: Streaming Pattern Discovery in Multiple Time-Series. In: VLDB 2005, pp. 497–708 (2005)
9. Chakrabarti, K., Mehrotra, S.: Local Dimensionality Reduction: A New Approach to Indexing High Dimensional Spaces. In: VLDB 2000, pp. 89–100 (2000)

---

[3] http://cs.scu.edu.cn/~tangliang/papers/slice_tangliang_waim09.pdf

Complete and Equivalent Query Rewriting
Using Views

Jianguo Lu[1], Minghao Li[1], John Mylopoulos[2], and Kenneth Cheung[3],[*]

[1] School of Computer Science, University of Windsor
{jlu,li12355}@uwindsor.ca
[2] Department of Computer Science, University of Toronto
jm@cs.toronto.edu
[3] IBM Toronto Lab
kkcheung@ca.ibm.com

Abstract. Query rewriting using views is a technique for answering
a query that exploits a set of views instead of accessing the database
relations directly. There are two categories of rewritings, i.e., equivalent rewritings using materialized views applied in query optimization,
and maximally contained rewritings used mainly in data integration.
Although maximally contained rewritings are acceptable in data integration, there are cases where an equivalent rewriting is desired. More
importantly, the maximally contained rewriting is a union of contained
queries, many of which are redundant. This paper gives an efficient algorithm to find a complete and equivalent rewriting that is a single conjunctive query. We prove that the algorithm is guaranteed to find all
the complete and equivalent rewritings, and that the resulting rewriting
is guaranteed to be an equivalent one without additional containment
checking. We also show that our algorithm is much faster than others
through complexity analysis and experimentation.

1 Introduction

Query rewriting using views [1][8] amounts to replacing tables in a query using
view definitions. Different applications require different kinds of query rewritings.

In query optimization using materialized views [1][11], the rewriting should
be equivalent, but not required to be complete, i.e., it can use either base tables
or views. While in data integration [8][10] and other applications such as query
migration [7][9], the rewriting must be complete, i.e., it can refer to views only
since it can not access database tables directly. In this case, equivalent rewriting
does not always exist when there is an insufficient number of views. In order
to obtain the best rewriting result under this constraint, maximally-contained
rewriting is needed.

In maximally contained query rewriting algorithms [8], the output is a union
of all of contained rewritings that can be found, i.e., the maximally contained

---

[*] The views in this article are not intended to represent official opinions from IBM
Corporation or IBM Canada Ltd.

Q. Li et al. (Eds.): APWeb/WAIM 2009, LNCS 5446, pp. 641–646, 2009.

rewriting Q'_M is $Q_1 \cup Q'_2 \cup ... \cup Q'_n$, where $Q'_i, 1 \leq i \leq n$, is a conjunctive contained rewriting of query Q using views. Among all those contained conjunctive rewritings $Q'_1, ..., Q'_n$, there might be a single conjunctive rewriting Q'_i that is equivalent to query Q. Obviously, answering Q using Q'_i will be much more efficient than using Q'_M. Since Q'_i is already equivalent to Q, there is no need to evaluate all other contained rewritings in $Q'_1, ..., Q'_n$.

However, existing algorithms will include all of the contained conjunctive queries into the result, thereby rendering the evaluation of the rewriting inefficient. The problem is exacerbated if the query or the view set is large, which will result in a large number of contained rewritings.

A naive approach to eliminating the duplicate conjunctive rewritings is to conduct containment checking [2] on each conjunctive rewriting against the original query. This approach is not efficient since many non-equivalent conjunctive rewritings may be generated by the same view.

We propose a Tail Containment Mapping (hereafter TCM) rewriting algorithmwhich expands the idea in bucket algorithm [8]. It produces complete and equivalent rewritings (hereafter CE rewritings) for conjunctive queries and views. TCM bucket algorithm removes inappropriate views while forming the buckets, so that each produced rewriting in the end is automatically a CE rewriting without extra containment checking. Our experiment shows that our algorithm is much faster than the naive approach that removes redundant rewritings afterwards. The data in the experiment contains hundreds of relations, views, and queries that are collected from an e-commence application [7].

2   Equivalent and Complete Rewriting

A rewriting of a query using views is a query that refers some views and/or database tables. Given a query Q and a set of views Vs over the same database schema. A query Q' is a rewriting of Q using Vs, if Q' refers to one or more views in Vs. More specifically, Q' is a contained rewriting of Q, if $Q' \subseteq Q$; Q' is an equivalent rewriting of Q, if $Q' \subseteq Q$ and $Q \subseteq Q'$; Q' is a maximally-contained rewriting of Q if $Q' \subseteq Q$, and for any contained rewriting $Q"$ of Q, $Q" \subseteq Q'$; Q' is a complete rewriting if Q' refers to views only.

2.1   Expanding Maximally Contained Rewriting Algorithms

Bucket algorithm. Complete rewriting algorithms such as bucket algorithm [8] can find maximally contained rewritings, which implies that if there are equivalent rewritings, these algorithms can find them. The only problem is that these algorithms do not generate a single conjunctive query as the result; instead, it produces a union of all contained conjunctive queries.

The bucket algorithm proceeds in two stages. First, it constructs a bucket for each subgoal of the query. The subgoal is called the owner of the corresponding bucket. A view V is added to the bucket if one of the subgoals of V matches the owner subgoal. If there are multiple subgoal of V matching the owner subgoal of a bucket, view V is added to the bucket multiple times.

Example 1. *Given a query and four views as below :*
$Q(x,r) : -A(x,y), A(y,z), B(z,r).$
$V_1(y,z,r) : -A(y,z), B(z,r).$ $\qquad$ $V_2(x,y,z) : -A(x,y), A(y,z).$
$V_3(r,z) : -A(r,z), B(z,r).$ $\qquad$ $V_4(z,r,w) : -B(z,r), C(r,w).$
The buckets of subgoals $A(x,y)$, $A(y,z)$, *and* $B(z,r)$ *are:*

$A(x, y)$	$A(y, z)$	$B(z, r)$
$V_1(x,y,r')$	$V_1(y,z,r'')$	$V_1(y',z,r)$
$V_2(x,y,z')$	$V_2(x',y,z)$	$V_3(r,z)$
$V_3(x,y)$	$V_3(y,z)$	$V_4(z,r,w')$

When a view is put into the bucket, the distinguished variables of the view are renamed. All the distinguished variables selected from the view subgoal that matches the owner subgoal of the bucket are renamed to the corresponding query variables. Other distinguished variables are renamed to fresh variables (primed variables) so that they are not the same as other variables.

In the second stage, the algorithm performs Cartesian products between views in all the buckets. For each combination of views, create a candidate rewriting by joining them together. Then containment checking is applied on each candidate rewriting against the query, to make sure it is indeed contained in the original query. Based on the buckets in Example 1, there are 27 contained rewritings generated in the combination stage. To save space, we only list some of them below:

$Q'_1(x,r) : -V_1(x,y,r'), V_1(y,z,r).$ $\qquad$ $Q'_2(x,r) : -V_2(x,y,z'), V_1(y,z,r).$
$Q'_3(x,r) : -V_2(x,y,z), V_1(y',z,r).$ $\qquad$ $Q'_4(x,r) : -V_1(x,y,r'), V_2(x',y,z), V_1(y',z,r).$
$Q'_5(x,r) : -V_1(x,y,r'), V_3(r,z).$ $\qquad$ $Q'_6(x,r) : -V_2(x,y,z'), V_3(r,z).$
$Q'_7(x,r) : -V_2(x,y,z), V_4(z,r,w').$ $\qquad$ $Q'_8(x,r) : -V_1(x,y,r'), V_2(x',y,z), V_3(r,z).$

A naive approach to expanding the bucket algorithm. A naive approach of expanding the bucket algorithm for CE rewritings is to remove properly contained conjunctive queries from the rewritings of bucket algorithm. For each contained rewriting we can conduct the containment checking with Q. If the conjunctive rewriting contains Q, it is a CE rewriting. Otherwise it is not equivalent to Q, and will be eliminated. Among the 27 contained rewritings in Example 1, 25 rewritings need to be removed, and the CE rewritings are:

$$Q'_2(x,r) : -V_2(x,y,z'), V_1(y,z,r).$$
$$Q'_3(x,r) : -V_2(x,y,z), V_1(y',z,r).$$

In general, there may be a large number of contained rewritings that need to be removed when the sizes of the query and view set are large. Some of the produced rewritings are eliminated due to the same reason. In Example 1, all contained rewritings that refer V_3 or V_4 are eliminated in the end. Hence, if we can remove V_3 and V_4 before generating the rewritings, the algorithm will be much more efficient.

A closer inspection on the example reveals that there are at least three cases when a view should not be used in CE rewriting. One is when the view contains an extra table. For example, V_4 introduces a subgoal $C(r, w)$ that can not be mapped to any subgoal in Q. Hence V_4 can not be included in the bucket. Another case is when a view introduces extra constraints on variables. For example, in V_3 there is another join condition on the first attribute of A and the second attribute of B. In this case, in the expansion of the rewritings that refer V_3, such as Q_5', the two subgoals A and B from V_3 will not be able to be mapped to any subgoal in Q. A third case is that V_1 can not be used in the bucket $A(x, y)$, even though it can be used in another bucket $A(y, z)$ in the same query. This will be explained further in Section 3.2.

Based on the observations above, we can see that views V_3 and V_4 should not be used to generate CE rewriting of Q. Therefore, we should eliminate them from the buckets before rewritings are constructed. In addition, V_1 should not be added to the first bucket.

3   TCM Algorithm

We introduce the TCM algorithm that generates CE rewritings by screening the views to be used in the buckets, so that the rewritings produced in the combination stage are automatically CE rewriting without extra containment checking.

Definition 1. *(TCM mapping) Given queries V and Q, and a mapping from the variables in V to the variables in Q is called a TCM mapping if each subgoal of V can be mapped to one of the subgoals in Q.*

Similar to containment mapping [2], TCM mapping requires that all relations referred by V also occurs in Q, and all join conditions in V are retained in Q. These two conditions guarantee that query Q has the same or stricter conditions than the query V. What differs from the containment mapping is that TCM mapping dose not require the mapping between the head variables of V and Q.

Example 2. *V_4 does not have a TCM mapping to Q, since the subgoal $C(r, w)$ can not be mapped to any subgoal in Q. V_3 does not have a TCM mapping to Q either because the variable r in $A(r, z)$ and $B(z, r)$ can not be mapped to two different variables in Q.*

Theorem 1. *If Q' is a CE rewriting of Q, each view referred by Q' must have a TCM mapping to Q.*

Intuitively speaking, Theorem 1 shows that views that do not have TCM mapping to a query Q can never be used to generate equivalent rewritings of Q, no matter how they are combined with other views. Based on this, we develop the first rule of our algorithm:

Rule 1. *If a view dose not have a TCM mapping to the given query, remove it before running the bucket algorithm.*

Example 3. *By applying Rule 1, the resulting intermediate result of the Bucket algorithm is as follows:*

$A(x, y)$	$A(y, z)$	$B(z, r)$
$V_1(x, y, r')$	$V_1(y, z, r'')$	$V_1(y', z, r)$
$V_2(x, y, z')$	$V_2(x', y, z)$	

Using Rule 1 only is not enough to guarantee the resulting rewritings are all CE rewritings. For example, in the combination $\{V_1(x, y, r'), V_1(y, z, r''), V_1(y', z, r)\}$ from the bucket in Example 3, the following rewriting can be generated:

$$Q_1'(x, r) : -V_1(x, y, r'), V_1(y, z, r''), V_1(y', z, r).$$

After simplification, Q_1' becomes $Q_1'(x, r) : -V_1(x, y, r'), V_1(y, z, r)$, whose expansion is:

$$Q_1'^{exp}(x, r) : -A(x, y), B(y, r'), A(y, z), B(z, r).$$

Q_1' is a properly contained rewriting, i.e. $Q_1' \subset Q$, because in the expansion of Q_1', subgoal $B(y, r')$ can not be mapped to any subgoal of Q. Although V_1 has a TCM mapping to Q, it is mapped to the subgoal $A(y, z)$, not $A(x, y)$. Hence we need to refine the TCM mapping to Bucket-TCM mapping as below.

Definition 2. *(Bucket-TCM mapping) Given a bucket (or subgoal) $A(x_1, \ldots, x_n)$ in query Q, and a subgoal $A(y_1, \ldots, y_n)$ in a view V. A mapping θ is a Bucket-TCM mapping from V to Q wrt $A(x_1, \ldots, x_n)$ if*

- *$A(y_1, \ldots, y_n)$ is mapped to the subgoal $A(x_1, \ldots, x_n)$; and*
- *All other subgoals in V are mapped to some subgoals in Q.*

Different from the TCM mapping, Bucket-TCM is relevant to one particular subgoal. With this definition we can see that V_1 does not have a Bucket-TCM mapping to Q wrt the bucket $A(x, y)$, hence V_1 can't be added to this bucket. Generalizing from this observation, we developed the following theorem:

Theorem 2. *Given a rewriting $Q' : -V_1, \ldots, V_n$ of query $Q : -A_1, \ldots, A_n$, which is generated using a bucket-based algorithm. Each view V_i, $1 \leq i \leq n$, is selected from a the A_i. Q' is a CE rewriting iff each V_i has a Bucket-TCM mapping to Q wrt A_i.*

Based on Theorem 2, we develop the second rule for our algorithm:

Rule 2. *When adding a view V to a bucket A of query Q, V must have a Bucket-TCM mapping to Q wrt A.*

Example 4. *For the buckets in Example 3, $V_1(x, y, r')$ in the bucket of $A(x, y)$ does not have a Bucket-TCM mapping to Q. Hence V_1 should not be added to the bucket of $A(x, y)$. The resulting buckets are changed as below:*

$A(x, y)$	$A(y, z)$	$B(z, r)$
$V_2(x, y, z')$	$V_1(y, z, r'')$	$V_1(y', z, r)$
	$V_2(x', y, z)$	

In the combination stage of the bucket algorithm, two rewritings are generated as below, which are both CE rewritings.

$$Q_2'(x, r) : -V_2(x, y, z'), V_1(y, z, r).$$
$$Q_3'(x, r) : -V_2(x, y, z), V_1(y', z, r).$$

4   Conclusions

We identifies the problem of Complete and Equivalent (CE) rewriting, and propose an efficient algorithm to generate the CE rewritings. The algorithm can find all the CE rewritings. Moreover, the rewriting is guaranteed to be equivalent without additional containment checking. We also conduct an experiment that shows performance gain of our algorithm.

Acknowledgements

We would like to thank Calisto Zuzarte and Xiaoyan Qian from IBM for extensive discussions on the equivalent rewriting problem.

References

1. Chaudhuri, S., Krishnamurthy, R., Potamianos, S., Shim, K.: Optimizing queries with materialized views. In: ICDE 1995, pp. 190–200 (1995)
2. Chandra, A.K., Merlin, P.M.: Optimal implementation of conjunctive queries in relational data bases. In: Proceedings of the ninth annual ACM symposium on Theory of computing (STOC), pp. 77–90 (1977)
3. Chirkova, R., Li, C.: Materializing views with minimal size to answer queries. In: PODS 2003, pp. 38–48 (2003)
4. Florescu, D., Levy, A.Y., Suciu, D., Yagoub, K.: Optimization of run-time management of data intensive web-sites. In: VLDB, pp. 627–638 (1999)
5. Gou, G., Kormilitsin, M., Chirkova, R.: Query evaluation using overlapping views: completeness and efficiency. In: SIGMOD 2006, pp. 37–48 (2006)
6. Halvey, A.Y.: Answering queries using views: A survey. The VLDB Journal 10(4), 270–294 (2001)
7. Lau, T., Lu, J., Mylopoulos, J., Kontogiannis, K.: Migrating E-commerce Database Applications to an Enterprise Java Environment. In: Information Systems Frontiers, vol. 5(2). Kluwer Academic Publishers, Dordrecht (2003)
8. Levy, A.Y., Rajaraman, A., Ordille, J.J.: Querying heterogeneous information sources using source descriptions. In: VLDB 1996, pp. 251–262 (1996)
9. Lu, J.: Reengineering of database applications to EJB based architecture. In: Pidduck, A.B., Mylopoulos, J., Woo, C.C., Ozsu, M.T. (eds.) CAiSE 2002. LNCS, vol. 2348, p. 361. Springer, Heidelberg (2002)
10. Pottinger, R., Halevy, A.: Minicon: A scalable algorithm for answering queries using views. The VLDB Journal 10(2-3), 182–198 (2001)
11. Zaharioudakis, M., Cochrane, R., Lapis, G., Pirahesh, H., Urata, M.: Answering complex sql queries using automatic summary tables. In: SIGMOD 2000, pp. 105–116 (2000)

Grubber: Allowing End-Users to Develop XML-Based Wrappers for Web Data Sources[*]

Shaohua Yang[1,2], Guiling Wang[1], and Yanbo Han[1]

[1] Institute of Computing Technology, Chinese Academy of Sciences
[2] Graduate University of Chinese Academy of Sciences
Beijing, 100190, P.R. China
{yangshaohua,wangguiling}@software.ict.ac.cn, yhan@ict.ac.cn

Abstract. There exist numerous online data sources on the Web. It is desirable to facilitate end-users to build XML-based wrappers from the data sources for further composition and reuse. This paper describes Grubber, a tool that allows end-users to develop XML-based wrappers from these data sources with just a few mouse clicks and keystrokes. An active learning algorithm was proposed and implemented to reduce end-users' effort. Experimental results on real-world sites show that the algorithm can achieve a high degree of effectiveness. Compared with other similar tools, Grubber includes a number of usability improvements to lower the barrier of usage and we believe it is suitable for mass end-users to build situational applications.

Keywords: XML, Web data sources, Data extraction, Wrapper.

1 Introduction

Recently, a new paradigm of software development prevails, which allows end-users to build situational applications through composing online data sources. Situational applications refer to software created for a small group of users with specific needs [1]. Some mashup creation tools like Yahoo! Pipes [2] can be seen as the typical tools to enable this paradigm. However, web data is usually presented in HTML pages, which are not appropriate for programmable composition and reuse. If end-users were able to wrap these data resources into XML resources, they can develop situational applications more efficiently by reusing the full extent of web resources.

In this paper, we present a tool called Grubber to enable end-users to develop XML-based wrappers from web data sources. An XML-based wrapper is a program that extracts data from HTML pages and outputs XML documents. To reduce end-users' effort, an active learning algorithm is proposed and experimental results on real sites show that the algorithm achieves a high performance. In addition, a number of usability improvements have been made to Grubber to lower the barrier of usage.

---

[*] Supported by the National Natural Science Foundation of China under Grant No.60573117 and Grant No.70673098, and the National Basic Research Program of China under Grant No.2007CB310804.

Q. Li et al. (Eds.): APWeb/WAIM 2009, LNCS 5446, pp. 647–652, 2009.
© Springer-Verlag Berlin Heidelberg 2009

2 Related Work

In the past few years, many wrapper generation approaches have been proposed [3]. These approaches can be categorized into manual, supervised and un-supervised. Manual approaches require programming expertise. Un-supervised methods [4, 5] can achieve high automation degree. But it is not suitable for situational applications, for end-users usually need to decide to extract the interesting data for their special use. Supervised methods are more suitable. Thesher [6] and the system by Utku Irmak [7] are two excellence supervised tools. But they are not target for situational applications.

Some mashup creation tools provide ability to extract data from web pages. Yahoo! Pipes [2] provides two modules (*Fetch Page* and *Regex*) to extract data from web pages. But the two modules require knowledge of HTML and regular expression. Marmite [8] provides an easy way to extract data. However, at current version (0.2), Marmite just supports extracting links, prices, time and tables from web pages.

3 Grubber – The Development Tool

3.1 Data Model

In this study, a data model is a collection of concepts for describing data embedded in web pages from the same data source. It defines data relationships, data semantics and constraints. We use *type* (or *schema*) as an abstraction to model the data embedded in web pages. A *type* is defined recursively as follows [4]:

    i.   The *atomic type* (or *attribute*) T, represents a set of simple data values, such as strings and numbers.

    ii.   If $T_1, \ldots, T_n$ are types, then their ordered list $< T_1, \ldots, T_n >$ is also a type, which is called as a *tuple type*.

    iii.   If T is a type, then the set $\{T\}$ is also a type, which is called as a *list type*.

The attribute and tuple type usually has a label to define the data semantic. We also use the term value to denote an instance of a type. From the above definition, we can see that one type may nest in another, which is similar to XML schema. Given a data model populated with values, we can easily generate an XML document.

3.2 A Running Scenario

Grubber is built as an extension of Firefox. The user clicks the "Discovered sources" button to begin the wrapper creation process, which comprises 4 steps (see Fig.1):

Step 1 (*initializing*): The user provides a name for the wrapper, e.g., "Amazon".

Step 2 (*labeling*): The user selects the interesting data (e.g., text strings, images, links) presented in the web page and assigns each data value a label which will be mapped to an attribute of the target data model. For a multiple-valued attribute, the user just needs to label one of them. The output of this step is a set of "is-a" labeling statements, e.g., "the text string '$29.67' is an instance of the '*price*' attribute". After labeling at least one instance value for each attribute, the user enters the next step.

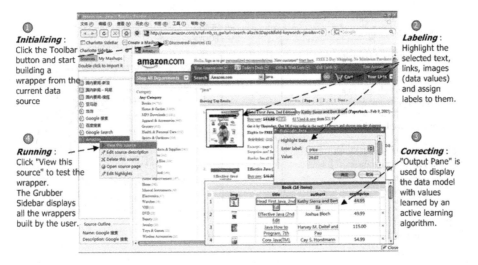

Initializing : Click the Toolbar button and start building a wrapper from the current data source

Labeling : Highlight the selected text, links, images (data values) and assign labels to them.

Running : Click "View this source" to test the wrapper. The Grubber Sidebar displays all the wrappers built by the user.

Correcting : "Output Pane" is used to display the data model with values learned by an active learning algorithm.

Fig. 1. A screenshot taken from Grubber, representing a scenario of building XML-based wrappers from web data sources with just a few mouse selections and keystrokes

Step 3 (*correcting*): An active learning algorithm is employed in the background to learn the data model and identify data values. The algorithm takes a set of statements and the DOM tree of the current page as input, and outputs a data model populated with values, which is displayed in the "Output Pane" in the right bottom corner. Each column represents an attribute or a tuple type. The user can correct misidentified values using "adding" or "removing" operations, which will be translated into new statements. After each correction, Grubber runs the active learning algorithm using the new set of statements. When the user is satisfied with the data model and values, the wrapper construction process is finished.

Step 4 (*testing*): After the above three steps, a XML-based wrapper is created and the user can test the wrapper by clicking the "View this source" menu item in the sidebar. The wrapper starts the following procedure when it is invoked: (1) construct a HTTP query, send it to the target data source and receive a responding document; (2) extract data values which confirm the defined data model from the document; (3) generate and response an XML document containing the extracted data values.

3.3 The Active Learning Algorithm

Fig.2a shows the modules of the proposed algorithm. The whole algorithm takes a DOM tree and a set of statements as input, and outputs a data model with populated data values. Before describing the processes of the algorithm, we first introduce two observations used in the algorithm and our similarity measure of two subtrees:

- **O1:** If two subtrees hold values of the same type, they have the same path.
- **O2:** If two subtrees s_1 and s_2 both hold values of some type t, and the subtree s_3 does not hold any data of t, we have: $Sim(s_1, s_2) > Sim(s_1, s_3)$ and $Sim(s_1, s_2) > Sim(s_2, s_3)$ where $Sim(s_1, s_2)$ denotes the similarity of s_1 and s_2.

We use cosine similarity to measure the similarity of two subtrees s_1 and s_2:

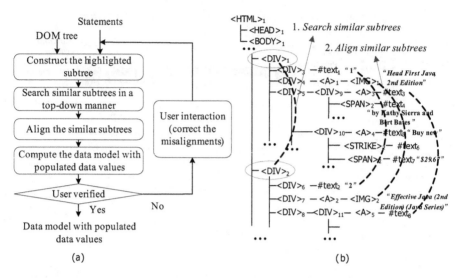

Fig. 2. (a) Modules of the active learning algorithm. (b) An example DOM tree, each node has a name with a suffix to differentiate with others.

$$Sim(s_1, s_2) = \frac{< x_1 \bullet x_2 >}{\|x_1\| \|x_2\|} \qquad \text{where: } x_1 \text{ and } x_2 \text{ are feature vectors of } s_1 \text{ and } s_2, \ \| x_1 \| \| x_2 \| = (x_1^T x_1 x_2^T x_2)^{1/2} \qquad (1)$$

A feature vector is used to represent a subtree's structure and visual layout. A feature can be a tag, an attribute, a text string, etc. In Grubber, we use the following types of features: (a) Tag features: tag name, attributes; (b) Layout features: left position, width, height of tags; (c) Text features: links, content, common text, length, isNum, isStartNum, isURL, isEmail, isPrice, isTime, isUppercase, isLowercase, etc.

We use binary-valued features, that means if a tree possesses the ith feature then the value of the ith vector component is 1 (otherwise, 0). The product $\langle x_1 \bullet x_2 \rangle$ is the number of common features of s_1 and s_2, and $\| x_1 \| \| x_2 \|$ is the geometric mean of the two vectors. Thus, $Sim(s_1, s_2)$ is between 0 and 1.

Now we give a brief introduction to the sub-processes of the algorithm as follows:

Construct the highlighted subtree. A highlighted subtree is a minimum subtree of the given DOM tree that contains all the highlighted nodes and their ancestors. In this process, we first identify the highlighted nodes using the "is-a" statements of the given statements, and then search for their ancestors by traversing the tree upward.

Search similar subtrees. This process is performed by traversing the highlighted subtree in a top-down manner. When visiting a subtree $s=T(r)$ (we use $T(r)$ to denote a subtree rooted at r), we compute a set of similar subtrees containing s as follows: (a) Find a set of candidate subtrees using observation **O1**. (b) Filter out those subtrees from the set whose similarities to s are less than a threshold θ. θ is initially given a default number and can be adjusted by user interaction. If there are missing subtrees, the user can add them by highlighting and θ will be decreased to accept the missing subtrees. Contrarily, θ is increased when removing incorrectly identified subtrees. The threshold adjustment strategy is based on the observation **O2**. In our scenario, a set of similar subtrees $\{T(\text{DIV}_1), T(\text{DIV}_2), \cdots\}$ (see Fig.2b) are found in this process.

Align the similar subtrees. The main idea of similar subtree alignment (see Fig.2b) is using the occurrence order and frequency of features. A feature is a skeleton feature if it occurs once and only once in every similar subtree. We firstly align those nodes that possess the same skeleton features and use them as boundary separators. And then align the nodes occurring between these separators recursively.

Compute the data model with populated data values. In this process, attributes in the same similar subtree form a tuple type and the responding data vales form tuple instances. In some cases, data values may be presented as part of strings, e.g., "Kathy Sierra and Bert Bates" is part of the node "#text$_4$". To extract such values, patterns (regular expressions, e.g. "by (\w+)") are generated from the highlighted instances.

Table 1. Experimental results on real web sites

No.	Site	A	C_1	Acc_1	O	C_2	Acc_2
1	Google	9	9	100.0%	0	9	100.0%
2	Yahoo!	8	7	87.5%	1	8	100.0%
3	Yahoo! Travel	6	6	100.0%	0	6	100.0%
4	AOL-Movie	8	7	87.5%	1	8	100.0%
5	Chow	4	4	100.0%	0	4	100.0%
6	IMDB-Tops	8	8	100.0%	0	8	100.0%
7	Sparknotes	4	4	100.0%	0	4	100.0%
8	IRS-Gov	3	3	100.0%	0	3	100.0%
9	Amazon	6	6	100.0%	0	6	100.0%
10	BlackBook	7	7	100.0%	0	7	100.0%
11	Dig	6	6	100.0%	0	6	100.0%
12	Indeed	6	6	100.0%	0	6	100.0%
13	Bankrate	8	8	100.0%	0	8	100.0%
14	Yahoo! NBA	5	4	80.0%	1	5	100.0%
15	ESPN	3	2	66.7%	1	3	100.0%
16	NBA-Rank	4	0	0.0%	0	0	0.0%
17	SportNews	4	4	100.0%	0	4	100.0%
18	Ubid	8	8	100.0%	0	8	100.0%
19	NSF-Search	10	10	100.0%	0	10	100.0%
20	Quote	4	4	100.0%	0	4	100.0%
21	IMDB-Movie	17	16	94.1%	2	17	100.0%
22	NSF	10	9	90.0%	1	10	100.0%
23	Weather	11	11	100.0%	0	11	100.0%
24	Gocurrency	1	1	100.0%	0	1	100.0%
25	UCC	1	1	100.0%	0	1	100.0%
Total	–	161	151	93.8%	7	157	97.5%

4 Evaluation

(1) Empirical Evaluation

We randomly chose 25 popular sites from Google Directory [9] and built wrappers from them. The evaluation procedure comprised two phases:

In the labeling phase, we labeled one value for each attribute and used 0.75 as the default threshold. Column A denotes the number of attributes for each site. Without any correcting operations, we counted the number of attributes whose data values are **completely** and **correctly** identified. The results are placed in Column C_1 and the accuracy ratios are placed in Column Acc_1. We gained an average accuracy of 93.8%.

In the correcting phase, Column O shows the number of operations applied for each site. We place the results after correcting in Column C_2 and Acc_2, which have the same meaning to the Column C_1 and Acc_1. We got an average accuracy of 97.5% in this phase. Compared with EXALG [4] (80% accuracy) and related work by Zhao H.K. et al [5] (92% accuracy), our approach achieves a higher accuracy with the help of end-users. They are both currently the best related works in the literature.

(2) Usability Discussion

By examining Grubber and other related tools, we carry out the following common aspects that affect usability and give a qualitative analysis of the usability of Grubber.

Sample web pages gathering. For one data source, only one sample web page is needed for labeling and the user provides sample web pages implicitly. Instead of collecting multiple sample pages manually, our approach can save much effort.

Labeling interface and effort. We provide a browser-based labeling environment, instead of exposing raw HTML code or DOM tree to end users. For multiple-valued types, Grubber usually only requires to label one value.

Data model (Schema) definition. The user just needs to provide the meaning of atom and tuple types. Grubber learns data relationships and complete data models.

User Expertise. Little user expertise is required in Grubber. The user doesn't need to write any code or master special IT knowledge like HTML, regular expression, etc.

5 Conclusion

This paper proposed Grubber, a tool allowing end-users to develop XML-based wrappers from web data sources. We put forward an active learning algorithm to reduce end-users' effort. Experimental results on popular sites indicate that the active learning algorithm can achieve high performance. In addition, Grubber includes a number of usability improvements to lower the barrier of usage and is more suitable for end-users to build situational applications.

References

1. Simmen, D.E., Altinel, M., Markl, V., et al.: Damia: data mashups for intranet applications. In: Proc. of International Conference on Management of Data, pp. 1171–1182 (2008)
2. Pipes: Rewire the web, http://pipes.yahoo.com/pipes/
3. Chang, C.H., Kayed, M., Girgis, M.R., Shaalan, K.: A Survey of Web Information Extraction Systems. IEEE Trans. Knowl. Data Eng. 18(10), 1411–1428 (2006)
4. Arasu, A., Garcia-Molina, H.: Extracting Structured Data from Web Pages. In: Proc. of the International Conference on Management of Data, pp. 337–348 (2003)
5. Zhao, H.K., Meng, W.Y., et al.: Mining templates from search result records of search engines. In: Proc. of SIGKDD 2007, pp. 884–893 (2007)
6. Hogue, A., Karger, D.R.: Thresher: automating the unwrapping of semantic content from the World Wide Web. In: Proc. of WWW 2005, pp. 86–95 (2005)
7. Irmak, U., Suel, T.: Interactive wrapper generation with minimal user effort. In: Proc. of the 15th International Conference on World Wide Web, pp. 553–563 (2006)
8. Wong, J., Hong, J.I.: Making mashups with marmite: towards end-user programming for the web. In: Proc. of CHI 2007, pp. 1435–1444 (2007)
9. Google Directory, http://www.google.com/dirhp

Performance Analysis of Group Handover Scheme for IEEE 802.16j-Enabled Vehicular Networks*

Lianhai Shan[1,2], Fuqiang Liu[1], and Kun Yang[2]

[1] Tongji University, Shanghai, 200092, China
[2] University of Essex, Colchester, CO4,3SQ, U.K.
joashan666@hotmail.com

Abstract. Multi-hop relay technology based on IEEE 802.16j has been intro-duced in vehicular networks, where a group of mobile stations (MSs) is within a mobile relay station (MRS). MRS is usually fixed on a large vehicle and it can help significantly reduce signaling overhead during group handover (GHO) process for every MSs in moving vehicles. This paper proposes a Markov chain-based analytical model to evaluate the overall performance of GHO for IEEE 802.16j-enabled vehicular networks in comparison with single handover (SHO). In particular, a channel borrowing mechanism is proposed to facilitate the effective operation of GHO. The simulations are conducted in terms of handover calls blocking probability using SHO and GHO schemes.

Keywords: vehicular networks, MRS, group handover, channel borrowing.

1 Introduction

Vehicular networks have recently attracted much attention from both academia and industries as not only a means to enable safe driving but also a platform to provide on-vehicle entertainment. As far as wireless communication technology is concerned, while the majority of the existing work on vehicular networks focuses on IEEE 802.11 as specified by wireless access vehicular environment (WAVE) [1], fairly recent research starts investigating into the potential application of IEEE 802.16 in vehicular networks [2]. IEEE 802.16j introduces the concept of mobile relay station (MRS), which relays data between base station (BS) and MS's [3]. MRS in vehicular networks can be fixed on a large vehicle such as a coach or a shuttle bus. The advent of MRS can not only enhance the signal strength but also significantly reduce signal-ing overhead by handing over multiple service connections simultaneously, i.e., group handover (GHO). A better network selection algorithm for group handover is pro-posed in [4], which may reduce the handover blocking and delay. Reference [5][6] proposed channel borrowing algorithm and adaptive QoS platform, which an acceptor cell that has no enough channels can borrow them from neighboring cell as long as this donor cell has some channels available after satisfying a minimum QoS level.

---

* This work was supported by the Ministry of Science and Technology (MOST) under the High-Tech 863 project (No.2007AA01Z239) and Tongji short-term Ph.D. visiting program.

This paper proposes a GHO scheme with channel borrowing. The number of channels allocated to each service flow in a BS is more than the minimal resource requested which is the minimal value to maintain its lowest level of QoS, so it can release some channels for borrowing when the BS have no enough free channels resource. In the rest of the paper, by GHO we mean GHO with the support of channel borrowing. The major contribution of this paper lies in its comprehensive modeling and evaluation of GHO for the first time to the best of our knowledge.

2 Preliminaries

As the typical application of GHO scheme is for vehicular networks, there are two possible communication situations: (1) a single MS connecting to a BS directly can hand over to a target BS individually; (2) a group of MSs connecting with BS through an intermediate MRS is considered as one single unit and hands over to the target BS in one go. This paper considers the situation where both situations coexist, as shown in Fig. 1.

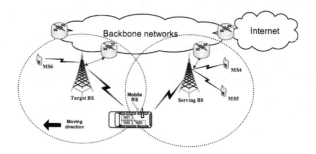

Fig. 1. System model

The proposed GHO scheme with channel borrowing can adjusts the grade of service (GoS) of the existing service connections and release some channels in target BS for coming GHO when the target BS don't have enough channels. B_{req} is the number of requested channels for handover, B_{avb} is available channels in the target BS. Then the BS can provide some borrowed channels from the existing service connections by reducing the existing connection's GoS. In order to guarantee the handover blocking probability (HBP), we must ensure $B_{req} \leq B_{avb} + B_{bor}$ by adjusting the GoS, where B_{bor} denotes the number of channels that should be borrowed from the existing connections.

In the proposed scheme, we use the following parameter definitions:
i_{BS} (j_{BS}) : number of rtPSs (nrtPSs) connecting with BS directly;
i_{RS} (j_{RS}) : number of rtPSs (nrtPSs) connecting with BS through a MRS;
b_R (b_N) : number of channels currently used by a rtPS (nrtPS);
b_{R_min} (b_{N_min}) : minimum number of channels used for maintaining a rtPS(nrtPS);
R_R (R_N) : number of channels released by a rtPS (nrtPS);

The maximum number of channels borrowed from other connections in the target BS can be calculated as

$$\max\{B_{bor}\}=(i_{BS}+i_{RS})\cdot(b_R-b_{R\_min})+(j_{BS}+j_{RS})\cdot(b_N-b_{N\_min}) \tag{1}$$

when $B_{bor}\leq j_{RS}(b_N-b_{N\_min})$, only nrtPS connections in the MRS releases some channels for borrowing. The rtPS in the MRS and others connections' channel in this BS will not be adjusted. The released channels of every nrtPS in MRS for GHO is calculated as

$$R_N=B_{bor}/j_{RS} \tag{2}$$

when $j_{RS}(b_N-b_{N\_min})\leq B_{bor}\leq j_{RS}(b_N-b_{N\_min})+j_{BS}(b_N-b_{N\_min})$, all nrtPS in the BS and MRS are needed to release channels. The released channels are then used for GHO, in which the released channels of every nrtPS can be calculated as

$$R_N=B_{bor}/(j_{RS}+j_{BS}) \tag{3}$$

when $j_{RS}(b_N-b_{N\_min})+j_{BS}(b_N-b_{N\_min})<B_{bor}<i_{RS}(b_R-b_{R\_min})+j_{RS}(b_N-b_{N\_min})$ $+j_{BS}(b_N-b_{N\_min})$, all of nrtPS in the BS and MRS are requested to decrease their channels to the minimum number of channels that can sustain a service at the low level of QoS, and the rtPS in this MRS are also requested to release channels, in which the released channels of every nrtPS and rtPS can be calculated as

$$R_N=B_N-B_{N\_min} \tag{4}$$

$$R_R=(B_{bor}-(j_{RS}+j_{BS})R_N)/i_{RS} \tag{5}$$

when $j_{RS}(b_N-b_{N\_min})+j_{BS}(b_N-b_{N\_min})+i_{RS}(b_R-b_{R\_min})<B_{bor}<\max\{B_{bor}\}$, all of nrtPS in the BS and MRS are requested to decrease to the minimum number of channels that can sustain a service just as equation (4) shown, and all of rtPS in the BS and MRS are also requested to releases some channels for GHO. The released channels of every rtPS for group handover MSs is calculated as

$$R_R=(B_{bor}-(j_{RS}+j_{BS})R_N)/(i_{RS}+i_{BS}) \tag{6}$$

Otherwise, because of user's differences, some users' nrtPS connections will be blocked to release more channels and guarantee the group MSs rtPS handover.

3 Analytical Model

In this section, we described analytical models for assessing the GHO scheme. We consider that all types of service calls arrival follow a Poisson process with rate λ, the channel holding time is an exponential distribution with service rate μ, the user dwelling time is assumed to be an exponential distribution with a mean of $1/\eta$, and the group MSs is composed of fixed multiple times as many as the single one.

In section II, we have described the system application scenario of GHO which is composed of two parts. Therefore, two types of handovers need to be considered respectively. The state of marked target BS is defined by three-dimensional Markov chain (i, j, t). where i is the sum of the number of rtPS; j is the number of nrtPS; t is the type of handover, and $t=0$ means single MS's call using GHO scheme and $t=1$ means group MSs' calls using GHO scheme. And we assume the number of group MSs in one moving vehicle is k times as many as the single one. B_s, B_g respectively denotes the sum of used channel by all of the single MS and group MSs. We represent the state of reference BS with a three-dimensional Markov chain with all state transitions shown in Fig. 2 and Fig. 3, where

$\lambda_{R\_sn}, \lambda_{R\_gn}$: new calls arrival rate of rtPS for single MS and group MSs;

$\lambda_{N\_sn}, \lambda_{N\_gn}$: new calls arrival rate of nrtPS for single MS and group MSs;

$\lambda_{R\_sho}, \lambda_{R\_gho}$: handover calls arrival rate of rtPS for SHO and GHO;

$\lambda_{N\_sho}, \lambda_{N\_gho}$: handover calls arrival rate of nrtPS for SHO and GHO;

$p(i_{BS}, j_{BS}, 0)$ and $p(i_{RS}, j_{RS}, 1)$ denote the probability of the system transition balance state single MS and group MSs, and that satisfies the regularity conditions such that

$$\sum_{(i_{BS}, j_{BS})\in C} p(i_{BS}, j_{BS}, 0) + \sum_{(i_{RS}, j_{RS})\in C} p(i_{RS}, j_{RS}, 1) = 1 \quad \text{and} \quad C = \{(i_{BS}, j_{BS}, i_{RS}, j_{RS}) \mid (i_{BS} + i_{RS})$$
$\cdot b_R + (j_{BS} + j_{RS}) \cdot b_N \le B\}$.

According to the system transition balance state equation and the regularity condition equation, then we use the SOR algorithm to get the value of each $p(i_{BS}, j_{BS}, 0)$ and $p(i_{RS}, j_{RS}, 1)$ [7]. And then we can get the values of HBP as follows:

The HBPs of rtPS is

$$P_{R\_ho} = \sum_{(i_{BS}, j_{BS})\in C_{R\_ho}} P_{BS}(i_{BS}, j_{BS}, 0) + \sum_{(i_{RS}, j_{RS})\in C'_{R\_ho}} P_{RS}(i_{RS}, j_{RS}, 1) \tag{7}$$

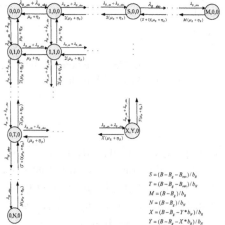

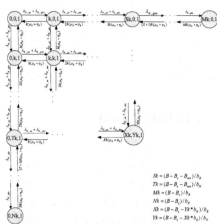

Fig. 2. Markov state transition diagram for a single MS using GHO scheme

Fig. 3. Markov state transition diagram for a group of MSs using GHO scheme

where $C'_{R\_ho} = \{(i,j) \,|\, i_{BS}b_{R\_min} + j_{BS}b_{N\_min} + i_{RS}b_{R\_min} + j_{RS}b_{N\_min} > B - b_{R\_min}\}$,

$C''_{R\_ho} = \{(i,j) \,|\, i_{BS}b_{R\_min} + j_{BS}b_{N\_min} + i_{RS}b_{R\_min} + j_{RS}b_{N\_min} > B - kb_{R\_min}\}$.

The HBPs of nrtPS is

$$P_{N\_ho} = \sum_{(i_{BS},j_{BS}) \in C'_{N\_n}} p_{BS}(i_{BS}, j_{BS}, 0) + \sum_{(i_{RS},j_{RS}) \in C''_{N\_n}} p_{RS}(i_{RS}, j_{RS}, 1) \tag{8}$$

where $C'_{N\_ho} = \{(i,j) \,|\, i_{BS}b_{R\_min} + j_{BS}b_{N\_min} + i_{RS}b_{R\_min} + j_{RS}b_{N\_min} > B - B_{rev\_rHO} - b_{N\_min}\}$,

$C''_{N\_ho} = \{(i,j) \,|\, i_{BS}b_{R\_min} + j_{BS}b_{N\_min} + i_{RS}b_{R\_min} + j_{RS}b_{N\_min} > B - B_{rev\_rHO} - kb_{N\_min}\}$.

In the actual handover process, the longest handover calls admission control (CAC) delay that can ensure the QoS is denoted as t_{max}. During the actual handover CAC process, we define T_{CAC_delay} denotes the time of handover CAC process; R_{nego_fail} denotes the HBP in this process; T_{cac} denotes the time of handover CAC process for each time; so the actual delay of handover CAC process can be denoted as:

$$T_{CAC\_delay} = T_{cac}/(1 - R_{nego\_fail}) \tag{9}$$

During the handover process, if the practical delay $T_{CAC\_delay} > t_{max}$, a handover dropping will happen. So we should reduce the HBP using the GHO scheme with channel borrowing.

4   Evaluation Results and Discussion

In SHO and GHO scheme simulation, we assume that the number of total channels is 35 [8], and the system traffic load is the same under the two different schemes. The channel number range occupied by one rtPS is from 2 to 4. The channel number range occupied by one nrtPS is from 1 to 3. And both rtPS and nrtPS in group handover have the following parameters: the new calls arrival rates of rtPS and nrtPS for single MS and group MSs in GHO are $\lambda_{R\_sn} = \lambda_{N\_sn} = \lambda/2$, $\lambda_{R\_gn} = \lambda_{N\_gn} = \lambda/2k$, where λ the calls arrival rate of rtPS and nrtPS is for SHO. The handover calls arrival rate for GHO initial value are respectively $0.3\lambda_{R\_sn}$, $0.3\lambda_{R\_gn}$ and it is 0.3λ for SHO. Both calls completion rate μ_R and μ_N are taken to be 0.5. The dwelling time is assumed to be exponentially distributed with mean values $\eta_R = \eta_N = 0.5$.

With the traffic loads increasing, we compared the performance of the proposed GHO scheme with SHO scheme. Fig. 4 shows the average HBP using the two different schemes. The GHO scheme improves handover efficiency with the rtPS and nrtPS traffic loads increasing. These results also prove that the GHO scheme reduce the HBP to ensure the HDP and delay for the group MSs in MRS's handover process. As we all know, the handover call blocking and dropping is more undesirable than new calls blocking. With settled traffic load and average number of group MSs, the results of HDP of GHO with vehicles moving at different velocity are shown Fig. 5. From the results, we can conclude that the decrease of HDP improves the connection QoS for group users in vehicular networks.

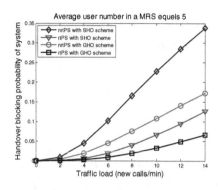

Fig. 4. Handover blocking probability   **Fig. 5.** Handover dropping probability

5 Conclusions

This paper has dealt with the performance analysis of GHO with channel borrowing for the IEEE 802.16j-enabled vehicular networks. Then we used the Markov chain model to conduct the performance analysis for SHO and GHO. The proposed GHO scheme reduces the HDP and handover delay to satisfy the the QoS in handover process. The analysis results show that GHO scheme has better overall system performance than the conventional SHO scheme. In the future, we will make further optimization in GHO process by considering dynamic two-level resource allocation.

References

1. White papers: Performance Evaluation of Safety Applications Over DSRC Vehicular Ad Hoc Networks (October 2004)
2. Yang, K., Ou, S., Chen, H.-H.: A Multi-hop Peer-Communication Protocol With Fairness Guarantee for IEEE 802.16-Based Vehicular Networks. IEEE Transactions on vehicular tchnonolgy 56(6), 3358–3370 (2007)
3. IEEE C802.16j-D5: Draft Amendment for Local and Metropolitan Area Networks, Part 16: Air Interface for Mobile Broadband Wireless Systems Multihop Relay Specification (2008)
4. Cai, X., Liu, F.: Network Selection for Group Handover in Multi-access Networks. In: Proc. IEEE ICC 2008, pp. 2164–2168. IEEE Press, Los Alamitos (2008)
5. Habib, I., Sherif, M., Naghshineh, M.: An adaptive quality of service channel borrowing algorithm for cellular networks. International Journal of Communication Systems 16(8), 759–777 (2003)
6. Wu, X., Yeung, K.L., Hu, J.: Efficient channel borrowing strategy for real-time services in multimedia wireless networks. IEEE Transactions on Vehicular Technology 49(4), 1273–1284 (2000)
7. Stojmenovic, I.: Handbook of Wireless Networks and Mobile Computing. John Wiley & Sons, Inc., Chichester (2002)
8. WiMAX Forum: WiMAX System Evaluation Methodology (January 2007)

Directly Identify Unexpected Instances in the Test Set by Entropy Maximization

Chaofeng Sha[1], Zhen Xu[1], Xiaoling Wang[2], and Aoying Zhou[2]

[1] Department of Computer Science and Engineering
Fudan University
Shanghai 200433, China
{cfsha,xzhen}@fudan.edu.cn

[2] Shanghai Key Laboratory of Trustworthy Computing
Institute of Massive Computing
East China Normal University
Shanghai 200062, China
{xlwang,ayzhou}@sei.ecnu.edu.cn

Abstract. In real applications, a few unexpected examples unavoidably exist in the process of classification, not belonging to any known class. How to classify these unexpected ones is attracting more and more attention. However, traditional classification techniques can't classify correctly unexpected instances, because the trained classifier has no knowledge about these. In this paper, we propose a novel entropy-based method to the problem. Finally, the experiments show that the proposed method outperforms previous work in the literature.

1   Introduction

For the present of unexpected instances in test set, the trained classifier can't correctly classify them. In practice, this phenomenon is very common. For example, the evolution of species may cause that many new unknown or undefined species emerge. If new species are classified to some known or predefined old species, it is obviously unsuitable, even leads to terrible consequences. Take SARS(Severe Acute Respiratory Syndrome) as an example, a kind of new disease having broken out in south China and spread to other countries. Initially, SARS was considered as a kind of common flu, and hence the patients got the incorrect treatment. Ultimately, When treating SARS as a new virus for treatment, this kind of new virus was completely overwhelmed. Currently, identifying unexpected instances[6] is an interesting topic in the data mining and machine learning communities.

With the absence of labeled unexpected (or negative) instances, traditional classification methods can't be applied directly. The crucial point of this problem is how to find or generate the negative instances and put them into the training data set for learning. Once negative instances are obtained, the traditional classification approach is directly for the rest task.

Q. Li et al. (Eds.): APWeb/WAIM 2009, LNCS 5446, pp. 659–664, 2009.

In order to solve the above problem, we present an entropy-based approach to identify unexpected instances hidden in the test set. Experimentally, The proposed technique outperforms existing ones.

The main contributions of this paper are as follows:

- This is the first work of using entropy maximization techniques to identify directly unexpected instances in test set.
- Besides text data, the proposed method also can deal with other types of data (e.g. nominal).
- Even when the proportion of unexpected instances is very small, the performance remains consistent.

The rest of this paper is organized as follows: Section 2 presents related research works and Section 3 briefly introduces the proposed approach. The experiments are shown in Section 4, followed by the conclusion in Section 5.

2   Related Work

PU learning(learning from Positive and Unlabeled examples) gets more and more focus in recent years. Now several different approaches are proposed to this problem. For example, the outstanding method LGN[6] firstly generates artificial negative instances based on the different distribution of equal words within training and test set respectively, and then use Naïve Bayesian classifier for the rest. Howerer, the assumption that every positive word has the same distribution is not easily satisfied in practice; Besides this, it only concerns text data, not other types.

Other approaches [1,5] also have similar shortcomings, e.g., when the percentage of unexpected instances is very small, these classifiers have very poor performances.

In this paper, however, a simple but effective approach is introduced. This new method directly classifies the unexpected instances hidden in test set using entropy, which firstly chooses negative instances and then realizes the rest classification. Besides text data, it also can apply to other types of data.

3   Our Approach

In this section, we propose the method named Naïve Bayesian classifier based on Entropy(NB-E) to find unexpected instances for classification.

3.1   Preliminaries

Entropy $H(x)$. The entropy of the probability distribution $P(x)$ on set $\{x_1, \cdots, x_n\}$ is defined as [4]:

$$H(x) = -\sum_{i=1}^{n} P(x_i) \log P(x_i)$$

where n is the number of known or predefined classes, and $P(x_i)$ is the class posterior probability of instance x belonging to the ith class.

The basic motivation is to use entropy as the direct measurement of confidence for correct classification, namely, the bigger the entropy H(x), the smaller the probability of correct classification, and vice versa. Therefore, the instances with the biggest entropy have nearly equal confidences to belong to any predefined class, i.e., the highest probability to be unexpected.

As an extreme example, there are three documents x_a, x_b and x_u belonging to three different classes respectively. After preprocess, the term vectors are $x_a \triangleright \{w_1, w_2\}$, $x_b \triangleright \{w_3, w_4\}$ and $x_u \triangleright \{w_5, w_6\}$. Here, x_a and x_b are training instances for two positive classes C_1 and C_2, and x_u as the negative example hidden in test set.

Then, C_1 and C_2 have the same prior probability, namely, $P(C_1) = P(C_2) = \dfrac{1}{2}$; Subsequently, by the naïve bayesian formula, the class conditional probabilities are as follows:

$$P(w_5|C_1) = P(w_6|C_1) = P(w_5|C_2) = P(w_6|C_2) = \frac{1}{6};$$

Then, the posterior probabilities of x_u are listed as follows:

$$P(C_1|x_u) = P(C_2|x_u) = \frac{1}{2}$$

Since $H(x_u)$ is biggest, the conclusion is made that x_u has higher probability of being unexpected or negative.

3.2 Naïve Bayesian Classifier Based on Entropy(NB-E)

As a highly effective classification technique, Naïve Bayesian classifier[2] is employed within this paper.

Based on the property of entropy, the proposed approach is presented in **Algorithm 1**, which mainly has the following four steps:

- From Line 3 to 5, TF-IDF technique is applied to do feature selection for document data;
- The entropy of the class posterior probabilities of any test instance, is computed and sorted by ascending order(Line 7 to 12). For any positive class, the top k instances are removed from test set and are labeled as the corresponding positive class, namely, every positive class has more k new instances for learning.(Line 14 to 17);
- From Line 19 to 24, by descending order, every instance is sorted based on the entropy. The biggest n instances chosen from test set are added into training set(labeled as "-"). At the same time, all positive training instances are viewed as one class (denoted as"+");
- From Line 25 to 31, there are only two different classes within training set. Finally, the multinomial Naïve Bayesian classifier is directly applied classification. Finally, the hidden unexpected examples are returned (Line 32).

Algorithm 1. NB-E()

Input: training set P, testing set U, negative percentage α, threshold
        parameter δ, choice factor n;

Output: unexpected instances set U_e;

1   $U_e = \phi$;

2   $C = \{c_1, \ldots, c_n\}$ is the set of classes appearing in the training set;

3   **if** P *is the document data* **then**

4   Using TF-IDF technique to extract representative attribute words for every class;

5   **end**

6   Build a naïve Bayesian classifier(NB) with the training set P;

7   **for** *each instance* $d_i \in U$ **do**

8   Use NB to classify d_i;

9       $H(d_i) = - \sum\limits_{c_j \in C} P(c_j|d_i) \log P(c_j|d_i)$;

10      $j \leftarrow \underset{c_j \in C}{\mathrm{argmax}}\, P(c_j|d_i)$;

11      $B_j = B_j \bigcup \{d_i\}$;

12   **end**

13   $k \leftarrow \lfloor \delta \times \underset{c_j \in C}{\mathrm{argmin}} |B_j| \rfloor$, $\delta \in (0,1)$ and the default value is 0.8;

14   **for** *each class* $c_i \in C$ **do**

15   Sort all the instances in B_i according to the ascending relation of the entropy of every instance's posterior probability;

16   Remove the first k instances from U and throw them into P as the instances of c_i;

17   **end**

18   Build a new Naïve Bayesian classifier(NB) with P; /* here, P is updated */

19   **for** *each instance* $d_i \in U$ **do**

20   Use NB to classify d_i;

21      $H(d_i) = - \sum\limits_{c_j \in C} P(c_j|d_i) \log P(c_j|d_i)$;

22   **end**

23   Rank the entropy of all instances in U by the descending order and choose the top n instances and insert into P as the instances of unexpected class "-";

24   Merge all positive classes in P and view them as a whole positive class "+";

25   Build a new Naïve Bayesian classifier(NB) with P; /* here, P is updated */

26   **for** *each instance* $d_i \in U$ **do**

27   Use NB to classify d_i;

28   **if** $P(-|d_i) > P(+|d_i)$ **then**

29         $U_e = U_e \bigcup \{d_i\}$;

30   **end**

31   **end**

32   **return** U_e;

---

4 Experimental Evaluation

4.1 Experimental Dataset

For the evaluation, we mainly use two kinds of representative data collections –
20 newsgroups[1] and uci letter.[2]

Here, several parameters are set as follows:

- α is the proportion of negative instances hidden in test set with respect to
  positive ones. In order to simulate various situations, α has 8 different values,
  namely, $\{5\%, 10\%, 15\%, 20\%, 40\%, 60\%, 80\%, 100\%\}$;
- n is the number of selected likely negative instances. As different cases, n
  has 2 rows of different values, i.e., $\{5, 10, 15, 20, 30, 30, 30, 30\}$(denoted as
  NB-E) and a constant 10(denoted as NB-E$^+$) respectively.

Additionally, as to the number of positive sub-classes, there are mainly 2 different
assumptions: 2 positive sub-classes existing in the training set and 3 ones.

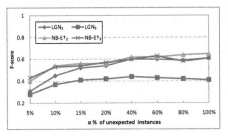
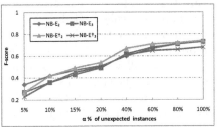

(a) F-score with different values for α in   (b) F-score with different values for α in
    document experiments   nominal experiments

Fig. 1. The comparison results with different values for α in 2 and 3-classes experiments

4.2 Experimental Results

As for document data, we mainly compare NB-E with LGN [6], which has the
best performance among former approaches, however, previous works are never
done for nominal data, hence no comparison here.

Fig.1 shows that as to document data, NB-E outperforms LGN, especially
when the number fo positive sub-classes increases; Both at document and nominal data, NB-E has consistent performances, without being influenced by the
number of positve sub-classes too.

Fig.2 shows that the values of parameter n has little influence on the final
results, only if not too big or too small; On the other hand, because the number attribute words of document data is greater than the one of nominal data,
document data have higher accuracy and F-score values.

---

[1] http://people.csail.mit.edu/jrennie/20Newsgroups
[2] http://archive.ics.uci.edu/ml/datasets.html

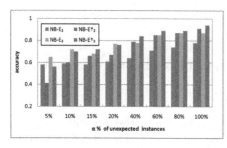
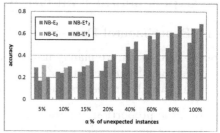

(a) *accuracy* with different values for n in document experiments

(b) *accuracy* with different values for n in nominal experiments

Fig. 2. The comparison results with different values for n in 2 and 3-classes experiments

5 Conclusion

To identify unexpected instances hidden in test set, we propose a novel entropy-based approach called NB-E, which is simple but very useful. Meanwhile, the experiments prove that this approach has excellent performances, such as high F-score and accuracy at document and nominal data.

There are still some places to be improved, however, e.g., much more datasets are evaluated; The cluster techniques are applied to improve the *accuracy* of the chosen negative instances; Besides these, many other classification methods are used together, inclusive of logic regression, SVM and ensemble methods, etc.

Acknowledgement. This work is supported by NSFC grants (No. 60673137 and No. 60773075), National Hi-Tech 863 program under grant 2008AA01Z1470967 and Shanghai Leading Academic Discipline Project(Project NumberB412).

References

1. Liu, B., Dai, Y., Li, X., Lee, W., Yu, S.: Building Text Classifiers Using Positive and Unlabeled Examples. In: IJCAI 2003 (2003)
2. Györfi, L., Györfi, Z., Vajda, I.: Bayesian decision with rejection. Problems of Control and Information Theory 8, 445–452 (1978)
3. Devroye, L., Györfi, L., Lugosi, G.: A Probabilistic Theory of Pattern Recognition. Springer, Heidelberg (1996)
4. Cover, T., Thomas, J.: Elements of information theory. Wiley Interscience, Hoboken (1991)
5. Li, X., Liu, B., Lee, W., Yu, S.: Text Classificaton by Labeling Words. In: AAAI 2004 (2004)
6. Li, X., Liu, B., Ng, S.: Learning to identify unexpected instances in the test set. In: IJCAI 2007 (2007)
7. Guo, Y., Greiner, R.: Optimistic active learning using mutual information. In: IJCAI 2007 (2007)

Decentralized Orchestration of BPEL Processes with Execution Consistency

Weihai Yu

Department of Computer Science
University of Tromsø, Norway
weihai@cs.uit.no

Abstract. Scalability, consistency and reliability are among the key requirements for orchestration of BPEL processes. We present a fully decentralized approach to orchestration of BPEL processes that is of continuation-passing style, where continuations, or the reminder of the executions, are passed along with asynchronous messages for process orchestration. Furthermore, we identify and address some consistency issues that are more challenging for decentralized orchestrations.

Keywords: Continuation-passing messaging, Decentralized orchestration, Dynamic process structure, Consistent orchestration.

1 Introduction

WS-BPEL [4], or simply BPEL, is the *de facto* standard for services composition based on the process technology. Individual web services are composed into BPEL processes. Today, orchestration of BPEL processes is typically carried out by central BPEL engines that are faced with scalability and reliability challenges. In view of these challenges, several research groups have proposed approaches to decentralized process orchestration. They are based on static definitions of processes for decentralization: they either pre-allocate resources in the distributed environment, even for the parts that will not be executed [1-3], or involve central repositories for process management tasks that cannot be properly planned in advance [7, 8].

Our approach is fully decentralized. It does not involve static process instantiation or a central repository. We further identify and address some consistency issues that are more challenging for decentralized orchestration. Our performance result shows high scalability of the approach [9].

2 Centralized and Decentralized Services Composition

In BPEL, *processes* and *composite services* are synonymous. A BPEL process consists of *activities* that are either basic or structured. Basic activities include activities for providing and invoking services, for assignments, etc. A *service* is provided through an invokable operation at a *site* with a receive activity. There are two types of invocations, a synchronous request-response invocation with an invoke activity and an asynchronous invocation with a send (an invoke without output parameter in

Q. Li et al. (Eds.): APWeb/WAIM 2009, LNCS 5446, pp. 665–670, 2009.

Process:
  flow(link(s_to_i),
    sequence(
      send(*i*, getInvoice),
      send(*i*, sendShippingPrice,
        in: shipping_price,
        target(s_to_i)),
      receive(invoiceCallback)
    ),
    invoke(*s*, arrangeShipping,
      out: shipping_price,
      source(s_to_i))
)

Service getInvoice at site *i* :
  sequence(
    receive(getInvoice),
    receive(sendShippingPrice,
      in: shipping_price),
    send(*engine*, invoiceCallback)
)

(a) Centralized

Process:
  flow(
    invoke(*i*, getInvoice),
    sequence(
      invoke(*s*, arrangeShipping,
        out: shipping_price),
      send(*i*, sendShippingPrice,
        in: shipping_price)
)
)

Service getInvoice at site *i* :
  sequence(
    receive(getInvoice),
    receive(sendShippingPrice,
      in: shipping_price),
    reply(getInvoice)
)

(b) Decentralized

Fig. 1. Centralized and decentralized compositions of example services

the BPEL standard) and a receive (callback) activity pair. A structured activity consists of a collection of constituent activities. Some structured activities specify the order in which the constituent activities are executed, such as in sequence (sequence) or in parallel (flow); some allow selective execution of one of its constituent activities based either on the incoming message (pick) or on the outcome of a Boolean expression (if-else).

Fig 1 shows an example process adapted from the purchase-order process in the BPEL standard specification [4]. The process consists of two parallel branches.

The standard BEPL specification assumes a central engine at which the executions of the activities are conducted. The process-aware data variables are shared by all activities in all branches. In Fig 1-a, the s_to_i link synchronizes two activities in the two branches so that send(*i*, sendShippingPrice), the target activity of the link, cannot be executed before invoke(*s*), the source activity of the link. This assures that the shipping_price value sent to the asynchronously running getInvoice service is the one returned by the arrangeShipping service.

Decentralized composition assumes no central engine. Activities are carried out at the locations determined at run time. Data are sent directly from its source, without the involvement of a central engine. In Fig 1-b, the second branch sends shipping_price directly after the return of the arrangeShipping service. The synchronization between the two branches is achieved by the message (rather than the link). Since the invoking branch (i.e., the first branch) does not interact with the getInvoice service, it does not have to invoke the service asynchronously. The getInvoice service now does not have to send a return message to the central engine.

3   Process Structure and Its Dynamics

BPEL processes are block structured extended with synchronizations between parallel branches. A block can be an (basic or structured) activity or a service body. Blocks can be arbitrarily nested.

We can represent a process as a directed graph where the nodes are blocks, and the directed edges are the order in which the blocks are executed. A block with constituent blocks can either be represented *unexpanded* with a single node, or *expanded* with nodes for the constituent blocks.

The graphical representation of a block must be *structurally sound* (also called symmetrical control structure in [5]). That is, it consists of a start node and an end node; every node is reachable from the start node; and the end node is reachable from every node. The execution of a process can generally be regarded as progressively marking the graph from the start node to the end node. A block for a selective structured activity is *undecided* if a selection of a constituent activity has not been made. If undecided blocks are represented as unexpanded nodes, all nodes will be marked when the execution terminates.

The graph of a process may change during the process execution. Fig 2 illustrates the changing structure of the example decentralized process. Fig 2-a is the process structure (fully expanded) at the start of an execution. In Fig 2-b, after the invocation of the external service getInvoice, the node for the invoke activity is replaced by the activities of the service body (except the first receive activity that is considered as having been executed after the invocation). Notice that the synchronization between the two branches (dashed line) becomes visible only after the invocation.

It is sometimes useful to graphically represent the remaining activities of a process at a certain point of its execution. A *continuation graph* during the execution of a process is the sub-graph with the nodes that have not been executed. For a partially executed block with parallel branches, we add a virtual start node preceding the first activities of each branch. A continuation graph of a block is also structurally sound. Fig 2-c is the continuation graph after the synchronization message is delivered. The flow node is the virtual node of the partially executed block with two parallel branches.

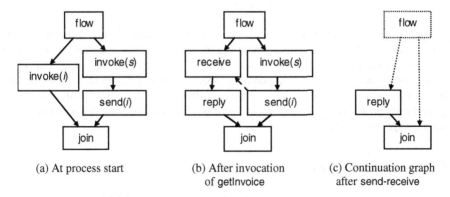

(a) At process start   (b) After invocation   (c) Continuation graph
                                  of getInvoice   after send-receive

Fig. 2. The changing process structure

4 Decentralized Execution with Continuation-Passing Messaging

Traditionally, a message for services orchestration contains only a basic activity. Information like activity execution order is maintained at the central engine. With continuation passing messaging, information like activity execution order is carried in messages in terms of continuations. The service sites can interpret the messages and conduct the execution of services without consulting a central engine. More specifically, a message contains a control activity, a continuation and an environment. The *control* activity is the activity to be carried out immediately. The *continuation* is a stack of activities that will be carried out after the control activity. This corresponds to the path from the current node to the end node in the continuation graph. The *environment* contains information of run-time status and process-aware data variables. Expressions can be evaluated within an environment. Fig 3 illustrates the messages for the orchestration of the example process starting at site *p*. The environment parts of messages are not shown in the figure.

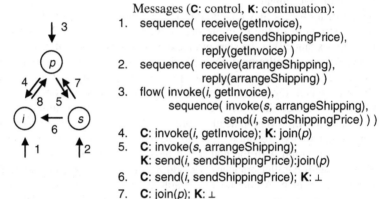

Messages (**C**: control, **K**: continuation):
1. sequence(receive(getInvoice),
            receive(sendShippingPrice),
            reply(getInvoice))
2. sequence(receive(arrangeShipping),
            reply(arrangeShipping))
3. flow(invoke(*i*, getInvoice),
       sequence(invoke(*s*, arrangeShipping),
                send(*i*, sendShippingPrice)))
4. **C**: invoke(*i*, getInvoice); **K**: join(*p*)
5. **C**: invoke(*s*, arrangeShipping);
   K: send(*i*, sendShippingPrice):join(*p*)
6. **C**: send(*i*, sendShippingPrice); **K**: ⊥
7. **C**: join(*p*); **K**: ⊥
8. **C**: join(*p*); **K**: ⊥

Fig. 3. Messages of the example process

This basic messaging scheme is also extended with process management and recovery support [9]. Process management is achieved through scope management. In a scope with multiple branches, the scope manager functions like the virtual start node in the continuation graph. An important property is that the current location of any branch can be obtained at its scope manager. Recovery support is achieved with a second continuation (called *failure continuation*) that is automatically generated during message interpretation.

5 Execution Consistency

We deal with two consistency issues for the orchestration of BPEL processes. Both are related with the dynamic nature of the process structure. Although none of the issues are specific to decentralized process orchestration, they are more challenging when only partial information about a process is available at branches.

5.1 Consistent Activity Pairing

When a message arrives at a site, it is possible that there is no matching receive activity awaiting. A naïve way to handle this situation is to throw a "service unavailable" exception, which, in the worst case, could lead to the rollback of the entire process. This could be inappropriate in some cases. In Fig 3, if message 6 arrives at site *i* before *i* reaches the corresponding receive activity, *i* should wait till the matching receive activity is in place. On the other hand, it is possible that a site is waiting for a message that will never be sent, because, for example, the other branch selects a constituent part other than the one containing the expected send activity. In this situation, the site should throw an exception rather than waiting for ever.

We refer to this issue as *consistent pairings of activities* for communication between parallel branches. The BPEL standard requires that pairings are declared at flow activities in terms of links, so that they are visible by static analysis tools. However, consistent activity pairing is a dynamic issue (due, for instance, to selective activities and late binding to services) that cannot be sufficiently addressed based on static analysis. The BPEL standard therefore further assumes that the established correlations should enable the central engine to handle this, though the specification does not mandate any specific form of handling. Activity pairing in a distributed setting is closely related to consistency issues in process choreography. [6] gives a consistency criteria for potentially successful interactions between partner processes.

To facilitate dynamic support for consistent activity pairing, we register additional information in the environment parts of messages. This includes potential pairings visible at the start of multiple branches. When a site receives a message with a potential pairing activity of that site, it waits for that pairing activity if it has not been available yet. When a site executes a selective activity like pick and if-else and excludes a constituent part that contains one of the pairing activities of a potential pairing, it informs the corresponding branch (via the scope manager) to invalidate the registered pairing. An attempt to execute an activity of an invalidated pairing will result in an exception.

5.2 Consistent Updates

The BPEL standard allows concurrent updates of process-aware variables at different parallel branches. This may leave the values of variables non-deterministic.

Deterministic variable values require *consistent updates*. Basically, if two parallel branches update the same variable, the updates must be synchronized with either a message or a link. Generally with multiple parallel branches, for every data variable, there is exactly one update path across the branches.

This seemingly obvious requirement for consistent update of process-aware variables has been surprisingly overlooked. The BPEL standard leaves this as an issue to be addressed at the application level. The closest related work seems to be $ADEPT_{flex}$ [5] where it states that "different branches are not allowed to have write access to the same data element, unless they are synchronized". We do not know of any work on consistent update that takes into account the dynamic nature of process structuring during execution.

To support consistent updates, we associate every variable with an *update token* and devise a mechanism that guarantees that, at any time, there is at most one branch holding the token of a variable. Only the branch holding the token has the right to update the variable. Our mechanism has the property that, in normal cases, tokens are passed around with orchestration messages and no extra communication for coordination is needed.

6 Conclusion

Our contributions presented in this paper are twofold. First, we present decentralized composition and orchestration of services with changing structures. It addresses the scalability and reliability constraints of the centralized approach, because there is no central performance bottleneck and point of failure. It does not unnecessarily pre-allocate resources or involve central repositories as in most of the existing decentralized approaches. The continuation-passing approach is novel for distributed computing. Second we identify and address some consistency issues that take into account the dynamic nature of process structures at run time.

References

[1] Benatallah, B., Dumas, M., Sheng, Q.Z.: Facilitating the rapid development and scalable orchestration of composite Web services. Distributed and Parallel Databases 17(1), 5–37 (2005)

[2] Muth, P., Wodtke, D., Weißenfels, J., Dittrich, A.K., Weikum, G.: From centralized workflow specification to distributed workflow execution. Journal of Intelligent Information Systems 10(2), 159–184 (1998)

[3] Nanda, M.G., Satish, C., Vivek, S.: Decentralizing execution of composite web services. In: Proceedings of the 19th annual ACM SIGPLAN conference on Object-oriented programming, systems, languages, and applications, Vancouver, BC, Canada, pp. 170–187 (October 2004)

[4] OASIS, Services Business Process Execution Language, Version 2.0 (2007), http://docs.oasis-open.org/wsbpel/2.0/wsbpel-v2.0.pdf

[5] Reichert, M., Dadam, P.: ADEPTflex-Supporting Dynamic Changes of Workflows Without Losing Control. Journal of Intelligent Information Systems 10(2), 93–129 (1998)

[6] Rinderle, S., Wombacher, A., Reichert, M.: Evolution of process choreographies in DYCHOR. In: Meersman, R., Tari, Z. (eds.) OTM 2006. LNCS, vol. 4275, pp. 273–290. Springer, Heidelberg (2006)

[7] Schuler, C., Weber, R., Schuldt, H., Schek, H.-J.: Scalable Peer-to-Peer Process Management - The OSIRIS Approach. In: Proceedings of IEEE International Conference on Web Services (ICWS 2004), pp. 26–34 (2004)

[8] Ye, X.: Towards a Reliable Distributed Web Service Execution Engine. In: Proceedings of IEEE International Conference on Web Services (ICWS 2006), pp. 595–602 (2006)

[9] Yu, W.: Scalable services orchestration with continuation-passing messaging. In: Proceedings of INTENSIVE 2009, Valensia, Span, April 21-25 (to appear, 2009)

Let Wiki Be More Powerful and Suitable for Enterprises with External Program

Yang Zhao[1,3], Yong Zhang[2,3], and Chunxiao Xing[2,3]

[1] Department of Computer Science and Technology, Tsinghua University,
Beijing, P.R. China
[2] Research Institute of Information Technology, Tsinghua University, Beijing, P.R. China
[3] Tsinghua National Laboratory for Information Science and Technology
zy.dennis@gmail.com
{zhangyong05,xingcx}@tsinghua.edu.cn

Abstract. Wiki technology is used popularly not only on Internet but also in intranet of enterprises. However with the rapid growth of pages in wikis, lots of useful information may be hidden in them and difficult to be grasped. If a wiki has some functional modules to catch and take a good use of the information, it will become more powerful and suitable for enterprises and benefit the enterprises a lot. We choose the method of external program as better solution and implement it. It can empirically help wiki have some computational capabilities, monitoring mechanism and better collaboration by distilling, processing and centralizing the wiki information.

Keywords: Wiki, External Program, Computational Capabilities, Monitoring Mechanism, Collaboration.

1 Introduction

Wiki technology is used popularly not only on Internet but also in intranet of enterprises now. In late 2006 IBM Research deployed a wiki which currently contains about 400 project proposals [1]. Intel also has company-wide internal wiki, called Intelpedia, which has grown to tens of thousands of users and over 8000 pages in the six months since its launch [2].

Wiki provides a very collaborative, efficient and concise web environment to enterprises. However, with the rapid growth of pages in wikis, lots of useful information may be hidden in them and difficult to be grasped. If a wiki has some functional modules to catch and take a good use of the information, it will become more powerful and suitable for enterprises. In this paper, comparing to other ways, we present a use of external program to meet this need. The external program can help wiki have some computational capabilities, monitoring mechanism and better collaboration by distilling, processing and centralizing the wiki information.

Section 2 comes up with the idea of making wiki more powerful and suitable for enterprise, and discusses how to do and what to do. In Section 3, we provide three possible scenarios which are suitable for using external program. Section 4, 5 describe implementation and conclusion.

Q. Li et al. (Eds.): APWeb/WAIM 2009, LNCS 5446, pp. 671–676, 2009.

2 Deal With the Wiki's Information

We think wiki should have some more functional modules which are computational capabilities, monitoring mechanism and better collaboration. These works are originally done by artificial way which waste a lot of work and time and may have some blind spot of management. If we do something to meet this need, no matter they are internal or external, will make the wiki more powerful and suitable for enterprise and will benefit the enterprises a lot.

2.1 How to Do?

Wiki software is a type of collaborative software allowing web pages to be created and edited using a common web browser. It is usually implemented as an application server that runs on one or more web servers which is divided into presentation layer and logical layer. The content is stored in a file system, and changes to the content are stored in a relational database management system.

Fig. 1 shows three methods we present to make wiki stronger. We think the third one is obviously better than the other two.

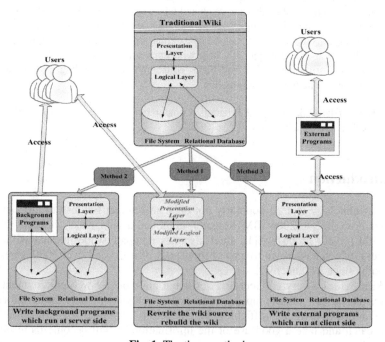

Fig. 1. The three methods

Method 1 is to rewrite the wiki source and rebuild the wiki.

This method solves the question from the ground floor. It could be a very effective way, but also has some problems. Firstly, it demands the developers to go down to the wiki's source to understand its architecture and relevant functions before modifying. The modification needed is difficult and time-consuming. Secondly, it is not very safe because the core code may be modified wrongly by accident and some main functional modules may break down. Besides, should we rebuild the wiki from a pure content management system to an integrated multi-functional content management system? It may disobey the wiki's original intention.

Method 2 is to write background programs which run at server side.

This method aims to use programs to extract the needed information from file system and database of the wiki and do some further work. We think it is possible but not a good one because it is hard to analyze the wiki's data file and database, and it is even harder to trace a page with its original file to its "current looking" after adding lots of changes saved in the database.

Method 3 is to write external programs which run at client side.

We think this method is obviously the best and most lightweight way to deal with the problem. Firstly, its implement is quite simple and the HTML data is easy to be analyzed. Secondly, external program fetches the information directly from the client browser, so everyone accessible could use the program personally to serve himself, and users can even build a wiki external programs library by sharing the programs. Thirdly, it is impossible to affect running of the wiki server and hardly do harm to the system. Fourthly, the people who use the program have the same permission rights as they login the wiki from web browser. Fifthly, there are not many differences when they reuse an external program to another wiki. The only change is to deal with the HTML pattern differences between the two wikis. At last, it maintains wiki's feature of being a content management system and divides the function more clearly.

So we apply this method for the following work.

2.2 What to Do?

External programs' critical steps are distilling, processing and centralizing.

First thing we need to know is that what information we should distill from the wiki to benefit enterprises. As we described before, computational capabilities, monitoring mechanism and better collaboration are the best supplement to a wiki. That means we could distill the data of which needs automatic calculation, of which helps the different levels' managers do the supervision well, of which can go a step further to coordinate the time and event better.

The flow of an external program is as follow. It should pass the login authentication of the target wiki first. Distilling is to retrieve the information we want to process. It includes finding the target wiki pages, fetching the HTML code of these pages, and removing the useless information such as HTML tags and some junk information. Processing is to deal with the distilled data according to different enterprise needs. Centralizing is to clean up the results and put them together, which is processed by the Processing step.

3   Possible Scenarios in Enterprise

In this Section, we will provide three possible scenarios can be well done by external program: data calculation, daily reports checking and activity coordination which stand for computational capabilities, monitoring mechanism and better collaboration mentioned before.

3.1 Data Calculation: Let Wiki Have More Computational Capabilities

It is familiar to see a flat table full of data in a wiki, especially in enterprises. Because the table on the web pages (including wiki pages) is hard to be converted into an Office Excel sheet, it is troublesome to calculate them by hand or by copying-and-pasting for a lot of times. If we have an external program, we can extract the numbers from the wiki and process them by the way we want.

There is much data written in the wiki pages needed to be calculated, and because its fixed format, it's easy for external programs to analyze. This case takes place especially in the companies who usually record lots of structured data to the intranet built by wiki technology.

3.2 Daily Reports Checking: Let Wiki Have Monitoring Mechanism

Some manager in enterprise always asks their employees to write something about their work on their own wiki page daily. It is too much work for the manager or his secretary to check all these daily reports which may be a necessary measure of employees' performance. The work can be well done by an external program.

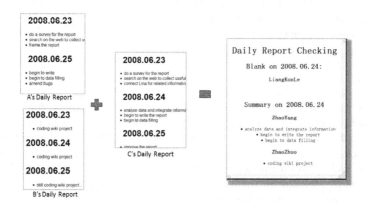

Fig. 2. Monitoring on the Daily Report of wiki

For example, everybody in an enterprise has its own daily report page which can be accessed by using http://wiki'sURL/${name}/DailyReport. So by using a list of the employees' names, the external program, after it extracts and processes the useful data, can tell who do not write the daily report that day and centralize the others' to one file by day which shows as Fig. 2.

Like daily report, a lot of things need to be supervised can be done by external programs after making rules to employees on how to write wiki pages with a predefined structure. It is without question a powerful enhancement on management.

3.3 Activity Coordination: Let Wiki Have Better Collaboration

Wiki is known as a collaborative tool, but we think it can be more collaborative with the help of external programs.

We take activity coordination as an example. Arranging meetings, especially tele-conferences, it's really troublesome for the manager. The more people are involved, the more difficulty to coordinate everyone's time. With wiki, the manager can establish one single schedule page to let everybody involved fill the busy time. This is much better than using a lot of emails or phone calls to finally determine a proper time. However the disadvantage of this method is that if there are more than two meetings in a week which involve not quite different group of people, someone will fill the meeting schedule at least twice.

Considering some companies may ask employees to maintain a personal schedule with the same pattern, as Fig. 3 shows, it is effective to use external program to process all the personal schedule of the people who are on the meeting list. There will be a result calculated by the program telling the proper time to hold the meeting.

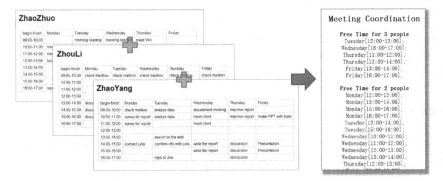

Fig. 3. Coordination of the Personal Schedule of wiki

4 Implementation

The three cases mentioned in Section 3 have already implemented in the Wiki of Web and Software Technology R&D Center of THU (http://166.111.134.22/west) and it works well. The wiki is built with MoinMoin [3].

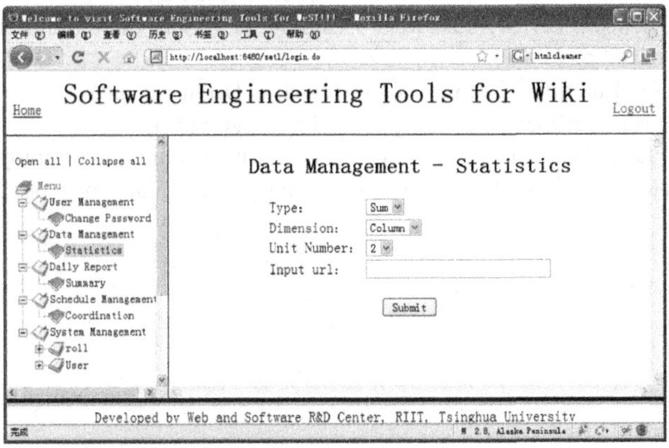

Fig. 4. Implementation

We implement the external program idea with a web application, as Fig. 4 shows. In the server side of the application we use Jakarta Commons httpclient package [4] provided by Apache to simulate a http client. Then it logins the system and gets the content of the target page. We use htmlcleaner 2.1 package [5] or Java method to extract the useful data of the target page's htmlscript. Based on the data we distill, we can do some further work as needed. Though the method is simple, it can let the program be really strong and can help the enterprise a lot.

5 Conclusion

Wiki technology is used popularly not only on Internet but also in intranet of enterprises. If a wiki has some functional modules to catch and take a good use of the information, it will become more powerful and suitable for enterprises and benefit the enterprises a lot. We choose the method of external program as better solution and implement it. It can empirically help wiki have some computational capabilities, monitoring mechanism and better collaboration by distilling, processing and centralizing the wiki information.

Acknowledgement

This work is supported by the Key Technologies R&D Program of China under Grant No. 2006BAH02A12, and the National High-tech R&D Program of China under Grant No.2006AA010101.

References

1. Ding, X.: Visualizing an Enterprise Wiki. In: CHI 2007, San Jose, California, USA, April 28 – May 3 (2007)
2. Organizational Uses of Wiki Technology, part of the Proceedings of Wikimania 2006 (2006), http://wikimania2006.wikimedia.org/wiki/Proceedings:KL1
3. McAfee, A.: Wikis at Dresdner Kleinwort Wasserstein (2006)
4. Argyris, C., Schoen, D.: Organisational Learning: A Theory of Action Perspective. Addison-Wesley, Reading (1978)
5. Hinchcliffe, D.: Enterprise 2.0: Lively Conversation Driving change, http://blogs.zdnet.com/Hinchcliffe/?p=186/
6. Hasan, H., Pfaff, C.C.: The Wiki: an environment to revolutionise employees' interaction with corporate knowledge. In: OZCHI 2006, Sydney, Australia, November 20-24 (2006)
7. http://moinmo.in/

DepRank: A Probabilistic Measure of Dependence via Heterogeneous Links

Pei Li[1,2], Bo Hu [1,2], Hongyan Liu[3], Jun He[1,2], and Xiaoyong Du[1,2]

[1] Key Labs of Data Engineering and Knowledge Engineering, Ministry of Education, China
[2] School of Information, Renmin University of China, Beijing, China
{lp,erichu2006,hejun,duyong}@ruc.edu.cn
[3] Department of Management Science and Engineering, Tsinghua University, Beijing, China
hyliu@tsinghua.edu.cn

Abstract. Dependence is a common relationship between objects. Many works have paid their attentions on dependence, but many of them mainly focus on constructing or exploiting dependence graphs on some specific domain. In this paper, we give a generic definition of dependence and make equivalence to the situation of graphs, and propose an algorithm called DepRank, which quantifies the dependence degree of a node pair based on probabilities of all dependence paths. Empirical study shows that DepRank method has reasonable results and maintains a well balance between accuracy and efficiency.

Keywords: link graph, probabilistic measure, object dependence.

1 Introduction

Dependence is a common feature of relationship that extensively exists. Generally, the relationship between objects can be modeled as a graph, with nodes and edges corresponding to objects and links respectively. Dependence is measured based on description of this graph; before that, we give a probability-theoretic definition of dependence, which says "dependence degree of object A on object B is the probability of B determining A". Supposing dependence degree of object A on B is $Dep(A \leftarrow B)$, it describes the degree of determinant and can be utilized by many other applications.

In this paper, we propose an effective method called "*DepRank*" to perform a probabilistic measure of dependence. Based on the view of random walk theory [4], we define "dependence path", which is a path without duplicate nodes. Hence, the probability of B determining A can be deduced as the sum of dependence values contributed by all dependence paths. Explicitly description is given in Section 3.

Taking a material-supply network among different companies shown in Figure 1 as an example, the directed edge from C to A means company C supplies materials to company A (so A has a dependence on C), and analogously for other edges. Our De-pRank method measures the dependence of company A on company B by computing the sum of dependence values on each path from B to A that visits a node no more than once. DepRank method will computes $Dep(A \leftarrow B)$ as Figure 2 shows.

Q. Li et al. (Eds.): APWeb/WAIM 2009, LNCS 5446, pp. 677–682, 2009.
© Springer-Verlag Berlin Heidelberg 2009

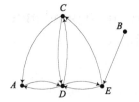

$$Dep(A \leftarrow B) = Dep(E \leftarrow B) \cdot Dep(C \leftarrow E) \cdot (Dep(A \leftarrow C)$$
$$+ Dep(D \leftarrow C) \cdot Dep(A \leftarrow D))$$
$$+ Dep(E \leftarrow B) \cdot Dep(D \leftarrow E) \cdot (Dep(A \leftarrow D)$$
$$+ Dep(C \leftarrow D) \cdot Dep(A \leftarrow C))$$
$$= \frac{1}{2} \cdot \frac{1}{2} \cdot (\frac{1}{2} + \frac{1}{3} \cdot \frac{1}{2}) + \frac{1}{2} \cdot \frac{1}{3} \cdot (\frac{1}{2} + \frac{1}{2} \cdot \frac{1}{2}) = \frac{7}{24}$$

Fig. 1. A material-supply network   **Fig. 2.** A computation of $Dep(A \leftarrow B)$

2 Related Works

By surveying related studies in recent years, these studies about dependence can be summarized as two types: (1) studies focusing on how to construct or exploit dependence graphs; (2) studies trying to assess the dependence between objects.

For the first type, lots of works have been done in some special domains [1, 2]. For the second type, we mainly introduce Memon et al. [3], who propose a method to discover dependence in terrorist networks based on shortest paths. The dependence of u on v is measured in Equation (1), where $occurence(u, v)$ is the number of times that node u uses node v in the communication in shortest paths, and $path(u, w)$ is the number of shortest paths between u and v, and $d(u, v)$ is the geodesic distance between u and v.

$$C_{dep(u,v)} = (1 + \sum_{w=1}^{n} \frac{occurence(u,v)}{d(u,v) \cdot path(u,w)}) \bigg/ (n+1) , \ u \neq v \neq w \qquad (1)$$

A comparison between Memon et al.'s method and DepRank will be given in experiments. The search of dependence paths on a graph in DepRank algorithm can be induced as a depth-first search (DFS) problem introduced in [5]. Besides, DepRank algorithm is based on link graph. Its survey can be found in [6].

3 DepRank Method

A Probability-Theoretic Definition. Since our goal is to provide a numeric measure of dependence, we first clarify our intuitions about dependence.

Intuition 1. The dependence of A on B is related to the extent of B determining A. The more extent of B determining A, the higher dependence degree of A on B.

Intuition 2. The maximum dependence of A on B is reached when $A = B$.

Furthermore, we assume the dependence of an object on *itself* is 1, which means the dependence degree of arbitrary pair of objects is between 0 and 1. From the view of semantics, dependence is the state of being determined, influenced or controlled by something else (see in [7]). For the material-supply network among companies shown in Figure 1, both company C and company D supply materials to company A, and we introduce *determinant*(C, A) to capture the probability of C determining A.

Definition 1. For two objects *A* and *B*, the dependence of *A* on *B* is defined as the probability of *B* determining *A*.

Equivalence to Dependence Paths in Graph. We propose a measure of dependence via heterogeneous links in a graph. Definition 1 we have given is a generic definition and does not rely on any specific domain. If we want to use this definition in practice, we should map it to a practical representation.

Most researchers consider random walk from the view of random surfers, such as works done in [8]. In our work, we exploit it from the view of nodes. Given a graph $G(V, E)$, supposing there is a node *A*, it has identical probability to be visited by surfers from adjacent nodes connected by incoming edges of *A* on the next step, and possibility of *zero* by surfers from other nodes. Applying this motivation to dependence measuring, we model random walk paths that make contributions to determinant probability as *dependence paths*. As Figure 3(a) shows, given a graph $G(V, E)$, supposing $S(A)$ is the set of all direct neighbors of node *A* connected by incoming edges and node $B \in S(A)$, for $S(A) \neq \{ \}$, $Dep(A \leftarrow B)$ can be calculated by

$$Dep(A \leftarrow B) = determinant(B, A) = 1/|S(A)| \tag{2}$$

where $|S(A)|$ is the number of incoming edges. For $S(A) = \{ \}$, $Dep(A \leftarrow B) = 0$.

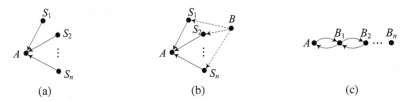

(a)   (b)   (c)

Fig. 3. (a) Direct neighbors. (b) Indirect neighbors. (c) A node pair. (d) A linear structure.

In Figure 1, $determinant(C, A)$ is 1/2, which means $Dep(A \leftarrow C) = 1/2$. Note that Equation (2) is only effective on direct neighbors of *A* connected by incoming edges. For other cases, we employ Theorem 1. Before that, dependence path $path(B \rightarrow A)$ is denoted as a random walk path from *B* to *A* without duplicate nodes. The reason why visiting duplicate nodes is not permitted is due to eliminating dependency cycles in travel. Naturally, $Dep(path(B \rightarrow A))$ is defined to be the dependence value of this path. Assuming $path(B \rightarrow A) = B \rightarrow N_1 \rightarrow ... \rightarrow N_n \rightarrow A$, we can get

$$Dep(path(B \rightarrow A)) = Dep(N_1 \leftarrow B) \cdot (\prod_{i=1}^{n-1} Dep(N_{i+1} \leftarrow N_i)) \cdot Dep(A \leftarrow N_n) \tag{3}$$

Theorem 1. Given a directed and unweighted graph $G(V, E)$, supposing $S(A)$ is the set of all direct neighbors of node *A* connected by incoming edges and node $B \notin S(A)$, and *m* is the number of different $path(B \rightarrow A)$, we can estimate $Dep(A \leftarrow B)$ by

$$Dep(A \leftarrow B) = \sum_{i=1}^{m} Dep((path_{B \rightarrow A})_i) \tag{4}$$

Proof. We use a forward propagation model to simulate determinant probability of *B* to *A* as shown in Figure 3(b). Backward propagation is not suggested, because duplicate nodes in a path produce cycles, and cycles make no contribution to influence the determinant of B to A. The influence of B propagates along the walk path starting from B, and we use all possible paths from B to A without duplicate nodes to estimate the probability of B determining A in the propagation process.

Equation (4) can be expanded by Equation (3) and every facient in Equation (3) can be computed by Equation (2). As an example shown in Figure 1, there are 4 different $path(B{\rightarrow}A)$ and a expansion of $Dep(A{\leftarrow}B)$ has been given in Figure 2.

We now discuss a special case. For a linear structure shown in Figure 3(c), $Dep(A{\leftarrow}B_i) = 0.5^{i-1}$ based on Theorem 1, where $1 \le i \le n$. This result is reasonable too, in which dependence degree decreases along with the increase of distance.

DepRank Algorithm. We propose DepRank, a method that can quantify dependence degree of a node pair effectively. A notice here is the search of dependence paths on a graph in DepRank algorithm, which can be induced as a depth-first search (DFS).

Major steps of DepRank algorithm are given as follows.

Algorithm 1. (**DepRank Algorithm**)
Input: A directed and unweighted graph $G(V, E)$; $A{\leftarrow}B$.
Output: $Dep(A{\leftarrow}B)$.
Method:
1. If $A = B$, return 1 and exit.
2. Obtain $S(A)$, a node set of all direct neighbors of node A connected by incoming edges.
3. If $B \in S(A)$, return $1/|S(A)|$ and exit.
4. Else if $B \notin S(A)$, (Theorem 1)
   (1) Obtain all possible paths from B to A without duplicate nodes by $DFS(B, A)$.
   (2) For every path $path(B{\rightarrow}A)$, compute $Dep(path(B{\rightarrow}A))$ using Equation (3).
   (3) Return the sum of all $Dep(path(B{\rightarrow}A))$ values and exit.

For $DFS(B, A)$ in Algorithm 1, we write the pseudocode as follows.

Subroutine. (**$DFS(B, A)$: Depth-First Search from B to A**)
Input: A directed graph $G(V, E)$; (B, A).
Output: A list of all possible $path(B{\rightarrow}A)$.
Method:
```
    a static list L
    a global list set LS
    dfs (B)
        add B to the end of L
        for each node C connected by a outgoing edge of B
            if (C is A)
                add A to the end of L
                add a copy of L into LS
                remove A from L
            else if (L contains C)
                print info "C has been visited"
            else dfs (C)
        remove B from L
    return LS
```

The depth-first search from B to A can be viewed as a problem of enumerating all simple paths in Graph Theory [9]. If the graph is dense, the size of simple paths may be astonished. Algorithm 2 can deal with sparse graph efficiently, but if your target is a large dense graph, some restrictions should be added to avoid huge complexity, e.g., only those dependence paths with length $< n$ are considered, where n is a manageable number (we recommend 10).

4 Empirical Study

A Comparison on Terrorist Networks. We have introduced the work done by Memon et al. [3]. They model the terrorist network as an undirected and unweighted graph, which is also adaptable to DepRank method.

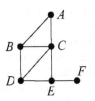

Fig. 4. A fictitious terrorist network

Table 1. Results of [3]

	A	B	C	D	E	F
A		0.3	0.7	0.2	0.3	0.2
B	0.2		0.4	0.4	0.3	0.2
C	0.2	0.2		0.2	0.4	0.2
D	0.2	0.3	0.3		0.3	0.2
E	0.2	0.2	0.5	0.3		0.2
F	0.2	0.2	0.35	0.25	1	

Table 2. Results of DepRank

	A	B	C	D	E	F
A	1	0.5	0.5	0.43	0.31	0.1
B	0.33	1	0.33	0.33	0.31	0.1
C	0.25	0.25	1	0.25	0.25	0.13
D	0.29	0.33	0.33	1	0.33	0.15
E	0.2	0.31	0.33	0.33	1	0.33
F	0.2	0.31	0.5	0.49	1	1

A comparison of these two methods is conducted on a fictitious terrorist network shown in Figure 4. Dependence scores computed by the method in [3] (Equation 1) and DepRank are shown in Table 1 and Table 2, respectively. Although these two methods have different theoretical models, they get similar results. It is hard to evaluate whose method is more accurate without performing extensively user studies; it lies on how the dependence is defined. It seems that dependence is not well defined in [3], while we give a generic definition of dependence and make equivalence to probabilities of dependence paths in graph.

Tests on Synthetic Datasets. To evaluate the scalability of DepRank, a set of synthetic graphs are generated. Graphs $G(V, E)$ with different scales on $|V|$ and $|E|$ are generated and can be divided into two types: (1) graphs with a fixed edge number per node on average and varying sizes of node; (2) graphs with a fixed node size and varying number of edges. The first type is for performance test on nodes, while the second is for performance test on edges. The conjunctions of edges and nodes are random.

Figure 5 and Figure 6 show the performance of DepRank algorithm on different node sizes and different edge sizes. Every experiment is repeated 100 times by choosing random node pairs for DepRank computation. For Figure 5 we assume $|E|/|V|$ is 5 on average, and for Figure 6, we assume $|V| = 500$ with varying edge sizes. The maximum length of a dependence path is set to be 10. Figure 5 shows time cost increases linearly along with $|V|$ when $|E|/|V|$ is fixed, which means the performance of DepRank is linear correlative to the scale of graphs with a roughly similar $|E|/|V|$. When $|V|$ is fixed in Figure 6, the performance of DepRank algorithm increases

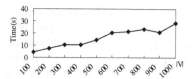   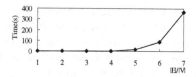

Fig. 5. Performances on different node sizes   **Fig. 6.** Performances on different edge sizes

rapidly while $|E|/|V|$ is rising, which means $|E|/|V|$ influences performance of DepRank remarkably. So, we suggest that when $|E|/|V| > 5$, only the dependence paths with a length below some manageable number are considered.

5 Conclusions

In this paper, we give a generic definition of dependence and then map it to the graph model. Considering a material-supply network, DepRank method can describe how much company *A* depends on another company *B* and which companies that company *A* is most dependent on. The results of computation can make a suggestion to manager of company *A* by avoiding risks of being determined by other companies.

Empirical study performed in the latter part of this paper shows that DepRank method has reasonable results compared with other congener methods. In future works, more applications based on our method need to be developed.

Acknowledgments. This work was supported in part by the National Natural Science Foundation of China under Grant No. 70871068, 70621061, 70890083, 60873017, 60573092 and 60496325.

References

1. Chen, L., Tong, T., Zhao, H.: Considering dependence among genes and markers for false discovery control in eQTL mapping. Bioinformatics 24(18), 2015–2022 (2008)
2. Liu, C., Chen, C., Han, J., Yu, P.S.: GPLAG: detection of software plagiarism by program dependence graph analysis. In: KDD 2006, pp. 872–881 (2006)
3. Memon, N., Hicks, D.L., Larsen, H.L.: How investigative data mining can help intelligence agencies to discover dependence of nodes in terrorist networks. In: Alhajj, R., Gao, H., Li, X., Li, J., Zaïane, O.R. (eds.) ADMA 2007. LNCS, vol. 4632, pp. 430–441. Springer, Heidelberg (2007)
4. Coppersmith, D., Feige, U., Shearer, J.B.: Random Walks on Regular and Irregular Graphs. SIAM J. Discrete Math (SIAMDM) 9(2), 301–308 (1996)
5. Cormen, T.H., Leiserson, C.E., Rivest, R.L., Stein, C.: Introduction to Algorithms, 2nd edn., pp. 540–549. MIT Press and McGraw-Hill (2001)
6. Getoor, L., Diehl, C.P.: Link mining: a survey. SIGKDD Explorations (SIGKDD) 7(2), 3–12 (2005)
7. The Free Dictionary, http://www.thefreedictionary.com/dependence
8. Jeh, G., Widom, J.: SimRank: A measure of structural-context similarity. In: SIGKDD 2002, pp. 538–543 (2002)
9. Rubin, F.: Enumerating all simple paths in a graph. IEEE Transactions on Circuits and Systems 25(8), 641–642 (1978)

Author Index